# Terrestrial Vertebrates of Pennsylvania

# Terrestrial Vertebrates of Pennsylvania

## A Complete Guide to Species of Conservation Concern

*Edited by* Michael A. Steele
Margaret C. Brittingham
Timothy J. Maret
Joseph F. Merritt

THE JOHNS HOPKINS UNIVERSITY PRESS | BALTIMORE

Printed in the United States of America on acid-free paper
9 8 7 6 5 4 3 2 1

The Johns Hopkins University Press
2715 North Charles Street
Baltimore, Maryland 21218-4363
www.press.jhu.edu

Library of Congress Cataloging-in-Publication Data

Terrestrial vertebrates of Pennsylvania : a complete guide to species of conservation concern / edited by Michael A. Steele . . . [et al.].
p. cm.
Includes bibliographical references and index.
ISBN-13: 978-0-8018-9544-9 (hardcover : alk. paper)
ISBN-10: 0-8018-9544-8 (hardcover : alk. paper)
1. Vertebrates—Pennsylvania. 2. Endangered species—Pennsylvania. 3. Wildlife conservation—Pennsylvania. I. Steele, Michael A.
QL606.52.U6T47 2010
596.16809748—dc22 2009048739

A catalog record for this book is available from the British Library.

*Special discounts are available for bulk purchases of this book. For more information, please contact Special Sales at 410-516-6936 or specialsales@press.jhu.edu.*

The Johns Hopkins University Press uses environmentally friendly book materials, including recycled text paper that is composed of at least 30 percent post-consumer waste, whenever possible. All of our book papers are acid-free, and our jackets and covers are printed on paper with recycled content.

# Contents

# Foreword

PENNSYLVANIANS HAVE MANY WONDERFUL OPPORTUNITIES TO CONNECT with wildlife. They represent an important aspect of the quality of life that many citizens of the commonwealth frequently enjoy. However, many of the species that are the source of this enjoyment face a difficult and challenging future. This book provides critical information concerning the basic biology of these terrestrial vertebrates, the requirements for their continued sustainability, and a plan for their future research and management.

The impetus for this book came from the development of the State Wildlife Action Plan that identified many species that need our attention. The book is truly a collaborative effort by all those who seek to share their interest and expertise to further the conservation of these important elements of our natural resources. To date, we have made considerable progress in managing these species, but much remains to be done. As you read, I urge you to consider the similarity in the habitats used by many of these species. Sound habitat management, above all else, will be the key for many of these species. Through a synergistic approach to habitat stewardship, we can achieve the greatest results with the fewest resources. To that end, the greater understanding of management requirements resulting from this book will allow us to be proactive, rather than reactive in our efforts to conserve and manage. Armed with this information, we must take on the difficult challenges necessary to prevent further degradation of these species' habitats so that we can ensure the future of these species for many generations to come.

Although the book focuses on species within Pennsylvania, there is no doubt of its important implications for a broad regional approach to conservation. Many of the species occur well beyond the boundaries of the Keystone State and are the focus of conservation concern in many other states. A regional understanding of the needs of these terrestrial vertebrates will pay greater dividends, and will certainly assist us in ensuring a better future for these species.

Some may view this book as a text for students or as a guide for land managers. It will certainly assist professional biologists in their quest for knowledge. However, I would offer that this publication is a tremendous asset for any wildlife enthusiast. It provides an incredible amount of information on species that are important to all of us, but it also provides a wealth of information on other species with which many of us may not be familiar. This book provides remarkable insights into our native wildlife species and those that have the greatest conservation needs. It provides yet another avenue that allows us to connect with wildlife in a meaningful way.

Carl G. Roe
Executive Director
Pennsylvania Game Commission

# Preface

WE ARE DEEPLY HONORED TO HAVE HAD THE OPPORTUNITY TO PREPARE this volume. Although it has required a substantial investment of time and energy, this project has been considerably rewarding, as we have had opportunity to interact with and learn from so many students, conservationists, resource managers, scientists, and naturalists alike. Six years have now passed since the book's inception and our first meeting with Lisa Williams (Pennsylvania Game Commission) to discuss how to capture the diverse efforts invested on behalf of Pennsylvania's rich biota over the past twenty-five years. It was 1985 when the first publication on *Species of Special Concern in Pennsylvania* was published by Genoways and Brenner. That volume, in the authors' own words, "represent[ed] the best estimation of the Pennsylvania Biological Survey of the status of the species of special concern in Pennsylvania as of 31 December 1982." Indeed, this was an impressive first attempt at pulling all available information together. In fact, unlike our contribution, which focuses only on the terrestrial vertebrates, Genoways and Brenner (1985) covered everything, from plants and invertebrates to the vertebrates—including the fish.

Soon after we began this project, it became evident that such a comprehensive approach to all species was no longer possible under one title. The sheer numbers of publications alone on the basic biology, conservation, and management of the 135 species reviewed here spoke to the need for a more focused approach. In the end, we decided to concentrate only on the terrestrial vertebrates: the reptiles, amphibians, birds, and mammals. Because these species share many of the same characteristics, habitats, conservation threats, and management issues, it seemed logical to concentrate on this assemblage of species of concern first. It is our hope that other authors will follow soon with updates on the fish, the plants, and the invertebrates.

Why focus on only the terrestrial vertebrates of Pennsylvania? At first glance, a volume that is largely centered on the distribution and biology of one state's biota may seem unnecessarily narrow and artificial. There are two reasons why we disagree with this assertion. First, and perhaps most important, it is at the state level where most species of conservation concern are first identified for protection and then subsequently researched and managed. As evidence of this, in 2001, the United States Congress created the State Wildlife Grants Program (administered by the U.S. Fish and Wildlife Service) to provide federal funds at the state level to reduce the decline of fish and wildlife species well before these species required federal attention under the Endangered Species Act. By 2005, each state was required to prepare a Comprehensive Wildlife Conservation Strategy (now referred to as Wildlife Action Plan) to be eligible for these funds. These federal requirements provided each state with an opportunity, as well as the funds, to take greater responsibility and leadership for the management of their nongame species. Indeed, this cooperative legislation reflected the primary role that each state has in managing their wildlife resources. In essence, it demonstrated that it is in fact at the local and state level where we best understand the biology and management of many of these species, and it is at the state level where many researchers work on many of the species.

A second reason for taking this Pennsylvania approach is that our state, we believe, is particularly important for the management and conservation of many of these vertebrates. Pennsylvania, with its diverse physiographic provinces, varied climate and habitats, and central location along major avian migratory pathways is, in essence, a crossroads for many of these vertebrate assemblages. Many of the rodents, shrews, and lagomorphs, for example, are relatively common boreal species whose geographic range extends to the southern Appalachians where their populations are highly fragmented. In Pennsylvania, populations of many vertebrates are likely important for maintaining gene flow between stable populations to the north and more precarious, isolated populations to the south.

Beyond the objectives outlined above, we were motivated by four additional goals. The first and foremost of these was to provide a useful manual for those personnel involved in the potential management and research of these species, not just in Pennsylvania, but elsewhere, across each species' geographic range. Second, we believe this volume—from the beginning overview of conservation biology, through the 135 species accounts, to the final chapter on new and emerging threats—has extraordinary educational value for students of all ages. To that end, we will seek the funds to distribute a copy of the book to all high schools in Pennsylvania. Third, it is our hope that this publication will encourage other states to produce similar formal publications on their Wildlife Action Plans and their species of concern. Although numerous agencies from many states regularly post information on their web pages, it is often prepared by a few individuals, usually limited in breadth and depth, and rarely peer-reviewed. Finally, throughout these pages we have taken great care to deliver a product that, although focused on the Pennsylvania scene, has broad utility beyond the state's borders. It is our hope that this volume will serve as an important resource for land managers, naturalists, scientists, and students across much of the eastern United States and Canada.

The book opens with three chapters that are intended to provide the context for the species accounts that follow. The first provides an overview of the, often misunderstood, field of conservation biology; it defines the discipline, illustrates the many kinds of problems it addresses, and provides a brief review of the most significant conservation issues facing vertebrates in the eastern United States. The second chapter focuses on the diverse habitats found in the state; it provides a physical context for understanding where these species live and why we are concerned about their future. The third chapter is a brief overview of Pennsylvania's recently completed Wildlife Action Plan, which served as the impetus for this project. It is in effect the guiding principle that the Pennsylvania Game Commission will use to manage and protect these species.

The species accounts are organized into three chapters: the amphibians and reptiles (chap. 4); the birds (chap. 5); and the mammals (chap. 6). Each begins with a brief introduction, followed by the species accounts, organized by their Wildlife Action Plan status, with the most imperiled species, those of *Immediate Concern*, first. These are followed by the species of *High-Level Concern*, *Responsibility* species, *Pennsylvania Vulnerable* species, and those designated for *Maintenance Concern* (see chap. 3).

The final list of 135 species for the both the Wildlife Action Plan and this publication were made by the Pennsylvania Game Commission, in close consultation with the Herpetological, Bird and Mammal Technical Committees of the Pennsylvania Biological Survey. Each of these committees is composed of professional biologists from across Pennsylvania and neighboring states. These committees also made the author assignments. The first draft of these accounts, completed in 2004–2005, served to guide the development of the Wildlife Action Plan. These accounts were then peer-reviewed, revised, and, within six months of completion of this manuscript, again updated by the authors to ensure the most current information. The final accounts were then condensed further, by eliminating much of the basic biology that can typically be found in field guides.

Each account is organized by the species' primary level of concern and a designation of Responsibility Concern for those species for which this category also applies (see chap. 3). The account then opens with a brief description of the species, a statement on why it was selected as a species (or subspecies) of concern, and a color photograph of the organism. This is followed by a discussion on the species' broader geographic range, a more focused examination of its distribution and abundance in Pennsylvania, and a detailed map of its state distribution. These maps are based on the most current data available, and when possible, reflect relative abundance, primary and secondary areas of occupancy, historic and current distributions, or other features relevant to the management and conservation of the species (e.g., bat hibernacula). State distributions are followed by brief discussions on the species' community and habitat associations and its behavior and ecology. The remaining three sections of each account focus on issues related to the species conservation: (1) primary threats, (2) management and conservation needs, and (3) monitoring and research needs.

Finally, we close the book with a brief review of several new and emerging issues in the conservation of terrestrial vertebrates and an appendix that summarizes the status of terrestrial vertebrates in other mid-Atlantic and northeastern states.

In closing, we are excited to see this project to fruition and hope it has its intended effect: to both generate interest in our valuable wildlife resources and serve as a guide for their research, management, and conservation. Whether one opens this book as an amateur naturalist, a seasoned scientist, or a beginning student hoping to learn just a little about Pennsylvania's terrestrial vertebrates, each we believe each can benefit in their own way. For the naturalist, there is detailed information on the state distribution and biology of their species, for researchers and managers there are specific guidelines on research, conservation, and management, and for even the beginning student, there is an opportunity to become more familiar with some of the fauna in her or his own backyard.

# Acknowledgments

WE ARE DEEPLY INDEBTED TO A VAST NUMBER OF AGENCY PERSONNEL, technical reviewers, volunteers, and other assistants, all of whom helped to make this project possible. Certainly, any attempt to name all those who have contributed would be futile. So instead we extend our appreciation to everyone and briefly recognize those that were most influential.

We first acknowledge the insight, leadership, and guidance of one individual in particular who was instrumental in the inception, design, and completion of this project—Lisa Williams of the Pennsylvania Game Commission. Much of the original plan for this project began with Lisa when she convinced us to focus on only the terrestrial vertebrates so that we could provide the greatest breadth and detailed information on this group of species. She then developed the approach we used, helped us to secure funding for the project, and provided invaluable assistance through the long arduous process. In short, the project simply would not have been possible without her leadership and steadfast guidance.

We are particularly grateful to the more than 150 scientists who served as reviewers for the species accounts and introductory chapters. Two reviewers offered their professional evaluation of each account and helped us to better focus the most critical and current information available on each species. We extend our deepest appreciation for their tireless efforts. Also behind the scenes were countless members of the scientific and wildlife management community who serve on the three Pennsylvania state technical mammal, bird, and herpetological committees that carefully evaluated and prioritized the species lists reviewed in this volume.

Range maps often provide only a rough idea of a species' distribution, even on a state or provincial level. The maps in this volume are designed to provide the most current information but also a level of detail that is rarely seen. This simply would not have been possible without the hard work, patience, and technical skills of Nick Bolgiano, who prepared all of the maps for the avian and mammal accounts. Nick devoted countless hours to preparing, revising, and then revising again many of these maps as we repeatedly updated the information for each account. Nick's talent and insight also allowed us to tailor each map to the kinds of information available for each species. The maps for reptiles and amphibians required a different approach because of the specific distributional data on hand for these species. They were designed and prepared by our herpetological editor, Dr. Timothy Maret, with much assistance from Danielle Bowers.

Shealyn Marino, research associate at Wilkes University, worked tirelessly to identify the finest photographs available and to secure the permission of the seventy-five or more photographers whose work is highlighted in this volume. To the photographers—we extend our deepest appreciation for sharing your fine works with our readers. These images are indeed a critical contribution that adds an immeasurable dimension to each account.

Numerous individuals assisted with the monumental task of editing this work of more than eighty-five authors. Andrea Lego, staff assistant in the School of Forest Resources at Penn State, provided invaluable assistance typing, editing, and standardizing the avian species accounts. Early versions of the manuscript were first carefully standardized by a devoted team of technical assistants at Wilkes University. We recognize the long hours and the impressive teamwork shown by Andrew Bartlow, Melissa Bugdal, MaryKathryn Hurst, Michael Kachmar, Richard Mebane, Jr., and Allison Otis. Melissa Bugdal in particular contributed significantly to all stages of the editing process. Copyeditor Andre Barnett at the Johns Hopkins University Press magically transformed the document into its final format. We appreciate her diligence and painstaking effort to scrutinize and finalize such a diversity of writing styles. Her efforts explain so much that is right about the document.

For their financial support the Pennsylvania Game Commission, the State Wildlife Grants Program, and the U.S. Fish and Wildlife Service who funded the initial project from which the species accounts were eventually adapted and the final publication. Patricia Barber, Daniel Brauning, and Lisa Williams assisted with the administration of these grants. Additional funding and assistance were provided by the Fenner Research Fund at Wilkes University and the expert administrative personnel in the Wilkes University Office of Sponsored Research: Anne Pelak, Julie Pius, and Amy Edwards.

We also extend our deepest gratitude to all individuals who have contributed to our knowledge of the biology and conservation of the species reviewed here. Since the last publication of this kind twenty-five years ago, countless scientific papers, theses, dissertations, and agency reports have been devoted to improving our knowledge of these species and our ability to manage and conserve them. The acquisition of such information is a slow process that was only possible through the hard work of nearly two generations of biologists without which we could not have produced this volume.

We thank Vincent J. Burke, senior acquisitions editor at the Johns Hopkins University Press, and his acquisitions assistants Jennifer Malat and Winnie Rodgers, for their insight, patience, and finely honed skills of supportive persistence that continually guided us through the publication process.

Finally, we thank our families for their encouragement, and patience during this long and evolving process. We could not have accomplished this without their support.

# Terrestrial Vertebrates of Pennsylvania

# 1

# Conservation of Terrestrial Vertebrates

## *An Overview and Synthesis*

Jeffrey Stratford
Michael A. Steele

### Introduction

Although only a small subset of our biological diversity, the terrestrial vertebrates are the most familiar and appreciated species for most humans. Despite this familiarity, the need to conserve these important biological resources, the various agencies involved in these endeavors, and the field of conservation biology, are all often misunderstood and underappreciated. We therefore begin this volume with a broad discussion of conservation biology, including important definitions and a discussion about why efforts to conserve these resources are so important. We give a history of conservation biology and environmental ethics. We provide an overview of the major threats to biodiversity and identify gaps in our knowledge, particularly concerning Pennsylvania vertebrates. We close the chapter with a discussion of the future of conservation biology and the role of natural history and education in conservation biology. Herein we make several references to Charles Darwin. Our motive is in part to pay homage to Charles Darwin's two hundredth birthday and the one hundred fiftieth anniversary of *On the Origin of Species*. We also seek to make the reader aware of the enormous challenges that face conservation biologists. Darwin may seem irrelevant to conservation biology, but he struggled with several questions that conservation biologists still grapple with today: What are the consequences of inbreeding? What causes rarity? Why do organisms go extinct? How should we define species? How do organisms cope with change? Although the tools we use to understand ecology and evolution have changed, Darwin's charge remains. Conservation biology not only must answer these basic questions but also must address societal challenges, such as the evaluation of ecosystems and species (see Sutherland et al. 2000 for a recent and comprehensive list of questions for conservation biologists). These challenges must be faced, or we will see countless species slip through our fingers into the oblivion of extinction.

## Background

### *What Conservation Biology Is*

The basic goals of conservation biology are to minimize or eliminate species extinctions and preserve ecosystem function. At first glance, then, conservation biology would seem no different than applied ecology and evolution. However, there are two other essential aspects of conservation biology in addition to the natural sciences: (1) institutional entities and (2) environmental ethics. Institutional entities are the societal constructs put in place to regulate human behavior; they include environmental law and conservation organizations. Environmental ethics is the "how" of conservation biology—how we value nature and species. If any one of these elements is missing, the goals of conservation biology are likely to fail. For instance, ecological sampling tools have indicated that tropical rain forests are filled with millions of species of invertebrates. Despite the potential usefulness of these taxa to humankind, the wholesale destruction of rain forests continues and nearly countless numbers of species go extinct even before their existence has been documented. Most fisheries are overexploited, but there is little public outcry. In these cases, the science is relatively clear, but public concern is generally deficient. There is little legislative protection, and the budget of conservation projects is meager compared with commercial interests. Indeed, the lack of funding for biodiversity might be considered an axiom of conservation biology. In other cases, the conservation "will" is present, but the science is lacking. Examples of scientific ignorance include our lack of understanding of how some species will respond to climate change. Indeed, we are still ignorant of the natural history of most species on this planet, and without such knowledge, the application of conservation measures is often futile.

### *What Conservation Biology Is Not*

A distinction should be drawn between conservation biology and animal rights. The focus of animal rights is the individual and individual suffering, while the focus of conservation is the population and ecosystem, particularly of native species. Take, for example, the management of feral cats in Cape May, New Jersey. Animal rights groups, such as Alley Cat Allies, favor trap-neuter-return (TNR) programs in which feral cats are returned to the wild once they are vaccinated and neutered (www.feralfeline.org/tnr.html). Wildlife advocates instead favor the removal of feral cats because their densities are negatively associated with avian diversity and abundance (Lepczyk et al. 2004b, Sims et al. 2008).

Another common misconception of conservation biology is that it is preservationism and antibusiness or antitechnology. Indeed, economic development, as measured by per capita income, may be associated with society's need for stronger environmental safeguards. Technology is essential to conservation biology as a tool for scientific inquiry as well as increasing economic efficiency and reducing stress on natural resources. Conservation biology acknowledges that economies are part of the system, regardless of whether they are considered natural or artificial. What conservation biology can do is predict the impacts of different economic models.

### *What Is Biodiversity?*

Natural communities can be parameterized in several ways. The simplest is species richness, a count of the number of species. Species richness can be assessed directly for easily detectable organisms such as trees. For more elusive taxa, species richness must be estimated using sampling. Species richness ignores the relative abundance or evenness of individual species. Diversity takes into account both species richness and evenness. Species richness and diversity are typically measured for a defined area and for particular taxa (e.g., birds, vertebrates, understory plants, fungi). Biodiversity is not an exact term and can refer to species richness or diversity and often refers to all taxa within an area. A biodiversity hot spot typically refers to an area with relatively high species compared with other areas and is sensitive to scale. Global biodiversity hot spots may include places such as Madagascar or the Amazon, but there are also hot spots within Pennsylvania.

Genetic diversity is another important element of natural communities. Genetic diversity can be measured within an individual, such as variation within gene families. Another measure is allelic diversity, which is the diversity of different forms of a gene at a particular locus. Heterozygosity is a commonly referred to measure by conservation biology and the proportion of the population that is heterozygous for a particular gene. These measures of genetic diversity are important to understand because genetic diversity is essential for a species to adapt to a changing world. Adaptation can occur through time as well as through space—a phenomenon that often causes populations in different areas to become genetically divergent. Genetic divergence is not only of scientific interest; there are regulatory consequences for populations that have diverged enough to distinguish them from other populations. These "distinct population segments" are protected under the U.S. Endangered Species Act (Fallon 2007).

### *What Is a Species?*

If populations are adapting to local environments or are becoming isolated from other populations, then populations may diverge to the point where they are recognized as different species. What exactly is meant by species? At what point must two populations diverge to be recognized as two distinct species? The species problem plagued Darwin. In a letter to his friend and confidant J. D. Hooker, Darwin writes: "After describing a set of forms, as distinct species, tearing up my M.S., & making them one species; tearing that up & making them separate, & then making them one again (which has happened to me) I have gnashed my teeth, cursed species, & asked what sin I had committed to be so punished" (www.darwinproject.ac.uk/darwin letters/calendar/entry-1532.html).

Darwin's frustration with barnacle systematics was caused by a lack of a consistent definition of species. In Darwin's time, a species was a population that shared particular elements of a phenotype. In other words, members of the same species, barring age and sexual differences, looked more or less the same. Minor variation was ignored although, as Darwin notes, minor differences can blur the differences between species. One of Darwin's great contributions was to focus on minor variations as being supremely important. This forced biologists to rethink the meaning of species. Today, different species concepts are used to define and delineate species. Historically, most people have used the biological species concept in which species are populations that are reproductively isolated from other populations. The biological species concept delineates a species but does not define it. As a response, the phylogenetic/evolutionary species concept was developed, which is a hybrid of the evolutionary species concept and the phylogenetic species concept (PSC). The evolutionary species concept defines a species as a population of organisms with its own evolutionary history and trajectory. The phylogenetic species concept delineates a species as a population with unique traits (synapomorphies) not shared with other populations. The phylogenetic/evolutionary species concept combines a definition and a method of delineation.

The choice of species concept is not trivial; because species are the units of evolution and the focus of conservation, it is essential for conservationists to have a clear idea of what is meant by species. There are implications for different species concepts, but as the types of data necessary to delineate species vary (genetic vs. ability to reproduce), so do the number of species within particular taxonomic levels, and the possible recognition of populations as unique species (Crother 1992, Zink 1996). For example, adoption of Ernst Mayr's biological species concept halved the number of avian species. And despite the advantages of the phylogenetic/evolutionary species concept, it is neither widely accepted nor widely taught.

## A Brief History of Conservation Biology and the Development of Environmental Ethics

It is nearly impossible to separate the history of conservation biology from the development of environmental ethics; the two are closely interdependent. A history of conservation without the accompanying change of ethics would conceal the motivation behind such events. Likewise, omission of conservation biology from the development of environmental ethics would make any changes in ethics appear inconsequential.

The development of environmental ethics in the United States may have been a response to different ecological crises (Cox 1993). These crises include deforestation and expansion of cities on the East Coast, large-scale deforestation on the West Coast, loss of soil in the Midwest (dust bowl era), and the post–World War II expansion of synthetic pesticides. American philosopher and historian J. B. Callicott et al. (1999) suggests three basic environmental ethical paradigms that developed in Western civilizations: (1) the Romantic-transcendental preservation ethic, (2) the resource conservation ethic, and the (3) evolutionary-ecological land ethic. Before these ethics emerged, the human view of nature was largely materialistic, and nature was to be exploited and feared. Some of this attitude stemmed from a particular Judeo-Christian philosophy, which places man with dominion over creation. The consequence of this worldview is exploitation of nature as a human right. However, out of the moonscape clear-cuts of the West Coast and the newly smoldering industrialized East Coast, a new environmental ethics emerged that would change the landscape of the American consciousness.

The ethical paradigm to first emerge in the mid-nineteenth century was the Romantic-transcendental preservation ethic. The central tenet of this ethic, espoused by Henry David Thoreau, Ralph Waldo Emerson, and John Muir, is that wild places are good for the human soul. This ethic was part of a larger movement that rejected materialism and maintained that human happiness was achieved through a communion with nature. For conservation, large undisturbed vistas and uncut forest tracts were essential. The preservation ethic also meant bringing nature to the urbanized world. With the campaigning of Thoreau, Muir, and many concerned citizens, the first national monuments and parks and the establishment of city parks like New York's Central Park were established in the United States.

As a compromise between preservationism, which called for elimination of all human interference, and, in contrast, the large-scale exploitation of natural resources, the resource conservation ethic arose. The founder of the resource conservation ethic was Pennsylvania native Gifford Pinchot. The United States lacked forestry schools, so after he graduated from Yale University, Pinchot trained as a forester in Germany. There he observed intensively managed forests, the techniques and philosophy for which he transported back to the United States. To Pinchot, forests were to be protected from overexploitation but still provide natural resources. Pinchot had a sympathetic ear in President Theodore Roosevelt, and national forests increased from 36 million acres to 148 million acres during Roosevelt's tenure (Cox 1993). The conservation ethic extended past trees to include animals. In 1903, Florida's Pelican Island Refuge became the first in the national wildlife refuge system (Cox 1993). Like forests, wading birds had been overexploited for their feathers, which were used to adorn hats.

Pinchot and Roosevelt also established the United States Forest Service of which Pinchot was the first director. The first American forestry school was established in 1900 by Pinchot at his alma mater, Yale University. Students in the forestry program were required to spend a summer at land donated by the Pinchot family in Milford, Pennsylvania, learning field methods.

One of these forestry students, Aldo Leopold, was to revolutionize our thinking about managing and thinking about wildlife. Leopold was a prolific scientist and writer, but perhaps his most important work was *A Sand County Almanac*. The evolutionary-ecological land ethic emphasized in this classic stressed the interdependence of organisms that have been evolved through the millennia. The land ethic complements the two human-centered ethical paradigms, with one based on stewardship of nature. Ecosystem health becomes the focus for its own sake while recognizing man's place in nature is still important.

Although each of these three ethics appears sequentially, we should not look at them as one replacing the other. We still see the preservation ethic manifested in our wilderness areas and wild and scenic rivers. The conservation ethic is apparent in our national forests, which provide some of the nation's timber as well as vast opportunities for camping, hunting, and fishing. The land ethic has permeated nearly all natural resource institutions and has been the impetus for the creation of several new organizations, including the Nature Conservancy. We have also seen university wildlife management programs diversify to include nongame animals and plants. Indeed, the *Journal of Wildlife Management* is now replete with articles on nongame animals and biodiversity issues.

## Valuing Biodiversity: Ecosystem Services

We can generalize how nature is valued into two systems: nature can have both inherent value and instrumental value. Inherent values do not come from the organism's or species' themselves. Instrumental values come from ecosystem services—the benefits that species or biological communities provide as a result of their function in nature. They include such benefits as seed dispersal, nutrient cycling, and pollination, among others (Sekercioglu 2006, Turner et al. 2007).

Many plants, for example, are dependent on animals, especially birds and mammals, for dispersal of their seeds and, thus, the establishment and regeneration of offspring. This process of animal-mediated dispersal is absolutely critical for some plant communities such as the oak forests of the eastern and central United States, where just a few common species (e.g., blue jays and eastern gray squirrels) are essential in the regeneration of oaks (Steele and Smallwood 2002, Steele et al. 2010). Forests without these keystone seed dispersers (Steele et al. 2005) would eventually lose their oaks. However, the importance of such vertebrates for plant dispersal extends to many other species as well. In Pennsylvania, it is estimated that eighty-three species of birds (47% of species belonging to 57.1% of bird families) contribute to dispersal of plants in some manner (Steele et al.

2010). Even the white-tailed deer, a common and often destructive herbivore, is an important long-distance dispersal agent of many herbaceous plants (Myers et al. 2004).

Other services provided by vertebrates in the eastern United States include reduction of plant pests by insectivorous birds (Marquis and Whelan 1994, Barber and Marquis 2009), pollination by hummingbirds, nutrient cycling, and possibly the control of pests (Sekercioglu 2006). Biodiversity may have an important ecosystem function in the control of disease transmission rates. The "dilution effect" refers to the observation that where there is a higher diversity of disease hosts, there is a lower incidence of disease. For humans, the two important diseases are West Nile virus and Lyme disease. Birds are important reservoirs for West Nile virus and where there is higher species richness of birds, there is a lower incidence of West Nile in humans (Swaddle and Calos 2008, Allan et al. 2009). Likewise, where there is a lower incidence of Lyme disease, there is higher species richness of small mammals, an important tick host (Ostfield and Keesing 2000).

The importance of ecosystem services has recently received a great deal of attention in the literature and has become an important strategy for the economic argument for conservation of vertebrates and other species as well. For example, a 1997 estimate of the mean value of ecological services was US$33 trillion (Costanza et al. 1997). More recently, US$57 billion was estimated for the value of wild insects in the United States alone (Losey and Vaughan 2006). Nature can be seen as a toolbox of species and genes that may have benefits directly or indirectly (Daily and Matson 2008). Conservation biology should not be about the best economic argument for species preservation, but it should advocate that there are numerous benefits that follow from conservation (Ghazoul 2007). There is also a more direct value of biodiversity in terms of hunting, fishing, and nature-related activities, such as bird-watching and nature photography. For instance, in 1992, bird-watching-related expenses in the United States were at least US$28 billion and this number is likely to have increased given the ever-expanding interest in birds, in nature watching, and in photography (Sekercioglu 2002).

Human-induced extinctions and population declines are likely to diminish or eliminate ecosystem services (Chapin et al. 2000, Dobson et al. 2006, Turner et al. 2007) and are likely to have economic consequences. Extinctions do occur naturally (Jablonski 2004), but the extinctions that concern conservation biologists are those caused by human activities—either directly (e.g., overharvesting) or indirectly (e.g., species introductions). Because biological communities are composed of interacting species, changes in the population of one species may ripple through the community and affect other species. Trophic cascades refer to the changes in trophic levels through changes in prey or predator abundance. Loss of keystone species may have more dramatic effects. For example, beavers create ponds that are used for many other wetland species. Ecological knowledge, therefore, is essential to be able to predict the consequences of population changes throughout the community.

## Primary Threats to Terrestrial Vertebrates

### *Habitat Loss and Fragmentation*

As the human population swells, land is converted for housing, agriculture, and places where material goods are produced, stored, and transported. This is not without consequence. In the United States (Flather et al. 1998, Wilcove et al. 1998) and elsewhere (Turner et al. 2007), the majority of endangered vertebrates are threatened by habitat loss and fragmentation. Habitat loss is the straightforward removal of required habitat. Habitat fragmentation, in contrast, is habitat loss accompanied by the fracturing of the remaining habitat. For example, the spread of agriculture into forested areas will often leave patches of forest surrounded by a matrix of crops. In the northeastern United States, habitat loss per se mostly takes the form of forest loss and wetland drainage, but habitat fragmentation and loss are usually concomitant.

There are numerous habitats that have been depleted by land conversion in Pennsylvania, including undisturbed sandy beaches, grasslands, rock habitat (e.g., talus slopes and caves), vernal ponds, and several types of barrens. Grasslands and early successional habitats, such as scrub, are being reduced, not only from direct land conversion alone but also from a lack of disturbance. Abandonment of farmland creates scrub habitat, but only temporarily, as the land converts to forests. Grasslands and scrub are maintained by mowing, clear-cutting, or fires. The loss of these temporary habitats is evidenced by the population declines of avian species associated with early successional habitats (Askins 1993, 2001). For example, Pennsylvania populations of northern bobwhite and golden-winged warblers have declined –7.59 percent and –9.73 percent annually from 1966 to 2007 (Sauer and Fallon 2008).

Habitat fragmentation has effects on biological communities beyond uniform habitat loss. Though the total summed area of remnant habitat may be equal to a single large remnant, we can expect fewer of the original species in the fragmented landscape. Fragmentation can affect organisms directly by altering the abiotic conditions (i.e., edge effects) or presenting novel environments that cause adverse behavioral responses. Increased temperature and decreased humidity are typically found along the edge of a forest patch where it joins nonforested habitat (Chen et al. 1995). The effect of the matrix on isolating forest patches is highly dependent on behavioral responses to the matrix (Blumstein et al. 2005). For example, a forest patch amid a sea of corn crops may be isolated to an amphibian but not to a songbird. In general, however, the effect of the matrix as a barrier increases as the matrix becomes increasingly dissimilar to the original vegetation. Adverse behavioral responses to novel environments, such as roads, are a part of a phenomenon called neophobia (Greenberg 1989). Many species of birds are less likely to cross open areas (Desrochers and Hannon 1997, Greenberg 1989, Moore et al. 2008), even species that are thought to be adaptable to human disturbance (Grubb and Bronson 2001). Gap-crossing is also an issue for North American mammals (Zollner 2000, Crooks 2002, Bakker and Van Vuren 2004) as well as amphibians (Trenham and Shaffer 2005). One solution to the problem of fragmentation is the construction of corridors between forest fragments. These strips of habitat allow movement of animals (and their genes) between larger patches. However, such connections may also facilitate the movement of disease (Cushman et al. 2008).

### *Urbanization and Urban Sprawl*

Urbanization is the conversion of landscapes to human dwellings, transportation systems, and manufacturing. Urbanization, as it spreads out from a core area, creates habitat fragmentation (Luck and Wu 2002) and is already responsible for endangering more plants and animals in the mainland United States than any other factor (Czech and Krausman 1997).

Urbanization is a form of habitat loss but a very different form, and there are several facets of urbanization that concern conservation biologists. (1) Urbanization has a much larger ecological footprint than the city limits (Pickett et al. 2001). Areas well outside the city limits are affected by the energy requirements of the city, including areas used to grow food, create power, and dispose of waste. (2) The flow of matter (water) and energy (sunlight) are highly modified in urban systems and are largely unavailable for biological communities. The result is largely barren areas and simplified vegetative communities, often dominated by invasive plants. Consequently, even a small amount of urbanization in an area can dramatically reduce avian species richness (Stratford and Robinson 2005), and many species that do persist in urban areas are introduced species (McKinney 2006, Olden et al. 2006). A survey of larval amphibians in Pennsylvania revealed a negative relationship between species richness and urbanization. Urban ponds in this study were more likely to be permanent and allow for the presence of predatory fish (Rubbo and Kiesecker 2005). (3) Urban centers often have higher pollution levels (Lepczyk et al. 2004a), and these pollutants may make their way into vertebrates (Ruiz et al. 2002, Roux and Marra 2007). (4) Urbanization is relatively permanent. Once urban centers form, there is little we can do to make amendments for conservation.

The suite of invasive species associated with urbanization tends to be the same species, regardless of geographic location. This "biotic homogenization" creates a less diverse community on a larger scale (McKinney 2006) at the expense of endemic species. Many of the invasive species found in urban areas of Pennsylvania (e.g., Norway rat [*Rattus norvegicus*], house sparrows [*Passer domesticus*], European starlings [*Sturnus vulgaris*]) are also found throughout the world. The effects of urbanization are pervasive enough to demand greater attention from conservation practitioners.

### *Pollution*

Urban areas are not the only source of pollutants; runoff from agricultural areas and atmospheric deposition of heavy metals are other important sources. Moreover, many persistent organic pollutants that have been banned are still affecting Pennsylvania vertebrates, including polychlorinated biphenyls (PCBs) and organochlorines (e.g., DDT). Fortunately, many of the direct negative effects of pollution on wildlife have largely been reduced with the elimination of the large-scale use of organophosphates, organochlorides, and carbamates, as well as the closing of smelters (see Stratford 2009 for review on contaminants and birds). Although the threat from contaminants is reduced, many vertebrates still appear to be at risk of heavy metal exposure, particularly in higher elevations (Rimmer et al. 2005) and aquatic systems (Evers et al. 2005, Rattner and McGowan 2007).

The indirect effects of contaminants may have a much larger effect on wildlife. Reductions of nontarget invertebrates, through insecticides, may reduce the prey base for insectivores, including songbirds, bats, shrews, amphibians, and other vertebrates. For example, the use of *Bacillus thuringiensis* (Bt) and diflubenzuron for gypsy moth outbreaks may also affect nontarget lepidopteron larvae (Rastall et al. 2003), which are important food sources for migrant songbirds (Cooper et al. 1990) and insectivorous mammals (Bellocq et al. 1992).

Acid rain still affects vertebrates, particularly at higher elevations in the northeastern United States. Like other pollutants, acid rain may affect some vertebrates directly, such as fish, but some of the important effects of acid rain are indirect. For example, acid rain is thought to decrease the availability of calcium in soils so arthropod prey contains less calcium for eggshell construction in birds (Hames et al. 2002).

### *Forest and Game Management: The Problems of Over- and Underexploitation*

Pennsylvania has more than 17 million acres of forest covering nearly 60 percent of the state (Pennsylvania Department of Agriculture). Roughly 12 percent of these forests are in state-owned forests. Because so much forest covers the state, management of these forests (on both public and private lands) can have significant effects on wildlife populations. Different strategies include the type of tree removal (clear-cutting vs. selective logging), the frequency of cutting, or the extent of areas where logging does not occur. Other aspects that influence forest structure include decisions to burn, burn frequency, the construction of roads, and how introduced species are dealt with. The use of pesticides on organisms such as gypsy moths can be looked at as a forest management issue.

Game management also has consequences for forests. In particular, management of deer populations affects forest structure directly (Liang and Seagle 2002, Holmes et al. 2008) and other organisms such as arthropods (Allombert et al. 2005) and songbirds indirectly (DeCalesta 1994). Overbrowsing by deer in Pennsylvania increases the colonization of invasive plants (Eschtruth and Battles 2009), which alters habitats for other vertebrates. Deer populations may also be affected by top predators (e.g., coyotes [*Canis latrans*]), which may also influence mesopredator abundance (Crooks and Soulé 1999). Mesopredators typically refers to mid-level mammalian predators such as foxes (*Vulpes* spp.), raccoons (*Procyon* spp.), possums (*Trichosurus* spp.), and skunks (*Mephitis* spp.). Mesopredators are efficient nest predators (Schmidt 2003) and respond positively to forest fragmentation (Crooks and Soulé 1999, Sinclair et al. 2005).

### *Invasive Species*

Invasive species have reshaped eastern North American forests from the canopy to the understory. In uplands, the American chestnut (*Castenea dentata*) has been rendered nearly extinct by the fungus *Cryphonectria parasitica*; in lowlands, the American elm (*Ulmus americana*) has been highly reduced by fungi (*Ophiostoma* spp.); and in highlands, the hemlock wooly adelgid (*Adelges tsugae*) has been decimating Pennsylvania's state tree, the eastern hemlock (*Tsuga canadensis*). Canopies in many areas are now covered in tree of heaven (*Ailanthus altissima*), Norway maples (*Acer platanoides*), and others. In the understory, plants such as garlic mustard (*Alliaria petiolata*) and Japanese knotweed (*Polygonum cuspidatum*) carpet large areas once occupied by native herbs. Such radical changes in structure and function are bound to affect vertebrate populations and communities. Devastating introductions can also take the form of epizootic diseases. The introduction of the West Nile virus, for example, has resulted in population declines of many birds (Kilpatrick et al. 2007, LaDeau et al. 2007) and mammals (Marra et al. 2004).

Czech and Krausman (1997) ranked causes of endangerment for plants and animals of the United States. They found interactions with introduced species ranked highest when Hawaii and Puerto Rico were included but ranked eighth when analysis was restricted to the mainland. This analysis is over a decade old, and since its publication, the United States has experienced the expansion of West Nile virus, the rapidly growing population of snakehead fish (*Channa argus*; Odenkirk and Owens 2007), and the introduction of earthworms (*Lumbricus* spp.) that alter understory vegetation (Bohlen et al. 2004) and many other introduced species. Cats have been present for decades, but as the landscape becomes increasingly (sub)urbanized, so does the spread of feral cats into the landscape (Lepzck et al. 2004b, Sims et al. 2009).

### *Climate Change*

Our species is pumping climate-altering chemicals into the atmosphere, including cooling aerosols and greenhouses gases such as methane and $CO_2$ that has been locked away for millions of years in the form of coal and oil. Given trajectories of climate change, based on greenhouse gas emissions, Pennsylvania will have warmer temperatures with longer droughts punctuated with larger precipitation events. This will change the plant species that live in the state. The vertebrates, of course, are likely to follow, but climate change also changes the types of species that are able to invade and establish in the state, including parasites and other pathogens. Species that are most susceptible appear to be species with narrow ecological niches or small geographic distributions (Thomas et al. 2004, Schwartz et al. 2006, Sekercioglu et al. 2008).

In the Northern Hemisphere, birds have already shown significant northward shifts in distribution (Hitch and Leberg 2007). In addition, climate change has already altered reproductive phenology in amphibians (Blaustein et al. 2001) and invertebrate prey (Jones et al. 2003). The negative effects of global warming will be compounded by other anthropogenic issues such as habitat fragmentation, invasive species, and introduced diseases. Species that are able to alter their distributions may be able to elude extinction, but less vagile species (i.e., those with reduced dispersal ability) may not be able to alter their distribution fast enough to track climate-induced habitat changes.

### *Synergies and Cascading Trophic Interactions*

It would be rare for any of the causes listed to work in isolation. For example, invasive plant species are often associated with disturbances that can come from edge effects accompanying forest fragmentation. One important lesson emerging from the science of ecology is that so many species interact, creating dependencies and feedbacks. Biodiversity, as its own entity, is a community parameter that influences community dynamics such as food webs and the abundance of organisms within trophic levels (Duffy et al. 2007, Bruno and Cardinale 2008).

Extinctions and changes in populations, therefore, do not occur in isolation. Instead, they potentially affect many other species. In trophic cascades, top predators affect the abundance of intermediate trophic levels and, in turn, affect the abundance of organisms lowest in the food chain. Predators, because of their rarity, are generally more prone to extinction than other organisms. However, the loss of predators can strongly influence ecosystems—often negatively. For example, the loss of top predators (e.g., coyotes) may result in the increase of smaller predators that are nest predators. This "mesopredator" release may in turn affect millions of birds (Robinson et al. 1995). Mitigating the loss of top predator may buffer against significant effects of these mesopredators (Finke and Denno 2004).

## The Future of Conservation

### *The Role of New Technologies in Conservation Biology*

There is a strong link between new technologies and new scientific knowledge. New technologies give us new tools to evaluate hypotheses that were previously untestable. Important technologies for conservation biology tend to expand the ends of the scales of our knowledge just as the microscope and the telescope did for cell biology and astronomy.

Advances in DNA sequencing have resulted in many new research opportunities. Genetic markers are DNA or RNA sequences that can be used to identify the breeding locations of migrant birds on their wintering grounds outside of North America (Clegg et al. 2003). Measures of heterozygocity reveal that inbreeding contributes to extinction of Rare Species (Spielman et al. 2004). DNA barcoding has been used to identify illegally traded species, morphospecies, and the presence of species from parts of organisms. DNA barcoding holds great promise for assessing biodiversity of elusive animals by screening soil and water for key DNA fragments (Valentini et al., in press).

A salient aspect of population ecology is understanding what habitats organisms use and the fitness and physiological consequences of habitat selection. On one end of the spectrum, satellite tracking (Berthold et al. 2001, Burger and Shaffer 2008) and monitoring technology (Bowlin et al. 2005) has miniaturized to the point that we can monitor the heart rate and body temperature of migrant songbirds (Cohn 1999, Wikelski et al. 2007). Cell phone technology is being developed to follow individual migrants from their wintering grounds to breeding grounds, allowing us to identify important wintering, breeding, and stopover areas. On a larger scale, radar can be used to identify important stopover habitats for large numbers of birds (Gauthreaux and Belser 2003, Gauthreaux et al. 2003, Schmaljohann et al. 2008).

These types of data are inherently spatial, and improving geographical information systems increases the precision and accuracy of identifying important areas for conservation (Naidoo et al. 2008). Computational advances, including the software (e.g., R, winBUGS) and statistical techniques (e.g., Bayesian statistics, mark-recapture), also increase our knowledge of ecological systems (Martin et al. 2007). Getting new data and knowledge to users and developers has blossomed with the growth of the Internet. Online databases such as NatureServe (www.natureserve.org/) and the Encyclopedia of Life (www.eol.org/index) allow rapid communication of species information as do online journals.

### *What about Natural History?*

A problem that plagued Charles Darwin was understanding and incorporating the factors that regulate populations and the causes of rarity into developing the hypothesis of evolution by natural selection. Darwin (1859) writes, "The causes which check the natural tendency of each species to increase are most obscure. Look at the most vigorous species; by as much as it swarms in numbers, by so much will it tend to increase still further. We know not exactly what the checks are even in a single instance." The toolbox of the modern population ecologist now includes many tools that were inaccessible to Darwin (e.g., PIT tags, satellite tracking), and population ecology is currently a flourishing science. An important advance for estimating populations is the ever-increasing sophistication of statistical models and the computational power of computers. However, computer models require data to estimate parameters such as population growth and population responses to external factors. Population models, particularly those that incorporate different management scenarios, require fecundity and survivorship estimates, and these can only be obtained from field studies that often require years of netting, trapping, counting, marking, and following animals (Beissinger et al. 2006).

At the lowest levels of biological organization, organisms must obey the laws of thermodynamics. At higher levels such as ecological interactions, there are few universals (Lawton 1999). Why should this be so? Evolution has produced an incredible array of adaptations in millions of species. Natural history is simply the science that studies this diversity (Futuyma 1998).

Natural history cannot be looked at as merely a hobby for amateurs or the pursuit of emeritus professors. Natural history is an essential cog in the wheel that drives conservation biology, ecology, and evolution forward. Therefore, conservation biology requires a steady recruitment of fresh, inquisitive minds that have that rare combination of seeking work under inhospitable conditions (e.g., heat, cold, mosquitoes, ticks) and understanding the rigors of science. Such minds can be fostered by bringing children into contact with the land and its inhabitants. Missing these opportunities will eventually disconnect humans from an emotional link with nature to the peril of many species, including our own.

2

# Wildlife Habitat

## *The Key to Abundance, Distribution, and Diversity*

Margaret C. Brittingham
Laurie J. Goodrich

### Introduction

Where animals are found and how abundant they are result from a number of complex factors, including their evolutionary history, physiology, morphology, behavior, and ecology—as well as the physical features of their habitat. In this chapter, we review the diversity of habitats, geographic location, and geography of Pennsylvania and discuss how these factors influence the distribution of terrestrial vertebrates, why some species are common and others are rare, and why Pennsylvania plays such an important role in the conservation of many of these species. We examine broad trends and relationships. A detailed analysis of terrestrial and aquatic habitats available to vertebrates can be found in Goodrich et al. (2002) and a summary of the physiographic provinces can be found in McWilliams and Brauning (2000) and Merritt (1987). We will not repeat these here but instead highlight the characteristics of the habitats, the physiographic provinces, and the geographic location of Pennsylvania that influence the distribution and abundance of species (fig. 2.1). We focus our examples on the terrestrial vertebrates of Conservation Concern. Detailed information on these species and their habitat requirements can be found in the individual species accounts, which follow the introductory chapters.

### Habitats for Terrestrial Vertebrates

Much of the information in the following sections was originally included in an interagency report on the status of wildlife habitat (Goodrich et al. 2002) and in a report on Pennsylvania's comprehensive wildlife conservation strategy (Williams 2007). Our goal here is to further refine and focus this information and, in so doing, provide a context for understanding the diversity and distribution of terrestrial vertebrates in Pennsylvania and neighboring states.

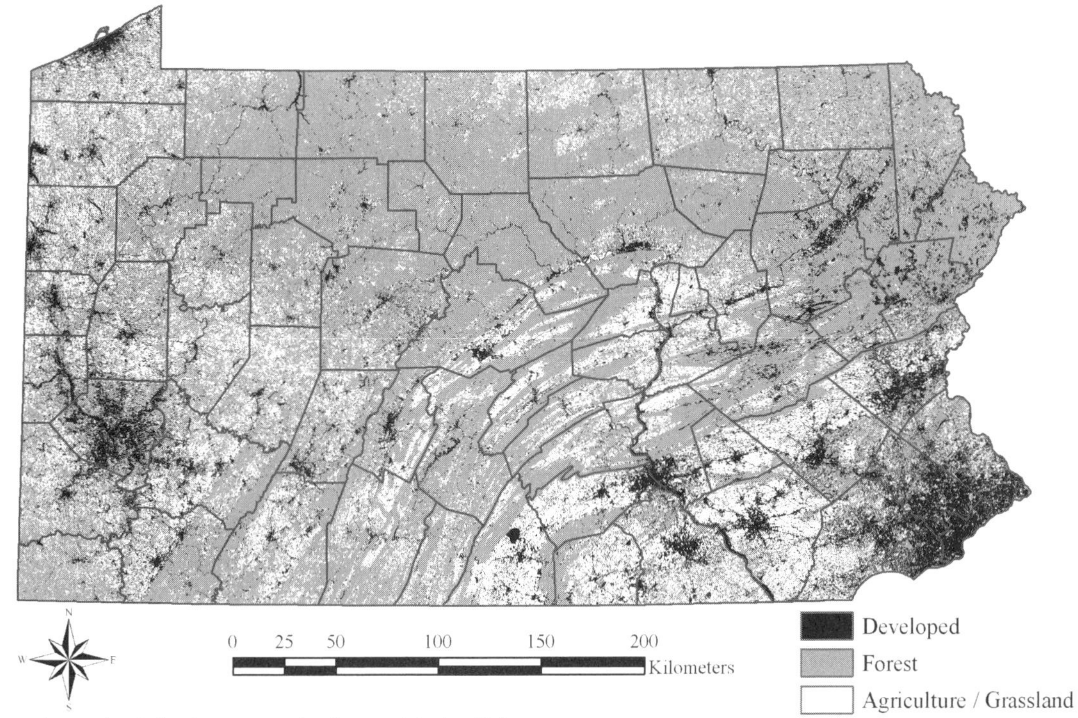


*Fig. 2.1.* Pennsylvania land cover. 1995 land cover data, Joe Bishop 2009.

## Forest Habitat

Forest habitat is the predominate habitat in Pennsylvania with more than 62 percent of the state in forest cover. Forest type and species composition vary with elevation, moisture, slope, and aspect. The varied terrain of Pennsylvania provides a high diversity of species and forest types (Fredrickson 1996, Fike 1999). Deciduous forest represents the vast majority of forest cover types found in Pennsylvania and comprises several natural community types, with oak-hickory and northern hardwoods making up most of Pennsylvania's upland forests (Alerich 1993).

Pennsylvania's extensive forests fill an important role in the conservation of forest wildlife within the northeastern United States (Rosenberg and Wells 1995). Many species that depend on large contiguous blocks of forest are still relatively abundant in Pennsylvania, although they may have declined in number. In other northeastern states with less forest cover, declines have been more extreme, and these species are much less abundant. These include many of the forest-dwelling Neotropical migrants, such as the wood thrush (*Hylocichla mustelina*) and scarlet tanager (*Piranga olivacea*), both of which are considered Pennsylvania Responsibility Species, because such a large proportion of the population breeds in Pennsylvania (Rosenberg and Wells 1995, Williams 2007).

A species that relies on large forested blocks for successful breeding and reaches its greatest abundance in large forested blocks is called a "forest-interior," or "area-sensitive," species. The diversity of area-sensitive birds is highest in the north-central and northeastern regions of the state where large blocks of forest occur and lowest in the southeastern and southwestern parts of the state where forest tends to be found in small woodlots and isolated patches; these patches result from habitat loss and forest fragmentation due to human development and agriculture (Goodrich et al. 2002, Brittingham and Goodrich 2010).

### *Coniferous and Mixed Forest*

Coniferous or mixed conifer-deciduous cover comprises 8.4 percent of Pennsylvania's land cover (Goodrich et al. 2002). Mixed forests include forests dominated by sugar maple (*Acer saccharum*), beech (*Fagus* spp.), birch (*Betula* spp.), spruce-fir (*Picea* spp./*Abies* spp.), and white pine-jack pine (*Pinus strobes*/*Pinus banksiana*). This cover type is found throughout the state with concentrations at high elevations and in the Pocono Plateau and in north-central regions of the state (Goodrich et al. 2002). The state is also dotted with plantations of Norway (*Picea abies*) and white spruce (*Picea glauca*), as well as red (*Pinus resinosa*), white, and scotch Pines (*Pinus sylvestris*) that were planted in the early and mid-twentieth century (Goodrich et al. 2002).

Although conifer forests make up a relatively small percentage of the total forest cover in the commonwealth, they are important for wildlife. Coniferous forests provide important habitat for breeding and wintering wildlife and add diversity to forest habitats. The northern flying squirrel (*Glaucomys sabrinus*) and red crossbill (*Loxia curvirostra*) are Species of Greatest Conservation Need whose abundance and distribution are closely tied to the abundance and distribution of coniferous forests across the state. Some birds, such as the black-throated green warbler (*Dendroica virens*) and blue-headed vireo (*Vireo solitarius*), preferentially nest in conifers (see species accounts). The distribution of fifteen birds that nest in association with conifers is concentrated in the northern regions of the state where the abundance of conifer cover is the greatest (Goodrich et al. 2002).

Eastern hemlock (*Tsuga canadensis*) is an important component of many conifer forest habitats, as well as a major component of remaining old growth forests (Davis 1996). It is a slow-growing, shade-tolerant late-successional conifer that provides a unique cover type (Orwig and Foster 1998). Several Species of Greatest Conservation Need, including the Acadian flycatcher (*Empidonax virescens*), blue-headed vireo, black-throated green warbler, and Blackburnian warbler (*Dendroica fusca*), depend on, or strongly prefer, hemlock habitats and are often found in highest densities in hemlock-associated habitat types (Benzinger 1994, Swartzentruber and Master 2003, Ross et al. 2004, Becker et al. 2008 ). In addition, the northern flying squirrel and water shrew (*Sorex palustris*), species of concern throughout the Northeast and mid-Atlantic, are closely associated with hemlock stands (Sciascia and Pehek 1995, Steele et al. 2004).

### *Old Growth Forests*

Strict age and size characteristics to define which forests are "old growth" are hard to establish because it varies with a number of factors, including forest type and soil conditions (Davis 1993). In general, old growth forests have been managed by nature without human interference for a prolonged period, have not suffered recent natural catastrophic disturbance that would reverse succession, and possess characteristics associated with forests that are usually classified as virgin (Davis 1993, Gross 1993). Old growth forests are often typified by large trees of advanced age, downed logs, an abundance of standing snags, and dense leaf litter. Today in Pennsylvania, old growth forests occur on less than 1 percent of the forest land and generally occur in small patches of less than 500 acres (195 ha) (Davis 1993, Goodrich et al. 2002).

Although there are no vertebrate species in Pennsylvania that are found only within old growth forests, many Species of Greatest Conservation Need are often closely associated with old growth or reach their highest abundance within old growth. This is either because of the structural features of the habitat or the abundance of conifer cones and other mast (Haney and Schaadt 1996). Examples of Species of Greatest Conservation Need that are often associated with old growth include birds such as the Blackburnian warbler, the winter wren (*Troglodytes troglodytes*), and the red crossbill. Northern flying squirrels are found in greater numbers in mature mixed-coniferous forest (Haney 1999, Mahan et al. 1999). The federally endangered Indiana bat (*Myotis sodalis*) uses large mature trees with loose bark or snags for roosting (Haney 1999), although recent studies show they will also use smaller trees for roosting (U.S. Fish and Wildlife Service 2007). Salamander populations increase in diversity and in abundance in mature forests, where the abundant woody debris provides the moist microhabitat they require (Haney 1999, Haney and Lydic 1999).

## Successional Habitats

Successional habitats in Pennsylvania may occur either as temporal or near-permanent habitat patches. Temporal thicket patches result primarily from farmland abandonment, reclamation and succession of reclaimed strip mines, forest clear-cutting, natural forest disturbances, and maintenance of shrub-wetlands by beaver activity.

Regenerating clear-cuts and reclaimed surface mines provide habitats that mimic natural shrub communities in structure and may be important to native thicket-associated species. These habitats are usually ephemeral, lasting generally five to ten years after disturbance, and they occur in a wider variety of forest types and at lower elevations. In addition to providing habitat for native priority species, these areas support additional early successional species that formerly may have been rare in this forested state. Depending on how they are managed, significant amounts of habitat for shrub-scrub species can also occur along power lines (Litvaitis et al. 1999). Near-permanent thicket patches occur in a few limited geographic areas as naturally occurring barrens. Barrens are discussed below under "Open Habitats."

Early successional forest or early successional habitat such as overgrown farmsteads, abandoned orchards, regenerating forests, and floodplain areas, are not easily quantified in current inventory and mapping methods. As a result, we do not have an accurate estimate for the total area of successional habitats. However, there is general agreement that this habitat type is currently in decline in Pennsylvania and the northeast United States (Askins 2000). From 1978 to 2002, the total acreage in Pennsylvania forestland remained stable, but the proportion in early successional stages (seedling, sapling, and nonstocked) declined from 21 percent to 12 percent (Alerich 1993, McWilliams et al. 1995). Although the aging of trees is a primary factor in the decline in successional habitats, factors such as highway and urban development, intensification of agriculture, and a reduction in farm abandonment all contributed to decreasing quantity and quality of this important habitat.

As the abundance of successional habitats has declined, a number of species associated with old fields and young forests have declined in numbers. Species of Greatest Conservation Need associated with young forests and old field habitat include birds such as the American woodcock (*Scolopax minor*), brown thrasher (*Toxostoma rufum*), golden-winged warbler (*Vermivora chrysoptera*), and yellow-breasted chat (*Icteria virens*), and mammals such as the least shrew (*Cryptotis parva*). Mammals such as the snowshoe hare (*Lepus americanus*) are associated with areas of dense vegetation particularly regenerating clear-cuts in the northern part of the state. A number of reptile species are often found in early successional habitats, including mountain earth snakes (*Virginia pulchra*), smooth green snakes (*Liochlorophis vernalis*), eastern fence lizards (*Sceloporus undulatus*), and northern coal skinks (*Plestiodon anthracinus anthracinus*).

## Farmlands, Grasslands, and Open Habitats

### *Farmlands*

Approximately 25 percent of Pennsylvania is in open herbaceous habitat with the majority of that maintained as farmland (Myers et al. 2000). Historically, Pennsylvania was a forested state, but as Pennsylvania was settled, much of the land was cleared for farming. The total amount of farmland peaked around 1900 with more than 65 percent of Pennsylvania in farmland (McWilliams and Brauning 2000). Since that time, there has been a steady decline in the amount of land devoted to farming and the number of farms. Both cropland and pastureland acreage have been reduced since the 1940s (Goodrich et al. 2002). The steepest decline in acreage occurred from the 1950s through late 1960s, a time when declines in farmland wildlife also were occurring. The loss of farmland habitat continues today, especially in southern Pennsylvania, where overall losses of agricultural land were estimated at as high as 37 percent in areas around Philadelphia between 1969 and 1992 (Goodrich et al. 2002).

Another change in recent decades has been a shift toward larger farms under intense mechanized production. Farmland that is harvested by frequent hay-mowing or crop-cutting leaves little opportunity for wildlife to find food and cover. The increased use of highly effective pesticides has reduced critical insect prey for grassland-associated wildlife. Because of the demands of modern equipment and economics, fewer brushy hedgerows and "odd areas" remain in today's farmland.

All of this has had adverse effects on farmland wildlife species, many of which thrived in a mix of open habitat, abandoned fields, hedgerows, and woods. For the eastern spotted skunk (*Spilogale putorius*), small farms and reverting farmland seem beneficial, but large-scale agriculture is associated with population declines (Polder 1968, Choate et al. 1974). Never common, this species may now be on the verge of extirpation within the state. Avian Species of Greatest Conservation Need associated with farmland include birds such as barn owls (*Tyto alba*), eastern meadowlarks (*Sturnella magna*), and grasshopper sparrows (*Ammodramus savannarum*). The northern bobwhite (*Colinus virginianus*) is a species that was closely tied to farmland and is now probably extirpated as a native breeder (see species accounts). In southeastern regions, only a small proportion of the area remains in agricultural habitat, and most of that is in small patches of less than 100 acres. Wildlife that require extensive grasslands (e.g., northern harrier [*Circus cyaneus*], short-eared owl [*Asio flammeus*]) are now limited by the availability of such habitats throughout this region.

### *Strip Mines*

Reclaimed surface mines provide extensive nonagricultural grassland habitat in Pennsylvania. A conservative estimate is that there are currently around 35,000 ha of reclaimed strip mines within a nine-county area in western Pennsylvania suitable as grassland bird habitat (Mattice et al. 2005). Statewide, the estimate would be much higher. Though once considered wastelands resulting from resource extraction, these sites can provide quality habitat for grassland-associated species. The acidic, nutrient-poor soils of reclaimed sites provide little potential for agricultural or timber production, and grasses and legumes tend to be the most successful and persistent vegetation types (Vogel 1981). These fields have a slow rate of ecological plant succession and provide habitat for grasshopper sparrows as well as many other grassland-associated birds (Bajema et al. 2001). Their suitability for grassland-associated species from other taxa (i.e., herptiles, invertebrates) has yet to be fully investigated. These anthropogenic grasslands are found primarily in the north-central and northwestern parts of the state where there are more than 300 surface mines (Yahner and Rohrbaugh 1996).

Grassland-dependent species associated with reclaimed surface mines include many of the same species associated with agricultural grasslands. However, reclaimed strip mines are somewhat unique in that they can provide large-scale grassland habitats for area-sensitive species. In western Pennsylvania, and particularly in Clarion County, northern harriers and short-eared owls, two species that require large blocks of grassland habitat, are using these extensive reclaimed strip mines.

Because of the extent of strip mines in Pennsylvania and declines in grassland species elsewhere, these habitats are regionally and even globally important for some grassland species. For example, nearly 9 percent of the global population of Henslow's sparrows (*Ammodramus henslowii*) breed in Pennsylvania, and the majority are breeding on reclaimed strip mines (see species account in chapter 5). Thus, Pennsylvania has a High Responsibility for this species.

### *Barrens*

Barrens habitats are naturally occurring open habitats that are often associated with dry or nutrient-poor soils (Goodrich et al. 2002). The vegetation tends to be a patchy mosaic ranging from woodland to shrubland to sparsely vegetated rocks interspersed with patches of grassy cover. Major types of barrens in Pennsylvania include shale barrens, eastern serpentine barrens, sand barrens, pitchpine scrub oak barrens, ridge-top scrub oak barrens, and mesic till barrens (Thorne et al. 1995, Fike 1999, Goodrich et al. 2002, Latham 2003). Depending on geographic location and soils, barrens habitat can provide suitable habitat for a number of priority species, including the whip-poor-will (*Caprimulgus vociferous*), prairie warbler (*Dendroica discolor*), golden-winged warbler, Appalachian cottontail (*Sylvilagus obscurus*), snowshoe hare, and timber rattlesnake (*Crotalus horridus*). These habitats are also home to a number of rare plant species (Latham 2003).

## Wetlands

Wetlands cover less than 2 percent of the land area in Pennsylvania (Tiner 1990). Natural wetlands are concentrated in previously glaciated counties of the northeastern and northwestern portions of the state, with more than 50 percent of the wetlands in the state occurring in these areas (Tiner 1990). The wetlands not only are most abundant here but also have their largest areal extent here. As a consequence, many of the Species of Greatest Conservation Need associated with wetland habitats are most likely to be found in these areas.

Most of Pennsylvania's wetlands (97%) are palustrine (bogs, fens, swamps, shallow pools). Emergent wetlands (marshes, meadows) and shrub swamps compose 10 to 20 percent of state wetlands. Large emergent wetlands, or undisturbed areas of small emergent wetlands mixed with fields, are needed to support many Species of Greatest Conservation Need, such as the American bittern (*Botaurus lentiginosus*), a State Endangered Species and Species of Regional Concern. This preference for large, undisturbed wetlands has made many species vulnerable to population losses, and has likely resulted in the disappearance of species from the commonwealth. The bog turtle (*Glyptemys muhlenbergii*) is a federally Threatened Species associated with emergent wetlands in the southeastern region, where wetland loss has reduced wetland wildlife abundance. Species of Greatest Conservation Need associated with shrub-scrub wetlands include birds like the olive-sided flycatcher (*Contopus cooperi*) and alder flycatcher (*Empidonax alnorum*). Chorus (*Pseudacris* spp.), leopard (*Lithobates* spp.), and northern cricket frogs (*Acris crepitans*), spotted (*Clemmys guttata*) and Blanding's turtles (*Emydoidea blandingii*), and eastern ribbon snakes (*Thamnophis sauritus*), all depend on a variety of wetland habitats. Wet meadows are important habitats for the eastern massasauga (*Sistrurus catenatus catenatus*), shorthead garter snake (*Thamnophis brachystoma*), and Kirtland's snake (*Clonophis kirtlandii*).

Thirty-six percent of Pennsylvania's wetlands are forested, and these wetlands are often the result of beaver activity. Peat bogs with floating vegetation mats are found in glaciated forested regions of northern Pennsylvania. Bogs are characterized by slow circulation of water and low rates of nutrient turnover. The yellow-bellied flycatcher (*Empidonax flaviventris*) is associated with glaciated bogs and swamps of northern Pennsylvania. Other Species of Greatest Conservation Need found here include the Canada warbler (*Wilsonia canadensis*), winter wren, American woodcock, and northern saw-whet owl (*Aegolius acadicus*).

Freshwater tidal wetlands occur in southeastern Pennsylvania. Brackish water reaches into Pennsylvania from the Delaware Bay at the far southeastern corner, allowing a small area of estuarine environment (Tiner 1990). Historically, species associated with this habitat include the least tern (*Sternula antillarum*) and black rail (*Laterallus jamaicensis*). Species of Greatest Conservation Need within this habitat include the marsh wren (*Cistothorus palustris*), common moorhen (*Gallinula chloropus*), and the eastern redbelly turtle (*Pseudemys rubriventris*).

## Streams, Rivers, and Lakes

Pennsylvania is host to 83,000 miles (133,575 km) of streams and rivers, which can be divided into eight primary drainage basins. We tend to think of streams as important habitats for fish, mussels, and aquatic invertebrates, but there are also a number of species we typically consider terrestrial vertebrates that are associated with streams. Species of Greatest Conservation Need associated with streams tend to fall into two groups. The first group is species that are associated with streams with high water quality, often headwater streams surrounded by forest cover. These species presumably declined in numbers or are limited at least in part because of poor water quality found in many streams. For example, the Louisiana waterthrush (*Seiurus motacilla*) is the only stream-dependent songbird in eastern North America. It feeds primarily on aquatic macroinvertebrates, which decline in number in acidified streams. Likewise, breeding waterthrush also decline in number on acidified streams (Mulvihill et al. 2008). Waterthrush breeding densities are lower on acidified streams and it takes almost double the length of an acidified stream to produce the same number of fledglings as a nonacidified stream (Mulvihill et al. 2008). Water shrews and river otters (*Lontra canadensis*) are mammalian Species of Greatest Conservation Need associated with high-quality streams. Because of their vagility, river otters can move from one watershed to another with little effort. Species such as water shrews, however, are likely to experience more permanent local extinctions when water quality declines. Amphibians and reptiles that use high-quality stream habitats include eastern hellbenders (*Cryptobranchus alleganiensis alleganiensis*), queen snakes (*Regina septemvittata*), and map turtles (*Graptemys geographica*). Wood turtles (*Glyptemys insculpta*) seldom venture far from the riparian areas along these streams.

The second group of Species of Greatest Conservation Need is a mixed group of species that use streams for feeding and often for a place to nest, but the reasons they are rare within the state are not directly associated with the quality of the stream habitat. For example, both osprey (*Pandion haliaetus*) and bald eagles (*Haliaeetus leucocephalus*) nest along many of our larger rivers and streams and forage within them. The causes for their declines were historically associated with indirect effects of pesticides on their reproductive output and direct persecution (see species accounts). Others, such as some of our colonially nesting waders, are on the northern edge of their geographic range in Pennsylvania (e.g., great egret [*Ardea alba*] and yellow-crowned night heron [*Nyctanassa violacea*]), and are naturally rare as a result.

Lake and pond habitats can be found throughout the state but are most abundant in the northwestern and northeastern parts of the state. There are no Species of Greatest Conservation Need associated exclusively with lakes and ponds, but many of the species associated with streams and rivers will also use larger lakes.

## Vernal Ponds

Another category of aquatic habitat within the state includes the temporary pools or vernal ponds located amid riparian and woodland habitats. Vernal ponds are found throughout the state, but there is limited information on specific locations. Vernal ponds are particularly important to amphibian populations as they provide ephemeral breeding sites that are free of predatory fish or other predators. Vernal ponds dry up in summer and only contain water during wetter months of the year. As a result of this periodic drying, wildlife needing water year-round are not able to survive. The dearth of fishes reduces predation on vulnerable amphibian eggs and young. Species of Greatest Conservation Need dependent on vernal ponds include a number of salamanders and frogs such as the Jefferson salamander (*Ambystoma jeffersonianum*), marbled salamander (*Ambystoma opacum*), the mountain chorus frog (*Pseudacris brachyphona*), and the eastern spadefoot (*Scaphiopus holbrookii*). Spotted turtles (*Clemmys guttata*) also frequently use vernal ponds.

## Urban and Suburban Habitats

In Pennsylvania, the largest concentrations of urban and suburban habitat are in the southeast and southwest. In the Ridge and Valley region, development occurs primarily in the valleys, frequently in areas previously devoted to agriculture. As an area becomes urbanized, we see shifts in the type of wildlife present. Urban and suburban areas tend to favor generalists over species that have very narrow habitat requirements. As a general rule, the diversity of wildlife is low in urban areas, but the abundance of wildlife may be very high. Species that can coexist with people often thrive in urban areas, and urban areas tend to have a much higher concentration of nonnative species (ex. pigeon [*Columba* spp.], house sparrow [*Passer domesticus*]), than rural areas. Amphibians and reptiles decline in both abundance and in diversity as native habitat is lost, and the remaining habitat is fragmented by roads and buildings that create major barriers to dispersal.

The amount of urban and suburban habitat in the commonwealth is increasing, but not all urban species are thriving. Some species that became dependent on human structures for nest sites are now declining as changes in building design and management make these structures no longer suitable for roost or nest sites. Chimney swifts (*Chaetura pelagic*), common nighthawks (*Chordeiles minor*), and Indiana bats are examples of Species of Greatest Conservation Need whose declines are attributed, at least in part, to a decline in availability of suitable roost or nest sites. However, the peregrine falcon (*Falco peregrinus*) is an urban success story as the species that formerly nested on cliffs adapted to nesting on buildings and bridges while exploiting the abundance of avian prey (primarily pigeons) in urban cities. Several reptile species, including shorthead garter snakes, find suitable habitat in vacant lots in urban areas.

Although we have not traditionally focused on urban and suburban areas as critical wildlife habitats, there is a growing awareness that if we are going to slow the accelerating loss of native species, we can no longer ignore these areas. Instead, we need to manage suburban and urban areas to provide habitat for the diversity of native plants and animals that occurred here historically and still depend on them for survival and reproduction (Tallamy 2007).

## Rock Habitats and Caves

The most essential feature of rocky habitat for wildlife is surface rock in the form of cliffs, ledges, outcrops, boulder fields, and caves. In Pennsylvania, the most common rock habitat types are caves and talus slopes (Goodrich et al. 2002). Man-made rock habitats, such as active and inactive mines and quarries, also provide habitat for wildlife.

There are four different cave types in Pennsylvania: (1) terrestrial solution caves, (2) aquatic solution caves, (3) tectonic caves, and (4) talus caves (Thorne et al. 1995). Solution caves occur in limestone bedrock and are the most common cave in the state, particularly in the Ridge and Valley Physiographic Region. A diverse invertebrate community can be found here, as well as several Species of Greatest Conservation Need, including the federally endangered Indiana bat, eastern small-footed myotis (*Myotis leibi*), and Allegheny woodrat (*Neotoma magister*; Goodrich et al. 2002).

Tectonic caves are formed by subsurface cracks in bedrock and may be associated with sandstone. They are usually dry and are also used by bats and woodrats. Talus caves are formed in boulder piles where openings occur between rocks. Many reptiles and small- to mid-sized mammals may use these for cover. Priority species, such as the Indiana bat, eastern small-footed bat, Allegheny woodrat, timber rattlesnake, and the northern copperhead (*Agkistrodon contortrix mokasen*), may be found using talus caves and talus slopes. Man-made caves, such as deep coal mine shafts or abandoned tunnels, also are inhabited by rock-associated wildlife and can be particularly important for hibernating bats.

Talus slopes, boulder- and rock-strewn regions of mountains or mountainsides, and ravines provide critical habitat for several priority species, including the rock shrew (*Sorex dispar*), rock vole (*Microtus chrotorrhinus*), Allegheny woodrat, eastern spotted skunk, timber rattlesnake, northern copperhead, eastern fence lizard, and northern coal skink. The complete distribution of talus slope habitat in the state is unknown at this time. However, the habitat type is closely associated with mountain ridges of the northern Ridge and Valley Physiographic Region and the southwest (Allegheny Mountains) portion of the Northern Plateau.

## Patterns and Trends

### *Generalists and Specialists*

Habitat can be thought of as the place where an animal lives, where it forages, where it is best able to escape predators, and where it reproduces. The habitat for a chickadee (*Poecile* spp.) might be a patch of deciduous woods while the habitat for a marsh wren would be an emergent wetland. Some species are generalists and can be found in a range of habitats. Others are specialists and will be found in a very narrow range of habitat types. As a general rule, all other things being equal, a species that is a generalist will have a broader range of ecological tolerances and, hence, be more abundant than a similar species that can be found in only a restricted range of habitats and ecological conditions. Thus, it follows—again, as a very general rule—Species of Greatest Conservation Need are often species that specialize on a narrow range of ecological and habitat conditions. An exception to this rule would be species that are locally rare because they are located on the edges of their geographic ranges (see chapter 1).

### *Habitat Abundance and Species Abundance*

For species that are habitat specialists, their abundance can be limited by the availability of the specific habitat type on which they depend. Thus, if a habitat is rare within the state, the species that specialize on that habitat type will also be rare. The best example of this as a group would be our wetland species. Wetlands are found in less than 2 percent of the state, and many are in poor condition. As a consequence, the species that depend on these wetlands are also rare. We can see this with species dependent on certain microhabitats such as birds dependent on hemlock or other conifers. Because this habitat type is rare, the species that depend on this habitat are also rare.

The converse is not always true. Just because a habitat is abundant does not mean that all species associated with it will also be abundant, because there may be other factors that have reduced population size below the carrying capacity of the habitat. For example, populations of many of our raptors were formerly reduced by pesticides that greatly reduced their reproductive output. The outcome was that those population sizes were well below the level that the habitat could support. For migrants that have both a winter and summer home, populations can be limited on the wintering grounds with the result that numbers are well below the level that could be supported by breeding habitat.

### *Geographic Location and Range Boundaries*

Wildlife communities associated with boreal forests to our north and those associated with ecosystems to our south intersect in Pennsylvania. As a consequence, in addition to species in which Pennsylvania is part of its core geographic range, Pennsylvania also hosts species that reach or approach the southern limit of their geographic range in Pennsylvania. Examples include birds like the yellow-bellied flycatcher and mammals such as the water shrew. Although the water shrew is reported from the Appalachian Mountains south of Pennsylvania, it is considered exceedingly rare, as it is reported from only three locations across West Virginia, Virginia, and North Carolina (Webster et al. 1985). But even in Pennsylvania, populations of water shrews are isolated and rare, which reflect both the distribution of the species' habitat in Pennsylvania (high-elevation montane streams) and its sensitivity to water quality.

In addition to the water shrew, there are several assemblages of birds and mammals associated with higher-elevation ridges, which have correspondingly cooler climates and more northerly vegetation, and are often associated with coniferous forest, a habitat type that is much more common in the boreal forests to our north. These high-elevation Appalachian forests extend from Pennsylvania to western North Carolina but are isolated and fragmented at higher-elevations across Pennsylvania and even more so to the south. In the case of the boreal small mammals (rodents and shrews)—some of which are considered relatively common species—both population densities and local abundances decline toward the southern terminus of the Appalachian Mountains, even where suitable habitat exists (Steele and Powell 1999). Thus, the condition and status of these Appalachian forests in Pennsylvania, as well as the species that rely on them, serve as an important indicator for the forests and species to the south.

A second group of species reaches the northern

edge of their distribution in southeastern and southwestern Pennsylvania. In this region, the winters tend to be relatively mild and the summers hot. The mountainous regions prevalent in the northern and central parts of the state do not extend into this region. The topography is generally flat with low rolling hills. Species that reach the northern edge of their geographic range here include birds such as the Kentucky warbler (*Oporornis formosus*), barn owl, summer tanager (*Piranga rubra*), and great egret; mammals such as the West Virginia water shrew and spotted skunk; and amphibians and reptiles such as the New Jersey chorus frog (*Pseudacris kalmi*), southern leopard frog (*Lithobates sphenocephalus utricularia*), rough green snakes (*Opheodrys aestivus*), and broadhead skink (*Plestiodon laticeps*).

Biologists creating state-level lists of Species of Greatest Conservation Need are posed with a dilemma when dealing with these species at the edges of their geographic range. They are generally listed as species of concern because they are rare within the state. However, they may be abundant throughout other areas of their geographic range. Thus, the question is whether these species should be listed, and how does their relative importance as a species of concern compare with other species. One argument for listing has been that range contractions are often noted at the edge of the geographic range, and thus, monitoring species in these locations can be used as an early warning for potential problems or changes (see discussion on geographic ranges in chapter 1).

### *Appalachian Mountains and Wildlife Distribution*

The Appalachian Mountains stretch across Pennsylvania as a series of parallel ridges that lie along a northeast-southwest direction and influence the distribution of wildlife in two important ways. These mountains form barriers to dispersal in some cases or conduits for geographic range expansions from the north in others. For amphibians and reptiles associated with lowland habitat, the mountains can serve as barriers to east-west dispersal (McCoy 1989). Consequently, we find a group of species that may be found to the east and west of the mountainous regions but not within them. Species of Greatest Conservation Need in this group include Fowler's toad (*Anaxyrus fowleri*), northern leopard frog (*Lithobates pipiens*), bog turtle, queen snake, and rough green snake. For others, such as the green salamander (*Aneides aeneus*) or mountain chorus frog, the mountains form an eastern barrier to dispersal, and the population is found to the west of the Allegheny Plateau. Species such as the eastern spadefoot toad (*Scaphiopus holbrookii*) are found to the east of Pennsylvania's mountainous regions. Finally, there are some species specifically adapted to mountainous habitats and restricted primarily to mountainous regions (McCoy 1989). These include reptiles such as the northern coal skink and timber rattlesnake and mammals such as the fisher (*Martes pennanti*) and northern flying squirrel.

The higher elevation of the ridge tops is associated with cooler climates and vegetation more common to the north. As a consequence, both mammals and birds that are more common to the north are able to expand their ranges southward into the commonwealth. The snowshoe hare and water shrew are examples of two mammalian Species of Greatest Conservation Need that show this pattern. These rather extensive ridge tops in Pennsylvania have historically represented a major corridor between the more extensive boreal forests to the north and the fragmented and isolated patches of high-elevation forests to the south.

The rocky ridges are generally forested, and the valleys between them are farmed or developed. The topography has played an important part in the importance of Pennsylvania to wildlife. The rocky ridges are generally not suitable for development, and, consequently, remain as relatively undisturbed large blocks of forest habitat. These provide habitats for many of Pennsylvania's forest specialists such as scarlet tanagers, broad-winged hawks (*Buteo platypterus*), and mammals such as the Allegheny woodrat and rock shrew.

The northeast-southwest ridges are also extremely important as corridors for migrating raptors and songbirds. The golden eagle (*Aquila chrysaetos*) is a Species of Greatest Conservation Need in Pennsylvania because of the high percentage of its population that migrates through the state during the spring and fall.

## Pennsylvania the Keystone State

In this chapter, we have reviewed the habitat types of Pennsylvania and their importance to Species of Greatest Conservation Need within the state and the broader region of the east. Pennsylvania is uniquely situated in the East to have both the northern limit for many species that are found primarily in the southern United States and the southern geographic range limit for many species that are much more common to our north and within the boreal forests of Canada. The northeast and southwest mountains serve as conduits for migration and dispersal for more vagile species and as barriers to dispersal for less mobile species. From a regional perspective, Pennsylvania is extremely important for supporting large populations of high-priority forest birds, while the large reclaimed strip mines provide valuable habitat for grassland species (Rosenberg and Wells 1995). Because of its location and diversity of habitats, Pennsylvania is home to an abundance of amphibian and reptile species. Pennsylvania is truly a keystone state in the East for wildlife because of its geographic location, topography, vegetation cover, and land use history.

3

# Pennsylvania's Wildlife Action Plan

Lisa Williams

## Introduction

Conservation of the nation's wildlife resources requires, above all else, a cooperative, nonadversarial approach at a regional and continental scale. This recognition led Congress to create the State Wildlife Grants program in 2001. The goal of the State Wildlife Grants program, which is appropriated annually, is to reduce the decline of fish and wildlife species at the state level before species require federal protection under the Endangered Species Act.

To receive funds under the State Wildlife Grants Program, each state in the nation was required to produce a Comprehensive Wildlife Conservation Strategy by October 2005. Now collectively referred to as Wildlife Action Plans, these plans were required to include seven elements: (1) a list of species of greatest conservation need; (2) habitat inventory; (3) threats analysis; (4) conservation actions and priorities for Species of Greatest Conservation Need; (5) monitoring and adaptive management recommendations for Species of Greatest Conservation Need; (6) plans for reviewing and revising the Wildlife Action Plans; and (7) recommendations for coordination and public outreach. This planning requirement was significant, and state agencies invested much time and personnel to accomplish the task.

Although the development of the state Wildlife Actions Plans was mandated by Congress, this effort represented more than a legislative mandate. Many states recognized this as an opportunity for state agencies to take a definitive leadership role in "nongame" conservation with the support of federal funds. Historically underfunded in their conservation, nonharvested fish and wildlife species represented the greatest conservation challenges for state agencies. Many populations were in decline. Detailed information on species' habitat and management requirements was often unavailable, and funding to implement conservation efforts focusing on nongame species was severely limited. The federal requirement to create a Wildlife Action Plan, coupled with federal funds to support development and implementation of these plans, created a historic opportunity for conservationists throughout the country.

## The Pennsylvania Wildlife Action Plan

> The people have a right to clean air, pure water, and to the preservation of the natural, scenic, historic, and aesthetic values of the environment. Pennsylvania's public natural resources are the common property of all the people, including generations yet to come. As trustees of these resources, the Commonwealth shall conserve and maintain them for the benefit of all the people.
>
> —*Article 1, Section 27 of the Pennsylvania Constitution*

As stated in Pennsylvania's constitution, the commonwealth has the responsibility to conserve and maintain its natural resources for the benefit of all citizens, including future generations. Implicit in this responsibility is the need to sustain the state's diverse fish and wildlife populations and ecosystems. The Pennsylvania Fish and Boat Commission has legal jurisdiction over fish, reptiles, amphibians, and aquatic invertebrates, while the Pennsylvania Game Commission has legal jurisdiction over wild birds and mammals. Together, these two agencies are responsible for managing all of Pennsylvania's fish and wildlife.

Funded primarily through hunting and angling license sales and fees, these two agencies have historically directed management efforts primarily at eighty-five species of game animals and sport fish. An additional twenty-seven species received targeted, albeit limited, management attention, and funding, because they were classified as state or federally Threatened or Endangered Species. Targeted management attention directed at the remaining 400 species of mammals, birds, fish, reptiles, and amphibians in the commonwealth was lacking. Although nongame species represent 75 percent of Pennsylvania's fish and wildlife, a comprehensive management program that ensured sustainable populations had yet to be developed before the creation of the State Wildlife Grants Program.

The Pennsylvania Wildlife Action Plan was an important milestone in the management of nongame wildlife in Pennsylvania. Before the Pennsylvania Wildlife Action Plan, no attempt had ever been made to analyze the conservation and management needs for the full suite of species of greatest conservation need in the commonwealth. For the first time in the histories of the Game and Fish and Boat commissions, the Pennsylvania Wildlife Action Plan presented an opportunity to develop a comprehensive strategy for managing the full spectrum of the commonwealth's nongame fish and wildlife resources.

## Development of the Pennsylvania Wildlife Action Plan

Recognizing the need for a comprehensive approach, the Pennsylvania Game Commission and Pennsylvania Fish and Boat Commission relied on technical experts to guide the Wildlife Action Plans planning process. An important partner in this process was the Pennsylvania Biological Survey. The Pennsylvania Biological Survey is a broadly based group of environmental science professionals serving on technical committees that focus on specific animal and plant taxa. Drawn from across the commonwealth, Pennsylvania Biological Survey membership comprises university researchers and professors, conservancy directors and specialists, curators at natural history museums, and biologists from state and federal agencies.

In January 2004, the Pennsylvania Game Commission and Pennsylvania Fish and Boat Commission initiated an expert-opinion process, whereby the commissions asked the various taxonomic technical committees of the Pennsylvania Biological Survey to identify species that were indicative of the diversity and health of the Commonwealth of Pennsylvania. With that request, Pennsylvania Biological Survey technical committees began developing the species lists that compose the Wildlife Action Plan—Priority Species. This technical input was invaluable: each Pennsylvania Biological Survey technical committee (Mammalogical, Ornithological, Herpetological, Fisheries) represents ten to thirty taxonomic experts, many of which have invested their entire careers with these species of concern. Technical committee findings, therefore, provided the best scientific assessment of a species' occurrence and status in the commonwealth.

Representing species in need of increased management attention at the species or habitat level, the Pennsylvania Wildlife Action Plan list includes Threatened and Endangered Species as well as other species with populations that are of concern in the commonwealth. In the case of birds, the Wildlife Action Plan included a number of relatively common species that the Ornithological Technical Committee considered important ecological indicators and, thus, critical species for monitoring, research, and management. Specifically, the commissions and their technical advisors considered several categories of species (table 3.1) and numerous sources of information (see www.wildlifeactionplans.org/pennsylvania.html) when identifying Wildlife Action Plan—Priority Species for Pennsylvania.

Beyond compiling the list of Wildlife Action Plan—Priority Species, the Pennsylvania Biological Survey technical committees developed detailed technical materials for each proposed species (i.e., identifying critical habitats, key threats, management recommendations, recovery, and monitoring recommendations). In addition, individual technical committee members, as well as a number of other colleagues, produced the individual species assessments, which were abbreviated, peer-reviewed, and standardized for presentation in this volume. In addition to the species accounts presented here, the Pennsylvania Wildlife Action Plan also included assessments of priority fish species and a statewide invertebrate assessment, the contents of which are beyond the scope of this volume.

*Table 3.1* Criteria and conservation categories considered by technical experts when identifying priority species for Pennsylvania's Wildlife Action Plan

- Globally Rare or Imperiled Species
- Federally listed Threatened and Endangered animals
- State-listed Threatened and Endangered animals
- Natural Heritage Program tracked and watchlist animal species
- Northeast Region Wildlife Species of Regional Conservation Concern
- Endemic species
- Responsibility Species (those species for which Pennsylvania supports core populations)
- Partners in Flight and All Bird Conservation priority species
- U.S. Fish and Wildlife Service's Migratory Birds of Management Concern
- Colonial waterbirds
- Forest interior breeding birds
- Shrubland successional breeding birds at risk
- Grassland breeding birds at risk
- Shorebirds with significant migratory concentrations
- Marshland breeding birds (e.g., rails, bitterns, sedge wren) at risk
- Species with small, localized "at-risk" populations
- Species with limited dispersal
- Species with fragmented or isolated populations
- Species of Special, or Conservation, Concern
- Sensitive aquatic species

## Guiding Principles of the Wildlife Action Plan in Pennsylvania

The Pennsylvania Wildlife Action Plan provided a statewide perspective on conservation by presenting recommendations at multiple levels: statewide management and planning needs, habitat-specific management needs, and species-specific information on threats, management, and monitoring needs. The Pennsylvania Wildlife Action Plan vision and purpose were based on five guiding principles that collectively resulted in the formulation of five conservation tiers used to prioritize the level of concern for each of the species reviewed in the Wildlife Action Plan (table 3.2). The principles used to develop these five conservation tiers are as follows.

1. *Conserving species at risk.* Species exhibiting warning signs today must be conserved before they become imperiled at the regional, national, or global level. Allowing species to become threatened or endangered results in the need for long-term and costly recovery efforts whose success may be questionable. Highly imperiled species require focused and immediate management attention, yet the needs of other declining species should not be overlooked. For this reason, the Wildlife Action Plan distinguishes between levels of imperilment, by denoting Immediate Concern, High-Level Concern, and Pennsylvania Vulnerable species.

2. *Keeping common species common.* Native wildlife species, both resident and migratory, must be retained in healthy numbers throughout their natural ranges to maintain their role in ecological processes. Common species may serve as important links in a food web, as indicators for habitats of interest or as suitable research or management proxies for more imperiled species. For these reasons, Wildlife Action Plan strategies and priorities strive to incorporate the needs of often-overlooked Maintenance Concern species and their associated habitats.

3. *Recognizing the unique role of Pennsylvania.* Pennsylvania straddles the border of many ecological systems and exhibits a diversity of physiographic provinces. As such, the commonwealth is home to a mixed variety of species from northern and southern climates, lowlands and uplands, grasslands and forests. Pennsylvania's ecosystems morph from marine and estuary environments in the southeast corner, through rocky mountain ridges and wide agricultural valleys in the central region, eastern deciduous forests of the central ridges to the northern forest of the Allegheny High Plateau, then end with the glaciated regions and Great Lakes in the northwest corner. Positioned at this ecological crossroads, Pennsylvania plays an important role in conserving many diverse species and habitats, both resident and migrant, both common and rare. For

*Table 3.2* Definitions for each of the five tiers of conservation for the priority species designated in Pennsylvania's Wildlife Action Plan

Conservation Tier 1: Immediate Concern
This tier comprises those species that are most at risk or are experiencing the most dramatic declines across their range. Immediate Concern species include globally Rare or Imperiled Species, nationally Rare or Imperiled Species, as well as those species in Pennsylvania or in the northeastern United States that are declining to the point of requiring federal listing in the near future.

Conservation Tier 2: High-Level Concern
This tier comprises nationally or regionally significant species that are vulnerable in Pennsylvania. High-Level Concern species include species with small, localized, and vulnerable populations, species with limited dispersal, species with fragmented or isolated populations, or species in need of additional research to determine status.

Conservation Tier 3: Responsibility Species
Responsibility Species are those species in which core populations occur in Pennsylvania or a significant proportion (>5%–10%) of their regional population occurs in Pennsylvania so that Pennsylvania has a high responsibility for conserving the species. This conservation tier includes species that may be relatively abundant or locally common and for which Pennsylvania serves as a population core, that is, a significant proportion of the species, population occurs in the commonwealth.

Conservation Tier 4: Pennsylvania Vulnerable
This tier comprises those species that are most at risk or are experiencing the most dramatic declines within the borders of the commonwealth but are not at risk at the regional, national, or global level.

Conservation Tier 5: Maintenance Concern
This conservation priority tier represents species that are fairly secure in Pennsylvania, but for which the Pennsylvania Biological Survey recommends some level of management attention. Many of these species, although still considered abundant and reasonably secure, have undergone recent declines that should be addressed. Species also were included in this tier if they serve as an indicator for high-quality habitat.

this reason, Wildlife Action Plan strategies and priorities incorporate the needs of Responsibility Concern species and their associated habitats.

4. *Voluntary partnerships for species, habitats, and people.* To be successful, a central premise of the Wildlife Action Plan is that the resources of public and private organizations throughout the commonwealth must be brought to bear on this effort. Human and capital resources must be combined, coordinated, and increased to achieve success in conserving fish and wildlife populations and their habitats. The Wildlife Action Plan's intended audience included decision makers, land managers, scientists, private landowners, and conservation organizations across the commonwealth, who collectively possess the ability to meet the Wildlife Action Plan's ambitious goals for fish and wildlife conservation. The power of conservation lies in the synergy that builds when diverse, committed partners work together toward a common goal.

5. *A comprehensive strategy.* Although the Wildlife Action Plan identified conservation priority species, single-species conservation is not generally the best approach to addressing conservation issues. That approach may be required in some cases, particularly when protecting highly endangered Species of Immediate Concern. A more practical approach for implementing conservation involves working at the habitat level, which will simultaneously benefit many species. For this reason, the Wildlife Action Plan strategies and priorities are presented at the levels of species, habitat, and species-suites so that the diverse stakeholders of the Wildlife Action Plan can find meaningful recommendations regardless of their scale and scope of interest. A comprehensive strategy is required before comprehensive results can be realized.

## Recognizing Responsibility Species

In keeping with the congressional intent of the State Wildlife Grants program, which is "endangered species prevention," special attention in the Wildlife Action Plan was given to species that are approaching the point of federal listing, are experiencing steep declines, or are being affected by widespread and pervasive threats. Attention to these species will enable resource managers to implement conservation and management measures at the state level before a species requires protection under the federal Endangered Species Act.

However, to limit the scope of the Wildlife Action Plan to rapidly declining or imperiled species would continue the inefficient pattern of "reactive" rather than "proactive" management. State Wildlife Grant funding provides resource managers with a unique opportunity to focus beyond imperiled species. That the Wildlife Action Plan is a requirement of each state also enabled staff to think about wildlife conservation across political borders. Therefore, significant effort was made in the course of the Wildlife Action Plan development to identify and emphasize the unique role and regional responsibility of Pennsylvania in conserving species of concern. An objective of highlighting Responsibility Species is to reach beyond rarity to achieve truly comprehensive and proactive management.

The overriding goal of the Wildlife Action Plan effort is to move toward proactive management of the species and habitats for which Pennsylvania has some regional, national, or global responsibility. This move from reactive to proactive management should result in increased conservation success on the ground as well as more efficient use of limited staff and funding resources.

For the purposes of the Wildlife Action Plan, Responsibility Species are defined as those species for which core populations occur in Pennsylvania or a significant proportion (>5%–10%) of the species' distribution occurs in Pennsylvania, and hence the commonwealth has a High Responsibility for conserving the species. The responsibility category includes species that may be relatively abundant or locally common and for which Pennsylvania serves as a "population core," that is, a significant proportion of the species' distribution occurs in the commonwealth.

A good example of a Responsibility Species is the shorthead garter snake (*Thamnophis brachystoma*). This relatively small *Thamnophis* species has one of the most restricted distributions of any snake species found in North America. It is endemic to the unglaciated portions of the upper Allegheny Plateau, with the bulk of its geographic range occurring in northwestern Pennsylvania and a small portion extending into southwestern New York. More than 90 percent of *T. brachystoma*'s geographic range occurs within Pennsylvania. As the primary stewards of this species, the long-term viability of *T. brachystoma* is the responsibility of Pennsylvania.

Responsibility Species in Pennsylvania include (1) those species for which Pennsylvania makes up a large portion of their global geographic range (e.g., eastern hellbender [*Cryptobranchus alleganiensis alleganiensis*]), (2) endemic species (e.g., shorthead garter snake), (3) migrating populations that rely heavily on Pennsylvania's migratory corridors (e.g., tundra swan [*Cygnus columbianus columbianus*]), (4) species that depend on large forest patches (e.g., wood thrush [*Hylocichla mustelina*]), and (5) species whose north/south gene flow would be disrupted if they were lost from Pennsylvania (e.g., snowshoe hare [*Lepus americanus*]).

Designation as a Responsibility Species implies that Pennsylvania has a high global, national, or regional responsibility for maintaining the species. Therefore, the main focus in managing these species is to ensure the continued viability of core populations, protect key habitats, and establish monitoring efforts as needed. It is anticipated that Responsibility Species, which are still currently abundant, can be protected through prudent attention to habitat management. Responsibility is distinct from status in that even a very common species with a Secure status could be a High Responsibility for Pennsylvania. The scarlet tanager (*Piranga olivacea*) and wood thrush (*Hylocichla mustelina*) are excellent examples of this. Rather than defining legal status, responsibility should be used as a key component in prioritizing conservation action. With limited funds and personnel, the typical conservation response is to focus on rarity, that is, which species are most endangered. However, prioritization of conservation actions based on each state's responsibility would likely result in more successful and cost-effective conservation over the long term.

## Comprehensive Strategies at Work

The northeastern states have several unique characteristics that aid in the conservation of fish and wildlife. Motivated stakeholders have led to the creation of multiple conservation organizations. Many private landowners retain an interest in wildlife viewing, feeding, hunting, or photography and are eager to manage their lands for wildlife. State resources are managed by multiple, strong conservation agencies that work in partnership with nongovernmental organizations.

The rolling hills, mountains, and sometimes steep valleys of the Northeast have resulted in large tracts of relatively undeveloped and, in some cases, perpetually protected public land. Several northeastern states have low population density relative to land area, and human populations are often concentrated in urban/suburban areas, leaving habitat available in more rural and remote areas. The region's mountains and streams have created large-scale connectivity for many terrestrial and aquatic species. This means that comprehensive conservation efforts spanning state lines can result in conservation benefits for multiple states. Together, these factors create unique opportunities for conservation in the Northeast.

Whereas many of the northeastern states enjoy similar opportunities for conservation, they also share many of the same threats to fish, wildlife, and habitat. The Wildlife Action Plans in the Northeast consistently identify threats that span the political boundaries separating states. Such large-scale threats as development and fragmentation of habitats, declining water quality and quantity, wildlife disease, and invasive species call for comprehensive strategies to be implemented at the regional, rather than state, level.

To this end, northeastern states have dedicated a percentage of annual State Wildlife Grants funds to support projects that fill regional conservation needs. Thus far, states have worked cooperatively to fund and implement two large-scale regional conservation needs projects: (1) a regional habitat classification system so that states across the region can identify, track, and map habitats of concern and (2) a performance monitoring framework that will enable states to track the success of their Wildlife Action Plan implementation efforts. Other regional conservation needs have been tentatively approved, including the creation of regional habitat cover maps; the establishment of a waterway connectivity assessment project; a study of impacts of invasive species on Wildlife Action Plan–Priority Species; the development of avian indicators and measures for monitoring; a regional initiative for biomass energy development for early succession species; the implementation of action plans for shrubland-dependent birds; and an investigation of the highly fatal white nose syndrome in bats. Additional projects and topics proposed for future years indicate that the regional conservation needs program is recognized as an important approach. The strength of the program is that it fosters a comprehensive multipartner approach to address large-scale multistate needs.

## The Need for Diverse Partnerships and Integrated Strategies

Although Wildlife Action Plans serve as blueprints for fish and wildlife conservation, they are not intended to replace existing conservation plans at the local, regional, or state level. In fact, complementary planning efforts will serve to assist in the implementation of statewide Wildlife Action Plans and vice versa. State and local conservation plans developed by nongovernmental partners have an important role to play in conserving the region's valuable fish and wildlife resources. The conservation and management strategies required for several hundred species are far too complex and variable to be treated in just one plan. Furthermore, implementation—that critical step in the process where a plan becomes an on-the-ground conservation action—must take place at state, county, township, and local levels. To be most effective, however, such local actions should be guided by an overall strategy: this type of statewide guidance is what the Wildlife Action Plans are designed to provide.

It is recognized that once statewide guidance is adopted it will be up to regional and local conservation partners to identify priority species and habitats that fall within their jurisdiction, set goals and objectives for their organization's involvement, identify local issues and opportunities, and develop strategies for implementing local conservation actions. It is also evident that as a consequence of working at different scales and in different contexts the plans of various partners may not always fully agree with that of the Wildlife Action Plans. Thus, it will be important over the next few years to resolve differences and adjust and refocus conservation objectives on a continuing basis as more information is acquired.

## The Need for Sustainable Wildlife Funding

The most significant impediment to implementing the conservation and management recommendations outlined in this volume is insufficient funding for conservation. Yet investing funds now to protect or restore wildlife populations is far more effective than waiting until populations reach critically low levels and need expensive "emergency room care" through the Endangered Species Act. Continued and adequate funding to support the implementation of state Wildlife Action Plans is crucial to achieve the goal of preventing wildlife from becoming endangered.

## Wildlife Action Plans in the Northeast

To access Wildlife Action Plan summaries and links to state Wildlife Action Plan Web sites, visit www.teaming.org.

# 4

# The Amphibians and Reptiles

## *Introduction*

Timothy J. Maret

Pennsylvania is home to seventy-seven species of amphibians and reptiles. One species of turtle (*Trachemys scripta* [the slider]) is introduced; the other seventy-six species are considered to be native to the commonwealth. Two additional species, the tiger salamander (*Ambystoma tigrinum*) and the smooth softshell turtle (*Apalone mutica*), have been extirpated from the state. Of the extant species, thirty-eight are amphibians (twenty-two salamanders and sixteen anurans) and thirty-nine are reptiles (four lizards, twenty-one snakes, and fourteen turtles). The diversity of species in Pennsylvania is comparable to that of surrounding states. West Virginia and Maryland each have eighty-eight species (not counting sea turtles that visit Maryland's shore), New Jersey has seventy-one, New York has sixty-nine, and Ohio has eighty-four. Almost half of the species in Pennsylvania (thirty-six) are considered to be of Conservation Concern and are formerly registered in the state's Wildlife Action Plan. This group includes five salamander species, nine anurans, three lizards, twelve snakes, and seven turtles. State endangered status has been granted to four reptiles and four amphibians, while one amphibian and one reptile are listed as state Threatened. Only one species, the bog turtle (*Glyptemys muhlenbergii*), is federally protected; its federal status is Threatened. The diversity of amphibians and reptiles is influenced by the high diversity of landforms, climates, and habitats found within Pennsylvania. Elevations range from sea level in the southeast corner of the state to 979 m at the top of Mount Davis in the Allegheny Mountains region of southwestern Pennsylvania. This range of elevations and climates allows for a diversity of species, including those typically associated with both northern (e.g., wood frogs [*Lithobates sylvaticus*]) and southern (e.g., broadhead skinks [*Plestiodon laticeps*]) regions. The tremendous diversity of terrestrial and aquatic environments within the state also provides a rich variety of habitats for amphibian and reptile species.

## Threats to Amphibians and Reptiles of Pennsylvania

The amphibians and reptiles of Pennsylvania face a variety of threats. Three primary threats that have greatly affected amphibian and reptile populations are (1) habitat destruction, fragmentation, and degradation; (2) exploitation in both wanton persecution and illegal collection; and (3) pollution and chemical contaminants. Other significant threats have emerged in recent years; these may well eventually surpass the previously mentioned threats in their effects. These emerging threats include (1) climate change, (2) emerging diseases, and (3) introduced species. These factors may interact synergistically, making their combined effects difficult to understand and predict.

Habitat destruction and fragmentation have increased dramatically in recent years, primarily due to unchecked urban and suburban sprawl. From 1982 to 2003, the rate of habitat loss due to sprawl increased from around 100 acres per day to 350 acres per day (Moyer 2003). Some habitat types, and therefore some species, have been hit much harder by habitat loss than others. Wetlands in particular have suffered tremendous losses over the past 100 years. Wetland-dependent species, such as the bog turtle and northern cricket frog (*Acris crepitans*), have suffered associated declines. The steadily increasing number of roadways necessary to accommodate increased development has fragmented and isolated remaining patches of habitat.

Roads are also deadly to individuals that attempt to cross them. The toll is particularly high on those species that must move between habitats, such as female turtles moving to upland habitats to lay eggs or amphibians returning to vernal ponds to breed. Road mortality to female turtles can be high enough to result in populations with male-biased sex ratios (Steen et al. 2006). Unless properly monitored and regulated, proposed energy development projects (e.g., industrial wind facilities and Marcellus shale natural gas development) have the potential to destroy, degrade, and fragment many relatively pristine habitats within the commonwealth.

A number of amphibian and reptile species have been, and continue to be, feared and indiscriminately killed by the general public. Unfounded dread of snakes, in particular, leads many people to kill snakes that they encounter. This is particularly devastating for venomous species, especially the timber rattlesnake (*Crotalus horridus*) and northern copperhead (*Agkistrodon contortrix mokasen*). Water snakes (e.g., the queen snake [*Regina septemvittata*]) and hellbenders (*Cryptobranchus alleganiensis*) have a history of persecution because they are perceived as negatively affecting sport fisheries (Anonymous 1934, 1935, 1936). Collection by hobbyists and for the commercial trade can also negatively affect amphibian and reptile populations. Bog (*C. muhlenbergii*), spotted (*Clemmys guttata*), and wood (*Glyptemys insculpta*) turtles are particularly threatened by illegal collection (Reed and Gibbons 2002). The effect of illegal collection on these and other species in the state is largely unknown but may be substantial.

The effects of pollution and chemical contaminants are most pronounced on those species that inhabit or use aquatic habitats. Acid mine drainage has rendered many streams unsuitable for habitation. Acid precipitation has acidified many streams and vernal ponds so that survival of aquatic amphibians is impaired (Rowe et al. 1992, Sadinski and Dunson 1992, Rowe and Dunson 1993). Siltation of streams due to various land use practices reduces their suitability as habitat for those species, such as the hellbender, which requires high water quality. Recent evidence indicates that a more sinister threat, endocrine disruption, can be caused by the application of many agricultural pesticides. These pesticides enter aquatic ecosystems, where, in addition to reducing growth and survival of amphibian larvae (Relyea and Diecks 2008), they cause sexual malformations that may affect reproduction and population viability (Hayes et al. 2006).

Probably the greatest future threat to our native amphibians and reptiles, and the greatest challenges to wildlife professionals, is climate change due to the accumulation of human-produced greenhouse gases in our atmosphere. Thomas et al. (2004) estimate, based on projected rates of warming, that between 18 percent and 35 percent of all terrestrial species will be "committed to extinction" by 2050. As temperatures in Pennsylvania warm, species distributions will likely shrink and shift northward. It is critical that resource managers begin making plans to meet the upcoming challenges posed by climate change.

The emergence of several diseases also poses a threat to Pennsylvania's native amphibians and reptiles. Perhaps the most notorious is *Batrachochytrium dendrobatidis*, the chytrid fungus that has decimated many amphibian populations worldwide (Berger et al. 1998). Populations of several amphibians in the United State have declined from chytrid fungus infections; however, other species appear to be resistant to infection. Although no species in Pennsylvania are known to be declining because of fungal pathogens, hellbenders, in at least one stream in northern Pennsylvania, have tested positive for the chytrid fungus (P. Petokas, personal communication). Viruses of the genus *Ranavirus* can infect and cause mortality in both amphibians and reptiles. *Ranavirus* infections have been implicated in die-offs of several amphibian and reptile species, including the deaths of box turtles in Pennsylvania (Johnson et al. 2008). Other potentially worrisome diseases include *Mycoplasma*, a bacterium known to cause respiratory infections in turtles, and *Saprolegnia*, a water mold capable of infecting amphibian embryos. Diseases often act synergistically with other environmental stressors. One example of such a synergism is that between chytrid infection and global climate change. The fungus thrives within a fairly narrow temperature range, and future warming will likely increase the area within Pennsylvania that is optimal for fungal growth. At the same time, the stresses of dealing with warmer temperatures may increase the susceptibility of amphibians to infection by the fungal pathogen. The combined result could be devastating to some amphibian species.

Invasive species, both plant and animal, may also threaten amphibian and reptile populations. Introduced fish and frogs have decimated native amphibian populations in parts of the western United States (Kats and Ferrer 2003), and the potential exists for similar situations here in Pennsylvania. For example, the northern snakehead (*Channa argus*), a fish that has successfully invaded the lower Potomac River, is a voracious predator that, should it take hold in Pennsylvania, could affect native populations. Invasive plants can have dramatic effects on the structure and function of native communities, altering them to the detriment of native amphibian and reptile populations. For example, when reed canary grass (*Phalaris arundinacea*) invades shallow emergent wetlands, it forms dense monoculture stands of little value to native wildlife species, including the federally Threatened bog turtle.

Of particular conservation concern are those species for which a significant portion of their geographic range and numbers exist within the boundaries of the commonwealth. The eventual fate of these species depends to a large extent on how these species fare within Pennsylvania. The citizens of Pennsylvania therefore have a responsibility to ensure that these species thrive within the commonwealth; otherwise, they may be lost forever. Primary among these species are the eastern hellbender, wood turtle, shorthead garter snake (*Thamnophis brachystoma*), and mountain earth snake (*Virginia pulchra*). At least two of these species, the eastern hellbender and wood turtle, have undergone substantial declines in recent years. The mountain earth snake is such a cryptic and secretive species that little is known of its present status. Developing conservation strategies for these species will be problematic. The causes of decline for the wood turtle and hellbender are multifaceted, and therefore, there is no single easily recognized solution to their decline. The long-term viability of these species will depend on halting their gradual decline. The Pennsylvania Fish and Boat Commission, the agency responsible for protecting amphibian and reptile species in Pennsylvania, prohibited the collection of all four of these species in 2008. However, because collecting is only a small part of the cause of decline of hellbenders and wood turtles, other conservation strategies will need to be developed and implemented. Protecting suitable amounts of high-quality habitat will be a key component of any successful conservation plan for these species.

## Conservation and Protection of Amphibians and Reptiles

A major challenge in conserving Pennsylvania's amphibians and reptiles is the paucity of information on their present status. For many species, the answer to how species are faring in the state is "we don't know." Little reliable long-term data exist on most species, so evidence of decline for many species is mostly anecdotal. This is not an easy situation to remedy. Many reptile and amphibian species are cryptic and spend much of their life under rocks and logs or even underground. Intensive surveys are needed for many species to determine their present status. However, surveys conducted over a short period of time provide little more than a snapshot of how many individuals were found within a particular area. These surveys need to be followed up with long-term monitoring to determine whether populations are stable, increasing, or decreasing. For amphibian species in particular, natural annual fluctuations in abundance can be considerable. Therefore, monitoring activities need to be conducted long enough to determine whether a decrease in abundance is evidence of decline or within the range of natural variability. Such monitoring activities will be costly and time consuming. Even intensive surveys may not be able to determine whether the most secretive species is present or absent from an area. For example, the presence of several species, including the eastern earth snake (*Virginia valeriae*) and eastern mud salamander (*Pseudotriton montanus montanus*), have not been confirmed in Pennsylvania in more than a decade.

A second challenge in conserving amphibian and reptile species is determining and addressing the causes of species declines. The causes of the precipitous decline of several species are poorly understood. For example, the northern leopard frog (*Lithobates pipiens*) and western chorus frog (*Pseudacris triseriata*) have virtually disappeared from large areas of their former geographic range. What makes the decline of these two species especially puzzling is that close relatives in the same genera (e.g., the pickerel frog [*Lithobates palustris*] and spring peeper [*Pseudacris crucifer*]) are thriving. Investigating the cause of decline will be critical to preserving remaining populations in the state.

A third, and larger, conservation challenge is that many species are showing evidence of decline but have not reached low enough numbers to be eligible for listing as Threatened or Endangered within the state. These species do not receive any formal protection other than a legal prohibition from collecting. Because collecting is usually not the major reason for decline, the likely scenario is that these species will continue to decline until they reach low enough numbers to be listed as Threatened or Endangered. By then, populations are usually small and fragmented and have lost much of their genetic variation, making them species extremely vulnerable to extinction. Historically, society has shown little interest in preserving species until their numbers have declined to dangerously low levels, and therefore, conservation measures have been instituted as last-ditch efforts to prevent extinction. A new management paradigm is needed in which intervention occurs well before species reach such dire conditions. Innovative approaches will be necessary to halt the decline of these still fairly common species while they still consist of viable populations.

Conserving the diversity of amphibians and reptiles native to Pennsylvania will entail addressing the present and future threats previously mentioned, and meeting the conservation challenges highlighted earlier. In addition, new unanticipated threats and challenges will almost certainly emerge. Meeting these challenges will require the combined efforts of dedicated researchers and resource managers. However, these efforts will be futile without the support of the general public. Unless concerned citizens take an active role in halting and reversing environmental destruction, conserving remaining habitats, and funding conservation initiatives, our rich wildlife heritage will gradually diminish.

## IMMEDIATE CONCERN

### Mountain Chorus Frog

Order: Anura
Family: Hylidae
*Pseudacris brachyphona*
(also *Responsibility Concern*)

The mountain chorus frog was selected as a Pennsylvania Species of Greatest Conservation Need because of drastic population declines throughout its geographic range in Pennsylvania, its restricted distribution in the state, and its uncertain status (fig. 4.1). Although mountain chorus frogs are not listed as Threatened or Endangered at the federal level or in Pennsylvania, their collection in Pennsylvania is prohibited (Pennsylvania Fish and Boat Commission 2008). The mountain chorus frog is listed as globally Secure (G5, NatureServe 2009).

#### *GEOGRAPHIC RANGE*

Mountain chorus frogs are found along the Appalachian Plateau from southwestern Pennsylvania southwest to central Alabama (Hoffman 1980, Mitchell and Pauley 2005).

#### *DISTRIBUTION AND RELATIVE ABUNDANCE IN PENNSYLVANIA*

In Pennsylvania, populations of mountain chorus frogs have been found in several counties in the western

*Fig. 4.1.* The Mountain Chorus Frog, *Pseudacris brachyphona*. Photo courtesy of Tom Diez.

portion of the state from Jefferson and Clearfield counties southwest to the West Virginia–Maryland border (fig. 4.2). The species appears to have declined dramatically in both distribution and abundance since the mid-1900s. At present, extant populations are known to exist in only a handful of locations in Greene and Westmoreland counties. Historical records seem to indicate that, in all probability, the mountain chorus frog has never been a common anuran in Pennsylvania. From the early 1960s to the mid-1970s, one of us (T. Diez) monitored several colonies of mountain chorus frogs in Westmoreland County. The number of breeding adults at each of the five sites seldom exceeded twenty-five individuals. Two of the sites were destroyed by road-building activity. More disturbing and puzzling is that, while the habitat and water quality at the other sites have changed little over the years, the frogs are no longer there. Sadly, this seems to be the case at many other historic sites in Pennsylvania. Similar declines have occurred in Maryland (Forester et al. 2003).

### *COMMUNITY TYPE/HABITAT USE*

Adult mountain chorus frogs are terrestrial and are found predominantly in forested habitats along slopes and hilltops (Forester et al. 2003). With the exception of the breeding season, it is unusual to find them near water (Hulse et al. 2001). Breeding occurs in a variety of shallow wetlands, including vernal ponds, roadside ditches, tire ruts, bogs, and springs (Mitchell and Pauley 2005).

### *LIFE HISTORY AND ECOLOGY*

Mountain chorus frogs are one of the least studied frog species in the United States (Hulse et al. 2001). Adult frogs emerge from hibernation in early spring and migrate to breeding habitats. In Pennsylvania, this emergence could occur from mid-March to late April. Once the breeding season is over, the adult frogs disperse through the woodlands and are rarely encountered. Like other species in the genus *Pseudacris*, their cryptic coloration and pattern make mountain chorus frogs difficult to find among the leaves and woodland shrubs. They probably hide under objects when inactive. Mountain chorus frogs feed on beetles, spiders, and other forest-floor invertebrates (Green and Pauley 1987). A number of larger vertebrates possibly feed on chorus frogs. Bullfrogs are a known predator (Barbour 1957). Adult frogs spend the winter hibernating underground (Mitchell and Pauley 2005).

In Pennsylvania, the breeding season begins somewhat later in the spring than for other *Pseudacris* species. Depending on spring temperatures, males can begin calling as early as the first week in April and may continue calling into mid-May (T. Diez, personal observation). Mountain chorus frogs prefer breeding habitats that have abundant vegetation along the edges. Once the spring chorus begins, the males are easily observed as they call from the vegetation along the edge of the water. At the height of the breeding season, males call incessantly. Males remain near the pools throughout the breeding season. Soon after the males begin their advertisement calls, the females arrive at the pools for breeding and egg laying. Small egg masses, containing ten to fifty eggs, are attached to submerged twigs and grasses (Hulse et al. 2001) or deposited on the bottom of the pool (Mitchell and Pauley 2005). The total number of eggs laid by each female varies from 300 to 1,500 (Hulse et al. 2001). The eggs hatch in three to ten days and tadpoles metamorphose within thirty to sixty-five days (Green and Pauley 1987, Hulse et al. 2001, Mitchell and Pauley 2005).

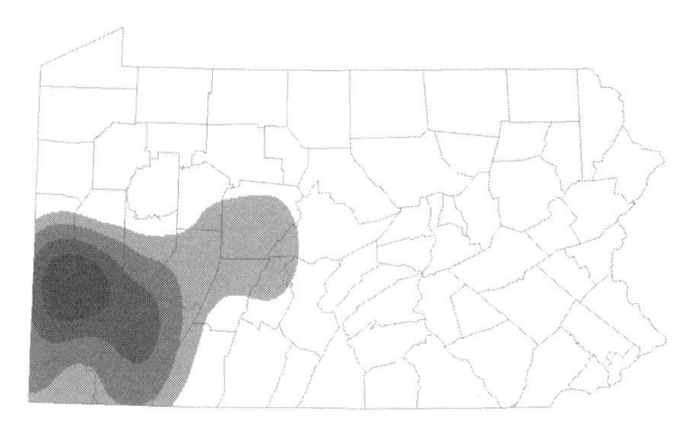

*Fig. 4.2.* Distribution of the Mountain Chorus Frog, *Pseudacris brachyphona*.

### *THREATS*

The cause of the rapid decline in this species in Pennsylvania is largely unknown, which is similar to the situation in Maryland (Forester et al. 2003). Habitat destruction and alteration from commercial forestry, urbanization, and recreational tourism have been cited as the most likely causes of the declines in Maryland (Forester et al. 2003). Mining activity and road improvement/construction have affected known breeding habitats in Pennsylvania. Loss of floodplain pool-breeding habitats is another threat (Murdock 1994). Other potential threats to mountain chorus frogs include acid precipitation (Rowe et al. 1992), chemical contamination (Semlitsch 2003), infectious diseases

(Carey et al. 2003), and habitat changes associated with global warming (Brooks 2004).

*CONSERVATION AND MANAGEMENT NEEDS*

Conservation of mountain chorus frogs will require protection of the forested hills where the adults live and the shallow wetlands where they breed. Current known breeding sites and the associated uplands should be given immediate protection.

*MONITORING AND RESEARCH NEEDS*

One difficulty in formulating a conservation plan for the mountain chorus frog is that there are very few known populations. To initiate a program designed to protect this species, we must first determine the location of remaining extant populations. Extensive surveys should be made in areas where these frogs once occurred and in surrounding areas with suitable habitat. Historical sites need to be monitored during the breeding season, when calling males are easy to locate. Any potential breeding sites should be carefully examined for the presence of tadpoles. These surveys should be followed by monitoring activities at selected sites to detect changes in population numbers and distribution.

Little is known about the biology of mountain wood frogs. Information is needed on many aspects of the biology of this species, including movement patterns, life history, and habitat requirements of adult frogs. Threats to the species need to be assessed and mitigated.

*Authors:* TOM DIEZ, TOM DIEZ NATURAL HISTORY PROGRAMS; TIMOTHY J. MARET, SHIPPENSBURG UNIVERSITY

## Eastern Hellbender

Order: Caudata
Family: Cryptobranchidae
*Cryptobranchus alleganiensis alleganiensis*
(also *Responsibility Concern*)

The eastern hellbender was selected as a Pennsylvania Species of Greatest Conservation Need because of population declines and its sensitivity to poor water quality (fig. 4.3). In addition, a significant proportion of the world's breeding population is found in Pennsylvania. Although eastern hellbenders are not listed as Threatened or Endangered at the federal level or in Pennsylvania, their collection in Pennsylvania is prohibited (Pennsylvania Fish and Boat Commission

Fig. 4.3. The Eastern Hellbender, *Cryptobranchus alleganiensis alleganiensis*. Photo courtesy of Tom Diez.

2008). Globally, the eastern hellbender is ranked G3/G4, Vulnerable/Apparently Secure (NatureServe 2009).

*GEOGRAPHIC RANGE*

The eastern hellbender's geographic range encompasses portions of the Ohio, Susquehanna, and Tennessee drainages in the east and east-central United States. A disjunct population occurs in Missouri. The species has been reliably reported from fifteen states east of the Mississippi River. Its distribution is peripheral in most of these states. It occurs in some of the northern-tier counties of Mississippi, Alabama, Georgia, and South Carolina. In North Carolina and Virginia, it is restricted to some western counties near the borders of Tennessee, Kentucky, and West Virginia. In Illinois and Indiana, the species is primarily restricted to counties bordering the Ohio River. In Ohio, it occurs primarily along the Ohio River and the southern portion of the Scioto River. New York populations are restricted to the Allegheny River and the southern portion of the Susquehanna River. Its largest area of contiguous distribution occurs in Kentucky and Tennessee.

*DISTRIBUTION AND RELATIVE ABUNDANCE IN PENNSYLVANIA*

In Pennsylvania, the geographic range of the hellbender includes the Ohio and Susquehanna drainages. The species is absent from the Delaware, Potomac, and Lake Erie drainages (fig. 4.4). A single specimen was reported from an eel weir in the upper Delaware River in 1990 but was undoubtedly an introduced individual. It appears that a significant decline in many populations of hellbenders has occurred within Pennsylvania. The Carnegie Museum collection lists hellbenders from nineteen streams within the state.

From fieldwork conducted in the 1990s, it appears that the animals have been extirpated from eleven (58 percent) of the streams (A. C. Hulse, personal observation). During the surveys in the 1990s, it was estimated that streams where hellbenders are known to exist have a total stream mileage of 819 linear miles and drain an area of approximately 6,050 square miles. Hellbenders were not found through the total stream areas but rather were restricted to regions of suitable habitat. The actual distribution of hellbenders sharply contrasts with the potential habitat for the species. The total stream mileage for streams where hellbenders could have existed within the state is 3,241 linear miles and drains an area of approximately 29,130 square miles. As a consequence, hellbenders occupy only 25 percent of the total linear mileage of streams available to them and only 20 percent of the total drainage area. In addition, many of the streams where hellbenders still exist appear to have declining populations.

*COMMUNITY TYPE/HABITAT USE*

Key habitat for adult hellbenders is moderate to large streams and rivers where there is an abundance of appropriate cover objects for the animals to take refuge under. Prime cover objects appear to be large flat rocks, primarily shale. In addition to the need for abundant rocky cover, streams must be of good quality with little or no pollution or siltation, as well as have water temperatures that remain cold or cool throughout the year. In addition, healthy and abundant populations of crayfish (the primary prey of hellbenders) must occur within the streams. Much less is known about the habitat requirements of larvae and juveniles, but it appears that they primarily use smaller rocks in riffle habitat as well as the interstitial spaces within the riffle zones (Nickerson et al. 2003).

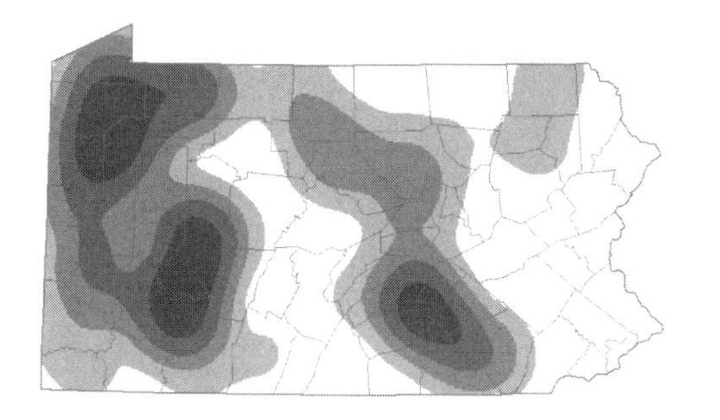

Fig. 4.4. Distribution of the Eastern Hellbender, *Cryptobranchus alleganiensis alleganiensis.*

*LIFE HISTORY AND ECOLOGY*

Eastern hellbenders prefer cool to cold, shallow, moderate-flowing areas with gravel or sandy bottom and an abundance of large, flat rock slabs. Such slabs serve as both shelter and foraging sites for the salamanders. Ultsch and Duke (1990) demonstrated that the hellbender's preference for cold, fast-flowing water is based on physiological requirements. Hellbenders have limited gas exchange ability and, consequently, are restricted to aquatic habitats where the oxygen content of the water is high.

Eastern hellbenders are probably active throughout the year. Specimens have been collected in every month but January, and there is no reason to believe that this entirely aquatic species should be inactive in the middle of winter. On a daily basis, activity is predominantly nocturnal. Bishop (1941) reported that this nocturnal pattern is disrupted during the breeding season when individuals may be found moving about at all hours. The author has also noted a dramatic increase in diurnal activity during the height of the mating season in late August and early September. Except during the breeding season, hellbenders are solitary creatures, with each individual occupying its own shelter. The occupant of a specific shelter site will actively defend the area against intruders (Hillis and Bellis 1971). However, during the fall mating season the author has seen as many as four individuals under a single rock.

For such large animals, eastern hellbenders have exceedingly small home ranges. In Pennsylvania, their geographic home range size has been estimated to average 346 $m^2$ (Hillis and Bellis 1971) and may be even smaller than that. Eastern hellbenders can achieve high densities in suitable habitat; in 1993, the author noted densities as great as one animal per linear meter of stream with biomass estimates as high as 500 kilograms per hectare (ca. 400 pounds per acre) of suitable stream. However, this was in a single stream and is not representative of population density throughout the species' geographic range in Pennsylvania.

Several authors have commented on the exceedingly skewed age distribution in this species, with large adults composing the majority of animals observed (Nickerson et al. 2003, Wheeler et al. 2003). This has also been my experience. The majority of animals collected are reproductively mature. This age distribution may be real or merely an artifact of collecting techniques in which adults are favored over other size and age categories. It is also likely that larvae occupy a different habitat than the adults (Nickerson et al. 2003).

In a long-lived species like the hellbender however, it is probable that larval and juvenile mortality is high and that recruitment into the adult population is very low. If this is the case, it has significant implications for management of the species.

Hellbenders feed predominantly on crayfish. Swanson (1948) suggested that in winter when crayfish are less active the animals feed more on fishes but that during the rest of the year crayfish constitute the majority of their diet. Hellbenders have often been unjustly accused by anglers of preying on trout, trout eggs, and other game fish. These accusations are unfounded. The few fish remains that have been reported are those of "rough" fish, with the majority being either minnows or suckers. The author has also occasionally found hellbenders to consume the following items: adult mudpuppies (*Necturus maculosus*), northern water snakes (*Nerodia sipedon*), brook lampreys (*Lampetra* sp.), and the larvae of Dobson flies (hellgrammites). Hellbenders consume eggs of their own species during the reproductive season.

Courtship and mating occur in late August and early September. Males construct shallow nest depressions under large, partially or completely embedded slabs of rock for the deposition of eggs. These depressions can be up to 30 cm by 50 cm in size. The author has also noted large cracks and crevices in boulders used as nest locations. When a receptive female enters the nest chamber, egg laying begins. Unlike all other Pennsylvania salamanders, hellbenders have external fertilization. Their eggs are laid in long, rosary-like strands. Once oviposition is complete the male chases the female away from the nest site and remains with the eggs. Males will occasionally consume some of the eggs, but the extent of this behavior and its effect on percent hatching success remains unknown. Clutch size for a Pennsylvania population (Kettle Creek) ranged from 235 to 478 eggs and was significantly correlated with female body size. Incubation varies from sixty to eighty-seven days. Larvae at hatching possess well-developed external gills, and fore and hind limb buds are present. The larval period lasts approximately two years. Bishop (1941) suggested that both sexes mature at five to six years of age, but this has not been verified.

### *THREATS*

Much of the key habitat for this species within the state has been dramatically degraded by one or more of the following: acid mine drainage, industrial pollution, domestic pollution, habitat alteration due to dam construction, and siltation. In addition, nonnative species of crayfish have been introduced into the state. Because the hellbender feeds primarily on crayfish, the introduction of alien species may affect its feeding patterns or success at capturing prey. It is unclear what the ultimate effect of introduced crayfish might have on the species. However it is likely that the effect will be negative.

Several threats within the state continue to influence the well-being of this species. Among them are acid mine drainage, pollution (industrial, agricultural, and domestic), potential habitat destruction by construction of impoundments, and potential highway construction along the margins of streams that hellbenders inhabit. Introduction or reintroduction of aquatic predators such as river otters may also pose a threat to hellbender populations. Several of the streams where river otters have been reintroduced by the Pennsylvania Game Commission contain viable populations of hellbenders. Stream "improvement for game fish habitat," such as the construction of makeshift rock barriers to form pools, can also negatively affect hellbender populations. Another threat of unknown magnitude is that of (illegal) hobbyist and commercial collecting for the pet trade.

### *CONSERVATION AND MANAGEMENT NEEDS*

Conservation of this species requires protecting their riverine habitats and maintaining high water quality in these habitats. Construction of dams or other impoundments on streams that support hellbender populations should be avoided, as should channel alteration and habitat alterations to enhance habitat for species of game fish. Care needs to be taken to avoid the introduction of potential predators (i.e., river otters) to streams where hellbenders exist until the potential effect of otters on hellbender populations has been studied. Road construction and disturbance of riparian vegetation along streams where hellbenders are known to occur should be minimized, and plans should be instituted for sediment control along these streams.

### *MONITORING AND RESEARCH NEEDS*

High-quality streams that support viable populations of hellbenders need immediate protection. Studies should be conducted of apparently recovered streams where hellbenders do not currently occur with an aim of reestablishing populations of hellbenders within these streams. In streams where hellbenders appear to be in decline, an attempt should be made to determine the causative agents responsible for the decline.

Recent survey efforts indicate that hellbender populations have declined or been extirpated in many streams. Additional surveys should be conducted both for the presence of hellbenders and for the size and age structure of the population, as well as determining the extent of suitable habitat in streams where hellbenders are located. Streams where hellbenders have been known to occur in the past should be resurveyed to determine the size of any present populations. Long-term research needs include long-term monitoring of select populations, including implantation of passive integrated transponder (PIT) tags into individuals for future identification. In addition, research is needed on the relationship between declines in hellbender populations and declining or changing species composition of crayfish in the same streams.

*Author:* ARTHUR C. HULSE, INDIANA UNIVERSITY OF PENNSYLVANIA (RETIRED)

## Green Salamander

Order: Caudata
Family: Plethodontidae
*Aneides aeneus*

The green salamander was selected as a Pennsylvania Species of Greatest Conservation Need because of its rarity, specialized habitat requirements, and extremely restricted distribution in Pennsylvania. The Pennsylvania Fish and Boat Commission lists it as a Threatened Species, and collecting this species is prohibited (Pennsylvania Code 2008; fig. 4.5). The green salamander is listed as Vulnerable/Apparently Secure (G3/G4) globally Secure (NatureServe 2009).

*Fig. 4.5.* The Green Salamander, *Aneides aeneus*. Photo courtesy of Rob Criswell.

### *GEOGRAPHIC RANGE*

The green salamander is distributed in a relatively narrow band from the Allegheny Mountains of extreme southwestern Pennsylvania southwesterly through the Appalachian Mountains to northern Alabama and northeastern Mississippi. The northern distribution also includes very southern Ohio, and small disjunct populations exist in southern Indiana, central Tennessee, and eastern West Virginia. A separate portion of the geographic range also exists in the Blue Ridge Escarpment, at the juncture of South Carolina, North Carolina, and Georgia (Petranka 1998, NatureServe 2005).

### *DISTRIBUTION AND RELATIVE ABUNDANCE IN PENNSYLVANIA*

The distribution of green salamanders in Pennsylvania is restricted to certain localized portions of Chestnut Ridge south of the Youghiogheny River in Fayette County. This represents a northerly extension of its distribution in West Virginia. Colonies in Pennsylvania are scattered and isolated, with individuals confined to suitable habitat and generally common but not abundant (fig. 4.6). At occupied sites, it is the most common salamander in suitable microhabitats.

### *COMMUNITY TYPE/HABITAT USE*

Across its geographic range, the green salamander inhabits shaded rock outcrops (sandstone, limestone, granite) and tree trunks or logs within humid forests, where it takes refuge in moist rock crevices, cracks in wood, underneath loose bark, or rarely under ground cover (Bishop 1947, Gordon 1952, Conant and Collins 1991, Wilson 2003). Green salamanders occur in several forest community types dominated by a variety of

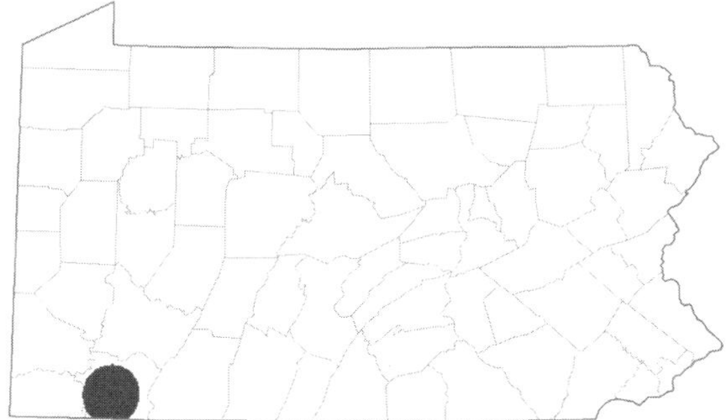

*Fig. 4.6.* Distribution of the Green Salamander, *Aneides aeneus*.

largely deciduous trees (Gordon 1952, Wilson 2003). In Pennsylvania, a mixture of tree species codominate sites, although oaks (*Quercus* spp.) and maples (*Acer* spp.) are usually common, eastern hemlock (*Tsuga canadensis*) is often present, and rosebay (*Rhododendron maximum*) typically grows on and adjacent to the outcrops (C. W. Bier, personal observation).

The green salamander is a habitat specialist adapted to use narrow crevices in rock outcrops and outcrop faces as its primary habitat. These crevices fall within a particular range of size, structure, and ambient conditions. Such crevices are narrow, shaded, damp but not wet, and do not vary widely in temperature (Gordon 1952). A Mississippi study showed the height of used crevices ranged between 30.5 cm and 396 cm (Cliburn and Porter 1987). Some moss or lichen growth is often present but not abundant. The use of tree bark, fallen logs, and ground environments is less well known but is reported in the literature (Pope 1928, Wilson 2003). The majority of published accounts for this species indicate that rock habitats are central to its distribution; however, recent literature suggests the importance of old growth forest habitats before widespread timber harvesting within parts of the geographic range (Wilson 2001, 2003). The occurrence of green salamanders beyond sandstone outcrops and boulders has not been observed in Pennsylvania to the author's knowledge.

### *LIFE HISTORY AND ECOLOGY*

The life history of the green salamander is closely linked to its niche as a dweller of aboveground crevices, the known majority of which are located in bedrock cliffs and massive blocks of stone. Crevices are used for shelter, foraging, establishing territories, travel, courtship and mating, egg laying and brooding, hatchling habitat, and hibernation (Gordon 1952, Cupp 1980, Davis 2004). Home ranges are on average quite small and extend just several meters; especially for females. However, occasional longer movements have been recorded (98 m), and some navigational capability beyond the usual territory has been suggested (Gordon 1952, 1959, 1961). During daylight, salamanders are often found near crevice openings and are visible with the aid of artificial illumination. Nocturnal activities (e.g., foraging) extend beyond crevices onto rock faces, and the occurrence of light precipitation further encourages this movement. Adults feed on terrestrial invertebrates, including mollusks, spiders, and insects, such as beetles, flies, and ants (Gordon 1952, Lee and Norden 1973). Little is reported on predators of green salamanders, although Gordon (1952) identifies the northern ringneck snake as a predator. Hibernation takes place in deep crevices, often communally. In Pennsylvania, high densities have been found around a presumed hibernating crevice on April 23 and October 20, with November 19 as the latest date any individuals were observed (C. W. Bier, personal observation).

Emergence from hibernating crevices occurs in late March and early April, with adults moving toward breeding crevices and males reacting territorially through physical confrontations (Cupp 1980). Mating has been recorded in both spring and autumn but more often in spring, beginning in later May and early June (Gordon 1952, Cupp 1971, Thompson and Taylor 1985, Canterbury and Pauley 1994). Breeding crevices in Pennsylvania are horizontal cul-de-sacs 10–20 cm deep (personal observation). Eggs are laid in early June in masses of ten to thirty and are attached to crevice ceilings several centimeters from the entrance (Gordon 1952, C. W. Bier, personal observation). The female remains close to her cluster for nearly three months (eighty-four to ninety-one days) until hatching. During this time, she acts defensively and protects the egg mass and recent hatchlings (Gordon 1952, C. W. Bier, personal observation). Larval stages occur within the egg, and hatchlings resemble miniature adults, 19.7 mm in average total length (Gordon 1952). Female green salamanders apparently breed biennially (Thompson and Taylor 1985, Canterbury and Pauley 1994).

### *THREATS*

The primary and immediate threats to the survival of the green salamander include habitat destruction and landscape fragmentation resulting in colony isolation. Sandstone quarrying, gas pipeline installation, and the mining of other bedrock types (coal and limestone) within the Pennsylvania geographic range of green salamanders are known to have directly damaged or possibly eliminated rock outcrop habitats (C. W. Bier, personal observation). Because the Pennsylvania range of this species has only recently been described (Bier 1985, Pennsylvania Natural Heritage Program 2005) earlier losses are undetermined.

Other threats involve more indirect effects on green salamander colonies and include sedimentation of crevices due to adjacent surface coal mining and the removal of trees and other vegetation through gas well development, road construction, forest practices, and rock climbing. The removal of shading vegetation alters the ambient environment of crevices, outcrop habitats, and surrounding forests, which results in warmer and drier conditions. Even-aged timber management is implicated in the extirpation of West Virginia colonies (Biggins 1987). Wilson (2003) suggests old growth forests, and those previously including large American chestnuts (*Castanea dentate*), were important to green salamanders for foraging. The loss of the associated structure has reduced habitat quality and caused the species' decline. The reduction of forest buffer zones around outcrops might also further serve to threaten colonies during times of drought.

Because green salamanders occur in a scattered distribution of isolated colonies, further isolation potentially threatens future viability through lack of gene flow and inability to recolonize habitats where populations have become extirpated (Larson et al. 1984). Genetic isolation has been linked to geographic distances between colonies (Johnson 2002). Isolation will increase as individual colonies are lost, thereby increasing intercolony distances of remaining colonies. Isolation is further aggravated by the creation of dispersal barriers within landscapes, such as highways and nonforest land cover. Severe declines have been recorded in vulnerable populations south of Pennsylvania, and several of these causal factors are suspect (Corser 2001).

Large-scale environmental impacts also potentially threaten green salamanders, including global warming (Donnelly and Crump 1998), acid deposition/precipitation (Biggins 1987, Wilson 2001), and the distribution of other toxins (Corser 2001). The Pennsylvania geographic range of this species is within a region of high acid deposition (Pennsylvania Department of Environmental Protection 2005). Although no information is known regarding this species, amphibians in general are known to be sensitive to habitat acidification, and the sandstone inhabited in Pennsylvania lacks buffering capacity. If global warming alters crevice and outcrop microclimates, the suitable parameters for green salamanders might be exceeded. Corser (2001) also suggests both epidemic pathogens and environmental pollutants as possible causal agents where populations have drastically declined in the Blue Ridge Escarpment of the Carolinas.

Overcollecting has been implicated in the decline of a colony of green salamanders in Maryland (E. Thompson, unpublished) and elsewhere (Corser 2001). Although there is no indication that this is an active threat in Pennsylvania, Hulse et al. (2001) warn of such a possibility.

### *CONSERVATION AND MANAGEMENT NEEDS*

All currently known colonies should be given protection and appropriate management. A conservation plan should be developed for green salamanders in the northern part of their global geographic range. Such a plan would address issues in this region, such as threats, strategies to address threats, adaptive management, and would act to link conservation and management in Pennsylvania, West Virginia, and Maryland. Rock outcrops and the surrounding forests must be protected at each location. Outcrop forests must be of significant size and maturity to provide protection of the local ambient climate and to serve as a buffer against other outside influences; for example, an invasive exotic plant (e.g., Oriental bittersweet [*Celastrus orbiculatus*]) could colonize and smother rock faces and crevice openings. The recommended forest buffer width of 100 feet suggested earlier in Pennsylvania (Shiffer et al. 1987) should be reevaluated and expanded. Old growth forest should be maintained surrounding outcrops as a component of core habitat (Wilson 2003), with additional forest as a buffer zone. Forested corridors should be designed and maintained between colonies as suitable dispersal routes, and the need for corridors to consist of old growth forest should be evaluated. The management of the green salamander should be a priority where it occurs on public land, for example, State Game Lands and State Forest.

### *MONITORING AND RESEARCH NEEDS*

Surveys are needed in areas of appropriate habitat to discover any unknown green salamander colonies. In Pennsylvania, green salamanders occur in few enough locations that regular detailed monitoring of each colony is feasible. Colony monitoring should include both assessments of each subpopulation (colony) as well as important parameters of suitable habitat. Monitoring should be designed to focus on critical issues that are perceived as threats. However, threats such as pathogens, as postulated by Corser (2001), should also be considered. Management decisions should be adjusted based on monitoring results, for example, increasing buffer size to address a related negative trend. Loss of genetic viability is also a potential threat that should be addressed through monitoring. Landscape-level monitoring should be included to reveal any active or future aspects of habitat loss and regional fragmentation. A potential necessary management scheme could be the translocation of individuals representing certain genotypes into substandard colony genomes. However, more research is needed to determine whether such actions are necessary.

Information is also needed regarding the use of forest habitats that surround occupied outcrops in Pennsylvania. Ecological descriptions of suitable outcrops, outcrop forests, and intercolony corridor forests are needed, including aspects of forest structure and condition. In addition, information is needed regarding which land uses act as barriers to dispersal. Mitigation options regarding immigration/dispersal corridors are needed. The genetic viability of this specialized species, existing as it does in isolated colonies, is significant. Information is needed on the vulnerability of green salamanders to loss of genetic variability and vigor. Related issues to be investigated include minimum viable population size and, eventually, an understanding of the dynamics of gene flow between colonies relative to long-term viability and management options. In general, an assessment of active and potential threats to the green salamander is needed, and this would be part of a comprehensive management plan for the species.

*Author:* CHARLES W. BIER, WESTERN PENNSYLVANIA CONSERVANCY

## Spotted Turtle

Order: Testudines
Family: Emydidae
*Clemmys guttata*
(also *Responsibility Concern*)

The spotted turtle was selected as a Pennsylvania Species of Greatest Conservation Need because of its uncertain status and evidence of population declines (fig. 4.7). In addition, the high risk of habitat destruction, spotty habitat distribution, illegal collection, and mortality from motor vehicles make this species extremely vulnerable to further population declines. Spotted turtles are not listed as Threatened or Endangered at the federal level or in Pennsylvania. Their collection in Pennsylvania is prohibited (Pennsylvania Fish and Boat Commission 2008). The global status of the spotted turtle is Secure (G5, NatureServe 2009).

*Fig. 4.7.* The Spotted Turtle, *Clemmys guttata*. Photo courtesy of Tom Diez.

### GEOGRAPHIC RANGE

Spotted turtles range from southern Canada and Maine south along the Atlantic Coastal Plain and Piedmont to northern Florida and west across Ohio, Indiana, and Michigan into northeastern Illinois (Ernst et al. 1994, Hulse et al. 2001). Although their geographic range is extensive, spotted turtle populations are generally small and often isolated.

### DISTRIBUTION AND RELATIVE ABUNDANCE IN PENNSYLVANIA

Spotted turtles are found throughout southeast and south-central Pennsylvania and in the far western part of the state (fig. 4.8). Present status and distribution, especially in the western part of the state, remains uncertain. Recent surveys have documented a number of populations throughout the state; however, many of these populations appear to be relatively small and are isolated from other populations.

### COMMUNITY TYPE/HABITAT USE

Spotted turtles use a variety of soft-bottomed aquatic habitats, including small streams, marshes, swamps, and vernal ponds (Ernst et al. 1994). They also spend considerable time on land during the spring and summer months (Ward et al. 1976). Commonly used terrestrial environments include upland woodlands and open habitats.

### LIFE HISTORY AND ECOLOGY

In early spring, spotted turtles emerge from hibernation and migrate from their hibernation sites to vernal ponds, marshes, bogs, or other wetlands to feed, bask, and mate (Perillo 1997, Litzgus and Brooks 2000, Joyal et al. 2001, Milam and Melvin 2001). During the spring and early summer, the turtles often disperse over land to other wetlands (Perillo 1997, Joyal et al. 2001, Milam and Melvin 2001). In late summer through early fall, spotted turtles often aestivate

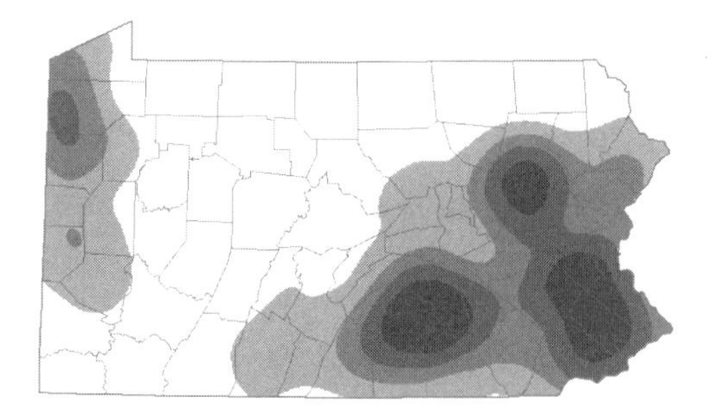

*Fig. 4.8.* Distribution of the Spotted Turtle, *Clemmys guttata.*

either on land or in water (Ward et al. 1976, Graham 1995, Haxton and Berrill 1999, Joyal et al. 2001). Aestivation sites on land range from areas along the perimeter of wetlands to areas considerably distant from water (Graham 1995, Haxton and Berrill 1999, Haxton and Berrill 2001, Joyal et al. 2001, Milam and Melvin 2001). In early fall, turtles return to hibernation sites, which range from sphagnum, blueberry, and forested permanent swamps, muskrat burrows, vernal ponds, and wet meadows, to the bottom of pools in streams (Ernst 1976, Ward et al. 1976, Graham 1995, Litzgus and Brooks 2000, Joyal et al. 2001, Milam and Melvin 2001). Spotted turtles have been found hibernating alone or in aggregations (Perillo 1997, Haxton and Berrill 1999, Joyal et al. 2001). Mean geographic home range size, including both aquatic and terrestrial habitats, varies from 0.52 ha (Ernst 1970) to 3.7 ha (Haxton and Berrill 1999). Milam and Melvin (2001) suggest that large home ranges may be necessary when turtles have to travel greater distances to acquire resources needed for reproduction and survival.

Spotted turtles are omnivores, feeding on aquatic invertebrates, amphibian larvae, fish, carrion, plants, and algae (Ernst 1976). A variety of avian and mammalian predators eat turtles and their eggs, especially raccoons (Ernst 1976, Ernst et al. 1994, Burke et al. 2000). Eggs and hatchling turtles are especially vulnerable to predation.

Reproductive maturity in both sexes is reached within seven to ten years (Ernst et al. 1994). Courtship and mating occur in the water in early spring, shortly after emergence from hibernation. In early summer, female turtles travel overland, sometimes considerable distances from wetlands, to open uplands to nest (Litzgus and Brooks 2000, Milam and Melvin 2001). Preferred nesting sites are open areas with well-drained soil (Hulse et al. 2001). After digging a cavity in the ground, females lay one to eight eggs (Ernst et al. 1994). In southeastern Pennsylvania, females usually lay three to five eggs (Ernst 1976). Sex determination is temperature dependent, with eggs incubated at higher temperatures tending to produce females. Hatchling turtles emerge either in the fall or in the following spring (Ernst et al. 1994).

*THREATS*

Destruction, degradation, and fragmentation of habitats are major threats facing spotted turtles since they frequently use isolated wetland habitats. Legal protection for these wetland habitats was affected by the U.S. Supreme Court decision in the case of *Solid Waste Agency of Northern Cook County v. U.S. Army Corps of Engineers* (No. 99-1178; SWANCC). This decision greatly reduced the protections afforded to isolated wetlands under the Clean Water Act (Gibbons 2003). Loss of these wetlands not only results in loss of habitat but also changes the spatial configuration of, and distance between, remaining ponds, thereby affecting overland movements and reducing the potential for recolonization of habitats (Semlitsch 2003).

Because spotted turtles may travel considerable distances on land, destruction of terrestrial habitats can also affect turtle populations, particularly if these habitats serve as corridors between wetland habitats or as aestivation sites. In addition, loss of proper nesting habitat can cause female turtles to travel farther overland in search of suitable conditions for nest construction. Several studies have recognized the importance of upland habitats surrounding wetlands to spotted turtles and have made recommendations for their preservation (e.g., Perillo 1997, Joyal et al. 2001, Milam and Melvin 2001, Semlitsch and Bodie 2003). A common proposal is to provide a protective buffer around wetlands. Buffer widths of 150-275 m have been recommended for a variety of aquatic and semiaquatic turtle species (Burke and Gibbons 1995, Bodie 2001, Buhlmann and Gibbons 2001). Perillo (1997) and Milam and Melvin (2001) recommended buffer widths of 200 m and 400 m, respectively, specifically for spotted turtles.

Highways and roads pose another threat to spotted turtles through isolation and fragmentation of habitats and by mortality from motor vehicles (Mitchell and Klemens 2002). Gibbs and Shriver (2002) created a model of the effects of road mortality on turtle populations and concluded that persistence of semiterrestrial turtles, such as spotted turtles are jeopardized by road densities characteristic of much of the eastern United States.

Collecting of turtles by hobbyists for pets or by

collectors for the commercial pet trade also threatens turtle populations. The extent of illegal collection for commercial trade is unknown. On the basis of adult survivorship, geographic range, and median monetary value in the retail trade, Reed and Gibbons (2002) rank spotted turtles as highly vulnerable to commercial harvest. Congdon et al. (1993) concluded that sustainable harvest is not possible for long-lived turtle species. Continued illegal collecting thus poses a serious threat to the persistence of this species in Pennsylvania.

Spotted turtles also face increased predation from mammals (particularly raccoons) and birds (Burke et al. 2000). Human disturbance, urbanization, and agriculture create conditions that cause predator populations to reach unnaturally high levels, resulting in high rates of predation, particularly on eggs and hatchlings (Mitchell and Klemens 2002).

*CONSERVATION AND MANAGEMENT NEEDS*

Conservation of this species will require protection of both aquatic and upland habitats, as well as the control or reduction of threats from overcollecting, predation, and road mortality. With the loss of federal protection previously provided by the Clean Water Act, increased state and local protections of isolated wetlands will be necessary. Sites of known current occurrence should be given high priority for protection.

*MONITORING AND RESEARCH NEEDS*

Recent surveys indicate the presence of spotted turtle populations throughout much of its historic stronghold in the south-central–southeastern part of the state. However, many of these populations appear to be small, and these surveys did not assess changes in abundance. These surveys should be continued and followed up with monitoring activities at selected sites to detect changes in population numbers and distribution. Circumstantial evidence hints at a substantial decline in recent years.

Research is needed on the numerous threats that face spotted turtles. Priority should be given to investigations of the effect of habitat destruction and fragmentation on spotted turtle dispersal, population structure, and population dynamics. Research is also needed on the effects of collecting, predation, and road mortality on different life stages of turtles, as well as an evaluation of methods to reduce mortality due to predators and roads.

*Author:* TIMOTHY J. MARET,
SHIPPENSBURG UNIVERSITY

## Blanding's Turtle

Order: Testudines
Family: Emydidae
*Emydoidea blandingii*

The Blanding's turtle was chosen as a Species of Greatest Conservation Need because of its extremely restricted distribution and poorly understood status (fig. 4.9). It is listed as a Candidate Species by the Pennsylvania Fish and Boat Commission, and collecting of this species is prohibited (Pennsylvania Code 2008). The global status of the Blanding's turtle is Apparently Secure (G4, NatureServe 2009).

*GEOGRAPHIC RANGE*

Blanding's turtles have a northern distribution, extending from southern Ontario in the east, westward into Minnesota, Iowa, and Nebraska. There are several disjunct populations in the northeast. One disjunct population is found in southeastern New York (Klemens 1993), and another larger population occurs from southwestern Maine and southeastern New Hampshire to northeastern Massachusetts. No recent localities have been found in Connecticut (Klemens 1993).

*DISTRIBUTION AND RELATIVE ABUNDANCE IN PENNSYLVANIA*

Historic records for the species are from Presque Isle in Erie County and Conneaut Lake and the nearby town of Linesville, both in Crawford County (fig. 4.10). Netting (1932) hypothesized that Blanding's turtles reached Conneaut Lake from Lake Erie by way of the Erie Canal. No specimens from Crawford County have been reported since 1906. This area has

Fig. 4.9. The Blanding's Turtle, *Emydoidea blandingii*. Photo courtesy of Tom Diez.

Fig. 4.10. Distribution of the Blanding's Turtle, *Emydoidea blandingii*.

undergone extensive environmental modification since the early 1930s. Among other things, the Shenango River was dammed and the Pymatuning swamp flooded to produce Pymatuning Reservoir in 1935, dramatically modifying much of the wetlands in the area. Specimens are sporadically found along the Lake Erie shore, especially in Presque Isle State Park.

### *COMMUNITY TYPE/HABITAT USE*

The Blanding's turtle is primarily an inhabitant of poorly drained lowlands. It is usually found in areas that are mosaics of marshes, wet meadows, ponds, and slow-moving streams. Ponds and other standing water are favored over streams. Recently, a detailed study of Blanding's turtles was conducted in southern Maine (Joyal et al. 2001). They found that the turtles prefer small (<0.4 ha) wetlands and that they use multiple wetlands throughout the year, including both temporary ponds and wetlands with longer hydroperiods. Although it has often been considered to be semiaquatic, recent studies indicate that it is primarily an aquatic species that makes occasional forays onto land (Ross and Anderson 1990, Rowe and Moll 1991). As a result, it appears that key habitat for Blanding's turtles is a landscape mosaic of wetlands interspersed among upland habitat.

Pappas and Brecke (1992) found that young turtles used shallow standing water in alder thickets and shallow eutrophic ponds that were bordered and invaded by tussocks of sedges. They postulated that habitat used by young turtles provided them with protection from predators as well as reducing potential competition with adults.

### *LIFE HISTORY AND ECOLOGY*

Blanding's turtles become active in the spring when water temperature is about 10°C. Activity reaches a peak in June, rapidly drops off in July, and then remains at a low level. Terrestrial activity occurs primarily in the spring of the year with the turtles being mainly aquatic during the rest of the year (Rowe and Moll 1991). Blanding's turtles are diurnal and generally exhibit a bimodal activity pattern with a peak occurring from mid- to late morning and a second peak in the late afternoon (Rowe and Moll 1991). Although nesting may occasionally continue after sunset, virtually all nesting activity is initiated in the afternoon (Congdon et al. 1983).

Turtles enter hibernation between the middle of September and the middle of November, depending on geographic area and local climatic conditions (Ross and Anderson 1990, Rowe and Moll 1991). Hibernation tends to occur in the same pond where summer activity takes place. During hibernation, turtles partially bury themselves in the soft bottom sediments of the ponds.

The Blanding's turtle is primarily an aquatic carnivore (Rowe 1992). Only 12 percent of their diet was composed of aquatic vegetation, and much of that may have been ingested incidental to the capture of animal prey. Mollusks, mostly snails, make up the majority of prey contained in their stomachs. Other major food categories are crayfish and aquatic insects. None of the food items recorded by Rowe (1992) were of terrestrial origin, giving added credence to the fact that these turtles are truly aquatic.

Blanding's turtles mate in the spring and summer (Graham and Doyle 1977). Mating occurs in the water, and not much courtship activity is involved (Ernst and Barbour 1972). The nesting season typically runs from late May through the end of June (Congdon et al. 1983, Linck et al. 1989, Rowe and Moll 1991). Females will move varying distances from their home pond to lay eggs. Nests have been located within 2 m of ponds and at distances of more than 1 km away from ponds (Congdon et al. 1983). Nests are located in sandy loam soil and tend to be in areas where grasses and sedge tussocks are common. However, a Massachusetts population showed a marked preference for laying eggs in a cornfield (Linck et al. 1989). Blanding's turtles most likely lay a single clutch a year. Clutch size varies from location to location. In Massachusetts, it ranged from nine to sixteen eggs (DePari et al. 1987). In Michigan, clutch size varied from three to fifteen eggs (Congdon et al. 1983), and in Ontario, a clutch ranged from six to eleven (MacCulloch and Weller 1988). Incubation lasts 73 to 106 days (Congdon et al. 1983). Average size of hatchlings from a Michigan population was 35.3 mm CL (Congdon and Van Loben Sels 1991).

Predation on nests in some populations is high. Congdon et al. (1983) reported that an average of 63 percent of all nests were destroyed by predators with a maximum predation rate of 93 percent and a minimum rate of 42 percent. One hundred percent of the nests studied by Ross and Anderson (1990) in Wisconsin were destroyed before hatching.

*THREATS*

Primary threats to Blanding's turtles in Pennsylvania are habitat destruction and the lack of information on the distribution of the species within the state.

*CONSERVATION AND MANAGEMENT NEEDS*

The major priority is to maintain adequate habitat for Blanding's turtles if and when they are located in the state. Joyal et al. (2001) suggest that the best strategy is to protect small wetlands and that conserving them in groups rather than as isolated entities is the best approach. Also, based on the extensive use of upland habitat by Blanding's turtles, terrestrial buffer zones of at least 500 m should be maintained around wetlands turtles use. Appropriate resource managers at all sites where populations are discovered should be made aware of the existence of the populations. Whenever possible, management for Blanding's turtles should be incorporated into appropriate resource management plans. Special attention should be given to the location of nesting sites, and attempts should be made to control predation by raccoon and other carnivores on the eggs of Blanding's turtles.

*MONITORING AND RESEARCH NEEDS*

Little is known concerning the biology or distribution of Blanding's turtles in Pennsylvania. There is an immediate need for programs to survey appropriate habitat both for the presence of Blanding's turtles and for the size of the populations on Presque Isle, as well as to determine the extent of suitable habitat where turtles are located. In addition, historic locations and their vicinity should be resurveyed for Blanding's turtles.

Once the location of populations is established through surveys, the populations should be subjected to intensive long-term monitoring. The sites designated for long-term monitoring should, whenever possible, be situated on public lands. Long-term studies should include mark-recapture procedures and should specifically target population size and structure as well as reproductive activity and success within the populations.

*Author:* ARTHUR C. HULSE, INDIANA UNIVERSITY OF PENNSYLVANIA (RETIRED)

## Wood Turtle

Order: Testudines
Family: Emydidae
*Glyptemys insculpta*
(also *Responsibility Concern*)

The wood turtle was chosen as a Species of Greatest Conservation Need because of its uncertain status, threat from illegal collection and commercial trade, mortality from motor vehicles, and as a representative of the guild of species associated with riparian habitats (fig. 4.11). Because a substantial portion of the world's breeding population occurs within the state of Pennsylvania, the commonwealth has a significant stewardship responsibility for this species. Wood turtles are not listed as Threatened or Endangered either at the federal level or in Pennsylvania. Their collection in Pennsylvania is prohibited (Pennsylvania Fish and Boat Commission 2008). In 1992, the wood turtle was added to Appendix II of the Convention on International Trade in Endangered Species of Wild Fauna and Flora (Convention on International Trade in Endangered Species 2007), thus restricting international trade in this species. The global status of the wood turtle is Apparently Secure (G4, NatureServe 2009).

*GEOGRAPHIC RANGE*

Wood turtles occur in fragmented and disjunct populations in North America. The largest portion of their

*Fig. 4.11.* The Wood Turtle, *Glyptemys insculpta*. Photo courtesy of Roy Nagle.

distribution is centered in the Northeast, ranging from New Brunswick and Nova Scotia south to northern Virginia. The other major area of distribution occurs from southern Ontario and northern Michigan west to eastern Minnesota. Wood turtles are absent from Ohio, Indiana, and Illinois. In general, the distribution ranks among the most northerly of New World turtles.

### *DISTRIBUTION AND RELATIVE ABUNDANCE IN PENNSYLVANIA*

In Pennsylvania, wood turtles are reported to occur in many portions of the state, although the highest densities likely occur in the central Valley and Ridge Province where large streams and associated riparian and forested habitats remain less intensively developed than areas to the east and west (fig. 4.12). Wood turtles are absent from much of the Allegheny Front and the extreme southwestern corner of the state.

### *COMMUNITY TYPE/HABITAT USE*

Characteristic habitat for wood turtles includes large streams with clear water and moderate flow, associated riparian areas, and partially forested bottomlands. Wood turtles may also be found in a variety of other habitats, including shallow wetlands, swamps, small streams, and rivers. Typical summer habitats include alder thickets, grassy areas in or near wetlands, and forests where canopy openings allow some degree of insolation (Farrell and Graham 1991, Ross et al. 1991, Kaufmann 1992a, Saumure and Bider 1998, Tuttle and Carroll 2003, Walde et al. 2003). Wood turtles often occur in association with fertile and glacial soils and low-intensity agriculture.

### *LIFE HISTORY AND ECOLOGY*

Wood turtles use aquatic habitats to overwinter and at other times to hydrate, thermoregulate, mate, and avoid predators. During late spring and summer, these turtles spend much of their time on land and are subsequently considered Pennsylvania's most terrestrial turtle other than the eastern box turtle. Wood turtles are rarely found more than several hundred meters from water, however, and juveniles are often found closer to water than are adults (Harding 1991). Range lengths for adult turtles in central Pennsylvania average about 450 m (Kaufmann 1995) but equal 1 km for some individuals (R. D. Nagle, personal observation). Some individuals show remarkable fidelity to particular sites among seasons and years (Harding and Bloomer 1979, Quinn and Tate 1991, Kaufmann 1995, Arvisais et al. 2002, Tuttle and Carroll 2003). Wood turtles feed on a variety of plant and animal material, including berries, mushrooms, earthworms, snails, amphibian and insect larvae, and fish, bird, and mammal carrion (Harding and Bloomer 1979, Farrell and Graham 1991, Tuttle and Carroll 2003, Walde et al. 2003). Substantial variation among individuals in both food preferences and habitat use, combined with the potential longevity of individuals, suggest that learned behaviors play key roles in the ecology of wood turtles (Compton et al. 2002). Wood turtles live at least forty-six years in the wild (Ernst 2001b), but maximum longevity has not been documented because few long-term studies have been conducted on natural populations.

Social interactions increase substantially in the spring and in the late fall when turtles congregate in aquatic habitats. Males establish well-defined dominance hierarchies through aggressive physical confrontations (Barzilay 1980), and the larger, older males obtain the highest levels of mating success within their activity areas (Kaufmann 1995). Some mating occurs in spring although the peak mating season occurs during autumn (Harding and Bloomer 1979, Kaufmann 1995).

Female wood turtles mature at minimum ages of fourteen to eighteen years and produce a maximum of one clutch of eggs per year (Harding and Bloomer 1979, Farrell and Graham 1991, Brooks et al. 1992). Nesting occurs from late May through early July, with females seeking open-canopy, well-drained soils with sparse vegetation in which to nest (Harding and Bloomer 1979, Buech et al. 1997). Most nesting activity occurs late in the day, with females beginning nesting activities during early evening and completing nests after dusk. Clutch size averages nine to ten eggs (Harding and Bloomer 1979, Farrell and Graham 1991, Brooks et al. 1992), and unlike most other North American turtles, wood turtle embryos lack temperature-dependent sex determination (Bull et al. 1985). The incubation period

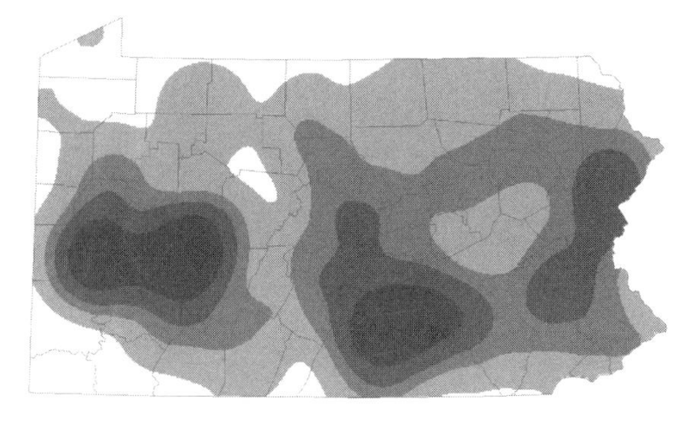

*Fig. 4.12.* Distribution of the Wood Turtle, *Glyptemys insculpta*.

is approximately sixty days at 25°C and decreases at higher temperatures (Ewert 1985). Hatchlings emerge from nests during autumn and migrate to aquatic habitats before the onset of winter.

*THREATS*

Throughout their geographic range, wood turtle populations have undergone substantial declines (Burke et al. 2000). Major factors associated with declines include habitat loss, habitat degradation, and the removal of animals from wild populations for pets (Harding and Bloomer 1979, Garber and Burger 1995, Ernst 2001a). Because much of their range overlaps a relatively dense human population, residential and commercial development probably have the greatest negative impacts. The requirement for both aquatic and terrestrial habitats and the propensity of wood turtles to make extensive movements on land appear to make population stability inconsistent with highly developed areas, particularly the presence of roads near riparian areas, which often result in direct mortality from automobiles. Wood turtles are rarely found in polluted waters or those with high sediment levels; their presence in an area may indicate a relatively healthy ecosystem. Wood turtles are highly prized in the commercial pet trade (Levell 2000), and illegal, commercial harvests have occurred in Pennsylvania in recent years and have resulted in prosecutions.

Turtles exhibit a unique suite of coevolved life history characteristics, including delayed sexual maturity, long reproductive life span, and low fecundity; such factors combine to make populations highly sensitive to changes in survivorship of adults and older juveniles (Crouse et al. 1987, Congdon et al. 1993, Congdon et al. 1994, Congdon et al. 2000). The presence of wood turtles in some areas should not be taken as evidence that populations in those areas are necessarily viable. Long life spans, long generation times, and relatively slow growth may contribute to the presence of turtles in a given area long after recruitment has ceased or populations reach levels below which sustainability or recovery is possible (Congdon et al. 1993, 1994). Acute impacts to wood turtles such as the loss of important nesting areas or unsustainable mortality of adults may remain undetectable until populations reach critical levels or become extirpated. Because wood turtles tend to occur in relatively low density, disjunct populations (Harding and Bloomer 1979, Harding 1991, Niederberger and Seidel 1999, Burke et al. 2000, Ernst 2001a), the loss of only a few adults in some areas may have substantial adverse impacts.

*CONSERVATION AND MANAGEMENT NEEDS*

Protection of essential habitats and consideration of the spatial ecology of wood turtles are vitally important to conserving this species. The establishment of buffer zones along streams would be of tremendous benefit to wood turtles (Arvisais et al. 2002). Although the level of protection would vary with habitat type and distance from streams, statutes such as those in Florida and Massachusetts, which delineate buffer zones of 100 feet around wetlands would protect many wood turtles, especially juveniles. Protection of most adult turtles would require larger buffer zones, such as the 275 m limit proposed for an assemblage of freshwater turtles in South Carolina (Burke and Gibbons 1995). Heavy equipment operations (e.g., logging) in and near riparian areas inhabited by wood turtles should be discouraged from April through October. Conversely, construction activities that affect streams (e.g., dredging, bridge construction) would be most detrimental to wood turtles during winter because individuals are concentrated in aquatic habitats, especially the deeper, slower pools that are often located near structures. Finally, wood turtles often use human-altered, disturbed sites for nesting, including railroad right-of-ways, power lines, and shale pits. Such areas are rarely considered biologically valuable or worthy of protection, but in some instances, their availability may be important for maintaining viable wood turtle populations. A comprehensive conservation strategy that addresses the protection of essential habitats and reduces road mortality is required to maintain viable wood turtle populations in Pennsylvania well into the future.

*MONITORING AND RESEARCH NEEDS*

The Pennsylvania Herpetological Atlas Project has documented the presence of wood turtles in fifty-seven of the sixty-seven counties in Pennsylvania, providing baseline information on their general distribution. Further monitoring is necessary to determine the relative abundance of wood turtles and population trends. Parameters to evaluate in relation to population stability include adult sex ratios and the ratio of juveniles to adults. Under relatively stable conditions, sex ratios of wood turtle populations are typically near 1 : 1 and juveniles comprise 20 percent to 30 percent of the population (Harding and Bloomer 1979, Kaufmann 1992a, Ernst 2001b, Walde et al. 2003). Sex ratios found to be strongly male-biased may indicate high levels of female mortality, resulting from nesting movements near open-canopy roadways (Aresco 2005). Low proportions of juveniles may indicate recruitment problems, such as the loss of important nesting areas. Management solutions to such problems may involve creation of new nesting habitat and reducing the appeal of areas near roads (e.g., by increasing canopy cover) or preventing access to such areas altogether.

Most studies of wood turtles have focused on geographic home range size, habitat use, and behavior (Quinn and Tate 1991, Kaufmann 1992a, 1992b, 1995, Arvisais et al. 2002, Compton et al. 2002, Tuttle and Carroll 2003). A few studies have estimated minimum age at maturity (Harding and Bloomer 1979, Ross et al. 1991, Brooks et al. 1992), but no long-term studies have documented other important life history characteristics, such as age-specific fecundity and age-specific survivorship. Such data are critical for a complete understanding of the effects of changes in survivorship and recruitment on population stability. Nesting habitats and nesting ecology have also received relatively little attention. Some researchers have suggested that agriculture may benefit wood turtles by providing canopy openings and increased edge habitats (Harding and Bloomer 1979, Kaufmann 1992a), whereas others have documented negative impacts such as increased levels of shell injuries in agricultural landscapes (Saumure and Bider 1998). The relationships between wood turtle population dynamics and specific agricultural methods and seasonal practices merit further investigation. Other areas in need of research include the effects of water pollution and timber-harvesting practices on wood turtle populations, as well as methods to reduce road mortality.

*Author:* ROY D. NAGLE, JUNIATA COLLEGE

## Bog Turtle

Order: Testudines
Family: Emydidae
*Glyptemys muhlenbergii*
(also *Responsibility Concern*)

The bog turtle was selected as a Pennsylvania Species of Greatest Conservation Need because of population declines, the spotty distribution of appropriate habitat, and illegal collecting for the commercial trade (fig. 4.13). Bog turtles are listed as Endangered in Pennsylvania (Pennsylvania Code 2008). The bog turtle was given federal protection in November 1997 when the species was listed as a Threatened Species under the Endangered Species Act (U.S. Fish and Wildlife Service 1997). In 1992, the bog turtle was added to Appendix I of the Convention on International Trade in Endangered Species of Wild Fauna and Flora (Convention on International Trade in Endangered Species 2007), thus banning international trade in this species. The global status of the bog turtle is Apparently Secure (G4, NatureServe 2009).

*Fig. 4.13.* The Bog Turtle, *Glyptemys muhlenbergii*. Photo courtesy of James Drasher.

### *GEOGRAPHIC RANGE*

Two separate geographical populations of bog turtles are recognized (U.S. Fish and Wildlife Service 2001). The northern population exists within New York, Massachussetts, Connecticut, New Jersey, Pennsylvania, Delaware, and Maryland. A disjunct southern population, separated by 250 miles from the northern population, exists in Virginia, North Carolina, Tennessee, South Carolina, and Georgia, primarily in the Blue Ridge Province (Lee and Herman 1999). A significant portion of the northern population's geographic range is found in Pennsylvania. Pennsylvania and New Jersey contain the highest number of extant bog turtle sites within the northern population (U.S. Fish and Wildlife Service 2001).

### *DISTRIBUTION AND RELATIVE ABUNDANCE IN PENNSYLVANIA*

Bog turtles are limited in distribution to portions of fifteen southeastern and eastern counties and possibly other isolated areas in northwestern Pennsylvania (fig. 4.14). Fragmented populations are documented from Cumberland County eastward to the Delaware River and northward to Monroe County. A northwestern

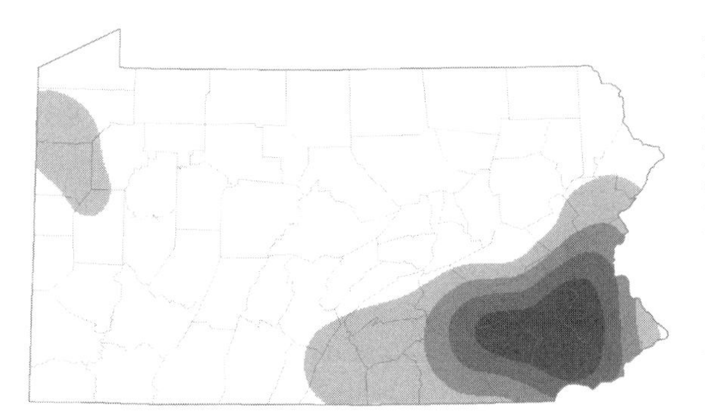

*Fig. 4.14.* Distribution of the Bog Turtle, *Clemmys muhlenbergii.*

population existed in Crawford and Mercer counties but is now considered historic or extirpated, with the last verified specimen captured in 1928 (C. Bier, personal communication). Although significant acreage of habitat exists, limited surveying has failed to confirm the northwestern population in recent years. Because of increased interest in the species and survey requirements when land development impacts are proposed, new occurrences of bog turtles continue to be discovered in Pennsylvania. Most of the new occurrences are presumed to have a low number of individuals and are effectively isolated from other colonies (C. A. Urban, personal communication).

### *COMMUNITY TYPE/HABITAT USE*

The bog turtle is a habitat specialist that relies on early successional, groundwater-driven emergent wetlands. The classic example of bog turtle habitat is a spring-fed emergent wetland meadow with dominant vegetation, consisting of sedges and other low herbs, often containing a scrub/shrub wetland component and with soft mud or soft soils. Herbaceous species typically encountered in Pennsylvania's bog turtle wetlands include sedges (*Carex* spp.), skunk cabbage (*Symplocarpus foetidus*), and cattail (*Typha latifolia*). Common woody species include red maple (*Acer rubrum*), silky dogwood (*Cornus amomum*), alders (*Alnus* spp.), and willows (*Salix* spp.). The preferred hydrologic characteristics typically encountered in Pennsylvania include persistent groundwater output and pockets of shallow surface water, which create saturated muddy conditions. During dry periods, the wet areas within suitable bog turtle habitat may be restricted to springheads.

Typically, wetlands that contain bog turtles are interspersed with a mosaic of wet and dry areas, often with subsurface flow. In addition, shallow rivulets (less than 10 cm deep), or pseudo-rivulets, are often present (U.S. Fish and Wildlife Service 2001). Persistent groundwater discharge is critically important to maintaining hydrology and saturated soils and may be linked to unique geological features. Bog turtle wetlands are usually situated above the floodplain. The third criterion of bog turtle habitat is suitable soils—saturated mud or mucky substrates a minimum of 7.5–10 cm deep—to facilitate burrowing. In Pennsylvania, the typical soil in a bog turtle wetland consists of a saturated mineral soil, for example, loam, and silt loam. In some areas, a true organic muck or peat may occur, or an organic surface layer may overlay a saturated mineral horizon. Portions of a wetland supporting bog turtles may lack one or more of these three required criteria. Recent radiotelemetry research has indicated that some individuals may spend considerable time during the active season within beaver ponds, stream systems, wooded swamps, and other habitats considered to be suboptimal (C. A. Urban, personal communication). These findings may be due to seasonal shifts in habitat use or in response to natural and anthropogenic habitat stresses.

### *LIFE HISTORY AND ECOLOGY*

Bog turtles in Pennsylvania typically emerge from hibernation in late March through April. The specific timing depends on local weather conditions. Early research suggested that bog turtles exhibited a bimodal activity cycle and aestivated during the hot, dry summer months (Ernst 1985). Recent telemetry research has indicated that bog turtles remain active and may not be visible because of dense vegetation in the summer and fall. Bog turtles are often associated with tussock sedge (*Carex stricta*) and other tussock-forming vegetation, which create a wide range of micro-atmospheric conditions allowing bog turtles to choose from a variety of temperature, solar, moisture, and humidity ranges. Bog turtles occasionally bask in the open; however, most individuals we have found are partly submerged in muck or shallow water with a portion of their carapace exposed or partially or totally obscured by living or dead vegetative matter. Bog turtles use surface runs and underground water-filled tunnels. Many of these features are likely started by small mammals and then enlarged by bog turtles. Other tunnels used by bog turtles result from subsurface water flow and groundwater discharges.

Ernst (1977) studied the geographic home range of bog turtles in Lancaster County, Pennsylvania, and found the mean home ranges of males and females to be 1.33 ha and 1.26 ha, respectively. In Maryland, Morrow et al. (2001) found that although males expand their geographic home range during the mating season, geographic home range sizes between males and females were not significantly different. Individual turtle home ranges varied from 0.003 ha to 3.12 ha, with considerable variation between years and sites. They suggest that the species' geographic home range may increase with decreasing habitat quality. Bog turtles do not defend a defined territory, but adult males almost always attack or threaten smaller males (Ernst et al. 1994). The probable life span of bog turtles is twenty-five to thirty-five years or more. In Pennsylvania, one (currently) living, wild female was recently documented at forty-eight years old (G. Gress, personal communication). Bog turtles are omnivorous and will eat insects, slugs, worms, frogs, salamanders, *Carex* seeds, Japanese beetles, berries, cattails, skunk cabbage, snails, and carrion (Nemuras 1967, Ernst 1985, Holub and Bloomer 1977 as cited in Ernst et al. 1994, T. Amitrone, personal communication). In Pennsylvania, bog turtles typically return to their hibernacula in October. They have been found hibernating in muskrat burrows, muddy rivulets, roots of vegetation, open marshes, and subterranean tunnels (Ernst et al. 1989, T. Amitrone, personal communication, G. Gress, personal communication, personal observation). Some bog turtles inhabit communal winter retreats and may exhibit site fidelity to their overwintering site (Ernst 1977, G. Gress, personal communication).

Breeding occurs from late April through early June (Barton and Price 1955). During mating, males exhibit aggressive behavior, often biting and chasing females. Copulation can occur on both land and in water. In Pennsylvania, bog turtles generally nest from June through early July. An average of three (range: one to six) eggs are deposited within a sedge hummock or sphagnum mat, or in soft soil above the water line (Ernst et al. 1994, Somers et al. 2000, Whitlock 2002). The female lays only one clutch of eggs per nesting season and may only nest once every two or three years (Somers et al. 2000). Heat and humidity are required for proper incubation of the eggs. In laboratory studies, Pennsylvania hatchlings emerged after a mean incubation period of fifty-five days (Zappalorti et al. 1995). Hatchlings in the wild typically emerge from mid-August through September and overwinter at or near the nest site. The growth rate of bog turtles is rapid for the first several years of growth. As the turtle matures, the rate of growth slows. Bog turtles are considered to be mature at an age of six to ten years (Ernst 1977).

### *THREATS*

In Pennsylvania, bog turtles have been lost from more than 50 percent of historical sites (Lee and Norden 1996), and possibly extirpated from the northwest part of the state. Primary factors in the species' decline in Pennsylvania are the loss of suitable habitat and illegal collection. Significant secondary factors are predation and the small, isolated nature of many remaining colonies.

Loss, fragmentation, and degradation of fragile wetland habitat are the primary threats to the northern population of bog turtles (U.S. Fish and Wildlife Service 2001). With the loss of family farms in southeastern Pennsylvania, wetlands that were formerly grazed or periodically mowed or burned are now subjected to natural succession that gradually transforms emergent wetlands into shaded scrub/shrub or forested swamps, unsuitable for the continued existence of a bog turtle colony. Smaller sedge meadow sites can all but disappear because of succession in less than twenty years (Lee and Norden 1996). Residential, commercial, and industrial development, as well as road construction and agricultural practices, have destroyed countless acres of bog turtle habitat in the past. Continued suburban sprawl and associated road construction adjacent to bog turtle areas result in increased road mortality, disturbances to surface and groundwater hydrology, changes to the vegetative community, and human contact. Construction of roads also degrades or eliminates dispersal corridors between areas of nearby suitable habitat. Replacement of more "movement friendly" bridges with culverts can present physical and possibly behavioral barriers to the species. Sedimentation and increased nutrients degrade habitats by decreasing microtopography (i.e., hummocks), accelerating succession, and changing the vegetative composition by decreasing diversity and increasing invasive species. Increased storm water runoff may result in higher water levels, increased pollutants (herbicides, oils, sediments), and increased scour to the wetlands (and adjacent streams). Groundwater withdrawal (for drinking water, agriculture) may dry up wetland habitats.

Invasive plants such as reed canarygrass (*Phalaris arundinacea*), common reed (*Phragmites australis*), purple loosestrife (*Lythrum salicaria*), and even cattail (*Typha latifolia*) can form dense homogeneous stands, eliminating or reducing basking and nesting areas and making movements more difficult. Establishment by alien woody species such as multiflora rose (*Rosa multiflora*) and succession by native species such as red maple (*Acer rubrum*) create shading and can contribute to higher transpiration rates, degrading the habitat for bog turtles. Mowing or cutting of vegetation in wetlands—legal in Pennsylvania—may degrade or eliminate the area as suitable habitat, and can physically harm or kill individual turtles or destroy their elevated nests.

Because of their rarity, small size, and handsome appearance, the demand for bog turtles as pets is a major reason for their decline in numbers (Hulse et al. 2001, U.S. Fish and Wildlife Service 2001). To reduce illegal collection, the locations of most bog turtle sites are not public knowledge. However, this may hamper conservation efforts, because potential local "watchdogs" are unaware of the species' presence.

Bury (1979) and Klemens (in U.S. Fish and Wildlife Service 2001) list a variety of potential predators. Although predation of bog turtles is difficult to measure, it is likely that predation from raccoons (*Procyon lotor*) and other mammals that forage along the edges of water bodies is higher in smaller, more linear systems where there is a greater edge-to-area ratio. Predation rates are also probably greater near agricultural and suburban areas because of the increased number of human "subsidized" predators (e.g., raccoons, opossums).

The bog turtle frequently occurs in small, isolated colonies. This makes the entire colony susceptible to inbreeding and vulnerable to predation, human collection, habitat loss, and localized pollution events (e.g., chemical spill). It also can give the perception that the species is more common, whereas the isolated colonies may be functionally extinct.

*CONSERVATION AND MANAGEMENT NEEDS*

An accepted conservation plan was prepared by Klemens for the northern population of the bog turtle (U.S. Fish and Wildlife Service 2001). Many of the recommended conservation and management priorities are currently being conducted in Pennsylvania. A few tasks, such as the development (and use) of standardized bog turtle survey protocols, have already been completed. The first Habitat Conservation Plan (HCP) for the bog turtle is currently under development in Pennsylvania. This particular plan strives to identify and conserve the best bog turtle sites within three watersheds in southeastern Pennsylvania and Delaware and to protect these sites in perpetuity. HCPs in other watersheds should be developed and implemented.

Several government agencies have direct or indirect roles that affect the bog turtle and its habitat. The U.S. Fish and Wildlife Service and Pennsylvania Fish and Boat Commission regulate the "take" (i.e., collection, harm, kill) of bog turtles, but neither agency regulates its habitat. The Pennsylvania Department of Environmental Protection and the U.S. Army Corps of Engineers both regulate work in wetlands and watercourses. However, neither agency can require an upland buffer nor prohibit mowing or cutting vegetation in a wetland unless a permit is obtained. As part of their permit process, a screening for potential bog turtle habitat must be conducted. The habitat screening and subsequent conservation and mitigation requirements can be powerful tools in bog turtle conservation. Both the Pennsylvania Department of Conservation and Natural Resources and Pennsylvania Game Commission manage lands containing bog turtle populations. The Pennsylvania Department of Transportation is continuously upgrading roads, replacing bridges and culverts, and conducting other maintenance activities within the geographic range of the bog turtle. Several nonprofit conservation organizations own or hold easements on bog turtle wetlands. In early 2006, Project Bog Turtle North, a nonprofit organization consisting of conservation groups and concerned individuals and companies, was formed. It is hoped that this group can work effectively with the aforementioned agencies to share resources to establish and implement conservation priorities for the long-term preservation of the species.

*MONITORING AND RESEARCH NEEDS*

Klemens (in U.S. Fish and Wildlife Service 2001) advises periodic monitoring (at least every five years) of known sites for population trends and for changes and threats to the habitat. Parameters to be monitored include, but are not limited to: population size and recruitment, succession, invasive plants, predation, conditions of hydrology, and changes to upland buffers. Periodic surveillance for poachers or environmental impacts (e.g., illegal filling or draining) should suffice on sites where access is limited or prohibited. However, at other sites, surveillance for these activities will need to be more frequent, if not continuous; neighborhood watch groups may be necessary (U.S. Fish and Wildlife Service 2001). It is unlikely that any metapopulations of bog turtles within Pennsylvania will persist through the twenty-first century without active management from conservation biologists, private landowners, regulatory agencies, and the general public.

Klemens (in U.S. Fish and Wildlife Service 2001) described research/survey priorities for the northern population of the bog turtle within an implementation schedule. We have summarized and modified his plans somewhat to be more specific for research and survey needs in Pennsylvania. In Pennsylvania, the most critical needs are to identify extant and historical sites in southeastern and eastern Pennsylvania, especially where developmental pressure is the highest. Areas to be surveyed should include riparian systems between known sites for other areas of habitat and for dispersal corridor suitability, other areas of suitable habitat within the watershed and watersheds adjacent to those with known populations, and historical sites and adjacent areas of suitable habitat in northwestern Pennsylvania. In addition, other research needs include conducting further studies on life history and ecology to improve our knowledge on population dynamics, movements (especially related to dispersal between colonies), genetics, predation rates, research on controlling invasive plant species, and preparing a predictive geographic information system model to assist in locating new potential bog turtle sites.

*Authors:* JAMES M. DRASHER, AQUA-TERRA ENVIRONMENTAL LTD.; THOMAS G. PLUTO, U.S. ARMY CORPS OF ENGINEERS

*Fig. 4.15.* The Kirtland's Snake, *Clonophis kirtlandii*. Photo courtesy of Tom Diez.

## Kirtland's Snake

Order: Squamata
Family: Colubridae
*Clonophis kirtlandii*

The Kirtland's snake was selected as a Pennsylvania Species of Greatest Conservation Need because of its rarity and limited distribution in Pennsylvania (fig. 4.15). Kirtland's snakes are listed as Endangered in Pennsylvania, and their collection is prohibited (Pennsylvania Code 2008). The global status of the Kirtland's snake is Imperiled (G2, NatureServe 2009).

### *GEOGRAPHIC RANGE*

The Kirtland's snake is considered a prairie peninsula species (Conant 1978). The bulk of its geographic range occurs in the north-central Midwest and extends from central Illinois eastward through Ohio and terminates with a large disjunct area in western Pennsylvania. The southern limit occurs in northern Kentucky, and the northernmost populations are located in southern Michigan (Wright and Wright 1957, Conant and Collins 1998, Hulse et al. 2001). It is considered a species of possible occurrence in Missouri (Johnson 1987) and Wisconsin (Hoy 1883).

### *DISTRIBUTION AND RELATIVE ABUNDANCE IN PENNSYLVANIA*

Kirtland's snakes are known historically from Allegheny, Butler, Forest, and Westmoreland counties in Pennsylvania (Hulse et al. 2001; fig. 4.16). Throughout the past forty years, repeated attempts to document this species at historic localities have failed. It is possible that the species is no longer extant in the commonwealth. Recent surveys suggest a decline throughout the entire geographic range of the species. Populations were once documented from more than 100 counties within eight states but have been observed in only a quarter of those throughout the past twenty-five years (Wilsmann and Sellers 1988).

### *COMMUNITY TYPE/HABITAT USE*

Kirtland's snakes are typically encountered in areas associated with crayfish burrows, including prairie wetlands, wet meadows, the grassy edges of creeks, streams, and ponds, and relatively open, wet woods (Conant 1943, Wright and Wright 1957, Minton 1972, Wilsmann and Sellers 1988, Phillips et al. 1999). However, along its geographic range periphery, it is frequently encountered in more wooded settings (Conant 1943). Kirtland's snakes are more commonly encountered in damp habitat remnants and vacant lots in urban settings than in undeveloped areas (Wright and Wright 1957, Smith 1961, Minton 1972, McCoy 1982,

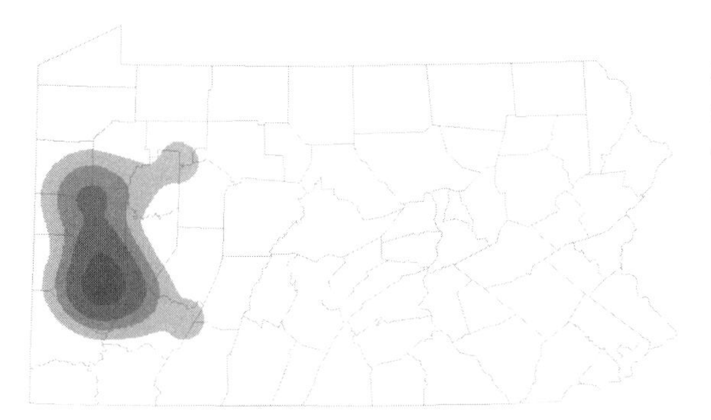

Fig. 4.16. Distribution of the Kirtland's Snake, *Clonophis kirtlandii*.

Phillips et al. 1999). Throughout the active season, Kirtland's snakes are nearly always associated with crayfish burrows. It is not known whether these burrows are also used for hibernation or whether relatively short seasonal movements occur. The relatively large number of individuals struck by vehicles suggests seasonal movements (Minton 1972).

*LIFE HISTORY AND ECOLOGY*

The Kirtland's snake is one of the rarest and most secretive snake species in North America, and consequently, little is known concerning its life history. A highly fossorial species, it seeks refuge in crayfish burrows or beneath logs and surface debris by day and typically reserves surface activity for the nocturnal hours or extremely mild and overcast days (Wright and Wright 1957, Phillips et al. 1999, Hulse et al. 2001). Its diet consists primarily of earthworms, leeches, and slugs (Atkinson 1901, Conant 1938, Minton 1944, Wright and Wright 1957, Tucker 1977, Phillips et al. 1999), but anecdotal reports of individuals feeding on insects and crayfish have also been reported (Bavetz 1993, Thurow 1993). Potential predators of Kirtland's snakes include other snakes, birds, carnivorous mammals, fish, and humans (Wilsmann and Sellers 1988, Harding 1997); however, the highly fossorial lifestyle likely limits predation. When encountered, individuals often become rigid and immobile, flattening their bodies, but retreat to nearby refugia when touched (Smith 1961, Conant and Collins 1998, Phillips et al. 1999, Hulse et al. 2001). The activity season for Kirtland's snakes has not been documented in Pennsylvania, but activity documented in Ohio ranges from March to October, with peak activity in April and May, likely coinciding with the breeding season (Conant 1938).

Females are courted by multiple males (Anton et al. 2003) shortly after individuals emerge from hibernation (Smith 1961, Minton 1972, Phillips et al. 1999, Hulse et al. 2001). Birth is related to temperature but typically occurs in August and September (Conant 1943, Wright and Wright 1957, Fitch 1970, Phillips et al. 1999); however, in unseasonably warm seasons, births may occur in late July (Minton 1972). The young are born live and typically average 7.3 in number; however, broods of four to fifteen offspring have been reported (Tucker 1976). Females in virtually all natracine snake species reproduce annually (Rossman et al. 1996). Because the Kirtland's snake is a natracine species, an annual female reproductive cycle is assumed.

*THREATS*

The decline of Kirtland's snakes has not gone unnoticed, and numerous researchers have speculated on potential causes and continued threats. Although many potential threats exist, the most eminent is habitat loss (Garman 1892, Minton 1972, McCoy 1982, Wilsmann and Sellers 1988, Harding 1997, Phillips et al. 1999). Direct habitat loss from residential and agricultural developments, and habitat alteration from wetland drainage and the destruction of native prairie marshlands, has been occurring on a relatively large scale since the settlement of the Midwest (Garman 1892).

Earthworms constitute the largest percentage of the diet of Kirtland's snakes (Harding 1997, Phillips et al. 1999), and a reduction of their numbers could pose a serious threat to remaining populations. Earthworm abundance, biomass, and biodiversity are negatively affected by soil pollutants (Spurgeon and Hopkin 1999), which can include herbicides, pesticides, oil contamination, and acid mine drainage. Competition with other earthworm specialists, such as shorthead garter snakes, Butler's garter snakes, eastern ribbon snakes, eastern garter snakes, northern brown snakes, and northern redbelly snakes, is another possible threat (Harding 1997), although direct evidence of interspecific competition is lacking.

Vehicular traffic can be a great source of mortality (Minton 1972, Bavetz 1993, Harding 1997). Land-management practices, such as prescribed burning and mowing, can negatively affect populations if performed when snakes are likely to be aboveground (Bavetz 1993). Because crayfish burrows are an integral component to the life history of Kirtland's snakes, their loss could also constitute a major threat (Tucker

1994, Phillips et al. 1999). Other threats include disease, long-term climatic changes, and removal from the wild by unscrupulous collectors (Harding 1997).

*CONSERVATION AND MANAGEMENT NEEDS*

Any sites where Kirtland's snakes are found should receive a high priority for protection. Conservation of this species will require protection of key habitats, as well as the control or reduction of threats from over-collecting and road mortality.

*MONITORING AND RESEARCH NEEDS*

Populations must first be identified before monitoring and management can be addressed. Therefore, the major need is for continued surveying of suitable habitats near historic sites. If Kirtland's snakes are found, a detailed life history examination should be conducted to learn specific population parameters and devise a comprehensive management and conservation plan. Other pressing research needs include an investigation of the threats to Kirtland's snake populations, including habitat loss and the ecological relationship between Kirtland's snakes and other earthworm specialists. Information garnered from these studies should culminate into a comprehensive management and conservation plan for Kirtland's snakes as well as a monitoring protocol.

*Author:* BEN JELLEN, WESTERN PENNSYLVANIA CONSERVANCY

## Timber Rattlesnake

Order: Squamata
Family: Viperidae
*Crotalus horridus*
(also *Responsibility Concern*)

The timber rattlesnake was selected as a Pennsylvania Species of Greatest Conservation Need because of concerns about illegal killing and collection, declines in abundance in some areas, its specialized habitat requirements, and as a representative of the guild of species associated with rocky habitats (fig. 4.17). Although the timber rattlesnake is a state Candidate Species (Pennsylvania Code 2008), rattlesnakes may be legally collected in Pennsylvania with the appropriate permit, with an annual limit of one snake (Pennsylvania Fish and Boat Commission 2008). The global status of the timber rattlesnake is Apparently Secure (G4, NatureServe 2009).

*Fig. 4.17.* The Timber Rattlesnake, *Crotalus horridus*. Photo courtesy of Howard Reinert.

*GEOGRAPHIC RANGE*

The timber rattlesnake is distributed widely from New Hampshire southward through the Appalachian Mountains to northern Florida and westward along the Gulf Coast to eastern Texas, Oklahoma, and Kansas. In the Midwest, it is found as far north as Minnesota and Wisconsin within the Mississippi River drainage and in southern Illinois, Indiana, and Ohio (Ernst and Ernst 2003). There has been a range-wide decline in the geographic distribution and abundance of the timber rattlesnake (Brown 1993). It has been extirpated from Maine and southern Ontario and is listed as an Endangered Species in Connecticut, Massachusetts, New Hampshire, New Jersey, Ohio, Vermont, and Virginia; a Threatened Species in Illinois, Indiana, New York, and Texas; a Candidate Species in Pennsylvania; and a Species of Concern in Minnesota, West Virginia, and Wisconsin (Ernst and Ernst 2003).

*DISTRIBUTION AND RELATIVE ABUNDANCE IN PENNSYLVANIA*

Before European settlement, the geographic range of the timber rattlesnake probably spanned most of the commonwealth. However, urbanization, agricultural development, and intense persecution have resulted in extensive range reduction and fragmentation (Martin 1982, Martin et al. 1990). The current distribution of timber rattlesnakes in Pennsylvania can be generally delineated on the basis of topographic relief and physiographic provinces (cf. Guilday 1985) (fig. 4.18). At present, this species is restricted to the forested Appalachian Mountain region that traverses the central portion of the commonwealth in a wide band from northeast to southwest. The leading ridge of the Appalachian Mountains (the Kittatinny Ridge) currently defines the southern extent of the range east of the Susquehanna River. Populations still persist east of the Kittatinny Ridge in the South Mountain region of Adams, Cumberland, and Franklin counties. Rattlesnakes are now absent from all of the Piedmont and Atlantic Coastal Plain provinces in the southeast. The western slope of the Allegheny High Plateau and Pittsburgh Plateau sections of the Appalachian Plateau Province defines the western edge of the range. The heavily forested north-central portion of the commonwealth harbors the largest populations (Martin et al. 1990).

### *COMMUNITY TYPE/HABITAT USE*

In Pennsylvania, timber rattlesnakes inhabit deciduous, hardwood forest (Reinert 1984a). A comparison of the distribution of rattlesnakes with forest cover within the commonwealth (Ferguson 1968, Brenner 1985) indicates that the most frequent occurrence and highest population densities are associated with oak forests. Suitable habitat in Pennsylvania consists of three major components: (1) foraging sites, (2) basking sites, and (3) overwintering sites (hibernacula). Foraging habitat typically has dense canopy cover (Reinert 1984b). Males and nongravid female rattlesnakes spend the bulk of their active season in such habitat. Forest openings, either as naturally occurring rock outcrops or human-created sites (road edges, pipelines, power lines, and timbered areas), are used for basking associated with shedding, digestion, and reproduction. Basking habitat has less than 20 percent overstory canopy closure (Reinert 1984b). Overwintering hibernacula are typically located on timbered, rocky hillsides having a southern or southeastern exposure. However,

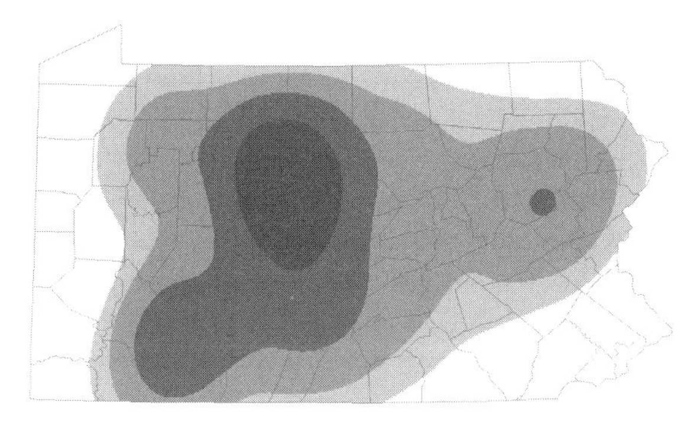

*Fig. 4.18.* Distribution of the Timber Rattlesnake, *Crotalus horridus*.

hibernacula are not restricted to such areas and have been discovered on north and west-facing slopes (H. K. Reinert, personal observation).

### *LIFE HISTORY AND ECOLOGY*

Timber rattlesnakes are ectothermic organisms with low energy requirements. They remain inactive during a six-month overwintering period. Overwintering aggregations of fifty or more adult rattlesnakes may occur, but most hibernacula in the commonwealth probably contain less than twenty adult snakes. Individuals establish a strong affinity for a particular hibernating site and will return to the same hibernaculum every winter, possibly for their entire lifetime. Rattlesnakes occupying the same hibernaculum have been found to share genetic similarity (Bushar et al. 1998).

During its active period (mid-April to mid-October), the timber rattlesnake functions as forest-floor predator on a variety of small mammals, which it captures from ambush sites adjacent to fallen logs (Reinert et al. 1984). Mice, voles, and chipmunks form the bulk of the diet (Surface 1906, Reinert et al. 1984, Clark 2002). Foraging animals in Pennsylvania range an average of 1.7 km from their hibernaculum over the summer season but may travel as far as 7 km (Reinert 1991). Complete season foraging ranges encompass an average of 105 ha for males and 50 ha for nongravid females (Reinert 1991, Reinert and Rupert 1999). In contrast to the rest of the population, gravid females occupy basking sites throughout most of the active season to maintain optimum conditions for embryo development. They do not feed during the later periods of gestation (Reinert et al. 1984), and their limited movements encompass areas averaging only 10 ha. Typically, they remain within 450 m of their hibernaculum; however, some may move more than 1 km to preferred basking areas (Reinert and Rupert 1999, Reinert 2005). Potential natural predators include hawks, owls, turkeys, coyotes, foxes, bears, weasels, black racers, and milk snakes.

Genetic studies indicate that timber rattlesnakes exhibit a metapopulation structure where each hibernaculum may represent a local population linked to nearby hibernacula through gene flow mediated by landscape features such as available basking sites (Bushar et al. 1998). Gene flow among populations may be easily disrupted by filter barriers such as roadways and unsuitable habitat. Consequently, genetic differentiation can be substantial even among nearby local populations (Bushar et al. 2005).

The timber rattlesnake is a long-lived (up to thirty years), late-maturing (five to nine years) species, with a low reproductive rate (Brown et al. 2005). Mating occurs from mid-July through October (Brown 1995, Coupe 2002). Males may travel extensively (more than 2 km a day) during the mating season to court distant females (Reinert and Zappalorti 1988a, Coupe 2002). Male-to-male combat during the mating season has been observed in Pennsylvania (J. Prowant, personal communication). Females store sperm and ovulation occurs after emergence from hibernation the following spring (Brown 1991). Birth occurs from late August through October. There are from three to fourteen young in a litter, with an average of eight (Galligan and Dunson 1979, Fitch 1985). Females typically produce broods once every three years (Brown 1993, Martin 1993). Birthing frequently occurs some distance (up to 1 km) from hibernacula, and neonates have been observed trailing adults to overwintering sites (Brown and MacLean 1983, Reinert and Zappalorti 1988b, Reinert 2005).

### *THREATS*

The major threats to existing populations of timber rattlesnakes include habitat loss, degradation, and fragmentation. Suburban sprawl and rural development cause habitat fragmentation and associated population isolation. Oil and gas exploration, construction of power lines, pipelines, and roadways, timbering, quarrying, and surface mining activity destroy, degrade, and fragment rattlesnake habitat.

Collecting and killing snakes also have negative effects on timber rattlesnakes. Organized and recreational rattlesnake hunting, as well as indiscriminant, wanton killing of incidentally encountered rattlesnakes can reduce population numbers. Improved trail and road access to backcountry areas may increase disturbance of rattlesnake populations by recreational and illegal collectors. Secondary paved roads, highways, and high human population density can cause high mortality in rattlesnake populations through accidental and deliberate causes.

### *CONSERVATION AND MANAGEMENT NEEDS*

The first step in ensuring the continued existence of timber rattlesnakes in Pennsylvania is to reduce the loss of snakes from existing populations. Small populations, low reproductive rates, and the relative ease of finding gravid females make this species extremely vulnerable to reckless killing and overcollecting. Changing the official status of this species from Candidate to Threatened would offer sustained and legitimate legal protection by closing the hunting season, prohibiting organized rattlesnake hunts, and instituting protective regulations. A second initiative that would provide substantial benefits to timber rattlesnake populations would be the complete protection for this species on all government-owned lands (State Game Lands, State Forest Lands, State Forest Natural Areas, State Parks, National Forests, and Military Reservations) through the joint cooperation of government agencies. To maintain viable, representative populations of this species, a minimum of eight intensively monitored and managed Timber Rattlesnake Management Areas should be established in the regions that currently contain extensive habitat and large populations.

To protect populations further, all private individuals/companies and government agencies undertaking projects within timber rattlesnake habitat should be required to perform a species impact review and assess the effect of the project on rattlesnake populations and habitat. This should include the protection of known hibernacula and the establishment of adequate buffer zones to maintain suitable foraging habitat.

Timber rattlesnakes are commonly viewed as dangerous animals that conflict directly with human habitation and human survival. Education programs that dispel these prevalent myths, elucidate the role of rattlesnakes as a beneficial and unique component of the forest ecosystem, and foster a greater tolerance for continued existence of rattlesnakes need to be developed and widely disseminated.

### *MONITORING AND RESEARCH NEEDS*

An assessment of the current distribution and occurrence of timber rattlesnakes is presently being performed (C. A. Urban, personal communication). This database should be used consistently to monitor potential impacts of development, logging, utility (pipeline, power line, oil exploration, cellular towers, windmills), and recreational projects (trails, parks, and campgrounds) upon timber rattlesnake habitat. All projects should be evaluated by biologists for the extent of their impacts and, if necessary, altered to reduce those impacts.

On the basis of the results of a statewide survey of timber rattlesnakes, small, isolated, and imperiled populations facing imminent extirpation should be identified. An assessment of the genetic structure of Pennsylvania populations is necessary to determine the extent of population isolation and its effect on gene flow and genetic diversity. Some research in this area has been completed (Bushar et al. 1998), and additional projects are under way (L. Bushar, personal communication). An intensive study to evaluate the acute effects of timbering operations on timber rattlesnake populations is also presently under way (H. K. Reinert, personal observation). This research should be extended to ascertain the long-term response of rattlesnake populations to timbered landscapes. In addition, experimental research should be undertaken to assess techniques to manage and improve habitat for timber rattlesnakes in Pennsylvania.

*Author:* HOWARD K. REINERT, THE COLLEGE OF NEW JERSEY

## Eastern Massasauga

Order: Squamata
Family: Viperidae
*Sistrurus catenatus catenatus*

The eastern massasauga was selected as a Pennsylvania Species of Greatest Conservation Need because it has suffered a major decline in abundance, is subject to illegal collecting, is globally rare, and has a restricted habitat (fig. 4.19). Eastern massasaugas are listed as Endangered in Pennsylvania, and their collection is prohibited (Pennsylvania Code 2008). In 1999, it became a candidate for listing as Threatened by the U.S. Fish and Wildlife Service. The global status of the eastern massasauga is Vulnerable/Apparently Secure (G3/G4, NatureServe 2009).

*Fig. 4.19.* The Eastern Massasauga, *Sistrurus catenatus catenatus*. Photo courtesy of Tom Diez.

### *GEOGRAPHIC RANGE*

The geographic range of the eastern massasauga extends from central New York and western Pennsylvania westward through Ohio, southwestern Ontario, Michigan, Indiana, Illinois, southern Wisconsin, extreme southeastern Minnesota, and eastern Iowa to central Missouri (Ernst and Ernst 2003).

### *DISTRIBUTION AND RELATIVE ABUNDANCE IN PENNSYLVANIA*

The eastern massasauga in Pennsylvania has experienced a massive reduction in its geographic distribution since Atkinson (1901) reported on its occurrence in the commonwealth more than 100 years ago. The historic geographic range of this species extended from Armstrong County northward through Butler, Lawrence, Mercer, Venango, and Crawford counties, and Pennsylvania populations were probably continuous with populations in Ohio (Atkinson and Netting 1927, Swanson 1930, Reinert 1985, Reinert and Bushar 1992). At present, extant populations are known to exist at only four geographically isolated sites in Butler and Venango counties (Jellen 2005; fig. 4.20).

### *COMMUNITY TYPE/HABITAT USE*

In Pennsylvania, the eastern massasauga is an inhabitant of fields, remnant prairies, and shallow, open wetlands. It occupies structurally different habitats on a seasonal basis (Reinert and Kodrich 1982). The close geographic proximity of these differing habitats is required for the survival of this species at any given location. Snakes overwinter submerged or partially submerged in mixed forb and rush-sedge marshes where crayfish burrows, mole tunnels, and other excavations allow them to reach the water table (Maple 1968, Reinert and Kodrich 1982, Seigel 1986). Following emergence from hibernation, they shift to slightly higher elevation, old field, or remnant prairie habitat dominated by grasses, goldernrods, and asters (Reinert

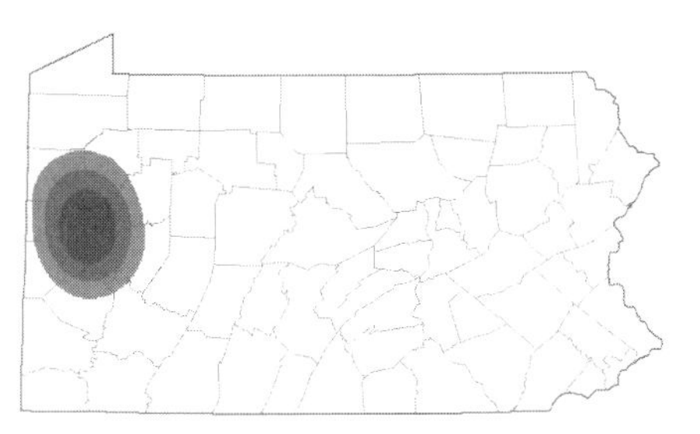

*Fig. 4.20.* Distribution of the Eastern Massasauga, *Sistrurus catenatus catenatus*.

and Kodrich 1982). These habitats frequently harbor an abundance of small mammal prey, which probably provide the major impetus for the observed habitat shift. Within this active season habitat, exposed areas with low or very sparse vegetation are often selected by gestating females for basking before parturition.

### *LIFE HISTORY AND ECOLOGY*

Emergence from hibernation occurs between late March and early April. Snakes rapidly move from the wetlands to surrounding old field and prairie habitats where they function as a predator on small mammals during the six- to seven-month active period. The return to wetland hibernacula occurs in mid- to late October (Reinert and Kodrich 1982). Radiotelemetry studies suggest a broad range of both inter- and intrapopulation variation exist in movement behavior (Reinert and Kodrich 1982, Weatherhead and Prior 1992, Johnson 2000). A summary of four telemetry studies from across the geographic range of the eastern massasauga suggests that total seasonal activity ranges of males and nongravid females average 25 ha (Johnson et al. 2000). Gravid females remain sedentary in their preferred habitat and have activity ranges of 2 ha or less (Reinert and Kodrich 1982, Johnson 2000).

The diet of adult eastern massasaugas consists predominantly of rodents (Keenlyne and Beer 1973). Meadow voles, short-tailed shrews, and a white-footed mouse were identified from scats of adult Pennsylvania specimens (Reinert 1978). Small snakes may be a significant dietary component of neonatal eastern massasaugas (Keenlyne and Beer 1973, Reinert 1978, Seigel 1986).

Mating has been observed in late July and August (Reinert 1981, Johnson 1995). Aggressive male-to-male interaction has been observed during the summer mating season in Illinois (Shepard et al. 2003). Females store sperm and ovulation occurs after emergence from hibernation the following spring (Aldridge et al. 2005). Parturition occurs in August or early September. In Pennsylvania, broods averaging seven young are produced once every two years (Reinert 1981). Maturity may be reached in two to three years in Pennsylvania (Reinert 1978), and the total life span may exceed twenty years.

### *THREATS*

Habitat loss is the major threat to eastern massasauga populations in the commonwealth. Wetland habitat required for overwintering has been drained, filled, or flooded, and the old field/remnant prairie habitat necessary for foraging and successful reproduction has been reduced by encroachment of forests and woody shrubs (Reinert and Bushar 1992, Jellen 2005). Several historic locations have been severely impacted by mining operations for oil, gas, and coal (Reinert and Bushar 1998). However, Swanson (1952) believed that oil wells created open habitat that was eventually beneficial to the species.

There is an increasing number of roads and higher traffic volume on many roads through rural regions surrounding the sites occupied by eastern massasaugas. The relatively small size and slow-moving nature of these snakes make them extremely vulnerable when crossing roadways. Reinert (1978) reported that of the five dead eastern massasaugas examined in Pennsylvania in 1976 and 1977, three were road kills, one was killed by a resident, and only one was due to natural predation. There are numerous historic reports of large numbers of eastern massasaugas being killed on oil leases, road construction sites, and when opportunistically encountered (Swanson 1952, P. Swanson personal communication). The eastern massasauga is routinely killed by the residents of most areas where it exists (H. K. Reinert, personal observation). The unnecessary removal of even a small number of snakes (especially gravid females) from imperiled populations could significantly impact population viability and long-term survival (Seigel and Sheil 1999).

### *CONSERVATION AND MANAGEMENT NEEDS*

The first crucial step in an effort to avoid the total extirpation of the eastern massasauga from the commonwealth is to immediately secure and manage significant acreage at the few remaining sites where extant populations persist. An intensive management program must be created for each site to improve habitat conditions and increase the amount of available habitat (Johnson et al. 2000). The management program should include removal of encroaching woody vegetation and maintenance of early successional stages (Johnson and Leopold 1998). All private individuals/companies and government agencies undertaking projects within the vicinity of potentially extant eastern massasauga populations should be required to perform a species impact review to assess the effect of the project on these snakes. Visible and tangible law enforcement programs should be instituted to enforce the prohibition of catching, taking, killing, or possession of eastern massasaugas. Rattlesnakes are commonly viewed as dangerous animals that conflict directly with human habitation and human survival. Education programs that dispel these prevalent myths, elucidate the role of eastern massasaugas in the wetland ecosystem, and foster a greater tolerance for continued existence of this species need to be developed and widely disseminated.

Genetic studies of eastern massasaugas indicate that this species exhibits a great deal of genetic differentiation even among geographically close populations (Gibbs et al. 1994, 1997, 1998). This implies that each population is genetically unique and that the maintenance of healthy genetic diversity for the species will require the maintenance of as many individual populations as possible.

*MONITORING AND RESEARCH NEEDS*

Intensive field surveys of historic sites are needed to assess the status of remaining populations. Remaining sites with extant populations and historic locations that still have the potential to harbor small populations of eastern massasaugas must be closely monitored for threats to the snakes and their habitat. Wetland hydrology also needs to be carefully monitored to maintain suitable hibernacula. The potential effects of suburban sprawl, road construction, agriculture, logging, mining (coal, gas, oil, clay, sand, and gravel), waterfowl management, and recreational projects (trails, parks, campgrounds) on eastern massasauga habitat should be evaluated by biologists.

Research is required to determine the spatial ecology, primary structural characteristics of preferred habitat, and genetic structure of the few remaining populations in Pennsylvania. Experimental assessments of the efficacy of habitat improvement are required for developing successful management programs.

*Author:* HOWARD K. REINERT, THE COLLEGE OF NEW JERSEY

HIGH-LEVEL CONCERN

## Northern Cricket Frog

Order: Anura
Family: Hylidae
*Acris crepitans*

The northern cricket frog was selected as a Pennsylvania Species of Greatest Conservation Need because of drastic population declines throughout its geographic range in Pennsylvania, restricted distribution in the state, uncertain status, and Pennsylvania's importance as the northern periphery of the its range (fig. 4.21). Although northern cricket frogs are not listed as Threatened or Endangered at the federal level or in Pennsylvania, their collection in Pennsylvania is prohibited (Pennsylvania Fish and Boat Commission 2008). The global status of the northern cricket frog is Secure (G5, NatureServe 2009).

*Fig. 4.21.* The Northern Cricket Frog, *Acris crepitans*. Photo courtesy of Richard Koval.

*GEOGRAPHIC RANGE*

The geographic range of the northern cricket frog in the eastern United States shows more of a southern affiliation. It occurs in greatest abundance from east Texas and the Florida Panhandle northeastward through the Gulf Coast states. The northern cricket frog's range constricts northward up, through the Coastal Plain states, entering southeast Pennsylvania and New Jersey, until it reaches the northernmost range in southeast New York (Conant and Collins 1991).

*DISTRIBUTION AND RELATIVE ABUNDANCE IN PENNSYLVANIA*

In Pennsylvania, the historical geographic range of the northern cricket frog appears to have been primarily in the southeast corner of the commonwealth. The majority of historical occurrences were from the southeast Coastal Plain Province. There were scattered reports of cricket frogs from the Piedmont and Valley and Ridge provinces, with the commonwealth's most northern historical report in Carbon County (McCoy 1982).

Although the northern cricket frog was geographically restricted and never abundant in Pennsylvania, current data suggest it is one of the commonwealth's rarest anurans. The northern cricket frog has drastically declined throughout its range in Pennsylvania. It appears to have disappeared throughout most of its southeast range and is declining in south-central Pennsylvania. Specific surveys in progress at historical locations have not confirmed the presence of northern cricket frogs.

There are few current verified records of the northern cricket frog in Pennsylvania. Occurrences of northern cricket frogs were confirmed during the Pennsylvania Herpetological Atlas Project from 1996 to 2002 (Hulse et al. 2001). Although this may reflect inadequate volunteer coverage throughout its geographic range, reports were only from Franklin County (T. Maret and T. Pluto, personal communication) and from Luzerne County (R. L. Koval, personal observation). The Franklin County observations are well within the species historical range, but the Luzerne County site is the commonwealth's most northern known occurrence. The current population of northern cricket frogs in Franklin County is reported to be of few individuals showing evidence of drastic decline (T. Pluto and T. Maret, personal communication). The Luzerne County location appears to be the commonwealth's stronghold for the northern cricket frog (fig. 4.22). The author has monitored a sustainable population since 1999, and as recent as 2005, has heard more than fifty singing males.

### *COMMUNITY TYPE/HABITAT USE*

Northern cricket frogs have an affinity for standing water with vegetated shores and edges, including impounded lakes, bogs, ponds, and vernal pools. They are usually found within standing water or along the shallow edges of wetlands. Northern cricket frogs reported in Franklin County were occasionally found in vernal pools and semipermanent ponds (T. Maret, personal communication) and found in somewhat larger restored wetlands (T. Pluto, personal communication). In Luzerne County, the habitat is much different, as they inhabit a large wetland complex consisting of a shallow glaciated lake and bog. The lake hosts sections of densely vegetated shorelines, emergent shrub margins, floating-leaf aquatic plants, and floating peat mats. The nearby bog exhibits an open water center with a concentric vegetative ring of bog plants quite different from the lake (R. L. Koval, personal observation).

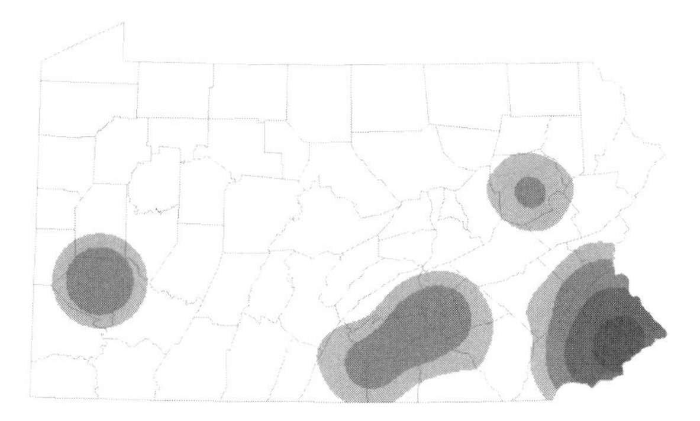

*Fig. 4.22.* Distribution of the Northern Cricket Frog, *Acris crepitans.*

### *LIFE HISTORY AND ECOLOGY*

The northern cricket frog's diet varies considerably, depending on its life stage. During its larval or tadpole stage, it consumes decaying vegetative matter. Adult northern cricket frogs consume varieties of small insects normally found within their aquatic environment. Smaller insects such as springtails, grasshoppers, damselflies, beetles, and flies are the primary food source (Labanick 1976).

Northern cricket frogs have been reported to be active year-round in the south, especially during mild winters. In Luzerne County, recently transformed juveniles have been observed in early October (R. L. Koval, personal observation). Activity depends on temperature, and this certainly varies throughout the commonwealth.

The northern cricket frog is one of the commonwealth's last species of frog to breed. Males usually are in full chorus by mid-June and may continue calling until August. Males call from emergent vegetation near the margins of wetlands, ponds, and lakes and have been observed calling from floating peat mats in bogs (R. L. Koval, personal observation).

Female northern cricket frogs may deposit up to 250 eggs onto submerged aquatic vegetation or on the bottoms of shallow-water bodies (Parmellee et al. 2002). The eggs are laid either in a small, filmy mass or singly. After forty to ninety days, the black-tipped tailed tadpoles emerge.

As colder weather approaches, northern cricket frogs may hibernate beneath cover objects found on land, under piles of leaves, or in muddy cracks and crevices along the edges of various wetlands.

### *THREATS*

A number of factors may have caused the decline of the northern cricket frog in Pennsylvania. The destruction of wetlands and vernal pools in Pennsylvania has eliminated, drained, cleared, and altered the breeding habitat of northern cricket frogs. Pollution of wetlands, ponds, lakes, and vernal pools from such contaminants as road salt, storm water runoff, phosphates, and nitrates, may also have contributed to recent declines. Aerial spraying of DDT, and other chlorinated hydrocarbon pesticides in the 1950s and 1960s, is thought to have contributed to the species' declines. Removal of aquatic vegetation along the margins of wetlands by mechanical and herbicidal control measures has reduced foraging habitat, and abusive motorized watercraft use has contributed to shoreline erosion, pollution, and disturbance of breeding habitats. Finally, the introduction of bullfrogs and predatory fish into many ponds and lakes may have reduced cricket frog numbers through predation on eggs, larvae, and adult frogs.

*CONSERVATION AND MANAGEMENT NEEDS*

Because of the rarity of the species in Pennsylvania, sites of known occurrences and any newly discovered sites warrant immediate protection. Protection of these breeding habitats should include vegetative buffer zones, restrictions on removal of aquatic vegetation, prohibition on the spraying of herbicidal and pesticide applicants, and no-wake zones along certain shorelines, especially on public waters. The protection of known breeding habitats of northern cricket frogs on private lands can be achieved through landowner conservation easements, land acquisitions, and landowner awareness.

*MONITORING AND RESEARCH NEEDS*

Continued monitoring of population densities and breeding success at the remaining locations of northern cricket frogs is critically important. Surveys at potential habitats, especially in the southeast and south-central sections of the state, are highly recommended. Extensive surveys need to be continued at nearby and adjacent wetlands to the Luzerne County site.

The most pressing research need is to determine the cause(s) of the dramatic decline in this species and develop strategies to stop and reverse the decline.

*Author:* RICHARD L. KOVAL, NORTH BRANCH LAND TRUST

## New Jersey Chorus Frog

Order: Anura
Family: Hylidae
*Pseudacris kalmi*

The New Jersey chorus frog was selected as a Pennsylvania Species of Greatest Conservation Need because of its low numbers and extremely limited distribution in Pennsylvania (fig. 4.23). New Jersey chorus frogs are listed as Endangered in Pennsylvania and are therefore protected by regulations that prohibit individuals from catching, taking, killing, and possessing them (Pennsylvania Code 2008). The global status of the New Jersey chorus frog is Secure (G5, NatureServe 2009).

*Fig. 4.23.* The New Jersey Chorus Frog, *Pseudacris kalmi*. Photo courtesy of Tom Diez.

*GEOGRAPHIC RANGE*

The New Jersey chorus frog is found on the Coastal Plain from Staten Island, New York, to the southern tip of the Delmarva Peninsula.

*DISTRIBUTION AND RELATIVE ABUNDANCE IN PENNSYLVANIA*

In Pennsylvania, the New Jersey chorus frog historically occurred in the Coastal Plain in extreme southeastern Pennsylvania. Pennsylvania is part of the northwestern extent of the geographic range for the New Jersey chorus frog, making it biogeographically and genetically important (Groves 1985a, Hulse et al. 2001). With the exception of a couple of fairly recent records, the records reported for this species in Pennsylvania are historic. The most recent record for Montgomery County is from 1985 (Pennsylvania Natural Heritage Program 2005). Three historic records are known from Bucks County with the last voucher specimen collected in 1951. A voucher specimen was collected from Montgomery County in 1942, and one record was reported by Groves (1985). Recent surveys for the New Jersey chorus frog during the Pennsylvania Herpetological Atlas could not confirm that populations

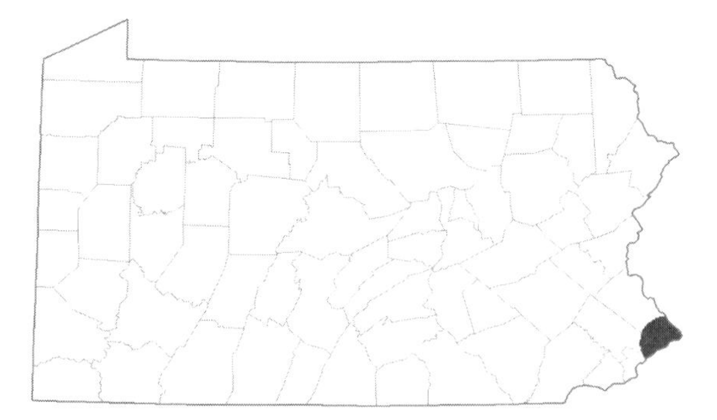

*Fig. 4.24* Distribution of the New Jersey Chorus Frog, *Pseudacris kalmi*.

still exist, but the surveys were concentrated in southern Bucks County (fig. 4.24). Because of heavy development pressures in these counties, suitable habitat may no longer be available. The species may be extirpated from the state (A. C. Hulse, personal communication).

### *COMMUNITY TYPE/HABITAT USE*

The New Jersey chorus frog is a member of the tree frog family (Hylidae), but it is not arboreal. Individuals are usually located on the ground and use a variety of wet habitats during breeding (Hulse et al. 2001). The chorus frog may use drier terrestrial areas during warmer months, where it may be found in leaf litter along the edges of swamps or marshes (Groves 1985a). After the breeding season, the New Jersey chorus frog may travel long distances from wetlands (Hulse et al. 2001). Because it is able to withstand freezing temperatures (Storey and Storey 1987), the New Jersey chorus frog hibernates mainly in terrestrial habitats under rocks, logs, ant mounds, or rodent burrows (Groves 1985a, Hulse et al. 2001). It may also hibernate with other amphibians and snakes (Groves 1985a).

### *LIFE HISTORY AND ECOLOGY*

Little information is available on prey preferences for this species. However, Whitaker (1971) showed that young western chorus frogs feed on mites and springtails usually found at pond edges. Adult frogs will eat spiders, snails, slugs, beetles, ants, and other terrestrial invertebrates (Whitaker 1971, Hulse et al. 2001). Adult chorus frogs are preyed on by snakes, particularly northern water snakes (*Nerodia sipedon*) and eastern ribbon snakes (*Thamnophis sauritus*), large spiders, and small mammals (Whitaker 1971, Groves 1985a). Salamander larvae, aquatic invertebrates, fish, and snapping turtles consume tadpoles (Groves 1985a).

Heavy rains trigger breeding, which usually takes place during late winter to early spring (early March to late May). The frogs use both permanent and temporarily inundated habitats, including forested swamps, marshes, wet meadows, floodplains, riparian corridors, ditches, and canals (Hulse et al. 2001). To attract the females to the pools, the males call both day and night (Hulse et al. 2001). Females usually enter breeding ponds in large numbers during a few nights of the breeding season (Whitaker 1971). The eggs are laid in shallow water in small groups, numbering from 500 to 1,500 eggs attached to sticks or aquatic vegetation (Wright and Wright 1949, Groves 1985a). Incubation, which may take three to thirteen days, is temperature dependent (Whitaker 1971). After hatching, the larvae will develop in thirty-five to fifty-five days and will emerge from water bodies in May or June (Groves 1985a, Hulse et al. 2001).

After the breeding period, sightings of the New Jersey chorus frog are scarce, and little is known about its geographic home range and habitat use (Hulse et al. 2001). Geographic home range information for the western chorus frog varied between 641 $m^2$ and 6,024 $m^2$ (Kramer 1974, Hulse et al. 2001).

### *THREATS*

The New Jersey chrorus frog is only found in the heavily developed southeast corner of the state. Therefore, habitat loss and fragmentation are the most imminent threats. Other threats to this species in Pennsylvania are largely unknown, but may include chemical contamination, increased mortality on roads, competition with other species (e.g., spring peepers), and disease.

### *CONSERVATION AND MANAGEMENT NEEDS*

Assuming that the species still exists in Pennsylvania, its conservation will require the protection of suitable wetland habitat. Heavy development pressures in southeastern Pennsylvania have caused extensive habitat destruction and alteration for this species.

### *MONITORING AND RESEARCH NEEDS*

Surveys are needed within its historic geographic range and the Coastal Plain physiographic province to determine if viable populations of New Jersey chorus frogs still occur in Pennsylvania. If frogs are found, the surveys should be followed up with monitoring activities at selected sites to detect changes in population numbers and distribution.

Little is known about the response of the New Jersey chorus frog to various land use practices (McLeod and Gates 1998). If extant populations of New Jersey chorus frogs are once again discovered in Pennsylvania, information will be needed to adequately identify and address specific threats and to develop conservation/management plans. Information is also lacking on interactions between the New Jersey chorus frog and the upland chorus frog (Groves 1985a, Hulse et al. 2001). To resolve this issue, additional research is needed throughout the geographic range of both species, with special emphasis placed on species overlap areas (Hulse et al. 2001). Finally, information is also needed regarding the habitat requirements of the New Jersey chorus frogs, including nonbreeding and hibernation habitats, migration corridors (Gibbons 2003), and home ranges.

*Author:* AURA STAUFFER, PENNSYLVANIA DEPARTMENT OF CONSERVATION AND NATURAL RESOURCES

## Eastern Spadefoot

Order: Anura
Family: Scaphiopodidae
*Scaphiopus holbrookii*

*Fig. 4.25.* The Eastern Spadefoot, *Scaphiopus holbrookii*. Photo courtesy of Tom Diez.

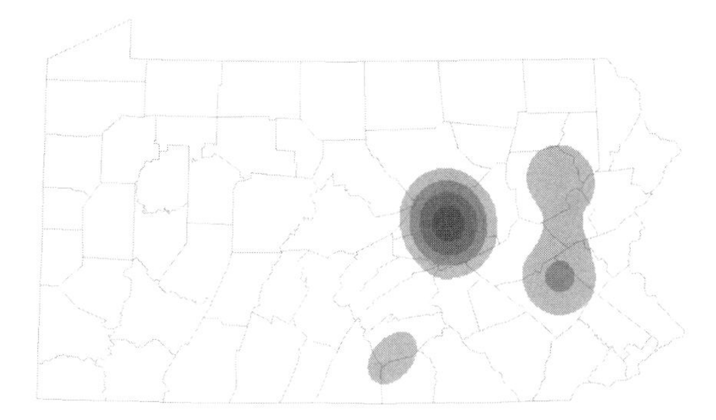

*Fig. 4.26.* Distribution of the Eastern Spadefoot, *Scaphiopus holbrookii*.

The eastern spadefoot was selected as a Pennsylvania Species of Greatest Conservation Need because of its rarity within the commonwealth (fig. 4.25). It is listed as an Endangered Species by the Pennsylvania Fish and Boat Commission, and collecting of this species is prohibited (Pennsylvania Code 2008). The global status of the eastern spadefoot is Secure (G5, NatureServe 2009).

### *GEOGRAPHIC RANGE*

The eastern spadefoot is the only spadefoot toad found east of the Mississippi River. The species ranges from southern New England to the Florida Keys and west to eastern Louisiana (NatureServe 2009).

### *DISTRIBUTION AND RELATIVE ABUNDANCE IN PENNSYLVANIA*

Only two extant breeding populations of spadefoots are known in Pennsylvania (Northumberland and Berks Counties), each of which represents a major watershed (Susquehanna and Delaware drainages, respectively). These populations appear to be disjunct. However, given the secretive nature of this animal, it could be present in isolated populations throughout the Susquehanna, Cumberland, and Delaware valleys (fig. 4.26). Anecdotal and historic records are known from the Cumberland Valley and several localities scattered throughout eastern Pennsylvania up through Monroe County.

### *COMMUNITY TYPE/HABITAT USE*

Key habitats for this species in Pennsylvania include temporary/ephemeral pools in low-lying areas, with sandy to loamy soils in agricultural areas and woodlands. Other important habitats include stream edges and floodplain wetlands. Studies are needed to determine specific habitat requirements in Pennsylvania (Hulse et al. 2001).

### *LIFE HISTORY AND ECOLOGY*

Given the fossorial nature of the eastern spadefoot and the paucity of records in Pennsylvania, little information on their biology in Pennsylvania exists. Most studies have occurred in the southern part of its geographic range. Eastern spadefoots prefer loose, friable soils, where they can use their hind feet for burrowing.

In the Northeast, spadefoots tend to be associated with sandy soils, although they have been documented as burrowing in silt loam and clay soils as well (Driver 1936, Pearson 1955). Jansen et al. (2001) discovered that spadefoots cannot burrow into sod, which has serious implications for this species when residential developments encroach on their habitat. The eastern spadefoot spends much of its time underground, with surface activity that depends on periods of heavy rainfall, particularly in the spring and midsummer, when they will exit their burrows to forage or breed (Pearson 1955). They have been documented burrowing to depths ranging from 5 cm to 2.5 m (Pearson 1955, Jansen et al. 2001). Pearson (1955) documented spadefoots spending 109 consecutive days in their burrows. Hulse et al. (2001) suggests that spadefoots may remain inactive in their burrows for more than 200 days out of a year. Although little data exist on eastern spadefoot hibernation, it is presumed that they hibernate below the frostline in these subterranean retreats (Hulse et al. 2001).

In Pennsylvania, spadefoots have been found to be active on the surface from April through September (Klemens 1993, Hulse et al. 2001). Eastern spadefoots are primarily nocturnal and are active during rain events, although they have also been known to emerge from their burrows on warm, humid, and overcast days (Bragg 1965, Mount 1975). Johnson (2003) found that spadefoots were often active on warm, humid evenings. Eastern spadefoots have very small home ranges (ca. 10 $m^2$), centered around their burrows. Although some toads exhibit high site fidelity to one burrow, many toads will have multiple (two to five) burrows within their geographic home range. Johnson's (2003) radiotelemetry study supported this site fidelity. It is currently unknown how far adult eastern spadefoots will travel to breeding habitats or how far they will disperse from breeding ponds.

Eastern spadefoots have no defined breeding season and are considered opportunistic, sporadic breeders. Reproduction occurs during, or within, one or two nights of heavy rain events and is concentrated at vernal pools, rain-filled depressions in farm fields, and along streams (Hulse et al. 2001). Breeding aggregations of spadefoot toads have been documented in temporary, or ephemeral, pools in Pennsylvania and West Virginia from April through September, often associated with tropical depressions, hurricanes, or rainfalls that exceed two inches in twenty-four hours (Green 1963, Hulse et al. 2001). Across the geographic range of the eastern spadefoot, breeding events have been known to lapse for several years (Hulse et al. 2001).

During a breeding event, male eastern spadefoots call from the perimeter of a temporary pool in shallow water. Chorusing males can be heard up to one mile away from the breeding pond. Spadefoots have also been documented calling from their burrows. Eastern spadefoots exhibit inguinal amplexus, a primitive form of mating behavior in the anurans in which males grasp females just anterior to the hind limbs, which is unlike other anuran males that grasp females behind the forelimbs. After mating, females lay their eggs in bands on vegetation in shallow water (Wright and Wright 1949, Hulse et al. 2001). Eastern spadefoots have accelerated larval development, similar to that seen in toads of the xeric Southwest. Spadefoot eggs hatch in twenty-four hours to seven days, and metamorphosis can occur within fourteen to sixty-three days, depending on ambient water temperature. Unlike the adults, the juvenile spadefoots are diurnal and will feed on small invertebrates and insects at the perimeter of the pool for two weeks before they disperse (Wright and Wright 1949, Hulse et al. 2001).

*THREATS*

Given that spadefoots occur in floodplains and valleys, they are threatened by habitat destruction from residential and industrial development, as well as habitat alteration and changes in water chemistry from agricultural practices (Jansen et al. 2001). The ephemeral habitats in which they breed often do not receive protection under the Clean Water Act (Gibbons 2003).

In Pennsylvania, current breeding populations occur in agricultural and in residential settings near Milton and the outskirts of Reading. It is likely that wetland habitats in this vicinity have been altered directly through development or indirectly through water degradation. Additional information is needed to identify and address adequately specific threats, such as development pressure, intensive agricultural practices, road mortality, and collection.

*CONSERVATION AND MANAGEMENT NEEDS*

Current information on this species in Pennsylvania is insufficient to develop a detailed conservation and management strategy. Specific needs include determination of site-specific threats, development of site management and monitoring plans for occupied habitats, vegetation and soil management of temporary pools and surrounding upland buffers, and the pursuit of conservation easements or direct acquisition for occupied sites.

*MONITORING AND RESEARCH NEEDS*

To determine the current size and geographic range of eastern spadefoot populations in Pennsylvania, surveys are needed to revisit the extant and historic records and to search for new locations around and between the currently disjunct populations. These surveys should be followed up with monitoring activities at selected sites to determine population status and detect changes in numbers.

Not much is known about the species' natural history in Pennsylvania. Information is needed regarding the habitat requirements, dispersal distances, geographic home range, and distance traveled to breeding sites. In addition, studies are needed to assess barriers to migration/dispersal and the efficacy of habitat corridors to other potential breeding populations.

*Authors:* CHRISTOPHER A. URBAN, PENNSYLVANIA FISH AND BOAT COMMISSION; KATHARINE L. GIPE, WESTERN PENNSYLVANIA CONSERVANCY

## Eastern Redbelly Turtle

Order: Testudinae
Family: Emydidae
*Pseudemys rubriventris*

The eastern redbelly turtle was selected as a Pennsylvania Species of Greatest Conservation Need because of its rarity and restricted distribution in Pennsylvania, loss of habitat, and the importance of Pennsylvania as the northern periphery of the turtle's geographic range (fig. 4.27). It is listed as a Threatened Species by the Pennsylvania Fish and Boat Commission, and collecting of this species is prohibited (Pennsylvania Code 2008). The global status of the eastern redbelly turtle is Secure (G5, NatureServe 2009).

*GEOGRAPHIC RANGE*

The redbelly turtle occurs along the Atlantic Coastal Plain and portions of the Piedmont from northern North Carolina to central New Jersey (Ernst et al. 1994). A small disjunct population occurs in Plymouth County, Massachusetts, and receives federal protection under the Endangered Species Act as a distinct population (U.S. Fish and Wildlife Service 1994).

*Fig. 4.27.* The Eastern Redbelly Turtle, *Pseudemys rubriventris*. Photo courtesy of James Drasher.

*DISTRIBUTION AND RELATIVE ABUNDANCE IN PENNSYLVANIA*

In Pennsylvania, the redbelly turtle occurs in the lower Delaware River watershed from Bucks County to the state line, the lower Susquehanna River watershed below Harrisburg, and at a few scattered locations in the Potomac River watershed (fig. 4.28). Survey data from recent years indicate this turtle's ability to colonize any appropriate habitats within its geographic range, although consistent populations remain primarily in the southeastern Coastal Plain. Like other charismatic turtle species, redbelly turtles are often picked up on roads or while nesting, and some historic and current locations may be the result of releases by humans (Conant 1951).

*COMMUNITY TYPE/HABITAT USE*

Eastern redbelly turtles in Pennsylvania rely on relatively deep waterbodies, such as moderate gradient rivers, reservoirs, ponds, and marshes, although transient turtles occur in faster-moving streams, shallow ponds or impoundments, and ditches (Hulse et al. 2001). In the industrial region around Philadelphia, redbelly turtles frequently use abandoned sandpit ponds and man-made lagoons or ditches. During the active season, redbelly turtles require abundant basking habitat and aquatic vegetation for feeding; winter habitat must have a soft bottom and sufficient depth for hibernation. Redbelly turtles seek out sandy or loamy sun-exposed soil for digging nests, usually within proximity (up to 250 m; Mitchell 1974) of the water. Agricultural and other open fields are frequently used for nesting, whereas shrubby or wooded areas seem to be

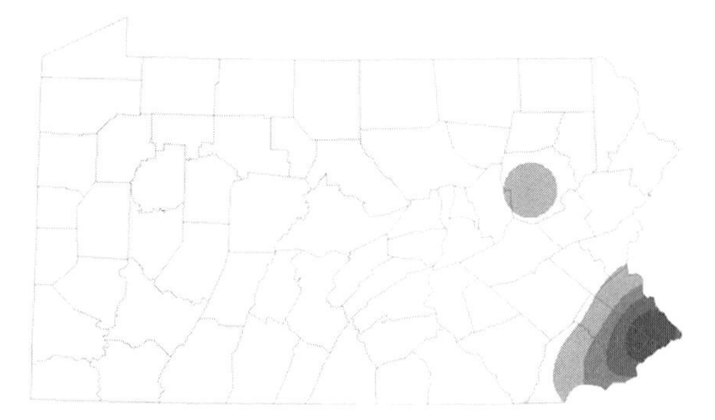

*Fig. 4.28.* Distribution of the Eastern Redbelly Turtle, *Pseudemys rubriventris.*

avoided (Swarth 2004). Swarth (2004) has documented nest sites typically within 225 m of wetland foraging habitat.

*LIFE HISTORY AND ECOLOGY*

Eastern redbelly turtle adults are thought to be primarily herbivorous, based on the difficulty of luring adults into baited traps (Ernst et al. 1994). Recent research in Maryland has confirmed that adult redbelly turtles feed almost exclusively on submerged aquatic vegetation and other wetland plants, while hatchlings and juveniles rely primarily on aquatic invertebrates for food (M. Fogel, J. Sage, and C. W. Swarth, unpublished data). The redbelly turtle is diurnal, emerging from hibernation in early to mid-April and becoming inactive again in mid- to late October. Redbelly turtles spend much time basking in the sun. They are quite shy, and one of the first turtles to drop into the water if alarmed. Hibernating turtles burrow in soft mud or rest on the substrate in deep water. Conant (1951) reported observations through the ice of several redbelly turtles resting on the bottom of a lake in Bucks County in December. Graham and Guimond (1995) observed and tracked redbelly turtles in the winter in a pond in Massachusetts; the individuals were in a depth of 1-3 m and lying exposed on the bottom.

There are no descriptions of redbelly turtle courtship or mating or the maturation age of females (Hulse et al. 2001). Females leave the water to lay eggs in June and July. Nesting season at Jug Bay, Maryland, is May 29 to July 23, with a peak from June 5 to 19 (Swarth 2004). Nests usually contain ten to twelve elliptical white eggs, 24-37 mm long. Swarth (2004) recorded a mean of twelve eggs and a maximum of twenty-two eggs in a nest and had evidence of double clutching. Young either hatch in late summer or fall or overwinter in the nest and emerge in the spring in April and May (Ernst et al. 1994, Swarth 2004). Both hatch periods have been observed in Pennsylvania populations. A study of emergence patterns in Maryland found that some nests exhibited a mixed emergence pattern of both fall and spring (E. Friebele and C. W. Swarth, unpublished data).

*THREATS*

Historically, the eastern redbelly turtle was regularly harvested for human consumption, leading to population declines in the Chesapeake Bay region in particular (Hulse et al. 2001). It is unknown to what extent the practice continues. In addition, it is unknown to what extent North American turtles are threatened by the pet trade or international trade, though anecdotal evidence points to these activities as potentially detrimental to most turtle populations.

In Pennsylvania, collection of redbelly turtles is prohibited except for legitimate scientific research. It is our opinion that commercial collection is not a significant threat to the redbelly in Pennsylvania because of the inaccessibility of many of the sites, the shyness of the turtles, and the low densities compared with other species. Primary threats to redbelly turtle populations in Pennsylvania are thought to be potential adverse impacts from industrial activities, such as pollution events, draining of former waste ponds, and river dredging; increased levels of nest predation due to development and habitat fragmentation; reduction in available nesting habitat from dense cover of invasive plant species, woody plant succession, and shoreline alteration (e.g., rip-rap); competition for basking and nesting sites by invasive red-eared sliders; isolation of populations through upland habitat alteration by development (U.S. Fish and Wildlife Service 1994); reservoir draining and dredging; road mortality to nesting females; and injury by motorboats (C. W. Swarth, unpublished data). Although aquatic habitats with redbelly turtles are provided significant protection in Pennsylvania, few regulations provide for identifying and protecting nesting habitat in the uplands.

*CONSERVATION AND MANAGEMENT NEEDS*

Efforts are needed to maintain the existing aquatic habitats of the redbelly turtle, including water quality protection, dredging limits, and controls of other direct disturbances. In addition, primary nesting sites need to be identified and managed for improved or continued nesting success of the turtles (including control of invasive and woody plant species). Landowners of potential nesting sites, such as farmers or recreational landowners, should be made aware of turtle nesting seasons and emergence. Access from the water to the nesting sites needs to be maintained as open space. Riverine habitats will be more difficult to manage than impoundments, given the extent of available habitats and the uncertainties involved in the habitat use of the turtles. Control of invasive red-eared sliders may be necessary in areas where they compete with redbelly turtles.

*MONITORING AND RESEARCH NEEDS*

Continued distributional surveys are needed to maintain accurate occurrence records and foster protection. Research is needed to document current levels of nesting success in redbelly turtle populations. Any efforts toward improving this success require follow-up monitoring studies. Several habitat enhancement projects have been implemented for the eastern redbelly turtle; the success and longevity of these structures and enhancements should be monitored, and the projects should be expanded to additional sites.

Other research needs include a population and demographics study at several established sites, studies of nesting success and associated threats, and identification of the level of interspecific competition with the aquatic nuisance species, red-eared slider. Finally, to protect eastern redbelly turtles better in riverine habitats, managers need to understand the habitat use and movement patterns of these turtles in the rivers of Pennsylvania.

*Authors:* KATHARINE L. GIPE, WESTERN PENNSYLVANIA CONSERVANCY; THOMAS G. PLUTO, U.S. ARMY CORPS OF ENGINEERS

## Northern Coal Skink

Order: Squamata
Family: Scincidae
*Plestiodon anthracinus anthracinus*
(also *Responsibility Concern*)

The northern coal skink was selected as a Pennsylvania Species of Greatest Conservation Need because of its uncertain status and restricted distribution (fig. 4.29). Northern coal skinks are not listed as Threatened or Endangered at the federal level or in Pennsylvania. Their collection in Pennsylvania is prohibited (Pennsylvania Fish and Boat Commission 2008). The

Fig. 4.29. The Northern Coal Skink, *Plestiodon anthracinus anthracinus*. Photo courtesy of Tom Diez.

global status of the northern coal skink is Secure (G5, NatureServe 2009).

*GEOGRAPHIC RANGE*

The northern coal skink has a fragmented distribution in the eastern United States. Its geographic range extends intermittently from northwestern New York in the vicinity of Lake Ontario to the mountains of northern Georgia and Alabama (Smith 1967). There is an isolated population in central Kentucky.

*DISTRIBUTION AND RELATIVE ABUNDANCE IN PENNSYLVANIA*

In Pennsylvania, the northern coal skink is primarily found on the Allegheny Plateau in the north-central and northwestern counties (fig. 4.30). A single locality for the species has been reported from Somerset County in south-central Pennsylvania. Recently, specimens were observed in south-central Pennsylvania in Adams and Franklin counties.

The coal skink is a rare lizard in Pennsylvania. Pennsylvania has the potential to play a significant role in the long-term survival of this subspecies, as approximately 25 percent to 30 percent of the geographic global range of the northern coal skink occurs within the commonwealth.

*COMMUNITY TYPE/HABITAT USE*

Key habitat for the northern coal skink is deciduous forest or mixed hardwood/coniferous forests with areas of open habitat where canopy cover is lacking. They tend to be found most frequently in open areas with an abundance of cover objects (i.e., rocks and logs) near forest margins. Clearings such as utility rights of way and open rocky hillsides along minor

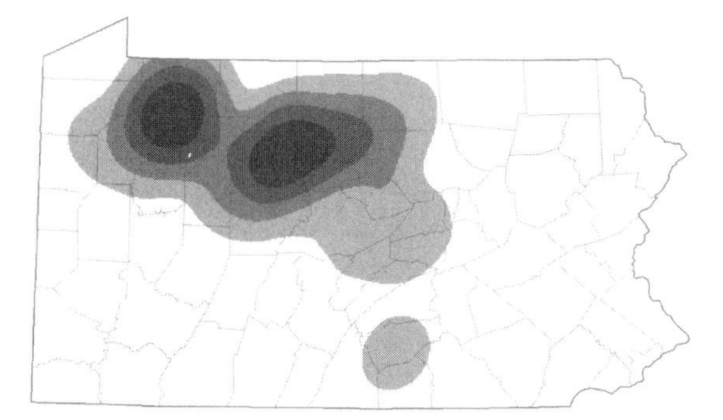

*Fig. 4.30.* Distribution of the Northern Coal Skink, *Plestiodon anthracinus anthracinus.*

roads account for the majority of habitat where coal skinks have been found. They do not appear to require any specialized seasonal habitat other than access to underground refugia used for hibernation.

### LIFE HISTORY AND ECOLOGY

Seasonal activity may commence as early as late March during unusually mild years, but usually does not begin until the first or second week of April. The lizards remain active until the end of September or beginning of October, depending on temperature. No information is available on hibernation in this species, but it is likely that they simply move downward in the rocky substrate until they are deep enough beneath the surface to avoid freezing temperatures.

The northern coal skink is a diurnally active terrestrial species that is seldom observed, because most activity takes place under leaf litter or cover objects, such as logs and rocks. Daily activity appears to occur from midmorning to late afternoon throughout the animal's activity season. Surface activity in populations appears to be extremely unpredictable. Sites known to harbor skink populations may yield seven or eight skinks in a matter of minutes on one day and then produce no animals after hours of intensive searching on other days with similar climatic conditions.

The northern coal skink is an active, wide-foraging species that generally searches for food under leaf litter, rocks, and logs. The diet reflects their foraging habits. Major food items are ground dwelling beetles, spiders, crickets, isopods, and centipedes.

Coal skinks readily autotomize their tails. This defense mechanism presumably reduces their chance of capture by potential predators. Damaged tails regenerate over several months. The incidence of broken tails in Pennsylvania coal skinks is exceedingly high. Seventy percent of males and 80 percent of females in the collection at the Carnegie Museum of Natural History possess broken tails.

Little is known about reproduction in the northern coal skink. Mating apparently occurs in May and early June (Clausen 1938). At this time, the throat and lower region of the head of mature males develops an orange reddish coloration. Egg laying occurs in late June and possibly early July. The eggs are usually laid in a shallow depression constructed by the female under a cover object. All nests the author observed have been under medium to large slabs of rock, although other species in the genus have been reported to lay their eggs in rotting logs and under other types of surface debris. Females generally remain in attendance of the nest. Clausen (1938) reported that an attendant female held in captivity would attempt to bite his finger if he got too close to her eggs. Clausen reported that his captive female refused food while attending the eggs but fed readily after the eggs had hatched.

The eggs are white when first laid but become stained a light brown or tan from contact with the substrate. The eggs are oblong and only slightly longer than they are wide. Clutch size in Pennsylvania varies from five to eleven eggs. The sample size from the region is too small to determine whether clutch size is correlated with size of female. Hatching depends on weather conditions. Eggs in nests discovered on July 27, 1990, had all hatched by August 12. A nest of eggs located on August 3, 1992, did not hatch until August 29. Summer 1992 was unusually cool, cloudy, and rainy, which probably delayed development that year.

### THREATS

Threats to the northern coal skink include mining and fossil fuel extraction, such as strip mines and gas and oil wells, clear cutting, highway construction and modification of highway shoulders, human development for residential and recreational housing, and collection of specimens by hobbyists and commercial collectors.

### CONSERVATION AND MANAGEMENT NEEDS

All sites where coal skinks exist should be entered into the Pennsylvania Natural Heritage Program database so that they can be readily recognized in response to development plans. Any development within regions where coal skinks occur should be conducted in a manner that would minimize impact on populations of the skinks.

*MONITORING AND RESEARCH NEEDS*

Surveys are needed to determine the present distribution and abundance of northern coal skinks in Pennsylvania, particularly in areas of the state with historic records of occurrence but no recent sightings and areas that have not been adequately surveyed. Once the location of populations is established through surveys, several populations should be selected for intensive long-term monitoring to determine population trends. The sites designated for long-term monitoring should be situated, whenever possible, on public lands. Long-term studies should include mark-recapture procedures and should specifically target population size and structure as well as reproductive activity and success within the populations.

*Author:* ARTHUR C. HULSE, INDIANA UNIVERSITY OF PENNSYLVANIA (RETIRED)

*Fig. 4.31.* The Broadhead Skink, *Plestiodon laticeps*. Photo courtesy of Tom Diez.

## Broadhead Skink

Order: Squamata
Family: Scincidae
*Plestiodon laticeps*

The broadhead skink was selected as a Pennsylvania Species of Greatest Conservation Need because of its extreme rarity and restricted geographic range within Pennsylvania (fig. 4.31). It is listed as a Candidate Species by the Pennsylvania Fish and Boat Commission, and collecting this species is prohibited (Pennsylvania Code 2008). The global status of the broadhead skink is Secure (G5, NatureServe 2009).

*GEOGRAPHIC RANGE*

Broadhead skinks range from southeastern Pennsylvania southward to northern Florida and west to east Texas and east Kansas. The broadhead skink reaches the northeastern limit of its geographic range in southern Pennsylvania and as such may be genetically different from more southerly populations.

*DISTRIBUTION AND RELATIVE ABUNDANCE IN PENNSYLVANIA*

That there are only two voucher specimens for the state attests to the broadhead skink's rarity in Pennsylvania. One was collected in West Chester, Chester County, in the early 1900s (fig. 4.32). The other specimen was just recently collected from the southern Susquehanna River in York County. It is doubtful that the West Chester population still exists because of the extreme modification of the land in that area. However, in the vicinity of the York County location, the area is relatively undisturbed and may support a population of broadhead skinks. Abundance and population trends for the species in Pennsylvania are unknown, in part, because of a lack of information on the precise geographic range of the species within the state.

*COMMUNITY TYPE/HABITAT USE*

Broadhead skinks are semiarboreal and are generally found in open deciduous woodlands and ecotonal areas in the vicinity of woodlands. Snags and fallen logs are often abundant in the habitat occupied by the skinks.

*LIFE HISTORY AND ECOLOGY*

The broadhead skink is so rare in Pennsylvania that nothing is known about its habits or reproduction in the state. As a consequence, material in this section relies on information published on studies in other parts of the skink's range. Herpetologists strongly agree that the broadhead skink is the most arboreal of all of the North American skinks. It is reported to frequent wooded areas where there is an abundance of dead, standing timber and large stumps and hollow logs (Mount 1975). They are often seen climbing high in trees (Smith 1961) and have been observed using abandoned woodpecker holes (Collins 1974).

Seasonal activity generally begins in March at lower latitudes (Vitt and Cooper 1985). Surface activity extends into September and early October in

*Fig. 4.32.* Distribution of the Broadhead Skink, *Plestiodon laticeps.*

northern Florida (Goin and Goin 1951). Broadhead skinks are diurnal. No information is available concerning geographic home range size. Males, for the most part, are nonterritorial; however, they will defend females during the reproductive season (Vitt and Cooper 1985).

Despite its arboreal tendencies, broadhead skinks spend much of their time foraging for food on the ground. The most detailed study of their diet and foraging behavior is Vitt's (1986). Skinks move over and through the leaf litter. They often burrow completely out of sight for several minutes before reemerging. Apparently, both visual and olfactory cues are used in prey location. Their diet is extremely varied. They feed on crickets and grasshoppers, beetles and their larvae, caterpillars, earwigs, snails, isopods, spiders, lizards and lizard eggs, as well as other items. The only foods that they have been shown to avoid are millipedes and velvet ants.

Vitt and Cooper (1985) have published the most detailed account of reproduction in broadhead skinks. Unless otherwise noted, all information comes from their paper on a South Carolina population. Mating generally occurs from May to early June. Males will engage in combat with any other male that they encounter. After breeding, males often remain with the female for three to four days, apparently to guard against the female mating with another male before she begins nest construction and egg laying (Cooper and Vitt 1986). When ready to oviposit, the female constructs a nest chamber in the rotted wood of hardwood logs. Females tend their eggs, although they will emerge to forage for food frequently enough during brooding to maintain their body weight. Clutch size varies from nine to eighteen eggs and averages 13.7 eggs. Egg number is not significantly correlated with female body size in this species. Incubation averages 48.7 days and neonates emerge from mid-July to early August. As in other skinks in Pennsylvania, the hatchlings of broadhead skinks possess brilliant blue tails.

*THREATS*

The main threats to this species are unknown. However, the greatest threat to the continued existence of this species in Pennsylvania is probably our lack of knowledge concerning its precise distribution within the state.

*CONSERVATION AND MANAGEMENT NEEDS*

Information is, at present, insufficient to devise a specific conservation or management plan for the species.

*MONITORING AND RESEARCH NEEDS*

The most pressing needs are to establish the size and structure of existing populations, to determine the extent of the geographic range along the lower Susquehanna River, and to search suitable habitat in the southern portions of York, Lancaster, and Chester counties for broadhead skinks. Assuming that populations are found, long-term monitoring efforts should be established to determine population trends.

*Author:* ARTHUR C. HULSE, INDIANA UNIVERSITY OF PENNSYLVANIA (RETIRED)

## Queen Snake

Order: Squamata

Family: Colubridae

*Regina septemvittata*

The queen snake was selected as a Pennsylvania Species of Greatest Conservation Need because of its limited distribution, poorly understood status, evidence of decline, vulnerability to water pollution, and the ongoing replacement of its native crayfish prey by more aggressive invasive crayfish (fig. 4.33). Queen snakes are not listed as Threatened or Endangered at the federal level or in Pennsylvania. Their collection in Pennsylvania is prohibited (Pennsylvania Fish and Boat Commission 2008). The global status of the queen snake is Secure (G5, NatureServe 2009).

*GEOGRAPHIC RANGE*

Queen snakes range widely throughout the eastern United States east of the Mississippi River (Gibbons and Dorcas 2004). There are disjunct western populations reported from central Arkansas. The northern-

*Fig. 4.33.* The Queen Snake, *Regina septemvittata*. Photo courtesy of R. T. Zappalorti.

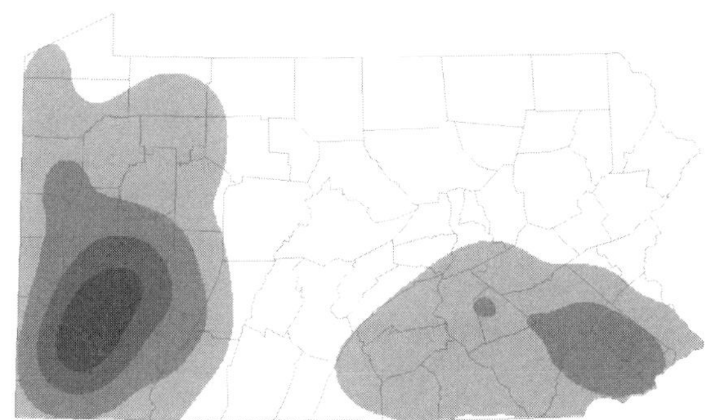

*Fig. 4.34.* Distribution of the Queen Snake, *Regina septemvittata*.

most extent of the geographic range is southern Ontario between Lake Michigan and Lake Erie.

### *DISTRIBUTION AND RELATIVE ABUNDANCE IN PENNSYLVANIA*

The queen snake has a discontinuous distribution in Pennsylvania that can be delineated relative to the physiographic regions of the commonwealth (see Guilday 1985). It occurs throughout the western third of the commonwealth on the Glaciated and Pittsburgh sections of the Appalachian Plateau and in the southeast from the Great Valley Section of the Valley and Ridge Province eastward through the Piedmont and Coastal Plain provinces (fig. 4.34). However, it is largely absent from the higher-elevation portions of the Allegheny High Plateau in north-central Pennsylvania, the Valley and Ridge Province in central Pennsylvania, and the Pocono and Glaciated sections of northeastern Pennsylvania.

Recent reports gained through the Pennsylvania Herpetological Atlas Project suggest that the queen snake still occurs throughout much of its historic geographic range within the commonwealth. However, it has been suggested that queen snakes are highly susceptible to the effects of water pollution and siltation and may be declining throughout much of Pennsylvania (McCoy 1982, Hulse et al. 2001).

### *COMMUNITY TYPE/HABITAT USE*

The preferred habitat of queen snakes in Pennsylvania has not been well determined. They are always found in proximity to water (Branson and Baker 1947). Because of their specialized diet, an abundant crayfish population is clearly necessary. Consequently, water quality suitable to maintain a stable population of aquatic benthic macroinvertebrates is a key aspect of the habitat. Range-wide, the descriptions of the species' habitat are fairly variable and include small, shallow streams, rivers, lakes, and marshes (Conant 1960, Ernst and Ernst 2003, Gibbons and Dorcas 2004). In many northern areas, queen snakes reportedly prefer shallow, rocky-bottomed streams, but in the southern portion of their geographic range, they have also been reported from sandy-bottomed streams and rivers (Gibbons and Dorcas 2004). In Berks County, Pennsylvania, queen snakes have been observed in small (<1 m wide) rivulets and large (>15 m wide) creeks (H. K. Reinert, personal observation). This species frequently basks on overhanging trees and shrubbery from which it will drop into the water when disturbed (Hulse et al. 2001, Ernst and Ernst 2003, H. K. Reinert, personal observation). Consequently, low, overhanging woody vegetation may be an important structural feature of inhabited streams. It is reported that the snake commonly hides under bankside rocks, logs, boards, and debris (Branson and Baker 1947, Hulse et al. 2001).

The hibernating habitat of queen snakes in Pennsylvania has yet to be clearly defined. It has been suggested that they use muskrat and crayfish borrows in stream banks (Hulse et al. 2001, Ernst and Ernst 2003).

### *LIFE HISTORY AND ECOLOGY*

The diet of queen snakes consists mostly of crayfish in the early stages of molting (Wood 1949, Brown 1979). Small fish are also eaten (Branson and Baker 1947, Wood 1949). Capture-recapture studies (Branson and Baker 1947, Ernst and Ernst 2003) suggest that resident snakes are very sedentary and make only limited movements (averaging approximately 25 m from capture locations), even over long periods (forty weeks).

Interestingly, none of the forty-nine specimens translocated by Branson and Baker (1947) to novel locations were ever recaptured. This suggests that either dispersal or high mortality occurred among these specimens.

Hulse et al. (2001) reported the active season of this species in western Pennsylvania to range from late March to mid-October. Reinert (1975) observed a queen snake outstretched on a branch overhanging a small stream in Berks County, Pennsylvania, on December 5, 1972. Fall aggregations of this species have been reported by several observers (Neill 1948, Wood 1949, Ernst and Ernst 2003).

The mating period in Pennsylvania is unknown but assumed to be in the spring. Ford (1982) described the courtship behaviors of captive specimens in the spring, and Minton (1972) observed spring copulation in the field in Indiana. Hulse et al. (2001) reported that birth occurs between August 1 and 20, and a Berks County female gave birth on August 26, 1963 (H. K. Reinert, personal observation). However, Swanson (1952) indicated that birth in Venango County occurred between early September and mid-October. In the Pennsylvania litters examined by Hulse et al. (2001), the number of young ranged from four to fifteen (mean = 9.7, SE = 0.60, $n$ = 20). Swanson (1952) reported a Venango County brood of eight, and a Berks County specimen gave birth to twelve young (H. K. Reinert, personal observation). Litter size correlates strongly with female body size (Hulse et al. 2001).

*THREATS*

The primary threat to the queen snake's survival is probably the degradation of stream water quality and the alteration of stream structure (Hulse et al. 2001). As a specialized feeder on aquatic prey, the suitability of its habitat is linked directly to water quality. Any alteration of streams that negatively affects aquatic invertebrates will negatively affect queen snake populations (McCoy 1982). In particular, siltation and acidification (through either precipitation or mine drainage) of streams may have negative effects on crayfish populations. Likewise, channelization and damming alters stream structure and may severely degrade habitat quality for queen snakes.

The effect of industrial and agricultural water pollution on snakes has not been studied in great detail. However, because of its trophic position, the queen snake could suffer from bioaccumulation of various chemical pollutants in surface water (e.g., pesticides). Fontenot et al. (1996) assayed a queen snake that contained significant quantities of polychlorinated biphenyls (PCBs).

As with most snakes, indiscriminant killing takes its toll on queen snakes. Water snakes in general have been heavily persecuted because of their suspected effect on sport fisheries (e.g., Anonymous 1934, 1935, 1936, Gibbons and Dorcas 2004), and the common public misconception that they are venomous (Gibbons and Dorcas 2004). Because few people can identify different snake species, the inoffensive, crayfish-eating queen snake is not spared. Ernst and Ernst (2003) report the shooting massacre of approximately 100 queen snakes at a single site in southeastern Pennsylvania.

*CONSERVATION AND MANAGEMENT NEEDS*

An assessment of the current distribution and occurrence of queen snakes is required to determine the actual status of this species within the commonwealth. Streams harboring extant queen snake populations should be assessed for present and potential habitat alteration and degradation, particularly changes in water quality. A conservation action plan should be developed for this species within the commonwealth.

*MONITORING AND RESEARCH NEEDS*

The distribution data collected in population surveys should be added to the Pennsylvania Natural Diversity Inventory (PNDI) database. The PNDI database should then be used consistently to monitor potential effects to queen snake habitat resulting from development, agriculture, utility, mining, and recreational projects. All projects that potentially affect habitat or populations should be examined before their initiation. Those found to have probable impacts should be evaluated by biologists for the extent of impact and, if necessary, altered to reduce those impacts. The effectiveness of this program at detecting potential impacts should be evaluated annually by assessing the number of projects examined and the number of projects altered.

Intensive field surveys of historic sites are clearly needed to assess the present status of queen snake populations. Basic field research is required to determine the spatial ecology, the primary characteristics of preferred habitat, and the genetic structure of populations in Pennsylvania. Research to evaluate the direct and indirect impacts of water pollution on queen snake and other aquatic snake species is also needed.

*Authors:* HOWARD K. REINERT AND ASHLEY P. PETTIT, THE COLLEGE OF NEW JERSEY

## Shorthead Garter Snake

Order: Squamata
Family: Colubridae
*Thamnophis brachystoma*
(also *Responsibility Concern*)

The shorthead garter snake was selected as a Pennsylvania Species of Greatest Conservation Need because of its uncertain status, limited distribution, and evidence of population declines (fig. 4.35). Because more than 90 percent of its geographic range occurs within Pennsylvania, the commonwealth is the primary steward of this species. Although shorthead garter snakes are not listed as Threatened or Endangered at the federal level or in Pennsylvania, their collection in Pennsylvania is prohibited (Pennsylvania Fish and Boat Commission 2008). The global status of the shorthead garter snake is Apparently Secure (G4, NatureServe 2009).

### *GEOGRAPHIC RANGE*

The shorthead garter snake has one of the most restricted distributions of any snake species found in North America, yet seems to be a habitat generalist. It is endemic to the unglaciated portions of the upper Allegheny Plateau, with the bulk of its geographic range occurring in northwestern Pennsylvania and a small portion extending into southwestern New York (Wright and Wright 1957, McCoy 1982, Rossman et al. 1996, Conant and Collins 1998 ).

### *DISTRIBUTION AND RELATIVE ABUNDANCE IN PENNSYLVANIA*

Shorthead garter snakes were historically found in Cameron, Clarion, Clearfield, Crawford, Elk, Forest, Jefferson, McKean, Mercer, Potter, Venango, and Warren counties. Introduced populations have become established in Allegheny, Butler, and Erie counties in Pennsylvania (Conant and Collins 1998), Chemung County in New York (Bothner 1986), and Mahoning County in Ohio (Novotony 1990; fig. 4.36).

The shorthead garter snake is locally abundant throughout its range (Conant 1950, Asplund 1963, McCoy 1982, Bothner 1986). However, its numbers drop off sharply at the periphery despite seemingly suitable habitat (McCoy 1982, Hulse et al. 2001). Current population densities are believed to be lower than historical densities, but stable (Bothner 1986).

### *COMMUNITY TYPE/HABITAT USE*

Key habitats for shorthead garter snakes consist primarily of open, wet areas, including meadows, old fields, wetlands, and pastures, nearly always in association with creeks, marshes, or streams (Swanson 1952, Klingener 1957, Asplund 1963, Rossman et al. 1996) and sometimes in dense woodlands (Hulse et al. 2001).

Seasonal movement is difficult to address in this species because of lack of data on movement, habitat usage, and geographic home range. However, throughout the active season, this species is rarely encountered more than a few hundred meters from water (Wright and Wright 1957, Hulse et al. 2001), which might indicate specific habitat requirements.

### *LIFE HISTORY AND ECOLOGY*

Individuals are gregarious, commonly encountered under cover objects, such as boards, stones, logs, and debris, or within clumps of grass and nearly always in proximity to water (Swanson 1952, Klingener 1957, Wright and Wright 1957, Asplund 1963). Its diet consists chiefly, if not exclusively, of earthworms (Ernst and Barbour 1989, Conant and Collins 1998, Hulse

*Fig. 4.35.* The Shorthead Garter Snake, *Thamnophis brachystoma*. Photo courtesy of Charlie Eichelberger.

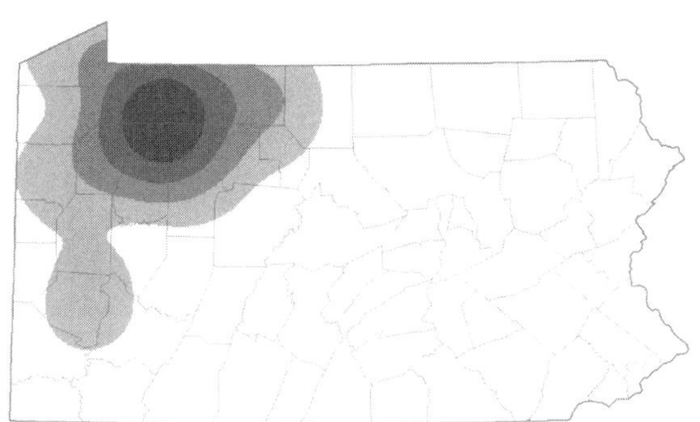

*Fig. 4.36.* Distribution of the Shorthead Garter Snake, *Thamnophis brachystoma*.

et al. 2001). The few reports of hibernacula both list westward-facing slopes within 60 m of a water source (Bothner 1963, Pisani 1967). Shorthead garter snakes have been reported to overwinter with eastern garter snakes, northern redbelly snakes, red-spotted newts, and spotted salamanders.

Mating occurs in the spring shortly after emergence from hibernation (Pisani 1967). Females briefly store sperm from these matings (Pisani and Bothner 1970) until ovulation in late May to early June (Pisani and Bothner 1970, Hulse et al. 2001). Birth is functionally related to temperature and typically occurs in August. However, in unseasonably warm seasons, it may occur in late July (Rossman et al. 1996, Hulse et al. 2001). Litter size is positively related to maternal body size and ranges from five to fourteen offspring (Hulse et al. 2001), with averages between 7.2 (Pisani and Bothner 1970) and 8.8 (Bothner 1986) reported. Females in virtually all populations reproduce annually (Rossman et al. 1996).

*THREATS*

Specific threats to the continued existence of the shorthead garter snake are yet to be determined. However, throughout the geographic range, habitat loss and increased competition with eastern garter snakes are speculated to be the primary threats (Bothner 1976, Bothner 1986, Rossman et al. 1996, Hulse et al. 2001).

Specifically, the chief causes of habitat loss are the encroachment of woody vegetation and human development. Succession on abandoned farms has also decreased the amount of suitable habitat (Bothner 1986). Because the species is an earthworm specialist (Ernst and Barbour 1989, Rossman et al. 1996, Conant and Collins 1998, Hulse et al. 2001) and earthworm abundance, biomass, and biodiversity are negatively affected by soil pollutants (Spurgeon and Hopkin 1999), oil contamination and acid mine drainage might also be contributing its decline and warrant examination.

Geographic range expansion of eastern garter snakes might also be a potential influence contributing to the decline of shorthead garter snakes. Bothner (1976) proposed that the eastern garter snake was gradually replacing the shorthead garter snake, particularly in the northern part of the geographic range. Competition for prey and overwintering and basking sites are potential factors contributing to population declines, although direct evidence is lacking.

*CONSERVATION AND MANAGEMENT NEEDS*

At this time, sufficient data are lacking that would accurately address, let alone identify, threats or propose conservation actions.

*MONITORING AND RESEARCH NEEDS*

Intensive monitoring is needed to delineate the current geographic range and abundance of shorthead garter snakes. Research priorities include determining demographic trends and threats, including habitat loss and the ecological relationship with eastern garter snakes. Studies should also be conducted to examine the spatial (geographic home range, seasonal movements, habitat usage and hibernation site fidelity), thermal (overwintering and gestating temperatures and basking propensity), and reproductive (female reproductive cycle, fecundity, age of sexual maturity and operative mating system) biology of the species. Information garnered from these studies should allow a comprehensive management and conservation plan and monitoring protocol to be designed.

*Author:* BEN JELLEN, WESTERN PENNSYLVANIA CONSERVANCY

## Eastern Ribbon Snake

Order: Squamata
Family: Colubridae
*Thamnophis sauritus*

The eastern ribbon snake was selected as a Pennsylvania Species of Greatest Conservation Need because of its uncertain status, evidence of population decline, and specialized food and habitat requirements (fig. 4.37). Although eastern ribbon snakes are not

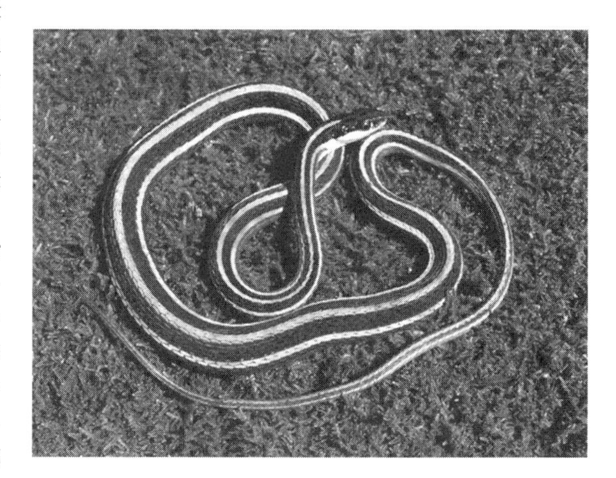

*Fig. 4.37.* The Eastern Ribbon Snake, *Thamnophis sauritus*. Photo courtesy of Tom Diez.

listed as Threatened or Endangered at the federal level or in Pennsylvania, their collection in Pennsylvania is prohibited (Pennsylvania Fish and Boat Commission 2008). The global status of the eastern ribbon snake is Secure (G5, NatureServe 2009).

*GEOGRAPHIC RANGE*

The eastern ribbon snake is found throughout the eastern United States. Two subspecies occur within Pennsylvania. The subspecies *Thamnophis sauritus sauritus* (eastern ribbon snake) occurs from southern Indiana through southern and eastern Pennsylvania, New Jersey, and southeastern New York, up into southern New Hampshire and south to the Florida panhandle. The subspecies *Thamnophis sauritus septentrionalis* (northern ribbon snake) occurs farther north into upstate New York, southern Ontario, and southern Maine, as well as western Pennsylvania and Indiana (Conant and Collins 1998). The separation of these two subspecies in Pennsylvania is still undetermined (Hulse et al. 2001).

*DISTRIBUTION AND RELATIVE ABUNDANCE IN PENNSYLVANIA*

Because of its fairly narrow habitat requirements, the eastern ribbon snake is localized and uncommon throughout much of Pennsylvania. Although found in all regions of the state, a number of counties across the state lack confirmed records of the species (McCoy 1982; fig. 4.38). Its increased rarity in Pennsylvania and other northeastern states is undoubtedly due to the historic destruction of wetlands.

*COMMUNITY TYPE/HABITAT USE*

The eastern ribbon snake is restricted to habitats associated with permanent water, where its prey is located. Ideal habitat is a freshwater marsh with standing water; dense growths of sedges, grasses, rushes, and emergent shrubs; and lots of frogs. Other habitats include bogs, the densely vegetated banks of streams and rivers, and ponds and lakes with lots of shoreline vegetation.

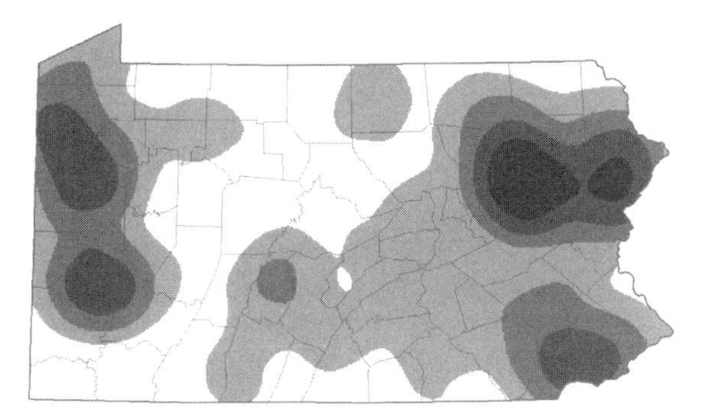

*Fig. 4.38.* Distribution of the Eastern Ribbon Snake, *Thamnophis sauritus.*

*LIFE HISTORY AND ECOLOGY*

The eastern ribbon snake primarily preys on small amphibians, secondarily on small fishes. With its large eyes, the ribbon snake is especially attuned to movements of its prey. Common prey species in Pennsylvania include wood frogs, small green frogs, red-backed and other salamanders, spring peepers, and American toads, as well as tadpoles and minnows (Rossman 1963). The dense herbaceous and shrubby cover found along the wet borders of its preferred habitats makes the ribbon snake difficult to detect and pursue. The three distinctive longitudinal stripes also make it difficult for a potential predator to focus on any reference point as the ribbon snake races through the grass and becomes a blur of continuous lines (Hulse et al. 2001). Ribbon snakes usually confine their activities to a relatively small area. Carpenter (1952) reported a maximum distance between captures of 278 m.

A ribbon snake often suns by draping its body around the branches of a shrub or small tree growing along the edge of the water. Occasionally, several ribbon snakes can be seen in the same branches. Ribbon snakes often share hibernation quarters in the spaces beneath loose boulders and smaller rocks that extend several feet beneath the surface of the ground (Lachner 1942). They emerge from hibernation at the end of April. Mating generally occurs in May, and the live young are born in late August or early September. From three to twenty-six young have been reported across the snake's geographic range (Ernst and Ernst 2003).

*THREATS*

The major threat to the survival of this uncommon species is the loss of wetlands (Ernst and Ernst 2003). It has been estimated that half of the nation's (and Pennsylvania's) wetlands have already been destroyed. Even though the remaining wetlands have acquired some legal protection, developers are still occasionally permitted to destroy them if they mitigate this loss by creating artificial wetlands elsewhere. In most cases, these man-made wetlands do not satisfy the requirements of the wildlife that once lived in the destroyed natural wetlands. The ribbon snake is a sensitive species, and the construction of homes, schools, shopping centers, and roads inevitably leads to a decrease in their numbers. Unable to adapt to changing conditions as easily as the garter snake or ring-necked snake, the ribbon snake soon disappears from suburban areas.

*CONSERVATION AND MANAGEMENT NEEDS*

Continued acquisition (either outright or through conservation easements) of the state's wetlands, including adequate buffer zones, is necessary to ensure the survival of this sensitive species. If a landowner expresses an interest and sensitivity to the needs of wildlife, it would be worthwhile to advise that landowner regarding various habitat management procedures (e.g., not mowing the edges of ponds, lakes, wetlands). Public land managers should be informed of the exact locations and the habitat requirements of these reptiles to prevent construction-maintenance projects that may jeopardize their survival.

*MONITORING AND RESEARCH NEEDS*

There is a great need to determine the extent of the ribbon snake's occurrence across the large areas of Pennsylvania where no records have been confirmed. Therefore, intensive surveys are needed to determine present distribution. Surveys should be taken throughout the historic sites, both past and current, to determine the species' present status and distribution. These surveys should be followed up with monitoring activities at selected sites to determine population status and detect changes in numbers.

*Author:* JOHN SERRAO

## Mountain Earth Snake

Order: Squamata
Family: Colubridae
*Virginia pulchra*
(also *Responsibility Concern*)

The mountain earth snake was selected as a Pennsylvania Species of Greatest Conservation Need because of its uncertain status, loss of habitat, and sporadic distribution. Because approximately 80 percent of the geographic global range of the species occurs within Pennsylvania, the survival of this species may depend on maintaining healthy populations within the state (fig. 4.39). Mountain earth snakes are not listed as Threatened or Endangered either at the federal level or in Pennsylvania. Their collection in Pennsylvania is prohibited (Pennsylvania Fish and Boat Commission 2008). The global status of the mountain earth snake is Secure (G5, NatureServe 2009).

Fig. 4.39. The Mountain Earth Snake, *Virginia pulchra*. Photo courtesy of Tom Diez.

*GEOGRAPHIC RANGE*

The mountain earth snake ranges from the Allegheny Mountains and Plateau of northern Pennsylvania southward through extreme western Maryland into adjacent West Virginia. Its geographic range barely extends into Virginia's Highland County.

*DISTRIBUTION AND RELATIVE ABUNDANCE IN PENNSYLVANIA*

In Pennsylvania, mountain earth snakes have a disjunct geographic range with three main centers of occurrence as follows: (1) along Laurel Ridge in Westmoreland, Somerset, and Fayette counties; (2) the Allegheny Plateau of Warren, Forest, Clarion, and Venango counties; and (3) the Allegheny Mountains of Potter, Cameron, Clinton, and Elk counties (fig. 4.40). The gap in its distribution in northern Pennsylvania may well be an artifact of survey intensity rather than of the biology of the animal, because individuals have recently been found in Clearfield County, approximately halfway between the two centers of distribution in the north. There is little information concerning the abundance of the species within Pennsylvania. There are recent records for the species from the Allegheny Plateau and Mountains, but there are no recent records of the species from the southern portion of its range in the vicinity of Laurel Ridge in Westmoreland, Somerset, and Fayette counties.

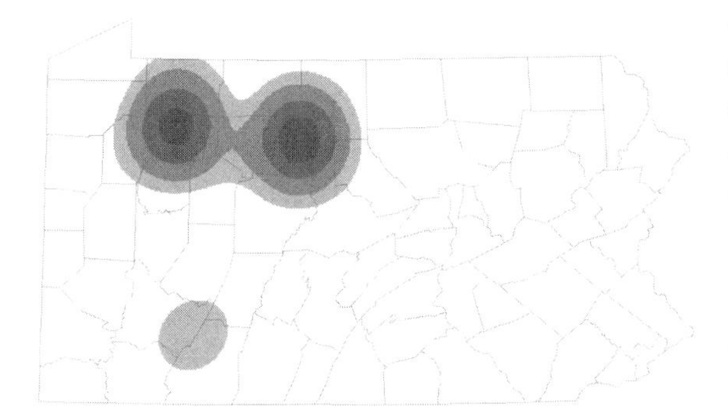

Fig. 4.40. Distribution of the Mountain Earth Snake, *Virginia pulchra*.

### *COMMUNITY TYPE/HABITAT USE*

Key habitat appears to be mountainous deciduous forest with intermingled clearing where cover objects such as rocks and logs are abundant. They are most commonly found on moderate to steep hillsides where there is an abundance of rocky cover and second-growth vegetation. In these areas, they are usually found under rocks. Clearings such as utility rights of way and open rocky hillsides along minor roads provide important habitats. They do not appear to require any specialized seasonal habitat other than access to underground refugia used for hibernation.

### *LIFE HISTORY AND ECOLOGY*

In Pennsylvania, mountain earth snakes begin to emerge from hibernation in late April but do not become common until mid-May. Activity remains fairly constant through July and then increases in August. It begins to taper off in September and ceases by mid-October. The snakes do not appear to have any well-developed daily activity cycle. Richmond (1954) reported them to be most commonly found following periods of heavy rain, but Bothner and Moore (1964) reported collecting a large series in Pennsylvania during fairly dry conditions. It is my experience that, while activity of mountain earth snakes in a given area is indeed sporadic, it cannot be adequately correlated with specific environmental conditions. Although they are usually found singly under cover objects, pairs have been observed sharing a refuge. In the laboratory, neonates tend to aggregate. Mountain earth snakes appear to feed exclusively on earthworms.

Nothing is known about when this species mates, but its relatively late emergence in the spring and continued fairly high activity in September suggest a fall mating period. Mountain earth snakes are viviparous. Parturition occurs later in mountain earth snakes than it does in most viviparous snakes in Pennsylvania. The earliest date for birth is August 19 (Richmond 1954), and the latest date for birth of a litter is September 30 (Bothner and Moore 1964). Most births appear to occur in late August and September. In a sample of specimens from the Carnegie Museum, litter size varied from four to eleven. Pisani (1971) reported a litter of fourteen. Litter size is not correlated with body size.

### *THREATS*

Threats to the mountain earth snake include but may not be limited to the following: mining and fossil fuel extraction, such as strip mines and gas and oil wells; clear-cutting; highway construction and modification of highway shoulders; human development for residential and recreational housing; and collection of specimens by hobbyists and commercial collectors.

### *CONSERVATION AND MANAGEMENT NEEDS*

All sites where mountain earth snakes exist should be entered into the Pennsylvania Natural Heritage Program database so that they can be readily recognized in response to development plans. Any development within regions where mountain earth snakes occur should be conducted in a manner that would minimize impact on populations of the skinks. In addition, resource managers of state and federal lands where mountain earth snakes are found should be alerted to the presence of the snake. Discussion of appropriate measures to ensure their continued existence on these lands should be discussed with land managers, and management plans should be implemented as soon as possible.

### *MONITORING AND RESEARCH NEEDS*

Surveys of appropriate and historic habitat are needed to determine the presence of mountain earth snakes and the size of their populations. Although survey work for this species is required throughout its geographic range in Pennsylvania, the southern portion of its range deserves particular attention, especially in the vicinity of Laurel Ridge. Once the locations of populations are established through surveys, several populations should be selected for intensive long-term monitoring. The sites designated for long-term monitoring should, whenever possible, be situated on public lands. Long-term studies should include mark-recapture procedures and should specifically target population size and structure as well as reproductive activity and success within the populations.

*Author:* ARTHUR C. HULSE, INDIANA UNIVERSITY OF PENNSYLVANIA (RETIRED)

**RESPONSIBILITY CONCERN**

## Jefferson Salamander

Order: Caudata
Family: Ambystomatidae
*Ambystoma jeffersonianum*

The Jefferson salamander was selected as a Pennsylvania Species of Greatest Conservation Need because of its uncertain status, evidence of possible population declines, and as a representative of the guild of species associated with vernal pond habitats (fig. 4.41). Jefferson salamanders are not listed as Threatened or Endangered at the federal level or in Pennsylvania. Their collection in Pennsylvania is prohibited (Pennsylvania Fish and Boat Commission 2008). The global status of the Jefferson salamander is Apparently Secure (G4, NatureServe 2009).

*GEOGRAPHIC RANGE*

Jefferson salamanders occur in deciduous forests from western New England south to Virginia and west through Kentucky and southern Indiana (Conant and Collins 1991). Along the northern periphery of its geographic range, Jefferson salamanders hybridize with blue-spotted salamanders, which often results in populations of polyploid gynogenetic females (Bogart and Klemens 1997).

Fig. 4.41. The Jefferson Salamander, *Ambystoma jeffersonianum*. Photo courtesy of Richard Koval.

*DISTRIBUTION AND RELATIVE ABUNDANCE IN PENNSYLVANIA*

Jefferson salamanders occur sporadically throughout the state (Hulse et al. 2001; fig. 4.42). Present status in Pennsylvania is largely unknown. At vernal ponds in south-central Pennsylvania, it is usually the least abundant of the three native ambystomatid salamanders.

*COMMUNITY TYPE/HABITAT USE*

Adult Jefferson salamanders are predominantly found in deciduous or mixed deciduous–coniferous forests near suitable breeding locations (Hulse et al. 2001). They have a strong affinity for upland forests (Petranka 1998). Although breeding occurs primarily in temporary or vernal ponds, Jefferson salamanders have been on occasion known to use permanent aquatic habitats (Douglas and Monroe 1981).

*LIFE HISTORY AND ECOLOGY*

Like other ambystomatid salamanders, adult Jefferson salamanders are seldom seen as they spend the majority of their lives underground. On occasion, they may be found under rocks and logs. Adults feed on a variety of terrestrial invertebrates. Predators on adult salamanders include owls, skunks, raccoons, shrews, and snakes (Petranka 1998). Jefferson salamander larvae feed on various aquatic invertebrates, tadpoles, and other salamander larvae. Larvae are gape-limited predators, so prey size increases as they grow. Larval Jefferson salamanders are more aggressive and eat a higher proportion of vertebrate prey (i.e., tadpoles and salamander larvae) than do other syntopic ambystomatid species and tend to inhabit deeper habitats within a pond. A number of larger aquatic predators, particularly insects, prey on salamander larvae.

Adult salamanders migrate to vernal ponds to breed in late winter/early spring. Most movement occurs on

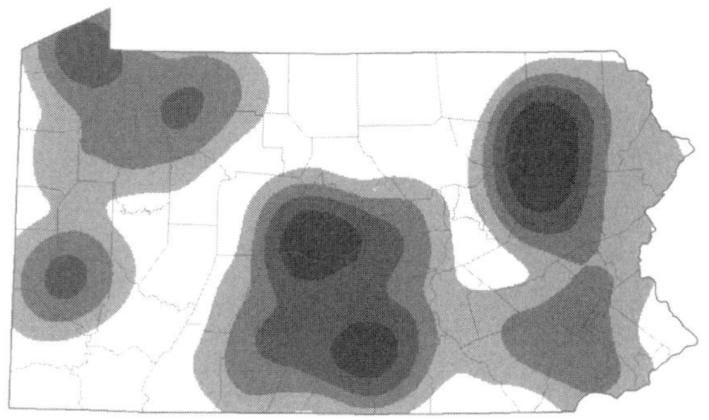

Fig. 4.42. Distribution of the Jefferson Salamander, *Ambystoma jeffersonianum*.

rainy nights, and males often arrive at ponds before females (Hulse et al. 2001). A rather elaborate courtship occurs underwater (Petranka 1998), during which the male deposits one or more spermatophores that the female picks up with her cloaca, thereby fertilizing her eggs. Females lay their eggs in small masses that they attach to submerged branches or aquatic vegetation. Females typically lay 150-300 eggs in groups of ten to twenty-five eggs (Hulse et al. 2001). In Pennsylvania, eggs usually hatch in April and May, and larvae metamorphose into terrestrial juveniles during the summer months. Age at sexual maturity and longevity are probably similar to that of the spotted salamander, *Ambystoma maculatum*, which reaches maturity in two to five years and can potentially live for more than twenty years (Petranka 1998).

*THREATS*

The primary threat facing Jefferson salamanders is habitat loss/destruction. Because of their complex life cycle, Jefferson salamanders require suitable terrestrial and aquatic habitats. The U.S. Supreme Court decision in *Solid Waste Agency of Northern Cook County v. U.S. Army Corps of Engineers* (No. 99-1178; SWANCC) greatly reduced the protections afforded to isolated wetlands, including vernal ponds, under the Clean Water Act (Gibbons 2003). Loss of isolated wetlands not only causes immediate loss of habitat but also changes the spatial configuration of, and distance between, remaining ponds, thereby affecting movements and recolonization processes (Semlitsch 2003).

Adequate terrestrial habitat around vernal ponds is also necessary for adult salamanders (Semlitsch 1998, Gibbons 2003). For ambystomatid salamanders, it has been recommended that a protected area, or buffer zone, extend 164 m (Semlitsch 1998) to 175 m (Faccio 2003) from vernal ponds to protect a majority of adults. Because Jefferson salamanders appear to move farther from ponds than other ambystomatid salamanders (occasionally in excess of 600 m; Petranka 1998), these recommendations may be conservative.

Another potential threat is habitat disturbance and mortality associated with timber harvest. Several studies have documented negative effects of timber harvest on ambystomatid salamanders (DeMaynadier and Hunter 1999, Naughton et al. 2000). However, Chazal and Niewiarowski (1998) found no difference in survival of recently metamorphosed *A. talpoideum* placed in forested and clear-cut habitats, which suggests that the process of timber harvest, rather than the associated habitat change, is detrimental to salamanders. In Pennsylvania, some protection to habitats is provided on state forest lands by regulations requiring a 30 m buffer of "no disturbance" habitat and an additional 30 m buffer of partial cut habitat (retaining 50 percent canopy cover) around vernal ponds. No such protection is presently required for habitats on private land.

Jefferson salamander larvae appear to be more sensitive to acidification (and associated toxicity of acid-soluble metals) than other salamander species, with considerable embryonic and larval mortality at pH 4.2–4.5 (Rowe at al. 1992, Sadinski and Dunson 1992). Rowe and Dunson (1993) found that the number of eggs laid by adult Jefferson salamanders in ponds in central Pennsylvania was positively correlated with pH, and Horne and Dunson (1994) found that low pH and its interaction with aluminum excluded Jefferson salamanders from some potential breeding ponds. Although progress has been made to reduce acid precipitation, it continues to be a problem in Pennsylvania (National Atmospheric Deposition Program 2004). Many temporary ponds in south-central Pennsylvania continue to have pH levels below 4.5.

Other potential threats to Jefferson salamanders include road mortality and reduced dispersal associated with increased development (Gibbs 1998), chemical contamination (Semlitsch 2003), infectious diseases (Carey et al. 2003), and habitat changes associated with global warming (Brooks 2004).

*CONSERVATION AND MANAGEMENT NEEDS*

Conservation of this species requires protection of vernal ponds and associated forested habitats. With the loss of federal protection previously provided by the Clean Water Act, increased state and local protections of isolated wetlands are necessary. Current sites of known occurrence should be given high priority for protection.

*MONITORING AND RESEARCH NEEDS*

The present status of Jefferson salamanders in Pennsylvania is largely unknown. Surveys are needed to determine distribution, particularly in areas of the state with historic records of occurrence but no recent sightings, as well as areas that have not been adequately surveyed. These surveys should be followed up with monitoring activities at selected sites to determine population status and detect changes in numbers. A recent project by the Western Pennsylvania

Conservancy to map and classify vernal ponds throughout the state is a useful step in identifying important habitats.

In addition to status assessment and monitoring, there are several immediate research needs. One is to determine the effects of habitat disturbance from logging on mortality and movements of adult salamanders, including salamanders' use of disturbed habitats following the cessation of logging activities. Another research priority is determining the effects of habitat loss and fragmentation on population viability and dispersal, particularly in relation to metapopulation dynamics. Research into the effects of chronic acidification of aquatic and terrestrial habitats on survivorship and population viability is also needed.

*Author:* TIMOTHY J. MARET, SHIPPENSBURG UNIVERSITY

**PENNSYLVANIA VULNERABLE**

## Upland Chorus Frog

Order: Anura
Family: Hylidae
*Pseudacris feriarum*

The upland chorus frog was selected as a Pennsylvania Species of Greatest Conservation Need because of recent sharp population declines (fig. 4.43). Although upland chorus frogs are not listed as Threatened or Endangered at the federal level or in Pennsylvania, their collection in Pennsylvania is prohibited (Pennsylvania Fish and Boat Commission 2008). The global status of the upland chorus frog is Secure (G5, NatureServe 2009).

*Fig. 4.43.* The Upland Chorus Frog, *Pseudacris feriarum*. Photo courtesy of Tom Diez.

### *GEOGRAPHIC RANGE*

The upland chorus frog is found from eastern Texas and Arkansas through the southern states (except Florida), northward through the Carolinas (except for the Coastal Plain), Virginia, Maryland, south-central and eastern Pennsylvania, New Jersey, to the southeastern tip of New York. Overall, the species is considered to be "stable" in parts of its geographic range, that is, midwestern and southern United States (International Union for the Conservation of Nature 2004). Regarding the upland chorus frog, the International Union for the Conservation of Nature states that "there are many secure populations that exist throughout the range."

### *DISTRIBUTION AND RELATIVE ABUNDANCE IN PENNSYLVANIA*

The upland chorus frog is found in southern, central, and southeastern Pennsylvania in a rough triangle from eastern Somerset County northeast to southern Lycoming County, southeast to Bucks County, with many areas lacking documented populations (fig. 4.44). The species is at the periphery of its geographic range in Pennsylvania. Long-term observations indicate that the species appears to be declining throughout its range within Pennsylvania and is no longer seen in areas where they were once abundant. Certain locations in Pennsylvania have relatively large populations of chorus frogs, while nearby areas of suitable habitat appear to be lacking them.

### *COMMUNITY TYPE/HABITAT USE*

Primary components of habitat include open palustrine emergent wetlands with areas of temporary shallow water, including ditches, for breeding in the spring. Adjacent forested uplands and grasslands are inhabited the rest of the year, including for hibernation. Some sites can be in agricultural or urban environments.

Skelly (1995, 1996) found that western chorus frog

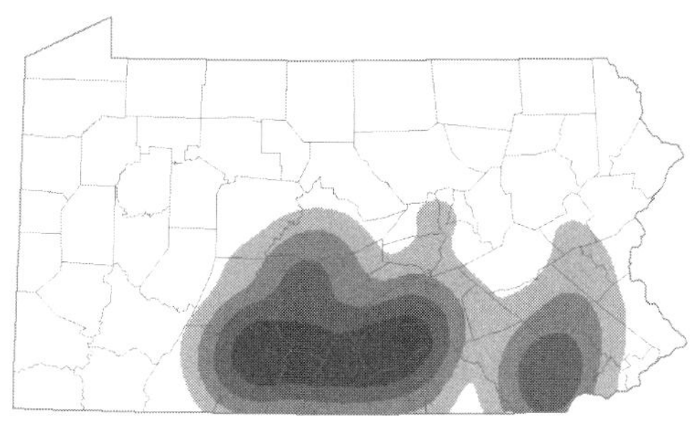

*Fig. 4.44.* Distribution of the Upland Chorus Frog, *Pseudacris feriarum*.

(*Pseudacris triseriata*) tadpoles were relatively more abundant in temporary ponds than those of the spring peeper, with spring peeper tadpoles tending to prefer less permanent bodies of water (he attributed this to lower predator densities). This is consistent with my observations of upland chorus frogs and spring peepers in a large restored/constructed wetland in Blair County. Calling chorus frog males (and noncalling individuals) were more prevalent in small, shallow, less permanent pools or areas of standing water.

Optimum habitat for the upland chorus frog is relatively common throughout its geographic range, although the habitat may be widely spaced. This type of habitat can be relatively easily restored and used by the species, if chorus frogs are still within the area. For example, two relatively large populations of breeding (i.e., calling) individuals colonized two restored wetland areas in Pennsylvania.

### *LIFE HISTORY AND ECOLOGY*

As with other anuran species, chorus frogs feed on a wide variety of invertebrates (Whitaker 1971). Christian (1982) found that larger western chorus frogs ate larger prey and tended to be more selective about their prey items. Larvae (tadpoles) feed on algae and organic detritus.

Various predators feed on adult chorus frogs. Documented predators include northern water snakes, eastern ribbon snakes, garter snakes, various fish species, and lycopid spiders (Whitaker 1971, Smith 1983). Various invertebrates and vertebrates also feed on the tadpoles, including diving beetle larvae, dragonfly nymphs, various fish species, salamander larvae, and adult red-spotted newts (Whitaker 1971, Smith 1983, Skelly 1996, Maret and Collins 1997, Bridges 2002).

Upland chorus frogs are probably short-lived. Platz and Lathrop (1993) found that most males of western and boreal (*Pseudacris maculata*) chorus frogs in breeding choruses from the Great Plains were two or three years old, with an occasional four-year-old individual. Smith (1987) also reported on low adult survivorship; his data suggested an adult survivorship of about 14 percent.

Comparatively little is known about chorus frogs outside of the breeding period. Individuals located outside the breeding season have been found in the leaf litter of woods or among dead vegetation of grasslands, relatively close to water (Whitaker 1971, Kramer 1973). The author observed an adult upland chorus frog active during the day in an upland field approximately 50 m from water (Blair County). Kramer (1973) reported that western chorus frogs generally remained within 100 m of their breeding pools, although two individuals in his study moved more than 200 m away. The home ranges of nine individuals varied from 641 $m^2$ to 6,024 $m^2$, with a mean of 2,117 $m^2$ and all included a breeding pool (Kramer 1974). Chorus frogs hibernate on land and are freeze tolerant (MacArthur and Dandy 1982, Storey and Storey 1986, Packard et al. 1998).

Chorus frogs are early breeders compared with other anuran species. Calling and breeding occurs in Pennsylvania from early March through April, depending on the weather. Primary breeding habitats (shallow areas of temporary standing water) are usually free from large aquatic predators (e.g., fish). Most calling occurs at dusk, but males may also call throughout the day. Small egg masses, 2.5-6 cm in diameter, containing 12-245 eggs, are laid usually attached to vegetation or within the open water. Female western chorus frogs lay multiple masses, totaling 440-752 eggs per female (Whitaker 1971). The time to hatching and metamorphosis is temperature dependent in chorus frogs. Whitaker (1971) noted that eggs hatched in eight to twenty-seven days and tadpoles metamorphosed between fifty-two and seventy days in Indiana. Growth rates of upland chorus frog tadpoles depend on food sources (Britson and Kissel 1996). If food resources are low or poor quality and the hydroperiod of the breeding pools is short, tadpoles may not survive to metamorphose.

### *THREATS*

Recent global amphibian declines have gathered much interest from the scientific community, and numerous authors have discussed the potential causes and synergistic effects (see Diana and Beasley 1998, Linder et al. 2003, Davidson 2004). The primary causes of the species' decline in Pennsylvania are speculative and may include various factors.

Habitat (wetland) loss has no doubt contributed to the species' decline in the past. Wetland losses continue; however, a net wetland gain is now occurring in Pennsylvania through wetland replacement associated with regulatory permitting requirements and government-sponsored conservation programs (Pennsylvania Department of Environmental Protection 2006). A more likely problem in Pennsylvania is fragmentation and isolation of breeding habitats from other such areas. For a short-lived species with apparently limited dispersal abilities, an isolated population could be extirpated with a drought of only two to three years (preventing local reproduction), coupled with no chance of recruitment from outside populations.

Other possible factors contributing to the species decline include chemical contaminants (pesticides, PCBs, metals), increases in ultraviolet radiation due to depletion of stratospheric ozone, increased parasitic trematode infections, and the chytrid fungus (*Batrachochytrium dendrobatidis*). The chytrid fungus has recently been verified in western chorus frogs from Colorado (Rittman et al. 2003).

*CONSERVATION AND MANAGEMENT NEEDS*

An overall conservation objective would be to conserve and protect as many major breeding areas and surrounding habitats as possible. Specific tasks cannot be proposed until surveys and research priorities are addressed.

*MONITORING AND RESEARCH NEEDS*

The need for information on the ecology, natural history, and current distribution of this species in Pennsylvania is evident. Other than distribution locations (Hulse et al. 2001), there are no recent references of research or of observations of the natural history of the species in Pennsylvania. There is little information on the upland chorus frog throughout its geographic range. In Pennsylvania, immediate needs include surveys to identify extant and historical sites throughout the species' range, as well as surveys of other sites of potential habitat throughout its range to detect new population occurrences.

Long-term needs include surveys of areas between known sites to assess for suitability as dispersal corridors, development of a monitoring protocol to record breeding and evaluate population changes at known sites, basic research on all facets of the biology and ecology of the species, and an investigation of the potential sympatry of the upland and New Jersey chorus frogs to determine similarities or differences in habitat and ecological requirements.

*Author:* THOMAS G. PLUTO, U.S. ARMY CORPS OF ENGINEERS

## Western Chorus Frog

Order: Anura
Family: Hylidae
*Pseudacris triseriata*

The western chorus frog was selected as a Pennsylvania Species of Greatest Conservation Need because of recent sharp population declines (fig. 4.45). Although western chorus frogs are not listed as Threatened or Endangered at the federal level or in Pennsylvania, their collection in Pennsylvania is prohibited (Pennsylvania Fish and Boat Commission 2008). The global status of the western chorus frog is Secure (G5, NatureServe 2009).

*Fig. 4.45.* The Western Chorus Frog, *Pseudacris triseriata*. Photo courtesy of Tom Diez.

*GEOGRAPHIC RANGE*

The western chorus frog ranges from eastern Oklahoma, Kansas, Nebraska, and southeastern South Dakota eastward to western Pennsylvania and farther northeast into northwestern New York, southern Ontario, and the extreme southern end of Quebec. Although still abundant throughout much of its geographic range, the species appears to be declining in many areas (International Union for the Conservation of Nature 2004). In the Midwest, the numbers of western chorus frogs appear to be declining in some areas, but the species may be expanding its range in other areas (Hemesath 1998, Minton 1998). On the basis of 1984-1995 surveys conducted in Wisconsin, Mossman et al. (1998) reported that, although western chorus frog populations were down, they were apparently rebounding somewhat after a decline during drought years.

*DISTRIBUTION AND RELATIVE ABUNDANCE IN PENNSYLVANIA*

The western chorus frog is distributed in the westernmost Pennsylvania counties, from Greene and western Fayette counties northward to Erie and Warren counties (fig. 4.46). The species is at the periphery of its geographic range in Pennsylvania. Long-term observations indicate that the species appears to be de-

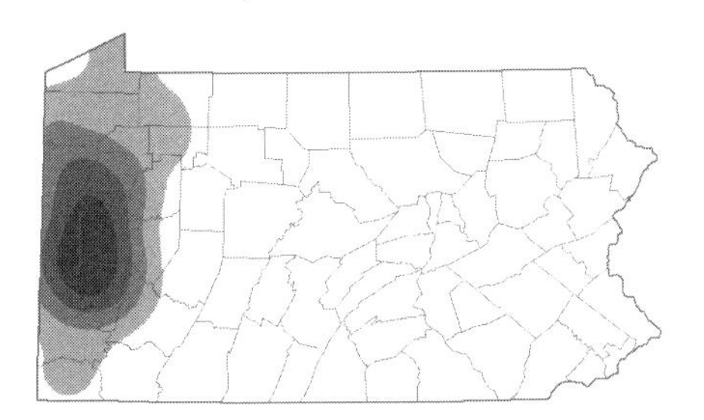

*Fig. 4.46.* Distribution of the Western Chorus Frog, *Pseudacris triseriata.*

clining substantially throughout its range in Pennsylvania and is no longer seen in areas where they were once abundant (T. Diez, personal communication).

### *COMMUNITY TYPE/HABITAT USE*

Primary components of habitat include open palustrine emergent wetlands with areas of temporary shallow water, including ditches, for breeding in the spring. Adjacent forested uplands and grasslands are inhabited the rest of the year, including for hibernation. Some sites can be in agricultural or urban environments.

Skelly (1995, 1996) found that western chorus frog tadpoles were relatively more abundant in temporary ponds than those of the spring peeper; chorus frog tadpoles tended to prefer less permanent bodies of water (he attributed this to lower predator densities).

Optimum habitat for chorus frogs (i.e., early successional palustrine emergent wetlands with temporary shallow-water habitat and adjacent forested or wooded areas) is relatively common throughout its geographic range, although the habitat may be widely spaced. As evidenced recently by two relatively large populations of breeding (i.e., calling) individuals of upland chorus frogs (*Pseudacris feriarum*) at two restored wetland areas in Pennsylvania, this type of habitat can be relatively easily restored and used by the species, if chorus frogs are still within the area.

### *LIFE HISTORY AND ECOLOGY*

As with other anuran species, chorus frogs feed on a variety of invertebrates (Whitaker 1971). Christian (1982) found that larger western chorus frogs ate larger prey and tended to be more selective about their prey items. Larvae (tadpoles) feed on algae and organic detritus.

A variety of predators feed on the adults. Documented predators include northern water snakes, eastern ribbon snakes, garter snakes, various fish species, and lycopid spiders (Whitaker 1971, Smith 1983). Various invertebrates and vertebrates also feed on the tadpoles, including diving beetle larvae, dragonfly nymphs, various fish species, salamander larvae, and adult red-spotted newts (Whitaker 1971, Smith 1983, Skelly 1996, Maret and Collins 1997, Bridges 2002).

The western chorus frog is a short-lived species. Platz and Lathrop (1993) found that most males in breeding choruses from the Great Plains were two or three years old, with an occasional four-year-old individual. Smith (1987) also reported on low adult survivorship; his data suggested an adult survivorship of about 14 percent.

Comparatively little is known about the species outside of the breeding period. Individuals located outside the breeding season have been found in the leaf litter of woods or among dead vegetation of grasslands, relatively close to water (Whitaker 1971, Kramer 1973). Petzing et al. (2002) reported finding an adult western chorus frog under slab rock in a dry creek bed in Illinois. Kramer (1973) reported that western chorus frogs generally remained within 100 m of their breeding pools, although two individuals in his study moved more than 200 m away. The home ranges of nine individuals varied from 641 $m^2$ to 6,024 $m^2$ with a mean of 2,117 $m^2$ and all included a breeding pool (Kramer 1974). Chorus frogs hibernate on land and are freeze tolerant (MacArthur and Dandy 1982, Storey and Storey 1986, Packard et al. 1998).

Chorus frogs are early breeders compared with other anuran species. Calling and breeding occurs in Pennsylvania from early March through April, depending on the weather. Primary breeding habitats (shallow areas of temporary standing water) are usually free from large aquatic predators (e.g., fish). Most calling occurs at dusk, but males may also call throughout the day. Small egg masses, 2.5-6 cm in diameter containing 12-245 eggs, are laid usually attached to vegetation or within the open water. Females lay multiple masses, totaling 440-752 eggs per female (Whitaker 1971). Times to hatching and to metamorphosis are temperature dependent. Whitaker (1971) noted that eggs hatched in eight to twenty-seven days, and tadpoles metamorphosed between fifty-two to seventy days in Indiana. Smith (1983) reported from northern Michigan that the minimum egg/larval development period was fifty-five days, with many tadpoles requiring greater than three months to complete development to metamorphosis. (Smith's population is possibly now considered to be the boreal chorus frog [*Pseudacris maculata*]) Growth rates of (upland) chorus frog tadpoles depend on food source (Britson and Kissel 1996). If food resources are low or poor quality and the hydroperiod of the breeding pools is short, then tadpoles may not survive to metamorphose.

*THREATS*

Although considerable research has been devoted to understanding the causes of amphibian declines across the globe (Diana and Beasley 1998, Linder et al. 2003, Davidson 2004), the specific factors contributing to the loss of the western chorus frog in Pennsylvania are still not clear. Loss of wetlands, formerly a problem, is now less of an issue because of regulatory requirements and various state-sponsored conservation initiatives that have resulted in a net gain of wetlands in some areas within the species' range (Pennsylvania Department of Environmental Protection 2006). In contrast, it appears that, in Pennsylvania, species-specific traits of the chorus frog (e.g., short life span and limited dispersal ability) increase the chances of isolation of breeding populations and local extinction, especially during droughts, which over time may contribute to increased fragmentation and contraction of the species' geographic range in the state. More information, however, is needed to better understand these potential threats.

Chemical contaminants, including pesticides, PCBs, and heavy metals, may also be negatively affecting chorus frog populations. Pathogens and parasites are also a concern. The chytrid fungus (*Batrachochytrium dendrobatidis*) has recently been verified in western chorus frogs from Colorado (Rittman et al. 2003). Trematode infections, which are known to cause abnormalities in a number of frog species, may be affecting chorus frogs. Finally, chorus frogs may be sensitive to increases in ultraviolet radiation due to depletion of stratospheric ozone.

*CONSERVATION AND MANAGEMENT NEEDS*

An overall conservation objective would be to conserve and protect as many major breeding areas and surrounding habitats as possible. Specific tasks cannot be proposed until surveys and research priorities are addressed.

*MONITORING AND RESEARCH NEEDS*

The need for information on the ecology, natural history, and current distribution of this species in Pennsylvania is evident. Other than distribution locations (Hulse et al. 2001), there are no recent references of research or of observations of the natural history of this species in Pennsylvania. Most scientific articles on western chorus frogs were from the Midwest. In Pennsylvania, immediate research/monitoring needs include surveys of extant and historical sites throughout the species' geographic range. Surveys should also include other sites of potential habitat to detect potential new population occurrences. Long-term needs include surveys of areas between known sites to assess for suitability as dispersal corridors, development of a monitoring protocol to record breeding and evaluate population changes at known sites, and basic research on all facets of the biology and ecology of the species.

*Author:* THOMAS G. PLUTO, U.S. ARMY CORPS OF ENGINEERS

## Southern Leopard Frog

Order: Anura
Family: Ranidae
*Lithobates sphenocephalus utricularia*

The southern leopard frog was selected as a Pennsylvania Species of Greatest Conservation Need because of its rarity and restricted distribution in Pennsylvania (fig. 4.47). It is listed as an Endangered Species by the Pennsylvania Fish and Boat Commission, and collecting of this species is prohibited (Pennsylvania Code 2008). The global status of the southern leopard frog is Secure (G5, NatureServe 2009).

*GEOGRAPHIC RANGE*

This species occurs throughout the Atlantic southern from Long Island and extreme southern New York

*Fig. 4.47.* The Southern Leopard Frog, *Lithobates sphenocephalus utricularia*. Photo courtesy of Tom Diez.

to Florida, including some of the lower Keys, west to eastern and central Texas, and north in the Mississippi Valley to southeast Kansas, Missouri, central Illinois, Iowa, Indiana, and extreme southern Ohio. Specimens from Florida have been designated as a separate race known as the Florida leopard frog or *Lithobates sphenocephalus sphenocephalus* (Conant and Collins 1998).

### *DISTRIBUTION AND RELATIVE ABUNDANCE IN PENNSYLVANIA*

In Pennsylvania, the southern leopard frog occurs in the Coastal Plain and Piedmont physiographic provinces in southeastern Pennsylvania in the lower Delaware Valley (fig. 4.48). Hulse et al. (2001) indicated that the ranges of the southern and northern leopard frogs in Pennsylvania are thought to be adjacent to each other, but do not overlap.

Only five extant and four historic records are known for this species in the state, and the records occur within a limited area of the Delaware Valley (Bucks, Chester, Delaware, and Montgomery counties). Historically, the southern leopard frog has also occurred in Philadelphia County, with the last report recorded in 1909 (Pennsylvania Natural Heritage Program 2005). The most recent records were reported during the Pennsylvania Herpetological Atlas in 2000 in Delaware County (Hulse et al. 2001).

### *COMMUNITY TYPE/HABITAT USE*

The southern leopard frog is nocturnal and semiaquatic. It occupies a variety of habitats, including marshes, ponds, wet meadows, and the edges of slow-moving rivers and streams; and it can also be found in brackish waters near coastal areas (Groves 1985b, Hulse et al. 2001). During periods of heavy rain, the frog ventures to virtually any permanent or semipermanent wet habitat (Groves 1985b).

*Fig. 4.48.* Distribution of the Southern Leopard Frog, *Lithobates sphenocephalus utricularia.*

### *LIFE HISTORY AND ECOLOGY*

Little information is available for the southern leopard frog in Pennsylvania (Groves 1985b). Most studies have occurred in other parts of its geographic range. Survey information from the Pennsylvania Herpetological Atlas recorded sightings of the frog as early as March 20 and as late as October 5, but it is likely that activity occurs from late February to early November. The frog probably hibernates at or below the water line in mud or under leaves (Hulse et al. 2001).

Adult frogs eat a varied diet consisting of both terrestrial and aquatic insects, snails, and terrestrial arthropods, while tadpoles consume algae, plant tissue, decaying plant matter, and aquatic invertebrates (Groves 1985b, Hulse et al. 2001, NatureServe 2009). The adults are preyed on by many vertebrate species, while a variety of vertebrates and aquatic invertebrates eat the tadpoles (Groves 1985b).

The peak month for breeding in Pennsylvania is April, but breeding may occur as early as February or March after a warm rain and milder spring weather (Groves 1985b). Males typically call to females while floating on top of shallow water near the edge of a water body. The eggs are laid in flattened masses, either by one individual or communally on aquatic vegetation at various depths in the water column (Groves 1985b, Hulse et al. 2001). Communal egg laying in shallow water is common during cold weather, while eggs are laid in deeper water during warmer periods (Caldwell 1986). Communal masses can contain hundreds to thousands of eggs, and the eggs will hatch between seven to twenty days, depending on the temperature (Groves 1985b, Hulse et al. 2001).

Observance of a Pennsylvania southern leopard tadpole in a laboratory setting showed that larval development lasted fifty to seventy-five days (Hulse et al. 2001). This result is comparable to tadpole development shown from studies in the south (Dundee and Rossman 1989). Ryan and Winne (2001) showed that small differences in hydroperiod length can dramatically affect juvenile recruitment by influencing the number of metamorphs and the length of the larval stage.

Males continue to call as late as October, but no egg laying has been observed in Pennsylvania at this time of the year (Groves 1985b). In northeastern Arkansas, egg laying occurred in late September after a heavy rain event (McCallum et al. 2004).

*THREATS*

Heavy development pressures have probably led to direct and indirect habitat alteration for this species in Pennsylvania. Increased traffic on roads near ponds is a significant threat in urban areas (Palis 1994). Pesticide exposure has been shown to cause deformities and to delay tadpole development (Bridges 2000).

*CONSERVATION AND MANAGEMENT NEEDS*

Surveys are needed in Pennsylvania to determine the current geographic range and extent of the southern leopard frog. After inventory surveys are conducted, specific threats to the species need to be identified and assessed, and a management plan developed.

*MONITORING AND RESEARCH NEEDS*

Intensive surveys are needed to determine the present status and distribution of this species. Survey efforts should focus on areas with historic records of occurrence but no recent sightings. These surveys should be followed up with monitoring activities at selected sites to detect changes in population numbers and distribution.

Not much is known about past and current community habitat types used by this species in Pennsylvania. The current records occur within a highly urbanized area near Philadelphia and the lower Delaware Valley. It is likely that wetland habitats in this vicinity have been altered directly through development or indirectly through water degradation. In addition, little is known about the natural history for this species in Pennsylvania.

*Author:* AURA STAUFFER, PENNSYLVANIA DEPARTMENT OF CONSERVATION AND NATURAL RESOURCES

## Rough Green Snake

Order: Squamata
Family: Colubridae
*Opheodrys aestivus*

The rough green snake was selected as a Pennsylvania Species of Greatest Conservation Need because of its rarity and limited distribution in Pennsylvania (fig. 4.49). Rough green snakes are listed as Endangered in Pennsylvania, and their collection is prohibited (Pennsylvania Code 2008). The global status of the rough green snake is Secure (G5, NatureServe 2009).

*Fig. 4.49.* The Rough Green Snake, *Opheodrys aestivus*. Photo courtesy of Tom Diez.

*GEOGRAPHIC RANGE*

The rough green snake is widely distributed in the eastern and central United States. Its geographic range extends from southern New Jersey south to the Florida Keys and west to eastern Kansas, Texas, and extreme northeastern Mexico.

*DISTRIBUTION AND RELATIVE ABUNDANCE IN PENNSYLVANIA*

The rough green snake is known from three locations within the commonwealth. All are located in southern-tier counties (fig. 4.50). A historic location is from Greene County near the town of Ninevah. Grobman (1984), however, dismisses this record on the basis that the nearest known locality to Ninevah is "110 miles to the southwest and 130 miles to the east." The second population is located in the vicinity of Horseshoe Camp, Chester County, and is extant. The Pennsylvania Herpetological Atlas Project recently received a voucher specimen that was found dead along a section of railroad track in Lancaster County.

*COMMUNITY TYPE/HABITAT USE*

Throughout most of its geographic range, the rough green snake is an inhabitant of riparian corridors and lakeshores where trees or woody shrubs dominate the vegetation. In Pennsylvania, all specimens have been collected in vegetation near streams. Plummer (1981) studying an Arkansas population, noted that 86 percent to 89 percent of all captures occurred within 3 m of water and that 97 percent were captured within 5 m of water. Goldsmith (1984) also noted their tendency to be found along the edges of bodies of water

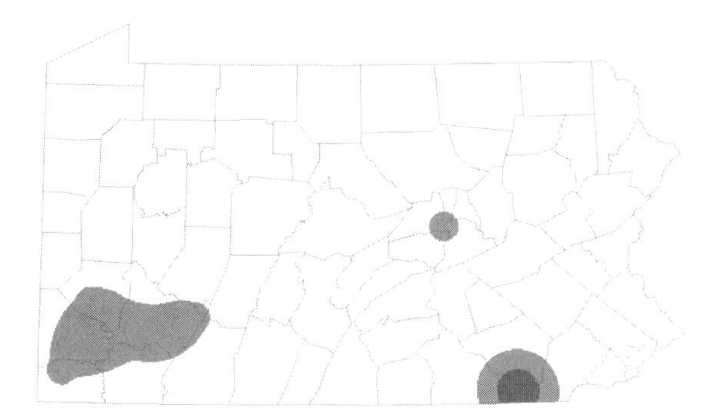

*Fig. 4.50.* Distribution of the Rough Green Snake, *Opheodrys aestivus*.

(both standing and flowing). The snakes do make occasional movements outside of these corridors, especially during nesting.

### *LIFE HISTORY AND ECOLOGY*

The rough green snake is extremely uncommon in Pennsylvania; as a result, information presented here comes from studies in other portions of the animal's geographic range. Rough green snakes are arboreal, usually found in trees and shrubs at a height of 1 to 3 m above the ground. Individuals are occasionally found on the ground, although these may be juveniles or females looking for oviposition sites (Goldsmith 1984).

The extent of seasonal activity in this species is variable, depending on geographic location. In West Virginia, they emerge in early May and cease activity by mid-October (Green and Pauley 1987). Rough green snakes are diurnal and spend most of the day slowly moving through the vegetation foraging for food (Plummer 1981). They often forage with head and anterior portion of their bodies elevated; intermittently stopping while foraging to move their head rhythmically from side to side. Collins (1974) suggested that it may be a form of cryptic behavior in that the snake's moving body would resemble a branch swaying in the breeze.

Rough green snakes have restricted home ranges. Plummer (1981) noted a mean movement of only 62 m (range 15 m to 247 m) from the point of original capture. In high-quality habitat, rough green snakes can reach very high densities, with up to 714 animals per hectare (Plummer 1985a).

Rough green snakes are almost exclusively insectivorous (Brown 1979, Plummer 1981). Main food items included in their diet are caterpillars, grasshoppers, crickets, spiders, dragonflies, and damselflies. A variety of other small insects are occasionally taken. Insects and spiders are generally small in relation to the size of the snake, and as a consequence, the animals have to feed at frequent intervals and consume numerous prey items to meet their nutritional and energetic requirements.

Mating occurs in both spring and fall. Oviposition has been reported from as early as June 17 (Guidry 1953) to as late as August 31 (Fitch 1970), but most egg laying appears to occur in late June and July (Goldsmith 1984). Incubation time is highly variable (thirty-four to ninety days) and may be a function of geographic location. It appears that northern populations have longer incubation periods (sixty-four to ninety days) than southern populations (thirty-four to fifty days). Females leave their arboreal environments to lay eggs on the ground or to move to specific nest trees. Eggs are laid in leaf litter, rotting logs and stumps, loose soil, trash piles, and cavities of living trees. Plummer and Snell (1988) demonstrated in the laboratory that females select nest sites on the basis of moisture content, preferring sites with higher moisture levels. Sites with higher moisture content produced larger and heavier young but did not affect hatching success. Communal nesting is known to occur in this species, with as many as seventy-four eggs found in a single oviposition site (Palmer and Braswell 1976). Clutch size varies from three to twelve eggs in Arkansas (Plummer 1990). The eggs are white and leathery. In Arkansas, males mature in two years and females mature in two to three years (Plummer 1985b). Time to maturity may be longer in the northern part of the geographic range.

### *THREATS*

Threats to this species are largely unknown; however, the biggest potential threat to this species would be the destruction of riparian habitat, because throughout most of their geographic range, rough green snakes are seldom found far from riparian vegetation. An additional threat is posed by our lack of information regarding the details of the species distribution within the commonwealth.

### *CONSERVATION AND MANAGEMENT NEEDS*

Riparian corridors, as well as a nonriparian buffer zone, should be immediately established at locations known to harbor rough green snakes.

*MONITORING AND RESEARCH NEEDS*

Long-term monitoring of rough green snakes at known locations is extremely important. A mark-recapture program (preferably using passive integrated transponder tags) should be instituted at both sites. In addition, the general region surrounding known locations should be intensely surveyed to determine whether additional populations or subpopulations exist.

The most pressing needs are to determine the size and structure of the existing populations of rough green snakes, along with a delineation of the habitat used by the snakes at these sites.

*Author:* ARTHUR C. HULSE, INDIANA UNIVERSITY OF PENNSYLVANIA (RETIRED)

## Eastern Earth Snake

Order: Squamata
Family: Colubridae
*Virginia valeriae valeriae*

The eastern earth snake was selected as a Pennsylvania Species of Greatest Conservation Need because of its extreme rarity and restricted distribution in Pennsylvania (fig. 4.51). Eastern earth snakes have not been reported within the commonwealth in more than fifty years. A recent survey effort was not successful at finding eastern earth snakes, and there are doubts as to whether the species still exists in the state. Their collection in Pennsylvania is prohibited (Pennsylvania Fish and Boat Commission 2008). The global status of the eastern earth snake is Secure (G5, NatureServe 2009).

Fig. 4.51. The Eastern Earth Snake, *Virginia valeriae valeriae*. Photo courtesy of Richard Valk.

*GEOGRAPHIC RANGE*

The eastern earth snake ranges from north-central New Jersey southward to northern Florida. It occurs westward to the Mississippi-Louisiana border and from there northward through central Tennessee and Kentucky to extreme southern Ohio and western West Virginia.

*DISTRIBUTION AND RELATIVE ABUNDANCE IN PENNSYLVANIA*

In Pennsylvania, the eastern earth snake's occurrence is restricted to the southeast Coastal Plain and adjacent Piedmont. Its western limit in the state roughly coincides with the edge of the Valley and Ridge Province. It has been reported from only four locations within the state (fig. 4.52). Roddy (1928) reported that an individual from Lancaster County was actually a misidentified northern brown snake; however, the species may occur in Lancaster County because the Chester County record is only a few miles from the Lancaster County border.

*COMMUNITY TYPE/HABITAT USE*

Key habitat is deciduous woodlands throughout most of the species' geographic range. In the south, they also occupy moist hammocks.

*LIFE HISTORY AND ECOLOGY*

Not much is known of the biology of this subspecies in Pennsylvania. As a result, the following discussion, and that on reproduction, will deal with information gathered in other parts of the animal's geographic range. The eastern earth snake is primarily an inhabitant of deciduous forests and adjacent open areas (Pisani and Collins 1971, Conant and Collins 1991). They are seldom encountered except during and after rains when they may be found under rocks, logs, and other surface debris. Nothing is known of the seasonal

Fig. 4.52. Distribution of the Eastern Earth Snake, *Virginia valeriae valeriae*.

activity of this subspecies. It has been found in Pennsylvania in May, June, and August. Eastern earth snakes appear to feed exclusively on earthworms (Blanchard 1923, Blem and Blem 1985). Like the mountain earth snake, the eastern earth snake is an inoffensive animal that never attempts to bite.

It is not known when courtship and mating occur. In Virginia, Blem and Blem (1985) noted that follicles began to enlarge in late March and early April and that by mid-May ovulation had occurred. Birth has been reported as occurring from early August (Blem and Blem 1985) to mid-September (Walker 1963). Litter size appears to vary from four to fourteen young. Blem and Blem (1985) reported an average litter size of 6.6 young from Virginia.

*THREATS*

Habitat fragmentation and development within the historic geographic range of the species constitute a great threat to the continued existence of the species within the commonwealth. However, one of the greatest threats to the species is our lack of understanding of its distribution and specific ecological requirements within Pennsylvania.

*CONSERVATION AND MANAGEMENT NEEDS*

All sites where eastern earth snakes are found to exist should be entered into the Pennsylvania Natural Heritage Program database so that they can be readily recognized in response to development plans. Any development within regions where earth snakes occur should be conducted in a manner that would minimize impact on populations of the snakes. Resource managers at sites where snake populations are discovered should be made aware of the existence of the populations. Whenever possible, management for eastern earth snakes should be incorporated into any resource management plans.

*MONITORING AND RESEARCH NEEDS*

Little is known concerning the status and biology of the eastern earth snake in Pennsylvania, and recent surveys have not detected any individuals. If future surveys do reveal the presence of eastern earth snakes, long-term monitoring of select populations should be instituted to determine population size and structure, as well as reproductive activity and population trends.

*Author:* ARTHUR C. HULSE, INDIANA UNIVERSITY OF PENNSYLVANIA (RETIRED)

### MAINTENANCE CONCERN

## Fowler's Toad

Order: Anura
Family: Bufonidae
*Anaxyrus fowleri*

The Fowler's toad was selected as a Pennsylvania Species of Greatest Conservation Need because of recent population declines in western Pennsylvania and uncertain status is eastern Pennsylvania (fig. 4.53). Fowler's toads are not listed as Threatened or Endangered either at the federal level or in Pennsylvania. Their collection in Pennsylvania is limited to one individual per day, with a possession limit of one individual (Pennsylvania Fish and Boat Commission 2008). The global status of the Fowler's toad is Secure (G5, NatureServe 2009).

*GEOGRAPHIC RANGE*

The Fowler's toad is restricted to the eastern United States, where it shows more of a southern and coastal affiliation. Its greatest distribution occurs primarily from East Texas and the Florida panhandle, continuing northeast from the Gulf Coast states through Arkansas, Indiana, and Ohio. The Fowler's toad's range continues northward up through the Coastal Plain states, entering the southern half of Pennsylvania. Its northern range extends from New Jersey, until it reaches the northernmost range in eastern Vermont (Barker and Caduto 1984, NatureServe 2009).

*Fig. 4.53.* The Fowler's Toad, *Anaxyrus fowleri*. Photo courtesy of Richard Koval.

### *DISTRIBUTION AND RELATIVE ABUNDANCE IN PENNSYLVANIA*

Records from historical data of the Fowler's toad in Pennsylvania indicate a distribution primarily along major river basins. The range of the Fowler's toad has been confined mostly to the southern two-thirds of Pennsylvania. Fowler's toads were reported along the valleys of the Ohio River basin in the counties of Allegheny, Beaver, and Lawrence; the north-central counties of Clinton, Clearfield, and Lycoming; low-lying areas in Cumberland and Fulton counties; along the Susquehanna River basin in Dauphin and Lancaster counties; and along the Delaware River in Bucks, Philadelphia, and Delaware (fig. 4.54). Fowler's toads north of their range were reported along the Lake Erie shore in the northwest, Venango County in the west, and Wyoming County in the northeast (McCoy 1982, Hulse et al. 2001).

Occurrences of Fowler's toads in western Pennsylvania have drastically declined over the past thirty years. They were reported to be so abundant along the Allegheny River near Pittsburgh that their choruses were deafening. What remains now is a small isolated river island population at Twelve Mile Island, north of Pittsburgh (T. Diez, personal communication). The few records from southeast Pennsylvania do not explain why they disappeared, which has caused great concern (A. Hulse, personal communication).

More recently, Fowler's toads have been confirmed at Presque Isle, Erie County, and Penn's Creek, Union County (T. Diez, personal communication). The Pennsylvania Herpetological Atlas Project also reported new discoveries of Fowler's toads in the northeast (A. Hulse, personal communication). Sustainable populations were discovered along the Delaware River in northeastern Monroe County and northward into Pike County (Serrao 2000, J. Serrao, personal communication) and along the Susquehanna River in Luzerne, Lackawanna, and Wyoming counties (R. L. Koval, personal observation). Surprisingly, Fowler's toads have not been confirmed along the Lehigh River basin from Carbon County northward.

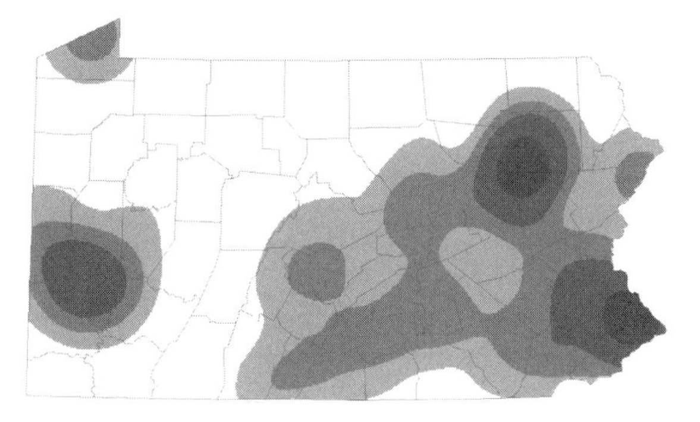

*Fig. 4.54.* Distribution of the Fowler's Toad, *Anaxyrus fowleri*.

The Fowler's toad is considered uncommon in Pennsylvania. Our current knowledge of this species shows drastic declines throughout its historical range, especially in western Pennsylvania, but it has expanded its range in the northeast.

### *COMMUNITY TYPE/HABITAT USE*

The Fowler's toad is a toad of river bottoms, lake edges, sandy places, and even urban gardens. They most often can be found not far from low-lying riparian corridors, shorelines, lakes and pond margins, and even in backyards (Oliver 1955). The Fowler's toad's affinity toward alluvium deposits of dry, gravelly, and sandy substrate restricts its distribution. High densities of Fowler's toads have been observed using flooded sand/gravel excavation pits along the Susquehanna River in Luzerne and Wyoming counties (R. L. Koval, personal observation). This species normally does not inhabit interior forests, mountain streams, swamps, or bogs, unlike that of the widespread American toad.

### *LIFE HISTORY AND ECOLOGY*

The diet of the Fowler's toad varies considerably, depending on its life stage. During the larval, or tadpole, stage, it consumes decaying vegetative matter. Newly transformed and adult Fowler's toads consume all types of invertebrates normally found throughout its terrestrial environment. Prey items include a variety of insects and invertebrate species.

Fowler's toads in Pennsylvania emerge from hibernation later than American toads. The Fowler's toad may require warmer weather to commence activity, usually by the month of May in Pennsylvania. They remain active throughout the summer and by October are seldom encountered (R. L. Koval, personal observation).

Fowler's toads become quite active on rainy nights and can be found along roads, railroad tracks, parking lots, and even in backyards. When breeding, they are found along the shorelines of lakes, rivers, quiet shallow back waters, ponds, ditches, vernal pools, and flooded meadows. Recently transformed Fowler's toads are often found along gravelly edges of river islands, riverbanks, lake shorelines, and sandy open areas (Klemens 1993, Hulse et al. 2001).

The Fowler's toad begins breeding in Pennsylvania

usually thirty to forty-five days later than the American toad. Males are usually in full chorus by late May and continue through June and early July. Female Fowler's toads deposit black stringy strands of approximately 8,000 eggs on the bottoms of shallow wetlands. The egg strands differ from the similar egg strands of American toads by lacking partitions and inner envelopes. The eggs hatch quickly in about three to seven days, depending on water temperature. The tiny black tadpoles develop in about forty to fifty-five days. Young Fowler's toads develop rapidly, and by August the following year, they are mature (Hulse et al. 2001).

Extreme weather variations may cause breeding seasons of Fowler's and American toads to overlap. An extended cold spell in early spring can delay the breeding cycles of American toads, while an unseasonably warm spring may stimulate early breeding of Fowler's toads. Both species congregating in the same breeding wetlands may be one cause of hybridization. Hybridization usually stops after the first generation as they are sterile and males usually cannot attract mates (Klemens 1993).

*THREATS*

Several factors may have caused the decline of the Fowler's toad in Pennsylvania. Destruction of breeding habitats, such as vernal pools, ponds, wetlands and river bottom wetlands, and pollution of breeding wetlands from contaminants, such as acid mine drainage, road salt, storm-water runoff, phosphates, and nitrates, are prime culprits. Aerial spraying of pesticides and herbicides near breeding wetlands; indiscriminate removal of vegetation along the margins of lakes, rivers, ponds, and vernal pools; construction of dams, dikes, and levees along rivers and streams; and the introduction of predatory game fish into ponds, lakes, and rivers may also have played a role in their decline. Channeling and dredging of rivers, and highway and road mortality near breeding habitats and travel corridors are also threats to Fowler's toad populations.

*CONSERVATION AND MANAGEMENT NEEDS*

Conservation of this species would be aided by protecting river corridors, wetlands, and vernal pools that provide breeding habitat; establishing and maintaining vegetative buffers along breeding wetlands; constructing "toad tunnels" along excessive road mortality areas; and protecting and conserving known breeding locations through conservation easements, land acquisitions, or landowner awareness.

*MONITORING AND RESEARCH NEEDS*

Monitoring and tracking occurrences of Fowler's toads in Pennsylvania will help understand the range and distribution of this species. Monitoring efforts should include both monitoring of known locations of Fowler's toads, especially during peak breeding seasons, and revisiting historical sites in western and southeastern Pennsylvania to verify whether populations are extant. Surveys are especially needed along the Susquehanna and Delaware rivers in the southeast section of the commonwealth and along the Allegheny River basin in the western part of the state. In addition to status assessment and monitoring, research needs to be conducted on the causes of decline in populations in the western part of the state.

*Author:* RICHARD L. KOVAL, NORTH BRANCH LAND TRUST

## Northern Leopard Frog

Order: Anura
Family: Ranidae
*Lithobates pipiens*

The northern leopard frog was selected as a Pennsylvania Species of Greatest Conservation Need. It has undergone a localized reduction in range and a considerable decline in abundance (fig. 4.55). Northern leopard frogs are not listed as Threatened or Endangered at the federal level or in Pennsylvania. Their collection in Pennsylvania is limited to one individual per day, with a possession limit of one individual (Pennsylvania Fish and Boat Commission 2008). The global status of the northern leopard frog is Secure (G5, NatureServe 2009).

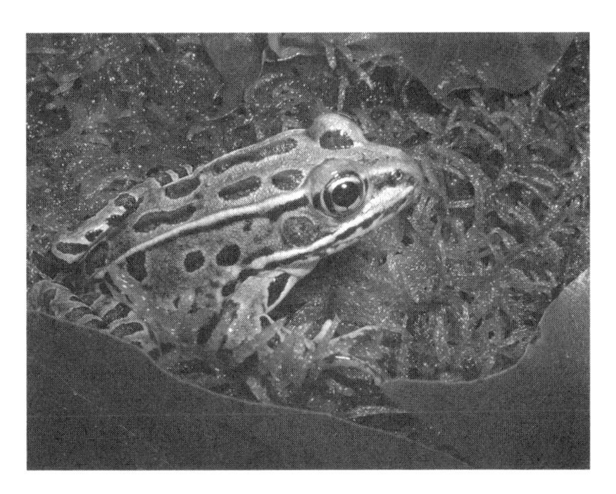

*Fig. 4.55.* The Northern Leopard Frog, *Lithobates pipiens*. Photo courtesy of Richard Koval.

### *GEOGRAPHIC RANGE*

The northern leopard frog ranges across most of northern North America (Hulse et al. 2001). This species is found from southern Quebec west to the western coast of Canada and south to Kentucky and New Mexico. The distribution in the western United States is somewhat spotty, as it has been introduced in many areas (Conant and Collins 1991).

### *DISTRIBUTION AND RELATIVE ABUNDANCE IN PENNSYLVANIA*

The northern leopard frog is found throughout Pennsylvania; however, it is most common in the western portion of the state (Hulse et al. 2001; fig. 4.56). This species used to be common in Pennsylvania and the Northeast, but in recent years, researchers have noticed its decline. The northern leopard frog is now absent from many areas where it once occurred and is present in low numbers in many portions of its range (Hulse et al. 2001). There is no good estimate of the current population in Pennsylvania and no obvious causes for continuing population declines. Although declines in Pennsylvania are not investigated in published reports, other studies in North America have reported similar trends. For example, Brodman et al. (2002) noted reductions in leopard frog populations from Jasper-Pulaski Fish and Wildlife Area in northern Indiana. Mossman et al. (1998) also reported declines in this species, as identified from the Wisconsin Frog and Toad Survey, which began in 1981. Similarly, local extinctions have been reported by Corn and Fogleman (1984) in Colorado and by Orchard (1992) in British Columbia.

### *COMMUNITY TYPE/HABITAT USE*

The northern leopard frog can be found in a variety of habitat types, including meadows, marshes, swamps, ponds, lakes, rivers, and streams. This species prefers wet meadows and fields and may be found far from water in summer months. They often use temporary ponds and wet meadows for breeding (Hulse et al. 2001). Leopard frogs are known to hibernate on pond bottoms in the winter (Emery et al. 1972).

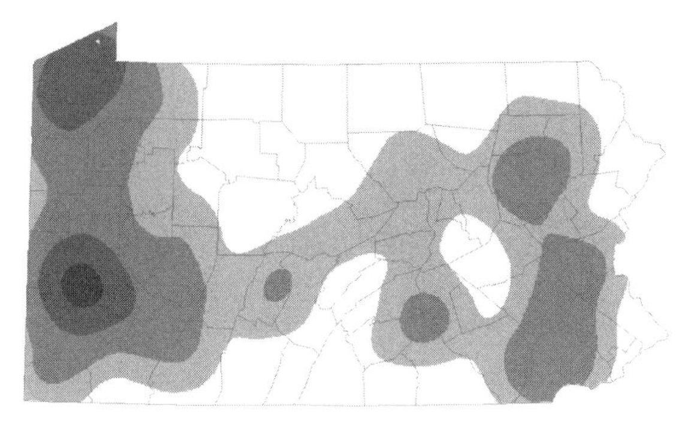

Fig. 4.56. Distribution of the Northern Leopard Frog, *Lithobates pipiens*.

### *LIFE HISTORY AND ECOLOGY*

Linzey (1967) explored the diets of northern leopard frogs in central New York throughout their entire activity period. Both mature and immature leopard frogs consumed mainly insect larvae, spiders, and adult insects. This frog may be active day or night but is most active at night during and after rainy periods (Dole 1965). Hulse et al. (2001) noted that northern leopard frogs usually hibernate from October to March in Pennsylvania. Emery et al. (1972) reported frogs hibernating in small mud pits in fairly shallow ponds in Ontario, Canada. Some activity was observed at overwintering sites; however, most movements were slow and generally a result of disturbance (Emery et al. 1972, Merrell 1977).

Northern leopard frogs commonly breed in temporary ponds, often in open habitats such as meadows and fields. Merrell (1977) notes that leopard frogs in Minnesota populations often emerged from hibernation in March and traveled to a new site for breeding in April. In the breeding season, males call from shallow, vegetated areas of the ponds. Conant and Collins (1991) describe the call as a rattling snore accompanied by clucking grunts. Females select a male, and they engage in amplexus. Females lay eggs in flattened masses containing 2,000 to 6,000 eggs. Hatching usually occurs within ten days but is dependent on temperature. Time to metamorphosis is also temperature dependent, but Merrell (1977) found that most juveniles in Minnesota had exited the pond by mid-July.

### *THREATS*

Amphibians have shown declines worldwide; a variety of factors may have contributed to the loss (Stuart et al. 2004). Habitat loss and fragmentation are major threats to amphibians in Pennsylvania. Pennsylvania ranks fifth in the nation in the amount of open space lost to development every day (Moyer 2003). Pennsylvania has also lost nearly half of its wetlands, which provide critical breeding habitat for many species of amphibians.

Increased road mortality as a result of habitat fragmentation is also a potential threat. Carr and Fahrig

(2001) compared the effects of road mortality on less mobile green frogs and more mobile leopard frogs. They found that northern leopard frogs may be more susceptible to road mortality because of their use of a variety of habitats and fairly long-distance movements between them. Linck (2000) also reported increased road mortality in a population of leopard frogs where breeding and overwintering habitats were separated by a roadway.

Water pollution and pesticides may negatively affect leopard frog populations. In a study on the effects of pesticides on amphibians, Relyea and Diecks (2008) treated five different amphibians with various pesticides. Northern leopard frogs did not show reduced survival rates as a result of pesticide exposure. However, this species did show reductions in growth as a result of exposure to pesticides.

Northern leopard frogs may be exposed to increased levels of ultraviolet (UV) radiation because of depleted stratospheric ozone. Diamond et al. (2002) examined the exposure of leopard frogs to UV radiation in wetlands in Minnesota and Wisconsin. They found that current levels are not likely to pose a significant risk to amphibians in these wetlands at the present time. However, environmental changes may increase these effects in the future. Palen et al. (2002) also examined this issue, specifically in wetlands in the Pacific Northwest. They suggest that amphibians in natural wetlands are somewhat protected by the dissolved organic matter in the water. However, they believe that amphibians may have been displaced from wetlands with high UV penetration in the past and are currently inhabiting wetlands with less UV exposure. Future changes in habitat quality and UV exposure rates may threaten this species.

*CONSERVATION AND MANAGEMENT NEEDS*

Protection of wetland habitat is critical for the survival of this and many other species of amphibians. Vernal pools are particularly important breeding habitat and currently receive little protection at the federal or state level. Adjacent wetlands, meadows, and forested habitat also must be maintained to protect this species. Further fragmentation and disturbance of northern leopard frog habitat may prevent this species from persisting at current locations.

*MONITORING AND RESEARCH NEEDS*

Statewide call surveys are critical to determine important breeding habitat for many species of amphibians in Pennsylvania. The Pennsylvania Herpetological Atlas is an important program that has increased knowledge of the distribution and abundance of amphibians and reptiles in Pennsylvania. This effort should be continued and expanded to include an intensive, statewide frog call survey. This method could be modeled after the Wisconsin Frog and Toad Survey that has been conducted since 1981 (Mossman et al. 1998). After initial surveys are completed throughout the state, more intensive monitoring should be conducted at critical breeding habitats.

Wetlands with noticeable reductions in leopard frog populations should be targeted for further study to determine possible causes for the declines. Wetlands with apparently stable populations should also be targeted for further study to determine what makes these habitats suitable for leopard frogs.

*Author:* KATRINA M. MORRIS, PENNSYLVANIA NATURAL HERITAGE PROGRAM

## Marbled Salamander

Order: Caudata
Family: Ambystomatidae
*Ambystoma opacum*

The marbled salamander was selected as a Pennsylvania Species of Greatest Conservation Need because of its uncertain status, evidence of possible population declines, and as a representative of the guild of species associated with vernal pond habitats (fig. 4.57). Marbled salamanders are not listed as Threatened or Endangered at the federal level or in Pennsylvania. Their collection in Pennsylvania is prohibited (Pennsylvania Fish and Boat Commission 2008). The global status of the marbled salamander is Secure (G5, NatureServe 2009).

Fig. 4.57. The Marbled Salamander, *Ambystoma opacum*. Photo courtesy of Andrew Hoffman.

### *GEOGRAPHIC RANGE*

Marbled salamanders occur throughout much of the eastern deciduous forest from southern New England south to northern Florida and west to eastern Illinois and Texas (Conant and Collins 1991). Several disjunct populations lie along the northern edges of the range (Conant and Collins 1991, Petranka 1998).

### *DISTRIBUTION AND RELATIVE ABUNDANCE IN PENNSYLVANIA*

In Pennsylvania, the range of marbled salamanders includes the southeast half of the state, extending southward into Maryland and eastward into New Jersey (Hulse et al. 2001; fig. 4.58). The species appears to be most abundant in south-central Pennsylvania but is relatively rare elsewhere. Because of its secretive nature, its present status in Pennsylvania remains uncertain, although recent surveys have documented a number of healthy populations.

### *COMMUNITY TYPE/HABITAT USE*

Adult marbled salamanders are found predominantly in mixed deciduous forests near suitable breeding locations (Petranka 1998), using both upland and floodplain forests (Petranka 1998, Hulse et al. 2001). In Pennsylvania, breeding occurs in temporary ponds (i.e., vernal ponds). These ponds fill with water in the winter or spring and often dry by midsummer. This periodic drying is critical, as it excludes predaceous fish that would otherwise prey on larval salamanders, which lack any natural defenses against fish predation (Kats et al. 1988).

### *LIFE HISTORY AND ECOLOGY*

Adult marbled salamanders are fossorial, spending the majority of time underground and thus are seldom seen. They are occasionally found under rocks and logs and often use rodent burrows that they aggressively defend against other salamanders (Smyers et al. 2002). Adults are generalized predators, feeding on a variety of forest arthropods and other invertebrates (Kenney and Burne 2000, Hulse et al. 2001). They, in turn, are preyed on by a variety of larger predators, including predatory birds, mammals, and snakes (Petranka 1998). Some protection against predators is provided by a milky skin secretion that is emitted by the salamanders when alarmed (DiGiovanni and Brodie 1981). Larval salamanders are also generalized predators, feeding on a large variety of aquatic invertebrates and occasionally on other amphibian larvae. Because they are gape limited, prey size increases as larvae grow larger. Small larvae feed mainly on crustacean zooplankton, adding larger prey as they grow. Larger larvae often consume caterpillars that fall into the water from overhanging trees. Larger aquatic predators, including dragonfly naiads and predaceous diving beetles, prey on larvae.

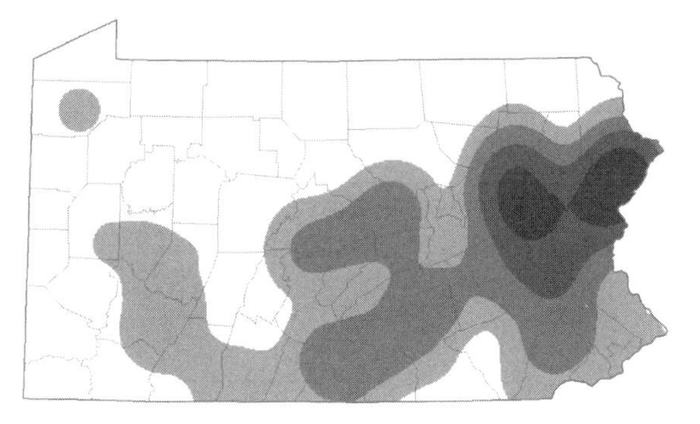

Fig. 4.58. Distribution of the Marbled Salamander, *Ambystoma opacum*.

Adults migrate to vernal ponds to mate during fall rains, with males usually arriving first. Courtship and mating occur on rainy nights in or near the dry or partially dry pond bed. A courting male deposits a spermatophore on the ground, and the female picks it up with her cloaca to fertilize her eggs. Females lay a clutch of around 100 eggs in shallow nests within the pond bed (Petranka 1998), often under rocks or logs. Most females brood their eggs until rains inundate their nests, at which time the eggs hatch. Larvae overwinter in the pond, metamorphosing into terrestrial juveniles during the late spring and early summer. Sexual maturity is reached in one or more years (Petranka 1998). Like other ambystomatid salamanders, adults are long lived and may potentially live more than twenty years.

### *THREATS*

The primary threat facing marbled salamanders is habitat loss/destruction. Because of their complex life cycle, the habitat needs of marbled salamanders include both terrestrial and aquatic environments. The U.S. Supreme Court decision in the case of *Solid Waste Agency of Northern Cook County v. U.S. Army Corps of Engineers* (No. 99-1178; SWANCC) greatly reduced the protections afforded to isolated wetlands, including vernal ponds, under the Clean Water Act (Gibbons 2003). Loss of isolated wetlands not only causes immediate loss of habitat but also changes the spatial configuration of, and distance between, remaining ponds, thereby affecting movements and recolonization processes (Semlitsch 2003).

Adequate terrestrial habitat around vernal ponds is also required for adult salamanders (Semlitsch 1998, Gibbons 2003). For ambystomatid salamanders, it has been recommended that a protected area or buffer zone extend 164 m (Semlitsch 1998) to 175 m (Faccio 2003) from vernal ponds.

Another potential threat is habitat disturbance and mortality associated with timber harvest. Several studies have documented negative effects of timber harvest on ambystomatid salamanders (DeMaynadier and Hunter 1999, Naughton et al. 2000). Chazal and Niewiarowski (1998) found no difference in survival of recently metamorphosed mole salamanders (*Ambystoma talpoideum*) placed in forested and clear-cut habitats, suggesting the process of timber harvest rather than the associated habitat change is detrimental to metamorphosed salamanders. Maintaining an intact forest canopy around breeding ponds may also be important for providing shade to the ponds. In Pennsylvania, some protection to temporary pond habitats is provided on state forest lands by regulations requiring a 30-m buffer of "no disturbance" habitat and an additional 30-m buffer of partial cut habitat (retaining at least 50 percent canopy cover) around vernal ponds. No such protection is presently required for habitats on private land.

Other potential threats to marbled salamanders include road mortality and reduced dispersal associated with increased development (Gibbs 1998), acid precipitation (Rowe et al. 1992), chemical contamination (Semlitsch 2003), infectious diseases (Carey et al. 2003), and habitat changes associated with global warming (Brooks 2004).

### *CONSERVATION AND MANAGEMENT NEEDS*

Conservation of marbled salamanders requires protection of vernal ponds and associated forested upland habitats. With the loss of federal protection previously provided by the Clean Water Act, increased state and local protections of isolated wetlands are necessary. Current sites of known occurrence should be given high priority for protection.

### *MONITORING AND RESEARCH NEEDS*

Recent surveys suggest that marbled salamander populations are secure. However, these surveys did not assess changes in abundance. These surveys should be followed up with monitoring activities at selected sites to detect changes in population numbers and distribution.

In addition to monitoring, several research priorities are as follows: (1) Determine the effects of habitat disturbance from logging on mortality and movements of adult salamanders, including the use by salamanders of disturbed habitats following the cessation of logging activities. (2) Determine the effects of habitat loss and fragmentation on population viability and dispersal, particularly in relation to metapopulation dynamics. (3) Research the effects of acid precipitation and chemical pollutants on survivorship and population viability is also needed.

*Author:* TIMOTHY J. MARET, SHIPPENSBURG UNIVERSITY

## Four-toed Salamander

Order: Caudata
Family: Plethodontidae
*Hemidactylium scutatum*

The four-toed salamander was selected as a Pennsylvania Species of Greatest Conservation Need because of its spotty distribution and uncertain status (fig. 4.59). Although four-toed salamanders are not listed as Threatened or Endangered at the federal level or in Pennsylvania, their collection in Pennsylvania is prohibited (Pennsylvania Fish and Boat Commission 2008). The global status of the four-toed salamander is Secure (G5, NatureServe 2009).

### *GEOGRAPHIC RANGE*

The four-toed salamanders are found throughout eastern North America. They range from western Florida into Alabama, north to Wisconsin, throughout the northeastern states, and into Nova Scotia. Their greatest abundance occurs in the northern states (Conant and Collins 1991).

*Fig. 4.59.* The Four-toed Salamander, *Hemidactylium scutatum*. Photo courtesy of Andrew Hoffman.

### DISTRIBUTION AND RELATIVE ABUNDANCE IN PENNSYLVANIA

Four-toed salamanders range widely across Pennsylvania, but their distribution is restricted to fewer than 100 known locations (McCoy 1982; fig. 4.60). During the Pennsylvania Herpetological Atlas Project from 1994 to 2002, few reports noted this species (Hulse et al. 2001). The species is seldom found in abundance anywhere in the commonwealth, and most reports are individual occurrences. For this reason, the four-toed salamander is considered uncommon in Pennsylvania. However, during a hot and dry July 2001 field visit at a reliable four-toed salamander location in Luzerne County, thirty-three adults were found in less than thirty minutes (R. L. Koval, personal observation).

### COMMUNITY TYPE/HABITAT USE

Four-toed salamanders are very much a woodland salamander, inhabiting deciduous, coniferous, and mixed forests (Shaffer 1991). Their preferred habitat in Pennsylvania may be forested lands adjacent to bogs, swamps, fens, wet meadows, vernal pools, lakes, and ponds. Egg deposition and incubation occurs in wetland or vernal pool habitats. For this reason, the species is most often associated with forests that contain wetlands with shallow standing water. Four-toed salamanders typically reproduce in forested wetlands and vernal pools that contain an adequate amount of moss (*Sphagnum*) or other wetland moss species (Hulse et al. 2001).

### LIFE HISTORY AND ECOLOGY

Four-toed salamanders emerge from hibernation in late March or early April and may be active through October. Breeding occurs in the fall and the gravid females usually venture to breeding wetlands/vernal pools for oviposition as early as March (R. L. Koval, personal observation), but normally around May.

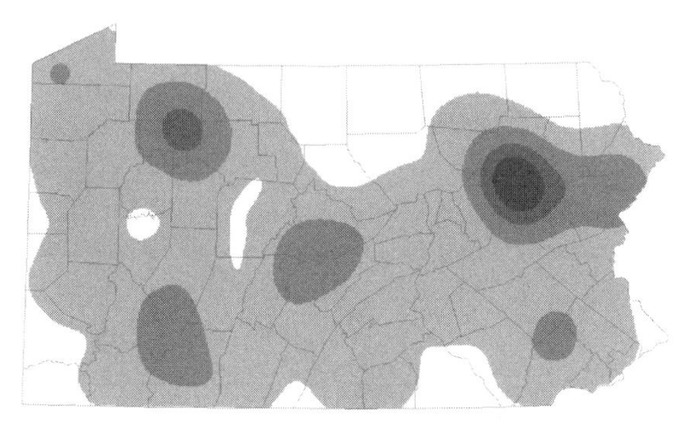

Fig. 4.60. Distribution of the Four-toed Salamander, *Hemidactylium scutatum*.

Four-toed salamanders exhibit antipredatory behaviors. They often turn into a tight coil when encountered, and their dorsal coloration resembles the tan-brown color of dead leaves and acts as camouflage. Sometimes they turn belly-up and expose their bright enamel white and black-blotched venter in an attempt to frighten away predators. Another effective method of defense used by four-toed salamanders is the ability to detach their tail. When the tail is grabbed by a predator, it breaks off and wiggles. The movement of the detached tail draws the attention of the predator, thus allowing the tailless salamander to escape. Four-toed salamanders have adapted well to this technique by exhibiting a constricted ring at the base of their tail, ready to detach.

The diet of the four-toed salamander is assumed to consist primarily of insects, such as beetles, spiders, isopods, slugs, and various small invertebrates (Stebbins and Cohen 1995).

Courtship in four-toed salamanders usually takes place in the fall and females appear to breed every other year. Males produce spermatophores that are engulfed by the female's cloaca after she is lured by the male. The following spring gravid females deposit between eight and thirty-eight eggs near or within forested wetlands/vernal pools. The clutch of sticky eggs attach to one another and to the vegetation, particularly sphagnum moss. Communal egg deposits are not uncommon and more than 1,000 eggs in a single nest have been reported, although this is not typical. Females guard the eggs from thirty to sixty-five days until hatching, at which time the larvae enter the water for development (J. J. Wilson, personal communication). Little is known of the biology of the aquatic larvae, which are assumed to consume benthic and planktonic organisms. Metamorphosis into terrestrial juvenile adults occurs in about 180 days (Klemens 1993, Petranka 1998).

### THREATS

The primary threat facing four-toed salamanders is loss of habitat. The habitat requirements of four-toed salamanders include both forested land and specific aquatic environments. Alteration and destruction of breeding wetlands are serious potential threats to this species. Pollutants and contaminants entering wetland ecosystems, such as acid mine drainage, road salt, storm-water runoff, nutrients, and pesticides, are threats to the aquatic larvae and the aquatic macroinvertebrates they prey on. Adequate terrestrial habitats along breeding wetlands are required for adults and metamorphosed young. Timber harvesting and vegetation removal along the margins of swamps, bogs, vernal pools, ponds, and lakes reduce available habitat and increase mortality of this species. Maintaining a shaded forest canopy of terrestrial habitat is important to this species' survival (Green 1997).

*CONSERVATION AND MANAGEMENT NEEDS*

Conservation of this species will require the protection of forested wetlands and vernal pools that provide breeding habitat and the adjacent terrestrial forests required by nonbreeding adults. Particular conservation measures should be given to protect wetlands and vernal pools that contain shallow water and a large amount of aquatic moss species. Pennsylvania has the opportunity and should consider becoming a "responsibility state" for the four-toed salamander.

*MONITORING AND RESEARCH NEEDS*

Surveys are needed to determine present status and distribution. Primary focus should be in areas that have not been adequately surveyed, particularly in western and southwestern Pennsylvania. Special attention should be given to locations with aquatic habitats that will support breeding and larval development.

Little is known about the effects of urban sprawl that encroaches on terrestrial habitats of nonbreeding adults, and the highly developed Pocono Region may be a research area for consideration. Little is also known of the diet of the four-toed salamander. Studies that include basic natural history elements for the species should be encouraged. Knowledge on the biology during larval development is also limited. Studies of this species' larval period should be encouraged.

*Author:* RICHARD L. KOVAL,
NORTH BRANCH LAND TRUST

## Common Map Turtle

Order: Testudines
Family: Emydidae
*Graptemys geographica*

The common map turtle was chosen as a Species of Greatest Conservation Need because of its restricted range, loss of habitat, mortality from motor vehicles, and its representation of the guild of species associated with riverine habitats (fig. 4.61). Common map turtles are not listed as Threatened or Endangered either at the federal level or in Pennsylvania. Their collection in Pennsylvania is limited to one individual per day, with a possession limit of one individual (Pennsylvania Fish and Boat Commission 2008). In 2006, the northern map turtle was added to Appendix III of the Convention on International Trade in Endangered Species of Wild Fauna and Flora (Convention on International Trade in Endangered Species 2007), thus regulating international trade in this species. The global status of the common map turtle is Secure (G5, NatureServe 2009).

*Fig. 4.61.* The Common Map Turtle, *Graptemys geographica*. Photo courtesy of Tom Pluto.

*GEOGRAPHIC RANGE*

Common map turtles range from southern Quebec and northern Vermont through the Saint Lawrence and Great Lakes drainages to eastern Minnesota (Ernst et al. 1994). They occur southward from the Great Lakes region to areas from northern Alabama to eastern Kansas. Isolated populations of common map turtles occur in the Delaware and Susquehanna river systems of Pennsylvania. Their range is the broadest and most northerly among *Graptemys* species.

*DISTRIBUTION AND RELATIVE ABUNDANCE IN PENNSYLVANIA*

In Pennsylvania, common map turtles are restricted to Erie County and portions of a few large river drainages. One population inhabits the Delaware River in areas of Northampton and Bucks counties and may have been introduced (Arndt and Potter 1973). Common map turtles occur in the Susquehanna River and some of its major tributaries, including the Juniata River. Although the Ohio River is included in historic range maps (Ernst et al. 1994), no localities for the entire river were reported during the recent Pennsylvania

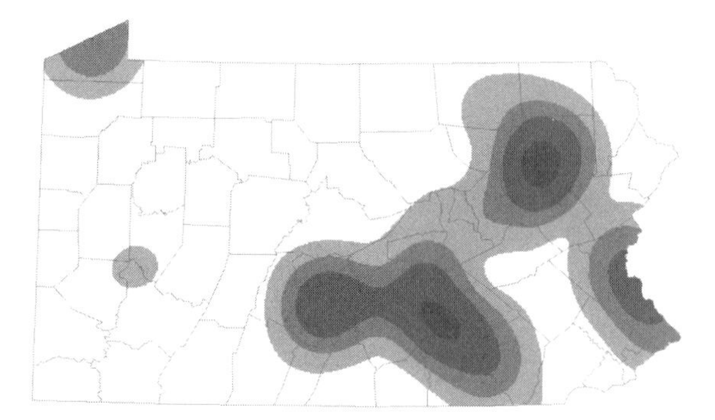

*Fig. 4.62.* Distribution of the Common Map Turtle, *Graptemys geographica.*

Herpetological Atlas Project. The relative abundance of common map turtles in Pennsylvania is poorly documented. However, on the basis of survey data, the highest densities appear to occur in portions of the Susquehanna and Juniata rivers (fig. 4.62).

*COMMUNITY TYPE/HABITAT USE*

Common map turtles occupy large rivers and lakes contiguous to rivers. Habitat characteristics appear to vary among areas of distribution, prompting some authors to suggest that food availability may be the major limiting factor, rather than any specific physical habitat component (Vogt 1981, Fuselier and Edds 1994). The presence of basking sites has been associated with the prevalence of common map turtles (Pluto and Bellis 1986) and other *Graptemys* species (Lindeman 1998), thus the absence of available structures, such as rocks and deadwood, may also limit populations. Adult common map turtles overwinter in the deeper, slower areas of rivers (Pluto and Bellis 1986). In Pennsylvania, common map turtles appear to be most successful in heterogeneous river habitats that contain numerous basking sites, some aquatic vegetation, ample levels of invertebrate prey, and variable depths and flow rates.

*LIFE HISTORY AND ECOLOGY*

Common map turtles are one of Pennsylvania's most aquatic turtles; other than females or hatchlings in nesting areas, individuals are rarely found in terrestrial environments. They are noted for frequent communal basking on sunny days and are sometimes observed stacked on top of one another on emergent logs or rocks. Smaller individuals generally bask and occupy habitats closer to shore compared with larger individuals (Pluto and Bellis 1986). Seasonal movements of individuals are substantial; in a study in the Raystown Branch of the Juniata River, range lengths of males averaged 2,115 m and those of females averaged 1,211 m (Pluto and Bellis 1988). The diet of common map turtles includes mussels, snails, crayfish, insect larvae, and fish carrion (Newman 1906, Vogt 1980, 1981).

Female common map turtles reach maturity at ten to fourteen years (Newman 1906, Vogt 1980). Nesting generally occurs from late May through early July and nesting activity is most concentrated during morning hours. Females have been observed to make extensive movements (>5,000 m upstream) during the nesting season (Pluto and Bellis 1988, R. D. Nagle, personal observation). Clutch size averages about ten eggs (White and Moll 1991, Nagle et al. 2004), but varies substantially with female body size. Some females produce two clutches per year. The incubation period is approximately sixty days and offspring sex is determined by incubation temperature (Bull and Vogt 1979). Most hatchlings delay emergence from nests until the following spring, with time from egg deposition to emergence averaging eleven months (Nagle et al. 2004).

*THREATS*

Among the greatest threats to common map turtles are death of adult turtles and loss of critical habitats. Because populations in Pennsylvania are isolated and restricted to a few river systems, such factors may place them at high risk of extirpation (Lovich 1995). Road mortality of nesting females can be substantial in some areas. Because their riverine habitats are often paralleled by roads and railroads, many nests are constructed in open canopy, sparsely vegetated areas containing disturbed soils and fill materials (Nagle et al. 2004). Although such habitats are rarely considered pristine or biologically valuable by land managers, recognizing their importance and ensuring their availability may be necessary to promote viable map turtle populations.

Other threats to common map turtles include habitat degradation and disturbance, water pollution, and invasive species. Anthropogenic removal of deadwood has contributed to declines of *Graptemys* species in the southern United States (Lindeman 1998), and such actions in Pennsylvania are likely to have negative effects. Continual anthropogenic disturbances to aquatic habitats (e.g., recreational and commercial boat traffic) may be detrimental to basking and exclude turtles from suitable habitats (Gordon and MacCulloch 1980). Primary prey for map turtles consist of filter-feeding benthic invertebrates, which can accumulate such pollutants as heavy metals and organochlorines in their tissues. Long-lived turtles that feed on such prey may accumulate high levels of contaminants over long time periods and transfer them to eggs and offspring (Nagle et al. 2001). The invasion of exotic zebra mussels (*Dreissena plymorpha*) is of concern because they displace native unionid clams, a prime food source of map turtles (Roche 2002).

*CONSERVATION AND MANAGEMENT NEEDS*

Long-lived turtles share a common suite of life history characteristics, including delayed maturity, long reproductive life spans, and relatively low fecundity; such factors combine to make population stability inconsistent with high loss of adults and older juveniles (Congdon et al. 1993, Congdon et al. 1994). Legal removal of turtles from wild populations may be detrimental, particularly for isolated populations or populations already reduced because of habitat degradation or road mortality. Thus, regulations to prohibit the collection and possession of common map turtles should be enacted within Pennsylvania. Before activities such as road construction along major rivers, open-canopy habitats should be carefully examined for nesting areas; such areas should be avoided where they are found. In areas with paved roads and substantial mortality of nesting females, management solutions such as the installation of fencing may be required. Because eggs, hatchlings, or nesting females are present at nesting areas during all months of the year, disturbances to such sites (e.g., heavy equipment operations) at any time may negatively impact common map turtle populations.

The highly aquatic nature and limited distribution of common map turtles in Pennsylvania make protection and management of specific riverine habitats a critical component of their conservation. Programs that improve water quality by reducing agricultural and industrial runoff and contaminant discharge are of substantial benefit to common map turtles. Management activities likely to be detrimental include channelizing and dredging rivers, removal of deadwood, and placement of rip-rap rock along riverbanks, which can exclude turtles from nesting areas.

*MONITORING AND RESEARCH NEEDS*

Surveys along major river systems are required to determine the present status of common map turtles in Pennsylvania. Spotting scope surveys appear to be the most effective means to determine the presence and densities of *Graptemys* species (Lindeman 1998). Surveys are also required to document major nesting areas for common map turtles in Pennsylvania. Some nesting areas may serve hundreds of females each year and such sites should be designated for protection. In areas where road mortality of nesting females is substantial, short (<1 m high) fencing should be installed to eliminate the ability of females to move onto roadways.

Despite its relatively broad distribution, the common map turtle remains one of the more understudied turtles of North America. Detailed studies are required in the areas of nesting ecology, nesting habitats, reproduction, and life histories, including age at maturity and age-specific survivorship and fecundity. Studies should also be undertaken to determine whether, because of their geographic isolation, populations inhabiting the Susquehanna and Delaware river systems have genetic uniqueness compared with populations from their major area of distribution.

*Author:* ROY D. NAGLE, JUNIATA COLLEGE

## Eastern Box Turtle

Order: Testudines
Family: Emydidae
*Terrapene carolina carolina*

The eastern box turtle was selected as a Pennsylvania Species of Greatest Conservation Need because of evidence of population declines, habitat loss, and high mortality among eggs and young turtles (fig. 4.63). In addition, illegal collecting and mortality from motor vehicles make this species extremely vulnerable to further population declines. Eastern box turtles are not listed as Threatened or Endangered at the federal level

*Fig. 4.63.* The Eastern Box Turtle, *Terrapene carolina carolina*. Photo courtesy of Tom Diez.

or in Pennsylvania. Their collection in Pennsylvania is prohibited (Pennsylvania Fish and Boat Commission 2008). In 1994, box turtles were added to Appendix II of the Convention on International Trade in Endangered Species of Wild Fauna and Flora (Convention on International Trade in Endangered Species 2007), thus restricting international trade in this species. The global status of the eastern box turtle is Secure (G5, NatureServe 2009).

### *GEOGRAPHIC RANGE*

The eastern box turtle ranges from Massachusetts to north-central Florida and westward from the eastern seaboard to western Illinois, Tennessee, and northeast Mississippi (Conant and Collins 1991). Six distinct subspecies are recognized in the eastern United States and are distributed from southern Maine south to the Florida Keys, and west to Michigan, eastern Kansas, Oklahoma, and Texas (Ernst et al. 1994). *Terrapene carolina carolina* is the only subspecies present in Pennsylvania.

### *DISTRIBUTION AND RELATIVE ABUNDANCE IN PENNSYLVANIA*

In Pennsylvania, the species occupies suitable habitat across most of the state with the exception of the northern portions of the northern-tier counties (Hulse et al. 2001; fig. 4.64). During field surveys, the eastern box turtle continues to be the most abundant terrestrial turtle species in the eastern part of the state (Pennsylvania Fish and Boat Commission 2003).

### *COMMUNITY TYPE/HABITAT USE*

Eastern box turtles are the only truly terrestrial turtles of the Northeast and occur in deciduous forests, old fields, ecotonal areas, and marshy areas. Box turtles appear to have strong microhabitat requirements for both temperature and humidity (Ernst et al. 1994), which may explain why during hot, dry periods eastern box turtles are often found in and near marshy areas and wetlands.

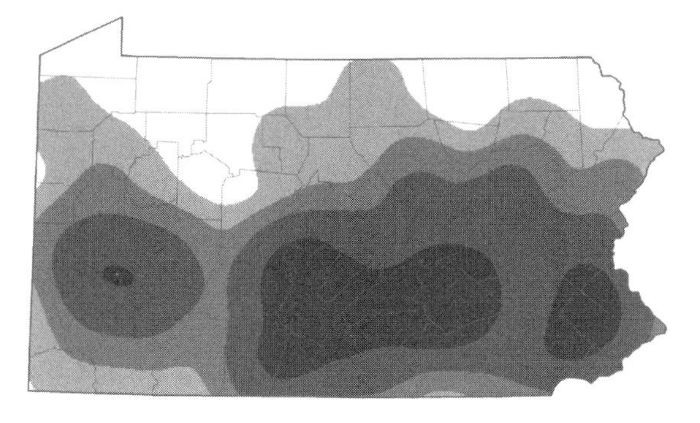

Fig. 4.64. Distribution of the Eastern Box Turtle, *Terrapene carolina carolina*.

### *LIFE HISTORY AND ECOLOGY*

In Pennsylvania, eastern box turtles generally emerge from hibernation in mid-April and are active until mid- to late October with activity peaking in June and early July (Hulse et al. 2001). Eastern box turtles are diurnal and are often seen in the middle of the day. They are omnivorous, but the young are chiefly carnivorous. Dietary items include snails, slugs, worms, insects, spiders, fungi, fruits, berries, and carrion (Hulse et al. 2001). They have been shown to be important vectors in seed dispersal as a variety of seeds from plants common to eastern deciduous forests have a higher germination rate after passing through the gut of eastern box turtles than noningested seeds (Braun and Brooks 1987). Predators include skunks, foxes, raccoons, crows, and snakes, and nest predation is common (Ernst et al. 1994).

The home range of a population of eastern box turtles in Pennsylvania is approximately 170 m linear distance (Strang 1983). Box turtle populations in Indiana are estimated to be similar to Pennsylvania's (Williams and Parker 1987). In Maryland, home range is estimated at 1.20 ha (2.92 acres) for males and 1.13 ha (2.75 acres) for females (Stickel 1989). Site fidelity to their home range area has been reported as high (Williams and Parker 1987), and overlapping home ranges are common as populations might consist of many transient individuals (Ernst et al. 1994). Box turtles have been known to travel up to 400 m (1,300 feet) in a linear distance outside of their home range (Hall et al. 1999). This tendency for home range overlap makes population size estimates difficult but does promote mixing of the gene pool (Kiester et al. 1982). Population size estimates vary and have been documented between about two and twenty-two turtles per hectare (Ernst et al. 1994).

Mating can occur at any time during their activity period, but in Pennsylvania, it peaks in early summer (June–July) and again in mid-fall (October; Hulse et al. 2001). Sexual maturity is reached in five to ten years of age for both sexes (Ernst et al. 1994). Mating partners generally occupy overlapping home ranges (Stickel 1989), and it is common for males to mate with multiple females or with the same female several times in their lifetime (Williams and Parker 1987). Females can retain viable sperm for several years after mating (Ewing 1943). Clutch size varies from one to eight eggs, but generally females will deposit four or five white, elliptical eggs in a nest and two clutches per year (Hulse et al. 2001). Suitable nest sites are often not found in a female's home range and are usually in an open or elevated patch of sandy or loamy soil (Ernst et al. 1994). In deciduous forests in central Pennsylvania, the author has observed female eastern box turtles constructing nests on residential septic sand mounds.

Eastern box turtles are long-lived and have been known to exceed 100 years of age in the wild (Ernst et al. 1994). Their population structure generally consists of a large percentage of adults more than twenty years old (Stickel 1978, Williams and Parker 1987, Hall et al. 1999, Henry 2003).

*THREATS*

Many actions threaten the long-term survival of eastern box turtles in Pennsylvania, but Pennsylvania Fish and Boat Commission (2003) recognizes habitat loss as the greatest potential threat. As denizens of upland forests, eastern box turtles do not receive many of the protections of other aquatic and wetland inhabiting turtles in the state. Development on uplands often does not require a permit or approval from resource regulating agencies. Terrestrial turtles are extremely vulnerable to road mortality (Gibbs and Shriver 2002). In Virginia, eastern box turtles accounted for 66 percent of 694 reptiles admitted to a wildlife rehabilitation center, and most of these cases were from trauma associated with automobile impact (Brown and Sleeman 2002). Other threats include illegal collection. Ernst et al. (1994) indicate that thousands of box turtles are illegally shipped overseas annually for the pet trade. Congdon et al. (1993) demonstrated that adult survival in long-lived turtles is so important that sustainable harvest is probably impossible. Increased mesopredator (raccoon, skunk, cat, dog) populations associated with urban and suburban sprawl are likely affecting box turtle populations throughout Pennsylvania.

*CONSERVATION AND MANAGEMENT NEEDS*

Recent studies indicate that eastern box turtle populations are declining throughout much of their range (Stickel 1978, Williams and Parker 1987, Hall et al. 1999). A thirty-year study in Maryland showed a reduction of about 50 percent from a population at an 11.7-ha (28.5 acre) site (Stickel 1978). From the early 1960s to the early 1980s, a population of eastern box turtles in Indiana dropped from about 5.7 turtles per hectare to 2.7 turtles per hectare (Williams and Parker 1987). A fifty-year study in Maryland found a more than 75 percent decline in eastern box turtles (Hall et al. 1999). Long-term population studies of eastern box turtles in Pennsylvania are lacking or unpublished (J. Drasher, personal communication).

For the immediate future, the protection of this species will be largely dictated by local interests as Pennsylvania local governments permit and regulate upland development. Even though it is one of the most commonly encountered turtles in Pennsylvania, the threats to this species are poorly understood by the general public, so education and outreach are recommended.

*MONITORING AND RESEARCH NEEDS*

In-depth population studies in Pennsylvania need to be conducted to determine the status of this species in the state. From these studies, conservation and management needs, including a listing for special protection if necessary, can be developed. Priority should be given to projects that provide information on population sizes and age structure, estimate population size in the commonwealth, identify characteristics of successfully breeding populations, determine key habitat characteristics, and develop best management practices that minimize land development effects on this species.

Much of the biology and natural history of the eastern box turtle is well documented and understood. The distribution of eastern box turtles across Pennsylvania is also well documented, largely because of studies conducted by Pennsylvania Fish and Boat Commission and the Pennsylvania Herpetological Atlas Project.

*Author:* JOSEPH J. WILSON,
WILSON ECOLOGICAL CONSULTING

## Eastern Fence Lizard

Order: Squamata
Family: Scincidae
*Sceloporus undulatus*

The eastern fence lizard was selected as a Pennsylvania Species of Greatest Conservation Need because of its spotty distribution, uncertain status, and evidence of possible population declines (fig. 4.65). Eastern fence lizards are not listed as Threatened or Endangered at the federal level or in Pennsylvania. Their collection in Pennsylvania is prohibited (Pennsylvania Fish and Boat Commission 2008). The global status

Fig. 4.65. The Eastern Fence Lizard, *Sceloporus undulates*. Photo courtesy of Andrew Hoffman.

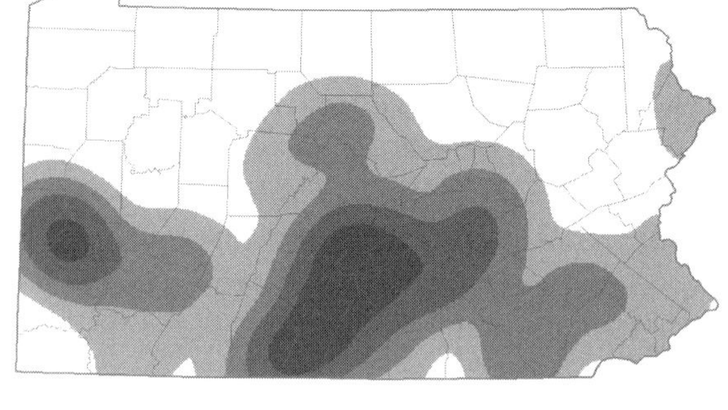

Fig. 4.66. Distribution of the Eastern Fence Lizard, *Sceloporus undulates*.

of the eastern fence lizard is Secure (G5, NatureServe 2009).

*GEOGRAPHIC RANGE*

The eastern fence lizard possesses the largest geographic range of any species of lizard in the United States. In the east it ranges from central New Jersey southward to central Florida. Its range extends westward to northwestern Arizona and central Wyoming. The eastern fence lizard appears to be common or abundant throughout much of its range, although there is little information available for the species at the periphery of its range. Ballinger and Watts (1995) reported dramatic declines in the species in the Arapaho Prairie of Nebraska that they attributed to changing land use patterns.

*DISTRIBUTION AND RELATIVE ABUNDANCE IN PENNSYLVANIA*

In Pennsylvania, the eastern fence lizard is found in scattered localities throughout the southern half of the state (fig. 4.66). It reaches the northern limit of the eastern portion of its range in Pennsylvania, and as a result, populations within the state may be genetically different from populations farther south or to the extreme west. The eastern fence lizard is a rare lizard in Pennsylvania. Its primary distribution lies in the southern half of the commonwealth, with the majority of sightings occurring in the Valley and Ridge Province. To attest to its rarity, Pennsylvania Herpetological Atlas volunteers reported only seventy-eight observations of this species. Fence lizards have been observed by volunteers in only 31 of the 700-plus U.S. Geological Survey 7.5-minute quadrangles in Pennsylvania. The lack of success in locating fence lizards attests to the scarcity of the animals because, unlike other native lizards, fence lizards are not secretive and can frequently be found perched on rocks or trees.

*COMMUNITY TYPE/HABITAT USE*

In Pennsylvania, the eastern fence lizard is generally found in open habitat within forests (i.e., rock slides, quarry faces, clearing due to clear-cutting or fire, rocky outcrops along waterways) and in the shale barrens of the south-central portion of the state (Huntingdon, Bedford, and Fulton counties). These locations provide them with ample basking sites, as well as refuge sites from surface cover objects and trees.

*LIFE HISTORY AND ECOLOGY*

The eastern fence lizard's seasonal activity may begin as early as mid-March during unusually warm years but generally does not begin until the end of March or early April. Adults usually cease activity by late August, but the young-of-the-year remain active until late September or early October. The eastern fence lizard is a diurnally active species that may be readily observed in regions where it occurs as it spends most of its active time perched on elevated rocks or logs. Activity usually occurs from midmorning to late afternoon. During hot spells, activity may become bimodal with a peak during midmorning and a second period of activity in late afternoon. Surface activity depends on the lizard's ability to maintain adequate body temperature. As a result, eastern fence lizards are seldom active on cool, cloudy days or when high winds greatly increase convective heat loss. Hibernation apparently occurs in deep cracks and crevices or caves within the area of their summer activity.

Eastern fence lizards are sit-and-wait predators whose primary mode of foraging is to perch immobile on a rock or log that has a clear view of the surrounding area. When an appropriate prey item comes within attack range, the lizard rushes from its perch to the prey and captures it. They feed on a wide variety of actively moving prey, such as ants, beetles, spiders, caterpillars, butterflies, and grasshoppers. Eastern fence lizards are highly territorial. Although animals are on their perches, they are also monitoring their territory.

No detailed studies of reproduction have been carried out on eastern fence lizards from Pennsylvania; however, Tinkle and Ballinger (1972) have studied this species in Hocking County, Ohio. It is likely that their reproductive behavior is not significantly different from Pennsylvania populations. In Ohio, courtship and mating occur shortly after animals emerge from hibernation in April. Males actively display to attract females and warn off other males from their territory (Cooper and Burns 1987). In Ohio, females generally lay two clutches of eggs a season. The first clutch is laid in mid- to late May and the second toward the end of June (Tinkle and Ballinger 1972). In Ohio, average clutch size was 11.8 eggs. Larger females had a tendency to lay larger clutches of eggs (Tinkle and Ballinger 1972), and the first clutch of the season is usually larger than the second clutch. Ohio hatchlings emerge from their eggs from the third week of August to the first week of October, after a natural incubation period of approximately eighty-five days. Those individuals that hatch late in the season exhibit little growth before they enter hibernation.

The wide geographic and climatic range of the species and its relative ease of study have made it an important species for examining the effects of geography and climate on life history parameters and physiological adaptations (Niewiaroski 1995, Adolph and Porter 1996, Angilletta 2001a, 2001b). Because populations of eastern fence lizards in Pennsylvania exist at the extreme northeastern limit of the species range, they potentially play an important role in future geographical studies of the species.

Angilletta et al. (2004) recently examined the effects of clinal variation in body size of *Sceloporus undulatus* with regard to life history traits. Eastern fence lizards exhibit clinal variation in size, with those populations from more northern latitudes achieving a larger body size than more southern conspecifics. They found that northern females delay maturation to a larger body size thus increasing clutch size. They also noted higher rates of juvenile survival from more northern populations.

*THREATS*

Threats to the eastern fence lizard in Pennsylvania include habitat modification due to changing land use patterns. However, the biggest threat within the commonwealth is our lack of information concerning the current range of the species.

*CONSERVATION AND MANAGEMENT NEEDS*

All sites where fence lizards are found to exist should be entered into the Pennsylvania Natural Heritage Program database so that they can be readily recognized in response to development plans. Any development within regions where fence lizards occur should be conducted in a manner that would minimize impact on populations of the lizards.

*MONITORING AND RESEARCH NEEDS*

Little is known concerning the biology of the eastern fence lizard in Pennsylvania. Primary research needs include surveying appropriate habitat for the presence of fence lizards and for the size of their populations, determining the extent of suitable habitat where eastern fence lizard are located, and resurveying locations where eastern fence lizard have been known to occur in the past to determine whether they still exist. Once the locations of populations are established through surveys, several populations should be selected for intensive long-term monitoring. The sites designated for long-term monitoring should be, whenever possible, situated on public lands. Long-term studies should include mark-recapture procedures and should specifically target population size and structure, as well as reproductive activity and success within the populations.

*Author:* ARTHUR C. HULSE, INDIANA UNIVERSITY OF PENNSYLVANIA (RETIRED)

## Smooth Green Snake

Order: Squamata
Family: Colubridae
*Liochlorophis vernalis*

The smooth green snake was selected as a Pennsylvania Species of Greatest Conservation Need because populations appear to be declining in Pennsylvania and throughout much of its range (fig. 4.67). Although smooth green snakes are not listed as Threatened or Endangered either at the federal level or in Pennsylvania, their collection in Pennsylvania is prohibited (Pennsylvania Fish and Boat Commission 2008). The

*Fig. 4.67.* The Smooth Green Snake, *Liochlorophis vernalis*. Photo courtesy of Tom Diez.

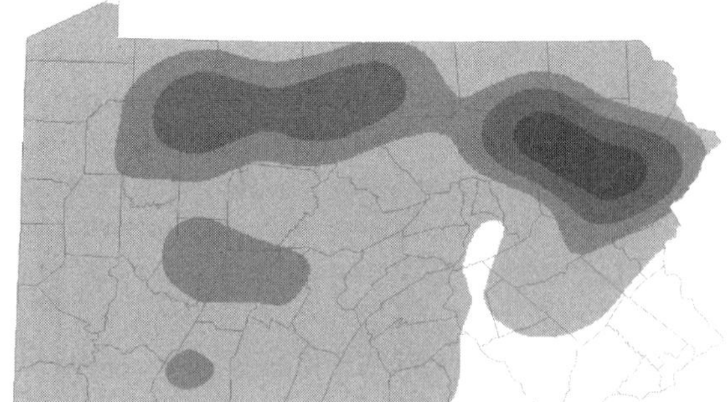

*Fig. 4.68.* Distribution of the Smooth Green Snake, *Liochlorophis vernalis*.

global status of the smooth green snake is Secure (G5, NatureServe 2009).

*GEOGRAPHIC RANGE*

Smooth green snakes range from Nova Scotia west to southeastern Saskatchewan and south through New England to northern Virginia. West of the Great Lakes states the species is present in a scattered distribution as far south as Texas and northern Central America and as far west as western New Mexico and eastern Utah (Conant and Collins 1991). New distribution records are routinely recorded for this species, particularly in the South, the West, and the Midwest (Black and Bragg 1968, Worthington 1973, Blahnick and Cochran 1994, Casper 1996, Akre and Robinson 2003).

*DISTRIBUTION AND RELATIVE ABUNDANCE IN PENNSYLVANIA*

Smooth green snakes, although widespread in distribution across Pennsylvania, are rarely found in abundance. The species is absent from the Piedmont and Atlantic Coastal Plain in the southeastern portion of the commonwealth. They are most abundant in the Glaciated Low Plateau, Glaciated High Plateau, and Pocono Plateau ecoregions of Pennsylvania, which account for more than one-half of known occurrences of this species in the state (Serrao 2000, Hulse et al. 2001; fig. 4.68).

Smooth green snake populations appear to be declining in Pennsylvania and throughout much of its range (Klemens 1993, Hulse et al. 2001). As recently as twenty years ago, this snake was a common inhabitant of old fields and open areas in central Pennsylvania, but it is seldom seen in these same locations now (J. J. Wilson, personal observation). Pesticides used in both home landscaping and agriculture may, in part, be contributing to the demise of this insectivorous species (Klemens 1993).

*COMMUNITY TYPE/HABITAT USE*

Smooth green snakes can be found in a variety of upland habitats, but they are most commonly found in old fields, pastures, and forest clearings. Forest clearings associated with secondary township roads and logging roads offer ideal habitat. Unlike the rough green snake, which is arboreal, smooth green snakes are seldom associated with woody vegetation but can occasionally be found climbing on low-lying, shrubby vegetation. Typically, smooth green snakes can be found under cover objects in open herbaceous upland habitats. Undisturbed herbaceous habitats are rare in Pennsylvania. Most herbaceous areas are used for agriculture, livestock grazing, or recreational purposes. Key habitats that once were used for moderate to light impact (i.e., livestock grazing) are now being converted to other human uses like residential areas. Primary and secondary habitat for this species is decreasing throughout the state but the adoption of "Green Space" and "Open Space" initiatives by local municipalities may help preserve some important habitat for smooth green snakes. It appears this species benefits from roadside disturbance. State and township roadways provide and maintain the grassy open habitat required by this species. The eggs of smooth green snakes are often found under cover objects along this habitat type.

*LIFE HISTORY AND ECOLOGY*

Smooth green snakes feed almost entirely on arthropods, favoring spiders, caterpillars, crickets, and grasshoppers (Conant and Collins 1991; Hulse et al. 2001).

The analysis of stomach contents from twenty-two Pennsylvania specimens confirms the preference for these items but also included snails, a slug, and a salamander (Surface 1906). This species is diurnally active from early April to November in Pennsylvania (Hulse et al. 2001), but the extent of its home range has not been documented.

Smooth green snakes often hibernate communally in underground retreats. In many parts of its range, hibernation frequently occurs in ant mounds (Hulse et al. 2001, Enrst and Ernst 2003). In Pennsylvania (Mercer County), two were found hibernating with other snakes and salamanders in a gravel bank at a depth of 80 cm (Lachner 1942).

Although the time of mating varies throughout its range, in Pennsylvania, copulation has been observed in early April (Hulse et al. 2001, Ernst and Ernst 2003). Incubation time is variable and can range from about four days to more than thirty days. On average, females deposit about seven eggs under cover objects in late July or August (Grobman 1989). Communal oviposition is well documented in this species (Cook 1964, Gregory 1975, Stuart 2002).

*THREATS*

Many human actions threaten smooth green snake populations in Pennsylvania. Poor livestock practices that degrade habitat can lead to a decline in prey items and cover objects, leaving the species more vulnerable to predation and disease. Recreational activities that result in habitat destruction (i.e., all-terrain vehicles) can have similar negative effects on the species. The overuse of insecticides for both agricultural and residential purpose has likely led to a decline in prey populations. Similarly, herbicide use and maintenance activities that remove cover objects along right of ways may negatively affect smooth green snakes. The direct loss of habitat due to development and sprawl is possibly the greatest threat to smooth green snake populations in Pennsylvania.

*CONSERVATION AND MANAGEMENT NEEDS*

Anecdotal evidence suggests that smooth green snakes are declining in number and distribution across Pennsylvania. Data need to be collected to determine the extent to which smooth green snakes are decreasing in abundance across the state. In addition, management practices to minimize the effects of development and land use activities on this species need to be developed and implemented.

*MONITORING AND RESEARCH NEEDS*

Building on work initiated in the mid-1990s by the Pennsylvania Herpetological Atlas Project, a monitoring program should be developed for this species. This monitoring program should continue to document the presence of this species throughout Pennsylvania and monitor a select number of populations in each physiographic region where smooth green snakes are known to exist.

The biology and life history of the smooth green snake are well documented, but detailed studies from Pennsylvania are lacking. Future research in Pennsylvania should focus on population studies that provide information on activity range, behavior and habitat preference, identification of key habitats that require protection, and identification of best management practices that minimize land development effects on this species.

*Author:* JOSEPH J. WILSON, WILSON ECOLOGICAL CONSULTING

## Northern Copperhead

Order: Squamata
Family: Viperidae
*Agkistrodon contortrix mokasen*
(also *Responsibility Concern*)

The northern copperhead was selected as a Pennsylvania Species of Greatest Conservation Need because of evidence of a decline in numbers (fig. 4.69). Pennsylvania may function as a stronghold for the continued survival of this species in the northeastern United States. Northern copperheads may be legally collected in Pennsylvania with the appropriate permit, with an annual limit of one snake (Pennsylvania Fish and Boat

*Fig. 4.69.* The Northern Copperhead, *Agkistrodon contortrix mokasen*. Photo courtesy of R. T. Zappalorti.

Commission 2008). The global status of the northern copperhead is Secure (G5, NatureServe 2009).

### *GEOGRAPHIC RANGE*

The northern limit of the distribution of the northern copperhead extends from western Massachusetts, southeastern New York, across northern New Jersey, Pennsylvania, and the southern extremes of Ohio, Indiana, and Illinois. The range continues south into northern Alabama, Georgia, and South Carolina (Conant and Collins 1998).

### *DISTRIBUTION AND RELATIVE ABUNDANCE IN PENNSYLVANIA*

The northern copperhead occurs throughout Pennsylvania, with the exception of the counties along the state's northern border, corresponding to the southern glaciation limit (fig. 4.70). However, northern copperheads have been discovered in river valleys of the glaciated region (Smith 1945a, McCoy 1982). It has been suggested that the northeastern distribution limit reflects the physiological constraints of the species and its preference for warmer habitat (Reinert 1984a). The northern copperhead has been recorded at altitudes of up to 2,500 feet (Smith 1945a).

The statewide abundance of the northern copperhead has not been determined. Recent reports gained through the Pennsylvania Herpetological Atlas Project suggest that this species still occurs throughout much of its historic range (A. Hulse, personal communication). However, this venomous snake suffers from long and persistent persecution, and as early as 1906, it was believed that its numbers were declining in some parts of the commonwealth (Stone 1906). The species has been extirpated from Philadelphia but still persists within twenty miles of Center City. Extant populations reportedly also still occur in areas of Chester, Bucks, and Montgomery counties in southeastern Pennsylvania (Gloyd and Conant 1990). Small colonies possibly persist in Pittsburgh, where they are offered refuge within rock ravines (Gloyd and Conant 1990).

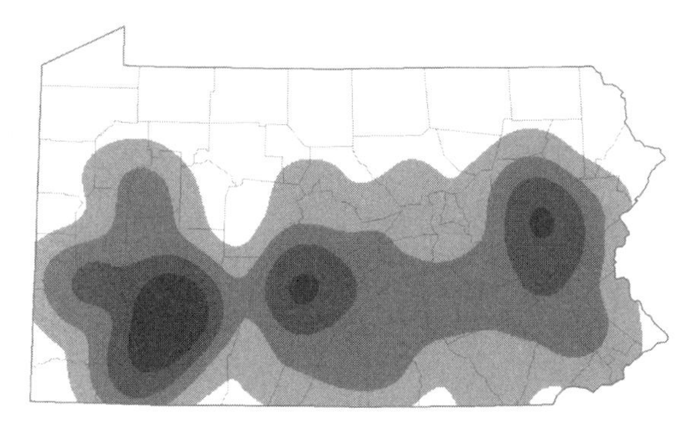

*Fig. 4.70.* Distribution of the Northern Copperhead, *Agkistrodon contortrix mokasen.*

### *COMMUNITY TYPE/HABITAT USE*

Northern copperheads inhabit areas of second-growth, deciduous hardwood forest found throughout Pennsylvania (Reinert 1984a). However, this snake prefers open, rocky areas over forested habitat, using sites with low-surface vegetation, reduced amounts of shade, and soil temperatures elevated by sunlight (Reinert 1984a). Areas of high rock density, little canopy cover, and little to no surface vegetation are selected by gravid females and are thus assumed to be important to the physiological demands of their reproductive condition (Reinert 1984b).

Overwintering hibernacula are typically rocky dens, to which copperheads return yearly (Gloyd and Conant 1990). Such sites may be shared with other species of snakes, such as timber rattlesnakes (*Crotalus horridus*), black rat snakes (*Pantherophis alleghaniensis*), and northern black racers (*Coluber constrictor constrictor*). However, it seems that the northern copperhead and timber rattlesnake usually hibernate in separate dens (Gloyd and Conant 1990).

### *LIFE HISTORY AND ECOLOGY*

Northern copperheads emerge from hibernation in April and are active until October, with peak activity occurring between May and September (Gloyd and Conant 1990, Ernst and Ernst 2003). The species demonstrates diurnal activity during the autumn and spring but becomes nocturnal as temperatures rise during the summer (Sanders and Jacob 1981). Seasonal movements from rocky hibernation sites to lowland regions in the summer have been indicated for this species in Pennsylvania (Swanson 1952). In Virginia, male copperheads moved an average of 22.3 m per day, and traveled an average of 2,962 m during the active season. Females moved 10.4 m per day and 1,580 m for the active season (Petersen 1995). Males had an average activity range of 8.07 ha, while females exhibited an activity range of 3.93 ha (Petersen 1995). Gravid females remain at birthing dens to meet thermoregulatory needs and thus had smaller home ranges (Fitch 1999). Excluding days when snakes did not move average distances traveled per day (in meters) were eighteen, twelve, and twelve for males, nongravid females, and gravid females, respectively (Fitch and Shirer 1971).

In Pennsylvania, northern copperheads include mice, voles, shrews, young opossums, sparrows, cicadas and moth larvae, salamanders, and even other snakes in their diet (Surface 1906). They may also feed on young gray squirrels, young cottontail rabbits, and chipmunks (Ernst and Ernst 2003). The northern copperhead is a sit-and-wait predator that uses crypsis to ambush prey (Fitch 1999). It has been suggested that young copperheads use their brightly colored tails as caudal lures to attract frogs and small mammal prey (Ditmars 1907, Carpenter and Gillingham 1990).

Female copperheads reproduce biennially, and most are mature at three years of age (Fitch 1960, Fitch and Shirer 1971). Courtship, copulation, and male-male combat occur in two distinctive reproductive periods, from March to April and from September to early October (Schuett 1982). After copulation, females store sperm through winter until ovulation occurs in the spring (Schuett and Gillingham 1986). During the summer and before giving birth, gravid females are known to aggregate in groups at rocky sites for basking and protection from predators (Fitch 1960, Fitch and Shirer 1971, Gloyd and Conant 1990, Fitch 2002). The females give birth anywhere between July and October, with most births occurring in September (Ernst and Ernst 2003). Gloyd and Conant (1990) give sizes of three copperhead broods from eastern Pennsylvania as numbering three, seven, and ten offspring. Hulse et al. (2001) reported broods ranging from four to nine with an average of 5.7.

### *THREATS*

The primary threats to populations of northern copperheads include habitat loss and indiscriminant killing. Habitat may be lost through rural and suburban development. Unlike timber rattlesnakes, copperheads use warmer, more open habitats (Reinert 1984a), and it is quite possible that the maturation of Pennsylvania forest may also reduce habitat suitability in some areas. Copperheads are commonly considered by the lay public to be more dangerous than rattlesnakes. This is often attributed to their lack of warning when disturbed or before striking humans (Surface 1906).

### *CONSERVATION AND MANAGEMENT NEEDS*

The first step in ensuring the survival of viable populations of northern copperheads in Pennsylvania is to reduce the loss of snakes from existing populations. Collecting and indiscriminant killing should be prohibited by closing the season and setting the possession limit for this species at zero.

A second initiative that would provide substantial benefits to copperhead populations would be the complete protection for this species on all government-owned lands (State Game Lands, State Forest Lands, State Forest Natural Areas, State Parks, National Forests, and Military Reservations) through the joint cooperation of government agencies.

Northern copperheads are commonly viewed as highly dangerous animals that conflict directly with human habitation and human survival. Education programs that dispel these prevalent myths, elucidate the role of these snakes in natural ecosystems, and foster a greater tolerance for continued existence of this species need to be developed and widely disseminated.

### *MONITORING AND RESEARCH NEEDS*

An assessment of the current distribution and occurrence of the northern copperhead is required to determine the current status of this species within the commonwealth and to identify key habitats and imperiled populations. This information should be added to the Pennsylvania Natural Diversity Inventory (PNDI) database. The PNDI database should then be used consistently to monitor potential effects of development, logging, utility (pipeline, powerline, oil exploration, cellular towers, windmills), and recreational projects (trails, parks, campgrounds) on copperhead habitat. All projects that potentially influence habitat or populations should be examined before their initiation. Those found to have probable impacts should be evaluated by biologists and, if necessary, altered to reduce those effects. The effectiveness of this program at detecting potential impacts should be evaluated annually by assessing the number of projects examined and the number of projects altered.

Ecological field studies are also required to determine the spatial ecology, the characteristics of active-season habitat and hibernacula, and the genetic structure of copperhead populations in Pennsylvania.

*Authors:* JOSEPH AGUGLIARO AND HOWARD K. REINERT, THE COLLEGE OF NEW JERSEY

## Eastern Hognose Snake

Order: Squamata
Family: Colubridae
*Heterodon platirhinos*

The eastern hognose snake was selected as a Pennsylvania Species of Greatest Conservation Need because of its restricted habitat, limited distribution, and unknown status in Pennsylvania (fig. 4.71). In addition, eastern hognose snakes are habitat specialists and are threatened by illegal collecting. Eastern hognose snakes are not listed as Threatened or Endangered at the federal level or in Pennsylvania. Their collection in Pennsylvania is prohibited (Pennsylvania Fish and Boat Commission 2008). The global status of the eastern hognose snake is Secure (G5, NatureServe 2009).

### *GEOGRAPHIC RANGE*

Eastern hognose snakes range from southern Florida north to southern Massachusetts and New Hampshire, west through Ontario, Minnesota, and southeastern South Dakota, and south to the Gulf Coast and central Texas (Conant and Collins 1991, Hulse et al. 2001, Ernst and Ernst 2003).

### *DISTRIBUTION AND RELATIVE ABUNDANCE IN PENNSYLVANIA*

In Pennsylvania, the species is distributed in the eastern, south-central, and southwestern portion of the state with a historic record from Erie County (Hulse et al. 2001; fig. 4.72). It is seldom found anywhere in abundance but is locally common in prime habitat throughout its geographic range (Ernst and Ernst 2003). No studies have been performed assessing the species abundance in Pennsylvania.

*Fig. 4.71.* The Eastern Hognose Snake, *Heterodon platirhinos*. Photo courtesy of Tom Diez.

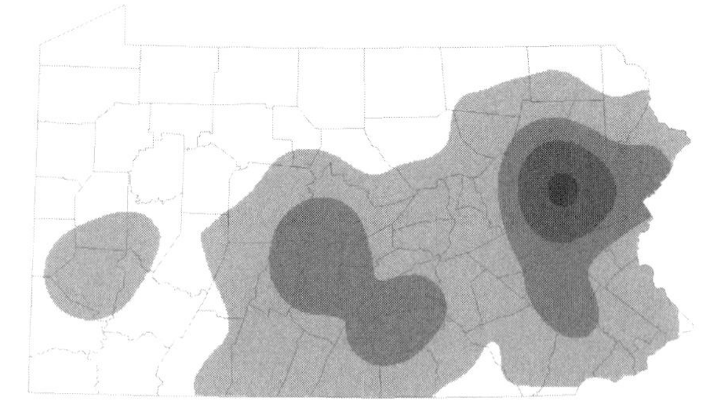

*Fig. 4.72.* Distrubution of the Eastern Hognose Snake, *Heterodon platirhinos*.

### *COMMUNITY TYPE/HABITAT USE*

Eastern hognose snakes are found in both forest and grassland, often in association with a watercourse, and are most always associated with sandy soils (Hulse et al. 2001, Ernst and Ernst 2003). In Pennsylvania, eastern hognose snakes are usually found along sandy floodplains or on sandstone ridges and mountainous areas (Hulse et al. 2001). The species is known to occupy mammal burrows and is seldom seen under cover objects (Hulse et al. 2001, Ernst and Ernst 2003). In the New England states, the species is associated with pine forests (Michener and Lazell 1989).

### *LIFE HISTORY AND ECOLOGY*

In Pennsylvania, eastern hognose snakes usually are active from late April through early October (Hulse et al. 2001). The species is diurnal, and individuals tend to restrict their movements to within a large home range of about 50 ha (Plummer and Mills 2000). Peak activity is usually in midmorning and late afternoon (Hulse et al. 2001). Inactivity for extended periods of time has been recorded for individuals during hot summer months (Plummer and Mills 2000).

The diet of eastern hognose snakes mostly comprises amphibians, especially toads, but also frogs, invertebrates, reptiles, birds, and mammals (Edgren 1955, Mills 1993, Hulse et al. 2001, Ernst and Ernst 2003). Hognose snakes are preyed on by crows, hawks, and owls (Edgren 1955). Eastern hognose snakes may be best known for feigning death behavior when threatened, in which they roll over on their dorsum as if dead. This antipredatory defense mechanism is commonly exhibited during encounters with humans. Copulating pairs have been observed feigning death together while continuing to copulate (Plummer 1996).

The species is also known to flatten its head and hiss when cornered.

Little is known about the mating period of eastern hognose snakes, but it is assumed to be in the spring (Hulse et al. 2001). Males are known to trail females during the breeding season (Plummer and Mills 1996). Eggs tend to be deposited in June (Edgren 1955). Clutch size has been shown to range from four to sixty-one eggs (Edgren 1955), but in Pennsylvania, the average clutch size is about fifteen eggs (Hulse et al. 2001). Growth is rapid in juveniles but slows in adults (Ernst and Ernst 2003). Longevity is known to be a maximum of about eight years (Snider and Bowler 1992).

### *THREATS*

Habitat destruction and human-related disturbances are likely the greatest threat to eastern hognose snakes. As a semifossorial species associated with floodplains, the eastern hognose snake is threatened by contaminated watercourses throughout much of its range in Pennsylvania. Populations associated with sandstone mountainous areas are less threatened by contamination sources and more threatened by timber harvesting and quarrying. Illegal collection may also be threatening this species in Pennsylvania. Accidental death by motor vehicle is likely another threat to the species.

### *CONSERVATION AND MANAGEMENT NEEDS*

The eastern hognose snake is in need of a conservation and management plan. Until further information is learned about the distribution, abundance, and life history characteristics of eastern hognose snakes in Pennsylvania, it will continue to be managed as all other Nonthreatened, Endangered, or Candidate reptiles in the state (Pennsylvania Fish and Boat Commission 2008). Education and outreach efforts should target local governments where land uses, in particular development on uplands, are permitted and regulated. Outreach and collaboration with the Pennsylvania Department of Conservation, Bureau of Forestry, and Pennsylvania Department of Environmental Protection, Bureau of Mining and Reclamation, may be warranted to curtail impacts on this species associated with timber harvest and mining, especially in areas where populations are known to occur.

### *MONITORING AND RESEARCH NEEDS*

The status of eastern hognose snakes in Pennsylvania is poorly understood. Intensive surveys are needed to determine present distribution, particularly in areas of the state with historic records of occurrence and with suitable habitat that have not been adequately surveyed. These surveys should be followed up with monitoring activities at selected sites to determine population status and to detect changes in numbers.

Much of the life history of the eastern hognose snake is poorly understood. This is particularly true for populations in the mid-Atlantic region. Future research in the commonwealth should include population studies on selected sites to determine home range, density, and related ecological measures, as well as life history studies associated with mating, reproduction, sexual maturity, and related traits.

*Author:* JOSEPH J. WILSON,
WILSON ECOLOGICAL CONSULTING

# 5

# The Birds
## *Introduction*

Margaret C. Brittingham

Pennsylvania is home to a diversity of birds, including breeders, winter residents, and migrants that pass through Pennsylvania on travels between breeding and wintering grounds. In Pennsylvania, the official state list of birds is 401 species (Pennsylvania Ornithological Records Committee 2000). Of these, 250 species occur every year either as residents or migrants. Approximately 200 species have been known to breed in Pennsylvania, and approximately 186 species regularly breed in the state (Gross 1997, McWilliams and Brauning 2000).

The passenger pigeon (*Ectopistes migratorius*) was once a common breeder in Pennsylvania forests but is now extinct. The Carolina parakeet (*Conuropsis carolinensis*) is also extinct. It was known to have occurred in Pennsylvania, but there are no official breeding records. Five other species that formerly bred in the state are considered Extirpated from the state. These include (1) the greater prairie chicken (*Tympanuchus cupido*), (2) piping plover (*Charadrius melodus*), (3) olive-sided flycatcher (*Contopus cooperi*), (4) Bewick's wren (*Thryomanes bewickii*), and (5) Bachman's sparrow (*Aimophila aestivalis*). Two of these, the piping plover and olive-sided flycatcher, are on the current list of Species of Greatest Conservation Need. The eskimo curlew (*Numenius borealis*) and brown-headed nuthatch (*Sitta pusilla*) were never breeders in the state, but were reported as migrants or visitors to the state. Both of these are now extirpated from the state, and the eskimo curlew may be extinct.

Eighteen species of birds are currently listed as Threatened or Endangered in Pennsylvania. The threats birds face and the reasons for their declines have changed over time. In the 1880s, when passenger pigeons were still a common sight in the forests of the commonwealth, overexploitation, resulting from unregulated hunting and shooting along with widespread clearing of the forest, were the major culprits. By the mid-1900s, laws and regulations protected birds, and overexploitation was no longer an issue, with the exception of some species of raptors that still lacked adequate protection. However, a new threat had emerged. Pesticide contamination, particularly from chlorinated hydrocarbons such as DDT, had become a major threat to many of the state's raptors and fish-eating birds. DDT was directly linked to declines in bald eagles (*Haliaeetus leucocephalus*), osprey (*Pandion haliaetus*), and peregrine falcons (*Falco peregrinus*). In the 1970s, DDT was banned in the United States, and this action, in conjunction with reintroduction programs initiated around the state in subsequent years, led to the recovery of many species that pesticides had negatively affected (Brauning and Hassinger 2009).

Today, the major threats to birds are habitat loss, degradation, and fragmentation. These are threats facing birds across the state and are much harder to address and more difficult to reverse than threats birds have faced in the past. Species in wetland habitats are particularly vulnerable. This is a habitat type that was always uncommon in Pennsylvania, and more than 50 percent of the original wetlands have been lost, and remaining wetlands are often degraded (Dahl 1990). Fourteen (78%) of the eighteen species included on the list of Threatened and Endangered Species breed in wetland habitats (Brauning and Hassinger 2009). Grassland and open habitats are another uncommon habitat type that support a number of rare species. Four species on the Threatened and Endangered Species list are associated with grasslands and open habitats (Brauning and Hassinger 2009). Forest fragmentation is a current ongoing threat to many of our area-sensitive forest songbirds (see chapter 7, "Critical and Emerging Issues in the Conservation of Terrestrial Vertebrates").

Seventy-eight species are currently listed as bird Species of Greatest Conservation Need in Pennsylvania. The list includes federal- and state-level Threatened and Endangered Species but goes well beyond that. It includes species like the eastern meadowlark (*Sturnella magna*) and whip-poor-will (*Caprimulgus vociferous*) that are not rare enough to be listed as Threatened within the state but have shown long-term population declines alerting us that these are species that need attention now. It includes species like blue-headed vireo (*Vireo solitarius*) and Acadian flycatcher (*Empidonax virescens*) that are representatives of "at risk" habitats, in this case hemlock forests.

Nine species of birds are listed as Pennsylvania Responsibility Species. These are species in which a significant portion of the population breeds in or migrates through Pennsylvania, and consequently, Pennsylvania is responsible for that species at regional and global levels. Five of the nine species breed within mature forest habitat, reflecting the importance of the large blocks of contiguous forest that occur in Pennsylvania today. Two species, the blue-winged warbler (*Vermivora pinus*) and golden-winged warbler (*Vermivora chrysoptera*), breed in old field and early successional habitat. Henslow's sparrow (*Ammodramus henslowii*) is associated with large reclaimed strip mines, reflecting the importance of this relatively new habitat type to grassland birds. The tundra swan (*Cygnus columbianus*) is the only nonbreeding migrant of the group. It is listed as a Responsibility Species because a major portion of the population stages in Pennsylvania during migration, and their ultimate success is dependent on wetland habitats they use while here. Some Responsibility Species, like the wood thrush (*Hylocichla mustelina*), are undergoing range-wide declines, making Pennsylvania's responsibility to these species even more critical.

Because birds for the most part are diurnal and vocal, and because there is an extensive network of skilled volunteers involved in surveys of birds, we know more about the distribution and abundance of birds than we do for other vertebrate groups. Exceptions include nocturnal species and wetland species. These groups are not well surveyed by current survey methods, and thus our level of information is much lower for these species.

Conservation of avian populations in the state involves multiple steps. Large-scale monitoring programs such as the Breeding Bird Survey and Pennsylvania Breeding Bird Atlases, in combination with surveys targeted at specific groups of species or specific locations, are used to identify species or groups of declining species. Research hypotheses are generated, and research is conducted to determine the cause or causes of declines. On the basis of research results, potential solutions or ways to mitigate declines are proposed. These can range from habitat management and manipulation, to direct protection, regulations, and legislation. Monitoring and evaluating occur throughout the process and changes and adjustments are made as we learn more about the causes and consequences of declines, the effectiveness of attempted solutions, and the tools we have to implement them.

The species accounts that follow are grouped by Wildlife Action Plan status and organized taxonomically within groups. Range maps were developed using a variety of data sources, with a particular emphasis on records reported in the first and second Pennsylvania Breeding Bird Atlases (Brauning 1992a, Mulvihill and Brauning 2009). Population trend graphs from Breeding Bird Survey data are presented when available (Sauer et al. 2008). Collectively, these species accounts provide a picture of the range of conservation issues facing birds in Pennsylvania, the Northeast, and, in many cases, birds around the world. They also provide summaries of research and monitoring needs and give guidelines for conservation and management actions for reversing these declines. We hope that these will be used to develop creative strategies to reverse declining trends of birds in all habitat types across the commonwealth.

## IMMEDIATE CONCERN

## Northern Bobwhite

Order: Galliformes
Family: Odontophoridae
*Colinus virginianus*

The northern bobwhite is a well-known, popular game species (fig. 5.1). It was selected as a Species of Greatest Conservation Need because it has undergone precipitous population declines in Pennsylvania and throughout most of its geographic range. Because of their precarious status in Pennsylvania, the Pennsylvania Biological Survey ranks them as a Candidate at Risk. Bobwhites are classified as Critically Imperiled (S1) in Pennsylvania, but global populations are considered Secure (G5, NatureServe 2009). Bobwhites are listed as a Species of Greatest Conservation Need by all northeastern states except Maine, New Hampshire, and Vermont. Bobwhites are listed as a game species in Pennsylvania, but the season has remained closed since 1982 in thirteen southeastern and south-central counties (Anon 1982). This restriction has been transferred to the contemporary wildlife management units encompassing those counties.

*Fig. 5.1.* The Northern Bobwhite, *Colinus virginianus*. Photo courtesy of Jacob W. Dingel III.

### *GEOGRAPHIC RANGE*

Bobwhites reside in farmlands, grasslands, and brushy areas from central and eastern North America to southern Mexico, with northern fringes extending into Wisconsin, Michigan, Ontario, Pennsylvania, New York, and New England (Brennan 1999). In eastern North America, good bobwhite habitat is characterized by well-interspersed woodland, brushland, grassland, and cultivated land (Leopold 1933).

The early bobwhites in Pennsylvania were likely of the subspecies *Colinus virginianus marilandicus*, whose historical geographic range was New England south to Maryland, Delaware, and central Virginia, and west to Pennsylvania. This race has been described as "large, brightly colored, [and] reddish" (Aldrich 1946, 494). Those that moved into western Pennsylvania from the Ohio Valley during the early to mid-1800s (Christy 1926) were probably of the subspecies *Colinus virginianus mexicanus*, whose range is westward of *C. v. marilandicus* (Aldrich 1946, Bolgiano 2000). This race has been described as "medium-sized to large, medium-toned, [and] grayish" (Aldrich 1946, 494). These two subspecies are often grouped with the nominate *Colinus virginianus virginianus* of the Atlantic seaboard (Johnsgard 1988a, Brennan 1999, Madge and McGowan 2002).

### *DISTRIBUTION AND RELATIVE ABUNDANCE IN PENNSYLVANIA*

In presettlement Pennsylvania, bobwhites were present in the lower Delaware and Susquehanna valleys and probably in the lower Great, or Cumberland, Valley as well. As forests were converted to agriculture, bobwhites moved into the upper parts of these valleys, into the Ohio Valley, much of the Ridge and Valley Province, and some agricultural regions of the Allegheny Plateau. By the twentieth century, bobwhites were no longer common, but even so, their statewide population numbered approximately a half-million during the 1920s to 1935. South-central Pennsylvania was the region with the highest density, particularly after farms were abandoned (fig. 5.2). Severe winters of 1935–1936 and 1944–1945 killed up to 90 percent of the state's bobwhites. Bobwhite numbers increased again during

1958–1970 coinciding with federal set-aside programs that increased the amount of available habitat in agricultural areas. With the ending of these programs and subsequent "fencerow-to-fencerow" farming, bobwhite populations declined precipitously in Pennsylvania and elsewhere, as documented by Christmas Bird Count and Breeding Bird Survey data (fig. 5.3). Today, if naturally reproducing bobwhites exist in Pennsylvania, they are most likely to be found in the southern parts of Franklin, York, Lancaster, and Chester counties. Bobwhites found elsewhere are likely to be released birds (Latham and Studholme 1952, Church et al. 1993, Brennan 1999, Bolgiano 2000, Sauer et al. 2004).

### *COMMUNITY TYPE/HABITAT USE*

Bobwhites require successional mosaics within their home ranges, which they occupy year-round. A typical geographic home range is about 10–60 ha but can be larger in poor habitat. Their preferred nesting cover in eastern North America consists of scattered shrubs and briars, interspersed with moderately dense herbaceous or grassy vegetation. Too much grass or woody vegetation is undesired. Nests tend to be concentrated on idle land—especially if it is about two years past agricultural use—in pastures, hay fields, roadsides, or fencerows, but not where these occur in isolated patches. This habitat requirement means that bobwhites have a relatively short window of nesting opportunity during the plant succession cycle. After about seven to ten years, too much woody vegetation has grown and a disturbance, such as fire or clearing by humans, is needed to maintain the successional mosaic (Klimstra and Roseberry 1975, Roseberry and Klimstra 1984, Brennan 1999, Taylor et al. 1999).

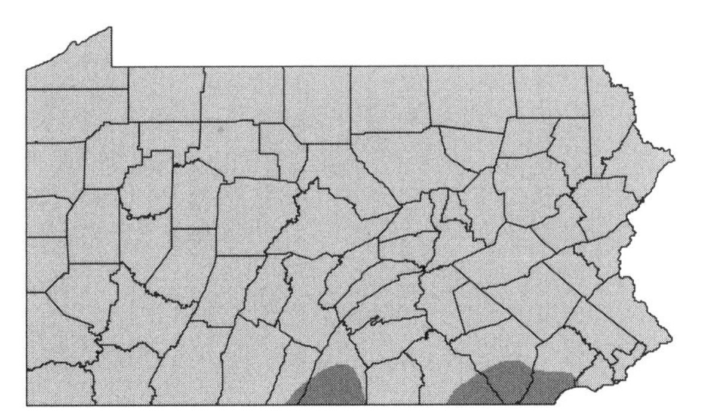

*Fig. 5.2.* Distribution of the Northern Bobwhite, *Colinus virginianus*. Dark shading = last known wild populations; light shading = pen-reared birds.

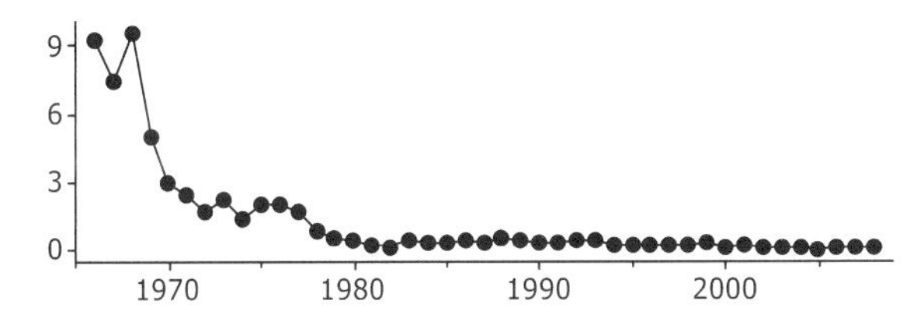


*Fig. 5.3.* Northern Bobwhite, *Colinus virginianus*, population trends from the Breeding Bird Survey.

### *LIFE HISTORY AND ECOLOGY*

In the fall, bobwhites band together in coveys of about fourteen birds to retain sufficient heat within their nightly sleeping ring. Their preferred winter habitat consists of wooded areas with dense understory used for cover and resting, accessible agricultural grains or weed seeds, and low, grassy cover for roosting. Coveys break up during March and April when pairing commences the reproduction cycle. Food includes seeds, small fruits, and insects when available (Latham and Studholme 1952, Rosene 1969, Roseberry and Klimstra 1984).

Bobwhites' life spans are short even under perfect conditions, in part because of their high metabolism. In the northern part of their range, annual survival may average about 17 percent, and it is unusual for a wild bird to live more than three years (Roseberry and Klimstra 1984, Guthery 2000). Severe winter weather may kill up to 90 percent of bobwhites, which occurred during the winters of 1855–1856, 1935–1936, and 1944–1945 (Latham and Studholme 1952, Bolgiano 2000). The main predators of Pennsylvania bobwhites are thought to be Cooper's hawks (*Accipiter cooperii*), house cats (*Felis catus*), red foxes (*Vulpes vulpes*), and gray foxes (*Urocyon cinereoargenteus*; Latham and Studholme 1952). The effect of predation and hunting on bobwhite populations may increase as habitat declines (Robel 1993).

Bobwhites begin nesting in late April to early May, with most of the hatch occurring during mid-June to mid-August. Nests are in clumps of grass stems. Nest success tends to be low, with hatching in about one-third of nests where eggs are laid. The primary causes of nest failure are predation and farming activities, especially mowing (Latham and Studholme 1952, Klimstra and Roseberry 1975, Roseberry and Klimstra 1984). Newly hatched bobwhites require a high-protein diet for rapid growth during their first few months. This insect diet can be most easily found in low-growing herbaceous vegetation, preferably where there is bare ground for easy movement of chicks and a plant canopy to shield the birds from predators (Rosene 1969, DeVos and Mueller 1993).

*THREATS*

The primary factors influencing the species' decline in Pennsylvania include changes in land use, principally intensive agriculture, development, and forest regrowth, with the result that the landscape no longer provides the necessary habitat mosaics within the required spatial scale. A lack of adequate nesting, brood-rearing, and winter cover is due to intensive mowing of hay fields, loss of fencerows, and clean farming practices. Small population size increases the risk of extinction, and severe winter weather may periodically kill up to 90 percent of populations. In addition, bobwhites may have lost adaptability because of interbreeding with imported bobwhites of southern races (Christy 1926, Todd 1940a, Latham and Studholme 1952, Bolgiano 2000).

*CONSERVATION AND MANAGEMENT NEEDS*

Hunting should remain closed in areas where naturally breeding bobwhite populations may be self-sustaining. Release of pen-raised bobwhites near these naturally breeding populations should be discouraged. Biologists of the Pennsylvania Game Commission should be in contact with the Southeast Quail Study Group, a group of wildlife managers working on bobwhite recovery. The maintenance and recovery of Pennsylvania's bobwhites will depend on the spatial linkage with bobwhites in Maryland and Delaware and the success of maintenance and recovery programs in those states. The Conservation Reserve Enhancement Program (CREP) is an opportunity for directed conservation effort. Multiple CREP fields in proximity could potentially become core habitat areas. Certain practices, such as not mowing during the nesting season and planting warm-season grasses and food plants, can enhance the effectiveness of conservation set-aside lands (Guthery 2000, Dimmick et al. 2002).

*MONITORING AND RESEARCH NEEDS*

Intensive monitoring is necessary to delineate the precise geographic range of naturally reproducing bobwhites in Pennsylvania, to monitor their population status over time, and to document the effectiveness of CREP to provide bobwhite habitat. In Pennsylvania, the most pressing research needs are to determine where wild bobwhites exist in the state, if there is sufficient probability of success for bobwhite restoration efforts, and the genetics of the wild bobwhites. The latter would help to determine the conservation value of Pennsylvania's bobwhites, to guide restoration efforts, and to document historical distributions.

*Author:* NICHOLAS C. BOLGIANO

## Piping Plover

Order: Charadriiformes
Family: Charadriidae
*Charadrius melodus*

The piping plover is a small, distinctly marked North American shorebird (fig. 5.4). It is considered a species of Immediate Concern in Pennsylvania, where it is listed by the Pennsylvania Biological Survey as an Extirpated Species. It is also listed as a Species of Greatest Conservation Need by all northeastern states where they occur or have historically occurred. Plovers breeding within Pennsylvania are part of the Great Lakes population, which is listed as federally Endangered by the United States Fish and Wildlife Service. The global population is considered Vulnerable (G3, NatureServe 2009).

*GEOGRAPHIC RANGE*

Piping plovers breed only in North America, along the Atlantic Coast from Newfoundland south to South Carolina, on the Great Lakes in the United States and Canada, and west from Minnesota and Manitoba to Alberta (Haig et al. 2005). The geographic range is divided into three geographic regions: (1) beaches of the Atlantic Coast, (2) shorelines of the Great Lakes, and (3) along alkali wetlands and major rivers of the northern Great Plains and Prairie Canada (Haig et al. 2005).

*Fig. 5.4.* The Piping Plover, *Charadrius melodus*. Photo courtesy of Glen Tepke.

Though declining, the northern Great Plains breeding population remains the largest (Ferland and Haig 2002). The Great Lakes population, in which Pennsylvania's historic nesting history lies, remains extremely imperiled. This population has almost tripled to 100 breeding individuals between 1991 and 2005 (Haig et al. 2005, Stucker and Cuthbert 2005). Although they are increasing in number, the Great Lakes population range has not expanded to narrow the current gap among the three breeding populations, which may potentially inhibit interregional gene flow (Haig and Oring 1985, 1988, Plissner and Haig 1997). Piping plovers from these three breeding populations winter in coastal areas of the United States from North Carolina to Texas. They also winter along the coast of eastern Mexico and on Caribbean islands from Barbados to Cuba and the Bahamas (Haig 1992). Critical habitat has been designated for the Great Lakes breeding population and for all three populations of piping plovers on the wintering grounds (U.S. Fish and Wildlife Service 2001b, 2001c).

### *DISTRIBUTION AND RELATIVE ABUNDANCE IN PENNSYLVANIA*

As many as fifteen pairs were documented on the outer shores of Presque Isle after the plovers were first confirmed breeding in 1911 (Todd 1940a; fig. 5.5). Piping plovers have not nested in Pennsylvania since the late 1950s (Stull et al. 1985), so this species is currently listed by Pennsylvania as an Extirpated Species. No recent (since 1980s) nesting has been documented on Lakes Erie or Ontario, but sightings continue and provide some hope for recolonization from the Great Lakes population (U.S. Fish and Wildlife Service 2003b).

Nonbreeding sightings at Presque Isle State Park have been documented in five of the past ten years (Pennsylvania Society for Ornithology Files 2005), with few of these sightings during the spring nesting season. The observation of three birds in April and May 2005 (McWilliams 2005) was the first breeding-season observations since 1992 (McWilliams 1992, McWilliams and Brauning 2000).

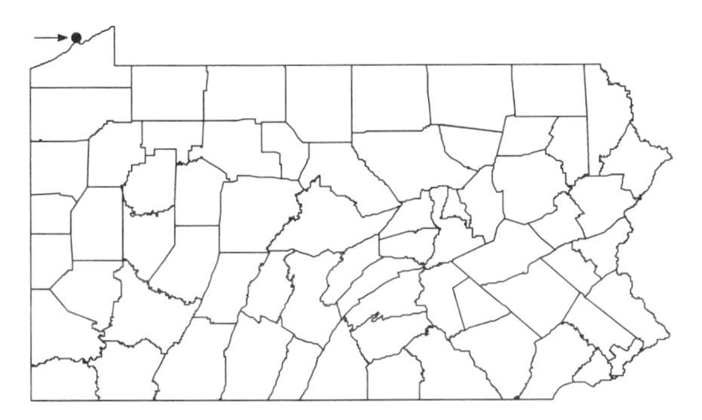

*Fig. 5.5.* Historical range of the Piping Plover, *Charadrius melodus.*

### *COMMUNITY TYPE/HABITAT USE*

The piping plover is a specialist restricted to beach habitats of rivers, lakes, and coastal dunes across its geographic range. The critical habitat resource of the piping plover in Pennsylvania is the tip of Presque Isle State Park known as "Gull Point." This site, now designated a Natural Area in the state park system (Department of Conservation and Natural Resources 1992), supports the best hope for nesting within Pennsylvania and the focus of conservation action. In addition, Price (2002) identified moderate potential for recolonization on beaches 9 and 10.

### *LIFE HISTORY AND ECOLOGY*

Piping plovers are migratory shorebirds that spend approximately three to four months a year on their breeding grounds. In the Great Lakes region, birds begin arriving on breeding grounds in late April, and most nests are initiated by mid- to late May (Pike 1985, Haas and Haas 2005). Eggs typically hatch from late May to late July (Lambert and Ratcliff 1981, Pike 1985). Although piping plovers typically produce one brood per year, they have produced two broods at some Atlantic Coast sites (Bottitta 1997) and in the Great Lakes (U.S. Fish and Wildlife Service 2003b). Breeding adults depart nesting grounds in the Great Lakes as early as mid-July, but the majority depart by mid-August (Wemmer 2000). Females typically depart before males. Juveniles usually depart a few weeks later than adults, and most disperse by late August.

Piping plovers feed primarily on exposed beach substrates by pecking for invertebrates. The diet generally consists of invertebrates, including insects, marine and freshwater worms, crustaceans, and mollusks (Cuthbert et al. 1999, Haig and Elliot-Smith 2004).

### *THREATS*

Shoreline development in the Great Lakes region and throughout the wintering grounds poses a threat to the Great Lakes population of piping plovers. The extirpation of piping plovers from Pennsylvania appears to be related to the expansion of beach recreation at Presque Isle State Park that altered the physical nature of beaches through direct and indirect effects.

Predation was identified as the primary cause of nest failure, and predators are suspected in the majority of disappearances of unfledged chicks. A diversity of actual and potential predators are implicated, including herring gull (*Larus argentatus*), ring-billed gull (*Larus delawarensis*), merlin (*Falco columbarius*), peregrine falcon (*Falco peregrinus*), great horned owl (*Bubo virginianus*), American crow (*Corvus brachyrhynchos*), red fox (*Vulpes vulpes*), coyote (*Canis latrans*), raccoon (*Procyon lotor*), striped skunk (*Mephitis mephitis*), domestic cat (*Felis catus*), and dog (*Canis familiaris*). Human developments near beaches attract increased numbers of predators, such as skunks and raccoons (U.S. Fish and Wildlife Service 1985).

Disturbance by humans and pets is another direct cause of mortality. Beach walking, bike riding, kite flying, fireworks, bonfires, horseback riding, kayaking, windsurfing, camping, and close-up photography are among the many activities that disturb piping plovers and disrupt normal behavior patterns (Howard et al. 1993). High pedestrian use may deter piping plovers from using nesting habitat (Burger 1991, 1994). In addition, home range size was greater on high-use beaches in Michigan (Haffner et al. 2009). Repeated flushing of birds from their nests by pedestrians exposes eggs to potentially lethal extremes in temperature (Bergstrom 1991). Pedestrians accompanied by pets present an even greater disturbance to breeding piping plovers (Pike 1985), as dogs frequently chase and attempt to capture adults and chicks (Lambert and Ratcliff 1979).

Endangered populations, by virtue of their small size and geographic isolation, are inherently at greater risk of extinction than larger populations (Caughley and Gunn 1996). Small, isolated populations are more likely to be destroyed by random environmental events than larger, widespread populations. Similarly, very small isolated populations are more strongly affected by demographic stochasticity, random changes in sex ratios, or ability to find mates (known as the "Allee effect"), which all influence population persistence (Haig et al. 2005). In an analysis of the Great Lakes population through 1999, up to 29 percent of adults remained unmated throughout the breeding season, suggesting that Allee effect may occur (Wemmer 2000).

Piping plovers may accumulate contaminants from point sources and nonpoint sources at breeding, migratory stopover, and wintering sites. Oiling also poses a potential threat to piping plovers migrating and breeding along Great Lakes waterways. The magnitude of threat that pollution presents to piping plover habitats and associated shorebirds is yet unknown.

### *CONSERVATION AND MANAGEMENT NEEDS*

The Great Lakes population of piping plover was listed as Endangered under the provisions of the U.S. Endangered Species Action (U.S. Fish and Wildlife Service 1988), and Critical Habitat was designated for the Great Lakes breeding population (U.S. Fish and Wildlife Service 2001b). The Great Lakes piping plover population has been assigned a 2C (high degree of threat and recovery potential) recovery priority (U.S. Fish and Wildlife Service 2002c). Conservation measures under way to protect the piping plover include recognition, research, protective management, requirements for federal protection, and prohibitions against certain practices. Thirty-five units (extending 500 m [1640 ft] inland) were designated along the Great Lakes shorelines of eight states, including an approximately 2.29-km (3.7-mile) section of Presque Isle State Park in Pennsylvania.

The Great Lakes basin has been identified as a refuge for a diversity of globally rare species and ecosystems (The Nature Conservancy 1994). Many piping plover breeding beaches within the Great Lakes harbor rare dune features or provide habitat for other species of special status, including some that are federally listed in other states. Adequately protecting Great Lakes piping plover breeding habitat may safeguard a significant proportion of shoreline biodiversity (Cuthbert et al. 1998). Given the imperiled nature of beach ecosystems, both within the Great Lakes region and along the Atlantic and Gulf coasts, an ecosystem approach to conservation will benefit both piping plovers and other inhabitants of coastal ecosystems. The Endangered Species Act provides for possible voluntary land acquisition and cooperation with the states and requires that recovery plans be developed for all listed species.

A long list of conservation management actions is presented in the federal recovery plan (U.S. Fish and Wildlife Service 2003b). To assess these activities and determine a course of action for the piping plover in Pennsylvania, the author proposes further collaboration with the team at Presque Isle State Park to assist recovery efforts. Conservation efforts should focus on continued monitoring, evaluation of suitable habitat at Gull Point and appropriate beaches, development of a rapid-response plan on detection of additional birds, and assessment of other factors that are deterring recolonization of breeding piping plovers.

Conservation education activities initiated by Presque Isle State Park include training an environmental education specialist in plover surveys and conservation and establishing a volunteer monitoring program. The park is currently working on developing a permanent wayside exhibit that would be located in the Gull Point Special Management Area. In addition, numerous presentations to the public and to park staff, signage to educate park visitors about the effects of dogs off the leash within the critical habitat area, and other conservation measure have been undertaken. Visitors are now educated and informed about the species, its recovery needs, and restrictions on dogs in the park. Also, Presque Isle State Park is working cooperatively with the U.S. Fish and Wildlife Service, U.S. Army Corps of Engineers, Department of Conservation and Natural Resource Bureau of State Parks, Cleveland Museum of Natural History, and the Pennsylvania Game Commission to develop a plan and approach to habitat enhancement and improvement on Gull Point.

*MONITORING AND RESEARCH NEEDS*

Sightings of piping plovers on Presque Isle State Park in 2005 (McWilliams 2005) raised a modest hope that restoration of the species may occur through dispersal from other Great Lakes sites. In response to those sightings, the Presque Isle State Park Resource Management Plan was updated to include pertinent research and a thorough contact list in the event of nesting plovers. Annual spring shorebird surveys were initiated in 2005 in the designated Critical Habitat Area to search for additional plover activity, and in 2009 were conducted by trained volunteers (K. Ryan, personal communication). A host of research questions have been posed in federal and state recoveries plans (U.S. Fish and Wildlife Service 2003b). With only a single historic and potential nesting location in Pennsylvania, research activities should be focused on local issues of practical significance to survival and nesting success at Presque Isle State Park, with careful attention to the social and cultural implications to the large number of humans who also use the beach. The initial focus of research should be to quantify factors (disturbance, predation) that might limit nesting success of piping plovers at its historic breeding site. Specific consideration also should be given to vegetative changes that are occurring on Gull Point due to invasion of native and exotic upland plants and loss of beach habitat. Considerations of human dimensions will be essential to successful implementation of conservation actions.

*Author:* DANIEL W. BRAUNING, PENNSYLVANIA GAME COMMISSION

## Upland Sandpiper

Order: Charadriiformes
Family: Scolopacidae
*Bartramia longicauda*

A medium-sized shorebird, the upland sandpiper was selected as a Species of Greatest Conservation Need because it has undergone precipitous declines in abundance, and is now a rare breeder restricted to a small number of relict populations in the state (fig. 5.6). In Pennsylvania, this species is listed as Threatened. It is on the Species of Greatest Conservation Need list of all northeastern states and is listed as Endangered by eight states. Consequently, it is considered a migratory bird of management concern in the Northeast. The global status of the upland sandpiper is Secure (G5, NatureServe 2009).

*GEOGRAPHIC RANGE*

The breeding range is centered on the Great Plains, from south-central Canada south to northern Oklahoma. It is found in lower densities from the Midwest east to the Appalachian Mountains, with smaller populations in Oregon, Alaska, and British Columbia,

*Fig. 5.6.* The Upland Sandpiper, *Bartramia longicauda*. Photo courtesy of David McNicholas.

and in the northeastern United States (Houston and Bowen 2001). This species' fortunes have been inextricably linked with changes in land use and grassland management across North America. Upland sandpipers have declined substantially in most areas since the nineteenth century, at which they were described as abundant birds across the Great Plains and the Midwest (Houston and Bowen 2001). In the Northeast, populations increased during the nineteenth century, when forests were felled and replaced with pastures and hay fields. During the twentieth century, a reversal of this trend, coupled with changes in grassland management, has resulted in substantial declines and local extirpations (Tate 1986). It spends the winter on grasslands in southern and eastern South America, primarily in Argentina.

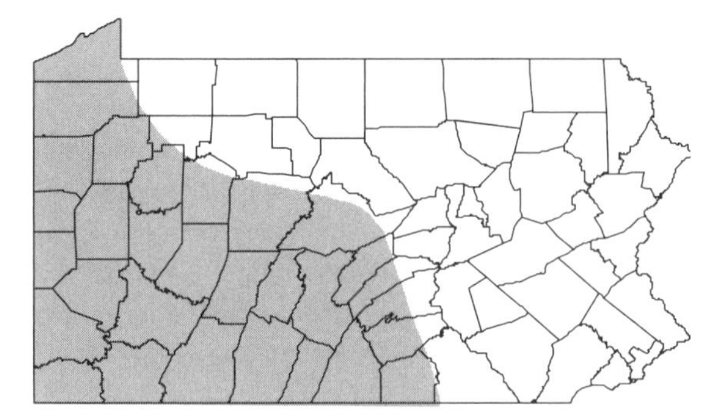

*Fig. 5.7.* Highest probability of occurrence of the Upland Sandpiper, *Bartramia longicauda*.

### *DISTRIBUTION AND RELATIVE ABUNDANCE IN PENNSYLVANIA*

During the first Breeding Bird Atlas (1983–1988), the upland sandpiper was found in fifty-four widely scattered atlas blocks and confirmed to breed in twenty-one (Brauning 1992a). The breeding locations were mainly in the Appalachian Plateau, Ridge and Valley, and Piedmont physiographic provinces, with concentrations in Erie, Crawford, Lawrence, and Butler counties, with a scattering of records southward between Westmoreland and Adams counties.

Once described as a common and delicious game bird (Warren 1890), upland sandpipers are now rare breeding birds in Pennsylvania (McWillians and Brauning 2000). Numbers were clearly lower in the 1930s than in the nineteenth century (Todd 1940a) and were much reduced by 1960. Since then, their numbers have dwindled still further. It has rarely been encountered on Breeding Bird Survey routes since 1980. It is restricted to the western half of the state, and by the early 2000s, breeding was only reported regularly from a handful of counties: Adams, Clarion, Crawford, Somerset, Verango, and Westmoreland (Fialkovich 2001, 2002) (fig. 5.7).

### *COMMUNITY TYPE/HABITAT USE*

The upland sandpiper was once an abundant bird of native mixed-grass prairies but has become increasingly dependent on nonnative grasslands as natural habitats disappeared. It prefers large fields with low to moderate forb and woody vegetation cover, moderate grass cover, moderate to high litter cover, and little bare ground. Fence posts or other display perches may be an important habitat component (Dechant et al. 2002). A mosaic of grassland cover heights and densities provides the ideal conditions on the breeding grounds: long vegetation for nesting and short vegetation for foraging. This mosaic was formerly typical in many farmland areas, where hay fields and grazed pastures were often found in proximity. Burning of native grassland has been shown to help maintain a suitable mosaic of grassland heights (Dechant et al. 2002). Although not strictly colonial, it often forms loose colonies in suitable habitat. On migration and before breeding, it often uses plowed or recently seeded arable fields (Houston and Bowen 2001). Most farmland in the northeastern United States is now unsuitable for this species, which more often relies on other marginal areas of grassland, such as airports (especially in Ohio, New Jersey, and New York), and reclaimed strip mines. This species is area sensitive (Horn et al. 2000); therefore, habitat fragmentation by land use changes could have detrimental effects on both habitat loss and reduction in habitat patch size.

### *LIFE HISTORY AND ECOLOGY*

This species depends on grasslands, both in the breeding season and in winter. It winters primarily on the pampas of Argentina, returning to the North American breeding grounds from late March onward, with the main arrival in Pennsylvania during mid-April, at which time flocks of passage migrants may be seen (McWilliams and Brauning 2000). Upland sandpipers leave the breeding grounds in late summer—with most having left by mid-August—although stragglers are occasionally seen into October.

During the breeding season, it occupies a range of grasslands, including meadows, reclaimed strip mines, and golf courses. Nests are usually located in stands of dense grass cover, 15–35 m tall at nest initiation. The young usually move to shorter vegetation, such as grazed pastures, for foraging, but may still use taller vegetation for shelter (Dechant et al. 2002). Grassland invertebrates, such as grasshoppers, crickets, and weevils, make up most of the diet; a few weed seeds are also taken (Houston and Bowen 2001).

Upland sandpipers nest in loose colonies, defending mates from neighbors not showing strong territoriality (Houston and Bowen 2001). The nests are usually initiated two weeks after returning to the breeding grounds; hence, eggs will be laid during May and June. A number of nest scrapes are made but only that chosen for the clutch of (usually) four eggs is lined. Adults become difficult to detect during the twenty-three- to twenty-four-day incubation period. The young are precocial and leave the nest almost immediately on hatching to find suitable foraging habitat in short grass (<10 cm), typically weed-rich grazed pastures. The young fly after around thirty days and become independent at that time.

*THREATS*

Changes in grassland management practices have been the primary cause of the long-term decline in numbers of this species across its geographic range. In Pennsylvania, losses of grassland and changes in grassland management are likely to have contributed to the long-term population decline. Hay-cutting patterns have also changed. At one time, hay was harvested once, either during June or July, but today it is cut two or three times to maximize yields, making the fledging of young impossible.

Between 1982 and 1997, more than 767,000 acres of pasture were lost to development in Pennsylvania (Goodrich et al. 2002). Grasslands have also been lost to woodland, through natural regeneration, and to row crops and "improved" pastures. Grassland patch sizes may have been reduced in some areas because of changes in land use. Traditionally managed hay fields, with only one cut per season, are now less common in Pennsylvania, although they are still found in areas farmed by the Amish. Wilhelm (1995) considered the Amish to be ideal stewards of upland sandpiper habitat in western Pennsylvania, although it should be noted that the species has been extirpated from the counties that hold the largest Amish communities.

Although there is no direct evidence of an effect on upland sandpiper populations, there can be little doubt that extensive pesticide use has greatly reduced numbers of the grassland invertebrates on which this species feeds. Upland sandpiper populations are also likely to be affected by changes in habitat extent and quality on the South American winter grounds. Loss of grasslands and heavy use of pesticides are suggested to have caused severe declines in upland sandpiper numbers wintering on Argentinean grasslands (Hooper 1997, Vickery et al. 2003).

*CONSERVATION AND MANAGEMENT NEEDS*

Retention of the upland sandpiper as a breeding bird in Pennsylvania may ultimately depend on increasing the extent of suitable habitat, by improving existing grasslands and possibly by creating new extensive grasslands in suitable areas. Both measures will require public funds to manage state-owned grasslands or to subsidize management on privately owned grasslands—possibly through the Conservation Reserve Enhancement Program (CREP) or through conservation easements. Management should be targeted, prioritizing areas close to existing upland sandpiper populations. This would include areas where hay fields, strip mines, and CREP fields are found alongside low to moderately grazed pastures. If possible, it would be useful to introduce a management practice through CREP, or another state or federally funded program, designed specifically for upland sandpipers. Such a practice should be made available in areas with contiguous grasslands of not less than 100 ha. Such fields should be managed to create heterogeneous vegetation structure by mowing to a height of 10–15 cm on a three-year rotation.

*MONITORING AND RESEARCH NEEDS*

The Breeding Bird Survey does not adequately monitor the population of this species in Pennsylvania. A long-term monitoring program in key areas is desirable. Such surveys need not be annual but should be repeated every four to five years at minimum. Surveys should be carried out either in the second half of May or during July (Hooper 1997), because adults can be difficult to detect between those times. The second Pennsylvania Breeding Birds Atlas should provide information on the current geographic range of this species within Pennsylvania. However, the atlas will not be sufficient to provide accurate population estimates; hence, detailed surveys of key areas of the state are required.

Most of the current literature on habitat management for this species is based on studies in the Great

Plains and Midwest and notably on prairie habitats. Suggestions for grassland management in Pennsylvania are therefore based on best guesses rather than specific research. Maintaining viable populations of this species in Pennsylvania may require targeted habitat management or habitat creation at a large scale. Management and habitat recreation/restoration should first be considered in areas of the commonwealth where this species still nests. A before-after control-impact experiment of grassland management would provide better information on which to base management advice.

*Author:* ANDY WILSON, PENNSYLVANIA STATE UNIVERSITY

## Short-eared Owl

Order: Strigiformes
Family: Strigidea
*Asio flammeus*

The short-eared owl is a small to medium-sized owl and is termed the rarest and most threatened species of owl in the northeastern United States (Melvin et al. 1989; fig. 5.8). It is considered a Species of Immediate Concern in Pennsylvania's Wildlife Action Plan and is listed as Endangered in Pennsylvania and New Jersey (Brauning et al. 1994). It is a Northeast Region Priority Species. Global populations are Secure (G5, NatureServe 2009).

### *GEOGRAPHIC RANGE*

The short-eared owl is primarily an Arctic species, nesting circumpolar in North America, Europe, Asia, and in scattered locations south to California, Colorado, Illinois, Missouri, Ohio, Virginia, and Utah. It is more numerous in western and central North America than in eastern North America (American Ornithologists' Union 1998). Pennsylvania is at the southern periphery of the short-eared owl's breeding range. The short-eared owl has never been a common breeder in the northeastern United States, but it has declined from historical nesting sites, and is now restricted to Vermont, New York, New Jersey, Massachusetts, and Pennsylvania (Tate 1992).

*Fig. 5.8.* The Short-eared Owl, *Asio flammeus*. Photo courtesy of Vic Berardi.

Short-eared owls winter mostly from the southern parts of most Canadian provinces south to southern Baja California, southern Mexico, the Gulf Coast, and Florida (American Ornithologists' Union 1998). Winter populations fluctuate annually, presumably in response to small mammal populations within the northern winter range.

### *DISTRIBUTION AND RELATIVE ABUNDANCE IN PENNSYLVANIA*

Listed as a state Endangered Species since compilation of the first list (Haas et al. 1985a), the short-eared owl has always been understood to be uncommon as a nesting bird within Pennsylvania. The number of confirmed short-eared owl nesting events in Pennsylvania since Audubon's time (early 1800s) until the first year of the first Breeding Bird Atlas in 1983 totaled less than ten (Brauning 1992a).

Breeding records were confirmed during the first atlas project at the Philadelphia Airport and on Clarion County surface mines (Buckwalter 1988, Brauning 1992a). Since the first atlas, many additional nest sites have been documented. An unusually high number of birds were found during the summer of 1997, when young were documented in Allegheny, Clarion, Jefferson, Lawrence, and Venango counties (McWilliams and Brauning 2000). Most of these reports, and most recent nesting events, appear to be confined to reclaimed surface mines. Since 1997, occasional summer sightings suggest that a small handful of pairs nest annually, with the most consistent records occurring on reclaimed surface mines in southern Clarion County. During the second Breeding Bird Atlas, there were three probable and one confirmed nesting attempts (Mulvihill and Brauning 2009; fig. 5.9). Populations traditionally fluctuate in response to the cycles of microtine rodents, their preferred food base, and declines of these resources in their northern nesting areas may account for years of influx in Pennsylvania.

In the nonbreeding season, groups of birds are regularly found across Pennsylvania ranging in size from single individuals to fairly large numbers (up to ca. 100 birds). They were annually found at some locations, but habitat at many of these areas has been lost to development or changes annually with agricultural practices.

*COMMUNITY TYPE/HABITAT USE*

Primary habitat during all seasons is large, unmowed grassy fields with minimal incursion of shrubs and trees. Most nesting sites in Pennsylvania have been open grasslands of over 100 ha. Broad expanses of open land with annual and perennial herbaceous vegetation for nesting and foraging are required. Habitat types frequently mentioned as suitable include fresh marshes, prairies, grassy plains, old fields, river valleys, meadows, savanna, and open woodland (Clark 1975, Holt and Melvin 1986, Holt and Leasure 1993). In general, any large area with tall grassy vegetation and some dry upland for nesting and that supports suitable prey, may be considered potential breeding habitat, although many such potential sites will not have breeding short-eared owls. Winter roosts have been reported in abandoned dumps, quarries, gravel pits, storage yards, stump piles, old fields, small evergreen groves, bayberry thickets, dunes, and open, abandoned cellars (Clark 1975, Bosakowski 1986).

*LIFE HISTORY AND ECOLOGY*

Although often seen foraging by day over open fields, short-eared owls are primarily crepuscular and nocturnal, active from early evening until dawn (Holt and Leasure 1993). Small rodents form most of the diet, but birds and insects are also taken. Voles are the mainstay of the diet, although shrews (*Sorex* spp., *Blarina* spp.), rabbits (*Sylvilagus* spp.), and even muskrats (*Ondatra zibethicus*) are taken. Microtine abundance dictates winter appearances of this predominantly northern breeder. In years of high vole (*Microtus* spp.) populations, short-eared owls remain to breed south of their traditional geographic range (Clark 1975).

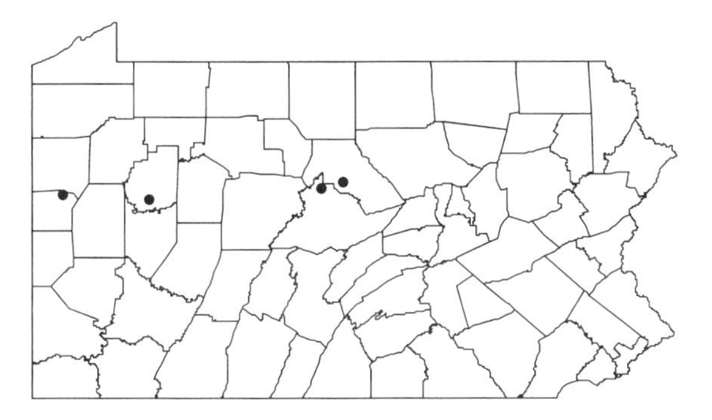

*Fig. 5.9.* Distribution of the Short-eared Owl, *Asio flammeus*, from Breeding Bird Atlas II.

The short-eared owl is unusual among the owls in that it nests on the ground, generally in a slight depression, often beside or beneath a bush or a clump of grass (Terres 1980). Many nests are near water, but generally are on dry upland sites (Pitelka et al. 1955, Holt and Melvin 1986, Tate and Melvin 1988, Combs and Melvin 1989). Sites on reclaimed surface mines in Pennsylvania appear to have no association with wetlands (Buckwalter 1988). The same nest site may be used in successive years.

Courtship generally begins in late March or early April. The breeding season is often reported to commence in relation to vole abundance, with a larger prey population yielding an earlier start to breeding activities (Randall 1925, Snyder and Hope 1938, Lockie 1955, Holt and Leasure 1993). Polygyny may result in two nests within one short-eared owl territory. Two broods are sometimes raised, and, if the nest is destroyed or depredated, the female may renest (Lockie 1955, Holt and Leasure 1993). Short-eared owls are opportunistic, moving into and breeding in areas with high rodent densities. This results in local population fluctuations.

Short-eared owls tend to congregate and roost communally in the winter (Banfield 1947, Craighead and Craighead 1956, Clark 1975), often in sheltered sites near hunting areas. Winter roost concentrations have been found to subsequently support nesting pairs in some instances.

*THREATS*

Habitat loss and degradation is the biggest problem. Short-eared owls have suffered, like many birds associated with grasslands, from a dramatic decline in agricultural acreage and changes in practices on remaining farms (Schwalbe and Schwalbe 1993, McWilliams and Brauning 2000). The species is declining in many parts of the geographic range because of loss and degradation of marshes, grasslands, and low-use pastures. This may be a result of development, changing land use patterns (e.g., farmlands to woodlands or to development), changing farming practices (e.g., hay fields to row crops), reforestation and natural succession, wetland loss, or a combination of these factors. Populations have declined because of reforestation of farmlands and fragmentation and development of coastal grasslands (Holt and Leasure 1993). Loss of open grasslands through natural succession reduces available hunting and breeding habitat.

Short-eared owls also are vulnerable to mammalian predators, populations of which have been augmented because of human-caused increases in food resources. As a ground-nesting bird, short-eared owl eggs and young may fall prey to various mammalian ground predators, such as foxes (*Vulpes* spp.), raccoons (*Procyon lotor*), and mustelids. Predation by the striped skunk (*Mephitis mephitis*) may have caused the extirpation of the short-eared owl as a breeding bird on Martha's Vineyard (Melvin et al. 1989). Domestic cats and dogs have been known to disturb owl nests (Tate 1991). The potential for an increase in the threat of predation or disturbance by domestic or feral cats and dogs may be high. Short-eared owls also have been shot by humans (Bent 1938, Clark 1975).

Prey abundance may be a limiting factor in the owl's distribution and breeding success (Melvin et al. 1989). This owl's reported reliance on microtine rodents emphasizes this specific component (Lockie 1955, Hagen 1969, Clark 1975).

Collisions with large, aerial radio antennas or high-tension guy wires, as well as vehicle collisions, can potentially cause substantial mortality. Clark (1975) reported mortality caused by collisions or entanglement with trains, cars, aircraft, farm machinery, and wire fencing.

### *CONSERVATION AND MANAGEMENT NEEDS*

From a habitat standpoint, land preservation efforts should be aimed at protecting large tracts of open habitat with low vegetation. In areas where short-eared owls nest, observation of hunting flights and territorial displays during the breeding season will help to delineate areas used by resident owls. Any adjacent similar habitat not used by the birds should be considered potential habitat. Areas of approximately 50 ha or larger, containing low, open grasslands or similar habitat with abundant small mammal populations, should also be considered as potential breeding or wintering habitat.

Management of suitable habitat includes maintaining large tracts of open grassland, such as reclaimed surface mines or potentially large abandoned fields. Protection of such habitat is crucial to the persistence of breeding and wintering short-eared owls in Pennsylvania. Loss of such habitat around the Philadelphia International Airport is probably irreversible. Restoration or new establishment of grasslands (e.g., from strip-mined areas) may offer potential new habitat.

Active maintenance of open habitat may benefit short-eared owls where natural succession into woody vegetation occurs. Any management practice used to maintain open habitat and inhibit the growth of woody vegetation, such as mowing or burning, might be used. However, these management practices must be employed outside the nesting season to avoid the destruction of nests, eggs, and young. In addition, care must be taken to allow for adequate buildup of the litter layer that provides habitat for microtine rodents. *Microtus* populations require adequate cover for several aspects of their ecology (Birney et al. 1976). Maintenance of an adequate prey base is essential because distribution and abundance seem to be tied to prey density (Lockie 1955, Clark 1975, Adair 1982, Melvin et al. 1989).

Conservation actions include preparing cooperative agreements with landowners of regularly occupied sites to ensure reduced disturbance during nesting season, promoting reclamation regulation that will establish suitable habitat on newly mined sites, and developing long-term management practices to sustain grassland habitat suitable for short-eared owls and other grassland specialists. Identification of potential sites for mine reclamation into grassland can potentially generate new nesting and wintering habitats.

### *MONITORING AND RESEARCH NEEDS*

Adequate monitoring procedures are needed to assess occupation and population parameters of historic sites. Many of the present monitoring techniques, such as walking large areas in an attempt to flush sitting females, are labor and time intensive and may disturb nesting. Furthermore, some procedures, such as counting birds observed from roadsides, may underestimate the number of individuals present. Short-eared owls are relatively inconspicuous and easily missed. Through diligent observation, the general area of a nesting territory can be ascertained. If one is fortunate enough to observe courtship, then it is likely that the pair will nest in the immediate vicinity. Development and implementation of a standardized population monitoring procedure should be a high research priority. Monitoring and habitat management efforts may be developed for this species in conjunction with other high-priority grassland obligate species, such as the Henslow's sparrow (*Ammodramus henslowii*).

Research needs for this species are closely tied to monitoring and management. Expanded efforts are needed to locate and study local breeding and wintering populations to accurately determine number of breeding and wintering birds; locate regularly occurring populations, which would facilitate long-term ecological studies; and determine more precisely the limiting factors and management needs of these populations. The location of regularly breeding and wintering populations would provide the data necessary for land protection efforts.

Mapping of breeding territories through observing territorial displays during the breeding season allows for a good estimate of territory size (Lockie 1955, Tate 1991). More data are needed concerning relationships between territory size and the abundance of small mammals to determine the amount of open habitat and the prey base needed to support a breeding pair. Such information would enhance the efficacy of land preservation efforts.

Research on the management of open habitat and its effect on prey populations is needed. The effect of such practices as burning, mowing, or plowing on small mammal populations must be considered. Any management for restoring or maintaining open grassland habitat must also manage for a sufficient prey base.

Habitat fragmentation, or isolation, may be important. The high mobility of the owl may allow it to use disjunct areas of habitat, but whether it will occupy an isolated patch of appropriately sized habitat and breed successfully is not known. Wildlife managers seeking to design preserves would need to know to what degree use of fragmented habitat is detrimental to the owl's territorial integrity and breeding success.

*Author:* DANIEL W. BRAUNING, PENNSYLVANIA GAME COMMISSION

## Olive-sided Flycatcher

Order: Passeriformes
Family: Tyrannidae
*Contopus cooperi*

The olive-sided flycatcher is a big boreal pewee similar to the greater pewee (*Contopus pertinax*) and dark pewee (*Contopus lugubris*) in size and habits (fig. 5.10). It is extirpated as a breeding species in Pennsylvania (Brauning et al. 1994, Gross 1998). The olive-sided flycatcher is currently listed as a Species of Conservation Concern by the United States Fish and Wildlife Service for the Bird Conservation Region 28, the Appalachian Mountains (U.S. Fish and Wildlife Service 2002c). Partners in Flight (PIF) considers it a priority species for conservation on the National Watch List and it is a priority for Pennsylvania (Carter et al. 1996). It has more recently been designated as a Watch List species by PIF (Rich et al. 2004). The global population is ranked G4, Apparently Secure (NatureServe 2009).

*Fig. 5.10.* The Olive-sided Flycatcher, *Contopus cooperi*. Photo courtesy of Ted Ardley Victoria, British Columbia, Canada.

### *GEOGRAPHIC RANGE*

The olive-sided flycatcher's breeding range is widespread across mountainous and northern parts of the continent. The breeding range stretches from Alaska and the Rocky Mountains (south as far as northern Baja California) across most of Canada and the northern United States east to the Maritime Provinces, New York, and northern New England. In the Northeast, it is found in the Adirondacks and Catskills of New York; most of Vermont and New Hampshire (Fitchel 1985, Robbins 1994); central and western Massachusetts sometimes in northwest Connecticut; and throughout most of Maine (American Ornithologists' Union 1998, Altman and Sallabanks 2000). This species formerly nested much farther south in the Appalachians than it is currently found (Simpson 1992, American Ornithologists' Union 1998, Altman and Sallabanks 2000). Its wintering ground includes the highlands of Central America south to Panama and northwestern South America (Fitzpatrick 1980, Ridgley and Tudor 1994, DeGraaf and Rappole 1995, Stotz et al. 1996, Altman and Sallabanks 2000).

### *DISTRIBUTION AND RELATIVE ABUNDANCE IN PENNSYLVANIA*

The olive-sided flycatcher is considered extirpated as a breeding bird with no confirmed records of nesting in Pennsylvania since the 1930s (Poole 1964, Gross 1992b, McWilliams and Brauning 2000). Sadly, the report by Hicks of a nest on Hemlock Island, Pymatuning Swamp, on June 15, 1932, is the last confirmed breeding record of this magnificent boreal bird in Pennsylvania (Todd 1940a). Previously, it was an uncommon and locally distributed breeding bird in northern and mountainous parts of the state (Todd 1940a, Warren 1890, E. L. Poole, unpublished manuscript; fig. 5.11). Nearly at the southern edge of its breeding ground in Pennsylvania, it has been found only in higher-elevation forests and wetlands, principally on the Allegheny Plateau but also in the Ridge and Valley Province. The first documented nesting in Pennsylvania was at a lumber yard five miles east of Hazleton, Luzerne County (Young 1896), in an area now decimated by open-pit mining. It also nested at nearby Delano in northeastern Schuylkill County. Its loud whistle was heard commonly in the summer near Lopez (Dutch Mountain), Sullivan County (Stone 1900)—an area broadly known as North Mountain. There were many records of nesting olive-sided flycatchers in both northern corners of the state: the Pocono Mountains (Carter 1904, Harlow 1913, Street 1954) and the northwestern counties, especially Warren (Simpson 1909, Todd 1940a). Olive-sided flycatchers were associated with old growth forest on North Mountain (Sullivan County) and Tionesta forest areas (McKean and Warren counties, Cope 1936); a summer breeder at Little Rouse Pond near Lopez, Sullivan County, and two individuals observed in July 1933 in the "virgin timber" of Tionesta. Dickey (in Bent 1942) found them in the upper portion of the Allegheny River in hemlock and white pine forests. Simpson found six nests over the years, 1904 to 1915, near Warren, mostly in what is now the Allegheny National Forest (ANF; R. B. Simpson, unpublished manuscript). These sites include freshly timbered sites and old forest, often near headwaters.

During the first Breeding Bird Atlas (1983–1989), there were no confirmed nestings, but some summering birds were observed (Gross 1992b). The locations where single summering olive-sided flycatchers have been reported reflect the former breeding range in the state. They include territorial males in a burn near Mehoopany, Wyoming County, in 1988 and 1989; in a forest near Arnot, Tioga County, in 1987; and in a spruce swamp in Lackawanna County in 1987 (Gross 1992b). Other summer birds have been observed at Black Moshannon State Park, Centre County (Breeding Bird Atlas and Pennsylvania Society for Ornithology Special Areas Project); a hemlock swamp in Cameron County; and Game Lands 166, Blair County (Pennsylvania Society for Ornithology Special Area Project). One recent intriguing report of a territorial olive-sided flycatcher came from the old growth hemlock—beech forest of Tionesta Scenic Area of ANF in 1993 (Haney and Schaadt 1994b, D. W. Brauning and D. A. Gross, personal observation). A singing bird persisted into June, was not detected for a period, and then apparently returned (or began singing again). The forest was more open than usual because of an outbreak of the defoliating caterpillar, elm spanworm (*Ennomos subsignarius*). Territorial birds also were observed in the old tornado blowdown area along Cherry Run in Tionesta Scenic Area, where many snags provided foraging opportunities for this flycatcher. There are opportunities for this species to reclaim its former haunts in Pennsylvania.

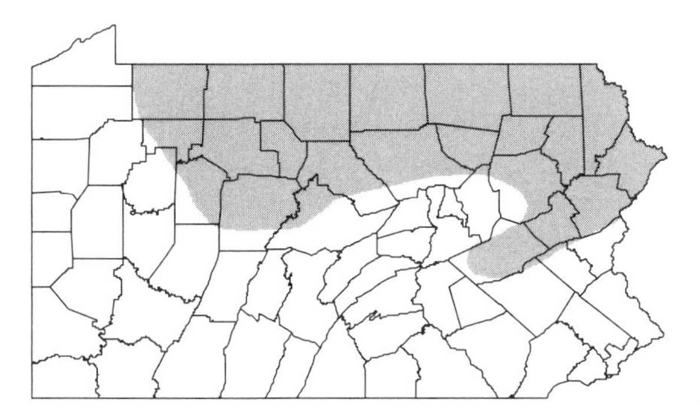

*Fig. 5.11.* Historical distribution of the Olive-sided Flycatcher, *Contopus cooperi*.

### *COMMUNITY TYPE/HABITAT USE*

The olive-sided flycatcher is one of the characteristic birds of the Canadian Life-Zone (Todd 1963). Within the boreal conifer forest and wetlands, it can be found in a diversity of habitats as long as there are prey and perches available (Altman and Sallabanks 2000). Olive-sided flycatchers tend to nest in forests with a low percentage of canopy or openings in otherwise continuous forests (DeGraaf and Rappole 1995, Altman and Sallabanks 2000). In Pennsylvania, it has been confined to higher elevations (more than 500 m, usually more than 600 m). This species is associated with old growth conifer forest but also with forest clearings, sphagnum bogs, burned over forest, swampy lake edges, and beaver meadows (Hall 1983b, Peterson 1988b, Gross 1992b, Haney and Schaadt 1996, Altman and Sallabanks 2000). Territories are characteristically dominated by conifers, including eastern hemlock (*Tsuga canadensis*), red spruce (*Picea rubens*), eastern tamarack (*Larix laricina*), and eastern white pine (*Pinus strobus*). They are most strongly associated with hemlocks in Pennsylvania.

In Canada, olive-sided flycatchers are characteristic

breeding birds of burned-over areas with shrubs and standing snags (Erskine 1977). In western North America, the olive-sided flycatcher is more common in some forests where there have been recent burns, compared with adjacent unburned plots (Altman and Sallabanks 2000). But the pattern in burned versus unburned forests is somewhat inconsistent, depending on the kind of forest. Some of the recent reports of oversummering birds in Pennsylvania have been in recent burned lands. In New York, fire suppression has reduced the availability of the open-forest habitat this species generally prefers in that state (Peterson 1988b). The habitat structure and prey availability may be confounding factors in understanding the relationship between this species and a fire cycle.

*LIFE HISTORY AND ECOLOGY*

Called "courageous" and "brave," the olive-sided flycatcher is one of the feistiest and most tyrannical of the Tyrannidae—prompting one admirer to call it "the peregrine of flycatchers" (Marshall 1988). Males declare their territory loudly from a conspicuous perch from dawn to dusk and defend their turf from many bird species, including raptors. They are particularly intolerant of potential nest predators, such as corvids and sciurids. They also are feisty in migration, making it difficult to interpret territorial behavior in migration or breeding.

Olive-sided flycatchers sally out from perches, often above the canopy or at the forest edge. Among North American flycatchers, it is the only species that forages exclusively by this method (Murphy 1989). They primarily forage from the top one-third of the canopy (Altman 1999b in Altman and Sallabanks 2000). Its diet comprises a wide variety of flying insects, particularly bees, flying ants, and wasps (*Hymenoptera*; Beale 1912, Bent 1942). They also consume flies (*Diptera*), bugs (*Hemiptera*), grasshoppers (*Orthoptera*), moths (*Lepidoptera*), and dragonflies (*Odonata*). Poor weather can severely hamper flycatchers' ability to find and capture their prey, delaying nesting or affecting nestling survival (Altman and Sallabanks 2000).

Olive-sided flycatchers migrate late in the spring and into early summer, so nesting season begins later than for most species; many are still migrating in early to mid-June (McWilliams and Brauning 2000). They have been observed as early as the first week of May, but most migrate through the state later in May, and some straggle through the state in the first and second weeks of June. Fall migration starts early, with some observed in the last week of July but more likely observed in the last week of August. Some travel through as late as early October. Migrants are most frequently observed at ridge-top hawk-watching sites. At one time, fairly large flocks could be observed in migration in Pymatuning Swamp (Sutton 1928a).

Pairs occupy a large territory for a bird of its size, up to 40 to 45 ha (Altman and Sallabanks 2000), but most territories are in the 10- to 20-ha-size range (Altman 1999 in Altman and Sallabanks 2000). Territories invariably include conifers but also maples (*Acer* spp.), aspens (*Populus* spp.), mountain ash (*Sorbus americana*), and other deciduous species. The loss of conifers from parts of its range has been linked to declines in population in parts of its range (LeGrand and Hall 1989, Altman and Sallabanks 2000). Pairs nest only once but will renest if the first attempt fails. Only the female broods the nestlings, but the males help feed the nestlings and fledglings. With its rigorous nest defense, this species is rarely a host for cowbird parasitism (Friedman 1963, Altman and Sallabanks 2000).

*THREATS*

The olive-sided flycatcher has been threatened by the destruction of conifer forests in Pennsylvania and elsewhere on its breeding grounds (Gross 1992b, Altman and Sallabanks 2000). Penn's Woods now contain many fewer conifers than in pretimbering days; most boreal conifer forests were cut before 1900. The populations reported by naturalists and ornithologists in the early twentieth century were only relicts of its former geographic range. The spruce forests of northern Pennsylvania, especially in the Pocono Mountains and North Mountain, were either destroyed or badly fragmented by the time they were visited by naturalists. These forests are still in recovery; red spruces are regenerating well on North Mountain and slowly reclaiming their original range. Hemlock woolly adelgid and other conifer pests greatly threaten conifer forest health in Pennsylvania (Department of Conservation and Natural Resources 2005, U.S. Department of Agriculture 2005). Logging is not a great threat to this species, which can thrive in cutover areas and forest edges (Altman and Sallabanks 2000). However, poor timbering practices, such as the lack of buffers around wetlands and riparian forests or not leaving some snags, conifers, and standing timber in cutover areas, are problematic for species such as the olive-sided flycatcher, which rely on trees for perching and nesting. Other factors, such as excessive deer browse and atmospheric acid deposition, may further stress these forests, slowing tree regeneration and decreasing vegetative diversity (Gross 2003). Little is known about the effects of pest control measures in forests on insectivorous birds such as the olive-sided flycatcher. The lack of a fire regime may also affect the chances for this species to nest in the state once again.

*CONSERVATION AND MANAGEMENT NEEDS*

The olive-sided flycatcher is one of the most conspicuous birds addressed in the boreal songbird initiative (Bird Studies Canada 2005, Blancher and Wells 2005). Although this initiative concentrates on the vast Canadian boreal forest, it has brought needed attention to boreal forests in general, including those of the Appalachians. Boreal forest persists in patches down the Appalachian Mountains, including Pennsylvania. Forest practices should be implemented that favor conifer regeneration and protect high-elevation wetlands and streamside habitat. These practices also would benefit other wildlife of conservation and recreational value, including snowshoe hare (*Lepus americanus*), northern flying squirrel (*Glaucomys sabrinus*), ruffed grouse (*Bonasa umbellus*), yellow-bellied flycatcher (*Empidonax flaviventris*), blackpoll warbler (*Dendroica striata*), and Canada warbler (*Wilsonia canadensis*). Low conifers provide thermal cover for wildlife in winter. As hemlocks are increasingly threatened by pests (Department of Conservation and Natural Resources and U.S. Department of Agriculture web sites), spruce deserves more attention as a conifer cover. Wetlands need more protection in the Pennsylvania highlands, an achievable goal where watershed and conservation organizations are active and agencies are important landowners. Headwater swamps and bogs should be given high priority for acquisition and protection by state agencies, watershed organizations, and land trusts.

There have been widespread declines in its populations throughout its geographic range (Sauer et al. 2005). The decline of this species in the center of its breeding range suggests that decline is occurring on its wintering ground or migration stopovers (Rappole and McDonald 1994), but there also may be issues concerning the health and extensiveness of its breeding habitat. One of the greatest threats may be in the South American and Central American highlands where this species spends its winters. Conservation of highland and foothill forests in Panama and northwest South America may be the most important component for managing this species. In addition to the olive-sided flycatcher, a suite of species that winter in this region are undergoing severe declines, including the golden-winged warbler (*Vermivora chrysoptera*) and cerulean warbler (*Dendroica cerulea*; Robbins et al. 1992, DeGraaf and Rappole 1995, Stotz et al. 1996). Therefore, a trans-hemispheric approach will be necessary for an effective conservation strategy for this species.

*MONITORING AND RESEARCH NEEDS*

The rarity of this species makes inventory and monitoring a difficult challenge because so little has been done to test monitoring. Off-road area searches in appropriate habitats are more likely to be fruitful than randomized surveys and the most likely way to find this species. Initial observations need to be followed by repeated visits to confirm nesting. Searches in burned-over areas, conifer forests, and high-elevation wetlands are most likely to be profitable. In particular, the following areas are advised for regular searches, perhaps as part of inventory and monitoring of birds in these locations, especially where hemlock or spruce forest and swamps are found: the Pymatuning region; Allegheny National Forest, especially the Tionesta scenic area; Sproul State Forest, especially the burn area; Black Moshannon State Park; North Mountain, including SGL 13, 57, 66, and Ricketts Glen State Park; Mehoopany Mountain, Wyoming County; Pocono Lake and Long Pond area, including Two Mile Run and Adam Swamp; Lackawanna State Forest, Delaware State Forest; and higher-elevation locations in Important Bird Areas. As a confirmation of absence, an audio-lure check (tape-playback) is a good way to ensure complete coverage of appropriate habitat, but members of *Contopus* (pewees) often do not vocalize in the nestling stage (D. A. Gross, personal observation).

If it still nested in the state, this conspicuous flycatcher could be detected and monitored by fairly standard methods like those used for monitoring birds by the U.S. Geological Survey, the Breeding Bird Atlas, and Important Bird Areas Project. Standard point count, area searches, and spot-mapping techniques should be fairly effective, especially with modern GPS technology (International Bird Census Committee 1970, Bibby et al. 1992, Ralph et al. 1993). The Mountain Birdwatch project that focuses on boreal Species of Conservation Concern (Lambert 2005) should be expanded into Pennsylvania, with this species as one of its targets. Training volunteers is key to monitoring rare species successfully as has been accomplished at Audubon birding workshops. This kind of training should be continued and expanded. This PIF Watch List species should be a priority species for the Nature Conservancy landscape conservation initiatives and for Pennsylvania Game Commission and Department of Conservation and Natural Resources land management in northern and highland Pennsylvania.

For Pennsylvania, the greatest need is to locate any breeding pairs and document the occurrence and location's attributes. Like other boreal species, its populations in the Appalachians are few and isolated, perhaps a regional metapopulation. On a regional basis, the genetic flow and colonization patterns in these eastern populations deserve study, perhaps of birds in migration. There is a lack of information on productivity and associated environmental attributes, including landscapes, management techniques, pest control, and silvicultural treatments (Altman and Sallabanks 2000). Its relationship with disturbance regimes caused by forest fire and beavers also is poorly known and deserves more study. The great mystery for this species is its wintering ground ecology and mortality as related to habitat, forest alterations, and management practices.

*Author:* DOUGLAS A. GROSS, PENNSYLVANIA GAME COMMISSION

## Loggerhead Shrike

Order: Passeriformes
Family: Laniidae
*Lanius ludovicianus*

The loggerhead shrike is a large-headed predatory passerine with a hooked bill (Yosef 1996; fig. 5.12). Pennsylvania has considered the loggerhead shrike as either an Extirpated or an Endangered Species for many years (Gill 1985, Brauning et al. 1994, Brauning and Siefkin 2005). This species is ranked Apparently Secure (G4) at the global level (NatureServe 2009). Loggerhead shrikes have been extirpated in most of the Northeast and are nearly gone from Minnesota, Michigan, and Wisconsin. This species is also currently undergoing serious declines in the southeast. It is listed as a Critical Recovery Species for the Appalachian Mountain Bird Conservation Region (Partners in Flight) and is listed as a Migratory Nongame Bird of Management Concern by the United States Fish and Wildlife Service (Yosef 1996).

*Fig. 5.12.* The Loggerhead Shrike, *Lanius ludovicianus*. Photo courtesy of Shawn Carey.

### *GEOGRAPHIC RANGE*

This species has an extensive breeding range, but the range is shrinking at an accelerating rate (Yosef 1996). The historic breeding range included most of the continental United States where suitable habitat was available. Loggerhead shrikes are year-round residents in the southern United States and Mexico and are migratory in the northern part of their range.

### *DISTRIBUTION AND RELATIVE ABUNDANCE IN PENNSYLVANIA*

This species historically bred throughout the Northeast, including Pennsylvania, but now only remnant and fragment populations remain (fig. 5.13). Loggerhead shrikes have bred primarily in western counties, including Greene, Allegheny, Huntingdon, Lawrence, Mercer, Crawford, and Erie, but mostly the latter three counties (Todd 1940a, Brauning et al. 1994). These were fairly small and localized populations, but it was considered locally common in Erie and Crawford counties (Warren 1890). After thirty-four years without documentation of nesting in the state, loggerhead shrikes were confirmed nesting in Adams County in 1992 (Brauning et al. 1994). This seems to be a short-time influx of shrikes into south-central counties that has not continued. In surveys conducted in June 2000, no breeding loggerhead shrikes were found in Franklin and Adams counties, areas suspected of still containing populations of this species (Brauning and Siefken 2005a).

In Pennsylvania, it is suspected that this species is essentially gone as a breeding species (Brauning and Siefken 2005a). Historically, this species bred throughout the Northeast, possibly due to clearing of virgin forests in the mid-1800s and replacement by open farmlands (Yosef 1996). The clearing of forests and newly open farmlands created habitat for early successional species, such as loggerhead shrikes. However, for

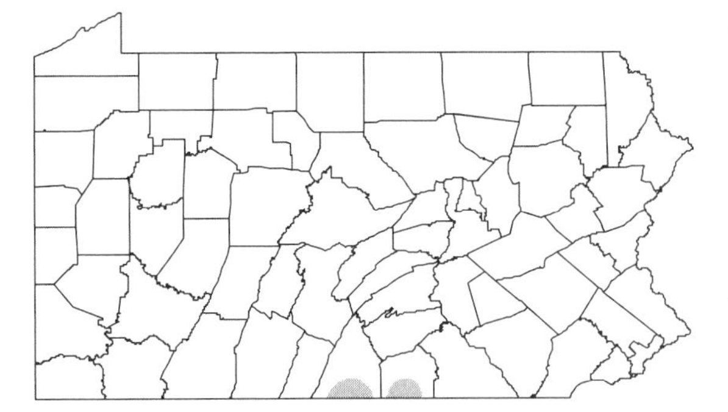

*Fig. 5.13.* Last confirmed nesting locations of the Loggerhead Shrike, *Lanius ludovicianus*.

unknown reasons, this species has almost completely disappeared from New England and almost the entire Northeast. The remaining reliable, although declining, population continues in southeastern Ontario, where active recovery efforts continue. Conversion of pasture farms back to row crop farms, woodlands, suburbs, and other development is possibly the causes of near extirpation of loggerhead shrikes in the Northeast. However, unoccupied suitable habitat remains in the Northeast where this species once bred (Yosef 1996). Future outlook for this species is not good in Pennsylvania and the Northeast, especially because loggerhead shrikes are continuing to decline farther south into Virginia, Maryland, and the Southeast. This species is still relatively common in the south-central states, and in the Southwest.

## *COMMUNITY TYPE/HABITAT USE*

Loggerhead shrikes are birds of open country with short grasses and forbs of low stature interspersed with bare ground and shrubs or small trees (Bent 1950, Yosef 1996, Dechant et al. 1998a). This species will use pastures with fencerows, old orchards, mowed roadsides, sagebrush (*Artemisia* spp.) desert, riparian areas, open woodlands, farmsteads, suburban areas, mowed road rights of way, abandoned railroad rights of way, cemeteries, golf courses, and reclaimed strip mines (Bent 1950, Yosef 1996, Dechant et al. 1998a). Breeding loggerhead shrikes usually settle near isolated trees or large shrubs and scattered trees or shrubs, particularly thorny or thick species, serve as nesting substrates, impaling stations, and hunting perches (Yosef 1996, Dechant et al. 1998a). Abundance of open habitat, foraging areas, and elevated perch sites has been found to be the most important factors in habitat suitability in the upper Midwest (Dechant et al. 1998a). It has been speculated that rainfall is a negative correlate with the nesting range of this species (Bent 1950), which explains the difficulty of this species to establish long-term viable populations in the moist landscape of the Northeast.

In much of this species' breeding range, hawthorns (*Crataegus* spp.), osage orange (*Maclura pomifera*), crabapples (*Malus* spp.), and red cedar (*Juniperus virginiana*) have been cited as important low tree types that loggerhead shrikes use as nesting substrates, impaling stations, and hunting perches (Warren 1890, Bent 1950, Dechant et al. 1998a, Brauning and Siefken 2005a). In Pennsylvania, this species historically bred in open country and prefer pasture and cropland with scattered trees and hedgerows, with hawthorns and crabapples as selected nesting trees (Brauning 1992a). In addition, red and white cedar (*Chamaecyparis thyoides*) were also regarded as important trees for nesting (Brauning 1992a). No information is known in habitat selection of loggerhead shrikes during migratory periods (Yosef 1996). However, during the wintering season, habitat selection does not seem to differ from summer habitat, although hay fields and idle pastures are heavily used. One study in Virginia showed that shrike individuals move from pastures to shrub and open forest habitats, which may provide more cover and food during periods of wet, cold weather (Blumton 1989).

## *LIFE HISTORY AND ECOLOGY*

In Pennsylvania, loggerhead shrikes are short-distance migrants (Yosef 1996). Special ecological requirements for this species include an abundant amount of prey items that include small birds, small rodents, and large insects (Yosef 1996, Dechant et al. 1998a). Small prey may be impaled before eating, but large invertebrates and vertebrates are always impaled before eating. Insects usually make up the majority of this species diet (Yosef 1996).

Compared with most passerines, loggerhead shrikes are early nesters (Yosef 1996). Second or even third broods are sometimes attempted after the initial brood has fledged. A few records have been documented of brood parasitism by brown-headed cowbirds, but generally, shrikes will chase cowbirds that approach a nest containing eggs (Yosef 1996). Territories of this species are usually about 6–9 ha in size (Dechant et al. 1998a). Loggerhead shrikes generally hold larger territories than other insectivorous passerines of similar size, possibly because of a function of specialized foraging behavior (Yosef 1996). Territories abandoned during the breeding season are usually not occupied until the next season, indicating that "floaters" are rare. In the eastern United States, territories include more land in pasture and less land in old field than expected at the macrohabitat level and include greater lengths of fences, utility wires, rights of way, and water sources. Another study has shown that territories within permanent pasture were more likely to be reoccupied the following year than territories located in areas where tall grasses overtook pastures by the end of the summer (Yosef 1996).

*THREATS*

The loggerhead shrike is in sharp decline in many areas of its breeding range, especially east of the Mississippi River (NatureServe 2005c). It is almost extirpated from the Northeast and north-central part of the United States and continues to decline precipitously in parts of the middle Atlantic and southeast (NatureServe 2005). Part of the decline can be attributed to reforestation and loss of open habitat, and thus return to presettlement conditions when loggerhead shrikes were probably absent from much of the heavily forested northern states (NatureServe 2005). However, the decline of this species has proceeded beyond what can be explained by sheer habitat loss, and much suitable habitat remains unoccupied in the northern states. In most of the breeding range, this species is declining, even in areas where open habitat is abundant. Possible reasons for the declines include pesticides, deteriorating winter habitat quality, and high predation pressure on roadside habitats. Locally, mortality from vehicle collisions may be significant, possibly because of low flights undertaken by shrikes for foraging opportunities (Yosef 1996, NatureServe 2005). Many of the perches that this species uses are concentrated along the highway. It has also been postulated that open habitats near human settlements attract feral cats and other predators, which may cause significant mortality during the breeding season (Yosef 1996). Pesticides and other contaminants may also have a negative effect on shrikes (Yosef 1996). However, contaminants' role in the species' decline remains unknown.

Loss of habitat on both the wintering grounds and breeding grounds still may be the primary factor in the species decline (Yosef 1996). Wintering areas along the Gulf Coast have lost much suitable habitat. In many areas of the breeding range, change in agricultural usage from hay fields and open pasture to intensive row crops has reduced the amount of available breeding habitat (Kruse and Smith 1992, Yosef 1996). In addition, hedgerows and shrubs have been eliminated in many areas, reducing the number of perches and breeding sites for this species (Grubb and Yosef 1994). The threat of suburban development remains high in many areas where open habitat is replaced by structural development (Yosef 1996).

*CONSERVATION AND MANAGEMENT NEEDS*

Although shrikes appear to be currently extirpated from Pennsylvania, measures should be undertaken to protect remaining suitable habitat of this species particularly in counties where the shrike formerly nested. This includes retaining old field and pasture habitat and providing potential perch sites and suitable nest trees criteria. Other management options include discouraging multiflora rose (*Rosa multiflora*) and encouraging other thorny shrubs, such as hawthorns or tall isolated shrubs (NatureServe 2005). In addition, adding hunting perches can enhance habitat (NatureServe 2005). Management of reclaimed strip mines for loggerhead shrikes should be encouraged in potential breeding areas. Biocides should be discouraged at potential breeding sites. Because loggerhead shrikes' reproductive success is relatively high, this species might be able to expand its existing population if the cause or causes of the declines are identified and eliminated (NatureServe 2005c). Reintroduction of loggerhead shrikes to the Northeast and upper Midwest has also been proposed and should be evaluated further.

In Pennsylvania, all historical and recent shrike population areas should be maintained and possibly expanded. This could provide larger areas that would support a population of loggerhead shrikes if future expansion of the northern Virginia population occurs. Continuing management of this species and possible reintroduction to these areas is feasible.

*MONITORING AND RESEARCH NEEDS*

In Pennsylvania, the search for breeding loggerhead shrikes should continue although this species has possibly not bred in the state for several years. Surveys of areas that historically (Franklin, Cumberland, Adams, and possibly York and Lancaster counties) held this species should be conducted, and intensive research and monitoring should occur if breeding pairs are located.

A region- or nation-wide study should be designed to determine causes of declines of this species. This study should be nationwide but site specific and should take place on the breeding grounds, wintering grounds, along the migratory route. This study should emphasize landscape analysis, biocide application in shrike habitats, mortality along roadsides, and habitat changes and preferences in all seasons, and it should identify causes of declines and ways to reverse declines. Ultimately, a long-term nationwide conservation recovery plan should be produced.

*Author:* FREDERICK C. SECHLER, JR., NEW YORK NATURAL HERITAGE PROGRAM, THE NATURE CONSERVANCY

## Sedge Wren

Order: Passeriformes
Family: Troglodytidae
*Cistothorus platensis*

The sedge wren is a small, nondescript wren (fig. 5.14). Global populations are considered Secure (G5), but breeding populations in Pennsylvania are Critically Imperiled (S1B, NatureServe 2009). The sedge wren is listed as Endangered in Pennsylvania.

### *GEOGRAPHIC RANGE*

Sedge wrens breed from Alberta across southern Canada to Quebec; south to northeastern Montana, the Dakotas, and eastern sections of Kansas and Nebraska to Missouri, and northeast from northwestern Kentucky to northwestern Vermont. Breeding and summering may also occur locally east of this geographic range to northeastern New Brunswick and south through eastern states to Virginia, West Virginia, and Arkansas and sporadically elsewhere (Herkert et al. 2001). Wintering occurs along the Atlantic Coast from Maryland through Florida and east into most of Texas (McWilliams and Brauning 2000). Some birds also winter in northeastern Mexico (Sibley 2000).

### *DISTRIBUTION AND RELATIVE ABUNDANCE IN PENNSYLVANIA*

The sedge wren is sparsely and sporadically distributed throughout Pennsylvania (McWilliams and Brauning 2000). It is a very rare breeder in the state. Judging its abundance is confounded by the unpredictable nature of its movements, and its low fidelity toward any particular location. During the first Pennsylvania Breeding Bird Atlas (BBA 1983–89), it was detected in only thirteen blocks (less than 1% of the total), with four "possible," six "probable," and three "confirmed" breeding records (Leberman 1992f; fig. 5.15). The most regular recent sites are in Crawford and Butler counties (McWilliams and Brauning 2000). A review of *Pennsylvania Birds* issues from 1996 to 2008 resulted in fifty sedge wren observations reported between May 1 and October 1. Many of these sightings were of single birds, and only eight reports indicated likely or confirmed breeding activity. It appears to be absent from much of its potential geographic range in the state, even where suitable habitat exists (McWilliams and Brauning 2000).

Documented historic breeding occurrences are limited but leave little doubt that its numbers have declined in Pennsylvania. Warren (1890, 311) considered it a "regular but apparently rare" summer resident, and reported nests, based on the observations of others, in Bucks, Crawford, and Lancaster counties, and on a Lehigh River Island. Sutton (1928a, 227) considered it a "rather rare and very local migrant and summer resident at Pymatuning Swamp and Conneaut Lake." Fricke (1930) located a colony and reported a nest at two Pymatuning Swamp locations in May 1930. Todd (1940) located additional nesting colonies in this region

*Fig. 5.14.* The Sedge Wren, *Cistothorus platensis*. Photo courtesy of Clint Murray.

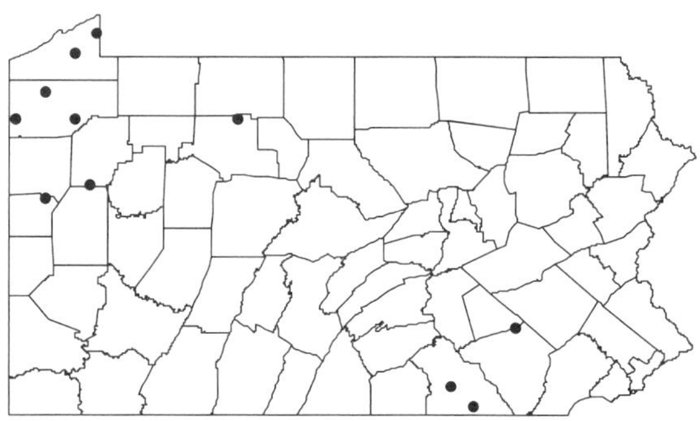

*Fig. 5.15.* Distribution of the Sedge Wren, *Cistothorus platensis*, from Breeding Bird Atlas I.

between 1930 and 1937 and thought the sedge wren was increasing in numbers. He also reported regular nesting activity on Presque Isle, Erie County, and noted possible breeding based on the presence of birds after the middle of May in Centre, Mercer, Somerset, and Washington counties. Poole (1964) added records for Luzerne and Monroe counties.

### *COMMUNITY TYPE/HABITAT USE*

The sedge wren nests in dense, tall growths of sedges and grasses in wet meadows, hay fields, retired croplands, and upland pond and lake margins, as well as in coastal, brackish marshes. Typical nesting habitat features scattered shrubs and an absence of standing water (Gibbs and Melvin 1992c). In Pennsylvania, it may also nest among clumps of sweetflag (*Acorus*), and one small colony used habitat more typical for marsh wrens during the first BBA (Leberman 1992f). It prefers dense, lush, and undisturbed hay fields and has responded favorably where hay lands have been idled as a result of enrollment in the Conservation Reserve Program in the Midwest (Dechant et al. 1998b). Sedge wrens are highly sensitive to habitat conditions and abandon sites that become too dry or too wet, or where shrubs become too prevalent (Gibbs and Melvin 1992c). Shrub densities are important in some locations. In Minnesota, average vegetative characteristics of sedge wren territories, expressed in stems per square meters, included 303 sedge stems, 16 forb stems, and 50 shrub stems, with a predominant vegetation height of 1.1 m. Wrens occurred in higher densities where shrub densities were moderate versus dense or sparse (Niemi and Hanowski 1984, Niemi 1985).

### *LIFE HISTORY AND ECOLOGY*

Sedge wrens are secretive and unpredictable in their movements. The first spring arrivals may appear in the last week of April but more typically during the first or second week of May. Migrants may continue to be seen until at least the first week in June. Migration occurs during September and early October, although sightings during July and August may represent early migrants, and two mid- to late November sightings have been reported (McWilliams and Brauning 2000). Information on the sedge wren's diet is limited but includes spiders and insects (Warren 1890, Howell 1932). The young are fed moths, spiders, mosquitoes, flies, grasshoppers, and bugs (Walkinshaw 1935).

Sedge wrens are typically colonial nesters but may also nest as lone pairs. Males establish territories used for courting, nesting, and much of their foraging and build multiple "dummy" nests that figure prominently in courtship and may serve as dormitories and decoys for predators. Nests are constructed of sedges and fine grasses woven into a globular mass, with the entrance positioned on the side. They are located in thick vegetation less than 0.5 m above the ground (Walkinshaw 1935). Nests with eggs have been found in Pennsylvania from late May through July, and birds occasionally set up territories in August (McWilliams and Brauning 2000). Sedge wrens are characterized by high mobility during the breeding season and low site tenacity between seasons (Burns 1982). At most breeding sites in Pennsylvania, nesting is not observed in subsequent years (McWilliams and Brauning 2000).

### *THREATS*

The greatest known threat to Pennsylvania's sedge wrens is loss or degradation of wetland and grassland habitat. The birds have also been affected by intensive mowing practices typical of modern agriculture (McWilliams and Brauning 2000). Sedge wrens use "drier" wetlands (wet meadows and margins) and grasslands in Pennsylvania. The state has lost more than 50 percent of its original wetlands, and much of the remaining acreage has become fragmented or reduced in quality and function (Tiner 1990). Those wetland types preferred by sedge wrens are the most easily drained or otherwise modified by agricultural or developmental interests.

### *CONSERVATION AND MANAGEMENT NEEDS*

Although all the factors that may contribute to the sedge wren's rarity in Pennsylvania have not been identified, it should benefit from the stringent protection of remaining wetlands, especially sedge meadows, and from implementation of agricultural programs, such as the Conservation Reserve Enhancement Program, that idle grasslands within the state. Such programs and safeguards should be strongly supported by the conservation community and government agencies. Conservation efforts should also include protecting sedge wren habitat from disturbances, such as mowing, burning, and grazing during the breeding season.

### *MONITORING AND RESEARCH NEEDS*

Where breeding sedge wrens are located, sites should be monitored for five consecutive years to determine degree of site fidelity, demographics, and changes in habitat characteristics. A comprehensive status survey of the sedge wren in Pennsylvania is desperately needed. All known locations where breeding activity has been confirmed during the past twenty years should be surveyed during three periods—early June, late June–early July, and late July, for three consecutive years. Habitat characteristics at surveyed sites should be evaluated, especially at the more reliable locations, in an effort to predict potential habitat elsewhere.

*Author:* ROBERT W. CRISWELL, PENNSYLVANIA GAME COMMISSION

### HIGH-LEVEL CONCERN

## American Bittern

Order: Ciconiformes
Family: Areidae
*Botaurus lentiginosus*

The American bittern is a cryptically colored brown heron (fig. 5.16). It is listed as a State Endangered Species, and the Pennsylvania breeding population is considered Critically Imperiled. As a result of population declines, American bitterns are listed as a Species of Greatest Conservation Need on all state lists in the Northeast, and American bitterns were listed as a migratory nongame bird of management concern by the United States Fish and Wildlife Service because of a downward trend in their population (U.S. Fish and Wildlife Service 1987). The bittern's global ranking is G4, Apparently Secure (NatureServe 2009).

### *GEOGRAPHIC RANGE*

The breeding range of the American bittern extends across the mid–United States northward and across Canada. Breeding is discontinuous in the southern portion of its range. They are associated primarily with inland, freshwater emergent marshes; historically, their range may have shifted northward, tracking the wetlands created by retreating glaciers (Gibbs et al. 1992a). They winter in marshes along the Pacific and Atlantic coasts and in marshes in the southern United States.

Fig. 5.16. The American Bittern, *Botaurus lentiginosus*. Photo courtesy of Alan D. Wilson, naturespicsonline.com.

### *DISTRIBUTION AND RELATIVE ABUNDANCE IN PENNSYLVANIA*

American bitterns are rare breeders in Pennsylvania. During the first atlas (1983–1988), they were reported in 53 of 4,928 atlas blocks, with 70 percent of the probable and confirmed records coming from the glaciated section of northwestern Pennsylvania (Brauning 1992a; fig. 5.17). They are regular breeders only in the largest wetlands in Crawford County, and since 1980, breeding has been confirmed in only eight counties (Crawford, Delaware, Erie, Lackawanna, Lawrence, Monroe, Potter, and Sullivan; McWilliams and Brauning 2000).

American bitterns were listed as Threatened in Pennsylvania from 1979 to 1997 and then downgraded to Endangered in 2000 (Genoways and Brenner 1985, McWilliams and Brauning 2000). They have a very patchy distribution and are more abundant on Breeding Bird Survey (BBS) routes in Canada than in the United States, suggesting a latitudinal trend in abundance (Gibbs et al. 1992a). Few quantitative data are available to assess past trends in abundance and distribution of American bitterns, but the general consensus is this species has declined (Gibbs and Melvin 1992a, McWilliams and Brauning 2000).

### *COMMUNITY TYPE/HABITAT USE*

In Pennsylvania, American bitterns prefer extensive freshwater wetlands, especially ones with dense stands

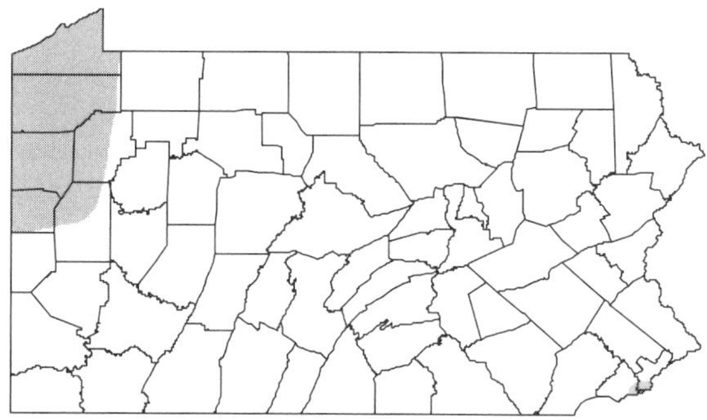

Fig. 5.17. Historical distribution of the American Bittern, *Botaurus lentiginosus*.

of cattails, spatterdock, bulrushes, and sedges interspersed with open water (Brauning 1992b). They occasionally use marsh areas adjacent to streams, bogs, or wet meadows (Brauning 1992b). They are more abundant on larger than on smaller wetlands and are considered to be area sensitive (Cashen and Brittingham 1998). They breed only in large marshes and are regular breeders only within the largest marshes in Crawford County (McWilliams and Brauning 2000). However, American bitterns will use smaller wetlands during migration (Cashen and Brittingham 1998, McWilliams and Brauning 2000). Gibbs et al. (1992a) suggest that they prefer impoundments and beaver-created wetlands over wetlands of glacial origin; however, there is no clear evidence of this trend in Pennsylvania. In Pennsylvania, sites regularly used for breeding, such as Conneaut and Geneva marsh (Crawford County), are large emergent marshes. In comparison to least bitterns, American bitterns use a wider variety of wetland cover types, less densely vegetated sites, shallower water depths, and exclusively freshwater habitats (Gibbs et al. 1992a). Wetlands cover 2.5 percent of the area in the state (Goodrich et al. 2002), and emergent wetlands, potential breeding habitat for American bitterns, make up about 10 percent to 20 percent of that 2 percent. Consequently, available breeding habitat is rare and often in small isolated patches.

### *LIFE HISTORY AND ECOLOGY*

American bitterns are cryptically colored solitary birds that breed in freshwater emergent marshes (Gibbs et al. 1992a). They are crepuscular and active primarily at dawn and dusk. They are carnivorous predators that feed along the shoreline and marsh edge and prey on insects, amphibians, crayfish, and small fish, relying on stealth to catch their prey (Gibbs et al. 1992a). American bitterns are migratory birds arriving in Pennsylvania from early April to mid-May (McWilliams and Brauning 2000). American bitterns breed from May through early July (McWilliams and Brauning 2000). They are thought to be monogamous or possibly polygynous (Gibbs et al. 1992a). They are presumed to be single brooded with the female building the nest, incubating the eggs, and caring for the young. The nest is made of sticks, grasses, sedges and placed among dense emergent vegetation or occasionally over dry ground in adjacent grassland areas (Ehrlich et al. 1988, Gibbs et al. 1992a). Dates of fall migration are not clear but appear to occur primarily in August to September, with stragglers reported into November (Wood 1979).

### *THREATS*

For American bitterns, as for many other species associated with wetland habitat, the primary threat is loss and degradation of emergent wetland habitat. In Pennsylvania, more than 50 percent of historic wetlands have been lost, and many of the remaining areas are degraded (Goodrich et al. 2002). Loss of emergent wetlands in Pennsylvania exceeds both the regional and national average (Tiner 1990, Goodrich et al. 2002). From 1982 to 1992, 91 percent of wetland loss in the Northeast was due to development (Goodrich et al. 2002). Degradation of wetland habitat by runoff, pollution, and acidic deposition may also be problems, but there is no research looking specifically at how these pollutants affect American bitterns. Runoff from agricultural chemicals may significantly and indirectly affect this species through its effect on prey populations of aquatic insects, crayfish, and amphibians (Gibbs et al. 1992a). Loss of emergent wetlands from conversion to lakes, ponds, and reservoirs has been a problem in the past and probably continues to be (Tiner 1990, Goodrich et al. 2002). Lowering of the water table as a result of suburban and urban development is a problem in other parts of the Northeast and may also be a problem in Pennsylvania (S. M. Melvin, personal communication).

### *CONSERVATION AND MANAGEMENT NEEDS*

A primary conservation and management need for American bitterns and other wetland-associated species is to minimize habitat loss and degradation and to increase amounts available through habitat restoration where possible. Creation and management of impoundments, specifically for American bitterns and associated wetland species, should also be attempted. In addition, we need to identify where American bitterns are currently breeding within the state and develop site-specific conservation and management plans for those sites.

### *MONITORING AND RESEARCH NEEDS*

Multispecies marsh bird survey protocols have been developed by Conway (2004). These surveys are designed to monitor both the birds and the habitat. Data are sent to a centralized location so that regional and national databases can be developed. The Pennsylvania Game Commission and other public agencies with wetland habitat on their land should consider developing and implementing a statewide inventory and monitoring program using the survey protocols. Wetland surveys conducted during the second Pennsylvania Breeding Bird Atlas will be useful in identifying wetlands with breeding bitterns. In addition, all historically occupied sites should be surveyed.

A primary research need is to determine the abundance and distribution of American bitterns in natural and restored wetlands. On sites where bitterns are present, biologists need to evaluate habitat requirements of American bitterns and develop a fine-scale habitat requirement model to better predict potential habitat suitability. Additional research at these sites could be conducted to characterize geographic home range size, extent to which bitterns use multiple wetlands, site fidelity, annual survivorship, life span, and age of first breeding. Much of this information could be collected through radio telemetry and banding studies. However, because of the very low abundance of American bitterns within the state, this is probably not a high priority at this time.

*Author:* MARGARET C. BRITTINGHAM, PENNSYLVANIA STATE UNIVERSITY

## Bald Eagle

Order: Falconiformes
Family: Accipitridae
*Haliaeetus leucocephalus*

*Fig. 5.18.* The Bald Eagle, *Haliaeetus leucocephalus*. Photo courtesy of Alan D. Wilson, naturespicsonline.com.

The bald eagle (*Haliaeetus leucocephalus*) is one of America's best-known wildlife symbols (fig. 5.18). It was recently removed from the federal list of Threatened and Endangered Species (U.S. Fish and Wildlife Service 2007a) but is currently listed by the Pennsylvania Game Commission as a Threatened species. Bald eagles are listed as a Species of Greatest Conservation Need by all Northeastern states. Global populations are Secure (G5, NatureServe 2009).

### *GEOGRAPHIC RANGE*

Bald eagles breed near seacoasts, rivers, and large lakes from central Alaska, northern Yukon, east across northern Canada, south to Baja California, the Gulf of Mexico, and Florida; the species winters generally throughout the breeding range most frequently from southern Canada southward (American Ornithologists' Union 1983). Historically, bald eagles bred in a broad band across much of Canada, but were restricted to coastal areas and major river systems in the United States. Following the drastic decline in the 1900s, the species was considered Threatened or Endangered throughout its range, except in Alaska (Stallmaster 1987). Bald eagles occur widely throughout North America in migration.

### *DISTRIBUTION AND RELATIVE ABUNDANCE IN PENNSYLVANIA*

As recently as 1985, the only nesting bald eagles in Pennsylvania occurred in Crawford County with only three or four pairs experiencing sporadic nesting success (Haas et al. 1985b). In recent decades, the breeding population has increased, spreading to major rivers and lakes. Man-made lakes have produced shallow flat-water habitat that this species uses for foraging. The bald eagle is now distributed throughout Pennsylvania in suitable habitat, and its recovery is one of the great success stories of modern conservation. Bald eagle nests have occurred in forty-six of Pennsylvania's sixty-seven counties between 1996 and 2008 with the greatest concentration in Pennsylvania's major river corridors in the northeast and southeast regions and the wetland complex in the northwestern corner (Gross 2009) (fig. 5.19). Results of the Second Pennsyl-

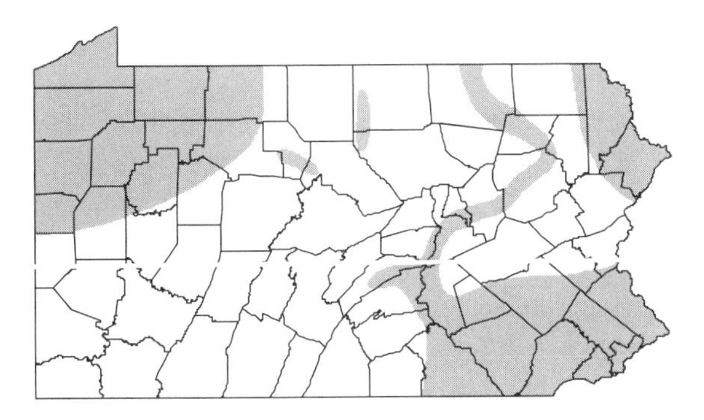

*Fig. 5.19.* Highest probability of occurrence of the Bald Eagle, *Haliaeetus leucocephalus*.

vania Breeding Bird Atlas indicate an increase of 800 percent in distribution (from 52 blocks to 423 blocks) has occurred since the first atlas (Mulvihill and Brauning 2009). In 2007, Philadelphia County's first active bald eagle nest in more than 200 years was discovered (Beer 2007).

Pennsylvania never had a substantial bald eagle population because of a general lack of water bodies capable of supporting breeding pairs, but the statewide population has increased in the past two decades so that the species was recently upgraded to Threatened rather than Endangered. Approximately 132 territorial nesting pairs of bald eagles were observed in Pennsylvania in 2007; some 151 eagles fledged in 2007 (Gross 2008). Eagle nesting in the state has increased as the Chesapeake population has recovered, and some of the breeding pairs are likely a product of this expanding population. Other eagles are a product of eagle chicks released from artificial nest platforms in a reintroduction technique called hacking. Pennsylvania's bald eagle reintroduction program began in 1983 (Mitchell and Brady 1986) and continued through 1989. Over several years, the Pennsylvania Game Commission (PGC) released about eighty-eight nestling eagles obtained from Saskatchewan at hack towers on Haldeman Island, Dauphin County, and Shohola Falls, Pike County, and several others at foster nests (Brauning and McWilliams 2000, Brauning 2002b).

Annual midwinter bald eagle survey results are reported in PGC bald eagle breeding and wintering reports. The latest results were for 2008 when 183 eagles were observed (101 adults and 82 immatures); Lancaster, Crawford, Huntingdon, Wayne, and Pike counties accounted for many of the eagles observed (Gross 2008). In prior years, wintering bald eagles observed during the annual January observation period varied from 125 in 2004, 118 in 2003, 114 in 2002, to 105 in 2001 and 76 in 2005. In addition to the counties mentioned for 2008, Armstrong, York, and Warren counties had large numbers of eagles in some winters (Brauning 2002b, 2003, Gross 2004, 2005, 2006). After nesting, southern bald eagles, especially immatures, often move north and travel widely until migrating south to breed (U.S. Fish and Wildlife Service 2007c).

### *COMMUNITY TYPE/HABITAT USE*

Critical elements of bald eagle habitat include water with an adequate prey base, open water during the winter, presence of roost trees with a view of the water, and large trees for nesting within 1,500 m (1 mile) of water (U.S. Fish and Wildlife Service 2007c). Large impoundments, including water supply, flood control, and hydroelectric reservoirs, such as Conowingo Pond and Allegheny Reservoir; large lakes and wetland complexes, such as Pymatuning-Conneaut; and large river systems, particularly the Delaware and Susquehanna, are widely used in Pennsylvania. Lakes and reservoirs greater than 8 ha (20 acres) located away from human disturbance are preferred (Peterson 1986, Brauning and Hassinger 2000). Although bald eagles were considered a wilderness bird attributed to wariness from decades of persecution, current Pennsylvania eagles are much more tolerant of human presence and disturbance (Brauning 2002a). As the breeding population grows, eagles are choosing nest sites close to roads, parking lots, and other human activity centers and are found next to small streams in wooded areas (D. Gross, personal communication).

Nest trees are generally taller than the surrounding canopy (Stallmaster 1987), and in Pennsylvania, white pine (*Pinus strobus*) and sycamore (*Platanus occidentalis*) are chosen preferentially (McWilliams and Brauning 2000). Large trees with easily approachable branches strong enough to support a nest located close to water and in an open discontinuous canopy are more important than tree species (Peterson 1986).

Wintering birds concentrate near open water, often below dams and on unfrozen rivers or lakes, especially the Delaware River, Lackawaxen River, and the lower Susquehanna where both waterfowl and fish are abundant (McWilliams and Brauning 2000).

### *LIFE HISTORY AND ECOLOGY*

Bald eagles in Pennsylvania are migratory, although birds are present virtually all year. Bald eagles are among the earliest raptors to migrate in the fall. Migration begins in mid- to late August, and migrants are present through December (McWilliams and Brauning 2000). Adults migrate earlier than immatures. During the winter months, northern bald eagles move south and congregate around open water to feed on waterfowl and fish.

Bald eagles mate for life; however, when either member of the pair dies, the surviving eagle remains on the nesting territory to recruit a replacement. Pairs are territorial, but in areas with abundant prey nests may be less than 800 m (0.5 mile) apart (Andrew and Mosher 1982, Peterson 1986).

During the winter (late December to February), breeding birds in Pennsylvania choose nest sites in tall trees within 1,500 m of water to which they return year after year, adding to the structure each season (Leberman 1992a). Many eagle pairs build alternate nests within their territory and may move between nests in different years (U.S. Fish and Wildlife Service 2007c). By late June or July, the young fledge and are as large as the adults when they leave the nest. Fledglings and young nonbreeding birds roam widely and may move well north of their natal site during late summer. Prey availability during the breeding season, defined as the presence of fish in adequate abundance located in shallow water accessible to eagles, is a limiting factor to reproductive success (Peterson 1986).

The Pennsylvania Game Commission monitors the success of all known bald eagle nests in Pennsylvania; in 2005, active nests were known in twenty-eight counties, some 42 percent of the counties in Pennsylvania. The most recent results available are for the 2005 nesting season (Gross 2006). In 2005, some ninety-nine active nests were monitored, of which 74 percent were successful. These ninety-nine nests produced 118 fledgling eagles, which yielded 1.2 fledglings per active nest and 1.6 fledglings per successful nest. The Pennsylvania Game Commission uses the metrics of 50 percent nest success and 0.7 young per nest as a measure of a stable population. Bald eagle nest success in Pennsylvania from 1996 through 2005 varied from 93 percent to 67 percent with an average of about 80 percent (Gross 2006).

Pennsylvania's bald eagle nesting population increased by 28 percent from the 2004 to 2005 breeding seasons; this growth represents an increase in both population and active nests in new areas of the state. It is very likely that there are active nests not known to the Pennsylvania Game Commission because of the difficulty of finding nests in remote areas or in thickly vegetated forested habitats.

Pennsylvania does not publish the causes of nest failure for individual nests, but weather-related nest destruction, nests falling from broken branches, and human disturbance are among the causes of nest failure. New Jersey Department of Environmental Protection reports that PCB (polychlorinated biphenyls) contamination is the likely cause of some nest failure on lower Delaware River bald eagle nests (Clark et al. 1998, Smith and Clark 2006). PCB contamination may cause failure for bald eagle nests on the lower Delaware River in Pennsylvania.

### *THREATS*

The draft Pennsylvania Recovery and Management Plan for the bald eagle (*Haliaeetus leucocephalus*) identifies several threats to bald eagles in Pennsylvania (Brauning and Hassinger 2000). Shooting eagles, although not as prevalent as in the past, continues to be a concern. The Pennsylvania Game Commission knows of six bald eagles shot in Pennsylvania in the past eight years (Gross 2006). Habitat loss is perhaps the most serious limiting factor. Water-based recreation and development of waterfront property are the greatest threats to continued growth of the population. Human disturbance of nesting pairs, especially during incubation, remains a major issue throughout the state (Brauning 1992b, Brauning and Hassinger 2000). Logging and land clearing for agriculture can produce habitat loss, nest abandonment, or nest failure. As a high-order predator, bald eagles are susceptible to environmental contaminants such, as direct poisoning by organochlorine pesticides, and bioaccumulation of heavy metals, such as mercury and lead. In watersheds with known contamination of PCBs, such as in the lower Delaware River, PCB contamination may affect reproductive success (Clark et al. 1998). Eggshell thinning attributable to DDE, a metabolite of DDT, may affect nest success because eagles may spend part of the year in areas with high background levels of DDT. Diseases and other natural factors may produce mortality, but there is no good evidence that any of these are limiting factors in the bald eagle (Byrd et al. 1990).

### *CONSERVATION AND MANAGEMENT NEEDS*

The National Bald Eagle Management Guidelines (U.S. Fish and Wildlife Service 2007c) and the draft Pennsylvania Recovery and Management Plan for the bald eagle (Brauning and Hassinger 2000) provide a framework for conserving and managing bald eagles in Pennsylvania. The actions required or recommended include an annual program to locate and protect existing nest sites and to locate potential areas for population expansion. Both the federal and state agencies require the protection of active eagle nests and winter roosts of greater than ten birds by buffer zones. The Pennsylvania Game Commission recommends a primary buffer of 400 m (0.25 mile) diameter around nests where virtually all human activity is restricted between February and early September and a secondary buffer of 800 m (0.5-mile) diameter to restrict habitat alterations based on site-specific review. United States Fish and Wildlife Service requires a buffer of 100 m (330 feet) for most human activities not visible from the nest and a buffer of 200 m (660 feet) for activities visible from the nest to limit disturbance at active bald eagle nests and roosts. The Pennsylvania Game Commission should develop cooperative agreements with private landowners to identify locations of active nests and roost sites, determine acceptable buffers, and monitor possible threats. In addition, it is important to foster cooperation with other state agencies to identify locations of active nests and roost sites, determine acceptable buffers, and monitor possible threats. Currently, the Pennsylvania Game Commission does not routinely test for the presence of environmental contaminants from chicks, eggs, or eagles at failed nests or birds found dead. A limited program of testing to determine whether environmental contaminants are present and could be contributing to mortality is suggested. Pennsylvania Game Commission and federal authorities are committed to prosecute illegal killings by actively pursuing the apprehension of eagle killers and offering rewards for information leading to their arrest and conviction. This should be continued along with preventative law enforcement. Preventative law enforcement involves news coverage of shooting cases to inform the public of serious treatment of such cases.

*MONITORING AND RESEARCH NEEDS*

Pennsylvania Game Commission staff and volunteers conduct annual statewide midwinter and breeding season surveys. The surveys currently include provisions to monitor nesting success and chick production at occupied nesting territories to determine reproductive success, to determine the geographic expansion of the population, and to identify the effect of habitat alterations. An addition to the current program is suggested to evaluate the role of environmental contaminants at failed nests. The Pennsylvania Game Commission should continue to establish and conduct standardized routes for the national midwinter eagle survey to identify the density and distribution of eagles and habitats used. In addition, they should recruit more volunteers to expand the geographic coverage of the midwinter bald eagle survey and to identify, protect, and manage potential and active nesting territories and foraging areas for active nest territories.

Research needs identified in the draft Pennsylvania Recovery and Management Plan for the bald eagle (*H. leucocephalus*; Brauning and Hassinger 2000) include documenting essential habitat characteristics of nesting, roosting, and foraging habitats and characterizing active nest sites in terms of species of nest tree, nest height, proximity to water, and other attributes that may affect reproductive success. Research to evaluate the effectiveness of efforts to protect known breeding, roosting, and foraging habitat for eagles and to identify, inventory, and monitor unoccupied potential eagle habitat is also recommended. The Pennsylvania Game Commission should seek cooperation from the Pennsylvania Fish and Boat Commission to determine whether the fish component of the prey base is adequate for eagles in potential habitats. In some locations, it may be necessary to determine whether barbiturate poisoning of eagles from feeding on improperly disposed of euthanized pets placed at landfills is occurring. If in the future the state is populated by invasive, exotic submerged aquatic vegetation, such as *Hydrilla*, it may be appropriate to evaluate for potential avian vacuolar myelinopathy (AVM) disease as a source of bald eagle mortality (Robert Ross, U.S. Geological Survey, unpublished data).

*Authors:* ROB BLYE AND DOUGLAS P. KIBBE, URS CORPORATION

## Northern Harrier

Order: Falconiformes
Family: Accipitridae
*Circus cyaneus*

Northern harriers are medium sized, long tailed, long winged, and sexually dimorphic hawks (fig. 5.20). This species is categorized as Secure (G5) at the global level. However, it is considered Vulnerable to Critically Imperiled as a breeder in twenty-nine states, including Pennsylvania (NatureServe 2009). This species is listed as a Species of Greatest Conservation Need because

*Fig. 5.20.* The Northern Harrier, *Circus cyaneus*. Photo courtesy of Michael Brown.

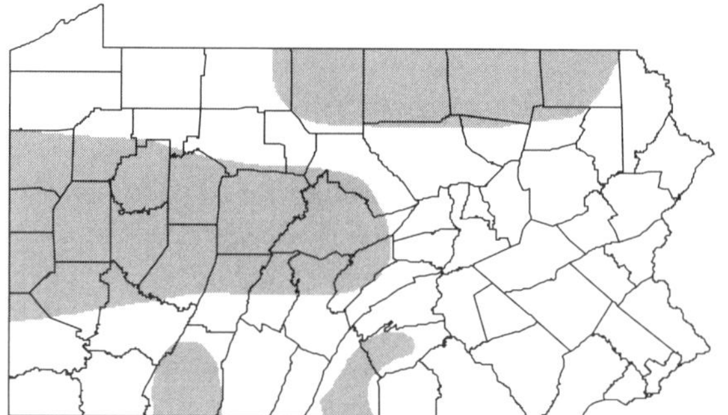

*Fig. 5.21.* Highest probability of occurrence of the Northern Harrier, *Circus cyaneus*.

of perceived population declines and a loss of nesting habitat.

### *GEOGRAPHIC RANGE*

Northern harriers occur throughout north-temperate North America and throughout Europe and Asia (MacWhirter and Bildstein 1996). In North America, this species is widely but locally distributed, breeding from north Alaska and Canada, primarily south of tundra, to north Baja Peninsula, Mexico, north New Mexico, north Texas, Kansas, Iowa, Wisconsin, Michigan, Pennsylvania, southeast Virginia, and probably in northeast North Carolina (MacWhirter and Bildstein 1996). The wintering range of this species includes the southern tier of the United States south to Panama and rarely to Venezuela and Columbia (NatureServe 2004).

### *DISTRIBUTION AND RELATIVE ABUNDANCE IN PENNSYLVANIA*

Northern harriers occur with irregular frequency and abundance throughout the state of Pennsylvania (Goodrich 1992a). During the first Pennsylvania Breeding Bird Atlas, this species was recorded in all regions of the state. However, during the 1980s, breeding northern harriers were rare. Breeding northern harriers were more common in the north and Northwest, especially where reclaimed strip mines or extensive open wetlands occur (fig. 5.21). Northern harriers were much less common in the Southeast, where urban and suburban sprawl is a limiting factor (Goodrich 1992a). Today, northern harriers are most common in extensive reclaimed strip mines in the western part of Pennsylvania, extensive grasslands scattered throughout the state, large open wetlands in the northern part of Pennsylvania, and open marshes in southeastern part of Pennsylvania. There are indications that this species may be declining in Pennsylvania (Goodrich 1992a).

The total estimated global population of northern harriers is 1.3 million with 35 percent of the total population occurring in the United States and Canada (Rich et al. 2004). It is unknown what proportion of the population of northern harriers actually breeds in Pennsylvania.

### *COMMUNITY TYPE/HABITAT USE*

Throughout North America, northern harriers prefer open wetlands as key habitats for breeding, including marshy meadows, wet lightly grazed pastures, open bogs, freshwater and brackish marshes, and riparian woodland (MacWhirter and Bildstein 1996). Across its range, the species nests in old fields, dry uplands, upland prairies, mesic grasslands, croplands, and cold desert shrub-steppe. In Pennsylvania, the Northeast, and the Midwest, northern harriers mostly breed in open wetlands. However, this species also seems to use large reclaimed open strip mines for breeding in Pennsylvania (Goodrich 1992a). Abundance of northern harriers depends on the size of the undisturbed habitat in wetland and upland areas and density of thick vegetation growth (MacWhirter and Bildstein 1996).

Little data exist on habitat usage during spring and fall migration (MacWhirter and Bildstein 1996). One study showed fledglings that were migrating southeast from central Wisconsin seemed to prefer freshwater marshes during stopover. This species has been seen in both open wetlands and uplands during migration. In winter, a variety of open habitats dominated by herbaceous cover are used. The densest population is typically associated with large tracts of undisturbed habitats dominated by thick vegetation growth (Apfelbaum and Seelbach 1983). In the eastern United States, the limit of the species' breeding and winter range coincides with dense deciduous and coniferous forest, which it does not occupy (MacWhirter and Bildstein 1996). However, this species will breed in open wetlands within contiguous forests (Sechler 2004).

*LIFE HISTORY AND ECOLOGY*

Northern harriers are solitary short- to long-distance migrants with an exceptionally long (about three months) fall passage between mid-August and mid-November (MacWhirter and Bildstein 1996). Northern harriers prey on small and medium-sized mammals, primarily rodents, birds, reptiles, and frogs (MacWhirter and Bildstein 1996). It appears that frequency of use of certain habitats is related to a combination of prey biomass and vegetative cover. Males tend to prefer more open habitats than females due to the smaller home ranges of females. In addition, males prefer more birds as prey compared to females (MacWhirter and Bildstein 1996). Females also tend to hunt more in taller and denser vegetation than males do (Bildstein 1987).

This species is not particularly territorial during the breeding season with the exception of close proximity to the nest, where both sexes are intolerant of conspecifics, chiefly those of the same sex and unrelated fledglings of either sex (MacWhirter and Bildstein 1996). This species is generally monogamous during the breeding season, but polygyny is common with a well-structured hierarchical harem of two to five females. Northern harriers seem to be the only raptor exhibiting polygynous behavior on a regular basis (MacWhirter and Bildstein 1996).

The pairing of males and females generally occurs on the breeding grounds; males arriving on the breeding grounds between five to ten days before females. Nest building occurs between April to early June in the Northeast, earlier dates in the Southeast, and is completed within several days to two weeks (MacWhirter and Bildstein 1996). This species generally nests from April through July and is single brooded per breeding season; however, renesting may occur if the nest is destroyed or deserted during egg laying (Bildstein and Gollop 1988, MacWhirter and Bildstein 1996). Northern harriers generally leave for the wintering grounds between August and November (Saunders 1913, Bent 1961). This species' site fidelity is somewhat high, as northern harriers may return to the same general area to breed as the previous year (Hammerstrom 1969).

*THREATS*

The loss of open wetlands to draining, flooding, and peat mining, and the loss of extensive grasslands to suburban development, intensive row crop farming, and reforestation have resulted in declines for this species in Pennsylvania (NatureServe 2004). This may especially be a problem in private lands, where draining of bogs and open wetlands eliminates available habitat for this species. Damming of wetlands by beavers may temporarily make habitat unsuitable, but eventually, abandonment of these wetlands leaves suitable habitat for northern harriers. Large reclaimed strip mines becoming reforested will likely be detrimental to this species in the state (Sechler 2004). Logging next to open wetlands in northern Pennsylvania may be a disturbance to breeding northern harriers and may cause this species to abandon nests. Pesticide and herbicide usage in open wetlands and grasslands used by this species could severely affect breeding success for this species.

*CONSERVATION AND MANAGEMENT NEEDS*

Because northern harriers rely on both upland and wetland type habitats, measures should be taken to conserve and manage both these types of areas for this species. Across the species' range in North America, steps should be taken to preserve native grasslands (Johnson 1996). Collaboration with ranching and farming interests to maintain native rangeland and pastureland should be done (Johnson 1996). In addition, the protection of grasslands through conservation easements, land purchases, and implementation of farm programs, such as the Conservation Reserve Enhancement Program (CREP) that assist farmers with conservation of wildlife habitat, should all be high priorities. Steps should also be taken to help protect wetlands from drainage through conservation easements, land purchases, tax incentives, management agreements, restoration, and continuation of the Wetland Reserve Program and further enforcement of wetland protection laws under the Clean Water Act (Dechant et al. 2003). A mosaic of grasslands and wetlands should be maintained, while some units should be treated to halt succession (Dechant et al. 2003). Other management recommendations include periodic mowing, burning, or grazing to maintain the two- to five-year-old accumulations of residual vegetation preferred by northern harriers. Management techniques, such as mowing, burning, or grazing, are recommended every three to five years to maintain habitat for small mammal prey. Disturbing nesting areas should be avoided during the breeding season, especially from April to July. In the east, mowing dates should be moved ahead until after breeding season of grassland birds is finished. Artificial flooding of wetland habitats should not be conducted in optimal northern harrier habitat. In addition, northern harrier populations and populations of their prey follow similar patterns of fluctuation (Dechant et al. 2003) so management activities should also be developed to provide suitable habitat for their prey.

In Pennsylvania, many wetland habitats that contain breeding northern harriers should be protected with a forested buffer and from succession and artificial flooding. Large reclaimed strip mines should be protected and managed for early successional/grassland-dependent bird species, such as northern harriers. Forest managers in Pennsylvania should be made aware of habitat preferences of northern harriers, and encouraged to take appropriate steps to protect their breeding habitat.

*MONITORING AND RESEARCH NEEDS*

Aerial photograph interpretation should be conducted in counties such as Potter, McKean, Wyoming, Tioga, Bradford, Susquehanna, Wayne, Pike, Carbon, and Monroe specifically to identify potential wetland habitat/community types that could contain nesting northern harriers. These areas should be identified as potential habitat, and efforts made to conduct thorough surveys throughout the breeding season to determine whether northern harriers nest there. Open wetlands in all types of landscapes, from heavily disturbed to extremely remote, should be surveyed for this species and other species that depend on this type of habitat. This survey effort should occur from May to July. In addition, areas that are not covered by the Breeding Bird Atlas that have large grassland or reclaimed strip mines should be surveyed and monitored for nesting northern harriers and other grassland-dependent species. Data should be analyzed to determine habitat preferences and nesting success across landscape types. Threats to sites, predation rates, and nesting behavior should also be recorded. The data should identify where the gaps of knowledge are in determining the next steps for conservation actions.

Many aspects of this species' habitat ecology are still not well understood. Research should be conducted to determine the size of preferred breeding habitats in Pennsylvania and how much fragmentation affects habitat use. Related studies should also be conducted to determine the sensitivity of this species to disturbances during the breeding season and to monitor its breeding success. In open grasslands and reclaimed strip mines, habitat studies should be initiated to determine vertical structure of preferred nesting and foraging areas. Assessing the effect of predation on reproduction and monitoring habitat loss and the effects of environmental contaminants on populations is also needed (MacWhirter and Bildstein 1996).

*Author:* FREDERICK C. SECHLER, JR., NEW YORK NATURAL HERITAGE PROGRAM

## Peregrine Falcon

Order: Falconiformes
Family: Accipitridea
*Falco peregrinus*

Peregrine falcons are medium-sized, fast-flying powerful birds of prey (fig. 5.22). They were selected as a Wildlife Action Plan species of High-Level Concern and are listed as Endangered by the Pennsylvania Game Commission. They were delisted from the federal Endangered Species list in 1999 and currently receive post delisting monitoring oversight by the United States Fish and Wildlife Service (U.S. Fish and Wildlife Service 1998, 1999). Global populations are Apparently Secure (G4, NatureServe 2009).

*GEOGRAPHIC RANGE*

The peregrine falcon is a cosmopolitan species, breeding on all of the world's continents except Antarctica (Brown and Amadon 1968). Of the nineteen geographical forms or subspecies, two historically occurred in Pennsylvania. Pennsylvania's nesting population was part of the anatum race (*Falco peregrinus anatum*) found in scattered locations of the eastern and western United States (Cade et al.1988). The Arctic subspecies (*Falco peregrinus tundrius*) nests in the Canadian high Arctic and in Greenland and winters in South America. This subspecies passes through the state in low numbers during spring and fall migration. Follow-

*Fig. 5.22.* The Peregrine Falcon, *Falco peregrines*. Photo courtesy of Glen Tepke.

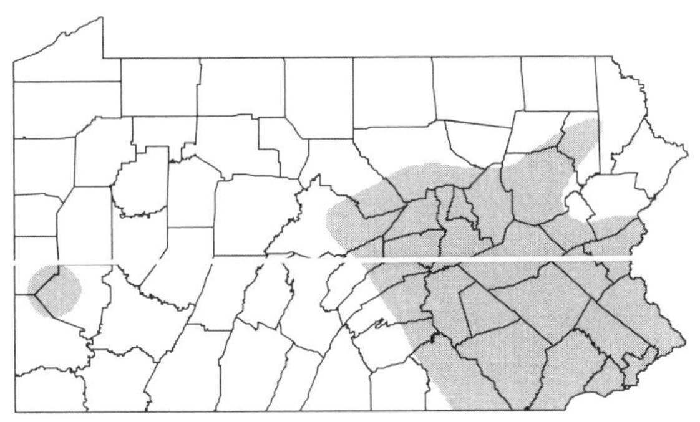

*Fig. 5.23.* Highest probability of occurrence of the Peregrine Falcon, *Falco peregrines*.

ing extirpation in the 1960s, peregrines were restored to eastern North America through a massive reintroduction effort involving the release of thousands of juveniles (Berger et al. 1969, Barclay and Cade 1983, Barclay 1988).

### *DISTRIBUTION AND RELATIVE ABUNDANCE IN PENNSYLVANIA*

The maximum historic breeding population of peregrines nesting in Pennsylvania appears to have been forty to fifty pairs (Hickey 1942, Poole 1964, Rice 1969). Evidence from the early decades of the twentieth century indicated that many, if not most, of the eyries (nest sites) were regularly occupied. Hickey's extensive surveys of the species in the 1930s documented routine occupancy at many sites (Hickey 1942). The population crashed during the 1950s and disappeared from the state as a nesting species by the early 1960s (Hickey 1969). Reintroduction programs and the banning of the insecticide DDT with support from the United States Fish and Wildlife Service's Endangered Species Act paved the way for recovery (Hickey and Anderson 1968, Cade et al 1988). Currently, the reintroduced population of diverse genetic origin is found nesting in more than twenty-four locations across Pennsylvania, with most sites east of the central mountains (fig. 5.23, McMorris and Brauning 2009).

### *COMMUNITY TYPE/HABITAT USE*

Peregrines require large open areas for hunting and tall and inaccessible ledges for nesting. They are rarely found in contiguously forested areas. Peregrines traditionally nested on high cliffs overlooking major river systems of central and eastern Pennsylvania (e.g., Susquehanna and Delaware rivers). Resident adult peregrines appeared to spend much of the year near nest sites, vacating nesting territories for only a few months in the middle of winter (Poole 1964). This contrasts sharply with the migratory habits of the Arctic subspecies. Because the species was reestablished in the 1980s, at least some birds nesting on bridges and buildings remain on their territory year-round as permanent residents.

Peregrines do not build nests but lay their clutch in a shallow indentation in the ground scratched out with their talons. They are also known to use nests built by other cliff-nesting species, most notably the common raven (*Corvus corax*). Nests are typically placed on high, inaccessible locations, often near large bodies of water. Nesting sites must be steep enough to afford the falcons protection from potential mammalian nest predators. Cliffs or structures that are topographically varied with recesses and overhangs for shelter provide more locations for the nest scrape. Nest sites must also have ledges large enough to accommodate the clutch and brood (Ratcliffe 1993). Historic nest sites included urban settings in Europe and Pennsylvania (Groskin 1947, 1952).

Modern urban nest locations include tall buildings and bridges (McMorris and Brauning 2009). Nests on bridges often are placed within enclosed steel beams accessible through holes of various sizes. Typical locations on buildings are ledges and small rooftops with a southern or western exposure. Like cliff sites, nesting birds use only existing substrate for nesting material, sometimes including feces of rock pigeons (*Columba livia*).

During migration, these falcons may be found anywhere across Pennsylvania but are observed most commonly along ridges, such as Kittatinny Mountain, with other migrating raptors. Nesting birds that remain year-round on their territory may be found in proximity to nesting territories. Cliff-nesting pairs appear to move into nearby urban areas where other winter residents occasionally may be found.

### *LIFE HISTORY AND ECOLOGY*

Peregrines feed almost exclusively on live birds captured in flight. The majority of prey items range in size from 50 to 500 g. However, there is no apparent lower limit to the size of potential prey with peregrines taking small birds, such as chimney swifts (*Chaetura pelagica*), and birds as large as black grouse (*Tetrao tetrix*) at 1,250 to 1,400 g in the United Kingdom (Ratcliffe 1993). A wide range of species has been documented at Pennsylvania nest sites, with northern flicker (*Colaptes auratus*), blue jay (*Cyanocitta cristata*), and rock pigeon (*Columba livia*) frequently encountered (Brauning 1988). Little is known of specific prey preferences at historic eyries or whether food availability limits nest-site selection or nesting success. "Duck Hawk," a historic name of the eastern peregrine falcon, suggests a popular diet. However, the peregrine is clearly an opportunistic species, taking advantage of appropriately sized prey as they become locally available.

Historic sites originally were along major rivers in forested regions, where most hunting activity would have been confined to open areas along the river or above the forest canopy. Peregrine populations probably benefited from land clearing and agricultural development in the nineteenth century because of the increased availability of certain prey and additional open foraging areas.

Migratory patterns of the tundra race (*F. p. tundrius*) are well established along eastern coastal borders (outside of Pennsylvania) and the central ridges (for example, Kittatinny Ridge) within the state during the fall migration. The fall migration typically occurs in September and October. Spring migration follows a more inland course on less fixed lines, resulting in widespread sightings during late April and May. Post-breeding dispersal of young appears to occur about six to eight weeks after fledging, in unpredictable directions (Brauning 2004).

Peregrines generally return to the same nest site each year and remain paired with the same mate until one of the pair dies or is driven off. Some pairs in Pennsylvania begin egg laying as early as late February. Clutches found in June or July are assumed to have replaced earlier failed nesting attempts, not second broods. Adults have remained active at nest sites for more than twelve years in Pennsylvania.

Productivity ranges considerably among nest sites and across years. Average production of young at buildings with established pairs was 2.8 young per year, considerably higher than established bridge-nesting pairs, 1.5 young per site until 2004. Productivity on bridges has increased considerably as a result of intensive management but remains below replacement levels (McMorris and Brauning 2009). Continued poor reproductive success of extant pairs, including those nesting on cliffs, raises questions concerning the viability of Pennsylvania's population.

Patterns of juveniles assessed through satellite-telemetry studies indicate that some juveniles spend their first winter in Central or South America, while others may remain in North America (Canadian Peregrine Foundation 2000; Falcon Track Project 2000). Pennsylvania's own studies have demonstrated that dispersal from the nest may occur as early as five weeks from fledging (Brauning 2004, McMorris and Brauning 2004), and that subsequent observations of young at nest sites may reflect visits following dispersal wanderings.

### *THREATS*

Although the current population represents a substantial recovery, several factors continue to place the species in jeopardy in Pennsylvania. Reproductive success has been poor at some sites due to disturbance, disease, and predation. An additional potential threat is environmental toxins. That most active nests are on man-made structures frequently brings them into conflict with routine human activities. Routine bridge maintenance or inspection activities and larger bridge renovations can be a problem, particularly during incubation and fledging periods. Nests on buildings have similar problems due to maintenance of rooftop equipment, window washing, and other related activities. Rock climbers may disturb birds nesting on traditional cliffs (Lanier and Joseph 1989). Competition with common raven, another cliff-nesting species, in combination with disturbance by rock climbers may affect nesting success at natural sites (Brambilla et al. 2004).

Urban hazards continue to result in juvenile mortality at both bridge and building nest sites. Unmodified nest sites on bridges and buildings are often inadequate to support natural development of juveniles. Automobile strikes are another documented cause of mortality on bridges, particularly when nests are placed above the roadway. As a result of these risks, poor fledging success contributes considerably to reduced productivity on most bridges. Recent experience with survivorship of juveniles on cliffs suggests that productivity on these sites may be no better, for apparently different reasons, possibly predation (McMorris and Brauning 2005). Respiratory fungal infection (*Aspergillosis*) was found to be a factor in the loss of some young over the fifteen years of monitoring, most likely as a secondary factor (Brauning and Dooley 1991).

A potential and certainly historic threat to hatching success, nestlings, and adults is environmental toxic contamination (Hickey and Anderson 1968). Current attention is on PCBs (polychlorinated biphenyls), lead, and flame-retardants (Lindberg et al. 2004). Recent contaminant analyses of Delaware Bay peregrine falcon eggs in New Jersey indicate that DDE levels are sufficiently elevated to produce eggshell thinning (K. E. Clark, personal communication). The United States Fish and Wildlife Service and the New Jersey Division of Fish, Game, and Wildlife are investigating environmental contaminants, including PCB and DDT metabolites. Exposure to avicides through secondary contact with rock pigeons and European sstarlings (*Sturnus vulgarus*) is a potential concern.

*CONSERVATION AND MANAGEMENT NEEDS*

The following management activities have a demonstrated track record of boosting reproductive rates and promoting an expanding peregrine population. Anthropogenic nest sites must be modified to provide a secure location for development and fledging of the young by placing wooden nest trays on bridges and buildings near existing nest sites or near prospective sites. At three weeks of age, young may be moved from existing, inadequate nest sites to a more secure nest box within about 50 yards of the active nest. This procedure immediately improves fledging success and can be used to coerce adults into using nest trays during subsequent years. Nest boxes should be placed at potential locations during early winter.

Medical treatment should be administered to young to enable them to overcome life-threatening infection during critical developmental periods. Nestlings found with visible signs of *Trichomoniasis*- or *Aspergillosis*-type illnesses should be pulled from nest sites, if possible, for a six- to eight-day intensive treatment period. If young respond to treatment, all efforts should be made to return young to the nest site.

Young should be banded with United States Fish and Wildlife Service bands and individual identifying color bands. All accessible nestlings in Pennsylvania have been color-banded since 1992. Observations of color-marked birds are useful for determining population levels, turnover, and survival of individual birds and contribute to population models that predict the rate of recovery of this endangered species. Urban-nesting falcons provide unique opportunities to see color bands and individually identify birds. This provides a much higher rate of "band recovery" than experienced by many species.

Because nests on buildings and bridges are vulnerable to disturbance, a memorandum of understanding should be developed for each agency overseeing structures with nesting peregrines. Agreements should regulate access to bridges by maintenance and inspection personnel during incubation and nestling periods (March–July). Disturbance of nest sites before incubation and of incubating pairs should be severely limited and restricted to less than fifteen minutes per incident. Nest sites should not be approached within two weeks of projected fledging dates to avoid prefledging the young. Desertion of nests, egg failure, and premature fledging are potential results of excessive disturbance. Peregrine management will require cooperation with institutions not normally involved in wildlife-related activities. Nesting pairs, or boxes, are currently on bridges managed by a variety of agencies and organizations and an even greater number may yet become involved as new nest sites are discovered.

All historic sites should be evaluated for suitability at ten-year intervals. Sites left unoccupied for the past forty years may have become overgrown by vegetation, potentially making them unsuitable for nesting peregrines. Sites may be improved to restore their suitability for nesting falcons. The presence of cliff-nesting competitors (ravens, vultures) should be noted. Great horned owl (*Bubo virginianus*) populations near the cliff should also be estimated. There is potential for serious conflicts between recreational rock climbers and peregrines nesting at traditional sites. This is particularly the case at popular climbing sites, such as the Delaware Water Gap. If pairs are observed prospecting for a nest site at historic eyries, a suitable nesting area should be closed to rock climbing until nesting has been completed or it is determined that the site was not used. Protection of wild peregrine populations is essential to continued restoration. Led by the Pennsylvania Game Commission Bureau of Wildlife Protection, all illegal takings of wild peregrine falcons should be prosecuted to the full extent of the law.

Hacking was the primary method of reestablishing peregrine falcon populations during the 1970s and 1980s and can be credited with restoring this species in the eastern United States. Additional releases during the 1990s contributed to the expanding population (Brauning 2008). However, early hacking failed at historic natural sites in Pennsylvania because of great horned owl predation of fledglings. The growing peregrine falcon population suggests that recruitment is not now the limiting factor. The population is dynamic, with young moving widely across the eastern United States. Young peregrines have routinely dispersed long distances (hundreds of miles) from natal areas to establish nest sites (Barclay 1995, Brauning 1998b). Recruitment from urbanized populations to natural cliff sites appears to be slow but has occurred in several cases (McMorris and Brauning 2005). For these reasons, further release or hacking of young peregrines is not recommended.

*MONITORING AND RESEARCH NEEDS*

Annual surveys should be implemented to determine the number of nesting pairs in natural and man-made situations. Reproductive success at all known sites should be carefully monitored, particularly during the fledging stage. All known active nest sites should be assessed each year to monitor nesting success. Monitoring should be conducted to determine the initiation of incubation and to monitor fledging success. Nests should be visited once each year to determine hatching success and to band young. Other developmental data can be gathered at the same time as banding, and young may be checked for disease. Each year a sample of suitable historic eyries, and likely bridge and building sites, should be surveyed during March and early April to determine the presence of peregrine falcons at potential nest sites. All suitable historic nest sites should be checked at least once every five years.

Existing monitoring programs at hawk-watching stations across the state are adequate to track population trends of nonbreeding, primarily migratory *P. f. tundrius*. Wintering individuals occur at widely scattered locations, often including urban areas. Monitoring birds observed during winter may lead to new nest sites.

Ongoing research is needed to evaluate the effects of environmental contaminants on nesting peregrine falcons through analysis of unhatched eggs and deceased young or adults. Continued low reproductive outputs at some sites suggest that this should be a priority. Unhatched eggs and eggshell fragments should be collected when young are banded or treated for disease. In addition, sources of potential contaminants may be evaluated by determining food items found at nest sites. Similarly, migratory patterns of progeny of Pennsylvania nests and of nesting pairs should be determined to assess potential exposure to contaminants and other hazards. If contaminant problems or toxic levels of contaminants are found in eggs or in deceased young or adults, a full study of the source of the environmental contamination should be initiated.

*Author:* DANIEL W. BRAUNING, PENNSYLVANIA GAME COMMISSION

## Virginia Rail

Order: Gruiformes

Family: Rallidae

*Rallus limicola*

Virginia rails are slender, long-billed wetland inhabitants (fig. 5.24). They were selected as a species of High-Level Concern in Pennsylvania because they are rare breeders in the state and depend on emergent wetlands, an at-risk habitat type in the state. The Virginia rail is classified as a migratory game bird. It is a Partners in Flight Priority IIA (high regional concern) species. Global populations are Secure (G5, NatureServe 2009).

*Fig. 5.24.* The Virginia Rail, *Rallus limicola*. Photo courtesy of Geoff Malosh.

*GEOGRAPHIC RANGE*

The Virginia rail breeding range in North America extends in a broad band across the northern United States into southern Canada (Conway 1995). Pennsylvania is on the southern border of the Virginia rail's contiguous breeding range in the East. Isolated breeding populations are found in the southeastern United States. In the West, the species ranges south into Mexico and Central America into South America. In the Northern Hemisphere, most populations are migratory, wintering on the southern coasts; however, south of the United States, populations are nonmigratory.

*DISTRIBUTION AND RELATIVE ABUNDANCE IN PENNSYLVANIA*

Although Virginia rails are found throughout Pennsylvania during the breeding season, the localized distribution of suitable emergent wetlands gives the species a scattered occurrence in the state. Less than 3 percent of the state's atlas blocks yielded Virginia rails during the first Breeding Bird Atlas. Western and northeastern portions of the state yielded the greatest number of records, a reflection of wetland prevalence in these areas (Schwalbe 1992). Currently, the highest probability of occurrence for Virginia rails is in the northwestern and northeastern parts of the state (fig. 5.25). High densities (e.g., ten to twenty-five pairs per hectare) of Virginia rails have been reported in other parts of its range (Zimmerman 1977), but estimates are difficult to obtain and are lacking from Pennsylvania. Conneaut Marsh is reported to support several hundred pairs (McWilliams and Brauning 2000). Populations are thought to be declining in the northeastern United States, but quantitative surveys of this secretive species are generally lacking. Pennsylvania provides little suitable habitat for this species and is on the boundary of its breeding range. Insufficient information exists to discuss population trends in Pennsylvania. Results of the second Pennsylvania Breeding Bird Atlas indicate an increase of 17 percent in distribution has occurred since the first atlas (Mulvihill and Brauning 2009). However, these results may reflect an increased effort in surveying wetland birds during the second atlas rather than a true increase in numbers. General trends based on roadside Breeding Bird Surveys in the northeastern United States indicate a decline in the population, but specific surveys are lacking.

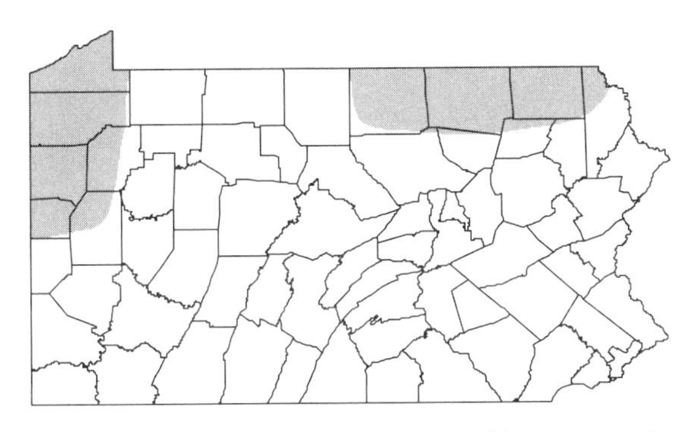

*Fig. 5.25.* Highest probability of occurrence of the Virginia Rail, *Rallus limicola*.

*COMMUNITY TYPE/HABITAT USE*

Virginia rails prefer dense emergent freshwater wetlands with an interspersion of mudflats and open water (Manci and Rusch 1988). Water depths up to 15 cm appear to be preferred, although deeper wetlands may also be occupied (Sayer and Rundle 1984). Cattails and sedge marshes are the normal habitat. Wetlands with trees or shrubs, which would offer perches for predators, appear to be avoided. During the winter season, most of the North American population winters in freshwater, brackish, and salt marshes of coastal southern United States.

*LIFE HISTORY AND ECOLOGY*

Virginia rails are skulking summer residents of Pennsylvania's emergent wetlands. More often heard than seen, they are easily overlooked. Surveyors frequently use taped calls to elicit responses. Territorial birds may respond strongly to both Virginia rail and sora (*Porzona carolina*) playbacks (Kibbe 1985d). Virginia rails probe and pick for prey, consuming invertebrates throughout the year, with some seeds and other vegetative matter during the winter months. Mucky substrates, which provide invertebrate fauna with shallow standing water throughout the summer, are preferred (Gibbs et al. 1991).

Although a few may overwinter in the larger marshes, most Virginia rails abandon Pennsylvania during the winter months for the coastal marshes of the southeastern United States. By mid-March, many males have returned to breed. Females arrive about a week later. Courtship (including courtship feeding, preening, and calling) and territorial behavior precede pairing. By early May, nesting has started at Pennsylvania's latitude. Nests often include a ramp and are placed in dense marsh, often over water less than a foot deep. Laying may commence before the nest is completed. Dummy nests are often constructed in the bird's territory.

*THREATS*

Pairs breeding in marginal habitats may be subject to greater predation. This may be particularly true in areas where feral house cats are allowed to roam. Increased residential development of areas surrounding wetlands may make these areas hunting grounds for feral cats. The threat of local extirpation of breeding populations is greatest in areas where wetlands occur in small pockets or are subject to degradation from acid mine drainage or other pollution. Protecting major wetland complexes, managing water levels, and providing buffer areas around wetland areas are all steps to protect habitat for this and other wetland species.

*CONSERVATION AND MANAGEMENT NEEDS*

Virginia rails are currently subject to fall hunting as a migratory game bird throughout their range, but harvests are generally believed to be small. Few hunters target the species in Pennsylvania, and most of the birds taken are incidental to other hunting efforts. Conservation needs of the Virginia rail parallel many other wetland species; protection and enhancement of the large emergent wetlands is critical for this species. Because few of these large complexes occur in the state, many smaller, drier wetland areas are occupied sporadically. These smaller, drier areas are more likely to be grazed by cattle and thus may offer less protection to breeding pairs and produce fewer young.

*MONITORING AND RESEARCH NEEDS*

At present, there is no way to assess qualitatively the change in this species' status. No survey has been conducted, even at a local level, to determine the breeding density. Development of a standardized survey procedure is an immediate need (Conway 2005, Conway and Nadeau 2006). Additional research into the best way to survey this secretive species is needed. Once a survey protocol is established, surveys need to be conducted on two levels: (1) statewide, to canvass the entire state identifying sites for further monitoring, and (2) localized searches of the principal wetland areas of Crawford, Butler, and Mercer counties to establish the current baseline population level in the historic stronghold of the species for long-term monitoring.

Use of taped playback calls is recommended as a survey technique, but other factors (e.g., time of day or night, weather influences, peak of calling period) that may influence the success of surveys need to be carefully determined (Conway 2005, Conway and Nadeau 2006). Surveys should be used to identify areas of greatest importance to the state's breeding population. As with other wetland birds, further research is needed to establish the species' nesting requirements and conservation actions, which may enhance breeding and survival. These research needs include studies on nest site selection, minimum wetland size, nesting density, factors affecting nesting success (e.g., predation, pollution, disturbance), and specific habitat attributes to successful breeding attempts.

*Author:* DOUGLAS P. KIBBE, URS CORPORATION

## Black Tern

Order: Charadriiformes
Family: Laridae
*Chlidonias niger*

The black tern is a distinctive dark tern of freshwater marshes (fig. 5.26). It is listed as a state Endangered Species because it is an extremely rare breeder in the state. The breeding population within the state is Critically Imperiled. The black tern is a Northeast Region Priority Species and is listed as an Endangered Species by Maine, Vermont, New York, and Pennsylvania. Globally, it is considered Apparently Secure (G4, NatureServe 2009).

*GEOGRAPHIC RANGE*

In North America, black terns breed from the southern Northwest Territories through southern Canada, south to central California, Colorado, Nebraska, and Iowa, and east to Pennsylvania, New York, Vermont, and Maine. They winter along both coasts, from Panama to South America (Novak 1992). The black tern has experienced a significant continental decline but re-

*Fig. 5.26.* The Black Tern, *Chlidonias niger*. Photo courtesy of Alan D. Wilson, naturespicsonline.com.

mains widespread and fairly common in much of the Prairie Province region and parts of the north-central and western United States (Novak 1992, McWilliams and Brauning 2000).

### *DISTRIBUTION AND RELATIVE ABUNDANCE IN PENNSYLVANIA*

The black tern is an uncommon to rare migrant across Pennsylvania, except in the Ridge and Valley and High Plateau provinces, where it is irregular. As a breeding bird, it is on the brink of disappearing from the state. It nests regularly only in Crawford County and has bred at Presque Isle, Erie County (McWilliams and Brauning 2000; fig. 5.27). During the first Pennsylvania Breeding Bird Atlas (BBA) project period (1983–1989) breeding was confirmed in only four atlas blocks, including the first breeding record for Presque Isle in more than twenty years. No breeding site contained more than a few birds, and probably no individual nesting area was occupied during every year of the project (Leberman 1992b). A nesting attempt at Presque Isle in 1993 is believed to have been unsuccessful, and no nesting was confirmed in Crawford County from 1992 to 1995. Up to three pairs were present at Hartstown, Crawford County in 1996, and at least two young were fledged (McWilliams and Brauning 2000). Since that time, a pair was confirmed nesting at Presque Isle (McWilliams 2005) but not again at Hartstown (R. Mulvihill, second Pennsylvania Breeding Bird Atlas, personal communication). Nesting was confirmed in 1997 and 1998 (Teats et al. 1998, Sandeen et al. 1999), and between 1999 and 2004, birds were observed each year by Pymatuning Laboratory of Ecology personnel, although in small numbers (no more than three birds per year) and without confirmation of nesting (A. Bledsoe, personal communication). During the second Breeding Bird Atlas (2004–2008), there were only eight records with only one confirmed (Mulvihill and Brauning 2009).

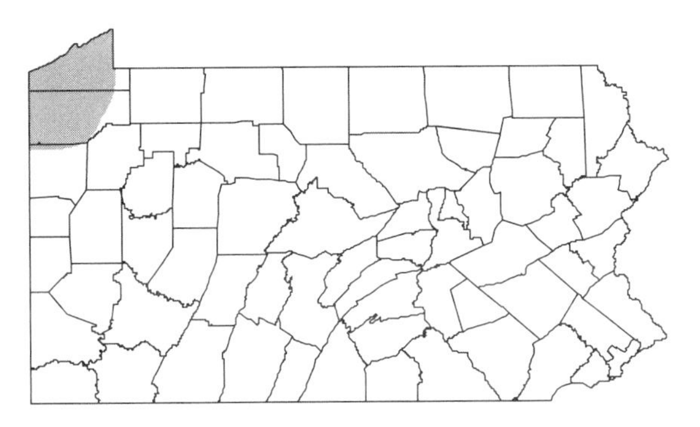

*Fig. 5.27.* Historical occurrence of the Black Tern, *Chlidonias niger*.

Historically, Sutton (1928a, 59) considered the black tern a "fairly regular and sometimes abundant transient visitant" near Conneaut Lake. Although breeding was suspected in 1926, the only nests he reported were two located about 1910. The marshes that resulted from the creation of Pymatuning Lake provided ideal black tern habitat (Leberman 1992b). Nesting was confirmed there in 1934, when a colony of about fifty pairs was located northwest of Linesville (Trimble 1940). On May 21, 1940, Grimm (1952) observed at least 500 terns at Hartstown, most of them migrants, and estimated twenty-five to thirty pairs were nesting there in June. Black terns were also present during the breeding season at the Smith Marsh and Conneaut Marsh, near Conneaut Lake. Black terns formerly nested in marshes near the entrance of Presque Isle, Erie County, where fifteen nests were found in 1958 (Stull et al. 1985).

### *COMMUNITY TYPE/HABITAT USE*

The black tern nests in freshwater wetlands and the shallow areas of ponds, lakes, prairie sloughs, and impoundments, preferring habitats with emergent vegetation interspersed with about 25 percent to 50 percent open water, stable water levels throughout the nesting season, and abundant nest substrates. Although they have been observed on wetlands less than 6 ha in size, black terns prefer wetlands of at least 20 ha. Smaller wetlands may be used if they are part of a large wetland complex (Novak 1992, Zimmerman et al. 2002). In Crawford County, black tern habitat includes cattails, spatterdock, water lily, bulrushes, and various grasses and sedges (Leberman 1992b). Vegetation densities appear to be more important than plant species (Cuthbert 1954).

### *LIFE HISTORY AND ECOLOGY*

Black terns generally arrive in Pennsylvania beginning the last week of April, and migration continues through the state until the first week of June. Although there is no apparent peak in the migration, most birds pass through the state during May. The fall migration period occurs between the last week in July and the second week in September (McWilliams and Brauning 2000). Black terns are generally insectivorous but vary their diets, depending on prey availability. Insects may compose as much as 94 percent of the diet, but arachnids and crustaceans may also be taken (Cuthbert 1954), and fish can make up to 40 percent of food items delivered to chicks (Goodwin 1960). Black terns forage on the wing over open water, wetlands, meadows, and cultivated fields, either flying in low circles or hovering (Goodwin 1960, Bent 1963).

Black terns are considered semicolonial nesters, breeding in loose colonies numbering up to several hundred individuals or as individual pairs, but typically in clusters of eleven to fifty nests (Dunn and Agro 1995, Zimmerman et al. 2002). Nests examined in Crawford County in 1997 and 1998 were built on floating logs, or floating mats of vegetation and muddy debris, and located in beds of spatterdock (Teats et al. 1998, Sandeen et al. 1999). Nests are also built on muskrat (*Ondatra zibethicus*) feeding platforms and inactive lodges. The nest typically consists of a small mass of aquatic vegetation, usually 2–6 cm high, with a cuplike bowl (Novak 1992) sitting 2–20 cm above the waterline and placed in water 0.05–1.2 m deep. They are generally placed adjacent to or within 0.5–2 m of open water (Zimmerman et al. 2002). Nests with eggs were located in Crawford County between May 18 and July 28 (Bush 1989, Teats et al. 1998, Sandeen et al. 1999). The latter date is the latest ever reported in Pennsylvania and probably represents a second nesting attempt (Teats et al. 1998). Black terns are vigorous nest defenders and will aggressively harass and attack intruders and perceived threats, including humans (Cuthbert 1954). A full clutch usually consists of three eggs (Dunn and Agro 1995). Eggs hatch in twenty to twenty-three days. The chicks are capable of walking, running, and swimming when they are two days old and fledge when they are approximately twenty-one days old (Goodwin 1960). Nest success (percentage of nests where at least one egg was hatched successfully) varies from 13 percent to 72 percent and is typically 30 percent to 50 percent (Dunn and Agro 1995).

*THREATS*

Loss of breeding habitat is considered a major factor in the decline of black tern numbers throughout its range (Novak 1992). The decline in Pennsylvania, where the birds are on the periphery of their breeding distribution, may be an artifact of the general continental reduction. With suitable habitat available, the factors limiting Pennsylvania's terns may be external to the state. Subtle changes, such as increases in the extent of vegetative cover and loss of open water, may also threaten the suitability of existing habitats for black terns. Other potential threats include flooding of nests due to heavy rains, human disturbance, predation, and contaminants (Novak 1992).

*CONSERVATION AND MANAGEMENT NEEDS*

Management is often necessary to maintain suitable vegetation to open water ratios on larger wetlands. Black tern habitat in Pennsylvania should be managed for suitable ratios of open water to vegetation, with optimal interspersion. This may be accomplished through mechanical cutting, herbicide applications, water-level management (kept stable during nesting season), and muskrat management. A draft black tern recovery and management plan recommends establishing and maintaining five separate breeding colonies, consisting of a minimum of ten pairs each (Kibbe 1995c). Kibbe (1995c) and Sandeen et al. (1999) recommended placing informational signage at public access areas to discourage human disturbance; deploying artificial nesting platforms; stabilizing water levels; dredging a portion of the Hartstown Marsh to reduce unwanted vegetation; increasing the extent of open water; and possibly increasing the incidence of spatterdock, which is preferred for nesting.

*MONITORING AND RESEARCH NEEDS*

Active and inactive black tern nesting locations in Crawford and Erie counties where suitable habitat remains should be monitored annually for breeding activity and changes in habitat conditions. Where conditions become unfavorable, management should be directed to those areas to improve the suitability for black terns. Research should be considered to determine the feasibility of restoring inactive black tern sites in Crawford County to optimum habitat conditions.

*Author:* ROBERT W. CRISWELL

## Long-eared Owl

Order: Strigiformes
Family: Strigidae
*Asio otus*

The long-eared owl is a secretive and rare medium-sized owl with narrow ear tufts (fig. 5.28). It is listed in Pennsylvania as a Candidate—Undetermined Species because of the uncertainties about its population size and trends in the state (Brauning et al. 1994). The regional conservation picture looks bleak for this species (NatureServe 2005). It is possibly extirpated in Delaware, Maryland, and Rhode Island. It is considered Critically Imperiled in Ohio, West Virginia,

*Fig. 5.28.* The Long-eared Owl, *Asio otus*. Photo courtesy of Jerry Burke.

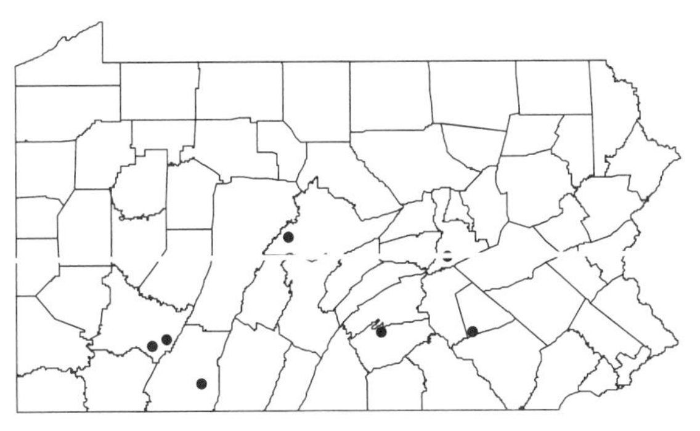

*Fig. 5.29.* Distribution of the Long-eared Owl, *Asio otus*, from Breeding Bird Atlas II.

Virginia, Kentucky, Illinois, and Connecticut. In neighboring New York, it is considered Vulnerable, and it is Threatened in New Jersey. In New England, it is considered Imperiled in Vermont, Massachusetts, and Maine. Global populations are considered Secure (G5, NatureServe 2009).

### *GEOGRAPHIC RANGE*

The long-eared owl has a circumboreal distribution with populations in North America, Europe, Asia, and isolated populations in central Africa, the Azores, and Canary Islands (Marks et al. 1994, Scott 1997, American Ornithologists' Union 1998, Konig et al. 1999).

### *DISTRIBUTION AND RELATIVE ABUNDANCE IN PENNSYLVANIA*

The long-eared owl is poorly known, but it is apparently rare and local in distribution as both a breeding and wintering species (Santner 1992a, McWilliams and Brauning 2000). It is almost at the southern edge of breeding ground in Pennsylvania, but is found in isolated locations in the Appalachians (Marks et al. 1994, American Ornithologists' Union 1998). The tendency for this species to roost in large numbers confuses the issue of abundance (giving the impression of abundance when it is absent from most of the state). Its status in the state has never been well understood. Because it was found in less than 1 percent of the atlas blocks, it is one of the rarest or most difficult-to-find breeding birds in Pennsylvania (fig. 5.29). According to Poole (1964), it was formerly much more common, so there is potential for nesting in other counties. At one time, this species was a resident of Cook Forest (Sutton 1928b). During the first Breeding Bird Atlas, nesting was confirmed in Tioga, Cumberland, Beaver, Lebanon, and Lancaster counties (Santner 1992a). Since that time, nests have been located at scattered locations, including state parks and private property in Columbia, Montour, Indiana, and Centre counties (M. Lanzone, personal communication, Pennsylvania Society for Ornithology Special Areas Project database).

### *COMMUNITY TYPE/HABITAT USE*

The long-eared owl is a raptor of a habitat mosaic where conifer woods mingle with field and meadow (Marks et al. 1994, Konig et al. 1999). They nest and roost in densely vegetated conifers in eastern North America and forage in adjacent fields or open woods. Sutton (1928a) found that this species was strongly associated with hemlocks (*Tsuga canadensis*) in Pennsylvania. In migration and on its wintering ground, the dense conifer vegetation is important apparently as camouflage and perhaps as thermal protection (Marks et al. 1994). It is strongly associated with coniferous woods, especially for wintering habitat, where owls fan out from roosting locations to hunt over open fields (Craighead and Craighead 1956). These conifers can be either native or ornamental exotics and include Austrian pine (*Pinus nigra*); Scotch pine (*Pinus sylvestris*); Virginia, or scrub pine (*Pinus virginiana*); red pine (*Pinus resinosa*); Norway spruce (*Picea abies*); and eastern red-cedar (*Juniperus virginiana*). In the West, it is commonly associated with riparian forests (Marks et al. 1994), but Pennsylvania's nesting records are generally in agricultural or park-like settings, where there is a combination of conifer woods and open fields (Santner 1992a, McWilliams and Brauning 2000, Pennsylvania Society for Ornithology Special Areas Project database, D. A. Gross, personal observation). It also inhabits conifer plantations, open mixed forests, and pine stands within forests (Marks et al. 1994). Short ground vegetation is an important component of its foraging habitat that allows the owls to find their rodent prey. The long-eared owl is often classified as a deep forest bird (Bent 1938, American Ornithologists' Union 1983), but this is misleading in that long-eared owls normally use these habitats for nesting and roosting only (Marks et al. 1994). Long-eared owls have nested in and at the edges of towns with fields nearby (D. A. Gross, personal observation).

### *LIFE HISTORY AND ECOLOGY*

The long-eared owl has the reputation of being more strictly nocturnal than other American owls, foraging chiefly between dusk and midnight (Bent 1938, Johnsgard 1988b). However, on occasion, long-eared owls will forage in daylight, either as a reaction to prolonged adverse weather conditions at night or in late spring when short nights force provisioning adults to hunt more frequently before dusk and after dawn (Scott 1997). The long-eared owl is associated with a mosaic of habitats that enable it to nest and forage in the same area. It forages for small rodents over open ground but typically nests and roosts in dense conifer stands (Marks et al. 1994). Although some researchers have considered it a prey specialist, more recent studies suggest that it is an opportunist that takes advantage of whatever small rodents and other prey are available, including a diversity of small mammals (including bats), lizards, snakes, and small birds (Marti 1976, Johnsgard 1988b, Marks et al. 1994, Sperring 2001).

Winter roosts are invariably in dense conifer stands where the owls usually perch almost invisibly next to the trunks of trees. In the eastern United States, most roosts are in densely vegetated conifer stands and can include seven to fifty owls (Bent 1938, Bosakowski 1982, Marks et al. 1994). The large numbers of owls at some roosts suggests year-to-year fidelity (Johnsgard 1988b). There is some suggestion that some pairs begin nesting at or around winter roost sites (Santner 1992a, Marks et al. 1994).

Pairs apparently bond over the winter, probably in conjunction with a communal roost. They nest in cavities or adopt the former stick nest of another bird, generally corvids or buteos. Resident long-eared owls begin nesting when many northern migrants are still on their wintering ground. Consequently, it is difficult to differentiate between the migrants and the breeders. Nesting is often complete before June (Santner 1992a). It is very easy for nesting birds to be overlooked.

### *THREATS*

Overall, nesting habitat may have declined in Pennsylvania since the nineteenth century. Concerns for this species are widespread in North America and Europe (NatureServe 2005, World Owl Trust 2005). There has been a downward trend noted in long-eared owls wintering in neighboring New Jersey, as measured by Christmas Count data (Bosakowski et al. 1989). There also has been a decline in numbers migrating south from Canada at Duluth, the center of its United States range, as indicated by long-term banding data (Marks et al. 1994). Because this is one of the most poorly understood owls of the Northern Hemisphere, there is much to be learned about threats to its populations. Some of the known threats are described below.

Direct persecution formerly occurred when long-eared owls were confused with great horned owls by those who killed that species for bounties (Santner 1992a, Ulrich 1997). Illegal taking may still be occurring, especially where it might be considered a predator of game birds (e.g. pheasants, grouse, bobwhites). Road mortality is an additional threat. Because it nests in mixed habitats and forages low over the ground, it is vulnerable to collisions with vehicles. Disturbance of nest sites and roosts is also a problem, as this species flushes so easily that it is a common victim of over-enthusiastic birders or incidental foot traffic through conifer groves (Sutton and Sutton 1994, D. A. Gross, personal observation). Disturbed nests are vulnerable to predation by crows (*Corvus* spp; Marks et al. 1994), and fledglings and "branchers" are sometimes killed or taken by people (Marks et al. 1994). Many roosts formerly occupied in Berks County are no longer active probably because of disturbance (Ulrich 1997).

There is little information concerning direct poisoning from pesticides and other contaminants. In Britain, direct mortality has been documented on owls that ate vertebrates that consumed seeds treated with Dieldrin (Ratcliffe 1980). Brodofacoum (poisoned bait for rodents), organochlorides, and PCBs have also been implicated in deaths, including secondary poisonings (Hedgal and Colvin 1988, Sheffield 1999). We know little about small mammal populations in the landscape where long-eared owls are most likely to be found. Reduced winter grass and forb cover could lead to reduced rodent populations in foraging areas of this species. The effects of rodenticides in reducing the prey base (in addition to poisoning) on any and all raptor populations is unknown but could be important, especially in an agricultural setting.

Decline in conifer cover in agricultural settings and forests due to modern farming practices has resulted in habitat loss. With more mechanized agricultural techniques, there has been a trend away from smaller farms with a mix of woodlots, hedgerows, and fields to larger fields. Therefore, there has probably been a steady decline in the availability of conifer cover in the landscape where this species might forage. Also, there has been a general decline in conifer component in the forests of the Allegheny Plateau (Lutz 1930, Whitney 1990). In addition, the larger great horned owl (*Bubo virginianus*) may compete for nesting and foraging areas and may also directly predate on the much smaller long-eared owl. Raccoons (*Procyon lotor*) have been found to be consequential predators of long-eared owl nests near water in southwest Idaho (Marks 1986). Cooper's (*Accipiter cooperi*), red-tailed (*Buteo jamaicensis*), and red-shouldered hawks (*Buteo lineatus*) are nest predators in California (Bloom 1994 in Marks et al. 1994).

West Nile disease may be having a significant effect on raptor populations, which is difficult to detect, especially in rare and secretive species. Owls, including this species, are known to be vulnerable to this disease (Gancz et al. 2004) and may be undergoing significant, but largely undetected, mortality in North America.

### *CONSERVATION AND MANAGEMENT NEEDS*

This species is one of the most poorly known birds in the state (Santner 1992a, Brauning et al. 1994, Gross 1998) and one of the greatest needs is a species-specific inventory to identify nesting locations deserving protection or conservation action. The Partners in Flight program can only "guesstimate" its population and has imprecise trends for this species (Rich et al. 2004). There is a great need to inventory this species using volunteer projects such as the second Pennsylvania Breeding Bird Atlas, Important Bird Area monitoring, Pennsylvania Society for Ornithology Special Areas Project, Pennsylvania Society for Ornithology Pennsylvania Birds county reports, and raptor surveys. The additional data collected will assist conservation biologists in assessing potential nesting and roosting areas through a gap analysis of known habitat attributes.

In Britain, there is active conservation for this and other owls. Land restoration that improves foraging areas and sheltered woods nearby is employed there (Scott 1997, World Owl Trust 2005). Proactive measures could be taken as part of wildlife and land conservation programs in appropriate habitat, including private lands. This might include planting shelter belts of conifers, placing posts and nest baskets, and creating more foraging rough-grass habitat near roosting sites.

### *MONITORING AND RESEARCH NEEDS*

Monitoring this species is a considerable challenge because it is secretive and generally silent, even on its breeding ground. It is one of the hemisphere's most enigmatic species and the most poorly known of the state's owls. Although statewide surveys identify some long-eared owl nest sites (Santner 1992a, Pennsylvania Society for Ornithology Special Area Project database), this species probably evades most observers unless a species-specific survey is conducted. Even night bird surveys that are successful with other species are not effective for this one. Call-broadcast surveys are notoriously effective for some species, but not for all species in all studies (Duncan and Duncan 1997, Francis and Bradstreet 1997, Takats et al. 2001). In Great Britain, area searches in appropriate locations are preferred over call-broadcast surveys (World Owl Trust 2005). A nest area search and survey protocol is employed that Pennsylvania would do well to duplicate if a breeding population were found and adopted for regular surveys (Hardey et al. 2006). However, nest surveys must be done cautiously because pairs are sensitive to disturbance (Mikkola 1983, Hardey et al. 2006). In a statewide survey for northern saw-whet owls (*Aegolius acadicus*), no long-eared owls were detected in 161 routes (eight stops each) in 2000 and 2001, but the survey was successful with other nontargeted night birds (Gross 2000). This suggests that long-eared owls are rare breeders, but the saw-whet survey did not target the habitat mosaic where long-eared owls typically nest and may have underrepresented this species.

Collection of habitat and geographical data are needed so the potential habitat and distribution of this species can be better understood. There is a need to establish a better search image and description of breeding habitat in the state for the sake of inventory and conservation initiatives. A protocol with training has been developed in Britain that might be adaptable here (World Owl Trust 2005, Hardey et al. 2006). With more information on nesting site locations, including choice of nest tree species and local configuration of breeding habitat, a better model could be developed for identifying breeding habitat and its subsequent protection. With an eye to possible conservation initiatives, we should also study possible habitat modifications that might enhance long-eared owl nesting potential (e.g., nest baskets, conifer plantings, grass and forb plantings).

Perhaps the most immediate research need is a species-specific inventory of nesting birds. There is a critical need to find the best inventory/monitoring methods for species. Little is known about migratory behavior and links between the few breeding pairs and more common winter roosts. Relationships between these breeding and winter populations need more study. The relationship between rodent populations and the occurrence of wintering and breeding populations of long-eared owls are not known, but deserve study. Rodent populations are probably a critical limiting factor for this species (Marks et al. 1994). Better understanding of diet could be easily achieved by studying pellets. There is a critical need to study fidelity to nest sites and dispersion of young from nesting grounds through radiotelemetry and banding. This study could be part of a larger-scale multistate program. Effects of rodenticides and other poisons in agricultural landscape on long-eared owls and other rodent predators are poorly known and deserve more study. Intensive banding and recapture projects may be necessary to better understand species demographics and gene flow necessary for conservation planning (Marks et al. 1994).

*Author:* DOUGLAS A. GROSS, PENNSYLVANIA GAME COMMISSION

## Marsh Wren

Order: Passeriformes
Family: Certhiidae
*Cistothorus palustris*

The marsh wren is a small wren associated with wetland habitats (fig. 5.30). Marsh wrens were selected as a Species of Greatest Conservation Need because they have a limited and spotty distribution in the state and have undergone population declines. State breeding populations are considered Vulnerable or Imperiled. Global populations are Secure (G5, NatureServe 2009).

### *GEOGRAPHIC RANGE*

The breeding range of marsh wrens is far-reaching, extending from northwestern Canada and Maine south to Southern California, Texas, and Florida. Within this range, however, marsh wren populations are highly fragmented and restricted to appropriate freshwater and saltwater marsh habitat (Kroodsma and Verner 1997).

*Fig. 5.30.* The Marsh Wren, *Cistothorus palustris*. Photo courtesy of Reinhard Geisler.

### *DISTRIBUTION AND RELATIVE ABUNDANCE IN PENNSYLVANIA*

The majority of Pennsylvania's marsh wrens can be found in the glaciated northwest, near the Lake Erie Shore, in the glaciated northeast, and in large marshes of Philadelphia and Delaware counties (fig. 5.31). The marsh wren is rare in that it has two distinctive forms that breed in Pennsylvania. The nominate race, *Cistothorus palustris palustris*, breeds along the Atlantic Coast from Rhode Island to Virginia. In Pennsylvania, it is found in the tidal marshes of the lower Delaware River. The second form, the prairie marsh wren (*Cistothorus palustris iliacus*) is the breeding marsh wren of the rest of Pennsylvania. This race is larger than *C. p. iliacus* and differs with its buffy chin, throat, and belly where *C. p. palustris* has white (Reid 1992).

The 3,000-acre Conneaut Marsh in Crawford County supports the state's largest population of marsh wrens, where surveys indicate a population of more than 500 territorial males (McWilliams and Brauning 2000). Marsh wrens can also be found scattered throughout Pennsylvania typically in marshes larger than 20 acres. During the first Pennsylvania Breeding Bird Atlas, marsh wrens were located in 77 of 4,928 atlas blocks (Reid 1992). In surveys of wetland birds conducted by the Pennsylvania Game Commission at seventy-three stations in twenty emergent wetlands of Crawford, Erie, Monroe, Pike, Tioga, and Wayne counties during 1992, 1993, 1996, and 1997, marsh wrens were detected during 23 percent of all

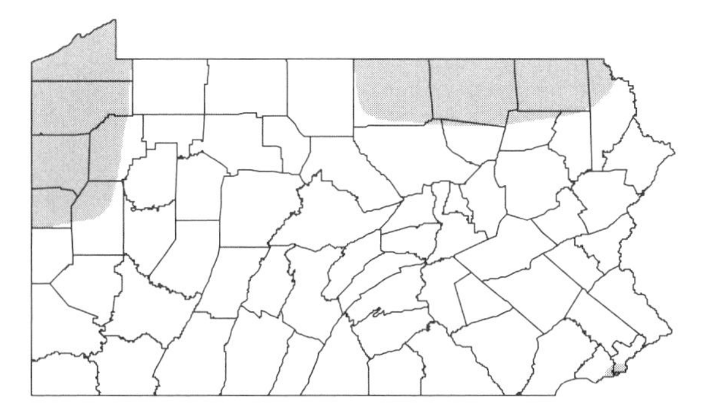

*Fig. 5.31.* Highest probability of occurrence of the Marsh Wren, *Cistothorus palustris*.

surveys, and the average relative abundance (birds per station) across all years for marsh wrens was 0.69 birds per survey (Brauning 1998a).

Loss of wetlands has resulted in significant declines of marsh wrens. With high developmental pressures, the coastal subspecies, *C. p. palustris*, is thought to be undergoing especially high population declines (D. A. Gross and R. Criswell, personal communication). In the late 1800s, marsh wrens were considered the most common breeding bird in marshes along the Delaware River ten to twelve miles south of Philadelphia. Because of developmental pressures, several historic breeding areas outside of Philadelphia have been destroyed. Examples of other historic breeding areas that no longer support marsh wrens include wetlands along the Delaware River in lower Bucks County and near New Galilee, Beaver County (McWilliams and Brauning 2000). Further evidence of decline is seen in a comparison between the first and second Pennsylvania Breeding Bird Atlas. In the second atlas, volunteers recorded marsh wrens in only fifty-four blocks compared with seventy-seven blocks during the first atlas.

### *COMMUNITY TYPE/HABITAT USE*

Throughout the year marsh wrens occupy a diversity of wetland habitats, including fresh, brackish, seasonal, semipermanent, or permanent wetlands, that contain emergent aquatic vegetation, such as cattail (*Typha* spp.), bulrush (*Scirpus* spp.), sedges (*Carex* spp.), and common reeds (*Phragmites* spp.). Marsh wrens are less commonly found in stands dominated by bluejoint (*Calamagrostis canadensis*) and reed canary grass (*Phalaris arundinacea*) because these plants are relatively short and have little stem strength. Marsh wrens may be found along the banks of slow-moving streams and rivers, where emergent vegetation has developed. They may also occupy restored wetlands (Zimmerman et al. 2002). Marsh wrens predominately occupy marshes larger than twenty acres (Cashen and Brittingham 1998) but may use wetlands smaller than 1 acre (Reid 1992). Marsh wrens may be confused with sedge wrens (*Cistothorus platensis*), which can be found in similar environments of Pennsylvania. The preferred habitats of the two species differ, however. Sedge wrens are typically found in wet meadows and grasslands while marsh wrens prefer cattail and bulrush marshes with an abundance of standing water.

### *LIFE HISTORY AND ECOLOGY*

Marsh wrens eat invertebrates, especially insects and spiders. They commonly forage by hopping and creeping near the marsh floor, gleaning insects from the stems and leaves of cattails, fallen bulrush, and other marsh vegetation. Marsh wrens have also been observed gleaning insects from at or just below the water surface and will occasionally sally into the air to catch a flying insect (Kroodsma and Verner 1997).

Marsh wrens are migratory in Pennsylvania. In Pennsylvania's Coastal Plain, marsh wrens typically arrive on their breeding grounds as early as the first week of April. In all other parts of the state, marsh wrens arrive the third or fourth week of April. Marsh wrens typically depart Pennsylvania before the third week of October (McWilliams and Brauning 2000). A section along the Delaware River near the town of Tinicum is the only region of Pennsylvania where, in some years, marsh wrens can be found until mid-January.

Marsh wrens are polygynous, with the percentage of males attracting more than one female varying greatly among populations. Males build numerous dome-shaped nests of cattail, sedge, or grass placed 1 to 3 feet over standing water. Prospective females will often inspect several nests before accepting a nest by lining it with feathers, strips of grass or sedge, cattail down, or other soft materials before laying eggs. A female may build a new nest, which is thought to be the more common behavior in New York (Kroodsma and Verner 1997).

### *THREATS*

Emergent wetland habitat loss is the largest threat to marsh wrens. From 1956 to 1979, Pennsylvania had a net loss of nearly 42,500 acres of emergent wetlands. This represents a 38 percent net loss of the state's emergent wetlands in seventeen years. Much of this net loss of emergent wetlands (64%) was due to changes to other vegetated wetland types (forested and shrub wetlands) not suitable to marsh wrens. The remaining net loss is attributed to urban development, conversion to farmland, channelization, and pond construction (Tiner 1990).

Another threat that may be a reason for decline is invasive plants, such as purple loosestrife (*Lythrum salicaria*) and multiflora rose (*Rosa multiflora*), which are replacing native wetland vegetation. These exotic species may create population sinks where marsh wrens may use wetlands containing a high proportion of exotics and have a much lower breeding success rate than marshes that primarily contain native vegetation.

*CONSERVATION AND MANAGEMENT NEEDS*

Statewide efforts should focus on protecting the state's approximately 28,000 acres of emergent wetlands. Emergent wetlands provide critical habitat for marsh wrens and more than half of Pennsylvania's avian species of concern.

*MONITORING AND RESEARCH NEEDS*

In Pennsylvania, regular statewide inventory and monitoring is needed to accurately describe the population trends of this species. Also needed is a detailed experimental and field investigation that explores the ecological relationship between invasive plant species, such as purple loosestrife (*L. salicaria*) and multiflora rose (*R. multiflora*), to determine the potential effect of these invasives on the breeding success of marsh wrens.

Intensive monitoring is necessary to delineate the precise range of marsh wrens in Pennsylvania, regularly track demographic and reproductive trends, and document patterns of habitat and space use by marsh wrens. Annual surveys of more than fifty emergent wetlands should be conducted annually where marsh wren populations have occurred since 1990. Additional reconnaissance surveys of emergent wetlands throughout the state should be conducted where potential marsh wren populations may occur but have not been documented in the past twenty years. At wetlands where breeding is confirmed, more intensive study of the breeding population is recommended wherein the status, behavior, and ecology of the species can be studied more closely.

*Author:* DANIEL P. MUMMERT, PENNSYLVANIA GAME COMMISSION

## Golden-winged Warbler

Order: Passeriformes
Family: Parulidae
*Vermivora chrysoptera*
(also *Responsibility Concern*)

The golden-winged warbler was selected as a species of High-Level Concern because of widespread declines throughout its range and because it is a Pennsylvania Responsibility Species with more than 8 percent of its global population breeding in Pennsylvania (fig. 5.32). It is a National Bird of Conservation Concern (U.S. Fish and Wildlife Service 2008) and Near Threatened (International Union for Conservation of Nature 2006). Global populations are Apparently Secure (G4, NatureServe 2009).

*GEOGRAPHIC RANGE*

The breeding range of the golden-winged warbler occurs entirely within Canada and the United States. The range generally has been shifting northward, coincident with or just before range expansion of the blue-winged warbler (*Vermivora pinus*; Confer 1992). On the basis of data from 1994 to 2003 (Sauer et al. 2006), the range may be separated into a Great Lakes–Upper Midwest region, an Appalachian Mountains region, and several peripheral locations in New England. Distribution of golden-winged warblers is generally fragmented throughout much of the breeding range. The winter range occurs in southern Central America and northern South America, including central Guatemala and northern Honduras southward to northern and western Venezuela and western Columbia (Confer 1992).

*Fig. 5.32.* The Golden-winged Warbler, *Vermivora chrysoptera*. Photo courtesy of Geoff Malosh.

*DISTRIBUTION AND RELATIVE ABUNDANCE IN PENNSYLVANIA*

The range of the golden-winged warbler in Pennsylvania extends from the southwest to the northeast, excluding the northwestern and southeastern regions of the state (fig. 5.33). Golden-winged warblers are locally and irregularly distributed within this range (McWilliams and Brauning 2000), probably because suitable habitat (early successional forest) is ephemeral and occurs only in isolated "pockets" across the landscape. Postsettlement populations of golden-winged warblers in the eastern United States likely peaked between the late 1700s and early 1900s in response to massive forest clearing and farm abandonment, then declined as regenerating forests matured and the rate of farm abandonment declined (Gill 1980, Confer 1992, Askins 2001, Lorimer 2001, Trani et al. 2001). Actual population sizes are unknown, but data from the North American Breeding Bird Survey indicate that golden-winged warblers have declined since 1966 at an annual rate of 2.5 percent within the breeding range as a whole, 8.6 percent in the Northeast (U.S. Fish and Wildlife Service Region 5), and 9.3 percent in Pennsylvania (Sauer et al. 2006; fig. 5.34). Since 1990, the annual decline in Pennsylvania has been 17.5 percent, suggesting that immediate action is required to conserve golden-winged warblers in the state.

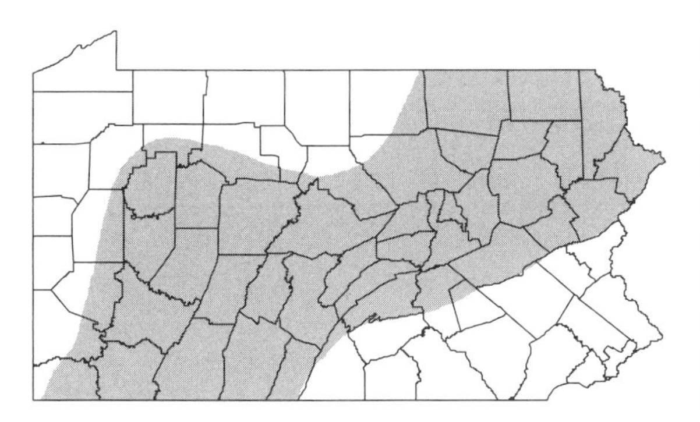

*Fig. 5.33.* Highest probability of occurrence of the Golden-winged Warbler, *Vermivora chrysoptera*.

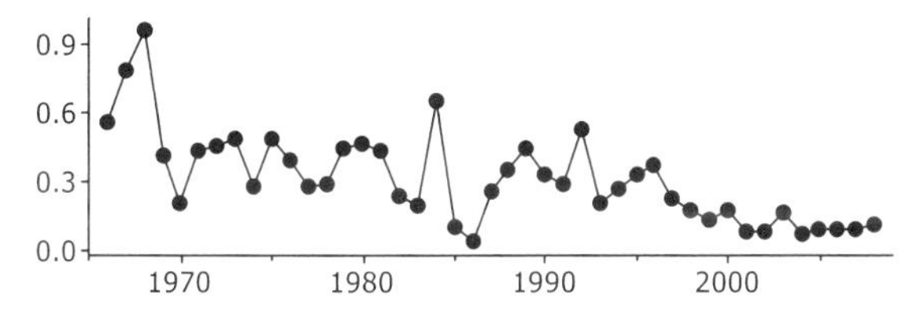


*Fig. 5.34.* Golden-winged Warbler, *Vermivora chrysoptera*, population trends from the Breeding Bird Survey.

*COMMUNITY TYPE/HABITAT USE*

Golden-winged warblers use a variety of early successional habitats during the breeding season, including pine-oak barrens, wetlands (e.g., bogs, beaver [*Castor canadensis*] meadows), recently burned or clear-cut forest, abandoned farmland, utility rights of way, and reclaimed or abandoned surface mines. Occasionally, golden-winged warblers may occur within relatively mature forest that has openings in the canopy and a dense shrub layer; examples include partial timber cuts, stands damaged by gypsy moths, and swamp forest (McWilliams and Brauning 2000). Although a wide variety of general habitat types are used, the species has specific microhabitat requirements. Most populations require a mosaic of shrubby thickets and herbaceous patches located along a forest edge. Occasional trees or snags (>5 m tall) throughout the shrubby areas are useful as song perches. The shrub layer (woody plants 1.5–3 m tall) in a typical territory consists of a dense thicket of species such as scrub oak (*Quercus ilicifolia*), dwarf chinkapin oak (*Quercus prinoides*), dogwood (*Cornus* spp.), American hazelnut (*Corylus americana*), alder (*Alnus* spp.), honeysuckle (*Lonicera* spp.), and aspen (*Populus tremuloides* and *P. grandidentata*). A shrub layer consisting mainly of stump sprouts, such as those resulting from cutting of oaks (*Quercus* spp.) and maples (*Acer* spp.), probably do not provide the type of shrub cover preferred by golden-winged warblers (Kubel 2005). Cutting aspen, however, is favorable because it results in dense root suckering (Roth and Lutz 2004). Patches of herbaceous vegetation are used for nest placement and consist predominantly of goldenrod, grasses, and scattered woody saplings (approximately 0.5 m tall); rough-stemmed goldenrod (*Solidago rugosa*) appears to be particularly important (Kubel 2005). Blueberry (*Vaccinium* spp.), sweetfern (*Comptonia peregrina*), mountain laurel (*Kalmia latifolia*), and ferns constitute a common plant community on disturbed sites in Pennsylvania, but such ground cover is rarely used for nesting.

*LIFE HISTORY AND ECOLOGY*

Males arrive on breeding grounds two to seven days before females (Ficken and Ficken 1968a, Confer 1992) and establish territories of about 0.4–2.6 ha (Ficken and Ficken 1968b, Murray and Gill 1976, Will 1986); size and boundaries of territories may be determined by population density, interactions with other males, and the spatial and structural characteristics of vegetation (Confer 1992). Males arrive in Pennsylvania as early as the last week of April or first week of May (Confer 1992, J. E. Kubel, personal observation). Pair formation occurs shortly after the arrival of females and nest building occurs within a few days thereafter (Confer 1992), but cold and wet weather may delay this process by as much as ten days (Will 1986). At one breeding area in central Pennsylvania, females arrived as early as May 8 and nest building commenced as early as May 15 (J. E. Kubel and R. H. Yahner, unpublished data).

Nest sites are usually characterized by open areas of forbs (especially goldenrod), grasses, and short (approximately 0.5 m) saplings that occur within or adjacent to shrub thickets near a forest edge (Confer 1992, Klaus and Buehler 2001, Kubel 2005). Nests are located on the ground and, although bulky, are well concealed by vegetation. Most nests incorporate one to four woody saplings approximately 0.5 m tall, which adults use to descend to the nest when feeding young (Will 1986, Kubel 2005).

Egg laying usually commences one to six days following completion of nest construction (Confer 1992, Kubel 2005). Golden-winged warblers sometimes initiate second or third nesting attempts after failure of previous attempts (Confer 1992, Kubel 2005), but there are no published reports of additional attempts subsequent to successful fledging of a brood.

Potential nest predators in Pennsylvania include snakes (e.g., eastern ratsnake [*Pantherophis alleghaniensis*], northern black racer [*Coluber constrictor constrictor*], eastern gartersnake [*Thamnophis sirtalis sirtalis*]), mammals (e.g., eastern chipmunk [*Tamias striatus*], striped skunk [*Mephitis mephitis*], Virginia opossum [*Didelphis virginiana*], common raccoon [*Procyon lotor*], red fox [*Vulpes vulpes*]), and birds (e.g., American crow [*Corvus brachyrhynchos*] and blue jay [*Cyanocitta cristata*]). The brown-headed cowbird (*Molothrus ater*) is the only known brood parasite. Golden-winged warblers and blue-winged warblers produce fertile hybrids by interspecific pairing or by extra-pair fertilization, which has resulted in development of mosaic hybrid zones, where backcrossing threatens persistence of genetically pure individuals (Gill 2004, Dabrowski et al. 2005).

### *THREATS*

Main threats to breeding populations of golden-winged warblers in Pennsylvania are: (1) low/declining availability of suitable habitat as a result of forest maturation, suppression of wildfire, infrequent use of even-aged forest management practices (e.g., clear-cutting) in areas suitable for colonization by golden-winged warblers, infrequent conversion of farmland to old field habitat, and incompatible types of land use (e.g., agriculture, commercial and residential development) in areas of potentially suitable habitat; and (2) continued introgressive hybridization and replacement of pure individuals or populations as a result of interspecific mating or extra-pair fertilization, backcrossing, and range expansion by the blue-winged warbler (Gill 1980, Confer 1992, 2006, Gill 1997, Confer et al. 2003). Whether genetically pure populations of golden-winged warblers still occur in Pennsylvania is not known, so the magnitude of this latter threat is not completely understood. Recent genetic studies (Gill 1997, Shapiro et al. 2004, Dabrowski et al. 2005) have shown that physical appearance (phenotype) alone is not sufficient to identify hybrids. Hence, genetic analysis now appears to be the only reliable way to identify pure individuals. Additional threats include brood parasitism by the brown-headed cowbird (Confer et al. 2003) and increased nest predation pressure from artificially inflated predator populations (Hogrefe et al. 1998).

### *CONSERVATION AND MANAGEMENT NEEDS*

The highest priorities for conservation of golden-winged warblers in Pennsylvania should be to identify, maintain, and enhance genetically pure populations. Whether pure populations still occur in Pennsylvania needs to be determined. If such populations exist and are identified, management strategies that ensure the viability of those populations (e.g., increasing habitat availability, preventing introgression of blue-winged warbler mtDNA) need to be developed. If genetically pure populations do not exist, then a conservation strategy needs to be decided on in consultation with the United States Fish and Wildlife Service, Partners in Flight, the Golden-winged Warbler Working Group, and other relevant organizations. Of particular importance would be whether to manage golden-winged and blue-winged warblers as a single species in Pennsylvania.

### *MONITORING AND RESEARCH NEEDS*

Golden-winged and blue-winged warblers (and hybrids) must be monitored over the long term to be aware of changes in range/distribution; to identify source and sink habitats/populations; to understand the factors contributing to the replacement of golden-winged warblers by blue-winged warblers; to detect colonization and abandonment of habitats/breeding areas; to determine population response to conservation action/inaction; to determine longevity of various habitat types created by particular techniques; and to assess overall values of particular conservation actions. Information gathered in pursuit of these objectives would be vital to adaptive management of the two species in Pennsylvania. Long-term monitoring plans should occur at the state level to maintain a general understanding of the status of golden-winged and blue-winged warblers in Pennsylvania. Monitoring should occur also at the site level to compile meaningful data sets that can be used to answer important questions about the ecology and long-term conservation of the species. Inclusion of genetic sampling during population monitoring would be extremely beneficial at any level.

Several lines of research are needed to identify suitable areas for long-term management and monitoring of breeding populations of golden-winged warblers in Pennsylvania. These include an extensive study that assesses population size and genetic purity of golden-winged and blue-winged warblers at all known breeding areas and identifies specific areas with the highest proportion of genetically pure golden-winged warblers; an intensive demographic study of multiple breeding populations of golden-winged warblers to identify the most productive breeding areas and habitat types (Kubel 2008); a field study that investigates what habitat conditions, if any, favor golden-winged warblers to the exclusion of blue-winged warblers and hybrids; a management study that experiments with different habitat creation/maintenance techniques (e.g., prescribed burning, clear-cutting) and monitors colonization and use of manipulated habitats by golden-winged and blue-winged warblers (and hybrids); and a study that identifies potential habitats (i.e., areas that do not currently support golden-winged warblers but could be converted to suitable habitat given proper management action) and investigates the feasibility of acquiring such habitats (if not already under state ownership).

*Author:* JACOB E. KUBEL, M.S., NATURAL HERITAGE AND ENDANGERED SPECIES PROGRAM, MASSACHUSETTS DIVISION OF FISHERIES AND WILDLIFE

## Cerulean Warbler

Order: Passeriformes
Family: Parulidae
*Dendroica cerulea*
(also *Responsibility Concern*)

The cerulean warbler (*Dendroica cerulea*) is a small, canopy-dwelling wood-warbler (fig. 5.35). In Pennsylvania it is listed as a Species of High-Level Concern. Because of severe population declines, it is currently being considered for listing as Threatened under the United States Endangered Species Act (U.S. Fish and Wildlife Service 2002a). It is considered a Species of Special Concern in Canada (McCracken 1993), and is a Partners in Flight Continental Watchlist species (Rich et al. 2004). It is included on the list of Species of Greatest Conservation Need throughout most of the Northeast. State and global populations are considered Apparently Secure (S4B, G4, NatureServe 2009).

### *GEOGRAPHIC RANGE*

Ceruleans breed in deciduous forests through much of eastern Northern America, from eastern North and South Dakota, southern Minnesota, Wisconsin, Ontario, and Quebec south to eastern Oklahoma and northeastern Texas, east through the central Gulf States to central Georgia (American Ornithologists' Union 1998). Within most of this range they are rather sparse and locally distributed, except in the core of their range in the central Appalachians from southwestern Pennsylvania south to the Cumberland Plateau regions of Kentucky and Tennessee, where they can be the most common forest warbler (Hamel 2000).

Ceruleans winter in the lower to middle elevations (800–2,200 m; Stotz et al. 1996) of the Andes of

*Fig. 5.35.* The Cerulean Warbler, *Dendroica cerulean*. Photo courtesy of Frode Jacobsen.

South America from Venezuela and Columbia south to eastern Peru and northern Bolivia. Although migration routes remain poorly known, en route ceruleans have been reported primarily from the Greater Antilles and Caribbean slopes of Central America (American Ornithologists' Union 1998, Hamel 2000).

### *DISTRIBUTION AND RELATIVE ABUNDANCE IN PENNSYLVANIA*

Cerulean warblers were reported from 17 percent of blocks in the first Pennsylvania Breeding Bird Atlas, but distribution was highly irregular. Fully 43 percent of the blocks documenting cerulean presence were located in seven southwestern counties, primarily in the Pittsburgh Plateau physiographic section. Elsewhere in the state, ceruleans were rather uncommon and local. Minor areas of concentration include the Lake Erie Coastal Plain and portions of the Allegheny, Susquehanna, and Delaware River valleys (Ickes 1992c, McWilliams and Brauning 2000). Their current distribution pattern is similar (fig. 5.36).

The overall range of the cerulean has been shifting northeastward over time (Hamel et al. 2004). Within Pennsylvania, this shift has been apparent as the species was once rare or absent from all but the southwestern part of the state (e.g., Harlow 1918, Todd 1940a), but seems to have expanded across much of the state since the Breeding Bird Surveys began. Whether this truly represents a northward shift in the species' range or a recolonization of maturing forests that have grown up since the extensive deforestation of the early 1900s is unclear. Breeding Bird Survey data indicate that the overall abundance of ceruleans has decreased significantly in Pennsylvania, at a rate of 3.0 percent per year (Sauer et al. 2005; fig. 5.37). However, that figure is probably driven by declines in the western third of the state; populations in central and eastern Pennsylvania have generally increased since the 1960s.

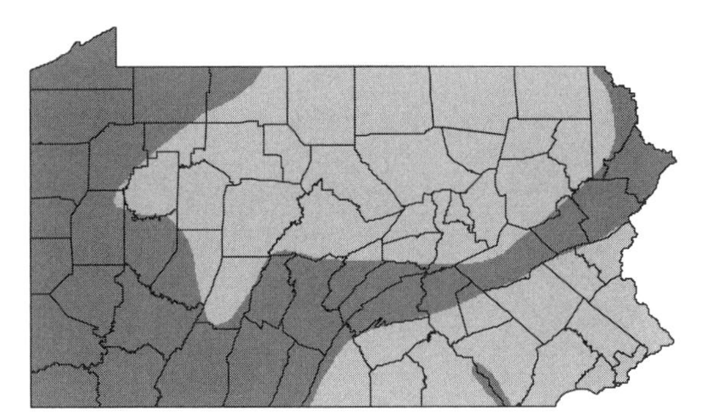

*Fig. 5.36.* Primary (darkened shading) and secondary (lighter shading) distribution of the Cerulean Warbler, *Dendroica cerulean*.

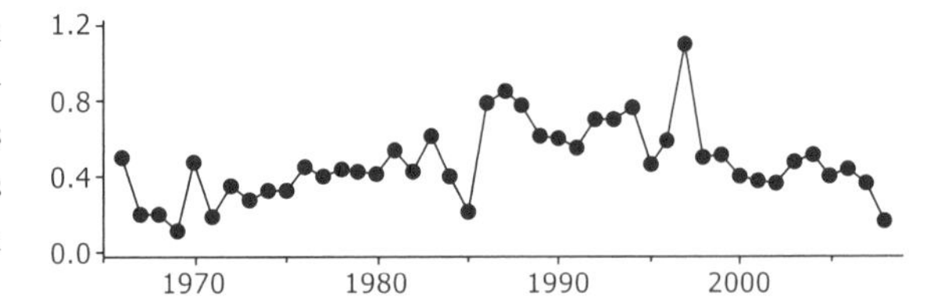


*Fig. 5.37.* Cerulean Warbler, *Dendroica cerulean*, population trends from the Breeding Bird Survey.

### *COMMUNITY TYPE/HABITAT USE*

The cerulean warbler breeds in extensive tracts of tall, mature deciduous forest. Considered to be an area-sensitive species, ceruleans avoid forest tracts below a certain size (Robbins et al. 1989, Weakland and Wood 2005). That tract size threshold varies regionally, from as small as 10 ha in Ontario to 1,600 ha in the Mississippi Alluvial Valley (Hamel 2000). Within large forest tracts, ceruleans breed in two general topographic locations: ridge tops and riparian corridors along waterways (Dettmers and Bart 1999, Rosenberg et al. 2000). Ridge-top sites are almost exclusively in oak-hickory forests; ceruleans appear to be rare or absent in northern hardwood or mixed hardwood-conifer types (Rosenberg et al. 2000, Stoleson 2004). Valley bottom habitats may include sycamore or elm-dominated riparian forests (McWilliams and Brauning 2000, Inman et al. 2002). In either habitat type, ceruleans favor areas of broken canopy, widely spaced large trees, and dense foliage 12–18 m up, often with an open understory (Jones and Robertson 2001, Inman et al. 2002). Structural complexity may be a critical attribute of high-quality habitat for this species; unlike many other area-sensitive birds, ceruleans seem to have an affinity for internal gaps or other openings and have been described as a disturbance-dependent bird of mature forests (Hunter et al. 2001, Rodewald 2004). Growing evidence suggests that certain types of timber management (e.g., partial harvests) may increase habitat suitability for ceruleans by increasing structural heterogeneity within even-aged forest stands (Rodewald and Yahner 2000, Stoleson 2004).

### *LIFE HISTORY AND ECOLOGY*

Cerulean warblers are insectivorous, catching prey primarily by gleaning from foliage high in the canopy (Hamel 2000). A study in Illinois found ceruleans to be extremely selective in foraging substrate, preferring hickories and silver maple but avoiding red maple (Gabbe et al. 2002). What little is known of their diet indicates they feed primarily on lepidopteran larvae and homopterans (Sample et al. 1993, Hamel 2000).

The cerulean warbler is a long-distance Nearctic-Neotropical migrant that spends less than five months of the year on the breeding grounds in Pennsylvania. Spring migration peaks in Pennsylvania during the middle two weeks of May (McWilliams and Brauning 2000). Fall migration is less well documented, as the birds tend to disappear quietly from breeding grounds by late July or early August; ceruleans are rarely reported after early September.

Although a socially monogamous species, ceruleans often appear to breed semicolonially, occurring in loose aggregations of territories while seemingly identical habitat nearby remains unoccupied (Rosenberg et al. 2000). Territory size can range from <0.4 ha to several hectares per pair, and average territory density varies considerably as well. Hamel (2000) reports the average density among 332 Breeding Bird Censuses as 43 pair per 100 ha. Current densities on the Allegheny High Plateau in northwestern Pennsylvania reach 76 pair per 100 ha in preferred habitat (S. H. Stoleson, unpublished data).

Ceruleans arrive on the breeding grounds about the time when oaks begin to leaf out. Males arrive first. Females arrive a week or more later, choose a mate, and begin building nests without male assistance. Ceruleans typically place their nests rather high in the lower canopy on the lowest horizontal branch of a large tree, often where a small side branch joins, and usually with open space below the nest. Nests are often located near a canopy gap (Oliarnyk and Robertson 1996, Rogers 2006). They are usually single brooded but will renest after an initial failure.

### *THREATS*

As an area-sensitive species, ceruleans are most threatened by the loss and fragmentation of its mature forest-breeding habitat, particularly in the densely populated southwestern Pittsburgh Plateau section, where historically the bird has been most abundant. Ceruleans are vulnerable to large-scale forest disturbances and fragmentation both directly through the loss of available habitat and indirectly through reduced densities in remaining forest (Robbins et al. 1989, Weakland and Wood 2005). Even natural disturbances affect populations; canopy damage from a large ice storm greatly diminished subsequent nesting success in an Ontario cerulean population (Jones et al. 2001). Actions or processes that reduce the structural complexity of forests are likely to also reduce their suitability for ceruleans. For example, ceruleans may be sensitive to the effects of deer overabundance on forest structure: in a controlled enclosure experiment, ceruleans occurred only in treatments with lower densities of deer (deCalesta 1994). In contrast, small-scale disturbances that increase structural complexity of forests, such as uneven-aged forestry techniques, may benefit ceruleans. Ross et al. (2001) documented successful breeding by ceruleans in partially harvested stands with as little as 12 $m^2$ basal area per hectare.

Although brown-headed cowbirds (*Molothrus ater*) parasitize cerulean nests, at least in some populations, they are unlikely to pose a significant threat, as ceruleans prefer extensively forested areas and nest very high (Hamel 2000). A Michigan study found relatively low rates of parasitism in ceruleans (<10%) in areas where understory-nesting hooded warblers (*Wilsonia citrina*) experienced rates in excess of 60 percent (Rogers 2006).

### *CONSERVATION AND MANAGEMENT NEEDS*

Although cerulean warbler populations appear to be increasing in the eastern half of Pennsylvania, those in the western half, with the bulk of the state's birds, are declining rapidly. Therefore, the primary conservation focus for Pennsylvania should be to reverse population declines and develop and maintain high-quality breeding habitat. Specific conservation strategies for ceruleans (synthesized from Rosenberg et al. 2000, Robertson and Rosenberg 2003, and Rich et al. 2004) include maintaining current levels of forest coverage within the state and minimizing fragmentation of remaining, large contiguous forest tracts. In addition, it is important to develop guidelines for timber management that promote structural complexity and maintain mature stands, especially at topographically appropriate sites. Long-range forest management plans are needed at as large a scale as possible to designate tracts that will be mature at each stage of the plan and to maintain connections between existing mature forest patches. Surveys should be conducted to identify important populations and sites on public and private land. Once identified, these areas should be targeted for management practices that protect or enhance populations.

*MONITORING AND RESEARCH NEEDS*

Populations of cerulean warblers have been assessed primarily through standard survey methods, such as the Breeding Bird Survey. However, such standard methods may be biased for forest-interior species like the cerulean (Rosenberg et al. 2000). Therefore, a targeted monitoring program should be designed and conducted to better track population trends of ceruleans. Also, because seemingly healthy populations elsewhere appear to function as sink populations (Jones et al. 2004, Rogers 2006), monitoring of nest success at selected sites across Pennsylvania should be conducted to gauge the health of cerulean warbler populations in the state. As forest management guidelines are developed, their effects on both cerulean abundance and productivity should be assessed to better understand how forest structure interacts with demography to determine habitat quality and to provide input on refinement of those guidelines.

The top research needs for successful conservation of the cerulean warbler have been discussed extensively by Hamel et al. (2004), Rich et al. (2004), and Rosenberg et al. (2000). Those most relevant to Pennsylvania are summarized here. As the greatest threats to ceruleans in Pennsylvania are the loss and degradation of its mature forest habitat, a better understanding is needed of the bird's need for and response to landscape configuration, patch size and shape, structural complexity, gaps, and specific floral elements (e.g., American chestnut). Researchers need to identify critical habitat components for ceruleans at multiple spatial scales.

In addition, we need to better determine how forest harvest and other management practices (such as oil and gas development), natural forest maturation, forest pests and pathogens, and effects of deer overabundance affect breeding habitat quality for ceruleans. Activities that alter forest structure and composition have great potential to affect ceruleans either positively or negatively, yet very little is known about how these activities influence habitat quality for ceruleans or other forest species.

The few populations studied appear to be functioning as population sinks (Jones et al. 2004, Rogers 2006), yet our understanding of the species' demography remains poor. Determining what factors drive population declines at the local and larger levels is vital for developing management strategies for reversing those declines.

*Authors:* SCOTT H. STOLESON, USDA FOREST SERVICE; FREDERICK C. SECHLER, JR., PENNSYLVANIA NATURAL HERITAGE PROGRAM, THE NATURE CONSERVANCY

## Prothonotary Warbler

Order: Passeriformes
Family: Parulidae
*Protonotaria citrea*

The prothonotary warbler is a large, plump, and short-tailed warbler that dwells in swampy forests, primarily in the southeastern United States (fig. 5.38). It was selected as a Species of High-Level Concern and is currently listed as a Candidate Rare Species in Pennsylvania because of its limited distribution within the state. It is a Partners in Flight Priority Species. Global populations are Secure (G5), and Pennsylvania breeding populations are considered Imperiled or Vulnerable (S2S3B, NatureServe 2009).

*GEOGRAPHIC RANGE*

During the breeding season, the prothonotary warbler is found in appropriate habitats throughout most of the southeastern United States. Within the northern half of the eastern United States, this species is found locally in scattered populations as far north as Connecticut and New York, with a small population ranging into extreme southwestern Ontario, Canada. In the Midwest, this species' breeding range has been expanding northward, especially in the Mississippi Valley between Wisconsin and Minnesota. During the nonbreeding season, the bulk of the population winters in the coastal lowlands of Panama, northern Venezuela, and northern Colombia, concentrated particularly in mangrove forests (Petit 1999).

*DISTRIBUTION AND RELATIVE ABUNDANCE IN PENNSYLVANIA*

A rare, but apparently increasing, species in Pennsylvania, the prothonotary warbler was detected in <1 percent of the state's atlas blocks in the first Breeding Bird Atlas (1983–1989; Leberman 1992e). During this effort, individuals were observed summering in most major river drainage basins in the state; however, many of these observations were of single males and may not represent breeders (Leberman 1992e). In general, in Pennsylvania this species is a rare and local breeder in the southeastern Piedmont and Glaciated northwest regions and is accidental elsewhere (McWilliams and Brauning 2000). Small populations may be regularly found in the Conneaut and Pymatuning Swamps of Crawford County, and at a few points along the lower Susquehanna River in Lancaster County (McWilliams and Brauning 2000; fig. 5.39).

*Fig. 5.38.* The Prothonotary Warbler, *Protonotaria citrea*. Photo courtesy of Geoff Malosh.

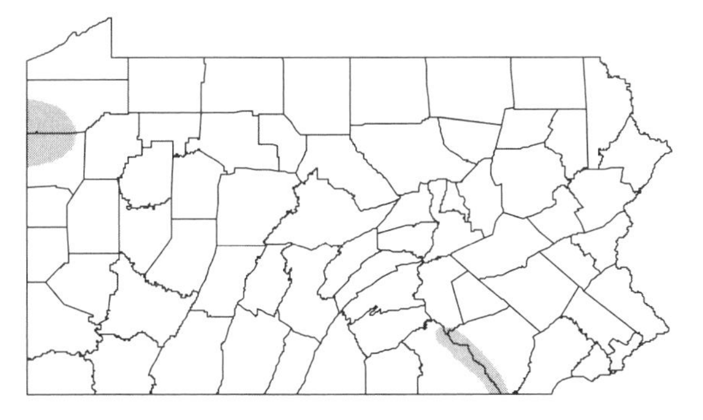

*Fig. 5.39.* Highest probability of occurrence of the Prothonotary Warbler, *Protonotaria citrea*.

### *COMMUNITY TYPE/HABITAT USE*

A species of forested wetlands, the prothonotary warbler breeds in a variety of habitats characterized by standing or slowly flowing water and some canopy cover, including seasonally flooded bottomland forests, bald cypress (*Taxodium distichum*) swamps, and river or lake edges (Petit 1999). Also important is the presence of low, rotting trees with shallow-nesting cavities, and moss for use as nesting material. Apparently area sensitive, this species requires relatively extensive habitat avoiding forests <100 ha and forested river borders that are <30 m wide (Kahl et al. 1985, Robbins et al. 1989). Within Pennsylvania, this species typically inhabits large swamps and wet woodlands bordering lakes, beaver (*Castor canadensis*) ponds, and forested river floodplains (McWilliams and Brauning 2000).

### *LIFE HISTORY AND ECOLOGY*

A medium distance, complete migrant, the prothonotary warbler is found very locally in Pennsylvania in forested wetlands between late April and early September (McWilliams and Brauning 2000). On the breeding grounds, territory size appears to vary by location and habitat type, but typically ranges between 0.5 and 2 ha, with the largest territory sizes observed in the more northerly populations (Petit 1999, Ontario Birds at Risk 2005). During the nonbreeding season, this species appears to be nonterritorial in its mangrove wintering habitat (Lefebvre et al. 1994).

The prothonotary warbler nests in tree cavities in wooded swamps or other flooded forest types (Petit 1999). Seasonally monogamous, this species primarily nests in natural cavities, abandoned woodpecker holes, or human-provided nest boxes that are located over or near standing water. Second broods are commonly attempted following successful first nests in southern populations, but are rarely attempted in northern populations (Petit 1999). Nests are vulnerable to predation, inundation from floods, destruction by competing cavity nesters, and even cowbird parasitism (Petit 1999). Reproductive output can be dramatically reduced by nest predation (Hoover 2006). Unlike most cavity nesters, the prothonotary warbler is a regular host of the brown-headed cowbird (*Molothrus ater*) with up to 50 percent of nests parasitized in some populations (Flaspohler 1996, Hoover 2003). Rates of brood parasitism are particularly high in moderately to highly fragmented landscapes (Hoover 2003). Cowbird parasitism primarily affects this warbler through egg removal, reduced hatching success, and reduced nestling survival (Petit 1989, Hoover 2003).

### *THREATS*

Like most wetland-associated species, habitat loss is probably the greatest threats to this species. In Pennsylvania, >50 percent of original wetland acreage has been lost statewide (Tiner 1990), and although rates of wetland destruction have declined, much of what habitat remains has been degraded. On the basis of sampling by the Pennsylvania Department of Environmental Protection, about 18 percent of stream miles sampled statewide were impaired. The primary causes of impairment of Pennsylvanian streams were agricultural and urban runoff, and acid-mine drainage, all of which have the potential to affect the prothonotary warbler by reducing the availability of its invertebrate prey (Goodrich et al. 2002).

Alteration or destruction of riparian forests may also affect habitat availability and suitability for this species. In the lower forty-eight states, only 10 percent of the original bottomland forests remain (Dickson et al. 1995), and >6 percent of the remaining forested swampland was drained or converted to other uses during the 1990s (Ontario Birds at Risk Program 1995). Continued habitat loss and fragmentation may lead to increases in nest predation or cowbird parasitism. Even partial harvesting of riparian forests can potentially affect this species by removing nest cavities and by affecting flooding regimes (Petit 1999). In addition, reduction of the area of remaining riparian forest, or reduction of the forest to a narrow buffer, may affect the suitability of this habitat for this somewhat area-sensitive species.

In some locations, competition for nest cavities can seriously affect settlement rates or nesting success. In northern parts of its range, house wrens (*Troglodytes aedon*) can be major competitors by making cavities unavailable or by taking over cavities formerly occupied by prothonotary warblers (Walkinshaw 1953, Petit 1989, Brush 1991). In general, competition and predation is a larger problem in cavities away from the water (Petit 1999). Rates of nest predation tend to be highest for those nests located over dry ground (including newly exposed mudflats) and water shallower than 60 cm, especially in areas where raccoons are abundant (Hoover 2006).

Contaminants and pollutants could also affect prothonotary warblers, but their importance is poorly studied. Direct contact with pesticides is probably limited on their breeding grounds. However, because pesticides, heavy metals, and other pollutants often accumulate in aquatic systems, this species may encounter these pollutants in their environment and diet. At a known polluted site on the breeding grounds, DDE, DDT, and mercury were all detected within prothonotary warbler adults, nestlings, and eggs (Reynolds et al. 2001). Other sources of pollution (e.g., agricultural runoff, acid rain, oil pollution) probably affect this species indirectly in some localities by degrading its breeding habitats.

*CONSERVATION AND MANAGEMENT NEEDS*

The prothonotary warbler requires mature forested habitat that lies in close association with water (preferably flooded) and contains adequate large dead or live trees that provide nesting cavities. These habitats have historically been destroyed in the state and remaining areas of such habitat have often been negatively affected by pollution or timber harvest. As such, the primary conservation need is to promote an increase in the abundance and distribution of this rare species across the state by maintaining, creating, or enhancing suitable breeding habitat. To accomplish this, the author recommends the following efforts (primarily from Sallabanks 1993). Enforce existing legislation to protect forested wetlands. Identify and inventory all sites with confirmed or potential prothonotary warbler habitat within the state. Target these sites for protection and management (particularly occupied sites). When possible, protect not only suitable forest habitat in, or directly adjacent to, water, but also adjacent areas of forest to create wide riparian buffers (>90 m) and large forest patches for this apparently area-sensitive species. If timber harvest is to be allowed in or near suitable habitat, minimize impacts by encouraging forest managers to leave large (>15 cm dbh) trees or snags that provide cavities and to leave enough trees in general to maintain a mostly closed canopy.

Current prothonotary warbler nests within the state need to be monitored to evaluate success rates and causes of failure. This information may be important to guide future management efforts. Efforts need to be initiated to add artificial nest boxes to occupied habitats and other appropriate sites. Nest boxes have been found to reduce predation and parasitism rates and can reduce problems related to competition for nest sites (Petit 1999). Addition of nest boxes may also increase local densities, or could be used to increase nest-site availability and encourage settlement in habitats where cavities are rare (e.g., in younger riparian forests; Twedt and Henne-Kerr 2001). See Mitchell (1988) for construction specifications and information on nest box placement. Currently, volunteers with the Bluebird Society of Pennsylvania install and monitor some nest boxes that are used by prothonotary warblers. Coordination of efforts and better data sharing with the Pennsylvania Game Commission could provide useful information for managing this species. At appropriate sites, water management practices should be considered to provide and maintain flooded forest habitat during the breeding season. The use of beaver reintroduction or management in some locations needs to be investigated as a means to create new potential habitat.

*MONITORING AND RESEARCH NEEDS*

Current monitoring efforts (i.e., Breeding Bird Survey, second Pennsylvania Breeding Bird Atlas) are adequate to monitor the general breeding population size, distribution, and trend for this species. Efforts are needed, however, to identify and inventory all forested wetland sites with confirmed or potential prothonotary warbler habitat within the state, and to monitor the breeding success and causes of failure of individuals currently breeding within the state. Efforts to monitor nesting attempts in the state are also critical to identify where management may be useful.

Although the breeding biology of this species has been well studied in several populations, the area requirements of this species probably is the least understood management parameter related to breeding habitat (Sallabanks 1993). Studies are needed to evaluate area sensitivity of this species across its range and to evaluate the influence of forest area or riparian buffer width on density, pairing success, and nesting productivity. Because nests of this species placed over water tend to suffer less impact from competitors and predators, there is some evidence that management of water levels (through levees and dams) may benefit prothonotary warbler populations (e.g., Hoover 2006), and may even be more beneficial than nest box supplementation (Brush 1991). This possible management option deserves additional investigation. During the nonbreeding season, it is important to evaluate population status, habitat use, and survival on the wintering grounds and during migration. Critical mangrove wintering habitats are currently suffering rapid destruction and degradation, yet almost nothing is known about the specific effects of these events on this species. In addition, little is known about habitat use and needs during migration (Petit 1999). Such information is critical to develop management strategies for this species, as it is possible that changes affecting these nonbreeding habitats are currently more important than issues on the breeding grounds.

*Authors:* CHRISTOPHER B. GOGUEN, PENNSYLVANIA STATE UNIVERSITY

## Henslow's Sparrow

Order: Passeriformes
Family: Emberizidae
*Ammodramus henslowii*
(also *Responsibility Concern*)

The Henslow's sparrow is a small, secretive grassland sparrow (fig. 5.40). It was selected as a Species of High-Level Concern because of range-wide declines in abundance and because it is a Pennsylvania Responsibility Species with approximately 9 percent of the global population breeding in Pennsylvania. State and global populations are Apparently Secure (S4B,G4, NatureServe 2009).

*GEOGRAPHIC RANGE*

Henslow's sparrows breed from South Dakota east across southern Ontario to New York and south to Kansas and east to North Carolina. They winter from South Carolina south to Florida and west across the Gulf states to Arkansas and Texas (McWilliams and Brauning 2000).

*DISTRIBUTION AND RELATIVE ABUNDANCE IN PENNSYLVANIA*

Henslow's sparrows are generally uncommon but breed widely in the state in a variety of grassland, pasture, field, and reclaimed surface mine habitats (fig. 5.41; McWilliams and Brauning 2000). These sparrows can be locally common in extensive grasslands of the western Appalachian Plateau provinces. They are most abundant and most regular in reclaimed surface mines in southwestern counties (McWilliams and Brauning 2000). North American Breeding Bird Survey data

*Fig. 5.40.* The Henslow's Sparrow, *Ammodramus henslowii*. Photo courtesy of Kent Nickell.

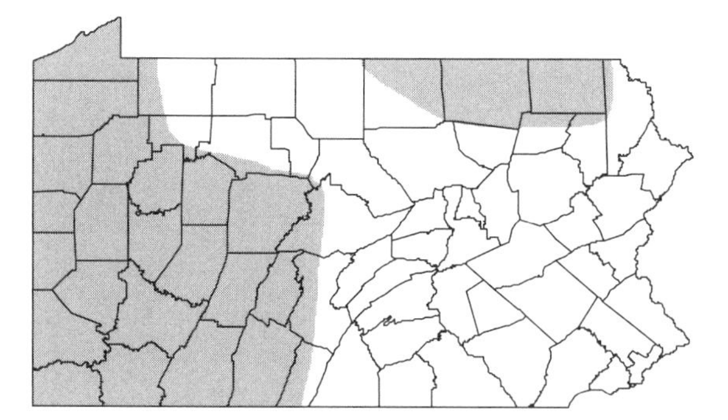

*Fig. 5.41.* Highest probability of occurrence of the Henslow's Sparrow, *Ammodramus henslowii.*

indicate a large and significant population decline for Henslow's sparrows (estimated with poor precision; Sauer et al. 2004) in North America (–8.1% per year) and in Pennsylvania (–2.7% per year) between 1966 and 2002 (Sauer et al. 2004).

In the eastern United States, and particularly in Pennsylvania, reclaimed bituminous coal fields are beneficial to grassland birds (Yahner and Rohrbaugh 1996). Widespread surface mining and subsequent reclamation in western Pennsylvania has resulted in an extensive patchwork of reclaimed sites among forests, woodlots, and agricultural fields. The acidic, nutrient-poor soils of reclaimed sites provide little potential for agricultural or timber production, and grasses and legumes tend to be the most successful and persistent vegetation types (Vogel 1981, W. G. Vogel, personal observation). These often undisturbed fields have a slow rate of ecological plant succession and are ideal for Henslow's sparrows, as well as compatible for many other grassland associated species (Bajema et al. 2001).

Mattice et al. (2005) estimated a total of 35,373 ha (95% CI = 26,758–46,870) of suitable reclaimed surface mine grassland habitat in their nine-county study area in Pennsylvania, and they estimated that 4,827 (95% CI = 2,734–8,211) singing males (or, roughly, 10,000 individuals) were present in this nine-county area. Diefenbach et al. (2007) modified this abundance estimate (based on the proportion of singing males likely to be present but missed during surveys) to 10,548 (95% CI = 3,906–15,107) singing males (or, roughly, 20,000 individuals) present in this nine-county area.

Pennsylvania's contribution to the global population of Henslow's sparrows is substantial. To place the estimate of Mattice et al. (2005) in context for conservation and management planning, we compared it to estimates for states reported in the 1996 Henslow's sparrow federal status assessment (Pruitt 1996). Although these estimates were based on different methods, the comparison is informative. Few states estimated populations exceeding several hundred birds, and only Oklahoma, Missouri, and Kansas reported populations in excess of 1,000 birds in known colonies or projected to occur in the matrix of natural and agricultural grasslands. In comparison to our estimates for Pennsylvania, Missouri is the only state with a larger population of Henslow's sparrows (Pruitt 1996).

Since publication of the Henslow's sparrow status assessment (Pruitt 1996), many studies have been initiated to evaluate the status of Henslow's sparrow populations. In some instances, significant new populations were identified. Notably, many of these have been on reclaimed surface mines, including locations in Indiana (Bajema et al. 2001), Illinois, and Ohio (Ingold and Murray 1998). These surveys indicate larger populations than expected from the 1996 assessment, suggesting that much of the extant population occurs on reclaimed surface mines.

K. Wentworth and M. C. Brittingham (personal communication) have investigated the use of agricultural lands in set-aside programs such as the Conservation Reserve Enhancement Program (CREP) by obligate grassland bird species and have found little use of these small parcels by Henslow's sparrows in the early years of the program. Perhaps as these sites become more established, there will be greater use by grassland birds. There is a need to estimate suitable habitat and sparrow abundance associated with hay fields, pastures, grassy fields, and other agriculturally related habitats (including set-asides) to compare with reclaimed surface mine habitats and inform conservation planning.

### *COMMUNITY TYPE/HABITAT USE*

Henslow's sparrow is a grassland species, characteristically found in prairies, wet meadows, neglected fields, and grassy swamps (Rising 1996, Herkert et al. 2002). Recent quantitative analyses of microhabitats (summarized in Herkert et al. 2002) have shown that litter density and depth, standing dead vegetation, and field size are important components of habitat. In Pennsylvania, they inhabit extensive grasslands with some perennial forbs or small shrubs used as perches. They may be found in meadows, uncut hay fields, and abandoned pastures, but most regularly and commonly are in grassy reclaimed surface mines (McWilliams and Brauning 2000). In general, Pennsylvania offers two contrasting, broad habitat types key to the maintenance of this species in the commonwealth: (1) agricultural lands, such as hay fields, pastures, and fallow fields, and (2) grasslands associated with reclaimed surface mines. Although the agricultural habitat types appear to be extensive (although declining) within Pennsylvania, several authors have concluded that most agricultural habitat essentially serves as a population sink (Bollinger et al. 1990, Kershner and Bollinger 1996, Rohrbaugh et al. 1999). Agricultural set-asides (e.g., Conservation Reserve Program lands) and prairie reserves provide reservoirs of grassland habitat that may help support remaining populations of some grassland bird species (Delisle and Savidge 1997, Koford 1999, Coppedge et al. 2001, Johnson and Igl 2001); yet, in Pennsylvania, preliminary results indicate these set-asides appear to be too small and isolated to serve as habitat for Henslow's sparrows (K. Wentworth and M. C. Brittingham, personal communication). However, the second broad habitat type, reclaimed surface mines, has inadvertently become the primary source of grassland bird habitat in Pennsylvania and in other areas. Recent studies have confirmed the existence of substantial grassland bird populations on reclaimed mines throughout the Midwest and the Northeast, which indicates these habitats may be important for conserving many grassland species (Yahner and Rohrbaugh 1996, Bajema et al. 2001).

J. A. Mattice (unpublished data) investigated landscape and patch-level habitat characteristics for Henslow's sparrows on reclaimed surface mines and found that patches generally needed to be greater than 30 ha for high probability of occupancy.

### *LIFE HISTORY AND ECOLOGY*

Pairs form on the breeding grounds immediately after spring arrival of females (Herkert et al. 2002). Henslow's sparrows typically place nest on the ground at the base of a clump of vegetation. Apparently, the female alone builds the cup-shaped nest of grass, lined with fine grass and sometimes hair. The nest is frequently overhung by vegetation to form a dome (Herkert et al. 2002). First nest attempts generally begin in early May in Pennsylvania, and Henslow's sparrows likely renest within a breeding season if a nest fails because of predation and likely attempt a second (or third) brood if the first is successful as it does in Kentucky (Herkert et al. 2002). Henslow's sparrows are occasionally parasitized by brown-headed cowbirds (*Molothrus ater*), and although it is considered an infrequent cowbird host, brood size is reduced when parasitized (Herkert et al. 2002). In general, little is known about the spatial and temporal patterns of nest success, causes of nest failure (i.e., nest predators), season-long productivity, or levels of cowbird parasitism for Henslow's sparrows in Pennsylvania.

### *THREATS*

The primary factors influencing species' decline in Pennsylvania include the changes in agricultural practices during the past fifty years that have made much agricultural habitat unsuitable for native grassland species (Warner 1994, Bolgiano 1999, 2000). Population declines are also due to the accelerating loss of agricultural areas and surface mine areas to urban sprawl (Vickery et al. 1999, P. D. Vickery, personal observation).

Reclamation practices, such as planting trees, as well as natural succession, can lead to an excessive density of woody vegetation on reclaimed surface mines, thus rendering them inappropriate for grassland species. Habitat protection for this species must include large tracts (greater than 20 ha) of grassland habitat and, if necessary, continual habitat management to prevent vegetation successional changes. Improved reclamation practices may be a threat to this species as well because a greater percentage of planted trees may become established, again rendering these areas unsuitable for obligate grassland species.

### *CONSERVATION AND MANAGEMENT NEEDS*

Identify the location and extent of suitable, high-quality reclaimed surface mine habitat for Henslow's sparrows across the nine-county bituminous coal area of Pennsylvania and coordinate protection and management of these locations. Mattice et al. (2005) estimated that roughly 35,000 ha of suitable habitat exist within this area. This assessment should not be done without considering other priority grassland species. Until a better understanding of population status and dynamics within these areas of Pennsylvania and a clearer understanding of Pennsylvania's regional/global responsibility for this species emerge, it seems reasonable to recommend maintaining the current estimated acreage of suitable habitat. Because land is currently being mined and subsequently reclaimed while reclaimed areas are simultaneously undergoing succession, these 35,000 ha will likely be secured within a shifting landscape mosaic, where not all 35,000 ha need protection as long as the acreage exists within the landscape. Some large blocks, however (such as the Piney Tract, Clarion County) should be preserved and managed. Finally, it is imperative to work with the Pennsylvania Department of Environmental Protection on the reclamation process. Specific areas that have a high probability of becoming important grassland bird habitat (e.g., open patches greater than 20 ha near other nonforested habitats) should not have trees planted as part of the reclamation process to increase the likelihood of providing quality grassland bird habitat.

*MONITORING AND RESEARCH NEEDS*

Until more information is gathered about density, demographic rates, and source habitats, it would be difficult to propose a fully specified monitoring strategy. However, efficient, statistically defensible methods of estimating grassland sparrow abundance over broad geographic areas have been developed by Diefenbach et al. (2007) and Mattice et al. (2005) that could be implemented as needed. This type of effort, while efficient and informative, is fairly time consuming and costly. It is therefore prudent to ensure that Breeding Bird Survey routes continue to be run throughout the state. These are the best data available for statewide population trends and continue to be valuable for most bird species, including the Henslow's sparrow. It may be valuable to augment the existing Breeding Bird Survey routes with additional routes in areas of Pennsylvania with grassland bird populations and relatively few active Breeding Bird Survey routes (such as nine-county bituminous coal area). In this manner, the less expensive, volunteer-based Breeding Bird Survey data could be used for general monitoring, and when a particular threshold (say, a decline of >5% per year) is reached, the more intensive (but more precise) monitoring methods of Diefenbach et al. (2007) and Mattice et al. (2005) could be triggered and implemented (Houser et al. 2006). The Pennsylvania Game Commission Conservation Reserve Enhancement Program Monitoring Survey should be continued and used to monitor use of Conservation Reserve Enhancement Program areas by Henslow's sparrows and other grassland birds and to better estimate the effect of Conservation Reserve Enhancement Program on grassland bird populations (Wilson 2009).

Not all patches of reclaimed surface mine habitat are equal to Henslow's sparrows because of such factors as patch size, vegetation characteristics, and landscape context. Because of this, and because we do not advocate maintaining *all* reclaimed surface mines as grassland bird habitat, it is imperative to identify and prioritize the best-quality areas. Mattice et al. (2005 and J. A. Mattice et al., unpublished data) have investigated and developed models explaining the relationship and relative importance of patch size, vegetation, and landscape context in terms of Henslow's sparrow patch occupancy and abundance. These models should be tested, verified, and then implemented to identify the most important high-quality patches of habitat available for conservation and management.

Little demographic data exist for the Henslow's sparrow in Pennsylvania in agricultural or in surface mine areas. Estimates suggest high sparrow abundance for much of the surface mine habitat in western Pennsylvania; however, empirical data on nest success, productivity, and annual survival estimates are needed in these habitats to assess their quality.

A better understanding of vegetation succession on reclaimed surface mines as it relates to grassland bird populations, especially in light of improved reclamation practices, is needed to evaluate and prescribe management actions to maintain surface mines as high-quality habitat. Populations of Henslow's sparrows among patches of reclaimed surface mine may be acting as a metapopulation with implications for conservation and management. Documenting movement patterns and exchange among patches are needed to understand these dynamics.

*Authors:* MATT R. MARSHALL, NATIONAL PARK SERVICE; DUANE R. DIEFENBACH, U.S. GEOLOGICAL SURVEY, PENNSYLVANIA COOPERATIVE FISH AND WILDLIFE RESEARCH UNIT

## Summer Tanager

Order: Passeriformes
Family: Cardinalidae
*Piranga rubra*

The summer tanager is a mid-sized, forest-dwelling songbird (fig. 5.42). A species at the northern edge of its range in Pennsylvania, the summer tanager has an extremely limited distribution in this state and is currently listed as a Candidate Rare Species in Pennsylvania. It is a Wildlife Action Plan species of High-Level Concern. Breeding populations in Pennsylvania are Vulnerable (S3B), and global populations are Secure (G5, Natureserve 2009).

*Fig. 5.42.* The Summer Tanager, *Piranga rubra*. Photo courtesy of David McNicholas.

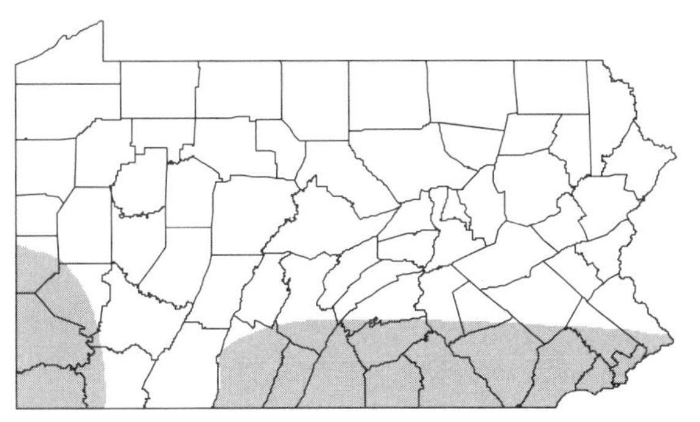

*Fig. 5.43.* Historical distribution of the Summer Tanager, *Piranga rubra*.

### *GEOGRAPHIC RANGE*

Summer tanagers breed throughout much of the southeastern United States, west across Texas, and into the Southwest (Price et al. 1995). They also breed south of the United States in northern Mexico (Robinson 1996). Summer tanagers winter in a variety of open woodlands and forests from central Mexico south through Central America to northern South America as far south as Bolivia and northern Brazil (Robinson 1996).

### *DISTRIBUTION AND RELATIVE ABUNDANCE IN PENNSYLVANIA*

The summer tanager was detected in only 1 percent of the state's atlas blocks in the first Breeding Bird Atlas (1983–1989) and confirmed breeding records were all limited to the three most southwestern counties (Ickes 1992f). During the second atlas, there were only three records (one confirmed) for this species, suggesting they are virtually extirpated from the state (Mulvihill and Brauning 2009). Within this limited portion of the state, this species was a rare but regular breeder in dry deciduous forests, with most other scattered detections throughout the state likely representing vagrant males (McWilliams and Brauning 2000). Historically, this species was most abundant as a breeder in the southeastern portion of the state (fig. 5.43). However, declines in abundance of this species from the adjacent Piedmont region of Maryland, perhaps in response to human-induced land use changes, may have led to a contraction of this species' range (Ickes 1992f). The single confirmed nesting of this species during the second Pennsylvania Breeding Bird Atlas occurred in Chester County in southeastern Pennsylvania

### *COMMUNITY TYPE/HABITAT USE*

Summer tanagers breed in a wide variety of deciduous and mixed-deciduous forest types. In the northern parts of its eastern range, this species prefers open deciduous woods often near large openings or edges. In the Southeast, it is often found in similar situations in dry, pine-oak forests. Western populations occupy riparian woodlands (Robinson 1996). Within Pennsylvania, the summer tanager is typically found in dry, upland forests, particularly open oak woodlands, although it may also be found in open parks and along road or utility cut edges (Ickes 1992f, McWilliams and Brauning 2000). At a local scale, summer tanagers have been found at higher densities along forest edges than in the interior (Kroodsma 1984). At a regional scale, however, occurrence of this species appears to be negatively correlated with the overall amount of fragmentation in a landscape (Rosenberg et al. 1999a).

### *LIFE HISTORY AND ECOLOGY*

A long distance, complete migrant, the summer tanager is an uncommon breeder at the very northern edge of its eastern range in Pennsylvania. Most sightings in the state occur from late April through September, although most nesting activity is restricted to May through July (McWilliams and Brauning 2000). Often described as a bee and wasp specialist, this species also takes a variety of other flying and nonflying insects. During the late summer, migration, and winter, fruit is often incorporated into the diet (Isler and Isler 1987).

The summer tanager breeds in a variety of open, deciduous woods. Seasonally monogamous, this species builds a shallow, and often flimsy, open nest (Bent 1958, Harrison 1975c). Two broods are typically raised in western populations (Rosenberg et al. 1991), but this trait does not appear to be as common in the East (Robinson 1996). Summer tanagers are parasitized by the brown-headed cowbird (*Molothrus ater*) throughout their range, and by the bronzed cowbird (*Molothrus aeneus*) in some southwestern populations. Parasitism rates vary substantially among populations but tend to be highest in highly fragmented, midwestern landscapes and lower in landscapes that contain larger forest tracts. Cowbird parasitism tends to reduce tanager productivity primarily through clutch reduction, presumably because of egg removal by female cowbirds, as tanagers are large enough to raise a cowbird with their own young (Robinson 1996).

*THREATS*

The greatest threat to this species is the continual loss and fragmentation of both breeding and wintering habitat. Unfortunately, the summer tanager's breeding range in Pennsylvania coincides relatively closely with the more densely populated portions of the state (i.e., the southeastern Atlantic Coastal Plain and piedmont uplands, and the southwestern Pittsburgh low plateau), where habitat loss and fragmentation are major problems. Fragmentation reduces the average size of the remaining forest patches, increases their isolation from other forest habitats, and increases the proportion of habitat close to an edge. Throughout its range, the likelihood of finding breeding summer tanagers in a landscape declined significantly with increasing fragmentation (Rosenberg et al. 1999a). Further, studies of the closely related scarlet tanager have indicated that increased fragmentation can also reduce pairing and nesting success (Robinson 1992, Roberts and Norment 1999). These possible effects have not been studied in summer tanagers but could represent an additional cost in fragmented landscapes.

An overabundance of white-tailed deer (*Odocoileus virginianus*) in the state has greatly limited forest regeneration and reduced the density of understory and subcanopy forest layers (Goodrich et al. 2002). These impacts could potentially benefit summer tanagers by creating the open habitats they prefer, or they could negatively affect the species by affecting the availability of foraging substrates. Specific studies are needed to evaluate the effects of this issue.

Lighted towers and buildings can be an important mortality source, particularly for night-migrating songbirds like the summer tanager (Goodrich et al. 2002). During spring and fall migration of a single year, nearly 300 tanagers were killed in collisions with a television tower in Florida (Stevenson and Anderson 1994). Given that thousands of new cellular and digital television towers are being built annually (Goodrich et al. 2002), tower collisions could represent a major mortality source for migrating tanagers.

*CONSERVATION AND MANAGEMENT NEEDS*

Given that Pennsylvania lies at the extreme northern edge of the summer tanager's range and currently supports only a tiny proportion of this species' total population, the persistence of this species in the state may depend on both the maintenance of high-quality breeding habitat within the state, as well as the maintenance of large areas of high-quality habitat in adjacent states to the south to support large source populations. The decline and near extirpation of the summer tanager from southeastern Pennsylvania may reflect not only habitat loss and degradation in this region, but also the decline of this species in adjacent eastern Maryland. Similarly, the apparent increase in occurrence of this species in extreme southwestern Pennsylvania may be the result of a healthy breeding population in adjacent West Virginia (Ickes 1992f).

To maintain and potentially increase the summer tanager population size within Pennsylvania, the author recommends that attempts to minimize fragmentation of remaining large contiguous forest tracts within the species potential range in Pennsylvania be made. Currently, fragmented landscapes in this range should be managed to maximize core habitat through reforestation and patch shape (i.e., circular rather than long and narrow) and to promote connectivity among patches using the protection or reforestation of corridors. Similar management activities should be encouraged in adjacent states to the south to maintain or increase potential sources of dispersers. Long-range forest management plans need to be developed at as large a scale as possible (perhaps even across state boundaries) to ensure that appropriate habitats will be available at each stage of the plan and to maintain connections between existing habitat patches.

*MONITORING AND RESEARCH NEEDS*

At a continental scale, current monitoring efforts (i.e., North American Breeding Bird Survey, Project Tanager) are adequate to monitor the breeding population size and distribution and to measure the population trend for this species broadly across its range. Although this species is too rare for these methods to be useful in Pennsylvania, the second Pennsylvania Breeding Bird Atlas represents an excellent opportunity to determine the current distribution and abundance of this species. This information could be used as a baseline for locating sites for additional local surveys and monitoring population status.

Currently, southwestern Pennsylvania appears to represent the stronghold for this species in the state, and large areas of potential habitat also appear to remain there. Management activities for summer tanagers within Pennsylvania should probably focus on protecting and enhancing these habitats to allow for future population increases and range expansion. Ultimately, however, tanager population growth in Pennsylvania may depend more on the health of populations in adjacent states than on specific management activities in-state.

Many basic aspects of the breeding biology of this species remain unknown (e.g., success rates, rates of double brooding, effects of cowbird parasitism), and very little is known about how habitat fragmentation affects breeding populations (Robinson 1996). Although populations in Pennsylvania are probably too small to support this type of research, larger populations in adjacent states could be studied, particularly because these may represent the sources of new individuals currently dispersing into Pennsylvania. These types of studies may be particularly important because the survival of this species in Pennsylvania, and its potential range expansion in the state, probably depend on the health and management of larger populations directly to the south.

Almost no information is available regarding this species during migration or on its tropical wintering grounds (Robinson 1996), yet it spends most of the year on these sites. A better knowledge of migration routes, winter distribution, and how forest fragmentation and disturbance might be affecting its survival during migration and on its wintering grounds is needed.

*Author:* CHRISTOPHER B. GOGUEN, PENNSYLVANIA STATE UNIVERSITY

## Dickcissel

Order: Passeriformes
Family: Cardinalidae
*Spiza americana*

The dickcissel is a grassland specialist whose core range coincides with North America's native prairies (fig. 5.44). Similar to most grassland specialists, the dickcissel has been in decline nationwide (Peterjohn and Sauer 1999). It was listed as a Species of High-Level Concern because of widespread declines throughout its range. In Pennsylvania, which is on the periphery of its breeding range, it is listed as Endangered. Global populations are considered Secure (G5, NatureServe 2009).

*GEOGRAPHIC RANGE*

The highest breeding density of dickcissels occurs from eastern Nebraska to eastern Iowa south to Arkansas and Oklahoma (Temple 2002). Dickcissels are regularly found as far north and west as Montana, south into Texas, and east to Pennsylvania and Georgia. However, dickcissels are known to be irruptive breeders, expanding their range one year and then disappearing again the following summer. During the nonbreeding season, dickcissels are communal with flocks of more than a million birds wintering in Venezuela (Temple 2002).

*DISTRIBUTION AND RELATIVE ABUNDANCE IN PENNSYLVANIA*

Historically, the dickcissel was locally abundant in Pennsylvania during the times of John James Audubon and Alexander Wilson but was an uncommon breeder by 1883 (Rhoads 1903). They stopped breeding along

*Fig. 5.44.* The Dickcissel, *Spiza americana*. Photo courtesy of Kent Nickell.

the Atlantic Coastal Plain in the late nineteenth century (Rhoads 1903, Gross 1921). They reappeared in the mid-1930s in southeastern Pennsylvania and elsewhere along the Atlantic Coast, but no stable population appeared (Stone 1928, Sharp 1934, Gross 1956) until the 1990s. Dickcissels were considered extirpated from Pennsylvania until a nest was located in Clarion County (Bell 1984). Dickcissels have nested regularly since then only in Adams and Cumberland counties in south-central Pennsylvania (D. W. Brauning, personal communication). Not enough birds are identified during the Breeding Bird Surveys to identify a trend in Pennsylvania.

The dickcissel is considered a rare and local breeder in Pennsylvania (McWilliams and Brauning 2000; fig. 5.45). During the first Pennsylvania Breeding Bird Atlas (1983–1989), dickcissels were reported in forty-five blocks, less than 1 percent of the total (Mulvihill 1992a). Most of these records came from an irruption in 1988 (Mulvihill 1988). Breeding males and females were found in the Cumberland Valley (Franklin County) from 1988 to 1996 (McWilliams and Brauning 2000) and continued there through 2005 (Brauning 2005). Wilhelm (1994) found three different males with multiple females nesting in Lawrence County. A pair nested in Snyder County during the summer of 2002, but they were absent in 2004 when the field was searched again (K. Wentworth and M. C. Brittingham, unpublished data). There were thirteen probable or confirmed reports of dickcissels during the second Breeding Bird Atlas (Mulvihill and Brauning 2009). Dickcissels are also found during spring and fall migration at bird feeders (McWilliams and Brauning 2000).

### *COMMUNITY TYPE/HABITAT USE*

Dickcissels are grassland specialists and nest primarily in prairies and old fields (Zimmerman 1971), but also in hay fields, pastures, waterways, hedgerows (Basore et al. 1986, Bryan and Best 1991), row crops (reduced tillage: Best et al. 1997; soybean: Wentworth 2001), and barley and wheat fields. Dickcissels prefer medium to high vegetation, with some forb coverage and moderate downed litter (Harmeson 1974, Finck 1984, Winter 1999). Dickcissel densities are higher on old fields than prairies (Zimmerman 1971).

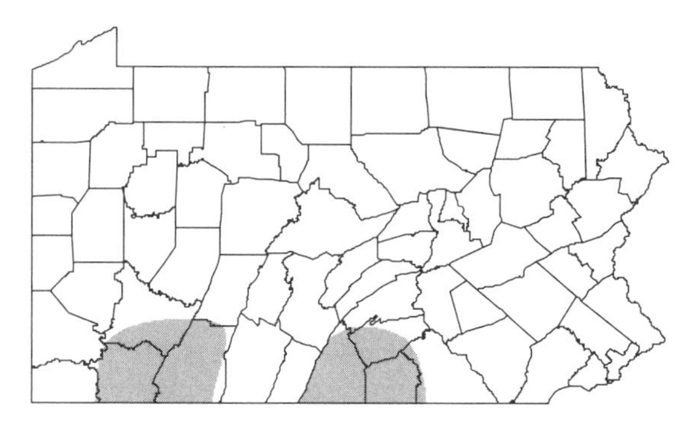

*Fig. 5.45.* Highest probability of occurrence of the Dickcissell, *Spiza americana*.

Conservation Reserve Program fields in the Midwest provide important habitat for dickcissels (Delisle and Savidge 1997, Horn et al. 2002). Field size has not been shown to be a factor in dickcissel abundance (Herkert 1994, Horn et al. 2002), but there are indications that larger fields may improve nest success (Winter and Faaborg 1998). This is important because Conservation Reserve Enhancement Program fields in Pennsylvania are relatively small (7.0 ha average; S. R. Klinger, personal communication). A dickcissel pair was found breeding in 2002 on a small Conservation Reserve Enhancement Program (CREP) field (<3 ha), in Snyder County, next to a house with horses, dogs, and kids (K. Wentworth and M. C. Brittingham, unpublished data). However, no dickcissels have been located during road surveys through agricultural regions in the twenty Conservation Reserve Enhancement Program counties in the Chesapeake Bay drainage (A. Wilson, unpublished data). Reclaimed strip mines in western Pennsylvania provide another possible habitat, but their use of this habitat type is patchy on reclaimed midwestern mines (Scott et al. 2002).

### *LIFE HISTORY AND ECOLOGY*

Dickcissels are long-distance migrants wintering in Central and South America. They arrive back on the breeding grounds in early May (Temple 2002). They are known for their erratic, seminomadic movements within their breeding range, which results in the birds' presence in some years and absence in others (Temple 2002). They are omnivorous during the breeding season and primarily granivorous on their wintering range, where they are often associated with crop damage (Basili and Temple 1999).

Dickcissels show resource-defense polygyny with the males with the best territories attracting multiple mates while those on poor-quality territories might not attract any (Zimmerman 1966). The cup-shaped nest is most often placed on forbs (nests are not attached to vegetation; Long et al. 1965); however, shrubs, grass, and standing litter are also used as a nesting substrate (Winter 1999). Dickcissel nests are heavily depredated and have a low nest success rate (proportion of nests where at least one young fledged) from 0 percent to 42.0 percent (Von Steen 1965, Zimmerman 1982, Hughes et al. 1999, Winter 1999). After nest failure, Walk et al. (2004) found that females emigrated from the area (62%), renested (32%), or ceased nesting (2%). Females will renest up to three times in a breeding season, four to fourteen days after the loss of a nest and travel 22 to 806 m from a lost nest. However, emigrating females have been located up to 32 km away though no nests were located (Walk et al. 2004). Females usually only raise one brood, but there have been examples of a female successfully raising two broods (Bollinger and Maddox 2000, Walk et al. 2004). Regardless of success or failure, dickcissel site fidelity is variable and often low with a range of 0 percent to 61 percent of banded birds returning to the same study area (Finck 1984, Zimmerman and Finck 1989, Walk et al. 2004).

*THREATS*

The greatest threat to dickcissels occurs on the wintering range where birds are often shot and poisoned (Basili and Temple 1999, Temple 2002). Because dickcissels roost communally in the nonbreeding season, one roost may have 30 percent of the global population (Basili and Temple 1999). During the breeding season, dickcissels in Pennsylvania are faced with the same threats that befall them in the Midwest—mowing of hay fields resulting in nest destruction, high predation rates, and habitat loss/fragmentation.

*CONSERVATION AND MANAGEMENT NEEDS*

Because the abundance of dickcissels in Pennsylvania is low and their occurrence erratic, it probably does not make sense to have a species-specific management program. Instead, needs of the dickcissel should be incorporated into a comprehensive management plan to increase the quantity and quality of grassland habitat in Pennsylvania. This includes encouraging the enrollment of fields into the Conservation Reserve Enhancement Program with a particular emphasis on Adams and Cumberland counties. Management of reclaimed strip mines for dickcissel and other grassland birds should also be encouraged.

*MONITORING AND RESEARCH NEEDS*

The Pennsylvania Game Commission Conservation Reserve Enhancement Program monitoring surveys (Wilson 2009) need to be continued, as these focused surveys are probably one of the best ways to identify when and where dickcissels are in Pennsylvania. Targeted surveys of fields where dickcissels have bred in the past ten years should also occur. Research should focus on determining the best way to manage agricultural fields and reclaimed strip mines for the guild of grassland obligates that currently breed in Pennsylvania.

*Author:* KEVIN WENTWORTH, PENNSYLVANIA STATE UNIVERSITY

**RESPONSIBILITY CONCERN**

## Tundra Swan

Order: Anseriformes

Family: Anatidae

*Cygnus columbianus columbianus*

The tundra swan is one of the few nonbreeders in Pennsylvania selected as a Species of Greatest Conservation Need. It was selected as a Responsibility Species (fig. 5.46) because of Pennsylvania's importance as a wintering area and staging area during spring and fall migration. Delaware is the only other northeastern state to include tundra swans as a Species of Greatest Conservation Need. The tundra swan is classified as a migratory game bird by the United States Fish and Wildlife Service, but there is currently no open hunting season in Pennsylvania.

*GEOGRAPHIC RANGE*

The breeding range of the tundra swan extends across the Arctic and sub-Arctic tundra regions of Alaska, the Yukon, the Northwest Territories, Nunavut, northeastern Manitoba, northern Ontario, and northern Quebec. Tundra swans have been differentiated into a western (WP) and eastern (EP) population

*Fig. 5.46.* The Tundra Swan, *Cygnus columbianus columbianus*. Photo courtesy of Rob Hanson.

for management purposes. This differentiation is based on migration or wintering affiliations and not on breeding or genetic differences (Eastern Population Tundra Swan Committee 2007). Analysis of band recoveries and neck collar observation data has found limited geographic overlap between these two population units (Limpert et al. 1991). The eastern population of tundra swans nest in tundra habitat from the north slope of Alaska across most of the Canadian Arctic and sub-Arctic regions (Bellrose 1980, Limpert et al. 1991).

### *DISTRIBUTION AND RELATIVE ABUNDANCE IN PENNSYLVANIA*

Eastern population tundra swans winter principally along the Atlantic Coast, with 98 percent of the wintering population generally occurring in three states: Maryland, North Carolina, and Virginia. The remainder winter in Pennsylvania, Delaware, New Jersey, New York, South Carolina, and southern Ontario. In Pennsylvania, the Lower Susquehanna River and Middle Creek Wildlife Management Area may winter 500 to 5,000 swans (Serie and Raftovich 2004). Lake Erie and Presque Isle winter up to 100 swans in some years, until freeze-up forces the bird south (McWilliams and Brauning 2000).

Tundra swans are a common to abundant spring and fall migrant through Pennsylvania (McWilliams and Brauning 2000). Migration occurs over a narrow band from northwest to southeast. Major staging areas are located along the lower Susquehanna River or at Middle Creek Wildlife Management Area and nearby agricultural areas (fig. 5.47).

The size of the eastern population of tundra swans has more than doubled since 1955 and currently averages about 100,000 (Serie et al. 2002). The annual midwinter waterfowl inventory conducted throughout the coterminous United States provides a reliable estimate of population size. The population is currently above the 80,000 goal established in both the Eastern Population Tundra Swan Management Plan (Eastern Population Tundra Swan Committee 2007) and North American Waterfowl Management Plan (U.S. Fish and Wildlife Service et al. 2004). Numbers in Pennsylvania are relatively small during the Mid-Winter Waterfowl Survey conducted each January (averaging 1,585 during the 1991–2005 period and ranging from 115 to 4,868) but build to their highest levels as the birds stage in the Susquehanna River Valley before spring migration. Average numbers of swans staging in Pennsylvania at any one time during spring migration are about 3,000–7,000 birds, but as many as 17,000 have been reported in some years. Because of staggered arrival and departure dates of individual swans, the overall number of swans stopping over in Pennsylvania during some portion of the spring migration, likely exceed these totals. In 2001, 27 percent of swans marked during winter in southern states were documented passing through Pennsylvania (Gregg 2002).

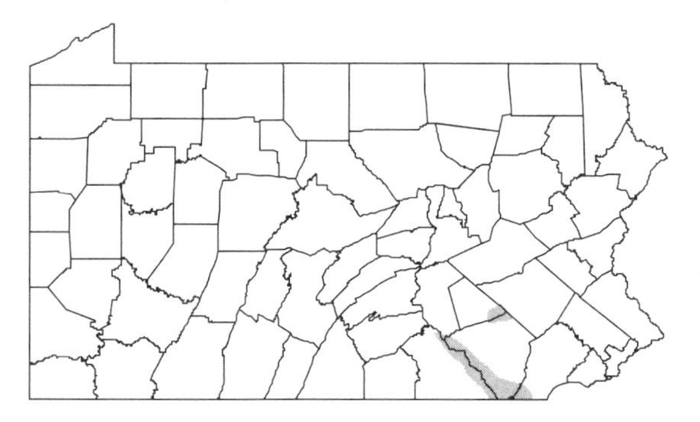

*Fig. 5.47.* Wintering distribution of the Tundra Swan, *Cygnus columbianus columbianus.*

### *COMMUNITY TYPE/HABITAT USE*

At every stage of their annual cycle, tundra swans are closely associated with wetlands. During summer, tundra ponds and lakes provide breeding and brood-rearing habitat. On the fall migration, swans primarily use large wetlands containing abundant submerged aquatic vegetation (SAV), the preferred food source during this season (Petrie et al. 2002, Badzinski 2003). In late fall or early winter, dietary preference shifts to agricultural grains, but wetlands continue to be important as roosting habitat. This pattern continues throughout the winter and much of the spring migration (Petrie et al. 2002). Thus, tundra swans wintering and staging in Pennsylvania require two distinct habitat types: (1) agricultural fields, primarily harvested corn and winter wheat for feeding habitat and (2) lacustrine and riverine wetlands for roosting habitat. Presence of submerged aquatic vegetation does not appear to be a major determinant of wetland use in Pennsylvania, but some feeding on submerged aquatic vegetation does occur at roost sites.

General patterns of feeding and roosting habitat use in Pennsylvania have been described by Gregg (2004). The most readily used agricultural fields are large; contain "pothole"-type wetlands, or sheet water; and have minimal human activity within them, though heavy foot and vehicle traffic may be present at roads or buildings on the perimeter of the field. Roosting sites, which are often shared with other species, include impoundments, quarry ponds, and shallow, slow-moving portions of the Susquehanna River (e.g., Conejohela Flats area). The amount of ice cover at roost sites is highly variable. As with feeding sites, potential disturbance activities are uncommon on the water body itself, but sometimes extensive in the adjacent uplands. For both feeding and roosting habitats, more detailed information needs to be obtained on site and landscape-level determinants of habitat quality and the interrelationships of these characteristics to maximize the effectiveness of habitat conservation and management for tundra swans in Pennsylvania (Gregg 2004).

*LIFE HISTORY AND ECOLOGY*

Tundra swans spend about 30 percent of the year on the breeding grounds, 20 percent in fall migration, and 25 percent each on wintering areas and in spring migration, with minor annual variations in the exact proportions, depending on weather conditions (Petrie and Wilcox 2003, Wilkins 2007). They are present in Pennsylvania during migration and winter periods. Fall migrants generally arrive during November, and variable numbers of swans overwinter in the Southeast from December through February. The number wintering in Pennsylvania is generally inversely related to the amount of ice and snow cover present and may vary somewhat over the course of a given winter as birds shift south or north in accordance with fluctuating weather conditions. Tundra swans show only low to moderate between-year fidelity to specific wintering sites, and there is no correlation between specific breeding and wintering areas (Gregg 2004, Wilkins 2007).

Spring migration is a crucial period in the annual cycle for tundra swans. Because they typically lose body mass over the winter and begin the spring migration with limited nutrient reserves (Bortner 1985), it is imperative that spring staging areas supply ample nutrition to fuel the long northward migration and provide needed reserves for successful reproduction (Petrie et al. 2002, Petrie and Wilcox 2003). The peak of spring migration through Pennsylvania usually occurs in late February or early March, but this timetable can be advanced (if springlike weather arrives early) or pushed back (if extensive snow and ice cover persist) by a few weeks. In general, the number of swans staging in Pennsylvania and the length of time individual birds remain in the state appear to be greatest in later springs (Gregg 2002), because birds are waiting for suitable staging habitats to thaw further north. Agricultural and wetland habitats in southeastern Pennsylvania serve as important migration habitat for the species during this critical portion of their annual cycle; in some years, they support a significant proportion of the entire eastern population during this period. In addition, these habitats support smaller numbers of wintering tundra swans. Consequently, this region is important to the long-term population health of the eastern population tundra swan.

In Pennsylvania, behavior is similar for both overwintering and migrating swans. A common daily pattern is one round trip to a feeding field in midmorning and another in late afternoon, with the remaining daylight hours and nocturnal periods spent resting on roost waterbodies, but variations on this theme are frequent. Swans may remain on roost sites for days at a time during inclement weather or spend entire nights afield feeding under a full moon. Movements between feeding and roosting sites in staging areas, as well as long-distance migratory flights, generally occur in flocks of five to fifty birds made up of one or more family groups (mated pairs plus young from the previous breeding season) and aggregations of subadult swans.

Because their breeding grounds are located thousands of miles from Pennsylvania, it is unclear whether the conditions swans encounter here have any direct effect on their reproductive performance. Nevertheless, Pennsylvania is one of the early stops on a three-month journey on which swans must acquire the necessary nutrient reserves to reproduce successfully once they reach the Arctic.

Tundra swans generally do not breed until four to five years of age. The vast extent and remote nature of the Arctic breeding grounds are believed to provide adequate habitat quantity and quality in this portion of the annual cycle to maintain the eastern population at desired levels (Eastern Population Tundra Swan Committee 2007).

*THREATS*

Although some direct mortality of tundra swans occurs in Pennsylvania because of illegal shooting (they are usually mistaken for snow geese [*Chen caerulescens*]), in-flight collisions with various obstacles, and disease, cumulative direct mortality in Pennsylvania is low, and of minimal concern for the maintenance of the overall population at desired levels. Of greater concern from a population standpoint is the incremental loss and degradation of agricultural (feeding) and wetland (roosting) habitats heavily used by swans wintering in and migrating through Pennsylvania, especially the lower Susquehanna Valley region. Feeding habitats are threatened primarily by residential and commercial development activities; conversion of farmland to nonfarm uses is increasing throughout Pennsylvania, and this trend is especially evident in the southeastern counties (Goodrich et al. 2002). Roost sites are generally more secure against direct habitat loss but remain threatened by uncontrolled human disturbance and by development of adjacent uplands. Wind energy development activities are a potential future threat because the presence of numerous large turbines in or near major staging areas and migration flight paths could increase collision mortality or impede use of key habitats by swans.

As habitat quantity and quality declines and disturbance increases, the energy expended by tundra swans relative to nutrients obtained can be expected to increase. This might not increase direct mortality levels in Pennsylvania but would result in a gradual decline in the body condition of a significant proportion of the population, which could negatively affect their annual survival and reproductive output and potentially lead to a decreasing population over the long term. It remains critical for long-term population maintenance that tundra swans be able to build up sufficient reserves for successful reproduction during the spring migration. This, in turn, requires that swans have access to sufficient food resources and resting habitat all along their migration routes, including staging areas in Pennsylvania.

### *CONSERVATION AND MANAGEMENT NEEDS*

The primary conservation need for tundra swans in Pennsylvania is the maintenance of sufficient quantities of high-quality feeding and roosting habitat through proactive identification and protection and management of key agricultural and wetland habitats in southeastern Pennsylvania, especially in Lebanon and Lancaster counties. Specific goals for the next five years should include securing 10,000 ha of known or potential feeding habitat through acquisition, easements, or land use planning/farmland preservation efforts; securing 200 ha of known or potential roosting habitat through acquisition, restoration, or enhancement; reducing and managing disturbance by restricting access to important habitats on public lands; developing educational materials and programs for private landowners and recreational user groups; and enacting regulations to prevent disturbance, if warranted.

Initial efforts should be focused on protecting known key feeding and roosting habitats. As additional information on important habitat features and disturbance effects is gained through research, habitat conservation efforts should be expanded to include additional areas (e.g., sites identified as potential habitat and buffer zones around use sites) as appropriate, and to implement habitat management techniques to create new or enhance existing habitat. State and local land use planning efforts should incorporate long-term swan habitat conservation as an explicit goal and discourage land use changes (e.g., inappropriately located wind farms) incompatible with this goal.

### *MONITORING AND RESEARCH NEEDS*

Monitoring and management of tundra swans is a cooperative effort between states in the Atlantic Flyway (including Pennsylvania) and the United States Fish and Wildlife Service. Population monitoring occurs through the Mid-Winter Waterfowl Surveys. The annual Mid-Winter Waterfowl Inventory conducted in early January provides a reliable estimate of tundra swan numbers on their wintering areas. Indices to productivity are derived from counts of gray-plumaged young and white-plumaged adults and subadults observed during fall and early winter. The proportion of young to adult is used to assess annual productivity. Additional monitoring of tundra swan feeding and roosting habitats is needed to assess disturbance and habitat losses, suggest appropriate conservation and management actions, and evaluate the effectiveness of these actions.

In addition to continuing and expanding monitoring efforts, a more thorough understanding of temporal and spatial habitat use patterns by tundra swans in Pennsylvania is needed to guide future habitat conservation efforts. Specific needs include GIS-based maps of known key habitats, catalogued and georeferenced to identify feeding and roosting sites known to local observers and managers or identified during prior research efforts. Research should be initiated to identify the relative importance and relationships of key habitat features. Although the general habitat preferences of migrating tundra swans are known, the specific habitat features that cause them to prefer certain feeding and roosting sites over other, apparently similar, habitats are not clearly understood. Quantitative habitat data should be collected and compared for swan use and nonuse sites to clarify the most important components of tundra swan habitat in Pennsylvania. This research may allow for identification of additional suitable habitat or provide insight into habitat management techniques that could be applied to enhance the amount and quality of habitat for migrating tundra swans in southeastern Pennsylvania.

Additional research is also needed on disturbance, identifying and quantifying the types and extent of disturbance at feeding and roosting sites, and assessing the effects of disturbance on distribution and body condition of wintering and migrating swans and on the optimal timing and methodology of population and habitat surveys for collecting needed data in the most cost-effective manner.

*Authors:* IAN D. GREGG, PENNSYLVANIA GAME COMMISSION; JOHN P. DUNN, PENNSYLVANIA GAME COMMISSION

## Wood Thrush

Order: Passeriformes
Family: Turdidae
*Hylocichla mustelina*

The wood thrush is a woodland thrush closely related to other thrushes in the genus *Catharus* (fig. 5.48). It was selected as a Pennsylvania Species of Greatest Conservation Need because of widespread declines and because it is a Pennsylvania Responsibility Species with more than 8.5 percent of the world population breeding in Pennsylvania (Rosenberg and Wells 1995). Although still common, the wood thrush is listed by Partners in Flight as a Priority Species because of significant long-term declines throughout its range (Rich et al. 2004). It is included on Species of Greatest Conservation Need lists in all northeastern states. Global populations are considered Secure (G5, NatureServe 2009).

*Fig. 5.48.* The Wood Thrush, *Hylocichla mustelina*. Photo courtesy of Lang Elliot.

### *GEOGRAPHIC RANGE*

The breeding range of the wood thrush extends from southern Canada, including New Brunswick, parts of Nova Scotia, southern Quebec, and Ontario, west to the Great Plains through north-central Minnesota, and from southeastern South Dakota, particularly along the Missouri River, south to eastern Texas and northern Florida (Roth et al. 1996). The wintering range of the wood thrush extends from southern Mexico to Panama.

### *DISTRIBUTION AND RELATIVE ABUNDANCE IN PENNSYLVANIA*

Extensive tracts of deciduous forest allow the wood thrush to breed throughout the state (fig. 5.49). In the first Pennsylvania Breeding Bird Atlas (PBBA), this species was found in almost every block surveyed (Brauning 1992a). Gap analysis indicates that 66 percent of the state provides potentially suitable breeding habitat (Myers et al. 2000). Wood thrushes nest in a variety of forest types and may be found even in wooded residential areas or urban parks (Roth et al. 1996). Along with other mid-Atlantic states, Pennsylvania contains a high density of breeding wood thrushes, with an average of 13.5 birds per Breeding Bird Survey route (five to twenty per route over much of the state), but numbers have been declining (Sauer et al. 2005; fig. 5.50).

### *COMMUNITY TYPE/HABITAT USE*

The wood thrush breeds in closed-canopy deciduous and mixed forest, often near water or in areas with

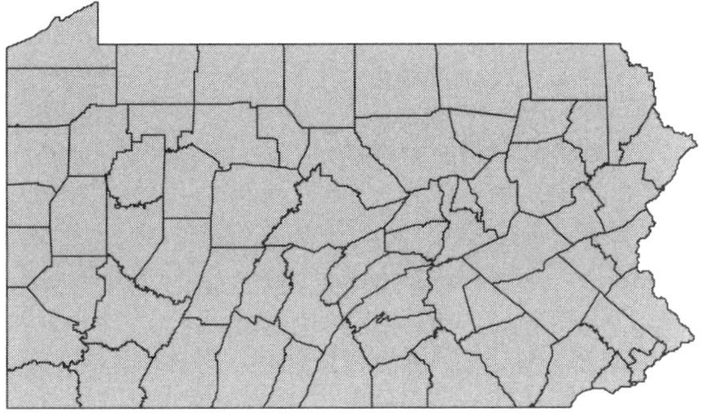

*Fig. 5.49.* Distribution of the Wood Thrush, *Hylocichla mustelina.*

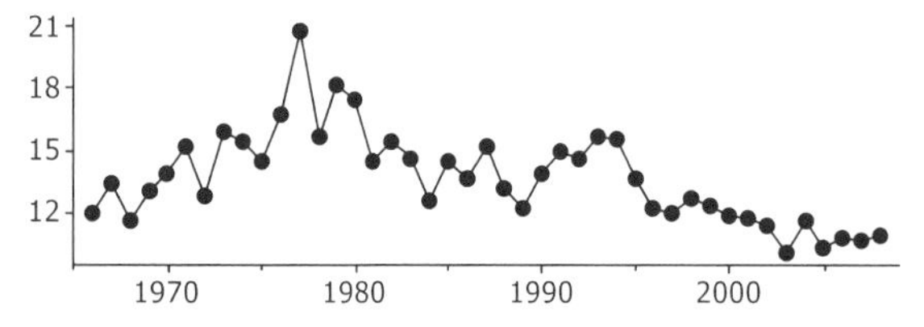


*Fig. 5.50.* Wood Thrush, *Hylocichla mustelina*, population trends from Breeding Bird Survey.

moist substrate (Bertin 1977). Key habitat characteristics include tall trees (>16 m in height), moderate midstory and understory, and a fairly open forest floor with moist decaying leaf litter (Roth et al. 1996). Mesic forest communities with oak (*Quercus* spp.), tulip poplar (*Tulipferia liriodendron*), American beech (*Fagus grandifolia*), or maple (*Acer* spp.) are all occupied (James et al. 1984). In Pennsylvania, nests may be built in areas with a high canopy and many trees while understory is important for nest sites. In some areas, spicebush (*Lindera benzoin*) provides a common nest substrate (Hoover and Brittingham 1998, Newell and Kostalos 2007), although growth characteristics are probably most important, and a wide variety of plant species are used, including southern arrowwood (*Viburnum dentata*), black gum (*Nyssa sylvatica*), flowering dogwood (*Cornus florida*), maples (*Acer* spp.), oaks (*Quercus* spp.), and eastern hemlock (*Tsuga canadensis*; Roth et al. 1996). Habitat use of juveniles during postfledging dispersal is characterized by dense understory and thick ground cover (Anders et al. 1998), and the presence of fruit may be important in habitat selection at this time (Vega Rivera et al. 1998). During molt, adults also move to areas with dense understory for protection from predators (Vega Rivera et al. 1999).

### *LIFE HISTORY AND ECOLOGY*

Foraging occurs primarily in leaf litter on the forest floor (Roth et al. 1996). Invertebrates taken include larval or adult insects (*Coleoptera*, *Diptera*, *Hymenoptera*, and *Lepidoptera*), millipedes, centipedes, and some snails (Holmes and Robinson 1988). Around migration, birds consume more fruit to build fat reserves (Martin et al. 1951). As a Neotropical migrant, wood thrushes migrate an average of 2,200 miles each fall and spring between breeding and wintering grounds (Yong and Moore 1994). Adults may live up to eight years and estimates of breeding site fidelity can be as high as 65 percent or more (Roth et al. 1996). Experienced breeders are more likely to show site fidelity (Brown and Roth 2004), and reproductive failure can lead to emigration (Roth and Johnson 1993). Mortality of young is high, and in one study, postfledging survival in the first eight weeks was 42 percent with highest mortality occurring the first week after fledging and again at three weeks when young become independent of their parents (Anders et al. 1997). Juvenile survival of wood thrushwood thrushes has not been found to be influenced by nestling condition (Anders et al. 1997, Brown and Roth 2004). Similar to other Neotropical migrants, natal philopatry is very low for wood thrushes (Roth et al. 1996), and recruits into a population often spend time exploring this area during postfledging dispersal in their first year of life (Brown and Roth 2004). Fluctuations in population growth rates may be linked to rodent populations and mast production (Schmidt et al. 2008).

Wood thrushes return to breed in Pennsylvania forests in mid-April. Pair bonds generally last one season and are typically monogamous. Extra-pair fertilizations are rare at 6 percent (Evans et al. 2008). The open-cup nest is placed from 1 m to 20 m in a tree or shrub, although the average height is 2–4 m (Roth et al. 1996, Hoover and Brittingham 1998, Artman and Downhower 2003). In Pennsylvania, about 20 percent of nests are parasitized by the brown-headed cowbird (*Molothrus ater*; Hoover and Brittingham 1993). Both parents care for young from the first nest until the female initiates incubation of a second clutch, at which time the male takes over sole care until independence. In final clutches, brood care is divided between the parents (Vega Rivera et al. 2000). Wood thrushes usually raise two broods annually (rarely three) and may continue re-nesting into August if predation occurs (Roth et al. 1996). In large forests, 30 percent to 60 percent of nests successfully fledge young (Roth et al. 1996), and a range of mammals, birds, and snakes have been identified as predators of wood thrush nests (Farnsworth and Simons 2000, Williams and Wood 2002).

### *THREATS*

Development and fragmentation of forests remain the primary threat to wood thrushes in the state. In addition to direct loss of suitable habitat, wood thrushes are less likely to breed in fragmented forest patches (Temple 1986, Boulinier et al. 1998), where nests experience high predation (Donavan et al. 1995, Hoover et al. 1995, Robinson et al. 1995). However, there may be some regional variation in the effects of forest fragmentation (Trine 1998, Friesen et al. 1999, Fauth 2001, Philips et al. 2005). Nest predation can increase along forest edges with increasing fragmentation and loss of forest in the surrounding landscape (Driscoll and Donavan 2004, Driscoll et al. 2005). Landscape-level changes influence the predator community (Chalfoun et al. 2002a), and recent work suggests low nest placement in dense understory along the forest edge may increase predation, possibly by making nests more accessible to ground foraging mammals or snakes while limiting effective nest defense (Newell and Kostalos 2007). In small fragments, wood thrushes are often unable to raise sufficient young to maintain the population, and such habitat is often occupied by inexperienced breeders showing lower subsequent site fidelity (Weinberg and Roth 1998). The long-term population dynamics of increased predation can lead to declining abundance (Roth and Johnson 1993). Brood parasitism by the brown-headed cowbird also poses a threat in fragmented forests surrounded by agricultural or housing developments (Rich et al. 1994, Ford et al. 2001, Philips et al. 2005). However, this remains a less significant threat in the more forested mid-Atlantic states (Hoover and Brittingham 1993). Evidence so far suggests that, in general, wood thrushes are not significantly affected by timber-harvesting techniques (Duguay et al. 2001, Robinson and Robinson 2001, Dellinger et al. 2007). Another potential threat to wood thrushes is acid ion deposition, which may limit distribution by affecting calcium-rich invertebrate prey necessary for breeding (Hames et al. 2002). A recent study has also found that forest songbirds in the Northeast, especially the wood thrush, show high levels of mercury contamination in their blood stream (D. Evers, unpublished data).

### *CONSERVATION AND MANAGEMENT NEEDS*

Because wood thrushes require extensive forest on both their breeding and wintering grounds, conservation efforts require protecting forest across several regions of the globe, as well as important stopover sites on migratory flyways. In Pennsylvania, specific management guidelines are needed to conserve a guild of forest-interior species, including wood thrushes, along with integrated long-term monitoring. Many large areas of source habitat are needed as part of a comprehensive strategy to conserve forest songbirds (Donovan et al. 1995, Robinson et al. 1995, Simons et al. 2000). Sink populations within 60 to 80 km may be linked to source populations (Tittler et al. 2006). Conservation efforts should attempt to protect and maintain extensive suitable large forest tracts. Priority should be given to forest >10,000 ha with >70 percent forest in a 5-km radius, whereas development should be focused in already fragmented areas. Maintaining ≥80 ha patches with ≥30 percent forest in the surrounding landscape may also be important (Hoover et al. 1995, Rosenberg et al. 2003, Driscoll and Donavan 2004). Roads and trails should be planned around the forest edge with the shape of the forest designed to conserve core areas, which may be one of the most important factors in wood thrush nest success (Rich et al. 1994, Hoover et al. 1995, Rosenberg et al. 2003, Driscoll et al. 2005). In urban and suburban areas, establishing 100-m buffer zones between residential development and forest may also be beneficial (Friesen et al. 1995, Philips et al. 2005). Agriculture or grassy areas adjacent to forests should be limited (Rich et al. 1994, Ford et al. 2001).

### *MONITORING AND RESEARCH NEEDS*

Continued monitoring with Breeding Bird Survey routes remains the most effective method to track long-term population change over large areas. Current point counts for the Pennsylvania Breeding Bird Atlas should provide more complete information on breeding density throughout the state while future work may be used to confirm Breeding Bird Survey results.

As breeding density does not necessarily equal reproductive success, coordination of long-term monitoring efforts to assess seasonal fecundity, although labor intensive, are probably most effective; monitoring nests initiated in May can provide a less labor-intensive measure of actual productivity (Underwood and Roth 2002). Development of a unified protocol for forest interior species of greatest conservation need could be used to identify important source populations for conservation within the state. This would allow adaptive management to conserve the most beneficial habitat for a range of forest breeding species, including wood thrushes. State monitoring should include participation in national programs, such as Breeding Biology Research and Monitoring Database (BBIRD; Martin et al. 1997), and Monitoring Avian Productivity and Survivorship Program (MAPS; DeSante et al. 2001).

The wood thrush has been extensively studied; however, such well-studied species may provide the best insight for effective large-scale conservation efforts. Human alteration of the natural environment remains complicated, and there are still questions to be answered about the declines of forest-interior Neotropical migrants. The relationship of breeding density to invertebrate food sources needs to be studied to identify the effects of soil pH and fragmentation.

Further work should look at the effect of high levels of mercury on the wood thrush. Although the effects of forest fragmentation on avian populations have been extensively studied, the underlying mechanism of a changing predator community is often ignored (Chalfoun et al. 2002b). Identification of local predators may provide a predictor of the effects of fragmentation on forest birds, and direct measures of reproductive success need to be correlated with predator assemblages in relation to fragmentation and landscape-level characteristics (Chalfoun et al. 2002a). GIS can be further used to improve the understanding of large-scale landscape and land use effects on forest bird conservation. Stopover habitat important to survival during migration, as well as the quantitative effect of collisions with man-made structures such as towers or windows, also need to be studied.

*Author:* FELICITY NEWELL, POWDERMILL AVIAN RESEARCH CENTER

*Fig. 5.51.* The Blue-winged Warbler, *Vermivora pinus*. Photo by Jacob W. Dingel III.

## Blue-winged Warbler

Order: Passeriformes
Family: Parulidae
*Vermivora pinus*

The blue-winged warbler was listed as a Species of Greatest Conservation Need because it is associated with a declining habitat type in Pennsylvania and because it is a Pennsylvania Responsibility Species with more than 5 percent of the global population breeding within the state (fig. 5.51). It is a Partners in Flight priority I (continental importance) species and a National Bird of Conservation Concern (U.S. Fish and Wildlife Service 2008). It is included on the lists of Species of Greatest Conservation Need in bordering northeastern states. Global populations are Secure (G5, NatureServe 2009).

### *GEOGRAPHIC RANGE*

The breeding range of the blue-winged warbler occurs entirely within Canada and the United States and extends throughout most of the eastern United States and into southern Canada. The range expanded northward and eastward for at least the past century, following massive forest clearing and high rates of farm abandonment in the northeastern United States during the late 1700s to mid-1900s (Gill 1980, Confer 1992, Askins 2001, Gill et al. 2001, Lorimer 2001, Trani et al. 2001). Range expansion continues in Michigan, Wisconsin, and central Ontario, presumably in response to the recent creation of early successional habitats, such as aspen clear-cuts and abandoned farmland (J. L. Confer, personal communication). Distribution of the blue-winged warbler is generally fragmented throughout much of the breeding range. The winter range occurs mainly from Mexicosouth along the Atlantic Slope of Mexico and Central America to central Panama; several winter records in the southern United States exist (Gill et al. 2001).

### *DISTRIBUTION AND RELATIVE ABUNDANCE IN PENNSYLVANIA*

Blue-winged warblers occur in areas of suitable habitat throughout most of Pennsylvania (Fig. 5.52), but abundance is greatest in western counties (Sauer et al. 2006). Historically, blue-winged warblers were rare in most regions of Pennsylvania (except the Southeast) until the 1930s when they moved up the Ohio River Valley and colonized western regions (Gill 1980, 1992). Through the late 1980s, blue-winged warblers were rare to uncommon in the Appalachian Mountain Section of the Ridge and Valley Physiographic Province and the Allegheny Front, Allegheny Mountain, eastern Pittsburgh Low Plateau, southern Deep Valleys, and High Plateau sections of the Appalachian Plateaus Physiographic Province (Gill 1992).

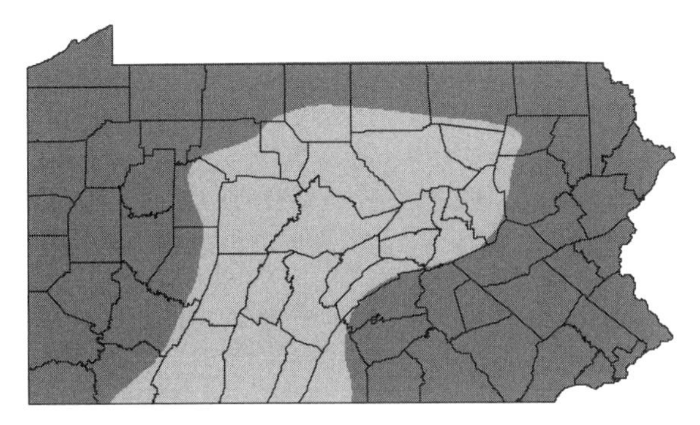

Fig. 5.52. Primary (darker shading) and secondary (lighter shading) distribution of the Blue-winged Warbler, *Vermivora pinus*.

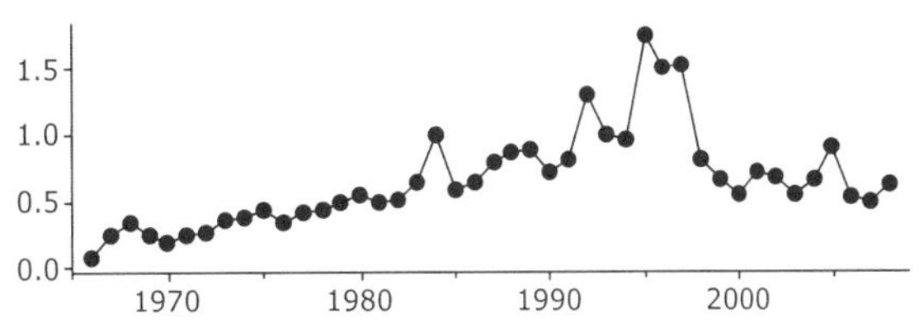


Fig. 5.53. The Blue-winged Warbler, *Vermivora pinus*, population trends from Breeding Bird Survey.

However, populations have been expanding into these regions recently (McWilliams and Brauning 2000, Sauer et al. 2006). Data from the North American Breeding Bird Survey suggest that populations increased in Pennsylvania by an annual rate of 6.1 percent during 1966–1994 but declined at the same rate during 1995–2005 (fig. 5.53; Sauer et al. 2006). Regional trends since 1966 include increases in abundance in northeastern, central, and south-central counties and declines in abundance in southeastern, northwestern, and extreme southwestern counties (Sauer et al. 2006).

### *COMMUNITY TYPE/HABITAT USE*

Blue-winged warblers use a variety of early to mid-successional habitat types during the breeding season, including pine-oak barrens, wetlands/bogs, recently burned or clear-cut forest, abandoned farmland, utility rights of way, and reclaimed or abandoned surface mines. Microhabitat requirements are similar to but generally less specific than those of the golden-winged warbler (*Vermivora chrysoptera*; Confer and Knapp 1981). Most populations require a mosaic of herbaceous patches and shrubby thickets located along a forest edge or within an open forest. Occasional trees or snags (>5 m tall) throughout the shrubby areas are useful as song perches. The shrub layer (woody plants 1.5–3 m tall) in a typical territory consists of a dense thicket that may include species such as scrub oak (*Quercus ilicifolia*), dogwood (*Cornus* spp.), American hazelnut (*Corylus americana*), alder (*Alnus* spp.), and aspen (*Populus tremuloides* and *Populus grandidentata*). Patches of herbaceous vegetation are used to place the nest and consist predominantly of goldenrod (*Solidago* spp), grasses, and scattered woody saplings (approximately 0.5 m tall).

### *LIFE HISTORY AND ECOLOGY*

Males arrive on breeding grounds two to nine days before females (Gill et al. 2001) and establish territories of 0.3–5.0 ha (Ficken and Ficken 1968b, Murray and Gill 1976, Confer and Knapp 1977, Canterbury et al. 1995b). Size and boundaries of territories may be determined by warbler density, interactions with other males, and the spatial and structural characteristics of vegetation (e.g., edges or rows of tall trees; Gill et al. 2001). Spring migration peaks in Pennsylvania the last week of April through the second week of May (Gill et al. 2001). Pair formation occurs shortly after the arrival of females and nest building occurs within a few days thereafter, but cold and wet weather may delay this process by as much as ten to twelve days (Will 1986, Gill et al. 2001).

The female appears to select the nest site, sometimes accompanied by the male (Gill et al. 2001). Nest sites are similar to those of the golden-winged warbler (Will 1986, Confer 1992, Gill et al. 2001, Klaus and Buehler 2001) and are characterized by open areas of forbs (especially goldenrod), grasses, and short saplings that occur within or adjacent to dense patches of shrubs near a forest edge or patch of trees. The nest is placed on or near the ground at the base of goldenrod or *Rubus* spp. or in tufts of grass or sedge (*Carex* spp.; Gill et al. 2001). Most nests are well concealed by vegetation and incorporate at least one woody sapling or dead goldenrod stem from the previous year, which adults use to descend to the nest when feeding young (Will 1986). Nests resemble those of the golden-winged warbler (Gill et al. 2001).

Blue-winged warblers will initiate additional nesting attempts following failure of a first attempt (Gill et al. 2001), but no published reports have documented additional attempts subsequent to successful fledging of a brood.

Potential nest predators in Pennsylvania include snakes, mammals (such as eastern chipmunk [*Tamias striatus*] and striped skunk [*Mephitis mephitis*]), and birds (such as American crows [*Corvus brachyrhynchos*] and blue jays [*Cyanocitta cristata*]). The brown-headed cowbird (*Molothrus ater*) is the only known brood parasite. Blue-winged warblers and golden-winged warblers produce fertile hybrids by interspecific pairing or by extra-pair fertilization, which has resulted in development of mosaic hybrid zones, where backcrossing threatens persistence of genetically pure individuals (Gill 2004, Dabrowski et al. 2005).

*THREATS*

The main threat to breeding populations of blue-winged warblers in Pennsylvania is low and declining availability of suitable habitat resulting from forest maturation, suppression of wildfire, infrequent use of even-aged forest management practices (e.g., clearcutting) in areas suitable for colonization by blue-winged warblers, infrequent conversion of farmland to old field habitat, and incompatible types of land use (e.g., agriculture, commercial, and residential development). A second threat, the magnitude of which is not completely understood, is introgressive hybridization and replacement of pure individuals or populations because of interspecific mating with or extra-pair fertilization by golden-winged warblers and backcrossing with hybrids (Gill 2004, Confer 2006). Although Gill (1997) concluded that hybridization appeared to threaten only golden-winged warblers (because of asymmetrical, nonreciprocal introgression), more recent evidence suggests that both species are threatened (Shapiro et al. 2004, Dabrowski et al. 2005). Brood parasitism by the brown-headed cowbird (Coker and Confer 1990, Canterbury et al. 1995a, 1995b, Gill et al. 2001) and artificially inflated populations of nest predators associated with urban and suburban sprawl (Hogrefe et al. 1998) may reduce reproductive success of blue-winged warblers.

*CONSERVATION AND MANAGEMENT NEEDS*

Because breeding ecology of blue-winged and golden-winged warblers and threats to the species are very similar (interrelated in the case of hybridization), conservation and management of blue-winged warblers in Pennsylvania should be coordinated with conservation and management of golden-winged warblers. However, the golden-winged warbler is considered a species of higher conservation priority, so efforts to conserve blue-winged warblers should be restricted to regions where golden-winged warblers do not occur (e.g., northwestern and southeastern Pennsylvania). Under that condition, the highest priorities for conservation of blue-winged warblers in Pennsylvania should be to identify, maintain, and enhance genetically pure populations. Whether pure populations still occur in Pennsylvania needs to be determined. If such populations exist and are identified, management strategies that ensure the viability of those populations (e.g., increasing habitat availability, preventing introgression of golden-winged warbler mtDNA) need to be developed. If genetically pure populations do not exist, then a conservation strategy needs to be decided on in consultation with the United States Fish and Wildlife Service, Partners in Flight, the Golden-winged Warbler Working Group, and other relevant organizations. Of particular importance would be whether to manage blue-winged and golden-winged warblers as a single species in Pennsylvania.

*MONITORING AND RESEARCH NEEDS*

Blue-winged and golden-winged warblers (and hybrids) must be monitored over the long term to be aware of changes in range/distribution; to identify source and sink habitats/populations; to better understand the factors contributing to the replacement of golden-winged warblers by blue-winged warblers; to detect colonization and abandonment of habitats/breeding areas; to determine population response to conservation action/inaction; to determine the longevity of various habitat types created by particular techniques; and assess overall values of particular conservation actions. Information gathered in pursuit of each of these objectives would be vital to adaptive management of the two species in Pennsylvania. Long-term monitoring plans should occur at the state level to maintain a general understanding of the status of blue-winged and golden-winged warblers in Pennsylvania. Monitoring should occur also at the site level to compile meaningful data sets that can be used to answer important questions about the ecology and long-term conservation of the species. Inclusion of genetic sampling during population monitoring would be extremely beneficial at any level.

Several lines of research are needed to identify suitable areas for long-term management and monitoring of breeding populations of blue-winged warblers in Pennsylvania. Such research needs are related to those for the golden-winged warbler and include: an extensive study that assesses population size and genetic purity of blue-winged and golden-winged warblers at all known breeding areas and identifies specific areas with the highest proportion of genetically pure blue-winged warblers; an intensive demographic study of multiple breeding populations of blue-winged warblers to identify the most productive breeding areas and habitat types; a field study that investigates what habitat conditions, if any, favor blue-winged warblers to the exclusion of golden-winged warblers and hybrids; a management study that experiments with different habitat creation/maintenance techniques (e.g., prescribed burning, clearcutting) and monitors colonization and use of manipulated habitats by blue-winged and golden-winged warblers (and hybrids); and a study that identifies potential habitats (i.e., areas that do not currently support blue-winged warblers but could be converted to suitable habitat given proper management action) and investigates the feasibility of acquiring such habitats (if not already under state ownership).

*Author:* JACOB E. KUBEL, M.S., MASSACHUSETTS DIVISION OF FISHERIES AND WILDLIFE

## Worm-eating Warbler

Order: Passeriformes
Family: Parulidae
*Helmitheros vermivorus*

The worm-eating warbler is a medium-sized wood warbler and is the only member of its genus (fig. 5.54). It is currently a Pennsylvania Audubon Watch List Species. Partners in Flight has listed this species as one of Moderately High Priority globally because of its dependence on large tracts of mature forest (Rosenberg 2003a). It is listed as a Pennsylvania Responsibility Species because 10 percent of the global population breeds in Pennsylvania.

*Fig. 5.54.* The Worm-eating Warbler, *Helmitheros vermivorus*. Photo courtesy of Bill Moses.

### *GEOGRAPHIC RANGE*

The worm-eating warbler breeds in eastern North America throughout most of the Appalachian region. Their population shows a pattern that ranges from eastern Massachusetts south to northern South Carolina and Georgia and extends west to eastern Texas and north to eastern Iowa. Winter populations are found in the Caribbean and Central America (Hanners and Patton 1998).

### *DISTRIBUTION AND RELATIVE ABUNDANCE IN PENNSYLVANIA*

Historically, the worm-eating warbler was likely less common in the state when a climax forest was dominate, because this forest type provides little understory (Brauning 1992b). However, the regenerating forests created after European settlers cleared lands would have allowed the worm-eating warbler to expand its range. By the end of the 1800s, it was considered to be found statewide, with the largest population in the southeastern counties (Warren 1890), especially Chester County and the lower Susquehanna Valley (Stone 1894). Breeding records from Todd (1940) in the western half of the state indicated that Clarion County was the northern limit of its breeding range, although it was observed in Crawford County.

Pennsylvania is considered an important breeding area for the worm-eating warbler, and the commonwealth may be responsible for a significant proportion of the global population (Rosenberg 2004). Current Breeding Bird Atlas data (Brauning 1992b) show that the worm-eating warbler was a confirmed breeder in many parts of the state, with the largest populations recorded in the Valley and Ridge Physiographic Province (fig. 5.55). It was also commonly found throughout the Southeast in areas with appropriate habitat, such as South Mountain in Cumberland, Adams, and Franklin counties, and Blue Mountain in Franklin County. Partners in Flight considers the Ohio Hills Physiographic Area, which includes Southwestern Pennsylvania, as supporting a high proportion of the world population of worm-eating warblers (approximately 8% of the total global population). The population in that region is considered large and stable (Rosenberg 2003a). The Northern Ridge and Valley Physiographic Area, which includes area from the Allegheny Front to the Kittatinny Ridge, holds about 5 percent of the total global population of worm-eating warblers (Rosenberg 2003b). Both of these areas are considered to have a high regional responsibility for this species.

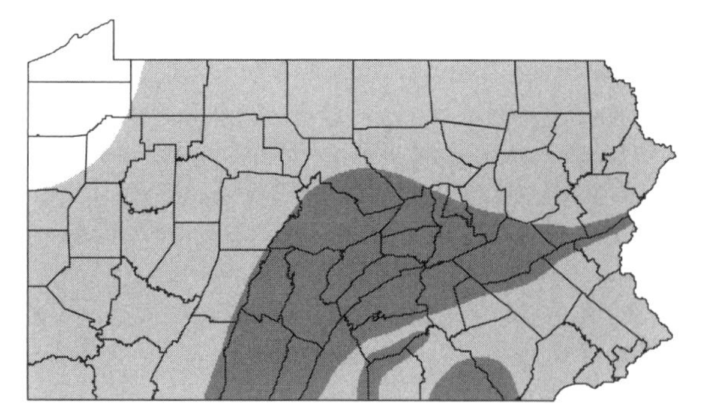

Fig. 5.55. Primary (darker shading) and secondary (lighter shading) of the Worm-eating Warbler, *Helmitheros vermivorus*.

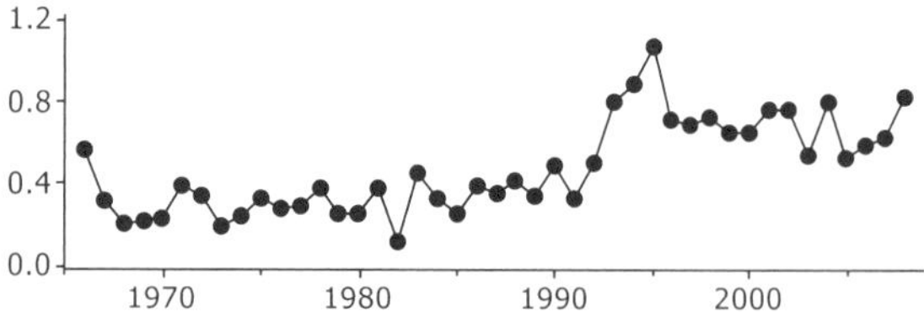


Fig. 5.56. Worm-eating Warbler, *Helmitheros vermivorus*, population trends from Breeding Bird Survey.

Although it is absent from most areas in the northwestern part of the state, the worm-eating warbler appears to have expanded its range into the northeastern and northern central areas of the state, such as Venango, McKean, and Potter counties during the last century. Records of this species have most conspicuously disappeared from the southwestern counties, where it was historically abundant (Todd 1940a), and may be due to increasing fragmentation in that part of the state (Brauning 1992b).

Worm-eating warbler populations in Pennsylvania have shown a nonsignificant decreasing trend from 1966 to 2003, according to Breeding Bird Survey records, which contrasts to a slight increasing trend across its entire range (Sauer et al. 2004; fig. 5.56). However, the data for the state trend are seen as deficient for an accurate analysis (see Sauer et al. 2004 for an explanation of trend data). Partners in Flight considers this species as having a stable population in Pennsylvania, and current goals are to maintain this population (which they have estimated at 46,000 individuals) at current levels (Rosenberg 2004).

*COMMUNITY TYPE/HABITAT USE*

The worm-eating warbler is most commonly found in large, mature tracts of deciduous or mixed forest types with a high percentage of canopy cover. Brauning (1992) noted that although the worm-eating warbler preferred large forested areas in Pennsylvania, it would use moderate-sized woodlots if other forested areas were nearby but would not use small or isolated woodlots. Although they can be found in a variety of forest types, they rarely breed in forest dominated by pine. They prefer areas of moderate to steep hillsides with dense patches of shrub cover, especially mountain laurel (*Kalmia latifolia*) and rhododendron (*Rhododendron maximum*). Although they have been found breeding in nearly flat areas, this is usually only in areas with no available slope habitat, such as along coastal sites (Hanners and Patton 1998).

*LIFE HISTORY AND ECOLOGY*

The worm-eating warbler is a ground-nesting species that prefers to locate its nest in dense shrub cover (typically mountain laurel or rhododendron) on a moderate to steep hillside. A typical nest site will be in a well-hidden location near the base of a sapling, against the roots of a shrub or sapling, or within a dense shrub cluster (Brauning 1992b, Hanners and Patton 1998) and will be constructed of many dead leaves, lined with moss stems and other fine material. Most nests in Pennsylvania are initiated during late May and late June (McWilliams and Brauning 2000), although data are limited. The worm-eating warbler raises one brood per year and is a common host of the brown-headed cowbird (*Molothrus ater*).

The worm-eating warbler forages for food in the understory, probing both live and dead leaves (Greenberg 1987, Hanners and Patton 1998) for arthropods, slugs, and spiders early in spring and for caterpillars later in the spring and in summer (Hanners and Patton 1998). Nestlings are fed a diet of caterpillars and other insects (Williams 2000).

*THREATS*

Because the worm-eating warbler is considered an area-sensitive species and is typically found at their highest densities in large tracts of unfragmented forest, a primary threat in Pennsylvania is the loss and fragmentation of large tracts of suitable habitat. Unfortunately, many of the areas that have seen the highest increase in housing units and associated infrastructure during the past ten years (Goodrich et al. 2002) are also areas where the worm-eating warbler has its highest breeding populations according to the Breeding Bird Atlas data. In particular, the continued increase in suburban and urban development around Philadelphia, New York, and Harrisburg may lead to the loss and fragmentation of habitat where the worm-eating warbler prefers to breed (Rosenberg 2003b); however, it is unknown how much of their key habitat is currently being threatened by sprawl or other development. The Chesapeake Bay Program reports that, in the Lower Susquehanna (York, Lancaster, and Adams counties), 11.6 percent of the forest in that area was lost from 1978 to 1992. They also report that the Harrisburg area lost 6.2 percent of its forest during the same time period. Brauning (1992) attributes the decline in breeding status in southwestern counties as potentially to the loss of large tracts of suitable habitat. Currently, only Greene County still has large areas of forested habitat.

*CONSERVATION AND MANAGEMENT NEEDS*

The worm-eating warbler relies on large areas of mature forest with a dense understory of woody vegetation for breeding. The Northern Ridge and Valley Physiographic Area, which includes area from the Allegheny Front to the Kittatinny Ridge, holds about 5 percent of the total global population of worm-eating warblers (Rosenberg 2003b). Both of these areas are considered to have a high regional responsibility for this species. Both the Breeding Bird Atlas (Brauning 1992b) and Breeding Bird Survey (Sauer et al. 1994) have identified areas that have high densities of worm-eating warblers, which are primarily in the Valley and Ridge Province. The Pennsylvania Gap Analysis has also identified potential areas of suitable habitat (Meyers et al. 2000), although these areas are not necessarily confirmed breeding areas. Partners in Flight has set its conservation target for this species in Pennsylvania as maintaining current population numbers. Therefore, to achieve this goal, those areas that contain the highest densities of worm-eating warblers should be identified positively and be protected from both habitat loss and fragmentation.

*MONITORING AND RESEARCH NEEDS*

Until more information is gathered about density, demographic rates, and source habitats, it would be difficult to propose a fully specified monitoring strategy. However, the best available data on statewide populations trends for most species, including the worm-eating warbler, comes from the Breeding Bird Survey. Additional routes are likely needed, especially where the population is the most dense, to ensure proper sampling and better statewide estimates of trend. In this manner, the cost-effective, standardized, volunteer-based Breeding Bird Survey data could be used for general monitoring, and when a particular threshold (say, a decline of >5% per year) is reached, more intensive monitoring methods or research projects could be triggered and implemented (e.g., Houser et al. 2006).

Limited demographic data are available on the worm-eating warbler in Pennsylvania. In particular, further investigation is needed to determine whether the areas that have the largest populations of this species may also be the most susceptible to habitat loss, fragmentation, or other threats. Little information also exists on how threats such as deer overbrowsing of preferred habitat may affect suitable nesting and foraging sites. Given the large population of white-tailed deer in the state, loss of understory habitat by overbrowsing may have a significant effect. Data on the effects of parasitism by the brown-headed cowbird within the state are sparse. Because other studies have shown that the number of host young from a parasitized nest decreases when cowbird young are present (Hanners and Patton 1998, DeCecco et al. 2000), this information is important. Finally, additional information on how different silvicultural practices on both a landscape and a smaller habitat scale affect the worm-eating warbler in Pennsylvania by altering habitat, predation, and parasitism rates is needed.

*Author:* JENNIFER A. DECECCO

## Louisiana Waterthrush

Order: Passeriformes
Family: Parulidae
*Seiurus motacilla*

The Louisiana waterthrush's ecological dependence on undisturbed and unpolluted forested headwater streams makes it a useful bioindicator, on the one hand, and a Species of Greatest Conservation Need, on the other (fig. 5.57). The waterthrush is a Northeast Region Priority Species and is included on state lists of Species of Greatest Conservation Need throughout the Northeast. Partners in Flight ranks the waterthrush as a Species of Regional Concern, regional stewardship, and continental stewardship, requiring management attention within the Appalachian Mountain bird conservation region (BCR 28), which includes virtually the entire range of the species within Pennsylvania. An estimated 44 percent of the global population of Louisiana waterthrush occurs within

*Fig. 5.57.* The Louisiana Watherthrush, *Seiurus motacilla*. Photo courtesy of Reinhard Geisler, Florida, www.reige.net/nature.

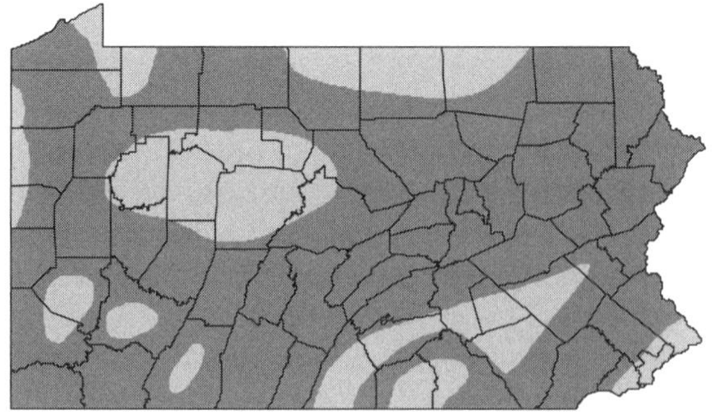

*Fig. 5.58.* Primary (darker shading) and secondary (lighter shading) distribution of the Louisiana Waterthrush, *Seiurus motacilla*.

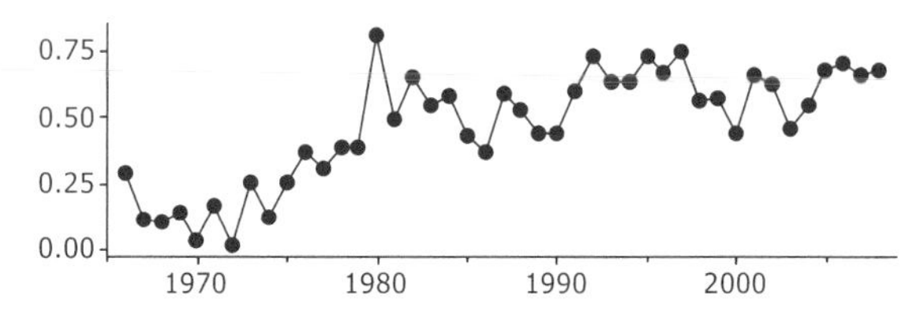


*Fig. 5.59.* Louisiana Waterthrush, *Seiurus motacilla*, population trends from Breeding Bird Survey.

this BCR. Within Pennsylvania, the waterthrush was selected as a Species of Greatest Conservation Need because it is a Responsibility Species and an indicator of high-quality streams and high-quality mature riparian forest habitat. Global populations are Secure (G5, Natureserve 2009).

### *GEOGRAPHIC RANGE*

Their range extends from southeast Minnesota and extreme eastern Nebraska, Kansas, Oklahoma, and Texas east to the Atlantic Coast and from extreme southern Maine, New Hampshire, Vermont, and northern New York south, as far as Alabama and Mississippi, excluding the Gulf and southeast Coastal Plain (Robinson 1995). The Louisiana waterthrush winters from New Mexico through Central America and in the West Indies (Ridgely and Gwynne 1989, Robinson 1995, Latta et al. 2006).

### *DISTRIBUTION AND RELATIVE ABUNDANCE IN PENNSYLVANIA*

The species is widely distributed at all elevations across Pennsylvania, where it occurred in 36 percent of survey blocks in the second Breeding Bird Atlas (compared with 26% in the first atlas), which were disproportionately found within the Valley and Ridge Physiographic Province (fig. 5.58; Mulvihill and Brauning 2009). This is likely the result of highly dissected topography and associated drainage patterns that provide numerous medium- to high-gradient forested headwater streams that are the preferred habitat of the species (Gross 1992a).

Currently, Breeding Bird Survey (BBS) data suggest the Louisiana waterthrush population in Pennsylvania appears stable with no significant trend in population size (Fig. 5.59). Detectability, using standard BBS protocol, however, is probably low due to early territorial establishment and specificity for a habitat that is not well sampled by BBS methods. Overall, the population is probably stable in Pennsylvania.

### *COMMUNITY TYPE/HABITAT USE*

Preferred habitat is small (first through third order) medium- to high-gradient unpolluted forested headwater streams having a benthic substrate of loose (i.e., not heavily embedded in sediments) gravel to cobble-sized rocks and characterized by well-developed, partially eroded banks or plentiful fallen trees with exposed root mats. Within such watersheds, areas with moderate to sparse undergrowth are preferred (Prosser and Brooks 1998). Surdick (1995) observed that most foraging sites had nearly three times the surface area of exposed rock than the territory average. This species sometimes will breed in bottomland forest on mud-bottomed streams but at lower densities than in upland areas (Graber et al. 1983, Walsh et al. 1999).

In the Pocono Region, this waterthrush is found primarily along the plateau perimeter where hemlock-dominated headwater streams tumble down steep ravines to surrounding lowlands. It is allopatric with the northern waterthrush that inhabits forested wetlands at higher elevations on the plateau (T. L. Master, unpublished data). The affinity for habitat dominated by eastern hemlock (*Tsuga canadensis*) does not appear to be as strong in central and western Pennsylvania where mixed deciduous trees more often compose the riparian forest (T. O'Connell, unpublished data; R. S. Mulvihill, unpublished data). In these deciduous settings, however, rhododendron and mountain laurel frequently make up a significant part of the riparian forest understory.

### *LIFE HISTORY AND ECOLOGY*

Louisiana waterthrushes characteristically teeter, or bob, as they wade into shallow water and walk over exposed rocks and fallen logs within and along stream banks searching for food. Their diet is mainly composed of aquatic macroinvertebrates, both nymphs and adults, but also can include small aquatic vertebrates, such as minnows, tadpoles, and salamanders.

The Louisiana waterthrush arrives earlier in Pennsylvania than any other breeding warbler, generally appearing during the last week of March and not later than the first few days of April (T. L. Master, East Stroudsburg University [ESU], unpublished data; R. S. Mulvihill, unpublished data). Males defend linear territories along medium- to high-gradient headwater streams. Contiguous territories, each about 250–300 m long, are typical for unimpacted forested headwater streams across Pennsylvania (R. S. Mulvihill, unpublished data; T. L. Master, unpublished data). Densities are usually lower (i.e., territories are longer, up to 1,000 m, and frequently disjunct) on environmentally degraded, for example, acidified streams (Mulvihill 1999, Mulvihill et al. 2008). Although generally monogamous, approximately 2 percent (4/283) of males were discovered to be polygynous in a study in western Pennsylvania (Mulvihill et al. 2002).

On average, three young fledged per successful nest over the course of a 1998 to 2000 study of waterthrushes on twenty-three streams in southwestern, central, and northeastern Pennsylvania (O'Connell et al. 2003). Predation was responsible for the failure (at both egg and nestling stages) of 30 percent of 222 nests monitored across Pennsylvania (O'Connell et al. 2003). Depredated nests containing full clutches or nestlings are usually replaced (as many as three times in the same season; R. S. Mulvihill, unpublished data). Double brooding was confirmed in this species for the first time at study sites in western Pennsylvania (Mulvihill et al. 2009).

The Louisiana waterthrush departs its Pennsylvania breeding grounds as early as mid-July, and most breeding waterthrushes have left Pennsylvania by early August (Curson et al. 1994, McWilliams and Brauning 2000). The species arrives on its middle American wintering grounds from late July to late August (Ridgely and Gwynne 1989, Stiles and Skutch 1989, Latta et al. 2006).

### *THREATS*

Pennsylvania contains approximately 45,000 miles of streams (Pennsylvania Department of Conservation and Natural Resources), three-quarters of which can be categorized as headwater streams preferred by this species (Nickens 2004). Population status of the species depends on water quality and integrity of surrounding riparian forest cover. Impacts that degrade water quality, reduce forest canopy cover, increase forest fragmentation, or modify natural headwater stream bank structure are the major threats to this species in Pennsylvania.

Acid mine drainage and acid deposition, in particular, are significant threats to the productivity of the species in Pennsylvania and throughout the Appalachian region (Mulvihill 1999, Mulvihill et al. 2008). Streams affected by acid mine drainage or acid deposition exhibit significantly reduced population density, smaller clutch size, and lower site fidelity by breeding adults compared with reference streams. Loss of eastern hemlock canopy cover due to hemlock woolly adelgid (*Adelges tsugae*) infestation may be an increasing threat along streams in eastern Pennsylvania (T. L. Master, unpublished data). Less than 5 percent of nests were parasitized by brown-headed cowbirds in an intensive three-year study (1998–2000) in Pennsylvania (O'Connell et al. 2003).

### *CONSERVATION AND MANAGEMENT NEEDS*

Preservation, protection, and restoration of unpolluted forested headwater stream environments would be the single most effective conservation measure for Louisiana waterthrush. In managed forests, a forested buffer at least 100 m wide along both sides of headwater streams will help preserve water quality (i.e., by reducing sedimentation). An undisturbed forested riparian buffer will also ensure the availability of peripheral foraging opportunities (e.g., in small tributaries and wetlands) for waterthrushes, which are especially important where the mainstream water quality is subpar (Mulvihill 1999).

This buffer will also promote the survival of young waterthrushes, which take shelter in dense riparian forest understory vegetation, up to 100 m from the main stream, during the critical postfledging period (S. C. Latta, unpublished data). This recommendation agrees with findings/results in Pennsylvania's Wildlife Action Plan (PA-WAP) that recommends a forested buffer of 100–200 m whenever possible to accommodate a range of wildlife. Linear, forested headwater stream corridors connected laterally to suitably sized forest fragments were identified in a northeastern Pennsylvania connectivity study as important to maintaining populations of this species (Fanok et al. 2008).

Human activities that lead to reduction or elimination of riparian forest cover (e.g., logging, second home development), substantial alteration of surface water flow or benthic habitat (e.g., sedimentation, filling, dredging), and degradation of water quality (e.g., acidification) in headwater watersheds should be avoided or discouraged whenever possible. Because the Louisiana waterthrush is an "umbrella" species within headwater watersheds, many other less observable taxa (e.g., freshwater mussels, nongame fish), important to overall biodiversity and ecosystem function, will undoubtedly benefit from conservation and management efforts aimed at this species.

*MONITORING AND RESEARCH NEEDS*

A targeted population monitoring or survey protocol is needed for this species. A simple, easily implemented monitoring protocol might consist of visits to roadside headwater stream crossings during the second to fourth weeks of April to record presence of singing males, using audio playback (e.g., Prosser and Brooks 1998). Traversing a standard length of stream reach (ca. 1,000 m) at each site, using song playback at ca. 250-m intervals, would provide more detailed density information. Occupancy of headwater stream reaches also can be accurately determined through careful visual inspection of rock and log surfaces within the wetted stream channels for the characteristic splay of Louisiana waterthrush (Mulvihill 1999).

Research on the effects of water quality degradation on food availability, prey selection, and territory size and density are needed to better inform conservation recommendations. In addition, more studies of possible effects of forest patch size and fragmentation on stream occupancy and reproductive success (Prosser and Brooks 1998, O'Connell et al. 2003) and the role of predators in affecting population size are needed.

Postbreeding behavior and habitat use before fall migration is virtually unknown and require investigation. This is the only period during which individuals expand their habitat use away from the immediate stream vicinity (S. C. Latta, unpublished data). Studies on migration and stopover ecology are likewise critically important to the long-term stability of populations. Detailed studies of wintering ecology, emphasizing habitat use, potential effects of habitat degradation on subsequent breeding success and population recruitment, niche-partitioning, and competition with tropical residents are also not well documented and are recommended (George 2004, Master et al. 2005). The species' reliance on riparian habitats in winter means that water- and habitat-quality differences throughout its Latin American wintering range are likely to have significant carry-over effects for breeding populations, and this deserves study.

*Authors:* TERRY L. MASTER, EAST STROUDSBURG UNIVERSITY OF PENNSYLVANIA; ROBERT S. MULVIHILL, POWDERMILL AVIAN RESEARCH CENTER, CARNEGIE MUSEUM OF NATURAL HISTORY; AND STEVEN C. LATTA, DEPARTMENT OF CONSERVATION AND FIELD RESEARCH, NATIONAL AVIARY

## Scarlet Tanager

Order: Passeriformes
Family: Cardinalidae
*Piranga olivacea*

The scarlet tanager is a mid-sized, forest-dwelling songbird common to abundant throughout most of Pennsylvania (fig. 5.60). It was selected as a Pennsylvania Responsibility Species because as much as 17 percent of the global population breeds in Pennsylvania (Rosenberg and Wells 1995). State and global populations are considered Secure (S5B,G5, NatureServe 2009).

*GEOGRAPHIC RANGE*

Breeding range broadly coincides with the eastern deciduous forest biome (Price et al. 1995). The scarlet tanager breeds north into southern Canada, as far west as southeastern Manitoba, and south into northern Georgia, Mississippi, Alabama, and Arkansas. The

*Fig. 5.60.* The Scarlet Tanager, *Piranga olivacea.* Photo by Jacob W. Dingel III.

western boundary of the range corresponds to the eastern edge of the Great Plains, although populations range westward into the Central Plains along forested river systems. The scarlet tanager winters in primary and secondary-growth forests from southern Central America through northern and western South America (Mowbray 1999).

### *DISTRIBUTION AND RELATIVE ABUNDANCE IN PENNSYLVANIA*

The scarlet tanager was detected in nearly 87 percent of the state's atlas blocks in the first Breeding Bird Atlas (1983–1989; Ickes 1992e). Pennsylvania plays a significant role in this species survival with 13 percent to 17 percent of its global nesting population occurring within its boundary (Rosenberg and Wells 1995, Moyer 2003). Within the state, scarlet tanagers are found in least abundance in areas with the least tree cover (Ickes 1992e). As such, this species is found throughout the state, but is most abundant in the northern forests and the central and southern mountains of the state, and least abundant in the heavily developed southeastern Piedmont region (McWilliams and Brauning 2000; fig. 5.61). North American Breeding Bird Survey (BBS) data indicate a relatively stable population trend within Pennsylvania from 1966 to 2003 (Sauer et al. 2004; fig. 5.62).

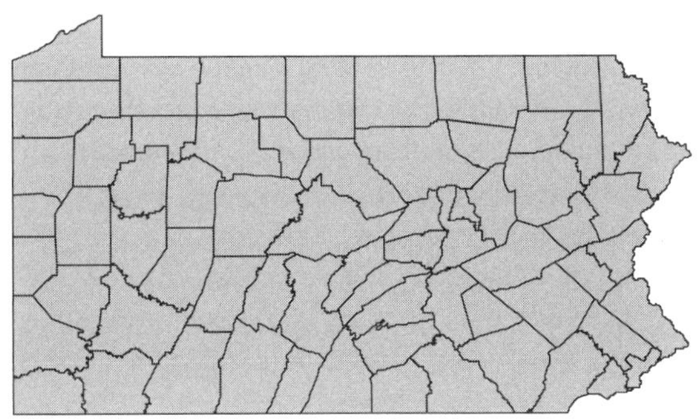

*Fig. 5.61.* Distribution of the Scarlet Tanager, *Piranga olivacea.*

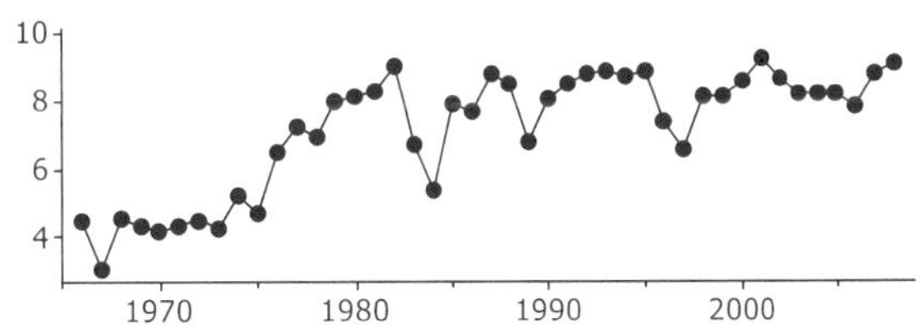


*Fig. 5.62.* Scarlet Tanager, *Piranga olivacea,* population trends from Breeding Bird Survey.

### *COMMUNITY TYPE/HABITAT USE*

Scarlet tanagers breed in a variety of mature deciduous and mixed deciduous forest types (Mowbray 1999). Within Pennsylvania, tanagers are common in oak or hickory woodlands, beech forests, mixed pine stands, and even pure, extensive stands of eastern hemlock (*Tsuga canadensis*; Ickes 1992e, McWilliams and Brauning 2000). This species can also be found in extensive plantings of shade trees and in parks, suburban areas, or cemeteries (Ickes 1992e). Scarlet tanagers appear to be less area sensitive than many forest birds and can be found in woodlots as small as 10 to 15 ha in some landscapes (Robbins et al. 1989, Roberts and Norments 1999). At a regional scale, however, tanager occurrence appears to be negatively correlated with the overall amount of fragmentation and the amount of edge within a landscape (Rosenberg et al. 1999a).

### *LIFE HISTORY AND ECOLOGY*

A long-distance complete migrant, the scarlet tanager is present in Pennsylvania between late April and October, although most nesting activity is restricted to May through July (McWilliams and Brauning 2000). During the breeding season, diet consists mostly of adult and larval insects and spiders. This species typically forages for prey within the forest midcanopy, capturing nonflying invertebrates from bark or foliage through gleaning. Scarlet tanagers also capture flying insects by hawking from a perch (Holmes and Robinson 1981). During the late summer, migration, and winter, fruit is often incorporated into the diet and may be especially important for fat deposition before migration (Rosenberg et al. 1999b).

In areas of sympatry, the scarlet tanager countersings in response to the song of the summer tanager (*Piranga rubra*) and maintains nonoverlapping territories with this species (Mowbray 1999). At a local scale, however, scarlet tanagers tend to be found in habitats with higher and denser canopy cover, a greater diversity of tree species, and a higher density of large trees than summer tanagers (Mowbray 1999).

The scarlet tanager breeds in a variety of mature forest types. Seasonally monogamous, this species builds a shallow, and often flimsy, open nest typically well hidden by foliage and far out from the trunk. Although studies of marked populations are lacking, this species appears to typically raise a single brood annually but will re-nest when early nests are lost (Mowbray 1999). Scarlet tanagers are parasitized by the brown-headed cowbird (*Molothrus ater*) throughout their range (Mowbray 1999). Parasitism rates vary substantially among populations, but tend to be highest in highly fragmented midwestern landscapes and lower in landscapes that contain larger forest tracts (Robinson et al. 1995). Cowbird parasitism tends to reduce tanager productivity primarily through clutch reduction, presumably due to egg removal by female cowbirds. Tanager eggs in parasitized nests are as likely to hatch as those in unparasitized nests, and the presence of a cowbird nestling does not appear to hinder the survival of tanager nestlings (Mowbray 1999).

### *THREATS*

The greatest threat to this species is the continual loss and fragmentation of both breeding and wintering habitat. Within Pennsylvania, loss of forest cover is primarily a problem in the more densely populated southeastern Atlantic Coastal Plain and Piedmont uplands and in the southwestern Pittsburgh low plateau. Increased forest fragmentation, however, is a problem throughout the state. The likelihood that a forest patch attracts scarlet tanagers is positively correlated with patch size but also with the amount of forest remaining in the surrounding landscape and the proximity of the forest patch to a large forest (Rosenberg et al. 1999b). As a result, in highly forested landscapes, including much of Pennsylvania, tanagers do not show area sensitivity and occur in virtually any forest patch, regardless of size. As regional forest cover is reduced, however, the probability of smaller or more isolated forest patches attracting tanagers decreases. Further, increased fragmentation can also reduce pairing and nesting success. In western New York, pairing success of male tanagers was highest in large (>1,000 ha) forest patches and fledging success increased significantly with increasing area of forest (Roberts and Norment 1999).

Although tanagers prefer mature forests over early successional forests, there is some evidence that changes that occur in very mature forests may negatively affect this species. Litwin and Smith (1992) reported a 50 percent decline in scarlet tanager density over a thirty-year period within a mature forest in New York and attributed this decline to the loss of vertical and horizontal heterogeneity associated with forest maturation. On the Appalachian plateau of Pennsylvania, however, there was no difference in the occurrence of tanagers between old-growth mixed forests and the surrounding younger forests (Haney 1999).

An overabundance of white-tailed deer (*Odocoileus virginianus*) in the state has greatly limited forest regeneration and reduced the density of understory and subcanopy forest layers (Goodrich et al. 2002). Because scarlet tanagers often forage in midcanopy levels, these changes could negatively affect this species; however, this threat has not been studied. Lighted towers and buildings may also be an important mortality source, particularly for night-migrating songbirds like the scarlet tanager (Goodrich et al. 2002). During the spring and fall migration of a single year, nearly 300 tanagers were killed in collisions with a television tower in Florida (Stevenson and Anderson 1994).

### *CONSERVATION AND MANAGEMENT NEEDS*

Pennsylvania currently supports a large and stable scarlet tanager population. As such, the primary conservation need, or responsibility, for Pennsylvania is to maintain the current population size and stable population trend within the state by maintaining abundant, high-quality breeding habitat. To accomplish this, it is important to maintain current levels of forest coverage within the state and minimize fragmentation of remaining large contiguous forest tracts. In addition, currently fragmented landscapes should be managed to maximize core habitat via reforestation and patch shape (i.e., circular rather than long and narrow) and to promote connectivity among patches using the protection or reforestation of corridors. Management efforts that promote a well-developed woody shrub and sapling layer should be implemented. Long-range forest management plans should be developed at as large a scale as possible to designate tracts that will be mature at each stage of the plan and to maintain connections between existing mature forest patches.

#### *MONITORING AND RESEARCH NEEDS*

Current monitoring efforts (i.e., Breeding Bird Survey, Pennsylvania Breeding Bird Atlas) are adequate to monitor the breeding population size and distribution and to measure the population trend for this species. Local survey efforts and studies, including nest searching and monitoring, would be useful in a variety of landscapes at various levels of fragmentation to better identify the features that constitute high-quality breeding habitat (i.e., source populations). These studies may also identify factors (e.g., particular predators, land uses, or forestry management practices) that negatively affect tanager populations; factors that may be targeted or corrected by future management efforts.

Given Pennsylvania's high responsibility for this species, it is important to determine the features associated with high-quality breeding habitat for tanagers so that managers can identify high-quality habitats (i.e., sources) for protection, and can initiate effective management efforts to improve the quality of fragmented landscapes for tanagers. Management activities for tanagers should also benefit many other forest interior species (Rosenberg et al. 1999b).

Activities that alter forest structure have the potential to affect tanagers, as well as many other songbirds, yet very little is known about how these activities influence habitat quality. Consequently, studies should be undertaken that better determine how forest harvest or management practices, natural forest maturation, and effects of deer overabundance affect breeding habitat quality for tanagers.

Almost no information is available regarding this species during migration or on its tropical wintering grounds (Mowbray 1999). Better knowledge of migration routes and a better understanding of how forest fragmentation and disturbance might be affecting its survival during migration and on its wintering grounds are needed.

*Author:* CHRISTOPHER B. GOGUEN, PENNSYLVANIA STATE UNIVERSITY

### PENNSYLVANIA VULNERABLE

## Green-winged Teal

Order: Anseriformes
Family: Anatidae
*Anas crecca (carolinensis)*

The green-winged teal is the smallest dabbling duck in North America (fig. 5.63). It has a globally Secure (G5) status due in part to its extensive world range (NatureServe 2009). It is recognized as a Species of Greatest Conservation Need within Pennsylvania because of its small breeding population size, which is considered Imperiled. Rhode Island is the only other northeastern state to currently list green-winged teals as a species of concern.

#### *GEOGRAPHIC RANGE*

The green-winged teal breeds mainly in Canada with smaller numbers in the northern states of the United States from Wyoming east of Massachusetts. Pennsylvania is at the southern edge of the species' North American range. Its Eurasian counterpart, *Anas crecca crecca,* breeds at similar latitudes across Europe and Asia, while a third subspecies, *Anas crecca nimia*, is restricted to the Aleutian Islands. During winter, it is found mainly to the south of the breeding range, although breeding and winter ranges overlap in the northern Rockies and Pacific Northwest. Some winter as far south as southern Mexico and the Caribbean Islands.

#### *DISTRIBUTION AND RELATIVE ABUNDANCE IN PENNSYLVANIA*

The species is most commonly found during spring and fall migration in Pennsylvania, with small numbers overwintering in the Coastal Plain, less often

*Fig. 5.63.* The Green-winged Teal, *Anas crecca (carolinensis)*. Image courtesy of Elaine R.Wilson, naturespicsonline.com.

elsewhere (McWilliams and Brauning 2000). The birds that pass through Pennsylvania are part of the Atlantic Flyway population (Gregg et al. 2003), which comprises a small proportion of the North American population of more than 3 million birds (K. Johnson 1995). The John Heinz National Wildlife Refuge at Tinicum is the most important site for this species in Pennsylvania, with up to 3,000 birds present during spring and fall migration (Cohen and Johnson 2004).

Pennsylvania supports a small breeding population of green-winged teals. Between 1983 and 1988 it was confirmed to have bred in only five blocks during the first Pennsylvania Breeding Bird Atlas (Brauning 1992a). Probable or possible breeding was recorded in an additional eighteen blocks, most of them in the north of the state, with concentrations in Crawford and Erie counties. The highest probability of breeding occurrence is currently in the northwest corner of the state (fig. 5.64). Given the secretive nature of this species during the breeding season and its preference for small wetlands, it could have been overlooked in some blocks. Even so, it is likely that the breeding population of this species in the state numbers tens rather than hundreds of pairs. This species has been listed as a breeding bird in three Pennsylvania Important Bird Area designations: (1) Marsh Creek Wetlands (12 or more pairs), (2) Conejohela Flats, and (3) Shohola Waterfowl Management Area (Cohen 2004a, 2004b, McNaught 2004).

There is no information on breeding population trends within Pennsylvania. There was a large increase in the number of teal harvested in the state between 1993 and 2002 (Gregg et al. 2003), during which time the continent-wide population increased from around 2 million birds to more than 2.5 million (U.S. Fish and Wildlife Service 2004b). Most of the green-winged teals are harvested during the fall migration period.

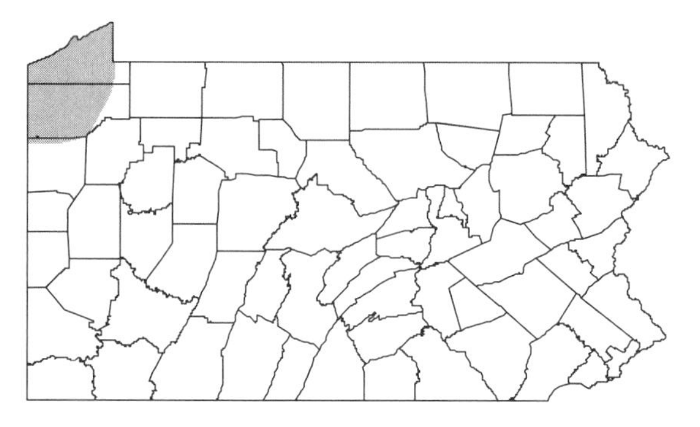

*Fig. 5.64.* Highest probability of occurrence of the Green-winged Teal, *Anas crecca (carolinensis)*

*COMMUNITY TYPE/HABITAT USE*

During the breeding season, green-winged teals prefer small, shallow wetlands with emergent vegetation (K. Johnson 1995). These include natural and artificial ponds, marshes, swamps, and beaver dams. They breed mainly in well-vegetated northern wetlands, often within forests or river deltas. Outside of the breeding season, it is found in a variety of wetland habitats, including brackish coastal waters. The largest flocks are found in extensive shallow wetlands, such as coastal marshes or river flood waters (K. Johnson 1995).

*LIFE HISTORY AND ECOLOGY*

The species is migratory, with birds returning to breeding areas from March to early May. Pair bonds are formed on the wintering grounds. Nests are typically within 20 m of water and well concealed within dense vegetation, often sedges, grasses, or shrub thickets. The clutch of six to nine eggs is laid from early May onward and incubated by the female only. The young are precocial and are led to water after hatching, where they are able to feed themselves on a diet of insect larvae taken on or close to the water surface. Adults also consume aquatic invertebrates, but rely most heavily on plant matter, especially seeds of aquatic vegetation. The young fledge after five to seven weeks.

Fall migration commences in midsummer in the north of the range, with more southerly populations leaving the breeding areas during September and October. During the fall and spring migrations, green-winged teals use a wide variety of wetland habitats, but generally prefer shallow freshwater or coastal marshes, as in the winter months. During winter it is gregarious, forming flocks of hundreds and often thousands of birds at favored sites.

*THREATS*

The loss of emergent wetland in Pennsylvania during the twentieth century is of concern for all species that rely on this habitat. Four of the five confirmed breeding records for this species in the Atlas of Breeding Birds in Pennsylvania (Brauning 1992a) were from state gamelands. The maintenance of high-quality emergent wetlands on state gamelands may therefore be crucial if the small breeding population of this species in Pennsylvania is to be sustained. Reduction in water quality through pollution, through agricultural runoff, or through acidification could all be detrimental to the ecology of the wetlands that this species relies on. There has been no research on whether water-quality issues have effects on green-winged teal breeding success, in Pennsylvania or elsewhere.

Green-winged teal populations are healthy across the continent and, indeed, have increased in recent years (U.S. Fish and Wildlife Services 2004). The current harvesting levels in Pennsylvania and surrounding states are not considered to be a threat to Atlantic Flyway populations of this species. It is not known where the birds that breed in Pennsylvania winter and, hence, an assessment of the effects of the hunting of these birds is not possible.

*CONSERVATION AND MANAGEMENT NEEDS*

Green-winged teals are found in good-quality emergent wetlands, a scarce habitat in Pennsylvania. Its status as a breeding bird in the state depends on the retention of this important habitat type but may also be influenced by changing fortunes of the large breeding populations farther north. The protection and enhancement of green-winged teal populations at a continental scale is therefore crucial to this species throughout the seasons in Pennsylvania.

*MONITORING AND RESEARCH NEEDS*

The population size of breeding green-winged teals in Pennsylvania is poorly documented. Data collected during the second Pennsylvanian Breeding Bird Atlas should provide some information on the current summer distribution of this species, but it may not adequately provide an estimate of population sizes. Targeted follow-up surveys in key areas are desirable.

Because the breeding population is small and fragmented in Pennsylvania, it would be difficult to carry out a research project, of any description, on this species within the state. Given that Pennsylvania supports a negligible proportion of the world population of this species, it should not be considered a strong candidate for research funds.

*Author:* ANDY WILSON, PENNSYLVANIA STATE UNIVERSITY

## Least Bittern

Order: Ciconiiformes
Family: Ardeidae
*Ixobrychus exilis*

The least bittern is the smallest North American member of the heron family (fig. 5.65). It has undergone precipitous declines in Pennsylvania and is listed as a state Endangered Species and a Pennsylvania Vulnerable Species. Its breeding population is considered Critically Imperiled (S1b, NatureServe 2009). Least bitterns are listed as a Species of Greatest Conservation Need on all northeastern state lists. The global population is considered Secure (G5, NatureServe 2009).

*Fig. 5.65.* The Least Bittern, *Ixobrychus exilis*. Image courtesy of Elaine R. Wilson, naturespicsonline.com.

*GEOGRAPHIC RANGE*

The least bittern nests primarily from the eastern Dakotas east across Michigan and southeastern Canada, and north along the coast to New Brunswick, and south to Texas and Florida. Scattered populations may be found in western states from Oregon south through Arizona. The species is found widely in Central and South America and the Caribbean (American Ornithologists' Union 1998). Northern populations are migratory to Central and South America, but breeding populations in the southern United States south are sedentary.

*DISTRIBUTION AND RELATIVE ABUNDANCE IN PENNSYLVANIA*

In Pennsylvania, the least bittern was listed as Threatened beginning in 1985 (Brauning et al. 1994) and downgraded to Endangered in 1999. This is a rare bird in Pennsylvania although easily overlooked and poorly monitored. It is very spottily distributed, found in 31 of 4,928 atlas blocks in the 1980s (Brauning 1992a) with similar results from the second atlas (Mulvihill and Brauning 2009). Pennsylvania is not peripheral to this species' geographic range, and it has a long history of occupation in extensive wetlands, primarily in southeastern and northwestern counties (fig. 5.66). The most reliable locations have been the Tinicum Marsh of Delaware and Philadelphia counties, on

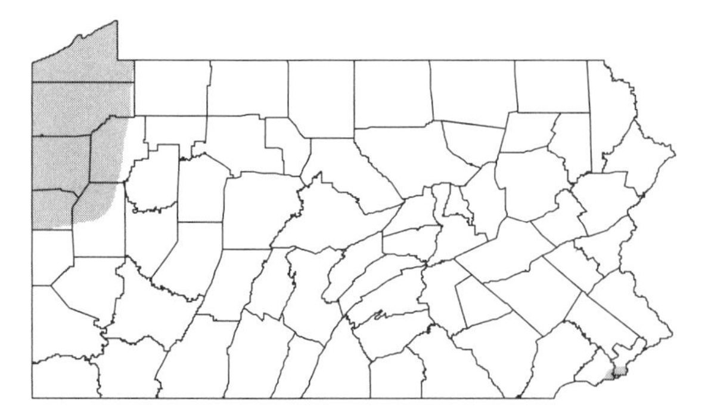

*Fig. 5.66.* Highest probability of occurrence of the Least Bittern, *Ixobrychus exilis*.

Presque Isle, and the Conneaut and Pymatuning marshes of Crawford County. Least bitterns may nest anywhere within the state in suitable habitat.

Least bitterns may be found in any of the larger tidal and freshwater marshes of Pennsylvania during the spring migration season, which ends in late April. Because of the inaccessibility of their habitat, they are seldom recorded during the fall migration period. Most have left the region by late September. Confirmed nesting has been documented recently from Berks, Chester, Crawford, Cumberland, Delaware, Erie, Indiana, Mercer, and Philadelphia counties and seasonal reports in another half-dozen counties (Brauning 1992a, McWilliams and Brauning 2000, F. Haas, personal communication).

Historically, they were considered very common in the Delaware River marshes as far north as Bucks County (Harlow 1918). A count of twenty-seven nests was made at Tinicum (Miller and Price 1959), but by 1992, the count was twelve pairs (Haas and Haas 1992) and six in 2001 (*fide* J. Miller 2001). There is clear indication that, at Tinicum (John Heinz National Wildlife Refuge), this species is declining severely. During the breeding season in the 1980s, least bitterns were found in only thirty-one (0.6%) breeding atlas blocks in Pennsylvania, but with the exception of Crawford and Erie, most counties had only one site that hosted least bitterns during the first atlas period (Brauning 1992a). Large sections of the state report no sightings. Studies of large wetlands in Crawford County since the first atlas project have consistently found a small number of least bitterns present during the breeding season (Brauning et al. 2002).

### *COMMUNITY TYPE/HABITAT USE*

Habitat used by least bitterns is palustrine and lacustrian wetlands dominated by tall emergent vegetation such as cattails (*Typha* spp.) interspersed with shrubs and open water (Gibbs et al. 1992b). Marshes with scattered bushes or other woody growth are preferred, but bitterns spend nearly all the diurnal period in dense, grasslike vegetation; open habitats, such as mats of emergent vegetation, are rarely used (Frederick et al. 1990). Least bitterns frequent the deepwater side of the marsh, generally hunting over open water. Some studies have shown them limited to wetlands larger than 5 ha (Brown and Dinsmore 1986), while others found territorial birds on smaller sites (Gibbs and Melvin 1992b). In Pennsylvania, small (<5 ha) wetlands along lake and pond shores are sometimes used for nesting.

Nesting usually occurs among dense, tall growths of emergent vegetation, particularly cattail, sedge (*Carex* spp.), or bulrush (*Scirpus* spp.) interspersed with some woody vegetation and open, freshwater (Weller 1961, Palmer 1962, Swift 1987, Frederick et al. 1990). Least bitterns are able to nest in pure cattail marshes over deep water by building their nests a meter or more above the water. They readily use artificial wetlands (Gibbs and Melvin 1992b), such as the predominately *Phragmites* wetland of the Glen Morgan Lake, a fly-ash basin in Berks County. Detailed description of vegetation for Pennsylvania nest sites is not available.

### *LIFE HISTORY AND ECOLOGY*

Least bitterns generally arrive in Pennsylvania in mid-April (Haas and Haas 2005) and commence nesting in early to mid-May in the Delaware estuary (Kibbe 1995a) and slightly later in other portions of Pennsylvania. They prefer to nest in emergent vegetation in wetter portions of marshlands, usually near or over open standing water. In addition to nests, the birds may build fishing platforms in areas of high prey densities (Gibbs et al. 1992b). Least bitterns can probably rear second broods (Kent 1951, Weller 1961), although this has not been demonstrated in Pennsylvania. Bitterns are opportunistic predators, eating fish, crustaceans, insects, amphibians, reptiles, small mammals, and birds, which they capture live. Stomach analysis (Palmer 1962) showed small fish to be the most abundant prey (40%) followed by dragonflies (21%), aquatic bugs (12%), and crustaceans (10%).

The following sections on threats and conservation draws heavily from the Gibbs and Melvin (1992b) *Birds of North America* account and the draft "Pennsylvania Recovery and Management Plan" (Kibbe 1995b).

*THREATS*

Historically, wetland losses in the northeastern states were primarily caused by draining, dredging, filling, flooding, pollution, acid rain, agricultural practices, siltation, and urbanization (Jorde et al. 1989). However, wetland losses were curtailed, in some cases reversed, through regulatory protection and through active restoration and mitigation efforts. At this time, the large emergent wetlands that support least bittern populations are currently protected by public ownership from direct threats.

Environmental contaminants and unnaturally high densities of predators, such as raccoons, are found in areas of fragmented or urbanized wetlands (Evers 1992). Pollution and environmental contaminants may impair reproductive capacity and predispose birds to disease in industrialized and agricultural portions of their range. Organochlorines, heavy metals, and PCBs have been found in many other species of herons, and some contaminants (DDE, dieldrin) have persisted in tissues of herons long after their use was banned in the early 1970s (Fleming et al. 1983). Acid rain could potentially reduce food supplies, particularly in nutrient-poor acidic wetlands in the northeastern portion of the state. In agricultural areas, siltation resulting from erosion and runoff containing insecticides may degrade nesting habitats and reduce food supplies.

Marshland invasion by purple loosestrife (*Lythrum salicaria*) and common reed (*Phragmites australis*) may alter and degrade habitats. In the southeastern counties, these invasive species pervade existing marsh habitats, and in the northwest, loosestrife is currently expanding. Natural changes in vegetative structure may result in unsuitable habitat for least bitterns, and impoundments and flooding of native wetlands may eliminate some habitat. Beaver activity across the northern counties creates considerable habitat over time, but in the short term may flood some suitable habitat.

In small, isolated, and urbanized wetlands, bitterns may be vulnerable to various generalist mammalian, avian, and reptilian predators. Sources of mortality of chicks and adults include predation by raptors, crows (*Corvus* spp.), raccoons (*Procyon lotor*), mink (*Mustela vison*), snakes, and snapping turtles (*Chelydra serpentina*; Bent 1926, Trautman 1940, Weller 1961, Hancock and Kushlan 1984). These mortality factors may account for some of the irregular occupation of small wetlands by bitterns.

Least bitterns are susceptible to fluctuations in water level during the breeding season that may occur as a result of beaver activity or eradication, water-level management, and longer-range changes in water levels on Lake Erie (Sandilands and Campbell 1988). Wetland management activities should consider impacts to this and other rare species.

*CONSERVATION AND MANAGEMENT NEEDS*

The most urgent management need is preservation, protection, and improvement of wetland habitats, particularly large (>5 ha), shallow wetlands with dense growth of robust, emergent vegetation. Wetlands bitterns use also need to be protected from chemical contamination, siltation, eutrophication, and other forms of pollution. Equal ratios of cover to open water are preferred, so wetland managers could benefit bitterns by periodically reversing vegetative succession while maintaining suitable habitats nearby to serve as alternate nesting areas during wetland manipulations (Gibbs and Melvin 1992b). Dense stands of cattail and bulrush, often eliminated with cutting, burning, or flooding treatments to improve waterfowl habitat, should be partially retained as habitat interspersed with open water. Maintaining stands of cattails in deep water (10–30 cm) is important because shallower water may reduce nesting potential and eliminate foraging sites for least bitterns (Weller 1961). Where littoral vegetation is scarce, moist-soil plant management can be used to reestablish and promote growth of dense stands of emergent vegetation. Deep-water pools should be maintained during drawdowns so that populations of small fish and dragonfly larvae, which make up the majority of the bittern's diet, are conserved for the following season. Water-level manipulation may also be needed to arrest succession in eutrophic situations. Liming and fertilizing dikes and adjacent fields can increase the productivity and raise the pH of nutrient-poor, acidic wetlands in the northeastern counties. Infestations of purple loosestrife can be controlled with herbicides, physical removal, and burning (Gibbs and Melvin 1992b). Site-specific conservation plans should be developed for occupied sites and restoration plans for sites with the potential to support least bitterns.

*MONITORING AND RESEARCH NEEDS*

One of the first monitoring actions should be to compile an inventory of wetland sites that have or could support least bitterns. Wetlands should be ranked by importance for birds and the threats to each site identified. Sites that can potentially support least bitterns should be monitored using the North American Marsh Bird Monitoring Protocol (Conway 2004). A sampling effort should be developed to assess adequately large historic populations, as well as satellite sites that probably support bitterns intermittently. In one study, the use of tape-recorded calls substantially increased the detection of least bitterns (Swift et al. 1988). Once baseline population estimates are established for historic and potential sites, habitat management efforts should be implemented to enhance or restore the dense emergent wetland vegetation least bitterns require. Responses by bittern populations should be incorporated into future wetland management activities on public sites.

The initial monitoring effort is under way to a limited degree through the second Breeding Bird Atlas project. The effectiveness of this survey by volunteers should be evaluated to determine the need for targeted efforts. The basic research need is to evaluate the response of least bitterns to wetland management practices. Such evaluation of this and other rare wetland birds will require considerable resources. Although the least bittern has been listed for more than five years as a State Endangered Species, surveys have been limited to multispecies efforts at high-priority locations, such as Conneaut Marsh (Brauning et al. 2002). The objective of these surveys was to detect a variety of wetland birds. Considerably greater effort will be necessary to determine relative abundance, population trends, reproductive success, and long-term viability of small local populations of least bitterns at this or other locations across the state, let alone initiate active conservation measures to ensure that this species continues to reside within Pennsylvania.

*Author:* DANIEL W. BRAUNING, PENNSYLVANIA GAME COMMISSION

## Great Egret

Order: Ciconiiformes
Family: Ardeidae
*Ardea alba*

This is the classic egret whose plumage is entirely white contrasting with a yellow beak and glossy black legs and feet (fig. 5.67). The great egret was initially listed as Threatened in 1990 and downgraded to Endangered status in 1999, as a result of its small nesting population confined to one major and one minor col-

*Fig. 5.67.* The Great Egret, *Ardea alba*. Photo courtesy of Joe Kosack, PGC Photo.

ony site. The great egret is listed as a Species of Greatest Conservation Need by seven other northeastern states. The global population is considered Secure (G5, NatureServe 2009).

### *GEOGRAPHIC RANGE*

The great egret breeds on every continent except Antarctica and is among the most cosmopolitan of all herons (Hancock and Kushlan 1984, McCrimmon et al. 2001). Its range blankets the eastern half of the North American continent from southern Canada to the Gulf Coast, from the Mississippi River east to the Atlantic seaboard and along rivers and large lakes in between (Palmer 1962, Hancock and Elliot 1978, McCrimmon et al. 2001). The species occurs primarily in isolated valleys in the western half of the continent.

Globally, the great egret population is stable if not increasing (Kushlan and Hafner 2000). In North America, it may be at its highest population level in history, having recovered from the effects of plume hunting following passage of the Migratory Bird Treaty Act in 1913 (Allen 1958, Ogden 1978). These increases are largely from the southeastern United States and have occurred despite continued degradation and drainage of wetlands.

### *DISTRIBUTION AND RELATIVE ABUNDANCE IN PENNSYLVANIA*

Historically, there were no great egrets nesting in Pennsylvania in the 1800s (Schutsky 1992a). It was recognized only as a postbreeding wanderer from the south. The bird was observed only once in western Pennsylvania from 1890 to 1925 (Todd 1940a). As the southern United States population grew, individuals were observed more frequently. The first two nests recorded in the state were observed in black willows (*Salix nigra*) in the freshwater tidal marsh in Tinicum Township along the Delaware River, Delaware County, in 1957. Nesting continued sporadically at this site through the 1950s (Miller and Price 1959, Miller 1979). Nesting was not recorded here in the 1960s but resumed in 1978 with three nests located at the Tinicum National Environmental Center, Philadelphia County (Miller 1979).

The great egret was reported from less than 1 percent of blocks during the first atlas. Confirmed records were located on Wade Island in Harrisburg, Rookery Island, in Washington Boro, Lancaster County, and along the Delaware River near Philadelphia. Peapatch Island in upper Delaware Bay may have been a likely source for the Philadelphia birds. A colony of up to eighteen nests, along with black-crowned night-heron (*Nycticorax nycticorax*) nests, was located on Mud Island in Philadelphia County in 1989 (McWilliams and Brauning 2000). The colony was abandoned in 1991 when dredge spoil was placed around the nest trees. Birds were observed building nests once again at Tinicum in 1997. Three nests were observed on Rookery Island in 1987 and 1988 amid a larger cattle egret (*Bulbucus ibis*) colony (McWilliams and Brauning 2000). By the mid-1990s and during the second atlas, the only remaining colonies were located on Wade Island in the Susquehanna River at Harrisburg, Dauphin County, and Kiwania Park in York (Fig. 5.68). The total number of nests on Wade Island has shown a generally increasing trend and rose to an all-time high of 197 in 2009 (C. Butchkoski, unpublished data). In recent years, one to three nests have also been observed at Kiwanis Park in York, Pennsylvania, as well.

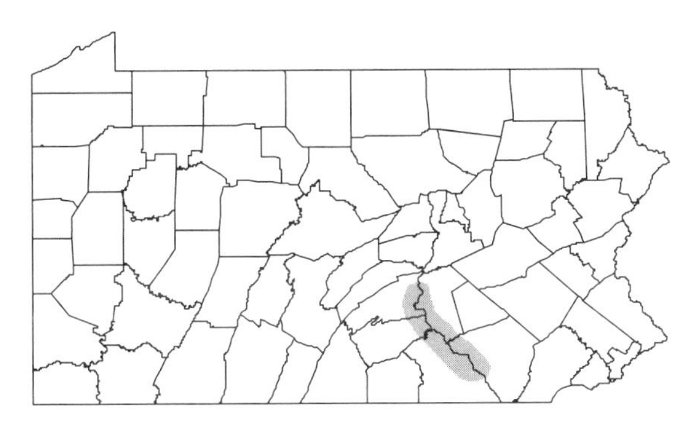

*Fig. 5.68.* Highest probability of occurrence of the Great Egret, *Ardea alba*.

### *COMMUNITY TYPE/HABITAT USE*

The great egret is a habitat generalist preferring open environments in salty, brackish, and freshwater (Hancock and Elliot 1978, McCrimmon et al. 2001). Freshwater marshes, riparian margins, lake shorelines, and especially drainage ditches are typically used inland from the coast (McCrimmon et al. 2001). On the Susquehanna River, the shallow margins of islands vegetated with water willow (*Justicia americana*) within 3 km of the colony are preferred foraging sites (Romano 2007). When the river level exceeds 3.5 to 4.0 feet, foraging occurs in surrounding freshwater habitats (Romano 2007).

### *LIFE HISTORY AND ECOLOGY*

Arrival along the Atlantic Coast generally occurs in early spring and in New Jersey as early as late February. Nests are located in both single and multispecies colonies, where they are often the first species to arrive. Birds begin returning during middle-late March at Wade Island (Romano 2007). Unpaired males establish territories after returning to nesting colonies. Great egrets often nest in mixed-species colonies as on Wade Island, where they nest with black-crowned night-herons and double-crested cormorants (*Phalacrocorax auritus*). Aggressive interactions with other species, which were not observed frequently on Wade Island (Romano 2007), may vary as a function of vegetational physiognomy. In a North Carolina heronry with high vegetative complexity, great egrets were more aggressive intraspecifically (McCrimmon 1978). In a New Jersey colony with less structurally complex vegetation, interspecific aggression was more prevalent (Burger 1978).

Small fish, invertebrates (especially crustaceans), and small vertebrates are favored food items. On Wade Island, their diet is primarily composed of small fish, rusty crayfish (*Orconectes rusticus*), and toad (*Bufo* spp.) tadpoles when available (Romano 2007). This species walks slowly (Rodgers 1983) and is primarily a diurnal predator that uses this behavior 60 to 90 percent of the time (Willard 1977, Master 1992b, Master et al. 1993).

Nesting activities begin on Wade Island in middle to late April. Nests are situated high (20–50 feet) in river birch (*Betula nigra*), silver maple (*Acer saccharinum*), green ash (*Fraxinus pennsylvanica*), American sycamore (*Platanus occidentalis*), and black willow trees (*Salix nigra*; Master 2001, 2004, Romano 2007). Individuals appear to reuse nests on Wade Island with repairs and additional sticks added throughout the nesting period

(Romano 2007). First-year mortality of 76 percent has been reported with a rate of 26 percent annually thereafter (Kahl 1963). Adults and fledglings undergo regional postbreeding dispersal away from the colony site, as do many species of herons and egrets (McCrimmon et al. 2001).

*THREATS*

Two primary threats to the existence of the great egret in Pennsylvania are synergistic. One is the continuing viability of Wade Island as a nesting site given ongoing erosion and loss of nesting trees. Exacerbating erosional effects are double-crested cormorants, which occupy high nesting sites preferred by great egrets. Their population has grown nearly exponentially since 1996.

*CONSERVATION AND MANAGEMENT NEEDS*

Of immediate concern is the protection of Wade Island from the continuing effects of erosion, especially along the northeastern corner of the island. The lingering threat of building the Dock Street Dam in Harrisburg remains a concern. Raised water levels would exacerbate erosion and reduce the extent of favored foraging habitat in water willow shallows along island margins (Master 2001, Romano 2007). The Pennsylvania Game Commission continually assesses numbers and potential effects of cormorants on the great egret nesting population (Master 2004) and discusses active cormorant population control options, agency participation in such options, public response to active control measures, and potential effectiveness.

*MONITORING AND RESEARCH NEEDS*

The number of nests of all species breeding on the island is monitored annually by the Pennsylvania Game Commission. This program should continue, perhaps with more consistency regarding the timing of the survey. Nesting activity at Kiwanis Park in York is being monitored (K. van Fleet, unpublished data), but other ecological attributes of this colony need to be investigated.

An effort was begun in spring 2004 to attract great egrets to an island of similar vegetative structure adjacent to Wade Island to provide an alternative nesting site and freedom from the activities of nesting cormorants, at least temporarily, using egret decoys. Relocation and attraction of herons to nesting sites using decoys has been successful on occasion (Dusi 1985, Crouch et al. 2002, Crozier and Gawlik 2003). The decoys on Wade Island proved ineffective through the summer 2008 (T. L. Master, unpublished data).

Baseline observations on habitat use and foraging behavior of great egrets and double-crested cormorants has been conducted (Romano 2007), but productivity data are needed to complete our understanding of the ecology of the Wade Island colony and to evaluate the degree to which cormorants are competing with and affecting the nesting population of Great Egrets (Master 2001, 2004, Romano 2007). Dead material, including eggshells, feathers and corpses, should be analyzed for contaminants, especially given the heavy agricultural and industrial runoff into this riparian system. All of this information is necessary for enlightened management of Pennsylvania's largest wading bird colony.

*Author:* TERRY L. MASTER, EAST STROUDSBURG UNIVERSITY

## Black-crowned Night-heron

Order: Ciconiiformes
Family: Ardeidae
*Nycticorax nycticorax*

The black-crowned night-heron is easily recognized by its hunched, stocky appearance, relatively short legs, and pied plumage (fig. 5.69). It is listed as an Endangered species in Pennsylvania as a result of a rapidly dwindling number of colony sites, subsequent concentration of remaining populations, continued threats from human activities, and potential competition with double-crested cormorants (*Phalacrocorax*

*Fig. 5.69.* The Black-crowned Night-heron, *Nycticorax nycticorax*. Image courtesy of Alan D. Wilson, naturespicsonline.com.

*auritus*) on Wade Island, location of the largest single colony. Black-crowned night-herons are listed as a Species of Greatest Conservation Need throughout most of the Northeast. They are considered globally Secure (G5, NatureServe 2009).

### *GEOGRAPHIC RANGE*

The black-crowned night-heron is one of the most widespread heron in the world and probably the most abundant although the cattle egret (*Bubulcus ibis*) may share that honor with it. The species occurs on every continent except Australia and Antarctica between 53° north and 50° south latitude (Hancock and Elliot 1978). Within this vast range, it is absent as a breeder only from many Pacific Islands, New Zealand, desert environments and high mountains (Johnson 1965). Because it is a notable migrant and undergoes post-breeding dispersal, individuals have even occurred in these areas on occasion. In North America, it is found throughout the contiguous United States, with the exception of the Pacific Northwest, and ranges northward into extreme southern Canada (Palmer 1962, Hancock and Elliot 1978). Throughout most of this range, the bird is a summer resident. It occurs as a permanent resident only along the Atlantic seaboard (a few individuals remain as far north as Long Island in winter), the Gulf and Pacific coasts and at inland sites in California (Hancock and Kushlan 1984). It has wintered in Pennsylvania at the John Heinz National Environmental Education Center (Tinicum National Wildlife Refuge) near Philadelphia (McWilliams and Brauning 2000).

### *DISTRIBUTION AND RELATIVE ABUNDANCE IN PENNSYLVANIA*

Current nesting activity is confined to four major colony sites and several smaller sites located primarily in Dauphin, Lancaster, Berks, and York counties with most birds (116 nests in 2009) occurring on Wade Island in the Susquehanna River (C. Butchkoski, unpublished data; fig. 5.70). Black-crowned night-herons were reported from 3 percent (152) of blocks during the first Pennsylvania Breeding Bird Atlas and only 2 percent (99) from the second atlas (Mulvihill and Brauning 2009). The core of their distribution remains similar and includes Lancaster, Dauphin, Cumberland, York, and adjacent southeastern counties (Brauning 1992a). Most of these locations probably represent nonbreeding birds and individuals foraging far from established colonies. A few may represent isolated individual nests and small colonies that are difficult to locate following leaf emergence. Confirmed nesting occurred in eighteen blocks during the first atlas and seventeen blocks during the second atlas, with almost all breeding sites confined to the lower Susquehanna River Valley and southeastern quarter of the state. Two colonies along the north branch of the Susquehanna River in Luzerne County and one in south-central Tioga County were not reported during the second atlas. There were few sightings and no confirmations in the western half of the state during both atlases.

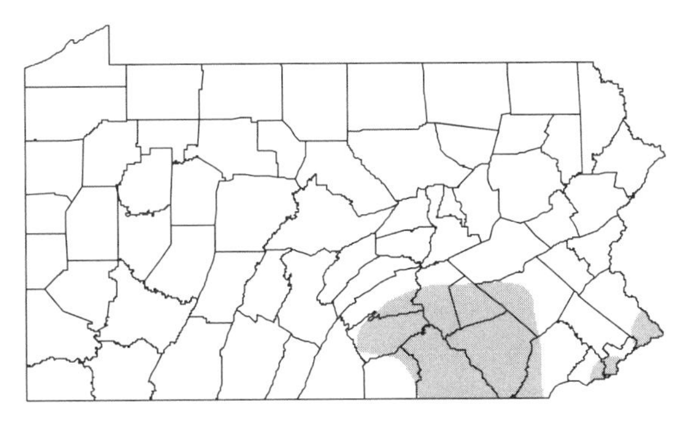

*Fig. 5.70.* Highest probability of occurrence of the Black-crowned Night-heron, *Nycticorax nycticorax.*

### *COMMUNITY TYPE HABITAT USE*

This species is found in fresh, brackish, and saltwater aquatic environments and in both remote wetlands and city parks. Therefore, habitat selection, both for foraging and reproduction, is so variable that it is difficult to characterize (Palmer 1962). Generally, shallow aquatic/terrestrial margins are preferred foraging sites. Shallow margins, vegetated with water willow (*Justicia americana*), surrounding islands, and especially the main shorelines, appear to be particularly important foraging areas for individuals on the Susquehanna River (Detwiler 2008).

### *LIFE HISTORY AND ECOLOGY*

Black-crowned night-herons are primarily crepuscular/nocturnal although they will feed diurnally during the nesting season when food demands are most urgent (Williams 1979, Fasola 1984). Their crepuscular/nocturnal habits may result from competition with diurnally active herons and egrets that sometimes harass them and reduce the amount and size of food they catch (Kushlan 1973, Watmough 1978).

Most often, they hunt from a crouched posture,

along a shoreline, waiting patiently to ambush prey. The black-crowned night-heron is an opportunistic forager, consuming small organisms, including fish, mollusks, crustaceans, and insects (Palmer 1962). Wolford and Boag (1971) and Hoffman (1978) reported stomach contents that included insects, crustaceans, fish, amphibians, nestlings, and mammals with insects, crustaceans, and fish predominating. Judging from the large number of exoskeletons observed, individuals on Wade Island appear to feed heavily on rusty crayfish (*Orconectes rusticus*; Detwiler 2008).

Black-crowned night-herons commence breeding at two years old and are single brooded but will renest if the first nest is destroyed (Palmer 1962). At Pennsylvania's latitude, nesting generally begins in late April/early May. Nest height varies greatly across the species' range from ground level to 160 feet (Davis 1993). Nesting colonies are typically located on islands, in swamps, or over water, suggesting that predator avoidance is a major consideration in site selection. Most nesting individuals are associated with other species in mixed heronries (Davis 1993).

On Wade Island in the Susquehanna River, most nests are located 10–30 feet above ground level, primarily in river birch (*Betula nigra*) and silver maple (*Acer saccharinum*) and secondarily in green ash (*Fraxinus pennsylvanica*), American sycamore (*Platanus occidentalis*), and black willow (*Salix nigra*; Master 2002, Detwiler 2008). Observations suggest that Wade Island birds reuse existing nests as observed in Massachusetts, where 86 percent of individuals in one colony occupied old nests (Davis 1993). The number of young fledged per nest averages 2.25 across North America. A value of 2.24 is given in Davis (1993) from Pennsylvania in 1932 with no specific reference reported. On Wade Island, relatively high nests (>4.8 m) fledged an average of 3.5 nestlings, while low nests fledged 2.6 nestlings (Detwiler 2008). First-year mortality is 61 percent, while adult mortality is approximately 31 percent throughout the species' range (Davis 1993). Erwin et al. (1996) found survival rates after leaving the colony to be only from 0.25 to 0.60 in the first two months based on two years of data in Virginia. Because of the two-day hatching interval, young are asynchronous, and thus, late hatchlings are often victims of siblicide. Youngsters moving about colonies beginning at four weeks old are not attacked by adults and will beg and be fed by any of the nesting adults (Davis 1993).

*THREATS*

During the twentieth century, black-crowned night-heron colonies suffered from a variety of human impacts, including egg collecting, hunting, breeding habitat loss, and eggshell thinning, resulting from pesticide contamination (Walsh et al. 1999). More recent, the species has conflicted with aquaculturists in Pennsylvania. Eighty were drowned at an aquaculture facility in Lebanon County in 1994, and ten were shot under a depredation permit as recently as 1995 (McWilliams and Brauning 2000). Unreported shooting likely continues, especially at private hatcheries. Habitat loss and general disturbance remain the major problems in Pennsylvania. All of the known colony sites are potentially threatened by a variety of human impacts, falling under the category of general disturbance. Some adverse effect from feral cats and dogs is also suspected. Foraging and migratory stopover sites are undoubtedly affected by continuing development, suburbanization, and degradation of wetland habitats. In addition to these general influences, there are site-specific effects as well. The colony at Wade Island, although in somewhat more pristine condition than the other three colony sites, probably has the greatest variety of potential threats (Master 2002, 2004, Detwiler 2008). Of most concern is the loss of nesting trees due to erosion, especially at the northeastern tip of the island, and potential competition from nesting double-crested cormorants whose nest numbers have risen nearly exponentially from 1 individual in 1996 to 120 in 2009 (PA Game Commission).

*CONSERVATION AND MANAGEMENT NEEDS*

All remaining colony sites should be afforded strict protection from habitat disturbance of any type, including that posed by feral cats and dogs. Habitat use patterns and identification of foraging sites at colony sites require attention (completed at Wade Island) to protect key feeding areas and migration stopover sites if any exist in the state. Fish hatcheries in the vicinity of existing colonies should have proper measures in place to discourage use of these facilities as foraging sites. All colonies should be regularly patrolled and identified by consistent, official signage describing the need to protect these unique sites. A Wade Island Cormorant Management Committee was established in 2004 to review annually developing and potential effects of double-crested cormorants on black-crowned night-heron nesting success. There remains a continuing need for erosion control to prevent loss of nesting trees at this site as well.

*MONITORING AND RESEARCH NEEDS*

Monitoring the number of nesting pairs at all colonies at a standard, effective time (following arrival of all individuals and before complete leaf emergence) each year is important because management decisions will be based on such population estimates. Observation of habitat use patterns around all remaining colony sites (completed at Wade Island), emphasizing distances traveled, identifying foraging sites, diet composition, productivity levels, adult survival rates and, competitive interactions with double-crested cormorants should be conducted. Identifying and protecting any existing key migratory stopover sites is also a high research priority.

*Author:* TERRY L. MASTER, EAST STROUDSBURG UNIVERSITY

## Yellow-crowned Night-heron

Order: Ciconiiformes
Family: Ardeidae
*Nycticorax violacea*

*Fig. 5.71.* The Yellow-crowned Night-heron, *Nycticorax violacea*. Photo courtesy of Christopher James Bohinski.

The yellow-crowned night-heron (*Nycticorax violacea*) is the most colorful and strikingly patterned heron in Pennsylvania (fig. 5.71). It was considered a Threatened species in Pennsylvania in 1990 and was downgraded to Endangered in 1999 because of an extremely low population level and a rapidly dwindling number of nesting sites. The yellow-crowned night-heron is also listed as Endangered in Delaware and Threatened in New Jersey. It is considered a Species of Greatest Conservation Need in seven northeastern states. Global populations are considered Secure (G5, NatureServe 2009).

*GEOGRAPHIC RANGE*

The breeding range of this species extends from southern Canada to the south Atlantic Coast of Brazil, west to Colorado and Baja California, east to the Atlantic Coast, and north to southern New England. It also breeds on many Caribbean Islands and the Galápagos Islands (Raffaele et al. 1998, Latta et al. 2006). Highest population densities occur along the Gulf Coast and inland in the southern Mississippi River Valley (Watts 1995). A northward range expansion in the United States and Canada along the Atlantic Coast and inland occurred from 1925 to the 1960s. The recovery reclaimed range that had been occupied in the 1800s (Watts 1995). Postbreeding dispersal extends this distribution somewhat north and west of the core breeding range (Palmer 1962, Hancock and Kushlan 1984, Raffaele et al. 1998). Northern individuals migrate south, but a few birds almost always remain near the northern distribution limit (Massachusetts) throughout the winter, especially along the Atlantic Coast (Palmer 1962).

*DISTRIBUTION AND RELATIVE ABUNDANCE IN PENNSYLVANIA*

There are few historic nesting records for Pennsylvania. Turnbull (1869) listed it as a straggler to the Philadelphia area. Warren (1890) stated that there had been no records of this species for twenty years. Several nests and a female with distended ovaries were found in Chester and Montgomery counties around the turn of the century (Stone 1894). During the mid-twentieth century, a nest was found at Ambler, Montgomery County, on April 20, 1946, and two nests were seen in Camp Hill, Cumberland County, in 1951, 1954, and

1955 (E. L. Poole, unpublished manuscript). A single pair also nested near Mount Joy, Lancaster County, in 1957, and a few nests occurred annually in Lancaster County along the Conestoga, Little Conestoga, and similar sycamore-lined streams during the 1970s and early 1980s (Amico et al. 1984). A nest was found in Chester County in 1973 (McWilliams and Brauning 2000), and Richards (1976) noted one the same year in Montgomery County. The increase in the number of nest sightings from the early 1900s through the early 1970s coincided with the northward range expansion of this species during the middle of the twentieth century (Watts 1995). More recently, ten nests were observed on a small island in the Susquehanna across from the Pennsylvania Governor's Residence in 1987 (McWilliams and Brauning 2000). Since then, nesting has occurred on Wade Island in 2001 (C. Butchkoski, unpublished data) and more regularly on the Conodoquinet Creek, in Kiwanis Park, York, and in the Belleview Park neighborhood of Harrisburg.

The number of atlas blocks where this species occurred dropped from twenty-one in the first atlas to only eight in the second atlas, all of them from the lower Susquehanna River Valley (fig. 5.72; Mulvihill and Brauning 2009). Missing from the second atlas were confirmed nesting records from the Conestoga and Little Conestoga Creeks in Lancaster County, four outlying "possible" records in this area (Schutsky 1992c), and records from counties in the western and southwestern portion of the state where occurrence has been very sporadic historically. Todd (1940) mentions a specimen taken along Ten-mile Creek near Waynesburg, Greene County, and Leberman (1976) lists a single bird observed at Crisp Pond, Powdermill Avian Research Center, on May 14, 1965.

At present, there are two known small colonies, one at Kiwanis Lake in York, where a small number of nests are integrated with more numerous black-crowned night-heron (*Nycticorax nycticorax*) nests and the other in the Bellview Park neighborhood of Harrisburg. There have also been reports of nesting along the Yellow Breeches Creek in northern York County.

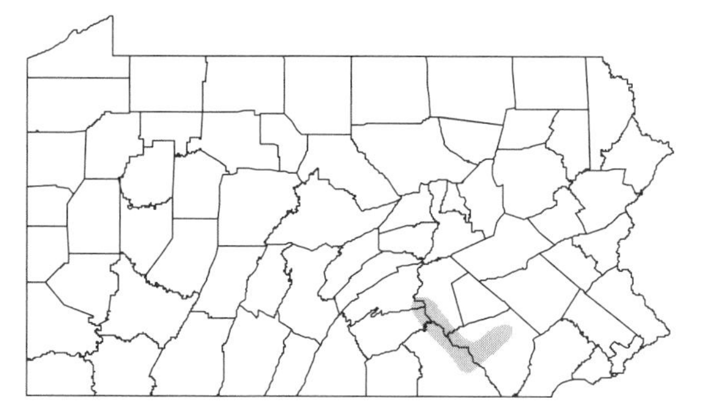

Fig. 5.72. Highest probability of occurrence of the Yellow-crowned Night-heron, *Nycticorax violacea*.

### *COMMUNITY TYPE/HABITAT USE*

Preferred foraging habitats include marine coastal areas (salt marshes, barrier islands, dredge spoil islands), as well as inland swamps, marshes, forested wetlands and riparian margins, the favored habitat in Pennsylvania along with suburban habitats with numerous large trees (Watts 1995, McWilliams and Brauning 2000).

### *LIFE HISTORY AND ECOLOGY*

The yellow-crowned night-heron walks slowly and deliberately when stalking prey with the body slightly hunched and head partially or completely withdrawn (Watts 1995). It is the most sedentary forager among the eight species of North American herons, spending 80 percent of its time in a stationary position (Rogers 1983). This species is primarily a solitary, crepuscular/nocturnal hunter in freshwater habitats (Hancock and Kushlan 1984). Shallow water margins are favored foraging sites but individuals have been seen feeding in plowed fields and on residential lawns (Mumford and Keller 1984, Wiltraut 1994). When breeding, choice of foraging habitat depends on proximity to suitable nesting substrate (Bentley 1994) and broadens considerably during postbreeding dispersal and migration (Watts 1995). Crustaceans are the mainstay of the diet, especially fiddler crabs (*Uca* spp.) in salt/brackish water environments (Riegner 1982) and crayfish (*Cambarus* spp.) in freshwater habitats (Niethammer and Kaiser 1983). Virtually all stomach content analyses and capture observations indicate crustaceans as the major dietary component, including those from nearby New York and New Jersey (91%; Howell 1932, Cottam and Uhler 1945, Riegner 1982). Fish, eels, amphibians, snakes, lizards, arthropods, and small mammals, including young rabbits, supplement the normal diet (Sutton 1967, Harris 1974, Riegner 1982, Wingate 1982, Niethammer and Kaiser 1983).

Migratory individuals from northern populations appear in the southern United States from early to mid-March, while those continuing north arrive from late March to mid-April (Sprunt 1954, Bull 1974, Crawford

1981, Bohlen 1989, Robbins 1991). After their arrival, breeding begins when temperatures permit crab emergence (Watts 1987). Presumably, the same scenario holds true in freshwaters with respect to crayfish availability. Individuals typically commence breeding at two years of age when adult plumage is attained, although rare instances of juveniles breeding have been reported (Wingate 1982).

This species is single brooded although pairs losing entire clutches will renest. In Virginia, renesting will not occur if the original nest is lost after June 15 (Watts 1987). Nesting occurs in mixed-species colonies although single-species colonies are probably more common and tend to be smaller and more loosely organized than in most other species, often consisting of a single initial nest, especially in inland situations (Palmer 1962, Watts 1995). In Pennsylvania, nests are typically located 30–80 feet up on a horizontal branch in tall shade trees, such as American sycamore (*Platanus occidentalis*), along riparian margins (Morrin1991, McWilliams and Brauning 2000). One characteristic common to many inland-nesting situations is an open understory below nests (Watts 1989, Laubhan and Reid 1991). Both sexes contribute to nest building and the same nest may be used for many years in succession (Watts 1995). There are no data on longevity or mortality rate (Watts 1995). Postbreeding dispersal of immatures occurs, but no information is available on distribution and ecology during this period (Watts 1995).

*THREATS*

The yellow-crowned night-heron was persecuted along with many other species of wading birds at the turn of the last century because they were hunted for plumes used for ladies hats during that period. Having survived those threats, the species expanded its range from 1925 through the 1970s, making it to Pennsylvania, where it remains a peripheral species at the northern edge of its inland range with a consistently small fluctuating population. All known colony sites, whether or not in use, are potentially threatened by a variety of human impacts, including degradation of streams and wetlands, human disturbance, disturbance from feral cats and dogs, and increasing suburbanization. Because of its small numbers and preference for crustaceans, it has never been persecuted at fish hatcheries like the black-crowned night-heron was in the past (McWilliams and Brauning 2000).

This species is adaptable to some degree of human interaction. A nesting colony was located above a campground near Chincoteague, Virginia, until two years ago (R. M. Erwin, unpublished data), and another was observed in a city park in downtown Charleston, South Carolina (T. L. Master, unpublished data). The long-used nesting sites along the Conestoga and Little Conestoga creeks in Lancaster County were constantly subjected to human activity as virtually all of them were within 100 m of houses, farms, and roads (Schutsky 1992c), and the species has been observed foraging on suburban lawns (Wiltraut 1994). This adaptability should facilitate any management efforts directed toward this species.

*CONSERVATION AND MANAGEMENT NEEDS*

The yellow-crowned night-heron is in urgent need of monitoring and protection; it is one of the rarest breeding birds in the entire state. A thorough, systematic search needs to be conducted to determine the extent of remaining nesting sites. Sites are difficult to locate because they are often under the canopy of tall shade trees such as the American sycamore. This is exemplified by an unconfirmed report of a nest in 2004 along a driveway in the Belleview Park area of Harrisburg (T. L. Master, unpublished data). It is a difficult species to develop a management plan for because the cause of population fluctuations is unknown and may simply be characteristic of a peripherally breeding species.

*MONITORING AND RESEARCH NEEDS*

The remaining colony site at Kiwanis Lake and Bellview Park and all recently used sites should be monitored and protected. A successful reintroduction of the species to Bermuda has been implemented and perhaps a similar program could be established in Pennsylvania in the future (Wingate 1982). Decoys, used successfully with some other species (Dusi 1985, Crouch et al. 2002), might aid in establishing new colonies, especially because inland colonies of this species often begin with a single nest (Watts 1995). Key research should be directed toward observation of habitat use patterns around colony sites, emphasizing distances traveled, identifying foraging sites, determining diet composition, calculating adult survival rates, and measuring and monitoring of productivity levels.

*Author:* TERRY L. MASTER,
EAST STROUDSBURG UNIVERSITY

## Osprey

Order: Falconiformes
Family: Accipitridae
*Pandion haliaetus*

Adult osprey are among our larger birds of prey (fig. 5.73). This species was declared Extirpated from Pennsylvania as a breeding bird in 1983 and was subsequently reintroduced. It is currently listed as Threatened in Pennsylvania. Global populations are Secure (G5, NatureServe 2009).

### *GEOGRAPHIC RANGE*

Osprey breed throughout North America from Alaska to maritime Canada south to Baja California and east to the states bordering the Gulf of Mexico and the Atlantic Coast of Florida. It also breeds throughout much of Europe, Asia, Africa, and Australia. Ospreys winter in the southern United States through Central and South America and in warm climates throughout the Old World (American Ornithologists' Union 1983).

### *DISTRIBUTION AND RELATIVE ABUNDANCE IN PENNSYLVANIA*

Historically, this species has never been common in Pennsylvania and has been known to nest in less than a quarter of the state's sixty-seven counties (Brauning 1992a, McWilliams and Brauning 2000). Declared Extirpated in the state in 1983 (Gill 1985), the species has made a remarkable recovery in the past two decades because of focused efforts to hack young birds into the wild and establish nesting sites in suitable areas of Pennsylvania (Schaadt and Rymon 1983). The state's current population now surpasses any known historical population level. Data from the United States Fish and Wildlife Service Breeding Bird Surveys indicate Osprey populations are increasing throughout the range in North America. Data from the second Pennsylvania Breeding Bird Atlas indicate an increase of 79 percent has occurred since the first atlas (Mulvihill and Brauning 2009). In 2004, there were sixty-five active nests (Brauning and Siefken 2005b). Osprey can currently be found nesting across the state with the highest probability of occurrence in locations where they were initially reintroduced (fig. 5.74).

*Fig. 5.73.* The Osprey, *Pandion haliaetus*. Photo courtesy of Vic Berardi.

### *COMMUNITY TYPE/HABITAT USE*

Because of its feeding habits, the osprey is closely associated with water. Across its range, it is most common on the coasts and along major rivers and lakes. It may occur wherever suitable foraging habitat is available. Because fish are only taken at or within a meter of the water surface, breeding osprey are found in areas where suitable shallow water for foraging (i.e., <2m in depth) remains open for a time period sufficient to raise their young to fledge. Nest sites free from mammalian predators are required but may be 10 to 20 km from the foraging areas (Poole et al. 2002).

Tall, isolated trees were historically used for nesting. Dead trees in open areas, on islands, or in standing water were preferred. Man-made structures (utility poles, large electric towers, buoys) now constitute 80 percent of the recent Pennsylvania nest sites (Brauning and Hassinger 2001). Nest sites are generally 3–4 km from foraging areas; however, birds regularly forage up to 14 km from nest sites (Hagan and Walters 1990). In Pennsylvania, almost all nest sites are close to water, either on structures in standing water, on an island, or overlooking a significant water body. Many sites are on small lakes and reservoirs associated with areas where the species was reintroduced or along

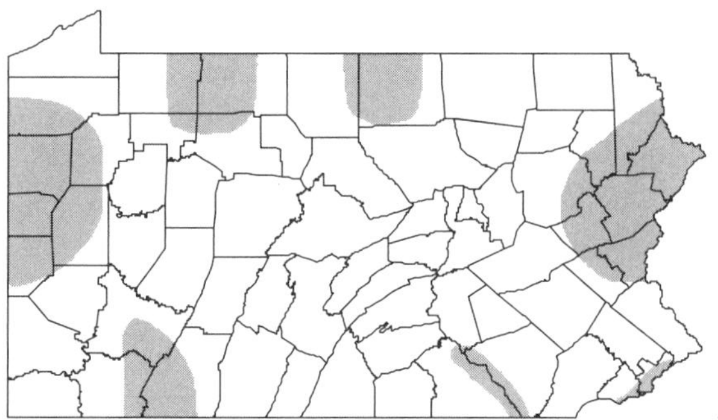

*Fig. 5.74.* Highest probability of occurrence of the Osprey, *Pandion haliaetus*.

Pennsylvania's major rivers, including the Delaware and Susquehanna. Presence of suitable shallow water for foraging is a critical element of successful nest sites.

*LIFE HISTORY AND ECOLOGY*

The osprey feeds exclusively on fish. Northern populations are highly migratory, annually leaving their breeding grounds in North America to winter in Central and South America. Birds typically reach sexual maturity in their third year of life, but many do not begin breeding until at least four years of age. Immature birds commonly spend the summer on the southern wintering grounds, not returning to their natal area until sexually mature at approximately two years old (Poole et al. 2002).

Ospreys typically arrive on breeding areas in mid-Atlantic states in late March and early April. The period from return to the nesting ground to fledging of the young encompasses 100 to 135 days. Egg laying is usually initiated in April within ten to thirty days after arrival on territory. Clutches will be replaced if destroyed. Ospreys begin dispersing from breeding territories in August, and most fall migration occurs in late August and in early September.

*THREATS*

Because of its specialized foraging behavior, osprey habitat is restricted to the vicinity of larger, slower rivers and lakes. These areas are also prime sites for a variety of human uses (e.g., vacation homes, industrial sites), which pose as potential disturbances to nests. Although undisturbed water bodies lacking human activity are considered the ideal nesting situation, human activity in the potential nesting site area year-round or on a relatively constant basis during the prenesting and nesting period, which allows osprey to become habituated to the activities is considered far preferable to irregular or occasional activities (Vana-Miller 1987). Osprey are frequently found nesting in areas subject to human activity (Spitzer 1989), and pairs habituate to human activities quickly (Poole et al. 2002).

Other potential threats to osprey include pesticides, which cause eggshell thinning; water-quality contamination, which impedes prey capture or reduces fish standing crop; seasonal human recreational use of nesting areas; and intentional shooting by misinformed individuals. Competition with bald eagles could be an emerging problem as their population increases dramatically as well.

*CONSERVATION AND MANAGEMENT NEEDS*

Marked increases in the population breeding in the state prompted upgrading of the species status from Endangered to Threatened in 1997 (McWilliams and Brauning 2000). This recovery was facilitated by improvements in water quality and successful conservation efforts (Brauning and Hassinger 2001). Because the state's current population now surpasses any known historical population level, the classification as Threatened seems problematic. A reason for the Threatened status may be that although their numbers are historically high, they are still clustered in relatively few sites. The Pennsylvania Game Commission recovery plan does not include a target population level to delist from Threatened in Pennsylvania. The reluctance of conservation groups to push for delisting the species is further compounded because a population level at which the species should be considered no longer Threatened has never been identified by the agencies responsible for maintaining the list. A population-level goal for delisting of species needs to be determined (the number of nesting pairs in state and their distribution).

Managers of state parks, wildlife management areas, state forests, and other public land agencies that have control of water bodies of sufficient size to attract nesting osprey pairs should be recruited to take an active role in protecting this species and its habitat, including placement of artificial nesting structures. Vana-Miller (1987) recommended that two potential nest sites be provided for each potential nesting pair. Such nesting areas require an adequate fish supply, suitable shallow water for foraging, early ice-out in time for birds to use foraging areas early in the breeding season, clear water, and little water surface obstruction in the littoral zone (Poole et al. 2002).

*MONITORING AND RESEARCH NEEDS*

Recent but now discontinued monitoring of nesting pairs indicates the Pennsylvania population is increasing at a comparable rate to the rest of the mid-Atlantic region. Local populations are stable or increasing. Funding and a protocol for accurately measuring the breeding population periodically should be established. The protocol should establish accuracy of counts to be conducted, frequency of these assessments, and justification of the level chosen for delisting. A campaign to involve the bird-watching public in the documentation of osprey nest sites is needed. This should include an osprey hotline, a dedicated compiler of osprey nest sightings, and a method of monitoring select sites to track nesting success.

Additional research is needed to establish the standing crop of fish necessary to ensure successful sustained breeding at an aquatic site. Failure to determine whether a site is capable of supporting a pair of nesting osprey may result in the expenditure of substantial investments of time, energy, and limited conservation resources on futile efforts. The effectiveness of conservation actions needs to be evaluated by means of a dedicated long-term program of monitoring to track population growth and nesting success at a variety of nest sites. Results will provide a basis for future habitat/nest placement and protection efforts.

*Authors:* DOUGLAS P. KIBBE, NORMADEAU ASSOCIATES, INC.; ROBERT W. BLYE, NORMADEAU ASSOCIATES, INC.

## Northern Goshawk

Order Falconiformes
Family: *Accipitridae*
*Accipiter gentilis*

This powerful bird of prey is the largest of the three accipiters found in Pennsylvania (fig. 5.75). The northern goshawk was selected as a Species of Greatest Conservation Need because it is a rare breeder within the state, and as a top predator that depends on forest habitat, it is thought to reflect the health of prey populations and forest habitat. The goshawk is listed as Candidate Rare in Pennsylvania. Global populations are Secure (G5, NatureServe 2009).

*Fig. 5.75.* The Northern Goshawk, *Accipiter gentilis*. Photo courtesy of Jim Martin.

### *GEOGRAPHIC RANGE*

The northern goshawk is the most widely distributed accipiter worldwide. Three subspecies are found in North America: (1) the Queen Charlotte goshawk (*Accipiter gentilis laingi*) is found in the Pacific Northwest, from southeast Alaska south to Oregon; (2) the Apache goshawk (*Accipiter gentilis apache*) inhabits the warmer forests of southern Arizona, southern New Mexico, and the Sierra Madre in Mexico; (3) while east of the continental divide, in southern Canada and the northern United States, *Accipiter gentilis atricapillus* is found in forested areas. Northern populations are migratory and irruptive, moving south in greater numbers during times of winter prey shortage (Squires and Reynolds 1997).

### *DISTRIBUTION AND RELATIVE ABUNDANCE IN PENNSYLVANIA*

Pennsylvania is at the southern limit of the breeding range of this species in eastern North America. Within the commonwealth, its breeding range is limited to large tracts of mature forests, and its status has been inextricably linked to them. Historical accounts suggest that this species was an established feature of the avifauna of northern Pennsylvania counties during the mid-nineteenth century, but deforestation, the extinction of the passenger pigeon *Ectopistes migratorius* (a key prey item), and sustained persecution resulted in the species' rarity by the turn of the twentieth century (McWilliams and Brauning 2000). Persecution continued until the middle of the twentieth century; bounties were paid in Pennsylvania until 1951, but the eventual cessation of persecution, coupled with regrowth and maturation of forests, saw this splendid raptor return to breed in most northern counties. Despite its improved fortunes, this has remained a scarce bird in Pennsylvania. During the 1983 to 1988 Pennsylvania

Breeding Bird Atlas survey period, it was reported from 120 atlas blocks, only 2 percent of the statewide total, and was confirmed to have bred in only forty-three blocks, most of them in the north of the state (Brauning 1992a). Kimmel and Yahner (1994) estimated nesting densities of only 1.2 pairs/100 $km^2$ in the northern forests of Pennsylvania. During the 1980s, the statewide population of northern goshawks was estimated at 150–200 territories, not all of which would have been active in any given year (McWilliams and Brauning 2000). Most of the suitable habitat for this species, as assessed by the Pennsylvania Gap Analysis Project (GAP), is in the northern half of the state, with some in suitably extensive forests within the Ridge and Valley provinces (Fig. 5.76; Myers et al. 2000).

Migratory and wintering status vary because of the irruptive nature of northern populations; larger numbers migrate south from breeding grounds in boreal Canada during food shortages there, caused by eight- to ten-year population cycles in ruffed grouse (*Bonasa umbellus*) and snowshoe hare (*Lepus americanus*) populations (Mueller et al. 1977). As a result of these irregular migratory patterns, autumn (mid-August to mid-December) counts of migrating birds vary considerably between years at Hawk Mountain (http://hawkmountain.org/cgi-bin/count/viewdate.cgi). Peak autumn migration time at Hawk Mountain is between mid-October and mid-November. Spring migrants occur rarely during the period from late February to mid-April (McWilliams and Brauning 2000). No more than twelve birds have been reported on the Audubon's Pennsylvania Christmas Bird Count during the years of 1998 to 2005 (http://audubon2.org/birds/cbc/hr/table.html), making it difficult to ascertain the status of this often secretive species outside of the breeding season. Pennsylvanian northern goshawks are not known to leave the state during the winter months.

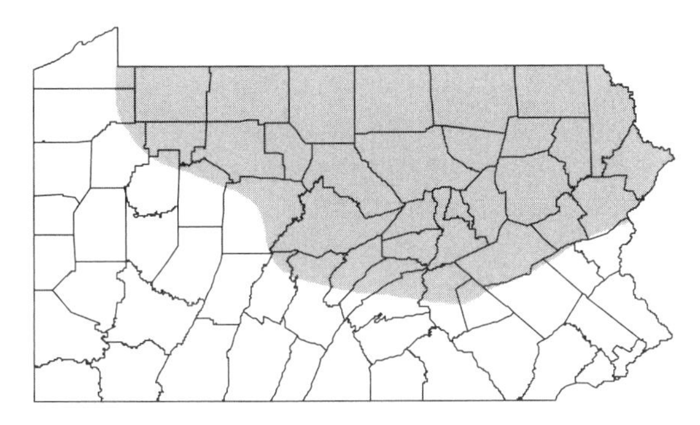

*Fig. 5.76.* Highest probability of occurrence of the Northern Goshawk, *Accipiter gentilis.*

### *COMMUNITY TYPE/HABITAT USE*

Through most of their large world range, northern goshawks are associated with old growth and mature forests with dense cover, open understories, and sparse ground cover during the breeding season (Squires and Reynolds 1997, Hardey et al. 2006). Although frequently associated with northern or upland coniferous forests, this species can be found in a wide range of forest types. Although there is little published information on habitat preferences within Pennsylvania, in neighboring northern New Jersey and southeastern New York, northern goshawks were found to nest preferentially in mature, mixed hardwood–hemlock stands (Speiser and Bosakowski 1987). The substantial stick nest is usually located in one of the largest trees within a forest stand. Speiser and Bosakowski (1987) found black birch (*Betula lenta*) and American beech (*Fagus grandifolia*) to be favored nest trees. Microclimate may be an important determinant of nest location; studies in some areas have shown that nests tend to be located on cooler north-facing slopes (Reynolds et al. 1982), and within stands with a high canopy density (Squires and Reynolds 1997). Indeed, a high canopy density may be the most consistent characteristics for nest location. Although northern goshawks are typically found in contiguous forests, they sometimes nest in relatively small stands within those forests, and nests are often located near small openings (Reynolds et al. 1982).

Most studies of habitat requirements during the breeding season concentrate on the habitat in the immediate vicinity of the nest and rather less is known about the foraging habitat required within the home range, which may extend over thousands of hectares (Iverson et al. 1996). Home ranges of 891–6,164 ha were estimated in a study in Minnesota (Boal et al. 2003). A study of seventeen radio-tagged male northern goshawks in Minnesota found foraging areas characteristically had high canopy and understory stem densities, high canopy closure, substantial shrub cover, and large amounts of woody debris (Boal et al. 2006). Winter home ranges are less well studied but are known to include forest edges, agricultural lands, and even suburbs (McWilliams and Brauning 2000).

### *LIFE HISTORY AND ECOLOGY*

The foraging habitats of the northern goshawk include the forested area in the immediate vicinity of the nest and a more extensive home range, which may include open areas and forest edges. Northern goshawks feed on small to medium-sized mammals and birds, along with small numbers of reptiles and insects (Squires and Reynolds 1997). It is believed that the passenger pigeon was an important food source historically, with the ruffed grouse (*Bonasa umbellus*) and snowshoe hare (*Lepus americanus*) being significant staples today in most northern populations. Mammalian prey accounted for more than 60 percent of the diet of breeding northern goshawks in a study in Minnesota (Smithers et al. 2005), predominantly red squirrel (*Tamiasciurus hudsonicus*), eastern chipmunk (*Tamias striatus*), and snowshoe hare, American crow (*Corvus brachyrhynchos*) and ruffed grouse were the dominant avian prey items. Predators of the northern goshawk include the great horned owl (*Bubo virginianus*) and the fisher (*Martes pennanti*). Increased predation of nest contents and of adult female northern goshawks by fishers caused a drop in nest success from 94 percent to 62 percent in a study on northern Wisconsin (Erdman et al. 1998).

Northern goshawks are monogamous, and adults generally return to nest in the same area each year (Squires and Reynolds 1997, Anonymous 1999). A substantial stick nest is built, sometimes over the top of an old nest, or unused nest of another raptor species (Anonymous 1999). The clutch of two to four (typically three) eggs is laid in late March to early May. Nest success across North America ranges between 44 percent and 94 percent, with between 2 and 2.8 young fledging per successful nest.

### *THREATS*

Although the extent of forest cover in the species' northern Pennsylvania breeding range is stable, there are concerns that deterioration of habitat quality within that area may affect this species. Invasive insects, such as the hemlock woolly adelgid (*Adelges tsugae*), emerald ash borer (*Agrilus planipennis*), and Asian longhorned beetle (*Anoplophora glabripennis*) are all having an effect on the composition of Pennsylvania's forests. These insect invasions may open up the canopy of mature forests where northern goshawks are found. While black birch is increasing, American beech and eastern hemlock (*Tsuga canadensis*), are decreasing within the state (Anon. 2004). Logging of mature forests is the key threat in several parts of this species' range in North America (Squires and Reynolds 1997, Anonymous 1999), but the extent of the risk from timber harvesting within Pennsylvania is difficult to ascertain. While this species requires mature forests for nesting habitat, some of its key prey species, notably the ruffed grouse, require early successional habitats; hence, the maturation of Pennsylvanian forests could result in the loss of key prey from some areas. Conversely, this predator is a potential prey item of the fisher, which has been reintroduced into northern Pennsylvania. This could potentially result in increased predation of nestlings and incubating females, as occurred in Wisconsin (Erdman et al. 1998).

Northern goshawks are susceptible to human disturbance during the nesting period, but the large tracts of relatively undisturbed forests within Pennsylvania should be favorable for this species. Although this species is popular among falconers, the legal taking of northern goshawks from the wild is not thought to have a detrimental effect on population levels (Erdman et al. 1998).

### *CONSERVATION AND MANAGEMENT NEEDS*

The retention of stands of mature forest with continuous canopy cover within the wider forest mosaic is crucial to this species. Although there may be aesthetical reasons for retaining stands of mature trees near roads or in other public areas, such stands may not be suitable for nesting northern goshawks because they tend to shun areas with frequent anthropogenic disturbance. Large deciduous trees within mixed forest are the preferred nesting tree and should be retained within known northern goshawk territories. Managing forests to enhance the populations of key prey species, such as the ruffed grouse and snowshoe hare, is also important.

Currently, the Nature Conservancy's Species Management Abstract for this species (Anonymous 1999) is our best guide to suitable forestry management practices, although these are largely based on the results of long-term studies in Arizona. It suggests that three nesting areas—stands of mature trees, each extending over 12 ha—are required per home range. In addition, three replacement nesting areas per home range should be in some phase of development to provide alternates to currently used sites. A post-fledging family area of approximately 170 ha should be managed to provide various forest conditions and prey habitat attributes. In addition, a foraging area of approximately 2,200 ha should be managed, incorporating larger forest openings and less canopy coverage than the post-fledging area. Forest regeneration should be sufficient to replace 10 percent of the forest every twenty years (Bassett et al. 1994, Graham et al. 1994), and timber-harvesting operations should leave a buffer of at least 8 ha around each nest (Reynolds et al. 1982). The ability to advise on specific management for Pennsylvania is somewhat hindered by the lack of published information about the foraging habitat and diets of northern goshawks within the state.

*MONITORING AND RESEARCH NEEDS*

Preliminary results from the second Pennsylvania Breeding Bird Atlas (PBBA) suggest that this species has declined in abundance in the state. Goshawks were confirmed or probable breeders in fifty-four blocks during the first atlas but only in forty-two during the second atlas (Mulvihill and Brauning 2009). After completing the second Pennsylvania Breeding Bird Atlas project, the data should be assessed to determine whether it is considered an accurate assessment of this species' status within Pennsylvania and should be used to guide any future efforts to target best forestry management practices for this species.

The Hawk Mountain Sanctuary's and Hawk Watch counts performed along Pennsylvania mountain ridges contribute to the study of the irruptive nature of the hawk's migration patterns. These count data are useful for assessing numbers or migratory tendency of northern populations of this species and may help guide assessments of status across its range in eastern North America.

State-specific information of nesting success, survival rates, and dispersion would greatly facilitate conservation efforts for this species. Such information is vital if this species is to be adequately monitored within the state. A within-state or regional study of nest location and habitat use would greatly enhance our ability to tailor forest management for this species. Radiotelemetry would be a useful way of assessing the foraging activities of northern goshawks during the breeding season.

The population of this impressive raptor within Pennsylvania is small and thinly scattered. Any assessment of the population status or conservation requirements of this species within Pennsylvania should not be made in isolation. There is little doubt that a regional assessment, particularly including neighboring New York state, would be more likely to yield sufficient sample sizes with which to assess the effects of large-scale processes, such as changes in forest structure through invasive species. A commitment to a long-term study of the ecological requirements, and population monitoring of this species within the mid-Atlantic/northeast United States region, is overdue. At present, we are too reliant on the results of studies of northern goshawk populations in Wisconsin, Minnesota, or even farther afield, for assessing the needs of this species within Pennsylvania. This species would greatly benefit from more collaborative research within the commonwealth.

*Authors:* ANDY WILSON, PENNSYLVANIA STATE UNIVERSITY; SUZANN RENSEL, CLEAR CREEK STATE PARK

## Golden Eagle

Order: Falconiformes
Family: Accipitridae
*Aquila chrysaetos*

Golden eagles inhabit a large range throughout western North America (fig. 5.77). A small population breeds in northeastern Canada, migrates southward through the Appalachian Mountains, and winters in the eastern United States. The golden eagle is listed as a Species of Greatest Conservation Need in Pennsylvania for two reasons: (1) it is a top predator and thus indicates habitat quality and (2) Pennsylvania contains significant migration corridors as evidenced by hawk watch sites, with the highest spring and fall counts in eastern North America. Global populations are considered Secure (G5, Natureserve 2009).

*GEOGRAPHIC RANGE*

Golden eagles are circumpolar, breeding across the northern portions of North America, Europe, Asia, and Northern Africa. In North America, breeding is widespread west of the Great Plains. The breeding population of eastern North America is confined to northern Quebec, Labrador, Ontario, and the Gaspe Peninsula. Kochert et al. (2002) depict the eastern and western populations as separate and distinct, but there

*Fig. 5.77.* The Golden Eagle, *Aquila chrysaetos*. Photo courtesy of David Brandes.

may be sporadic breeding pairs in Manitoba, southern Ontario, and southern Quebec. The eastern North America population winters from New York southward through the Appalachians, as well as scattered locations throughout the southeast United States (Milsap and Vana 1984, Wheeler 2003).

Except for reports from Tennessee, there are no known active golden eagle nests in the eastern United States (Lee and Spofford 1990). The last known successful nest in northeastern United States was found in Maine in 1984. The same pair continued nesting unsuccessfully until 1999 when they disappeared. Less recent but clearly documented nesting attempts occurred in both the Adirondacks of New York and mountainous areas of New Hampshire, but no success has been noted in these states since before 1970. In Tennessee, the Army Corps of Engineers reports that a nest produced six eaglets between 1993 and 2001. These Tennessee birds were reportedly released as part of a Georgia hacking program.

### *DISTRIBUTION AND RELATIVE ABUNDANCE IN PENNSYLVANIA*

Golden eagles may have historically nested in Pennsylvania. One nest is reported to have been opposite the confluence of the Pequa River with the Susquehanna River in York County before 1856 (Raub 1892). In addition, Van Fleet (1884) noted that the golden eagle was almost certainly a breeding bird of Clearfield County (Todd 1940c). Based on the lack of well-documented nesting records, it is likely that if they did indeed breed in Pennsylvania, it was in small numbers. There are recent summer records of golden eagles in Pennsylvania, but no evidence of breeding.

Golden eagles are observed during migration between the Allegheny Front and Blue Mountain (Kittatinny Ridge; fig. 5.78). Fall migrants enter the state in October and continue through late December. Peak flights generally occur from late October through mid-November along the Kittatinny Ridge, and from mid-November to early-December on the western ridges, such as Bald Eagle Mountain near State College. Most of the fall flight occurs to the north and west of the Kittatinny Ridge (K. Van Fleet 2001, personal observations).

The spring migration is shorter in duration, beginning in late February, peaking in early March, and tapering quickly through the end of March, although stragglers continue through April (D. W. Brauning, unpublished data from Tussey Mountain). The spring

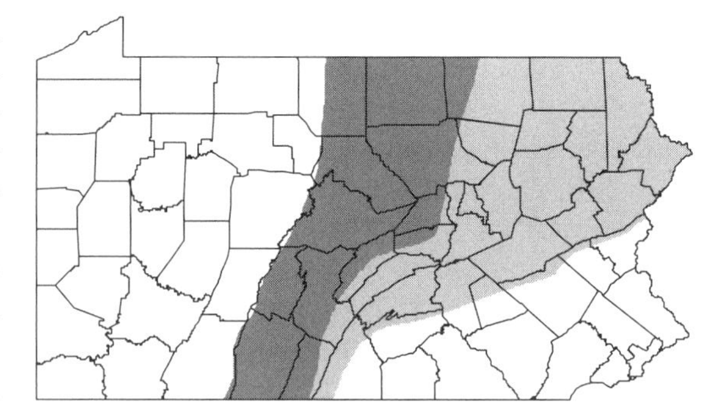

*Fig. 5.78.* Migration spring and fall (dark), fall only (light) of the Golden Eagle, *Aquila chrysaetos*.

flights are composed primarily of adult and subadult birds with few juveniles. Peak spring flights are equal in magnitude to the fall flights, which is unusual for diurnal raptors of Pennsylvania. Brandes (1998) and Brandes and Ombalski (2004) summarized available hawkwatch data and concluded that the spring passage through Pennsylvania is concentrated within a narrow corridor west of Harrisburg rather than across the state as in the fall (Fig. 5.78). Recent satellite data (Miller et al. 2007) support this contention of a narrow migration corridor through the western ridges. Satellite data show that golden eagles also migrate across the central northern-tier counties, such as Bradford, Clinton, Lycoming, Potter, Sullivan, and Tioga (Brodeur et al. 1996, Miller et al. 2007), where they may follow river canyons, escarpments, and other topographic features. There are no hawkwatch data from this area, so locations of concentrated migration are unknown.

A statistical analysis of fall migration data from 1974 to 2004 (Farmer et al. 2008) indicates strong upward trends in golden eagle counts at both Hawk Mountain and Waggoner's Gap over this period. Before this, counts had been declining steadily at Hawk Mountain from the mid-1940s through 1970. Recent counts at both these sites and spring counts from Tussey Mountain suggest that the population may have stabilized.

Recent data from Pennsylvania hawkwatches (compiled at hawkcount.org) show that 500–750 golden eagles are typically counted in the fall and 250–350 in the spring. Because there are many gaps in the migration monitoring network and some golden eagles winter north of Pennsylvania, we estimate the eastern population of golden eagles to be a few thousand individuals. Despite the high degree of uncertainty of this estimate, it can be concluded that a large fraction of the eastern North America golden eagle population migrates through Pennsylvania.

Golden eagles were found in Pennsylvania during eleven of the fifteen Christmas Bird Counts (CBC) from 1993 to 2008 (www.audubon.org/bird/cbc), although typically in small numbers. There are numerous historic records of goldens wintering in central Pennsylvania and Berks County (Wood 1973, Ulrich 1997). Pennsylvania appears to be north of the main wintering area, as records are more numerous to the south (e.g., Kentucky, Tennessee, Virginia, and West Virginia; Milsap and Vana 1984). As part of a study of wintering bald eagles along the upper Delaware River Valley, New York Department of Environmental Conservation has trapped and placed satellite transmitters on at least three golden eagles since 2000 (Nye 2004). Recent telemetry (Lanzone et al. 2007) and anecdotal information that golden eagles frequent deer dumps indicate that they are more numerous in Pennsylvania in winter than the CBC data indicate. This is likely due to their preference for remote areas away from human disturbance.

### *COMMUNITY TYPE/HABITAT USE*

Golden eagles are wary of humans and sensitive to disturbance. In western North America, they inhabit open country over a large breeding range. Despite their varied diet, these eagles are not well adapted to the extensive forest cover of Pennsylvania, except in remote areas where openings exist due to fire, agriculture, or wetlands. The nineteenth-century destruction of the commonwealth's forests to fuel the iron-making industry likely left a landscape more suitable to golden eagles than exists currently.

During migration through Pennsylvania, golden eagles inhabit rugged, mostly wooded terrain throughout the Allegheny Plateau and Valley and Ridge regions. There is little data on winter habitats because of the sparseness of records, although golden eagles tend to winter in semiopen mountainous regions (Wheeler 2003). Winter records are primarily from the Appalachian Plateau and Valley and Ridge regions, as well as the upper Delaware River. Recent satellite telemetry work (Lanzone et al. 2007) suggests that golden eagles winter in heavily wooded areas of Pennsylvania.

### *LIFE HISTORY AND ECOLOGY*

The resident (i.e., nonmigratory) populations of golden eagles breeding in western North America and Scotland have been well studied (Watson 1997, Kochert et al. 2002); however, little work has been done on the life history and ecology of migratory golden eagles of eastern North America. Two studies focused on golden eagles breeding in the Hudson Bay region (Morneau et al. 1994) and the migration cycle of these birds (Brodeur et al. 1996). Telemetry of eastern golden eagles is currently being conducted by T. Katzner and colleagues (see www.aviary.org) to better understand how terrain, weather, and habitat influence migration and wintering behavior and ecology.

Golden eagles are long-lived, with a life span in the wild of twenty to thirty years. They acquire a mate and nesting territory at age four or five and are believed to pair for life, although this has not been verified. Nonterritorial nonbreeding adults readily replace breeding adults that are killed. The pair bond is maintained year-round in resident populations, but the maintenance of pair bonds in migratory populations, such as those seen in Pennsylvania, is unknown.

A typical year for golden eagles breeding in Quebec includes nest building in late March to early April, laying in mid- to late April, hatching in late May to early June (about forty-five days after laying), fledging in late July to mid-August (about sixty to eighty days after hatching), migration from mid-October through late-November, wintering in the eastern United States from December through February, and migration back to Quebec from late February through early April (Morneau et al.1994, Brodeur et al. 1996, Miller et al. 2007).

Golden eagles hunt from a perch, from a soar, or using low contouring flight to surprise prey (Kochert et al. 2002). Across most of North America, they feed primarily on mammals, especially leporids (jackrabbits, hares) and sciurids (ground squirrels, marmots); however, they are known to take a variety of prey, particularly in times of scarcity of its major prey species. Many accounts exist of them working cooperatively in pairs to secure prey. Diet data are lacking for eastern North America golden eagles; however, available reports suggest that waterfowl and wading birds play an important role in the diet of the Quebec breeding population (Spofford 1971, Kochert et al. 2002). A pair formerly nesting in Maine consumed primarily bitterns and herons. Little is known about the diet of the eastern North America population outside of the breeding season, although waterfowl are also an important prey item for birds wintering in coastal regions of the eastern United States. During migration through Pennsylvania, golden eagles are occasionally seen with distended crops, hovering, or harassing vultures, indicating that they feed during migration. Maurice

Broun documented a golden eagle attacking and securing a red shouldered hawk (*Buteo lineatus*) at Hawk Mountain.

Carrion likely plays an important role in the winter diet, and there are many published accounts of golden eagles feeding on deer and sheep carcasses in the southeast United States, and they are known to frequent deer dumps in Pennsylvania. There is a published record of an immature golden eagle preying on turkey (*Cathartes aura*) and black vultures (*Coragyps atratus*) at Gettysburg National Battlefield, although it was unclear whether the vultures or their regurgitated meals were the intended prey. There is also a recent sight record from Waggoner's Gap of a golden eagle pursuing a vulture and forcing the vulture to regurgitate, with the regurgitated material then consumed by the eagle. At some Pennsylvania hawkwatch sites, including Waggoner's Gap, vultures are commonly observed to flap away from the ridge when a golden eagle approaches.

Golden eagles are known to prey on gallinaceous birds, including wild turkeys (*Meleagris gallopavo*), and published anecdotes suggest that wild turkeys may be an important prey item for golden eagles wintering in Pennsylvania. Turkey often feed in flocks in open fields near woodlands, making them vulnerable to attack by golden eagles.

### *THREATS*

Direct threats to golden eagles include electrocution on power lines, collisions with vehicles, poisoning due to feeding at carcasses containing lead shot, leg-hold traps set for fur-bearing mammals, and shooting. The installation of numerous wind turbines on ridge-top locations across the commonwealth is a cause of current concern, as it has been demonstrated at Altamont Pass, California, that such facilities can cause direct mortality of eagles and other raptor species. Golden eagles migrate during periods of low thermal lift, making them dependent on ridge updrafts that form at low altitude near sloping terrain. Little if any data exist to assess the effects on golden eagles of turbines located along ridge tops, which might include direct mortality, as well as indirect effects, such as avoidance of preferred habitat and increased energetic cost of migration. Land development in remote areas (including natural gas development), fire suppression, reforestation, and other changes in land use may have decreased and could continue to decrease the availability of suitable habitat for golden eagles in Pennsylvania.

### *CONSERVATION AND MANAGEMENT NEEDS*

Conservation of ridge-top migration routes and remote wintering habitat in Pennsylvania and neighboring states is needed to maintain stable populations of golden eagles in eastern North America. Development of new energy sources, such as wind and natural gas in remote wooded regions, has the potential to negatively affect golden eagles in Pennsylvania. Such activities should be managed with attention to these potential impacts.

For wind energy development, to minimize the risks of collision turbines should not be sited on major migration pathways (e.g., Allegheny Front, Bald Eagle Mountain, Tussey Mountain, Kittatinny Ridge). Wind energy projects have been constructed in some areas without sufficient data on golden eagles, which migrate well before (in spring) and well after (in fall) other raptor species.

### *MONITORING AND RESEARCH NEEDS*

Kochert and Steenhof (2002) recommend more effective monitoring of golden eagle populations. Pennsylvania is uniquely positioned to conduct monitoring due to the concentrated passage of the species during migration (Waggoner's Gap in fall and Tussey Mountain in spring consistently record the highest migration counts of golden eagles in eastern North America, and Hawk Mountain has the most extensive data record in existence). Periodic statistical analysis of trends in the count data can be used to identify potential population declines provided that consistent count methods and protocols are used at watch sites. At a minimum, standardized counts should be continued at the three sites mentioned earlier, as well as at Allegheny Front. The Hawk Migration Association of North America, Hawk Mountain Sanctuary, and HawkWatch International have developed the Raptor Population Index (RPI) project, which has published indices of trends in migratory raptor populations using count data and state-of-the-art statistical methods (Bildstein et al. 2008). If declining trends for golden eagles are identified, coordination of conservation efforts with other states and Canadian wildlife agencies will be necessary.

In addition to migration monitoring, a midwinter golden eagle survey in remote regions of the state is needed to augment CBC results, which are biased toward more populated areas. Deer carcass dumps equipped with motion-sensor cameras can be an effective method of obtaining data in remote areas (M. Lanzone, personal communication).

Little research has been conducted on the ecology of golden eagles of eastern North America, in part because of their rarity and remoteness of their breeding range. Continued satellite tracking programs (Lanzone et al. 2007, Miller et al. 2007) offer the ability to collect unprecedented data on golden eagle wintering behavior, migration route selection as related to weather and topography, and preferred habitats for foraging during migration and winter. Further development of quantitative migration models (Brandes and Ombalski 2004, Brandes et al. 2007) will allow predictions of locations of migration pathways where field data are lacking. The combination of satellite tracking data with ground-based count data and migration models will yield a much-improved understanding of golden eagle migration through Pennsylvania and result in more effective conservation and management.

Research is also needed to understand the effects of ridge-top wind energy development on migrating golden eagles and other raptors in Pennsylvania. The Pennsylvania Game Commission has a cooperative agreement with wind energy developers specifying minimum monitoring protocols for golden eagles and other raptors. However, these are not statistically valid before-after control-impact (BACI) studies, such as are necessary to understand impacts scientifically. Studies of raptor behavior near operating turbines are also needed to understand conditions resulting in collisions and develop mitigation strategies. Regional-scale research is also needed on the effects of wind energy facilities on migration patterns and wintering behavior.

*Authors:* DANIEL OMBALSKI, STATE COLLEGE BIRD CLUB; DAVID BRANDES, LAFAYETTE COLLEGE

## King Rail

Order: Gruiformes
Family: Rallidae
*Rallus elegans*

A secretive marsh dweller, the king rail is the largest of Pennsylvania's rail species (fig. 5.79). It was included as a Species of Greatest Conservation Need because it is a state Endangered Species with few recent breeding records. It is also listed as an Endangered Species in Connecticut and is listed as Threatened by Massachusetts and New York. Pennsylvania is located on the periphery of the king rail's breeding range, which contributes to its rarity. King rails are considered Apparently Secure (G4) globally (NatureServe 2009).

*Fig. 5.79.* The King Rail, *Rallus elegans*. Photo courtesy of Kent Nickell.

### *GEOGRAPHIC RANGE*

In North America, king rails breed in fresh and brackish wetlands from the Gulf Coast to southern Ontario and from the Atlantic Coast west to the 100th meridian. They winter in tidewater areas from Delaware to southeastern Georgia, through all of Florida, westward along the Gulf Coast into Texas, and north into seasonally flooded wetlands and the rice belt of the Mississippi Delta. Populations appear to be somewhat stable in much of the southern United States, but severe declines have been reported in the northern part of the range, including the southwestern shore of Lake Erie, the Western Shore of Maryland, and Delaware marshes (Meanley 1992).

### *DISTRIBUTION AND RELATIVE ABUNDANCE IN PENNSYLVANIA*

In Pennsylvania, the king rail is a rare breeder and migrant, with records scattered throughout much of the state (McWilliams and Brauning 2000). During the first Pennsylvania Breeding Bird Atlas period (1983–89), king rails were recorded in only five blocks and were confirmed breeding in Butler and Tioga counties and probably bred in Philadelphia County as well (Brauning 1992c). Additional breeding season records since 1960 are known only from Cambria, Crawford, Delaware, Lancaster, Mercer, and Somerset counties. Wetland nesting bird population studies conducted on larger Pennsylvania marshes since 1992 have failed to detect a single king rail (Brauning 1998a, Brauning et al. 2002). Birders have reported king rails intermittently from a section of the Conneaut Marsh, Crawford County, during the past several years (D. Brauning, personal communication) and the northwest corner of the state is the region of highest probability of occurrence (fig. 5.80). During the second Pennsylvania Breeding Bird Atlas, king rails were reported in only seven blocks and were not confirmed in any (Mulvihill and Brauning 2009).

The king rail has always been a rarity in Pennsylvania but formerly nested with some regularity in the tidal marshes of Delaware and Philadelphia counties, where it persisted until the early 1990s (McWilliams and Brauning 2000). In Crawford County, it nested near Hartstown in the 1920s–1930s period and was observed near Linesville each summer from 1934 to 1940 in the Pymatuning Lake environs, with a nest with eggs reported in 1935 (Trimble 1940). It has not been confirmed breeding there since (Grimm 1952). Pennsylvania is located on the periphery of the king rail's breeding range, and its present rarity here may be an artifact of the alarming declines of former metapopulations in Ohio and Delmarva.

### COMMUNITY TYPE/HABITAT USE

King rails inhabit freshwater marshes (tidal and nontidal), brackish tidal marshes, shrub swamps, and rice fields. Grasses, sedges, and rushes are important cover types, and cattails (*Typha*) appear to be an important habitat component throughout the bird's range (Meanley 1992). Habitat at a Butler County breeding site varied from standing water adjacent to a lake to a dense growth of alders and willows interspersed with clumps of dense cattails, skunk cabbage, grass tussocks, sedges, smartweed, and jewelweed (Wilhelm 1993).

### LIFE HISTORY AND ECOLOGY

Pennsylvania's king rails are migratory, and although there are relatively few recent spring records away from nesting locations, they range from March 30 to May 31. Little information is available on fall migration as well, with only a handful of post-1960 records available, ranging from September 1 to December 29 (McWilliams and Brauning 2000). King rails forage in shallow water in areas where they are concealed by vegetation or within a few steps of cover. Although king rails are omnivores, animal life is preferred over vegetation and, seasonally, accounted for 95 percent of the diet in spring, 90 percent in summer, 74 percent in fall, and 58 percent in winter, in birds collected from Arkansas rice fields. Crayfish are the most important food item in freshwater environs and composed 61 percent of the spring diet in Arkansas (Meanley 1992).

Nesting may begin as early as February in the southern part of the range but is initiated during May and June in northern sections (F. A. Reid et al. 1994). Males vigorously defend their nesting territories against inter- and intraspecific intruders. Nests are constructed of locally available vegetation and placed in a clump of grass or between several clumps or a sedge tussock, usually at sites where water depth is less than 25 cm. The base consists of wet, decaying vegetation, and the cup of dry grasses, are sedges, and rushes. A canopy and entrance ramp is usually included in the primary nest. Several additional brood nests are also built, usually without canopies (Meanley 1969, 1992). Young are semiprecocial and usually at least one hour old before leaving the nest. They are fed by the adults for their first three weeks, obtain at least 60 percent of their own food at four to six weeks, and are rarely fed by adults by seven to nine weeks (Meanley 1992). Nest success rates of 75 percent and 81 percent have been reported from Arkansas and Missouri, respectively (Meanley 1969, F. A. Reid et al. 1994).

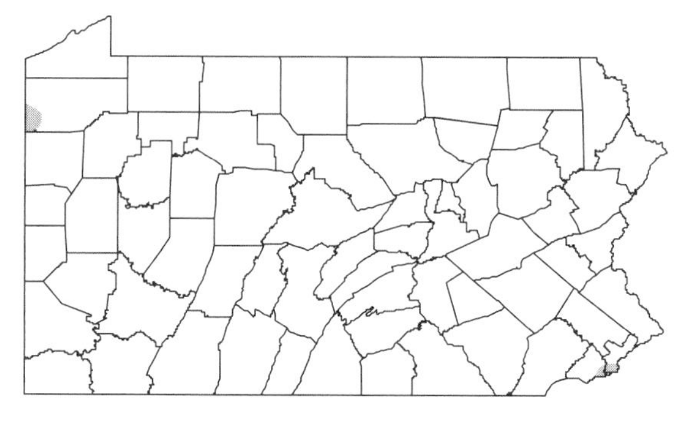

*Fig. 5.80.* Historical distribution of the King Rail, *Rallus elegans*.

### THREATS

Loss of wetlands is by far the most critical threat to king rail populations. In extreme southeastern Pennsylvania, where king rails formerly bred, nearly all tidal habitat in the Delaware Estuary has been filled or degraded (Goodrich et al. 2002), and more than 180 acres of nontidal emergent wetlands have been lost since the 1970s (Tiner 1990). Factors responsible for the king rail's reduction in the extensive marshes of Crawford County have not been identified, but the alarming decline of the Lake Erie population in Ohio, reported by Meanley (1992), may have also affected Pennsylvania's birds. Mammalian predators and fish crows (*Corvus ossifragus*) are known to prey on the eggs and young (Meanley 1969, F. A. Reid et al. 1994). A cat apparently captured and killed a juvenile in Tioga County during the first Breeding Bird Atlas (Goodrich 1989).

### *CONSERVATION AND MANAGEMENT*

Although all the factors that may be contributing to the king rail's rarity in Pennsylvania have not been identified, it should benefit from the stringent protection of remaining wetlands, and such efforts should be strongly supported by the conservation community.

### *MONITORING AND RESEARCH*

Where breeding king rails are located, sites should be monitored for five consecutive years to determine degree of site fidelity, demographics, and changes in habitat, and then site-specific conservation objectives should be developed. A comprehensive status survey of the king rail is needed. Specific effort, using taped playback calls, should be directed toward historic breeding locations, as well as other wetlands that appear to afford suitable habitat. It is possible that standardized marsh bird surveys may be missing it (Brauning 1998a).

*Author:* ROBERT W. CRISWELL, PENNSYLVANIA GAME COMMISSION

## Common Tern

Order: Charadriiformes
Family: Laridae
*Sterna hirundo*

The common tern is a bird of graceful flight and delicate features (fig. 5.81). In Pennsylvania, it is listed as a state Endangered Species with its last confirmed nesting attempt in 1995. Most of the Great Lake populations, which Pennsylvania's population is part of, are listed as a Conservation Concern. It is a Northeast Region Priority Species. Global populations are Secure (G5, NatureServe 2009).

Fig. 5.81. The Common Tern, *Sterna hirundo*. Photo courtesy of Glen Tepke.

### *GEOGRAPHIC RANGE*

The common tern is a widespread, colonial-nesting water bird found nesting across the Atlantic Coast and central Canada in North America and throughout temperate Europe and Asia (Nisbet 2002). It is highly migratory, wintering primarily in South America (American Ornithologists' Union 1998). It generally nests on islands of freshwater or salt water, close to the water's edge (fig. 5.82). Sandy beaches (such as Presque Isle State Park—Pennsylvania's only historic nesting location), and rocky maritime sites are used.

### *DISTRIBUTION AND RELATIVE ABUNDANCE IN PENNSYLVANIA*

Although it is found annually in migration across the state, conservation of this species in Pennsylvania is focused on the only historic nesting location—the eastern tip of Presque Isle in Lake Erie, now known as Gull Point. The Gull Point colony was active from 1927 to 1966, peaking to about 200 pairs in 1937 after receiving protection (Todd 1940a, Fingerhood 1992b). That population declined through the twentieth century to about fifteen nests in 1958 (Fingerhood 1992b) and then disappeared by 1967 (Stull et al. 1985). Two nesting attempts were documented there in 1995, but neither was successful (McWilliams 1995). This was the last documented nesting attempt through 2009. The species continues to be found commonly during spring and fall migration and rarely through the summer along the lakeshore and in Erie Bay around Presque Isle.

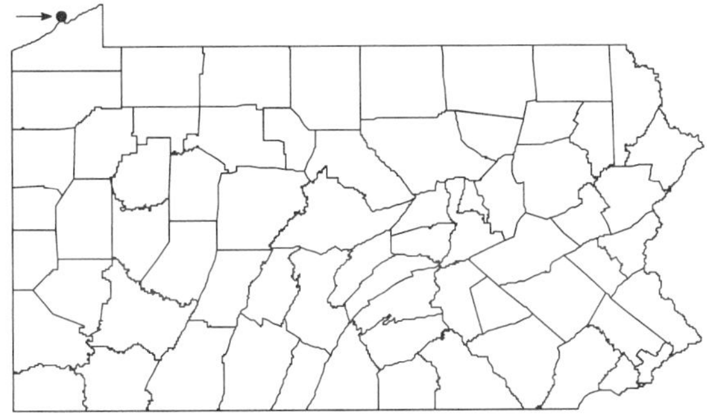

Fig. 5.82. Historical distribution of the Common Tern, *Sterna hirundo*.

*COMMUNITY TYPE/HABITAT USE*

The common tern is a specialist restricted to beach and island habitats of rivers and lakes and coastal areas. The critical habitat resource of the common tern in Pennsylvania is the tip of Presque Isle, known as Gull Point. This site, now designated a natural area in the state park system (Department of Conservation and Natural Resources 1992), supports the only historic nesting location and is the focus of conservation action for this species in Pennsylvania.

Common terns avoid areas dominated by woody vegetation (Nisbet 2002) and may show reduced nesting success when woody vegetation exceeds 30 percent (Cook-Haley and Millenbah 2002) but are more adaptable than some beach-nesting species. Artificial nest sites, such as dredge spoil islands, breakwaters, abandoned piers, bridge abutments, floating navigational platforms, and even gravel rooftops near water, are also employed when natural sites are not available (Nisbet 2002). This is occurring frequently around the Great Lakes and may be a restoration/expansion opportunity for this species.

*LIFE HISTORY AND ECOLOGY*

Common terns take small fish, crustaceans, and insects almost exclusively while flying. The terns are opportunistic in the feeding habits, responding to food abundance in both fish and flying insects, switching locations and techniques based on food availability (Nisbet 2002). Prey availability has not been evaluated specifically for the terns at Presque Isle.

Common terns typically are present around Presque Isle from mid-April through the end of June during spring migration and again from early August through the end of September in the fall (Haas and Haas 2005). The presence of birds during both periods suggests that, if suitable conditions existed, potential local breeders are available to colonize this site.

Males initially establish a territory and begin preparing a nest site. A small depression is found or excavated in sand or gravel substrate above the high-water line in an area with sparse vegetation. Individuals generally do not breed until age three. Eggs were discovered in 1995 on May 11 and July 8, reflecting both the start of nesting and a likely replacement clutch (McWilliams 1995). Pairs are normally present fifteen to twenty-five days before egg laying (Nesbit 2002), providing ample time to observe behavior before they establish a nest site.

*PRIMARY THREATS IN PENNSYLVANIA*

Historically, environmental contaminants (primarily DDT) contributed to reproductive failure that resulted in major population declines during the 1950s and 1960s. Common terns are considered highly sensitive to a range of environmental contaminants, from DDE to PCBs (Lorenzen et al. 1997, Nisbett 2002). These issues were worse before the 1980s, but renewed concern for PCBs indicates that contaminant issues continue.

The specific habitat requirements of nesting terns make these birds vulnerable to nest-site pressures. At a coarse scale, concerns can be summarized as direct disturbance and loss of habitat due to various factors and indirect anthropogenic effects on nesting colonies. Modern recreational activity often brings humans and terns into conflict at beach nest sites. Indirect anthropogenic effects play a significant role in tern colonies, including such issues as increased avian and mammalian predator populations, expansion of invasive vegetation into beach systems, and beach management. Predation by wild populations of mammals (fox and raccoons) and birds (gulls and corvids), as well as free-ranging pets (dogs and cats), undermines nest productivity.

Considerable change has occurred over the past forty years to the beach habitat on Presque Isle and the sand spit of Gull Point, which composes its tip. Breakwaters were established offshore of the beaches, and beach replenishment has been carried out by the United States Army Corp of Engineers. These changes, fluctuating lake levels, and other forces have changed the geological processes that sustain Gull Point and resulted in serious degradation of the habitat of Gull Point. The primary result is substantial revegetation by both native and exotic species, resulting in serious loss of habitat for beach-nesting species, including the common tern. Substantial populations of gulls year-round also present severe predation pressures on potential nesting terns (Cuthbert et al. 2003).

*CONSERVATION AND MANAGEMENT NEEDS*

The conservation challenges facing the common tern as a nesting species in Pennsylvania are probably better defined and understood than are the issues for most of the species of concern because they are focused on a single location. These challenges are formidable, but restoration of this species has reasonable chances of success if a systematic plan were undertaken to address known threats.

The first step in conservation is to formulate a recovery task force of interested and affected parties. This effort would contribute to the potential restoration of the federally Endangered piping plover (*Charadrius melodus*), recently observed in the Gull Point area. The primary objective of this group would be to draft a recovery task plan that would develop a brief list of conservation strategies needed to recover this species to Pennsylvania.

The critical nesting habitat in Pennsylvania, Gull Point, is designated as a Natural Area Management Unit within Presque Isle State Park (Department of Conservation and Natural Resources 1992). Highly restricted access from land and boat provides a high degree of protection from human disturbance. However, erosion and revegetation have affected the potential of this area to support terns and plovers. The historic nesting area, while always experiencing change due to a host of natural and anthropogenic factors, is now vegetated to a point where it is unlikely to be suitable for nesting terns without management. Potential nesting habitat for common terns is also designated as critical habitat for the federally endangered piping plover, so management actions proposed for piping plover recovery should be evaluated for their potential compatibility with tern recovery.

A primary management need is a contingency plan that would guide conservation actions on the report of breeding behavior or another nesting attempt. A quick response to active nesting behaviors will be essential to increase the likelihood of nesting success. This plan should include elements of attraction (decoys, broadcast sound recordings), habitat modification (increase or decrease in vegetation, as needed), predator control (trapping or shooting foxes, cats, owls, and gulls, electric fencing), and management of human disturbance (fences, signs, wardens).

*MONITORING AND RESEARCH NEEDS*

Modern nesting attempts by common terns on Presque Isle raise a modest hope of restoring the species through dispersal from other Great Lakes colonies. Current formal and informal shorebird monitoring efforts provide sightings, but monitoring must be sustained to evaluate the potential colonization by breeding pairs. Monitoring must quickly document nesting behaviors, identify potential nest sites, and evaluate the previously mentioned threats to nesting success. Habitat features at Gull Point should be evaluated to determine suitability of current conditions for common terns to colonize Presque Isle. Food availability should be evaluated to determine whether sufficient food resources exist to support nesting.

*Author:* DANIEL W. BRAUNING, PENNSYLVANIA GAME COMMISSION

## Yellow-bellied Flycatcher

Order: Passeriformes
Family: Tyrannidae
*Empidonax flaviventris*

The yellow-bellied flycatcher is a small tyrant flycatcher in the genus *Empidonax* (fig. 5.83). It is listed as Endangered by the Pennsylvania Game Commission and Pennsylvania Biological Survey. It was not listed by Gill (1985) as a Species of Special Concern, probably as an oversight, but it was listed by the Ornithological Technical Committee (OTC) as Threatened after the Pennsylvania Breeding Bird Atlas results brought this species back to our attention (Gross 1991, Gross 1992d, Brauning et al. 1994). In 2002, the Ornithological Technical Committee determined that it was rare and vulnerable enough to deserve Endangered classification in the state. It also is listed as a Partners in Flight (PIF) Stewardship species for the Northern Forest (Dunn and Blancher 2004). It is not listed by any other northeastern state, and its global population is considered Secure (G5, NatureServe 2009).

*GEOGRAPHIC RANGE*

The breeding range of the yellow-bellied flycatcher is restricted to the northern edge of the continental United States (south in mountains), forested Canada, and parts of eastern and central Alaska (Gross and Lowther 2001, Martin et al. 2006). It spends the winters in Mexico and Central America from the Veracruz area south to western Panama. It nests across boreal North America in Canada from the Maritime Provinces west to the Yukon River basin of central Alaska; the nesting range extends south into parts of Minnesota, Michigan, Wisconsin, New England, and New York, patchily into the Adirondacks, the Taconics, the Catskills, and down the Appalachians into North Carolina (Gross and Lowther 2001, Martin et al. 2006). At present, Pennsylvania is at the southern edge of its breeding range. There is a long history of yellow-bellied flycatchers nesting in isolated locations in the Appalachians south of the state (Simpson 1992). These locations represent remote, near-wilderness relicts of boreal habitat. It is unknown whether the Appalachian population,

*Fig. 5.83.* The Yellow-bellied Flycatcher, *Empidonax flaviventris*. Image courtesy of Alan D. Wilson, naturespicsonline.com.

including the mountains of New York and New England, is distinct from the more northerly population, but these pockets are rather isolated and may constitute metapopulations.

*DISTRIBUTION AND RELATIVE ABUNDANCE IN PENNSYLVANIA*

The "moss tyrant" is confined to cool, moist, shaded locations at high elevations in northern counties (fig. 5.84). The yellow-bellied flycatcher has been found scattered in conifer forests and wetlands in the following counties: Wyoming, Luzerne, Sullivan, Lycoming, Tioga, McKean, Elk, and Warren (Todd 1940a, Poole 1964, Gross 1991, Gross 1992d, 2002b, E. L. Poole, unpublished manuscript). It was formerly found in Monroe, Pike, Forest, Clearfield, and possibly in Bradford, Susquehanna, and Wayne counties (Todd 1940a, Poole 1964, E. L. Poole, unpublished manuscript). Some reports were of unmated singing males, so the breeding range is smaller in any year than the county list indicates. The Pocono Mountains were a stronghold

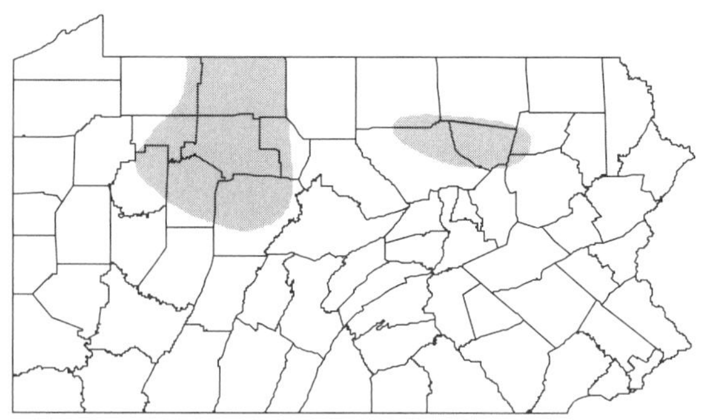

*Fig. 5.84.* Highest probability of occurrence of the Yellow-bellied Flycatcher, *Empidonax flaviventris*.

and frequently cited as the only breeding location in the state, but no nesting has been observed there for several decades (Gross 2002b, E. L. Poole, unpublished manuscript). It is among the rarest nesting species in the state. Sites are isolated from one another except for a cluster on North Mountain in Sullivan/Wyoming/Luzerne/Lycoming counties. Some locations only contain one or two pairs or include unmated, but persistently territorial, males.

*COMMUNITY TYPE/HABITAT USE*

In the words of Cottrille (Burt 2001): "Always, the yellow-bellied lives in the wildest and prettiest of places; places of tall spruce and tamarack, and moss, and shade; places that remind you of 'cathedrals.'" The yellow-bellied flycatcher is a characteristic breeding bird of North American boreal conifer forests and peatlands, including the Northern Forest Avifaunal Biome (Gross and Lowther 2001, Dunn and Blancher 2004, Rich et al. 2004). The "woodland waif" nests in cool, moist conifer or mixed forests, bogs, swamps, and muskegs; landscapes are often flat or poorly drained. In Pennsylvania, the yellow-bellied flycatcher nests in high-elevation cool conifer forests and wetlands. The average elevation of recent sites is 2,036 feet (range = 1,660 to 2,250). The lowest site was well shaded with a high density of fallen timber, providing a cool microclimate. Sites are typically cool, moist, and shady. They tend to be distant from roads and development in large-scale forests, all regularly used nesting Pennsylvania sites are over a mile from any human-occupied dwelling or paved road. Most territories in Pennsylvania are in "red spruce palustrine woodland" or "hemlock palustrine forest" (Fike 1999), also referred to as "boreal conifer swamps" or "northern conifer swamps" (Smith 1991). Range-wide, the dominant trees in nesting habitat usually include spruce (*Picea* spp.) or balsam fir (*Abies balsamea*) but may also be hemlock (*Tsuga* spp.), pine (*Pinus* spp.), or larch (*Larix* spp.; Bent 1942, Gross and Lowther 2001). Conifers may be mixed with as much as one-half deciduous species. Territories usually have a diversity of tree age, size, and foliage, with many snags and fallen timber. The vegetation of its breeding habitat usually is well stratified, with semiopen canopy, saplings and seedlings, and shrubs that provide high leaf density. The mid-story is often complex with highbush blueberry (*Vaccinium corymbosum*), mountain-holly (*Nemopanthus mucronatum*), rhododendron (*Rhododendron maximum*), swamp azalea (*Rhododendron viscosum*), mountain laurel (*Kalmia latifolia*), wild raisin (*Viburnum cassinoides*), red chokeberry (*Aronia arbutifolia*), and Labrador tea (*Ledum groenlandica*). Local ground cover usually includes characteristic plants of northern forests and wetlands such as goldthread (*Coptis trifida*), starflower (*Trientalis borealis*), bunchberry (*Cornus canadensis*), creeping snowberry (*Gaultheria hispidula*), and cinnamon fern (*Osmunda cinnamomea*).

All Pennsylvania yellow-bellied flycatcher territories are within large-scale forests, but small interior openings are tolerated. Pairs have nested within 100 feet of unimproved roads with low traffic and no right of way but have not been found near improved roads with wide rights of way. Pennsylvania locations have not had major human disturbance for more than seventy years (Mellon 1990, Gross 2002b). Like most Appalachian locations, Pennsylvania habitat approaches wilderness conditions. Although recent sites have been in forested wetlands, other habitats have been occupied, including thickets along streams and swampy areas (M. R. Simpson, unpublished manuscript), talus slopes, and rocky hillsides (cool, well-shaded microhabitats; Brunton and Crins 1975, Prescott 1987a), or dense deciduous understory (young cherry and birch) in second growth at high elevations (2,500 to 4,000 feet; Saunders 1927). It formerly was found in old growth conifer forests (Cope 1936).

In migration, yellow-bellied flycatchers visit various forests and thickets, particularly where there is a dense cover near the ground where they forage (Gross and Lowther 2001). On its wintering ground, it occupies forests, thickets, and pastures, where there is a dense mid-story and understory (Stiles and Skutch 1989, Gross and Lowther 2001). The vegetative structure of its winter ground habitat is similar to its breeding grounds.

### *LIFE HISTORY AND ECOLOGY*

Shy and reclusive, the yellow-bellied flycatcher is one of the most difficult species to study on its breeding ground, as nests are notoriously difficult to find (Harrison 1975b, Burt 2001). It nests in the state but migrates to wintering grounds in Latin America. Yellow-bellied flycatchers return to their nesting ground in late May to mid-June, with the males arriving first (D. A. Gross, personal observation). The timing of pair formation is somewhat prolonged, as some pairs are on the territory as early as May 20, while some individuals do not arrive until mid-June. Nests usually are built on the sides of mossy mounds, at the base of trees, in or under logs, or are well hidden in moss or ground herbs where the plumage of this bird blends with the habitat. Sometimes nests are found in tree roots or in the banks of small streams (Harrison 1975b, Burt 2001, Gross and Lowther 2001). Females seem to choose the nest sites as the males are watching nearby.

Pairs are apparently monogamous, but this has not been verified. In Pennsylvania, unmated males maintain territories throughout the breeding season (D. A. Gross, personal observation) and attempt to enter territories of pairs. This suggests a surplus of males and the potential for multiple sires of progeny. Year-to-year occupation of territories suggests that some pairs reform in successive years, but this has not been verified with marked birds. The yellow-bellied flycatcher is a typical, single-brooded Neotropical migrant that spends a relatively short time on its nesting ground, often less than ninety days, before returning south to its wintering ground, but some pairs attempt a second brood in Pennsylvania (DeGraaf and Rappole 1995, Gross and Lowther 2001). Nesting behavior has not been well studied and data are sparse for this species. Thirty-one nests have been discovered in Pennsylvania in recent years (Gross 2002b, D. A. Gross, personal observation).

### *THREATS*

Diminishment and fragmentation of conifer forests and wetlands has been the greatest threat to this species. The stronghold for this species was formerly the Pocono Mountains, but it has not nested there in more than seventy years (Street 1954, Gross 1991, 1992d, 2002, Mellon 1990, E. L. Poole, unpublished manuscript). Pocono bogs and swamps supporting yellow-bellied flycatchers have been destroyed, some in the late 1930s (Street 1954). Peat bog excavation, dam building for recreational lakes, vacation home development, and road-building played major roles in habitat destruction in the Pocono Mountains (J. F. Street, personal communication). Large scale timbering operations near Lopez certainly eliminated habitat in the North Mountain/Dutch Mountain region in the 1890s through the 1920s (Gross 2002a). Streamside habitat may have been eliminated in some parts of the Allegheny National Forest by timbering, road construction, and campsite development. Timbering in the early twentieth century of north-central forested counties probably fragmented the forests around the few locations of that region. Conifer forests have not yet recovered from the timber era of the late 1800s and early 1900s. This species has reestablished populations where spruce has returned to local dominance. In addition, climate change may also be threatening this and other bird species of northern distribution at the southern extent of their breeding range (Hitch and Leberg 2007).

*CONSERVATION AND MANAGEMENT NEEDS*

Forest practices should be implemented that favor conifer regeneration and protect high-elevation wetlands and streamside habitat. These practices also would benefit other wildlife of conservation and recreational value. Hemlock woolly adelgid and other conifer pests greatly threaten conifer forest health in Pennsylvania. Conifer forest and forested wetland species may be facing the same kinds of metapopulation extirpation effects as observed in fragmented temperate forest birds (Villard et al. 1992). The parts of the state that have potential for yellow-bellied flycatcher and other rare boreal species are heavily affected by acidic atmospheric deposition that may be affecting vegetative composition and prey availability for these insectivorous species (Sharpe and Drohan 1999). Conservation programs that address the habitat quality problems associated with atmospheric acid deposition could benefit this and other forest birds. Removal of understory vegetation from overbrowsing by white-tailed deer (*Odocoileus virginianus*) has the potential to decrease the available habitat of this species, as well as other species that live in Pennsylvania's forests.

*MONITORING AND RESEARCH NEEDS*

Nesting locations must be found before they can be protected, so inventory and monitoring potential nesting grounds are a necessary precedent to management. Species-specific surveys would be the most effective means of inventorying this species. Volunteer-based breeding bird surveys are potentially an effective way to find new locations for this species, as was demonstrated with the first Pennsylvania Breeding Bird Atlas (Gross 1991, 1992d). New nesting locations should be surveyed for population size, productivity, and habitat characteristics. Other programs that may be used to detect this species include the Important Bird Area monitoring (most recent breeding locations are in IBAs; Crossley 1999), the Cornell eBird and Birds in Forested Landscape projects (Cornell Laboratory of Ornithology 1997 and Web site), and the Mountain Birdwatch project that focuses on boreal Species of Conservation Concern (Vermont Institute of Natural Science 2005). When appropriate, area-specific surveys also could locate new sites and monitor known locations, including the county Natural Area Inventories, surveys in the Allegheny National Forest, and surveys conducted on Pennsylvania Game Commission lands. Species-specific surveys with a mapping component, following the International Bird Census Committee (IBCC 1970, or similar technique), should be conducted in appropriate habitat. As a confirmation of absence, tape-playback (audio-lure) is a good way to ensure complete coverage of appropriate habitat. Surveys for this species can be combined with surveys for species with similar habitat preferences, including olive-sided flycatcher (*Contopus cooperi*), blackpoll warbler (*Dendroica striata*), Swainson's thrush (*Catharus ustulatus*), Canada warbler (*Wilsonia canadensis*), and red crossbill (*Loxia curvirostra*). Productivity surveys should follow inventory searches or monitoring visits. As an indicator of boreal forest and a PIF Stewardship species for the Northern Forest, the yellow-bellied flycatcher should be a priority species for the Nature Conservancy landscape conservation initiatives in northern Pennsylvania (Rich et al. 2004).

The most immediate need is to identify locations not surveyed in the past twenty years and to field-check all recent locations for occupancy and productivity. It should be a priority to inventory other residents of the habitat to find associates. Data from undocumented sites should be collected and submitted to the Pennsylvania Natural Heritage Program database for better protection of sites. The Appalachian population may constitute a metapopulation with special adaptations. The source population for these sites is not known (nearby or in migration); the closest population to Pennsylvania is in the Catskills (Peterson 1988f, Gross and Lowther 2001). Many aspects of the behavior, ecology, and demographics of this species are poorly understood, including its vocal repertoire, territory size, reproduction, nest-site fidelity, dispersal patterns, colonization, landscape factors (such as fragmentation), source–sink patterns, migratory patterns, and responses to silviculture practices and other kinds of management (Gross and Lowther 2001).

*Author:* DOUGLAS A. GROSS, PENNSYLVANIA GAME COMMISSION

## Swainson's Thrush

Order: Passeriformes
Family: Turdidae
*Catharus ustulatus*

The Swainson's thrush is secretive and easily overlooked (fig. 5.85). It is the rarest nesting thrush of Pennsylvania, where it is near the southern extent of its breeding ground and considered a Candidate–Rare Species (Brauning et al. 1994). Global populations are considered Secure (G5, NatureServe 2009).

### *GEOGRAPHIC RANGE*

The breeding range of Swainson's thrush extends from the Canadian Maritime Provinces and New England west across Canada and the northern United States to Alaska south to California, but it is a rare and local breeder in the Appalachians (Mack and Yong 2000). The eastern (olive-backed) and western (russet-backed) subspecies groups may be distinctive enough to be considered separate species (Monroe and Sibley 1993). The eastern population of Swainson's thrush migrates primarily to western South America. The Appalachian subspecies spends the winter in Columbia and northeast Peru (Ramos and Warner 1980, Mack and Yong 2000).

*Fig. 5.85.* The Swainson's Thrush, *Catharus ustulatus*. Photo courtesy of Kent Nickell.

### *DISTRIBUTION AND RELATIVE ABUNDANCE IN PENNSYLVANIA*

The Swainson's thrush was found in less than 1 percent of the blocks in the first Pennsylvania Breeding Bird Atlas, making it one of the state's rarest breeding birds (Brauning 1992d). It is usually associated with high-elevation, conifer-dominated forests, mostly in the northern and western counties (Brauning 1992g, Sauer et al. 2005; fig. 5.86). Its breeding distribution is rather localized but much more widespread in the northwestern counties than elsewhere. The largest population in the state is probably in the 4,000-acre old growth hemlock–pine forest of the Tionesta Natural and Scenic Area in the Allegheny National Forest (Haney and Schaadt 1994a, 1994b, 1996). They are found throughout the Allegheny National Forest and in small numbers across the state line in New York's Allegheny State Park (Saunders 1938, Peterson 1988e). Not only is the Swainson's thrush rare, but it is also secretive and easily overlooked by observers.

The range and abundance of the Swainson's thrush has declined in the state. At one time, Swainson's thrushes were reportedly locally common to abundant in Pennsylvania's northern counties (Brauning 1992d, E. L. Poole, unpublished manuscript). It was once the most common thrush of Tamarack Swamp, Clinton County (Cope 1901), but has not been present in return visits since that time (Reimann 1947; Pennsylvania Society for Ornithology [POS] SAP database). It was also fairly abundant on North Mountain, Sullivan County, especially in the "virgin forests" (Dwight 1892, Cope 1936), but now they are only scattered and local (Conant 1990, Gross 2003, Sauer et al. 2005; PSO

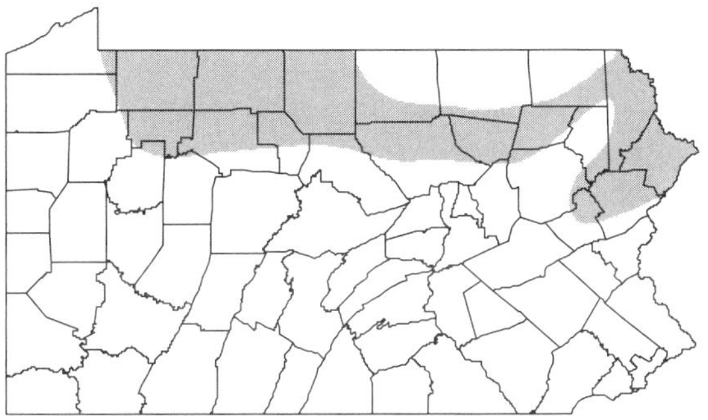

*Fig. 5.86.* Highest probability of occurrence of the Swainson's Thrush, *Catharus ustulatus*.

SAP database). It was rare and local in Warren County (R. B. Simpson, unpublished manuscript), where it was found nesting deep in the forest in Tionesta (Cope 1936). This species once was more common in the Appalachians south of the Mason-Dixon Line, where it was locally common in spruce-fir forests of the West Virginia mountains before timbering (Bent 1949, Hall 1984). By contrast, many of the more recent reports are of single birds singing on a territory.

### *COMMUNITY TYPE/HABITAT USE*

Continent-wide, this species often is associated with mixed forests, but it tends to inhabit cool, moist conifer-dominated woods in Pennsylvania and the rest of the Appalachians (Brauning 1992d, Simpson 1992, Mack and Yong 2000). Most territories are at elevations above 1,700 feet and are strongly associated with hemlock, but some have been found in spruce (Brauning 1992d, Gross 2003, D. A. Gross, personal observation). Swainson's thrush nests in dense conifer forests and wherever there is dense, tall understory vegetation in mixed or deciduous forests (DeGraaf and Rappole 1995). This thrush usually nests within the shrub layer of the forest (Bent 1949, Clark et al. 1983, Mack and Yong 2000), but it often nests in hemlocks in Pennsylvania (D. A. Gross personal observation). South of Pennsylvania, it is associated with spruce and fir in West Virginia and Virginia up to the timberline (Hall 1983b, Brauning 1992g, Simpson 1992, Buckelew and Hall 1994). In New York and New England, it favors spruce forests of the higher elevations but will tolerate mixed hemlock–northern hardwood forests of lower elevations (Kibbe 1985c, Peterson 1988e, Elkins 1994). In the Canadian Maritimes, it seems to prefer less open woods than the hermit thrush and can be quite common in conifer forests (Erskine 1977, 1992). Although a bird of fir and spruce forest in eastern Canada, it nests in successional aspen (*Populus* spp.) stands in western Canada and the United States (Erskine 1977, Mack and Yong 2000).

Although there are many mentions of mature forest as a primary breeding habitat, the Swainson's thrush also nests in dense stands of young trees. Common features of most Pennsylvania sites are small streams, seeps, and springs, where the thrushes forage (D. A. Gross, personal observation). Territories sometimes occur in gaps within an extensive forest, either natural (wetland, blowdown) or man-made (edge of clear-cut). Adults with dependent young have been observed foraging in blueberry thickets near nesting sites in postbreeding dispersal (D. A. Gross personal observation).

### *LIFE HISTORY AND ECOLOGY*

The Swainson's thrush spends the winter in South America and has one of the longest migrations of any songbird. Therefore, it is one of the latest species to arrive on its nesting ground (Mack and Yong 2000). In spring, the earliest migrants arrive in the fourth week of April, but most migrate through from the second to the fourth week of May, with the peak in late May (McWilliams and Brauning 2000). Fall migration can begin in early July but continues into the fourth week of October (Hall 1983a, McWilliams and Brauning 2000). The long migration can be explained by the large breeding range of this species, which stretches coast to coast far into Canada.

Males return to the nesting ground and establish territories from late May to mid-June. When it arrives on its breeding ground, this species is unusually quiet for a nesting songbird, so it may go undetected at first (Mack and Yong 2000). Generally, there is only one brood per season, but pairs may renest after nest failure. Pairs with dependent young have been observed in northern Pennsylvania in early August (Brauning 1992d; PSO SAP database, D. A. Gross, personal observation). Nests often are built in the shrub layer of the forest but also on conifer limbs and next to the trunk (Harrison 1975b, D. A. Gross personal observation).

### *THREATS*

Destruction and fragmentation of mature conifer forests are probably the biggest threat to the Swainson's thrush. Most sites occupied by Swainson's thrush are in large-scale, fairly unfragmented forest. The extent and maturity of the state's conifer forests have declined greatly since historic levels, especially on the plateaus (Whitney 1990). This species declined with the loss of the state's original forest (Brauning 1992d). When more widespread, Swainson's thrushes were found primarily in mature hemlock forest (Cope 1936). In the Appalachians, it was formerly more common in spruce-fir forests of the high peaks (Hall 1983b), so this kind of forest has the potential to be recolonized by this thrush if it increased in size. Although not strictly a conifer forest species, Swainson's thrush and its associates are threatened by conifer forest health. Tree diseases and pests diminish the ability of the forest to support wildlife. Hemlock woolly adelgid, elongate hemlock scale, fabrella needle blight, and other pests and diseases of eastern hemlock have the potential to destroy thousands of acres of conifer forest in Pennsylvania (Department of Conservation and Natural Resources Web site, U.S. Department of Agriculture Forest Service Web site) that are critical to the conservation of this and other conifer forest species. Healthy hemlocks are more likely to resist insect pests than those that are stressed by factors such as drought, *Armillaria* root rot, and other pests (U.S. Department of Agriculture Forest Service). There also is a lack of protection of riparian forests at higher elevations, particularly on private lands where professional foresters may not be involved with planning timber sales. Deer overbrowsing probably prevents regeneration and development of ground and shrub-level vegetation in forests where this species has the potential to occur. There also is a lack of protection for seeps and headwater tributaries on private and some public properties. More property owners should follow best management practices provided by forest professionals (Forested Wetlands Task Force 1993).

*CONSERVATION AND MANAGEMENT NEEDS*

Basic inventory of Swainson's thrush populations, often in remote forested areas, is necessary for its future conservation. A multiple species approach to conservation of its habitat should be achieved because conifer cover is important to a wide diversity of species. It co-occurs with Acadian flycatcher (*Empidonax virescens*), blue-headed vireo (*Vireo solitarius*), magnolia warbler (*Dendroica magnolia*), black-throated green warbler (*Dendroica virens*), Blackburnian warbler (*Dendroica fusca*), and other hemlock-associated species listed as Birds of Conservation Concern in Pennsylvania (Haney and Schaadt 1994a, 1994b; Pennsylvania Game Commission Web site). There is growing concern for this suite of species that is strongly associated with hemlocks (Tingley et al. 2002, Ross et al. 2004, second Pennsylvania Breeding Bird Atlas Web site), but it should be noted that some of these species also are associated with other kinds of forest types. Because boreal and conifer forests grow in patches along the Appalachian Mountains, including Pennsylvania, programs that promote conifer regeneration and release in appropriate habitat would aid this and other conifer-associated wildlife. As hemlocks are increasingly threatened by pests (Department of Conservation and Natural Resources and U.S. Department of Agriculture Web sites), spruce deserves more attention as a conifer cover. Because many Swainson's thrush locations are in headwaters of mountain streams, riparian forest protection will help protect the habitat of this species.

*MONITORING AND RESEARCH NEEDS*

The Swainson's thrush is one of the state's rarest forest songbirds, strongly associated with large, high-quality forests. It should be a goal to monitor the size and distribution of this species as a measure of forest health. The secretiveness and rarity of this species make it one of the more challenging songbirds to monitor or manage. Breeding populations are easily overlooked. However, it is a vocal thrush and can be detected and monitored by fairly standard methods, albeit with more special attention. Standard volunteer-based surveys are helpful for continued monitoring of this species where it is regularly found. These include Breeding Bird Surveys, the Pennsylvania Breeding Bird Atlas, PA IBA monitoring (Crossley 1999), and Cornell Laboratory's Birds in Forested Landscapes in appropriate habitat (Cornell Laboratory of Ornithology 1997). Where it is less rare (particularly in Northwestern counties), point counts and transects would be an effective survey method where there are clusters of thrushes. Because it is less detectable than most songbirds, a longer sample period (at least five minutes) and a double-observer, double-sampling, or removal model point count method would increase accuracy of the survey (Bart and Earnst 2002, Farnsworth et al. 2002, Thompson 2002). Point-count methodology has the potential to provide fairly robust abundance estimates for boreal birds (Toms et al. 2006). Where it is rare and locally distributed (north-central and northeastern counties), area searches are probably the most appropriate and effective survey method. Area searches can be supplemented by spot-mapping or territory-mapping using geographic positioning, especially for discrete populations in isolated habitat islands (International Bird Census Committee 1970, Bibby et al. 1992, Ralph et al. 1993). This would be particularly appropriate for public lands, where this species could serve as an indicator for quality conifer forest sites. Point count protocols could be used for monitoring a suite of conifer-associated species, including those already mentioned. For citizen-science projects, training the volunteers in situations like Audubon birding workshops is a key to successful monitoring of rare species and should continue in some form. Searches should be conducted for this species, particularly in conifer forests and wetlands in the ecological regions, where this species is a possible breeder. Off-road searches are more likely to be fruitful than road surveys. Evening surveys also are recommended in likely habitat. The Mountain Birdwatch project that focuses on boreal Species of Conservation Concern (Vermont Institute of Natural Science 2005) should be expanded into Pennsylvania to monitor some remote locations where this and other rare northern species occur.

Because little is known about the productivity, nesting ground fidelity, habitat preferences, postnesting dispersal, colonization pattern, and source–sink population dynamics of this species, these should be given priority for study. For better management, it would be critical to learn more about its reaction to silvicultural and other man-made disturbances. Research should be conducted on its survivorship and postnesting dispersal of young from nesting area. Little is known about the area sensitivity of this and other higher-elevation forest species in our state. The adverse effects of acid rain on the distribution of the similar wood thrush have been documented (Hames et al. 2002). There may be similar problems with Swainson's thrush as it occupies many locations considered affected by acidic atmospheric deposition. Hemlock stand health should be monitored in respect to occurrence and productivity of conifer-associated species such as this one. Because the Appalachian population of Swainson's thrush may be distinctive enough to be considered a subspecies, its systematics deserve study for conservation considerations.

*Author:* DOUGLAS A. GROSS, PENNSYLVANIA GAME COMMISSION, ORANGEVILLE

## Blackpoll Warbler

Order: Passeriformes
Family: Parulidae
*Dendroica striata*

The blackpoll warbler is a characteristic songbird of the boreal conifer forest of North America (fig. 5.87). It was selected as a Species of Greatest Conservation Need because of its limited distribution in Pennsylvania in high-elevation spruce forests and boreal wetlands. It was listed as an Endangered Species in Pennsylvania in 2005. Global populations are Secure (G5, NatureServe 2009).

### *GEOGRAPHIC RANGE*

The blackpoll warbler's breeding range extends coast to coast from the Canada Maritimes west across

*Fig. 5.87.* The Blackpoll Warbler, *Dendroica striata*. Photo courtesy of Reinhard Geisler, Florida, www.reige.net/nature.

Canada to Alaska (Hunt and Eliason 1999). In the northeastern United States, it inhabits higher-elevation conifer forests, where it can be one of the most common species. It is fairly localized in distribution in the southeastern part of its range extending south in the mountains of New England and New York, where it is found in the Adirondacks, the Taconics, and the Catskills (Peterson 1988a). The small Pennsylvania population is the southern extent of its breeding range (Gross 1994, Hunt and Eliason 1999). Blackpoll warblers spend the winter in northern South America east of the Andes (Ridgley and Tudor 1989, Hunt and Eliason 1999).

### *DISTRIBUTION AND RELATIVE ABUNDANCE IN PENNSYLVANIA*

In Pennsylvania, the range of the blackpoll warbler is confined to boreal conifer swamps and forests in the eastern part of the Allegheny Plateau known as North Mountain (Gross 1994, 1998, 2002a, 2002b, 2003, Davis et al. 1995; fig. 5.88). Most territories are found in Coalbed Swamp and nearby Tamarack Swamp of western Wyoming County (Game Lands 57), but some territorial males have been found in spruces within ten miles of this population center (Gross 2003). These additional observations seem to represent territorial unmated males, which seems to indicate the potential for this species to colonize more locations. As many as thirteen territories have been found in Coalbed Swamp, with additional birds in Tamarack Swamp. Summering males have been reported in the Pocono Mountains and elsewhere on North Mountain (Gross 1994). There is appropriate habitat in adjacent northern Luzerne and western Sullivan counties, as well as in the Pocono bioregion, including Monroe, Lackawanna, Carbon, Wayne, Pike, and eastern Luzerne counties. As a regularly singing bird, the blackpoll is generally well surveyed by Breeding Bird Survey routes, but it is

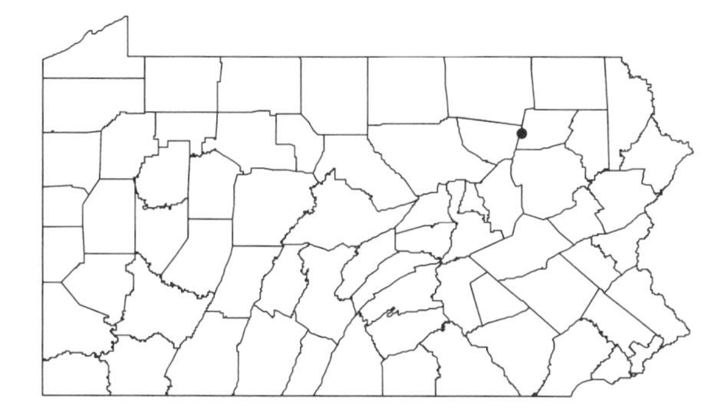

*Fig. 5.88.* Highest probability of occurrence of the Blackpoll Warbler, *Dendroica striata.*

not found on Pennsylvania routes—an indication of its rarity and spotty distribution (Sauer et al. 2005). Much of its habitat is off-road, so Breeding Bird Surveys probably underestimate populations in the mountain island and remote spruce forests in the southern part of its breeding range.

### *COMMUNITY TYPE/HABITAT USE*

On its breeding range, the blackpoll warbler is a denizen of boreal black spruce (*Picea mariana*) forest (Bent 1953, Erskine 1977), but in Pennsylvania, it is found in isolated red spruce (*Picea rubens*) forest (Gross 1994, 2002). Most territories in Pennsylvania are in "red spruce palustrine woodland" or "hemlock palustrine forest" (Fike 1999) also referred to as "boreal conifer swamps" or "northern conifer swamps" (Smith 1991). Territories in Pennsylvania include a diversity of vegetation, including ericaceous and lauraceous shrubs and deciduous trees, but are generally dominated by conifers, including red spruce, eastern hemlock (*Tsuga canadensis*), eastern white pine (*Pinus strobus*), and eastern tamarack (*Larix laricina*). Blackpolls also occupy territories in the uplands within and adjacent to the wetlands where conifers dominate. The populations in the southeastern part of its range tend to be in isolated "mountain islands" of spruce (*Picea* spp.) or balsam fir (*Abies balsamea*), including spruce bogs or conifer swamps (Walley 1989, Gross 1994). Their territories are characteristically in dense conifer stands, sometimes in dense regenerating deciduous clear-cut areas (Hunt and Eliason 1999).

During migration, blackpoll warblers stop over in a variety of forested, edge, and shrubby habitats. They are among those species that reached highest mean abundance in edge-dominated forests (forest-agricultural edge and suburban edge; Rodewald and Brittingham 2004).

### *LIFE HISTORY AND ECOLOGY*

Blackpoll warblers nest in conifer and mixed forests. Territory size seems to decline with increasing elevation (Morris 1979, Sabo 1980). Blackpoll warblers forage by gleaning insects and other arthropods from foliage and twigs and by hawking and hovering maneuvers (Morris 1979, Sabo 1980, Sabo and Holmes 1983).

This species claims the longest migration over water of any songbird in the world; its fall migration route crosses over the North Atlantic Ocean from the northeastern coast of United States to northeastern South America (Nisbet 1970, Nisbet et al. 1995). To accomplish this feat, blackpolls nearly double their body mass before flight and take advantage of shifts in wind direction that provide them the best route over water to their destination. Therefore, Pennsylvania is probably an important stopover habitat for this species.

Males generally arrive on the breeding ground before the females. However, the females choose the nest location and build the open cup nest, generally low in dense conifers. The two nests found in Pennsylvania were well concealed in low spruce seedlings (Gross 1994, D. Gross, personal observation). There is strong breeding site fidelity on part of the females. Territories in Coalbed Swamp are persistently filled year to year, but it is not known whether these are the same birds (D. Gross, personal observation).

### *THREATS*

The primary limiting factor for this species is the limited size and extent of boreal conifer forests. Development pressure in the Pocono Mountains, including a matrix of roads, fragments the habitat available to this and other conifer-dependent species. The blackpoll warbler has been threatened principally by habitat destruction in Pennsylvania and elsewhere on its breeding grounds (Gross 1994, Hunt and Eliason 1999). Most of Pennsylvania's boreal conifer forests were cut before 1900. The spruce forests of northern Pennsylvania, especially in the Pocono Mountains and North Mountain, were either destroyed or badly fragmented. With the Pocono Mountains, this occurred in the nineteenth century before North Mountain was deforested in the late nineteenth century and early twentieth century (Stone 1900, Taber 1970, Latham et al. 1996). These forests are still in recovery. Red spruces are regenerating well in some parts of the North Mountain forest, especially in the Dutch Mountain area, and blackpoll warblers seem to be adopting new locations in spruce and hemlock groves (Gross 2003). Often these conifer stands are associated with headwater wetlands. Poor timbering practices, such as the lack of buffers around wetlands and riparian forests, decrease conifer habitats that support species such as the blackpoll warbler. Other factors such as deer browse and atmospheric acid deposition, may further stress these forests, slowing regeneration and decreasing diversity of these habitats. Now, wind power development also threatens mountaintop forests. Programs for promoting conifer regeneration and release would help this and other conifer-related species. Boreal conifer species known to co-occur with blackpoll warbler include yellow-bellied flycatcher, olive-sided flycatcher, red-breasted nuthatch, winter wren, Swainson's thrush, Blackburnian warbler, northern waterthrush, Canada warbler, purple finch, and red crossbill. North of Pennsylvania, the Partners in Flight high priority Bicknell's thrush and rusty blackbird nests are in the same habitats (Dunn and Blancher 2004, Rich et al. 2004).

Migrant blackpoll warblers are fairly abundant passage migrants in Pennsylvania that take advantage of edge and forest habitats in the state, often moving along the treetops (McWilliams and Brauning 2000, Rodewald and Brittingham 2004). It is one of the many nocturnal migrant songbirds that have been documented in mortality studies of communication towers (Shire et al. 2000). Of the 230 species compiled by this summary report, the blackpoll warbler ranked as seventh among those suffering mortality. Over the weekend of October 8, 2005, more than 100 blackpoll warbler collided with human structures in the Quehanna boot camp, Clearfield County (D. Brauning and D. Gross, personal observations, birds identified by D. Gross). This mortality event was associated with an ana-type front in which precipitation and low overcast cloud ceiling followed after the passage of the front (which is atypical for the state's October weather; G. S. Young, Pennsylvania State University Meteorology Department, personal communication). Stalled fronts like this are dangerous for migrants that are brought very low in altitude, bringing them in contact with human structures that are sometimes made more attractive by lighting.

*CONSERVATION AND MANAGEMENT NEEDS*

Forest practices should be implemented that favor conifer regeneration and protect high-elevation wetlands and streamside habitat (Forest Wetlands Task Force 1993). These practices also would benefit other wildlife of conservation and recreational value. Tree pests and diseases are taking a heavy toll on conifer forests in Pennsylvania. Hemlock woolly adelgid and other conifer pests greatly threaten conifer forest health (DCNR Web site). Management practices to control these pests and improve conifer health should be implemented. The few blackpoll warblers found in the state are in large-scale forests, where there are numerous palustrine wetlands. Poorly planned development and timber practices decrease the size and quality of wetlands and riparian forests, even at high elevations. Accepted management practices for silviculture in riparian areas need to be followed, on both public and on private lands. Headwater swamps and bogs should be given high priority for acquisition and protection by state agencies, watershed organizations, and land trusts. The eastern part of the Allegheny Plateau, including the Dutch Mountain area, which has potential for blackpoll warbler and other rare boreal species is heavily affected by acidic atmospheric deposition that may be affecting vegetative composition and prey availability for these insectivorous species (Sharpe et al. 1999). Conservation programs that address the habitat quality problems associated with atmospheric acid deposition could benefit this and other forest birds. Deer overbrowsing also is impeding forest regeneration, especially on private lands on Dutch Mountain (D. A. Gross, personal observation).

To minimize accidental mortality from collision with tall man-made structures, the guidelines specified by the Federal Aviation Administration (2000) and United States Fish and Wildlife Service (2003) for communication towers and wind turbines should be followed.

*MONITORING AND RESEARCH NEEDS*

As one of the rarest breeding birds in Pennsylvania, the blackpoll warbler seems to be too rare and locally distributed to be surveyed by the Breeding Bird Survey, the standard method for monitoring songbirds, or other roadside methods. None of the known breeding sites are near a public road. Volunteer-based breeding bird surveys, such as the Breeding Bird Atlas and the Important Bird Areas project, have the potential to locate and monitor new locations. Locations should be monitored regularly using a spot-mapping or area search approach to mapping territorial birds (International Bird Census Committee 1970) using modern technology to improve a traditional technique (GPS units providing coordinates). There also is potential for sites to be adopted for Cornell's Birds in Forested

Landscapes (Cornell Laboratory of Ornithology 1997) and the Mountain Birdwatch project conducted by the Vermont Institute of Natural Science for boreal forest birds in mountain habitat islands (Vermont Institute of Natural Science 2005). Territories should be mapped for future reference and input into Pennsylvania Game Commission and Pennsylvania National Heritage Program databases.

Little is known about the demographics of the isolated nesting populations of these mountain island species. The productivity, territory size, and site fidelity of the Pennsylvania population should be studied. Using geographical information collected in the previously listed projects and historic records, researchers should find landscape and ecosystem attributes for areas where blackpoll warblers nest to better understand its distribution and potential conservation. Researchers need to identify habitat attributes and management techniques that improve the habitat for this species and others that inhabit the same ecosystem, including game and mammal species. Genetic flow and colonization patterns in the limited Pennsylvania populations also deserve study.

*Author:* DOUGLAS A. GROSS, PENNSYLVANIA GAME COMMISSION

## Red Crossbill

Order: Passeriformes
Family: Fringillidae
*Loxia curvirostra*

The red crossbill is a medium-sized stocky finch with a set of mandibles adapted for extracting seeds from cones (fig. 5.89). They are listed in Pennsylvania as a Candidate—Undetermined Species because of the uncertainties about its confusing taxonomy and its nomadic habits leading to an erratic breeding status (Gross in Brauning et al. 1994). In the eastern United States, the red crossbill is considered Critically Imperiled in Vermont, Virginia, Tennessee, Illinois, and Alabama and Vulnerable in New York and Maine (NatureServe 2005). The *percna* subspecies (*Loxia curvirostra percna)* is considered Endangered in Newfoundland (Committee on the Status of Endangered Wildlife in Canada 2004). These listings indicate a widespread concern for the populations of red crossbills, even in well-forested states with larger conifer forests. All of these populations are linked with populations in the western part of the continent by the nomadic habits of crossbills. The southern Appalachian populations are high priority species for United States Fish and Wildlife Birds of Conservation Concern (U.S. Fish and Wildlife 2002).

*Fig. 5.89.* The Red Crossbill, *Loxia curvirostra*. Image courtesy of Elaine R. Wilson, naturespicsonline.com.

### *GEOGRAPHIC RANGE*

Red crossbills have a circumboreal distribution with many races identified in North America, Europe, and Asia (Newton 1973, Clement et al. 1993, Groth 1996). In North America, it occurs irregularly in conifer forests across the northern part of the continent, south through the Rocky Mountains through Mexico to Nicaragua in the west and south through the Appalachians to North Carolina in the east (Atkisson 1996). Regionally, this species was reported as an abundant nester in some years in the spruce forests of the Adirondacks, which were timbered extensively in the late 1800s (Dickerman 1987, in Peterson 1988d). Their occurrence is driven by the availability of seeds in conifer cones.

### *DISTRIBUTION AND RELATIVE ABUNDANCE IN PENNSYLVANIA*

Red crossbills are rare and erratic in Pennsylvania and most of the northeastern United States (Fingerhood 1992a, Atkisson 1996). Historically, this species has nested (or attempted to nest) in Cameron, Carbon, Chester, Clinton, Clearfield, Luzerne, Lycoming, McKean, Monroe, Philadelphia, Sullivan, Tioga, Warren, and Wyoming counties (Conant 1990, Fingerhood 1992a; fig. 5.90). Recent sightings or attempted nestings include conifer forest sites near Lopez, Sullivan County; at Leonard Harrison State Park, Tioga County; and southern Sproul State Forest, Clinton County. Large-scale winter invasions include large flights of crossbills in Clarion, Dauphin, Cumberland,

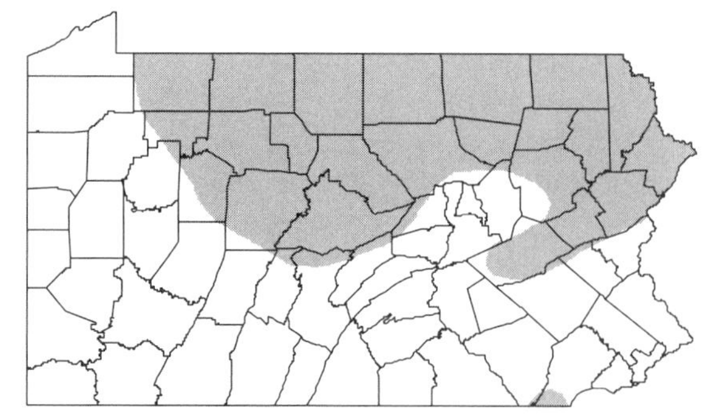

*Fig. 5.90.* Historical occurrence of the Red Crossbill, *Loxia curvirostra*.

and Luzerne counties (Pennsylvania Birds Local Notes; PSO Special Areas database). Historically, red crossbills once were much more common and widespread in the state (Fingerhood 1992a, Brauning et al. 1994), so current distribution or lack of it is somewhat misleading in consideration of the potential habitat and food sources for this species. Warren (1890) reported that crossbills bred regularly in Clinton, Clearfield, Luzerne, Lycoming, and Cameron counties, and probably in the higher mountainous regions in other parts of the state. Stone (1894) catalogued most of the same counties in the breeding range and added the Pocono Mountain and Tobyhanna areas of Monroe County.

Red crossbills are not just "northern mountain" birds in Pennsylvania and the northeastern United States (Fingerhood 1992a, Atkisson 1996). Early ornithologists in Pennsylvania considered the red crossbill a nesting species, even though many of their own experiences were largely confined to the southeast counties, including Philadelphia (Fingerhood 1992a). Crossbills were reported nesting in small cities and along rivers at fairly low elevations, including Pittston by the Susquehanna River and Williamsport and Renovo by the West Branch Susquehanna (Warren 1890, Todd 1940a). They also have nested in pine barrens of southwestern Chester County (Pennock 1912).

### COMMUNITY TYPE/HABITAT USE

Crossbills are denizens of conifer and mixed forests wherever cone crops are large enough to support the provisioning of young (Atkisson 1996). Seed cone availability is a critical factor for nesting (Benkman 1990), but in late autumn, crossbills rarely nest even when seeds are abundant (Benkman 1990, Hahn 1995, Groth 1996). Crossbills are primarily associated with extensive, mature forest but they also have a history of nesting in various kinds of scrub pine–oak barrens and artificial conifer plantings. Red crossbills inhabit various coniferous palustine forests, conifer-broadleaf palustrine forests, coniferous terrestrial woodlands, conifer-broadleaf terrestrial woodlands, as well as acidic glacial peatland complex and the mesic, serpentine, and ridge-top barrens complexes (Fingerhood 1992a, Fike 1999).

In Pennsylvania and the Appalachian Mountains, red crossbills have been associated with eastern white pine (*Pinus strobus*), red pine (*Pinus resinosa*), eastern hemlock (*Tsuga canadensis*), red spruce (*Picea rubens*), black spruce (*Picea mariana*), and white spruce (as a planting, *Picea glauca*; Griscom 1937, Dickerman 1987, Groth 1988, 1996, D. A. Gross, personal observation). In the Pocono Mountains, there is a mosaic of boreal conifer and mixed forests with pine–scrub oak till barrens that meet many requirements of this species because of their pine component (Latham et al. 1996). Red crossbills have nested at lower elevations in Pennsylvania and elsewhere in the Northeast where conifers were available (Austin 1968, Fingerhood 1992a, Atkisson 1996).

### LIFE HISTORY AND ECOLOGY

Crossbills are the ultimate conifer songbird. Their entire biology depends on the availability of conifer seeds (Newton 1973, Benkman 1993a, Atkisson 1996, Groth 1996). Crossbills live up to their name, because their mandibles are crossed for quick extraction of seeds from cones. The bills of different crossbill species and populations are more efficient at removing seeds from different kinds of conifers (Benkman 1987). Despite their unique adaptations, crossbills do not feed on conifer seeds exclusively (Austin 1968, Erskine 1977, Atkisson 1996). They also consume buds and seeds of other plants and a variety of insect matter. Crossbills also take grit material from roads making them vulnerable to vehicular collisions (Atkisson 1996, D. A. Gross, personal observation).

Red crossbills are among the most enigmatic of North American songbirds by habit and by taxonomy. Although long considered a biological species across much of the Northern Hemisphere, including much of Europe and Asia (Newton 1973, Clement et al. 1993, American Ornithologists' Union 1998), the red crossbill may constitute a suite of at least eight cryptic species in North America rather than a single species (Groth 1993b, Atkisson 1996, American Ornithologists' Union 1998, Parchman et al. 2006).

Ornithologists have long known of its nomadic habits. By their nature, crossbills occupy locations inconsistently and depend on conifer forest health and distribution, challenging conservation and management (Benkman 1993b, Atkisson 1996). Despite their flocking behavior, crossbills remain paired through most of the year even within their flocks (Atkisson 1996). Red crossbills are famous for their irruptive behavior, often in large flocks (Austin 1968). Food unavailability is almost certainly the trigger for irruptions in this and other boreal species (Bock and Lepthien 1976, Koenig 2001). Flocking is important for crossbills as foraging flocks attract other crossbills to highly profitable food sources (Benkman 1988).

The nesting behavior of red crossbills is irrevocably interwoven with the availability of cone crops, reacting opportunistically to seed availability and social factors but also regulated by photoperiod (Benkman 1990, Hahn 1995). Intake rates seem to provide the proximate cue for crossbills to initiate nesting (Benkman 1990). Nests are well concealed in dense cover of tree branches, usually in evergreen conifers (Atkisson 1996). There may be multiple clutches as the food supply permits.

*THREATS*

The greatest threat to this species is the decline in the size, diversity, and quality of conifer forests, especially mature and extensive forests. Red crossbills have been reported nesting in many parts of the state, but there is the best chance for conserving the habitat of this species in the larger-scale forests of the Appalachian Mountain and Allegheny Plateau regions. The small size and fragmentation of mature (including old growth) conifer forests is probably the biggest threat to red crossbills (Dickerman 1987, Benkman 1993b). Because seed production varies between tree species and in different regions, a diversity of cone-bearing trees is critical to the potential for species dependent on them (Benkman 1993a). The extent and maturity of the state's conifer forests have declined greatly since historic levels, especially on the plateaus (Lutz 1930, Whitney 1990). The large cone crops produced by old conifers provide an abundance of seeds necessary for a breeding population of red crossbills. The remaining conifer stands are smaller and more isolated than in the past, making it more difficult for crossbills and other seed-eating birds to find cone crops that can sustain the demands of provisioning young.

Smaller forest size also may make nesting more vulnerable to predation. Sharp-shinned hawks (*Accipiter striatus*) are known predators, but Cooper's hawks (*Accipiter cooperii*), merlins (*Falco columbarius*), and peregrine falcons (*Falco peregrinus*) also live in the same habitat (Atkisson 1996). Adults seem to react negatively to the presence of red squirrels (*Tamiasciurus hudsonicus*), blue jays (*Cyanocitta cristata*), and common ravens (*Corvus corax*) near their nests (D. Gross, personal observation), all of which are potential nest predators.

Forest health is another threat to the state's conifer birds, especially the strongly specialized crossbills. Tree diseases and pests diminish the ability of the forest to support wildlife. Hemlock woolly adelgid, elongate hemlock scale, fabrella needle blight, and other pests and diseases of eastern hemlock can potentially destroy thousands of acres of conifer forest in Pennsylvania (Department of Conservation and Natural Resources and U.S. Department of Agriculture Forest Service Web sites) that are critical to the conservation of this and other conifer forest species. Healthy hemlocks are more likely to resist insect pests than those that are stressed by factors such as drought, *Armillaria* root rot, and other pests (U.S. Department of Agriculture Forest Service). The lack of a fire regime in Pennsylvania's forests does not allow pine ecosystems, including barrens and early successional forests, to produce the kind of seed cone crop that would provide food for crossbills and the many other seed-eating species. There is a lack of recognition for the value of older conifers for food and shelter (including thermal cover) for a variety of wildlife species. Fire stimulates regrowth of conifers, including boreal black spruces in Maritime Canada (Committee on the status of Endangered Wildlife in Canada 2004).

*CONSERVATION AND MANAGEMENT NEEDS*

Because red crossbills are strongly associated with conifers, forest health may be a critical constraint for crossbill success. The forest must be healthy and mature enough for the trees to bear cones regularly. The size of the forest must be sufficient to produce cones in sufficient supply to nourish a viable population of crossbills and other seed-eating birds. A diversity of cone-bearing trees allows continued occupation of the species and subsequent nesting. Crossbills need an adequate amount of seeds to nest and raise young. Therefore, the size and health of the conifer forest is the key habitat component for this species. Proximity to other large, healthy conifer stands enhances the probability that crossbills will visit from other sites. Old growth conifers produce enough cones to sustain populations of crossbills and have higher densities of several species than other forests (Benkman 1993b, Haney and Schaadt 1996, Haney 1999). Because most conifers do not consistently produce seeds until they are twenty to thirty years old (Fowells 1965), a longer timber rotation cycle is required to maintain a forest that produces a seed crop. Isolated stands of white pines or other conifers attract crossbills long enough that they might nest, maintaining viable populations in our region. The lack of landscape-scale planning across ownership boundaries threatens all species that use large-scale forests. The lack of an old growth management plan in place for Pennsylvania's forests hurts all species that benefit from the attributes of mature forests (Haney and Schaadt 1996, Jenkins et al. 2004).

#### *MONITORING AND RESEARCH NEEDS*

The erratic behavior of this species makes it one of the more challenging songbirds to monitor or manage. Protecting nomadic species is particularly formidable because they may concentrate on small areas for brief periods and may not be present in critical areas for long periods, so those locations may not be recognized for their conservation value (Benkman 1993b). It is easy to overlook a nomadic breeding population because it sometimes appears and begins to nest when most songbirds are not nesting. However, popular birding activities and "citizen science" projects have great potential to detect crossbills. A two-pronged approach is needed for this conspicuous, vocal bird: volunteer- and agency-based surveys. Volunteer-based surveys that would be useful for detecting new locations for this species are Breeding Bird Surveys, the Pennsylvania Breeding Bird Atlas, Important Bird Area (several are located where this species formerly nested; Crossley 1999), Cornell Laboratory's Birds in Forested Landscapes in appropriate habitat (Cornell Laboratory of Ornithology 1997), and Mountain Birdwatch project that focuses on boreal Species of Conservation Concern (Vermont Institute of Natural Science 2005). Initial observations need to be followed by repeated visits to confirm nesting, sometimes at periods when other species are less apt to nest.

As for professional surveys, this species should be a priority for County Natural Area Inventories. Species-specific surveys should be conducted by experienced observers, perhaps as part of bird community or management area inventory. This is particularly appropriate in remote locations, especially old growth and other mature conifer forests, as well as pine–oak barrens. Surveys can be conducted in response to reported irruptions. In any of these surveys, recordings of the crossbills should be made to identify the "type" observed (Groth 1993a, 1996).

Red crossbills are among the most poorly studied and least understood birds in the state. We need field data on the occurrence, status (nesting or passage migrant), food plants, and habitat associations of each of the red crossbill "types" in the region. Field observers should make audio recordings of any red crossbills found in the state, especially nesting crossbills. These can be compared with recordings attributed to the various distinct populations (Groth 1996). Collecting these data should be coordinated with other researchers. Considering the strong connection this species has with conifer forests, habitat attributes of each site should be measured, including vegetative components, recent silviculture, and landscape attributes.

Red crossbill research should be integrated into any project concerning old growth or boreal conifer communities. The strong association of red crossbills with conifers invites many research projects, including studies on the timing and circumstances of migration, breeding, and molt, as well as productivity, multiple broodedness, age of maturation and first breeding, and association with food plants and competitors. Little is known about the interaction between conifer-related birds and other taxa of this ecosystem in the East, particularly with squirrels that are probably competitors and nest predators.

*Author:* DOUGLAS A. GROSS, PENNSYLVANIA GAME COMMISSION, ORANGEVILLE

## Pine Siskin

Order: Passeriformes
Family: Fringillidae
*Spinus pinus*

This small finch is an irruptive species that breeds in Pennsylvania irregularly (fig. 5.91). It was selected as a Pennsylvania Vulnerable species because of its restricted breeding population in Pennsylvania. Global populations are considered Secure (G5, NatureServe 2009).

#### *GEOGRAPHIC RANGE*

The pine siskin is a North American species breeding south into Mexico in the western half of the continent. In eastern North America, its occurrence south of the northern boreal forests of Canada, New York,

*Fig. 5.91.* The Pine Siskin, *Spinus pinus*. Photo courtesy of Geoff Malosh.

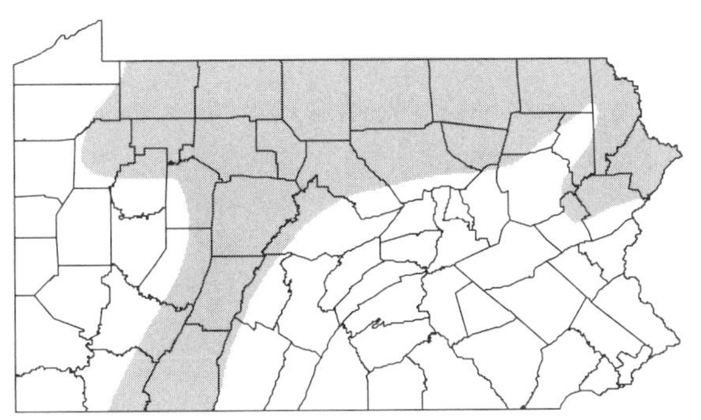

*Fig. 5.92.* Highest probability of occurrence of the Pine Siskin, *Carduelis pinus*.

and New England is erratic with incursions south into Pennsylvania and along the Appalachian Mountains (McNair 1988a, 1988b) occurring at unpredictable frequency, presumably in response to food scarcity in northern forests where they normally winter. Boreal forest populations in northern Canada are highly mobile, generally moving farther south in winter, and wandering widely in search of food concentrations.

### *DISTRIBUTION AND RELATIVE ABUNDANCE IN PENNSYLVANIA*

During invasion years, presumed to occur when northern cone crops are poor, this irruptic species may occur anywhere in Pennsylvania but is most likely to appear in mountainous areas harboring extensive conifer forests (Gross 1992d, McWilliams and Brauning 2000; fig. 5.92). Results of the second Pennsylvania Breeding Bird Atlas indicate a decrease of 18 percent in distribution has occurred since the first atlas (Mulvihill and Brauning 2009).

### *COMMUNITY TYPE/HABITAT USE*

The pine siskin is considered a bird of the northern boreal forest, preferring open stands of spruce and pine interspersed with birch and maple hardwood (Dawson 1997). Although these community types are considered key in the ecology of the species, use of a wide variety of other habitats, including alder swamps, weedy fallow areas, meadows, roadsides, and feeding stations occur. The pine siskin inhabits open coniferous boreal forest at all times of the year but is highly nomadic. During irruptions southward, believed to be food-shortage induced, the flocks may use a wide variety of habitats, including suburban feeding stations, alder swamps, birch stands, weedy fields, and parks. Suitable habitat for this species appears to be widespread in northern and mountainous portions of Pennsylvania (Gross 1992c, McWilliams and Brauning 2000).

### *LIFE HISTORY AND ECOLOGY*

Pine siskins are noted as a nomadic irruptic species (Bock and Lepthien 1976). Their peregrinations are assumed to be in response to food availability (Palmer 1968, Widrlechner and Dragula 1984). Siskins eat a wide variety of seeds and feed on both the ground and in treetops, depending on the season and type of seed material being consumed (Balph and Balph 1979). They will feed on conifer seeds and often feed with goldfinches on birch and alder seeds gleaned from the ground and in the catkins. Insects are also consumed as is some plant material (Jennings and Crawford 1983). Given the wide array of materials consumed, it is not surprising that the linkage between irruptic appearances and food availability is often obscure. Siskins, like many other winter finches, have a fondness for roadside salt (Bennetts and Hutto 1985). This propensity places them at risk during their incursions into northern urban areas.

Pine siskins are hardy birds and have been recorded breeding in February and March in Pennsylvania and New York (Yunick 1981, Gross 1992c). Nesting probably peaks in May but may continue into July or possibly August. Early season breeding often is in loose colonies in conifers (Peterson 1988c), but later nesting in a wide variety of deciduous trees and shrubs has been documented (Palmer 1968, Messineo 1985, Dawson 1997). Nests may be placed from 3 to 50 feet off the ground and are well concealed in foliage at the distal end of a branch. Territorial chases, courtship flight songs, and singing from elevated perches are circumstantial evidence of breeding (Dawson 1997).

There appears to be little fidelity to the nest site between years at this latitude in the eastern United States. Invasions are irregular and may occur every other year or less frequently. Some migratory (north/south) movement is believed to occur in most years, but the major influxes are believed to be associated with food shortages in the northern boreal forests where the majority of the birds normally occur. In Pennsylvania, pine siskins are a rare breeder after winter incursions (Gross 1992c).

*THREATS*

Pine siskins are known to be susceptible to organophosphate pesticide poisoning. Applications of pesticides to control insect outbreaks in boreal forest may affect this species during critical breeding periods. Because the male feeds the female throughout most of the reproductive cycle, and the female is solely responsible for the nest and nestlings, loss of either member of the pair will cause loss of the nest.

Outbreaks of disease around feeding stations, for example, salmonellosis, which is spread by contact with feces, have been identified as a potential threat to this ground-feeding species. Although some feeding stations practice periodic sterilization of feeders, most do not, and virtually none take measures to prevent possible buildup of pathogens in areas under the feeding station. The attraction of pine siskin to road salt is well established, and birds that visit Pennsylvania during the winter are likely to spend substantial amounts of time on roadsides foraging for grit and salt. Increasing traffic levels place these birds at significant risk from speeding vehicles. Broadcast spraying for control of boreal forest dwelling insects should be prohibited during years when incursions of boreal finches into Pennsylvania occur.

*CONSERVATION AND MANAGEMENT NEEDS*

Because of the nomadic habits of this species and its highly irregular breeding pattern in the state, conservation priorities seem premature. Additional study of breeding areas to determine features that appear to attract and induce breeding birds is needed. Feeding stations, particularly those offering thistle (niger) seed, may induce pairs to nest in areas where they normally might not (Siebenheller and Siebenheller 1987). Consequently, preservation of conifer stands is just one of probably many actions that may be appropriate to enhance breeding populations of this species.

*MONITORING AND RESEARCH NEEDS*

Tracking an irruptic nomadic species is a difficult task. Modern communication technology greatly facilitates and expedites gathering of avian species occurrence data. Requests for winter and early spring pine siskin sighting localities can be gathered by canvassing the Pennsylvania bird-watching community. Following up on these sightings to confirm probable nesting will be required to gather data on likely proximal causes of nesting activities. Considerable natural history investigation into the needs of this wandering species will be required before informed decisions can be made regarding the most appropriate management goals to set.

Habitats used for breeding in Pennsylvania need to be characterized to determine whether they share common habitat characteristics. A review of historical records from the first Breeding Bird Atlas and more recent records from the second atlas should be included. A review of banding records should also be conducted to ascertain whether long-term residency within the Pennsylvania population is indicated. Further study is required to determine what features induce pine siskins to occasionally breed in the state. As the popularity of bird feeding, particularly feeding of thistle (niger) seed increases, the potential to attract this boreal species to suburban areas will increase. Until such time as these factors can be clearly identified and quantified, action to enhance or protect critical habitat features attractive to these birds will be pure guesses.

*Author:* DOUGLAS P. KIBBE, URS CORPORATION

## MAINTENANCE CONCERN

## American Black Duck

Order: Anseriformes
Family: Anatidae
*Anas rubripes*

The American black duck was selected as a Species of Maintenance Concern because of long-term population declines throughout its range (fig. 5.93). It is currently included on the Species of Greatest Conservation Need list for every northeastern state. It is also listed as a Species of Moderate Priority under the Partners in Flight Watch List—Species Nesting in Pennsylvania (National Audubon Society 2005b). American

*Fig. 5.93.* The American Black Duck, *Anas rubripes*. Photo courtesy of Glen Tepke.

black ducks are not listed as Threatened or Endangered in any part of their range.

### *GEOGRAPHIC RANGE*

The breeding range of the black duck is extensive, covering the mid-Atlantic States, the Great Lakes States, New England, and eastern Canada from Manitoba to Newfoundland. Breeding densities are highest in Ontario, southern Quebec, Maine, and the Maritime Provinces. The black duck also has an extensive winter range. Some birds winter as far north as southern Ontario and some as far south as Florida. However, wintering densities are highest in the Atlantic coastal areas of New Jersey, Maryland, and Virginia.

### *DISTRIBUTION AND RELATIVE ABUNDANCE IN PENNSYLVANIA*

Historically, black ducks bred across Pennsylvania but were rare or irregular breeders and were most common in northeastern Pennsylvania (Warren 1890, Sutton 1928b). By the 1940s, perhaps in response to the impoundment of Pymatuning Swamp, black ducks were also common breeders in northwestern counties (Todd 1940a, Trimble 1940, Grimm 1952, McWilliams and Brauning 2000). Black ducks are currently widely distributed across the state but locally uncommon breeding birds with the highest concentration of breeders in the northeastern region of the state (Hartman and Dunn 1991, Brauning 1992a, McWilliams and Brauning 2000; fig. 5.94).

During migration, black ducks may be found anywhere in the state. Most black ducks found during migration or wintering in Pennsylvania would be birds that breed outside Pennsylvania. Black ducks are hardy, and during winter, some can be found in Pennsylvania wherever open water occurs. Most will be found in the southeastern portions of the commonwealth at places such as Middle Creek and the lower Susquehanna River. During mild winters, black ducks winter as far north in the commonwealth as Pymatuning and Presque Isle.

Although accurate estimates of total, range-wide abundance are not available, good estimates of relative abundance are. The longest series (1955–2004) of data comes from the Mid-Winter Waterfowl Survey (MWS). The MWS is conducted annually during January in the Atlantic and Mississippi flyways by the United States Fish and Wildlife Service and its cooperators, principally state and provincial wildlife agencies. Estimates derived from MWS have been used to monitor black duck population change over the life of the survey. MWS provides minimal estimates of wintering black ducks because it is an index, and not a complete census of wintering black ducks.

In Pennsylvania, the MWS reported an annual average of 1,903 black ducks wintering during the years of 2001 to 2004, only 14 percent of the 1955 to 1960 average of 13,555 (Serie and Raftovich 2004). The MWS data indicate a long-term decline in black ducks wintering in Pennsylvania. Christmas Bird Count (CBC) data provide additional information on black ducks wintering in Pennsylvania (National Audubon Society 2005a) and corroborate a decline in black ducks.

When considering the status of black ducks in Pennsylvania, resident breeding populations are more

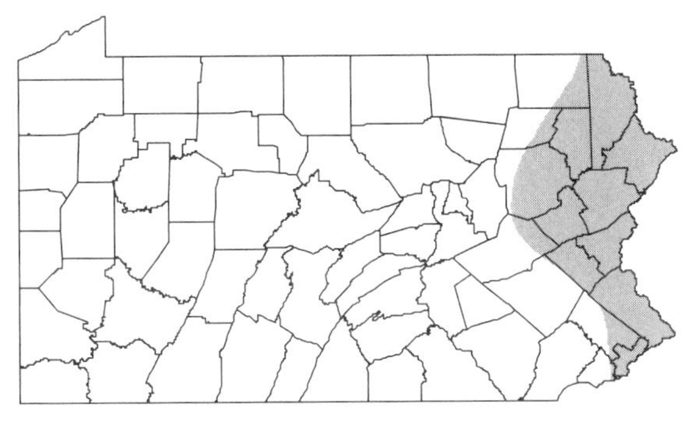

*Fig. 5.94.* Highest probability of occurrence of the American Black Duck, *Anas rubripes*.

relevant than wintering populations composed of birds from many areas. Estimates of black ducks breeding in Pennsylvania are available from the Atlantic Flyway Breeding Waterfowl Plot Survey (AFBWPS; Serie and Raftovich 2004). From1993 to 2003, an annual average of 2,468 pairs of black ducks was estimated to breed in Pennsylvania. Individual annual estimates vary considerably and, as with MWS estimates, should be viewed as indices rather than absolute counts.

Too few black ducks (zero to sixteen) are recorded in Pennsylvania during the Breeding Bird Survey (Sauer et al. 2005) to draw meaningful conclusions about breeding population trend. The AFBWPS data, while specific to breeding black ducks in Pennsylvania, are of a relatively short time series (1993–2003), are highly variable, and suggest no population trend. However, Pennsylvania Game Commission officials, arguably the most knowledgeable people about black ducks in Pennsylvania, believe that they have declined, particularly in the Northeast where breeding densities have been highest (K. Jacobs, J. Dunn, and I. Gregg, Pennsylvania Game Commission, personal communication). Preliminary results from the second Breeding Bird Atlas also suggest a significant decline in black ducks between the first (1984–89) and second atlas (Mulvihill and Brauning 2009).

### *COMMUNITY TYPE/HABITAT USE*

Throughout its life cycle, the success of the black duck has been linked closely to the availability of quality wetlands. It uses a wide range of wetland habitats for breeding. Inland in eastern Canada and the northeastern United States, black ducks nest in freshwater wetlands, including bogs, marshes with emergent vegetation, lakes, swamps, rivers and streams, and beaver flowages. Often, if not typically, these preferred wetlands are located in forested areas. Freshwater wetlands in forested regions make up most of the nesting habitat of black ducks in Pennsylvania.

During spring and fall migration, black ducks may be found anywhere in Pennsylvania but most likely on larger marshes and water bodies. Black ducks winter primarily along the Atlantic Coast from Maine to Florida in salt marshes and tidal estuaries, but in Pennsylvania, a few thousand can usually be found wintering wherever there is open water.

### *LIFE HISTORY AND ECOLOGY*

Black ducks form pairs during the late fall and winter. They migrate and nest early in the spring. Hens seek secluded cover near wetlands for nesting. Pair bonds dissolve after the hen begins incubation. The ducklings are precocial and leave the nest to follow the hen soon after hatching. Invertebrates, particularly insects, are important to young black ducks.

From the onset of egg laying, black ducks face considerable environmental risk, especially predation. Nesting success, the probability of at least one young leaving the nest, is not high, ranging from 21 percent to 80 percent, with most around 50 percent in several studies (cited by Longcore et al. 2000a). Once out of the nest, ducklings are also at risk. Ringelman and Longcore (1982) estimated duckling survival to be 42 percent in their southern Maine study. In the face of these odds, annual recruitment to black duck populations is not likely high. Results from banding studies indicate that mean survival rates for immature (first year) male and female black ducks are 64.9 percent and 57.9 percent, respectively, and 66.7 percent and 61.0 percent, respectively, for adult males and females (Francis et al. 1998). These rates equate to mean life spans of 2.4 years for males and 1.9 years for females.

Black ducks feed on a variety of seeds, tubers, roots, and other vegetable matter, which they obtain in shallow waters of fresh and saltwater wetlands. They also feed in flooded pastures and agricultural fields, particularly during migration, and often in the company of other ducks. Animal matter such as insect larvae, snails and mussels are also part of the black duck's diet, especially during the late winter and spring before nesting.

For more detailed information on the natural history of the black duck, see Bellrose (1976), Longcore et al. (2000a), and NatureServe (2005).

### *THREATS*

The long-term decline of black duck populations has stimulated much research into, and debate over, causes of the decline. Interactions with mallards (competition and hybridization), hunting, and habitat loss have all been implicated. Mallards have increased in the primary breeding range of the black ducks but are still outnumbered by black ducks. Recent thinking is that circumstantial evidence for competition with mallards exists but does not provide a clear explanation for the decline of black ducks (Conroy et al. 2002).

During the 1970s, some believed that overhunting was a major cause of the decline of black ducks. The issue went from controversial to litigious (Humane Society of the United States et al. 1982), and, in 1983, restrictive hunting regulations were implemented by federal and state wildlife agencies. Subsequent studies (Francis et al. 1998, Longcore et al. 2000b) suggested that, indeed, for some populations of black ducks, hunting mortality has been additive to other causes of mortality. Following restrictive regulations, black duck harvests declined range-wide and in Pennsylvania declined from an average of 11,100 per year from 1961 to 1965 to 6,600 per year from 1996 to 2000 (Serie and Raftovich 2004). In Quebec and the Maritimes, the breeding population increased following restrictive regulations. Regulations governing the hunting of black ducks generally remain restrictive, and appear to have helped slow the rate of population decline. Hunting is less of a factor in black duck population ecology today, and there appears to be little desire or need for further regulatory restrictions.

Habitat loss and degradation remain the most serious threats to the black duck throughout its range. In Pennsylvania, many of the same problems that resulted in the past loss and degradation of wetlands, remain threats today, making it unlikely that wetland trends soon will be reversed to the benefit of increased black duck populations. Although impounding and major drainage now occurs infrequently because of stricter environmental regulation, smaller-scale drainage and fill still occur. Development accompanying urban and suburban sprawl is the most serious threat to wetlands in Pennsylvania (Moyer 2004), especially in the rapidly developing northeastern part of the state. Invasive plant species affect wetlands across the state and are particularly insidious in the northwestern part of the state where Pennsylvania's largest marshes, including Presque Isle, Pymatuning, Hartstown, Geneva and Erie National Wildlife Refuge are located. Acidic precipitation and other forms of nonpoint source pollution remain threats to wetlands across the state.

*CONSERVATION AND MANAGEMENT NEEDS*

The black duck has received much attention from agencies and nongovernmental organizations concerned about its conservation status. The United States Fish and Wildlife Service formally lists it as a Game Bird Below Desired Condition. At least five major conservation plans address black duck conservation directly or indirectly, the most direct being the Black Duck Joint Venture (BDJV; Lepage and Bordage 2003) under the North American Waterfowl Management Plan (NAWMP; U.S. Fish and Wildlife Service 2004a).

The BDJV focuses mainly on surveys, banding, and research needs. The more general Eastern Habitat Joint Venture and the Atlantic Coast Joint Venture (ACJV) indirectly promote black duck conservation by focusing on the conservation of waterfowl/wetland habitats throughout the eastern two-thirds of the black duck's range. The ACJV identifies focus areas where waterfowl conservation efforts are to be concentrated in Pennsylvania, including the Northwest Focus Area and the Delaware Basin Focus Area, where breeding black ducks are most common. Also, in Pennsylvania, Audubon Pennsylvania's Important Bird Area (IBA) program (Crossley 1999) promotes conservation of numerous wetlands (e.g., Pymatuning-Hartstown Complex, Conneaut-Geneva Marsh, Cussewago Bottoms, Middle Creek Wildlife Management Area, and Shohola Waterfowl Management Area) important to black ducks in Pennsylvania.

Habitat protection is most important to maintaining a breeding population of black ducks in Pennsylvania. However, setting a habitat target specifically for black ducks in Pennsylvania is not recommended. Rather, the black duck's habitat needs should be included in comprehensive targets for all wetland-dependent species in Pennsylvania. The general habitat goals (protect the best of what remains, restore and improve degraded or impaired habitats) outlined in Pennsylvania's *Wildlife and Wild Places* (Moyer 2004) should be followed, and the specific acreage targets for wetlands presented in the Pennsylvania Waterfowl Management Plan 1991–2000 (Hartman and Dunn 1991) should be updated and followed. That plan's objectives are to protect 88,000 acres of critical waterfowl and wetland habitats from further loss and degradation, enhance 70,000 wetland acres, and create 11,000 acres of new wetlands. Qualitative goals and measures for Pennsylvania wetlands should also be adopted. To the extent feasible, working to protect habitat in cooperation with the NAWMP Joint Ventures programs is recommended. The Joint Ventures address breeding habitats for black ducks in Pennsylvania, as well as in their out-of-state wintering habitats.

A population objective for black ducks breeding in Pennsylvania should be established, but in the absence of an accurate census, an absolute numerical objective may be unrealistic. A more reasonable approach to establishing and monitoring a population objective would be to use an index figure, such as might be derived by modeling data from AFBWPS. Using MWS data to monitor black ducks breeding in Pennsylvania would not be effective, because most birds surveyed would be migrants from other areas.

*MONITORING AND RESEARCH NEEDS*

Considerable effort has gone into modeling black duck populations (Conroy et al. 2002). At the range-wide level, adaptive management has been applied successfully through a Black Duck Adaptive Harvest Management Working Group. However, given the accuracy and precision of population and habitat data for the relatively small population of black ducks in Pennsylvania, adaptive management for the species at the state level is not likely feasible and, therefore, not recommended.

The exhaustive amount of research done on black ducks (see Longcore et al. 2000a, Conroy et al. 2002) has not led to population recovery. Further research specific to black ducks in Pennsylvania is not recommended. However, continued surveys and monitoring are advised. The second Pennsylvania Breeding Bird Atlas will document the current distribution of black ducks breeding in Pennsylvania. AFBWPS should be continued to monitor numbers of black ducks breeding in Pennsylvania. Comprehensive assessments of current wetland and water quality in Pennsylvania should be made. Habitat assessments need not be specific to black ducks, because information gained through larger, species-comprehensive efforts can be applied to the black duck case. Range-wide research to develop more accurate population models and more effective means of monitoring habitats should be supported.

*Author:* JOHN TAUTIN

## Ruddy Duck

Order: Anseriformes
Family: Anatidae
*Oxyura jamaicensis*

The ruddy duck is a small diving duck with a long, stiff upturned tail (fig. 5.95). It is listed as a Species of Greatest Conservation Need in Pennsylvania because it is an extremely rare breeder, and breeders are associated with wetlands with dense emergent vegetation, a rare habitat type in Pennsylvania. It is also listed as a Species of Greatest Conservation Need by Maine, New York, and Maryland. Globally, the ruddy duck is Secure (G5, NatureServe 2009). The ruddy duck is hunted as a game species and is under the management regulations for hunted species.

*Fig. 5.95.* The Ruddy Duck, *Oxyura jamaicensis*. Photo courtesy of Glen Tepke.

*GEOGRAPHIC RANGE*

Ruddy ducks breed primarily in the prairie and parkland regions of Canada and north-central United States (Bellrose 1976, Brua 1999, 2002). They breed irregularly outside their usual range at scattered locations in the eastern United States, including Pennsylvania, New York, Ohio, Maryland, and Delaware (Brua 2002). Most of the Atlantic Flyway ruddy ducks winter from the Chesapeake Bay south to the Pamlico Sound, North Carolina (Bellrose 1976). They may also be found in extreme southeast Pennsylvania and south along the coast of North America and the Gulf of Mexico. Ruddy ducks are year-round residents of the desert southwest, Southern California, and parts of Mexico.

*DISTRIBUTION AND RELATIVE ABUNDANCE IN PENNSYLVANIA*

Pennsylvania is a major migratory flyway for ruddy ducks bound for wintering grounds in the Chesapeake Bay. Flocks of migrating ducks congregate in large numbers on the Delaware River in the eastern part of the state and Tinicum Marsh in the southeastern part of the state. They are regularly observed in winter (Fingerhood 1992c). In Pennsylvania between 1978 and 2007, an average of twenty ruddy ducks were reported on the Atlantic Flyway Midwinter Waterfowl Survey (MWS), which is conducted in mid-January (Jacobs et al. 2008). These are minimum estimates of the number of ruddy ducks wintering in Pennsylvania and are used as an index. Ruddy ducks are also reported annually on Christmas Bird Counts conducted across the state. From 1990 to 2008, a mean of 554 ruddy ducks were reported in Pennsylvania on these counts (Audubon 2008). In 2000, a high count of 506 individuals was recorded in Bucks County (American

Birds 2000). In summer, the ruddy duck is a casual visitor to Pennsylvania. Ruddy ducks were documented breeding at Lake Pymatuning between 1934 and 1940 (fig. 5.96). However, by 1952, there were no breeding records of ruddy ducks in Pennsylvania (Fingerhood 1992c). There have been three confirmed breeding records in the state since 1960 (McWilliams and Brauning 2000).

*COMMUNITY TYPE/HABITAT USE*

Ruddy ducks are found on lakes, ponds, freshwater marshes, and slow-moving rivers (McWilliams and Brauning 2000). They nest on freshwater marshes, sloughs, lakes, and ponds, where the water is surrounded by extensive emergent vegetation (Brua 2002). In winter, most of the Atlantic Flyway population winters in the upper Chesapeake Bay region in brackish estuarine bays (Brua 2002).

*LIFE HISTORY AND ECOLOGY*

The ruddy duck has a short, sturdy body adapted to life in the water. Its short legs are placed so far back on its body that it cannot walk well on land. It feeds on animal matter, including aquatic insects, crustaceans, zooplankton, snails, and other aquatic invertebrates (Bellrose 1976, Euliss et al. 1991, Brua 2002). Their primary food is midge larvae, chironomidae. They will also take aquatic vegetation and seeds. Ruddy ducks are opportunistic feeders and will shift their diet to include the most abundant foods in a given foraging area (Euliss et al. 1991).

The ruddy duck is a migratory species that travels in moderately sized flocks of five to fifteen individuals. This diurnal species will migrate mostly at night (Brua 2002). Winter flocks consist primarily of ruddy ducks, although sometimes they are associated with American coots (*Fulica americana*) and other diving ducks.

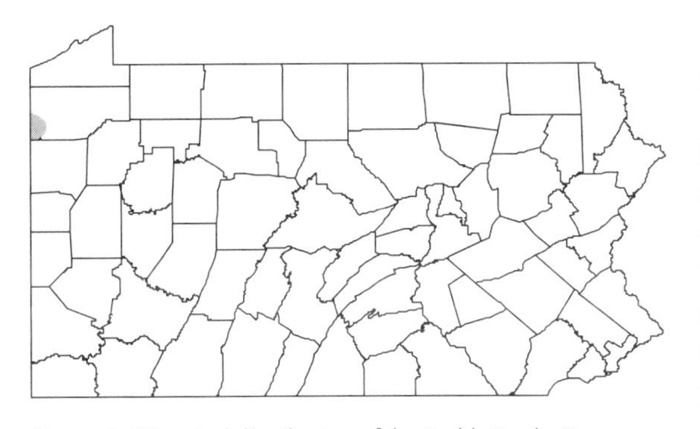

*Fig. 5.96.* Historical distribution of the Ruddy Duck, *Oxyura jamaicensis.*

In the breeding season, the ruddy duck is monogamous with a tendency toward small groups of breeding pairs or colonies. Nesting extends from May through July. The nest is built of cattails and other grasses and hidden among the dense vegetation of freshwater marshes. Sometimes a pair will use an abandoned nest and add materials to strengthen it. The precocial young leave the nest soon after hatching and are able to feed themselves within a short time. The female continues to tend to them after they have left the nest. A single clutch is common in the northern part of the range and a second clutch is often attempted in the southern part of the range.

Ruddy ducks engage in nest parasitism, both intra- and interspecific. A female ruddy duck will often lay her eggs in the nest of another female. Usually, these "dump" nests have a low hatching success, with complete loss sometimes commonplace. Chance of predation is increased in these "dump" nests as the number of eggs present may increase to twelve or more eggs. Greatest success occurs in nests that are initiated earlier in the season. Nests that are better concealed in vegetation and farther from habitat edges are also more successful. Few females have been observed to renest after the loss of their nest (Brua 1999).

*THREATS*

As with all of Pennsylvania's wetland birds, the primary threat to ruddy ducks is the loss and degradation of wetland habitat, particularly wetlands with dense emergent vegetation. Loss and degradation of wetlands within the prairie pothole region of North America continue to threaten breeders, and degradation of wintering habitat, particularly in the Chesapeake Bay Region, may affect wintering populations.

*CONSERVATION AND MANAGEMENT NEEDS*

Although the ruddy duck is primarily a migrant through Pennsylvania, there is evidence that the condition of migratory stopover sites has an effect on the breeding condition of these birds after reaching breeding grounds, which could have an effect on their reproductive success. Several management activities could be undertaken to support the current populations of this species. These actions include identifying and protecting concentrated migratory stopovers, identifying and protecting wetland habitats along migratory corridors, and wetland habitat restoration on public and private lands that emphasizes water resources and water quality. An important shift in water habitat management would include increased buffers between development and water resources with the goal of protecting water quality. Protection of wetlands and habitat restoration should be high priorities.

### MONITORING AND RESEARCH NEEDS

Wintering ruddy ducks are monitored by the Audubon annual Christmas Bird Count and the Atlantic Flyway Midwinter Waterfowl Survey. Because ruddy ducks are such rare breeders in Pennsylvania, there is no need to conduct species-specific surveys or research. Annual surveys of breeding birds associated with emergent wetlands in Pennsylvania would be useful for detecting when and where ruddy ducks breed in Pennsylvania in the future.

*Authors:* DARRYL SPEICHER, POCONO AVIAN RESEARCH CENTER; JACKIE SPEICHER, POCONO AVIAN RESEARCH CENTER

## Pied-billed Grebe

Order: Podicipediformes
Family: Podicipedidae
*Podilymbus podiceps*

The pied-billed grebe is the smallest of the grebes that occurs in Pennsylvania (fig. 5.97). It was selected as a Species of Maintenance Concern because of long-term declines in population size and its low abundance within the state. In Pennsylvania, it is listed as a Candidate Rare. It is included on the state lists of almost all the northeastern states and is considered a northeast region priority species. State breeding populations are considered Vulnerable (S3B), and global populations are Secure (G5, NatureServe 2009).

Fig. 5.97. The Pied-billed Grebe, *Podilymbus podiceps*. Image courtesy of Alan D. Wilson, naturespicsonline.com.

### GEOGRAPHIC RANGE

The pied-billed grebe breeds on ponds and wetlands across most of North America and south into Chile and Argentina in South America (Muller and Storer 1999). The breeding population along the Gulf Coast states is resident while the migratory population breeds from the prairie pothole area of the Great Plains north and eastward to the Maritime Provinces and New England. Breeding birds in North America winter in the Gulf Coast south into Central America.

### DISTRIBUTION AND RELATIVE ABUNDANCE IN PENNSYLVANIA

The pied-billed grebe may occur throughout Pennsylvania during migration, and there are summer records from thirty-six counties (Ickes 1992d). The principal breeding areas of the state historically have been the major marsh complexes of the southeast (i.e., Tinicum and other Delaware River Marshes) and the northwest (Pymatuning Marsh). Currently, the highest probability of occurrence is the northwestern corner of the state (fig. 5.98). Construction of reservoirs and ponds throughout Pennsylvania has provided water bodies suitable for foraging during migration, but these areas are infrequently used by birds for nesting if they lack the large emergent wetland complexes that pairs commonly use for breeding. If emergent wetlands develop, these man-made water bodies may be heavily used. A notable example is Glen Morgan Lake where more than 100 birds were found during July 1996 (McWilliams and Brauning 2000).

The pied-billed grebe is a fairly common migrant through the state and a rare breeder. During the breeding season, they are secretive and may easily elude casual observation. During the first Breeding Bird Atlas project, pied-billed grebes were found in less than 2 percent of the atlas blocks in the state. Tape surveys in other states (e.g., Vermont) have found large expanses of apparently suitable habitat unoccupied for unknown reasons (Kibbe 1985b). A similar situation appears to apply to Pennsylvania where large areas of the state are generally barren of pied-billed grebes; however, additional survey efforts using prerecorded tapes are recommended to confirm this. The species was once considered a regular breeder in the southeastern marshes but has become increasingly rare in the summer months and rarely breeds there now. Its status in the northwestern corner of the state as an uncommon breeder has remained more stable and it continues to be found in small numbers in the large wetlands of that region. Results of

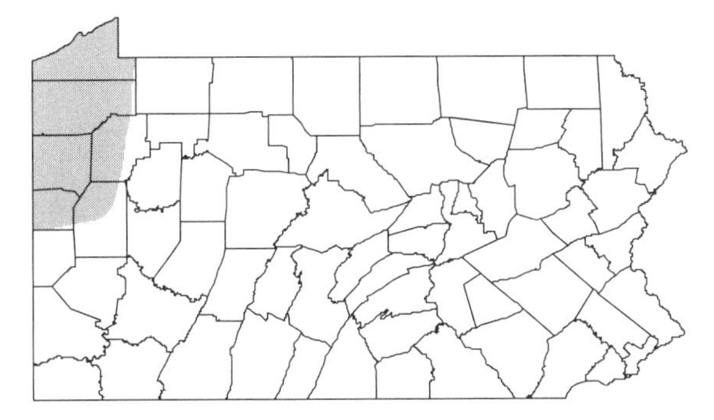

*Fig. 5.98.* Highest probability of occurrence of the Pied-billed Grebe, *Podilymbus podiceps*.

the second Pennsylvania Breeding Bird Atlas indicate a decrease of 28 percent in distribution has occurred since the first atlas (Mulvihill and Brauning 2009).

Migrant birds may be locally common throughout the fall until open water begins to freeze. Wintering birds are often reported from Lake Erie and the open water of major rivers but seldom spend the entire winter period within the confines of the state. Spring migration begins as soon as ponds become ice free and continues through mid-May (McWilliams and Brauning 2000).

### COMMUNITY TYPE/HABITAT USE

The pied-billed grebe depends on emergent wetlands throughout its life cycle, rarely going on dry land. The birds forage in and along the margins of these wetlands. A wide variety of wetland types may be used ranging from flooded agricultural fields and sewage ponds to pristine marshes. During the breeding season, they primarily use large extensive marshes with emergent and aquatic vegetation. They regularly breed in the large marshes in the glaciated northwestern section of the state. Greatest densities probably occur at all seasons in emergent marshes dissected by open water channels. In such habitats, the birds can skulk along the edges and disappear at will among the adjacent wetlands. During migration periods, pied-billed grebes can be found in open water, as well as the wetland habitats, which they prefer during the breeding season. During the nonbreeding season, birds continue to use these wetland communities although they occasionally may be seen in open water of canals, lakes and ponds, and brackish coastal bays.

### LIFE HISTORY AND ECOLOGY

Pied-billed grebes are opportunistic feeders, preying on a variety of aquatic organisms. Crayfish, aquatic insects, and small fish are the common food items. After reaching the breeding marsh, males may construct several nest starts/platforms, one of which is subsequently selected and occupied by the female. Although an area around the nest is defended against intruders, nests may be as close as 20 m apart, but the average is more than twice that distance. Pairs construct their floating nests in wetland vegetation usually in water greater than 25 cm deep to facilitate escape by diving. Additional platforms may be used for mating and brooding. Renesting occurs if the first clutch is destroyed and second broods are fairly common. The first nest site may be reused in such attempts. Nest failure may reach 50 percent because of weather, water fluctuation, wave action (including boat wakes), and spawning activities of carp (Muller and Storer 1999). Predation of eggs and young is another cause of reproductive failure.

### THREATS

Threats to the breeding population are largely unknown. Research into the factors that may reduce survival or restrict breeding in specific wetlands needs to be implemented. Although wetland protection regulations are already in place that provide reasonable protection for the large wetlands likely to be attractive to pied-billed grebes, wetland degradation through pollution, invasion by exotic species (plant or animal), or disturbance factors (e.g., boat wakes) that contribute to nesting failure may still be unchecked.

The species no longer uses the Delaware River marshes for breeding because of the level of existing disturbance, incursion of phragmites, and general degradation of these wetlands. Factors affecting the number of breeding birds in the remainder of the state are more difficult to assess. The species was probably always a rare breeder in central regions of the state because of the lack of large wetland complexes that it prefers. The wetlands of Crawford, Mercer, and Butler continue to support the majority of the breeding pairs that occur in the state. Leberman (in Ickes 1992d), indicated that the breeding population in northwestern Pennsylvania was smaller than it had been in previous decades, but survey data are lacking to quantify that assessment and no reasons have been proposed to explain the possible causes.

### CONSERVATION AND MANAGEMENT NEEDS

At present, there is no way to qualitatively assess the change in the status or this species or develop specific conservation needs. Recommended actions to resolve management needs include developing a survey protocol and implementing it statewide to identify areas of greatest importance to the state's breeding population. Research developed to understand factors contributing to success should be used to identify conservation measures that need to be implemented on a local or statewide level to enhance breeding populations.

*MONITORING AND RESEARCH NEEDS*

There has been no survey even at a local level to determine the breeding density. Adoption of a standardized survey procedure is an immediate need. Implementation of surveys of breeding areas needs to be undertaken as soon a standard survey protocol is selected. Use of taped playback calls is recommended as a survey technique (Conway 2005), but other factors (e.g., time of day or night, weather influences, peak of calling period) that may influence the success of surveys need to be carefully determined. Use of tape call playbacks to elicit better response rates from secretive marsh birds (Conway 2005, Conway and Gibbs 2005) is a well-established technique applicable to surveying of pied-billed grebes. Survey tapes and a set protocol establishing length of survey, timing (diurnal or nocturnal), seasonal, and weather constraints need to be developed. Type of access (bank vs. boat) and duration of survey also need to be established. The surveys, once implemented, will serve as a basis for further investigations into factors that may influence habitat choice, reproductive success, and population density. Once a survey protocol is established, surveys need to be conducted statewide to canvass the entire state identifying sites for further monitoring. In addition, standardized searches of the principal wetland areas of Crawford, Butler, and Mercer counties should be conducted to establish the current baseline population level in the historic stronghold of the species for long-term monitoring. Ultimately, management plans should be developed (in cooperation with public and private agencies) to increase productivity.

Further research is needed to establish the species' nesting requirements and conservation actions, which may be needed to enhance breeding. This includes studies on nest site selection, minimum wetland size used for breeding and the effect of size in successful density, nesting density and factors affecting it, nesting success and factors (e.g., predation, pollution, disturbance, boat wakes) that may affect it, and the importance of habitat and habitat attributes to successful breeding attempts.

*Author:* DOUGLAS P. KIBBE, URS CORPORATION

## Great Blue Heron

Order: Ciconiiformes
Family: Ardeidae
*Ardea herodias*

The great blue heron was selected as a Species of Greatest Conservation Need because it is a colonially nesting species that is vulnerable to disturbance and because it feeds in wetlands and riparian areas and is therefore sensitive to wetland loss and degradation (fig. 5.99). The great blue heron is globally Secure (G5, NatureServe 2009). Breeding populations in the state are ranked as Apparently Secure to Vulnerable, while nonbreeding populations are Apparently Secure. This species is included on the Species of Greatest Conservation Need list for eight northeastern states.

*GEOGRAPHIC RANGE*

The great blue heron breeding range extends from Alaska and southern-tier Canadian provinces across the entire United States to Central America.

*Fig. 5.99.* The Great Blue Heron, *Ardea herodias*. Photo courtesy of Michael Brown.

*DISTRIBUTION AND RELATIVE ABUNDANCE IN PENNSYLVANIA*

In Pennsylvania, breeding is concentrated in the glaciated northern counties of the Appalachian Plateaus Province and to a lesser extent in the southeastern corner of the state (fig. 5.100). Occurrence, however, especially after the breeding season, is statewide. In 1993, Pennsylvania inventoried 1,654 nests statewide (Pennsylvania Game Commission 2002). Assuming little change in total population over this two-decade period, Pennsylvania harbors only about 4 percent of the breeding population from midcontinent Canada to a portion of the Great Lakes and the entire eastern United States coast. Inland United States colonies (other than Pennsylvania and Illinois) are not included in this assessment. The Pennsylvania breeding population is not peripheral but rather is part of the core breeding population. Globally, Pennsylvania plays an important but not critical role in the survival of this species. When coastal and Great Lakes colonies are excluded from the picture, however, Pennsylvania supports a significant inland breeding population specializing in glaciated wetland habitats. Significant population increase trends were observed for the period between 1966 and 2003 in Pennsylvania (+2.58%, P = 0.0004, 82 routes) based on Breeding Bird Survey routes (Breeding Bird Survey 2003) (fig. 5.101). Within Pennsylvania, trends ranged from >1.5 percent increase per year for most of the state to 0.25 to 1.5 percent increase for the central and Pocono glaciated north to no change or loss in the Allegheny National Forest region (Breeding Bird Survey 2003).

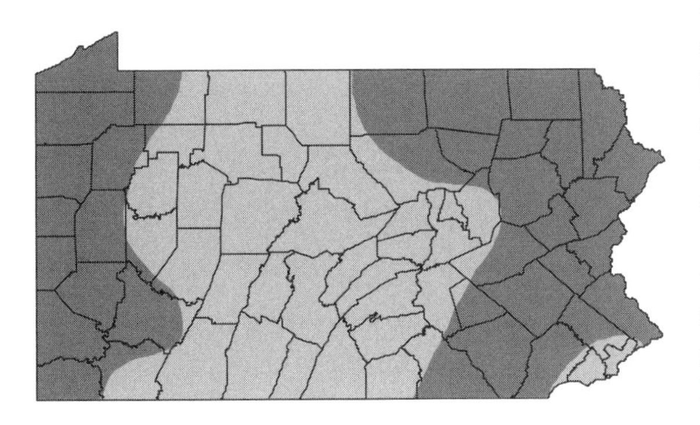

*Fig. 5.100.* Primary (darker shading) and secondary (lighter shading) distribution of the Great Blue Heron, *Ardea herodias*.

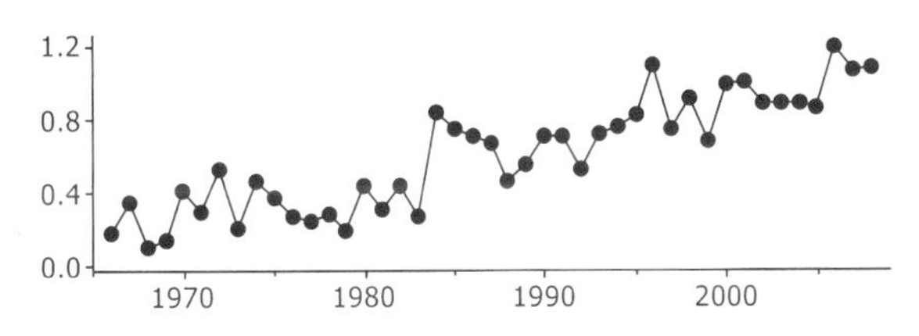


*Fig. 5.101.* Great Blue Heron, *Ardea herodias*, population trends from the Breeding Bird Survey.

*COMMUNITY TYPE/HABITAT USE*

Breeding/nesting habitats include mature hardwood or mixed hardwood/conifer forests, usually near rivers, wetlands, or lakes; palustrine forested wetlands; and lakes or emergent palustrine wetlands with wooded islands or snags. Foraging habitats during the nesting season include slow-moving or calm freshwaters of pond and lakeshores, streams, rivers, wet meadows, swamps, marshes, and fresh bays, occasionally fields and open deeper waters (DeGraaf and Rudis 1986, Butler 1992). Nonbreeding season foraging habitats are similar, but few individuals overwinter in northern Pennsylvania, where little open water exists. Greater use of tidal flats occurs during migration and coastal areas, estuaries, and open fields during winter months (DeGraaf and Rudis 1986).

Gap Analysis Project (GAP) shows limited primary and widespread secondary potential habitat for great blue herons in Pennsylvania (Myers et al. 2000).

*LIFE HISTORY AND ECOLOGY*

The great blue heron is primarily an aquatic carnivore feeding largely but not exclusively on fish. Other prey in freshwater habitats include arthropods, amphibians, and reptiles (Bent 1926, Kushlan 1978, Martinez-Vilalta and Motis 1992). Terrestrial prey occasionally taken include insects, amphibians, reptiles, rarely birds (as large as terns), and small mammals (DeGraaf and Rudis 1986, Martinez-Vilalta and Motis 1992). Some nineteen species of fish were identified at nesting colonies (dropped from nests) in Mercer County, the majority of which (64% of 286) were gizzard shad and common carp, as large as 36 cm total length (Ross 1990, R. M. Ross, unpublished data). Panfish composed 28 percent and sportfish (bass, walleye, and pickerel) only 3 percent.

Feeding is primarily crepuscular and diurnal, occasionally nocturnal, with mean distance of forays from nest in breeding adults 2–6 km, though distances up to 104 km have been recorded (Butler 1992). Solitary foraging is the rule, but feeding with conspecifics, other ciconiiformes, even cormorants occurs (Butler 1992). Adults defend both nests and feeding territories day and night (Bayer 1984, Butler 1992).

Northern populations of the great blue heron are

migratory (Henny 1972, Byrd 1978, Butler 1992). On the basis of banding of nestlings and subsequent recoveries, Henny (1972) estimated that nearly half of birds originating from colonies south of the Mason-Dixon line (to 34°N) migrate. Presumably, a majority of birds hatched north of this belt (including all from Pennsylvania) migrate. While many individuals overwinter in Pennsylvania where waterways are open (particularly southern counties), the available data suggest that many of these may not be Pennsylvania natals and that the majority of birds originating from Pennsylvania colonies migrate to southern United States, the Caribbean, and northern South America. Most birds from northern localities begin fall migration by late October (Butler 1992).

Great blue herons nest primarily in colonies, returning to the same colony, if undisturbed the previous year, from mid-February to April in Pennsylvania (Butler 1992, R. M. Ross, unpublished data). Nests are located typically near the tops of tall live trees in upland hardwood or riparian forests but may occur in wetland snags, shrubs, or even on the ground on islands (DeGraaf and Rudis 1986, Butler 1992). Nest and colony location is a response to both mammalian predator avoidance and food-source proximity (Butler 1992). All great blue heron colonies in Pennsylvania are monospecific, though owls or other raptors may nest in the same colony without apparent adverse interactions (Pennsylvania Game Commission 2002, R. M. Ross, unpublished data). In 2002, sixty colonies in Pennsylvania ranged from 1 to 383 nests and averaged 28 nests per colony (Brauning and Siefken 2003). Newly built nests are small thin platforms of sticks ~0.5 m in diameter but may grow in size both seasonally and annually, with weights up to 5 kg (Butler 1989). Dispersal in all directions occurs two to three months after nesting, with southward movement in autumn (Henny 1972). Maturity and first breeding occurs at two years of age; only one brood per year is produced (DeGraaf and Rudis 1986, Butler 1992). Annual mortality of banded birds ranged from 69 percent (first year) to 22 percent (third year and over; Henny 1972).

### *THREATS*

Reforestation, removal of organochlorine pesticides from agroindustry, and public environmental education/awareness have all contributed to the stable or increasing population trend observed in recent decades in Pennsylvania and across the continent. However, threats to the integrity of this species still occur inside the state, as well as on wintering grounds and during migration. Because nesting and feeding habitats are somewhat segregated at a local (sublandscape) scale, potential threats differ for the two aspects of life history.

Breeding habitats are abundant in Pennsylvania with 60 percent (17 million acres) of the state land mass in forested condition (Moyer 2003). This species is not area sensitive and thus can successfully use small forested tracts for breeding colonies. Colonies are often found near industries and commerce, though the likelihood of eventual disturbance and abandonment increases there. However, feeding habitats (wetlands and undisturbed riverine riparian stretches) are decreasing in size yearly. Contaminants in aquatic prey, such as DDE and other organochlorine pesticides, PCB, dioxins, and mercury may cause reproductive failure through eggshell thinning or other mechanisms, though no proof that this has occurred in great blue herons exists (Fitzner et al. 1988, Elliot et al. 1989, Butler 1992). Many of these contaminants persist or have increased in recent decades in aquatic habitats where herons feed.

Though mature forests in Pennsylvania are increasing, the greatest threat to nesting herons continues to be logging of forests without regard to the presence of heron colonies, especially on private lands where most colonies occur (Gill 1985, Schwalbe and Ross 1992). Of thirty-nine known owners of property on which great blue herons nest in Pennsylvania, only ten (26%) are public (Pennsylvania Game Commission 2002). Despite wildlife protection laws prohibiting the taking (or disturbance) of birds on nests, fines are insignificant to many developers or landowners and protection does not extend to the nonbreeding season (Ross 1990). Disturbance can easily result in the abandonment of an entire colony, especially early in the nesting cycle and when within 200–300 m of the colony (Ross 1990, Butler 1992). Disturbance from man-made structures or activity generally occurs only within 0.5 km (Butler 1992, Watts and Bradshaw 1994). Though splintering of disturbed colonies to locations nearby may occur in the same and subsequent years, these colonies often do not persist. Ross, however, documented the likely merger of the two largest great blue heronries in the state after abandonment of one in 1997: the following year the second heronry (11 km away) doubled in size to nearly 400 nests (Pennsylvania Game Commision 2002, R. M. Ross, unpublished data). This large colony persists but has gradually declined in size by 12 percent over the past five years (Brauning and Siefken 2004).

Threats to feeding habitat are related primarily to incremental loss of wetlands and undisturbed riparian stream reaches. Because this species appears to feed opportunistically on available prey, exotic and invasive prey species are not an issue. Continued loss of wetlands, however, would unfavorably affect food sources. Continued growth of Pennsylvania's rural human population may reduce the bird's access to wetland/riparian feeding habitat simply because of increased human encounters. Fish hatcheries are frequently visited by great blue herons, which may cause considerable losses if waterways are unprotected (Parkhurst et al. 1992). However, these losses can be prevented with the use of netting or electric fencing for exclusion (Mott and Flynt 1995; Pitt and Conover 1996). The remaining threats to feeding habitat relate to contaminants, which do not presently appear to adversely affect reproductive success. Acid mine drainage, however, resulting in fishless river segments and streams, continues to persist in the Susquehanna River West Branch, where great blue herons are largely absent (Schwalbe and Ross 1992).

*CONSERVATION AND MANAGEMENT NEEDS*

Protection methods for large breeding colonies in glaciated provinces need to be developed, especially with regard to heronries on private property. Easements, purchases, and public education need to be considered. To protect foraging habitat, wetland losses need to be curbed and buffers implemented both for large wetlands and riverine riparian areas.

*MONITORING AND RESEARCH NEEDS*

Existing survey methods, relying on volunteers, in addition to Pennsylvania Game Commision biologist field surveys, are adequate to monitor the breeding population of this large colonial breeder. Existing wetland survey methods used by the second *Atlas of Breeding Birds in Pennsylvania* should adequately quantify use of wetland habitats. Targeted riverine habitats can be surveyed by canoe. Effectiveness of conservation actions can be determined by nest count trends for implementation areas over a five-year period, Breeding Bird Survey data, or other heron encounter-rate data, such as canoe surveys and wetland surveys. Colony size or other identified habitat variables influencing colony longevity and growth can be used to alter conservation tactics where limited resources are available to protect important breeding areas.

Immediate research/survey needs for this species include a review of current population status and recent trends within the state. This information can be obtained by analyzing the existing nest count database (Pennsylvania Game Commission 2002) and by new field counts using standard counting techniques (Gibbs et al. 1988, Brucker 1992, Brauning and Siefken 2004, Dodd and Murphy 1995, Graham et al. 1996). Productivity estimates from new field data, in addition to trend inferences from these data, would help to identify problems in specific colonies or regions. The relationship between colony size and longevity should be examined using the existing database (Pennsylvania Game Commision 2002) to help in prioritizing colony protection measures. Other habitat-related factors can also be included in these analyses.

*Author:* ROBERT M. ROSS, UNITED STATES GEOLOGICAL SURVEY

## Sharp-shinned Hawk

Order: Falconiformes
Family: Accipitridae
*Accipiter striatus*

The sharp-shinned hawk is listed as a Species of Greatest Conservation Need because it is a rare breeder within the state and has shown apparent population declines in much of its range (fig. 5.102). Pennsylvania provides critical habitat for breeding, migrating, and wintering sharp-shins and may provide an important link from the prime nesting areas in the boreal forest to the more dispersed nesting population of the southern Appalachians. It is currently not considered Threatened or Endangered within the state. The sharp-shin is listed as a Species of Greatest Conservation Need

*Fig. 5.102.* The Sharp-shinned Hawk, *Accipiter striatus*. Photo courtesy of Shawn Carey.

in eight northeastern states. Partners in Flight ranks the sharp-shin as a species of High Regional Conservation Concern for the Appalachian Bird Conservation Region, which encompasses most of Pennsylvania. It is listed as globally Secure (G5, Natureserve 2009).

*GEOGRAPHIC RANGE*

The continental subspecies of the sharp-shin (*Accipiter striatus velox*) breeds throughout the boreal forest of Canada and northern United States and in eastern North America. Its range extends south within the eastern deciduous forest through the Appalachians into the northern region of the Gulf Coast states. Pennsylvania lies in the middle of the breeding range in the eastern United States and in the northern part of the wintering range (Bildstein and Meyer 2000). Eastern flyway wintering birds extend south from New England through the southern United States to Texas east to Florida and south into the Caribbean (Clark 1985, Goodrich and Smith 2008, N. Bolgiano, unpublished data). Pennsylvania may be a keystone state within the eastern range as it provides critical habitat for breeding, resident, wintering, and migrant birds. South of Pennsylvania nesting birds appear restricted primarily to the Appalachians (Bildstein and Meyer 2000).

*DISTRIBUTION AND RELATIVE ABUNDANCE IN PENNSYLVANIA*

Before 1900, sharp-shins were found regularly across the state (Goodrich 1992b). Persecution, as well as widespread forest cutting during the early 1900s, probably limited their range in Pennsylvania through midcentury. Today, nesting sharp-shins can be found throughout most of Pennsylvania although they are closely allied with higher elevations and the more forested regions of the state (Goodrich 1992b, McWilliams and Brauning 2000; fig. 5.103). Overall, the large home range size of this species results in this bird being rare across the state even where habitat is adequate. This species' preference for larger, undisturbed forests may preclude it from most urbanized or agricultural areas of the southeast, as well as regions near Pittsburgh where forests are highly fragmented (Goodrich et al. 2002). Prime sharp-shin habitat would lie in northern forests where larger patches of mixed conifer forest still exist.

Little information exists on the abundance of this species. The United States Fish and Wildlife Service Breeding Bird Survey does not monitor this species effectively (Sauer et al. 2004); however, the North

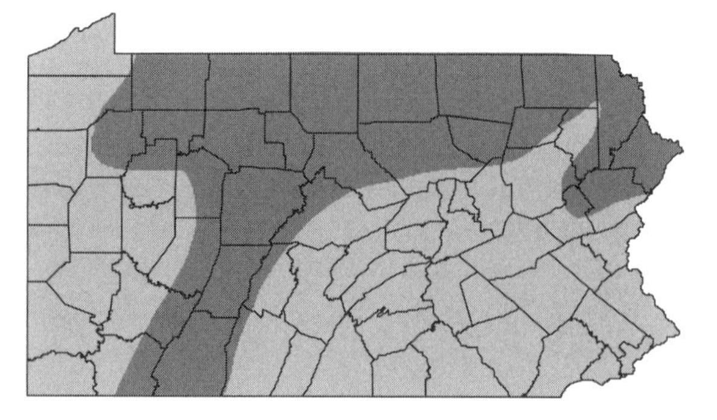

*Fig. 5.103.* Primary (darker shading) and secondary (lighter shading) distribution of the Sharp-shinned Hawk, *Accipiter striatus.*

American Landbird Conservation Plan estimates a global population of 1.1 million sharp-shins based on the Breeding Bird Survey data (Rich et al. 2004). Estimates on the breeding grounds are extremely suspect as all authors suggest regional population assessments are difficult because of the secretive nature of nesting birds and the widely dispersed distribution among nests (Bildstein and Meyer 2000). Migration counts at eastern watch sites plummeted in the late 1980s and early 1990s, particularly along the coast where juvenile birds concentrate (Panko 1990, Kerlinger 1992). Subsequent analyses of Christmas Bird Count data showed that wintering numbers were apparently increasing in northeastern states during the same period, and migration count declines could reflect an increase in "short-stopping" behavior (Duncan 1996, Viverette et al. 1996). Sightings from the Christmas Bird Count (www.audubon.org) and the Great Back Yard Bird Count (www.birdsource.org) imply that sharp-shins may winter in greater numbers in southeastern Pennsylvania, than in northern Pennsylvania and along the Coastal Plain to the east and south (Duncan 1996, Viverette et al. 1996)

The increasing abundance of bird feeders may be allowing greater numbers to winter in the northeastern states and eastern Canada as compared with previous years (Bolgiano 2005). Rebounds in migration counts during the late 1990s and thereafter, particularly at Cape May, suggest that at least some part of the observed decline may have been due to a real population change in northeastern birds (Bolgiano 2005). Wintering sharp-shins appear to be increasing or stable across the mid-Atlantic region (Bolgiano 1997). Recent sophisticated analyses of migration counts at seven eastern sites show no significant decline of –0.59 percent per year at Waggoner's Gap from 1990 to 2000 and a highly significant decline of –3.34 percent per year at

Hawk Mountain for the same period ($P < 0.01$; Farmer et al. 2008).

### COMMUNITY TYPE/HABITAT USE

Sharp-shinned hawks are widespread in mixed and coniferous northern forests in the breeding season, with many birds selecting larger forests blocks away from human habitation (Meyer 1987, Bildstein and Meyer 2000). The sharp-shinned hawk breeds primarily within conifer or mixed forest stands (Palmer 1988b, Wiggers and Kritz 1994, Trexel et al. 1999) but will also nest in a wide variety of deciduous stands (Platt 1976, Joy et al. 1994). Sharp-shinned hawks tend to nest in stands with higher densities of shorter and smaller trees and higher percentages of conifers than Cooper's hawks (Rosenfield et al. 1991, Trexel et al. 1999). In some studies, younger stands were selected presumably for their higher tree density (Joy 1990, Coleman and Bird 1999). In boreal forests, nests have been located near human activity areas in a large forest (Coleman and Bird 1999). However, in one Pennsylvania study, nests only were found in the continuous forests away from human activity areas, so further information on this aspect is needed. Nests were often placed within 140 m of a forest opening, possibly to facilitate foraging opportunities (Grimm and Yahner 1986).

Research on wintering and migration habitat use has been limited. Some birds may hunt more edge or suburban habitats during winter and have been noted to visit bird feeders regularly (Dunn and Tessaglia 1994). In migration, female sharp-shins apparently seek out larger forest patches in rural areas for resting and feeding en route (L. Goodrich, unpublished data). Birds migrating along the Kittatinny Ridge in Pennsylvania roosted in forest habitat exclusively, with juveniles roosting in the continuous forest of the Ridge for 25 percent of their roost nights. Females may be more apt to select large forests similar to nesting habitat, while the smaller males may be found in more transitional, suburban, or open habitats; however, few data are available on intraspecific differences (Meyer 1987).

### LIFE HISTORY AND ECOLOGY

Although highly secretive on nesting grounds, sharp-shinned hawks are regular and common migrants at migration watch sites continent-wide and are regularly observed on their nonbreeding grounds, particularly where bird feeders concentrate their prey (Davis 1992). Prey comprises small birds (>90%), particularly passerines, with an occasional insect or small mammal (Bildstein and Meyer 2000). The high dependence of boreal forest sharp-shins on songbirds has led to the suggestion that sharp-shin populations may cycle with spruce budworm specialist cycles (Bolgiano 2005). Banding return patterns suggest most birds do not survive beyond four to five years of age with a maximum-recorded age of thirteen (Keran 1981).

The sharp-shin is a partial migrant in Pennsylvania, thus nesting birds may include birds that migrate every year, birds that migrate some years, and birds that rarely migrate. In addition, Pennsylvania hosts birds that may winter within the state but nest farther to the north and birds that are in passage through the state on migration.

This species is presumed to be monogamous and pairs raise only one brood per year (Meyer 1987, Delannoy and Cruz 1988, Palmer 1988b). Some females may breed at one-year-old, but most will first breed at age two or older. In Pennsylvania, the sharp-shin builds its nest and lays eggs from May through mid-June and young fledge from July through September (McWilliams and Brauning 2000). Sharp-shinned hawks establish breeding territories with internest distances of 1–5 km (Reynolds and Wight 1978, Meyer 1987, Jacobs 1999).

### THREATS

Loss of habitat, declines in prey populations, collisions with stationary or moving objects, disturbance at nest sites, and shooting are some of the threats suggested for this species, although little is known about their regional influence on populations (Viverette et al. 1996, Bildstein and Meyer 2000). In Pennsylvania, loss or alteration of nesting habitat may be the largest threat. As forest fragmentation and increased residential sprawl continue statewide, the habitat for this species is increasingly reduced. Loss or rarity of undisturbed conifer stands or thicker forest types may also reduce nest site availability. An increase in the nonnative hemlock woolly adelgid (*Adelges tsugae*) and the associated decline of eastern hemlock (*Tsuga canaidensis*) populations also may reduce the availability of nesting sites in the southern counties. This species may benefit from forest cutting as it prefers thicker forests, although research on habitat use is limited.

Because sharp-shins prey predominately on songbirds, they are highly sensitive to bioaccumulation of environmental contaminants (Elliot and Martin 1994). They also have shown population-level declines in response to DDT use (Snyder et al. 1973, Bednarz et al.

1990) and may be sensitive to other pesticides such as organophosphates (Viverette et al. 1990). Because much of their diet comprises Neotropical migrant songbirds, sharp-shins will accumulate organochlorine compounds and other contaminants banned in North America because their primary prey winter in areas where such pesticides are still in use. Some contaminant load was evident in sharp-shins sampled in the mid-1990s (Elliot and Shutt 1993, Wood et al. 1996). Secondary poisoning from organophosphorus and carbamate compounds has been noted and could be a factor where forests are sprayed to control forest pests (Bildstein and Meyer 2000). Shooting and persecution may affect this species, although it is not suspected to have population-level effects (Bildstein and Meyer 2000). Bird band recoveries suggest a decrease of 7.5 percent to 3.7 percent of birds shot over the past thirty years (N. Bolgiano and L. Goodrich, unpublished data). Wintering or migrant sharp-shins are frequently victims of window strikes near bird feeders. Window strikes may be an increasing threat to this species as wintering birds frequent bird-feeding stations (Dunn and Tessaglia 1994) and as housing densities increase in some regions of Pennsylvania (Goodrich et al. 2002).

*CONSERVATION AND MANAGEMENT NEEDS*

Because the sharp-shin is sensitive to pesticide use, large-scale pest control programs should be limited within prime nesting areas. In addition, larger forests with conifer components could be targeted for protection from development to conserve habitats for this species and others (Myers et al. 2000, Goodrich et al. 2002). As forest fragmentation and alteration continues, conflicts with humans may increase and education on the role of predators in the ecosystem should be a continued statewide conservation priority. This species may do well in industrial forests that provide younger, thicker stands intermixed with older stands. Research on habitat use throughout the life cycle is needed to provide more informed conservation objectives. The lack of an effective monitoring program for forest-breeding raptors reduces the ability of state wildlife managers to conserve this species or to respond to possible declines and may be one of the most critical threats it currently faces across the region.

*MONITORING AND RESEARCH NEEDS*

Research is needed to determine the most effective and efficient method of monitoring Pennsylvania nesting populations. Current information suggests that nesting birds may be too difficult to monitor at the low densities at which they occur in Pennsylvania forests (Mosher et al. 1990), but other monitoring programs could be used effectively if additional research identifies that the source populations for wintering and migrating birds could be identified using band returns, stable isotopes, or other methods (Lott and Smith 2006, Goodrich and Smith 2008). Research using marked birds to define regions used by wintering Pennsylvania birds, as well as the propensity of Pennsylvania nesting birds to migrate or overwinter, is consistently needed. Winter and migration counts have been shown to be useful to derive regional trends (Bednarz et al. 1990, Bolgiano 2004, Farmer et al. 2008). Verification of state level trends from regional patterns still needs verification, but current analyses suggest it is possible (L. Goodrich personal observation).

Another monitoring goal would be to evaluate the use of breeding season surveys by volunteers to detect sharp-shins and other woodland raptors at a level that one could detect a population change (Rosenfield et al. 1988). As part of this evaluation, researchers would need to assess variations in detection probabilities and how the effort needed compared with other approaches of long-term monitoring. Breeding-season monitoring should be conducted using tape-call, the playback approach (Mosher et al. 1990) stratified by forest type across the state.

Additional research that will assist in the long-term conservation of this species includes assessing the nesting habitat requirements for this species (i.e. breeding forest size, territory and range size), the effect of fragmentation or suburbanization, mapping the breeding density or dispersion in different forest types across the state, and describing nest-site selection for birds in Pennsylvania and the importance of conifer forests to nesting birds. Because habitat use can change across life periods and locations, it is important to determine conservation threats beyond the nesting period. Defining the wintering range and migration pathways of Pennsylvania-nesting birds would allow better delineation of monitoring tools and identify potential threats to Pennsylvania birds. Assessing habitat use of this species during winter and migration periods and comparing males to females and adults to immature birds will help determine conservation needs during the nonbreeding periods. To assess contaminant levels and effects on fitness, researchers should monitor contaminant load in resident and migrant populations as an index to exposure of this species and its songbird prey. Periodic sampling of eggshell thickness and nest productivity for birds that are sampled for contaminant exposure could be used to assess how blood levels of contaminants relate to nesting success.

*Author:* LAURIE J. GOODRICH, HAWK MOUNTAIN SANCTUARY

## Red-shouldered Hawk

Order: Falconiformes
Family: Accipitridae
*Buteo lineatus*

The red-shouldered hawk was selected as a Species of Greatest Conservation Need because it is a rare breeder within the state and depends on extensive areas of riparian and bottomland forest (fig. 5.104). The red-shouldered hawk has been declining across its range throughout this century (Brauning 1992a, Partner In Flight 2004). Red-shouldered hawks are listed as a Species of Greatest Conservation need by eight northeastern states and are listed as Endangered by New Jersey. Global populations are considered Secure (G5, NatureServe 2009) .

### *GEOGRAPHIC RANGE*

The red-shouldered hawk occurs in North America, exclusive of the Rocky Mountain Region (Genoways and Brenner 1985). In the East, the red-shouldered hawk is found from the eastern edge of the Great Plains to the Atlantic Coast, with the largest populations in the Southeast, particularly on the Gulf Coast. From Maryland south through Florida, birds are not highly migratory and may be residents or move only a short distance (Wheeler 2003). A western population breeds west of Sierra Nevada and southwest Oregon to northern Baja California (Crocoll 1994). Only northernmost populations are migratory, and they winter from southern Wisconsin, Oklahoma, southern Ohio, and southern New England, south to the Gulf Coast and Mexico.

### *DISTRIBUTION AND RELATIVE ABUNDANCE IN PENNSYLVANIA*

In Pennsylvania, the red-shouldered hawk occurs as both a year-round resident and a migrant (Crocoll 1994). During the first Pennsylvania Breeding Bird Atlas in the late 1980s, individuals were recorded in 15 percent of the blocks across the state, although breeding was confirmed in only 3 percent of the blocks (Brauning 1992a). Statewide, the presence of this species is highly correlated with the proportion of total forest land, with the greatest concentration of breeding bird records coming from the Glaciated Section of northwest Pennsylvania and from the Allegheny High Plateau in the north-central portion of the state (fig. 5.105). These areas support large contiguous tracts of forest and provide relatively intact mature and lowland forest preferred by this species (Brauning 1992a). Although the present distribution of this species in the state is similar to what it was early in the century, low numbers exist throughout much of its historic range (Brauning 1992a).

### *COMMUNITY TYPE/HABITAT USE*

Throughout its range in North America and in Pennsylvania, breeding habitat for the red-shouldered hawk occurs in relatively extensive lowland, deciduous, or mixed forests, interspersed with small openings or marshes (Brauning 1992a). Hardwoods are preferred for nesting, which occurs in mature, contiguous hardwood or mixed hardwood forest (Kimmel and

*Fig. 5.104.* The Red-shouldered Hawk, *Buteo lineatus*. Photo courtesy of Bob Gress.

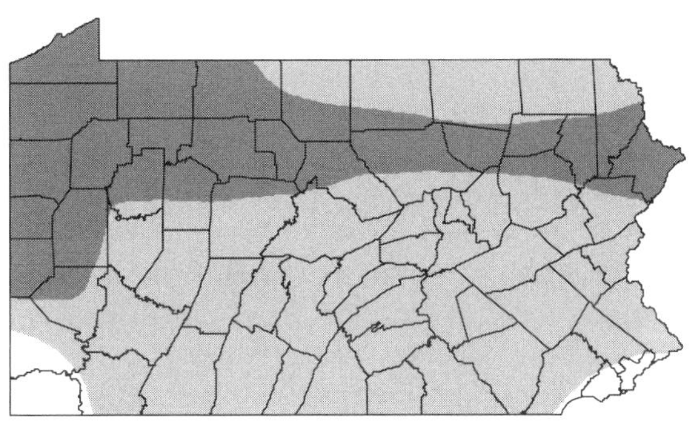

*Fig. 5.105.* Primary (darker shading) and secondary (lighter shading) distribution of the Red-shouldered Hawk, *Buteo lineatus*.

Fredrickson 1981, Titus and Mosher 1981) that typically has greater than 70 percent canopy closure (Titus and Mosher 1981, Titus 1984). Water is a critical element of red-shouldered hawk habitat, and breeding red-shouldered hawks are found in riparian and bottomland forests, as well as in upland forests adjacent to streams or other water sources (Crocoll 1994, U.S. Department of Agriculture-FS 2002). Nonbreeding habitat typically occurs in lowland areas near water (Palmer 1988a) and in level, open country (Bent 1937, Crocoll 1994).

*LIFE HISTORY AND ECOLOGY*

A species that prefers mature forest, the red-shouldered hawk is a highly territorial breeder (Cooper 1999), often using the same territory every year this hawk will refurbish an old nest or build a new one in the same area (Genoways and Brenner 1985, Crocoll 1994). They typically fly below the forest canopy, often gliding or swooping up to the nest. It hunts from a perch or flies low beneath the forest canopy and drops on its prey from close range (Palmer 1988a). Foraging also frequently occurs in open areas near water. The diet of the red-shouldered hawk consists of small mammals, frogs, crayfish, small reptiles, and birds, and large insects (U.S. Department of Agriculture-FS 2002), meadow voles and chipmunks are key prey in the East (Crocoll 1994). Nests are generally far from forest edges and occur in various deciduous trees and occasionally conifers (Palmer 1988a). Nesting areas are mostly found near some form of water (Titus and Mosher 1981, Crocoll and Parker 1989). Nests are often lined with bark, sprigs of evergreen, feathers, and down (Genoways and Brenner 1985, Crocoll 1994). Typical nest height is 11–15 m but can range from 1.5 m to 33.5 m (Ebbers 1989). The red-shouldered hawk frequently exhibits an aggressive response to disturbance, particularly from incubation through the nestling stage (Crocoll 1994).

Red-shouldered hawks are among the first hawks to return to their nesting grounds, which in Pennsylvania can occur in early to mid-March (Brauning 1992a). Eggs are usually laid in April or May (Genoways and Brenner 1985), and clutch-size ranges from one to six (Palmer 1988a); two to four eggs are the most common sizes throughout its range (Nature Conservancy 1999). The average incubation period is twenty-eight days (Brauning 1992a), and the nestling period lasts from five to six weeks (Crocoll and Parker 1989), with fledging occurring in June and July (Brauning 1992a).

*THREATS*

Habitat alteration and loss of habitat through development and forest management have been and probably will continue to be the greatest threat to viable populations of red-shouldered hawks in the state. Forest fragmentation can reduce habitat to unusable levels, which results in increased size and distance between territories or increased competition. However, disturbance-related effects vary somewhat based on landscape condition, as well as region, and some individuals have been reported to acclimate to human presence and achieve nesting success in urban landscapes (Crocoll 1994). Other threats to this species in the state include human disturbance, particularly during nest building and incubation, loss of wetlands, through development and draining (Partner in Flight 2003), and the introduction and spread of exotic insects and diseases, such as the hemlock woolly adelgid and beech bark disease complex, which can result in a long-term loss of forest cover (Partner in Flight 2003).

*CONSERVATION AND MANAGEMENT NEEDS*

Where known nests occur, they need to be protected. This should include protecting the nest site, as well as maintaining suitable habitat and landscape conditions preferred by this species within 0.5 miles (0.8 km) of the nest. Human disturbance and other activities within 0.5 miles (0.8 km) of active nests also need to be restricted during the nesting season (March 1 to August 15). The potential for recovery of this species in the state is good, but only if large tracks of mature forest are maintained. To provide habitat and to restore this species across Pennsylvania, private and public land managers will need to incorporate the needs of this species into forest and land management planning.

*MONITORING AND RESEARCH NEEDS*

Specific monitoring and research for this species is needed to accurately determine distribution and numbers across the state, to better identify home range size and movement patterns, to document landscape-level influences, to better assess effects from predation and competition, to better document migration patterns and winter habitat, to better assess nest productivity and success, and to evaluate existing management guidelines currently used to reduce effects from timber harvest and human disturbance. In addition, timber harvest adaptive management treatments that maintain or improve red-shouldered hawk habitat should be pursued and a landscape-level model that will help identify and facilitate management of red-shouldered hawk habitat should be developed.

*Authors:* SCOTT L. REITZ, USDA–FOREST SERVICE, ALLEGHENY NATIONAL FOREST; BRAD NELSON, USDA–FOREST SERVICE, ALLEGHENY NATIONAL FOREST

## Broad-winged Hawk

Order: Falconiformes
Family Accipitridae
*Buteo platypterus*

The broad-winged hawk is a small, crow-sized buteo (fig. 5.106). It was selected as a Species of Greatest Conservation Need because it has experienced declines within the core of its range and is sensitive to forest fragmentation. The state is an important migration corridor for broad-wings. Partners in Flight considers them a species of High Regional Concern and a Priority Species for Conservation Attention in Pennsylvania (Rosenberg 2004). Broad-wings are included on the Species of Greatest Conservation Need lists in seven northeastern states. Global populations are Secure (G5, NatureServe 2009).

### *GEOGRAPHIC RANGE*

Broad-wings breed in deciduous or mixed deciduous–coniferous forests from central Alberta, east across Canada to New Brunswick and Cape Breton Island, Nova Scotia, then south through east-central Texas and south and east along the Gulf Coast to northern Florida. Prime range occurs mainly east of the Mississippi River from Maryland north through southern Canada. Greatest densities occur in northern New York, Vermont, New Hampshire, and northward in spruce-hardwood forests (Robbins et al. 1986, Titus et al. 1989).

*Fig. 5.106.* The Broad-winged Hawk, *Buteo platypterus*. Photo courtesy of Vic Berardi.

Broad-wings winter commonly from southern Mexico, south through Middle America and South America to northern Peru and southern Brazil (Goodrich et al. 1996). They are regularly recorded in south Florida, Cuba, Haiti, Puerto Rico, and West Indies during winter months. A few recent records suggest some birds may linger into December in some northern states during mild weather. Pennsylvania harbored at least one broad-wing through early winter 2004 near Beltzville Lake, eastern Pennsylvania, and one in a prior winter (Miller 1994). Increasingly warmer winters due to climate change may allow greater numbers to winter farther north than found in the past.

### *DISTRIBUTION AND RELATIVE ABUNDANCE IN PENNSYLVANIA*

Broad-wings nest in larger forests throughout the state (McWilliams and Brauning 2000). They are likely rare within the highly fragmented forests of southeastern Pennsylvania or southwestern Pennsylvania, preferring forests of more than 40 acres, although larger patches may be needed where continuous forest is not found nearby (Grimm and Yahner 1986; L. Goodrich, personal observations). Pennsylvania provides abundant habitat for this species throughout the northern counties where forest remains extensive. The high proportion of fragmented forests throughout most of the southern part of the state probably limits nesting habitat, particularly around urban areas of Philadelphia and Pittsburgh (Goodrich et al. 2002). They prefer a mixed deciduous-conifer forest habitat and occur more commonly in northern counties where the northern hardwood forest predominates (Grimm and Yahner 1986, Goodrich et al. 2002). During the Pennsylvania Breeding Bird Atlas of the 1980s, broad-wings were sighted in every county, with the highest concentrations of confirmed nesting within the Pocono Plateau, Appalachians, and Alleghany Plateau regions (Senner and Goodrich 1992). In recent atlas efforts, the distribution appears similar (fig. 5.107).

The only estimate of the state population was derived by Partners in Flight using United States Fish and Wildlife Service Breeding Bird Survey data (Rosenberg 2004), resulting in an estimate of 36,000 birds in the state's forested habitat, with the largest numbers occurring in the Appalachian Mountains. The global population of broad-winged hawks had been estimated to

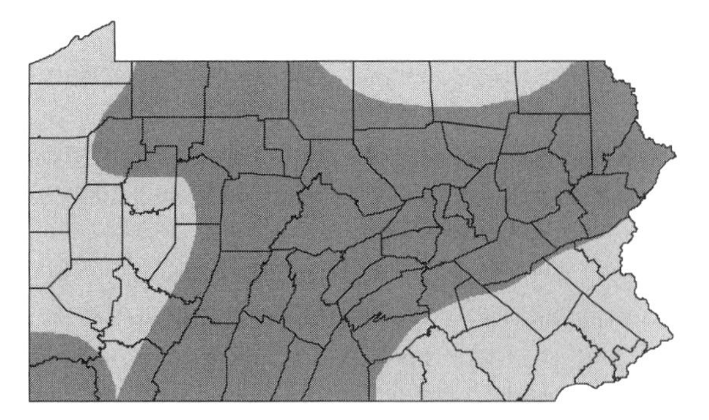

*Fig. 5.107.* Primary (darker shading) and secondary (lighter shading) distribution of the Broad-winged Hawk, *Buteo platypterus.*

exceed 1 million birds (Kirk and Hyslop 1998), but best estimates are derived from migration counts at Veracruz, Mexico, where biologists may count more than 90 percent of the global population of this complete migrant. From these counts, a global population estimate of 2 million to 3 million birds during postbreeding periods is most likely (Ruelas-Inzunza 2005). Viable breeders would number much fewer, as winter mortality can be high for young birds. A conservative global breeding population estimate might range from 1 to 1.75 million birds.

The broad-winged hawk was probably common in Penn's Woods before the 1800s but was considered rare by ornithologists during the early 1900s (Senner and Goodrich 1992). Widespread forest cutting in late 1800s and early 1900s undoubtedly reduced habitat for them until reforestation accelerated in the mid-1900s. Population trends appear complex, as regional populations may be exhibiting different trends. Analyses of Breeding Bird Survey data from 1966 through 1987 show a decrease at 1.1 percent per year in more-developed regions of northeastern United States (Titus et al. 1989). An analysis of Canadian Breeding Bird Survey trends from 1985 to 1994 reveal a significant decline of 3.33 percent (Kirk and Hyslop 1998).

Recent analyses of migration counts from Hawk Mountain and Waggoner's Gap, Pennsylvania, suggest a significant decline of 3.13 percent per year from 1974 to 2004 at Hawk Mountain in Pennsylvania and a nonsignificant long-term decline of 1.14 percent per year at Waggoner's. For the decade 1990 to 2000, broad-wing counts declined by 3.13 percent at Hawk Mountain where Waggoner's showed a significant 4.1 percent per year increase. Cape May, New Jersey, counts declined by a nonsignificant 1.42 percent per year for 1990 to 2000 while Lighthouse Point, Connecticut, counts declined significantly by 2.34 percent per year (Farmer et al. 2008).

Previous analyses have suggested that different subpopulations within the eastern flyway may use the coastal and mountain routes (Miller et al. 2002), and the difference in trends from the coast inland may represent different subpopulations showing different 1993trends. Waggoner's Gap likely draws birds from north-central Pennsylvania and Ontario, and Hawk Mountain may collect some northeastern Pennsylvania birds, as well as New England and eastern Canada birds. Cape May and Lighthouse Point may draw from a more easterly distribution. Further work is needed to identify the source populations for each count site, but the trends suggest some subpopulations may be declining. In most regions of Pennsylvania, the broad-wing is probably stable, but further monitoring is needed based on declines suggested by Pennsylvania migration counts.

*COMMUNITY TYPE/HABITAT USE*

Broad-wings nest in continuous or large deciduous or mixed-deciduous forests with openings and water source nearby (Goodrich et al. 1996). They often forage near small openings in the canopy (Crocoll 1984). Broad-wings will use younger forest stands than the red-shouldered hawk, but mature forests are commonly used as well (Titus and Mosher 1981). Some conifer component is preferred. No clear preference for nest tree species has been shown (Goodrich et al. 1996). They appear to avoid developed areas but may nest near dwellings on occasion and will forage along power-line corridors (Armstrong and Euler 1983). Broad-wings nest in deciduous or mixed deciduous–coniferous forest but place nests in a wide variety of trees (Goodrich et al. 1996). American chestnut (*Castanea dentata*) was formerly a typical nesting tree in the eastern United State and white (*Pinus strobus*) and red pine (*Pinus resinosa*) were used regularly in one Pennsylvania study (Grimm and Yahner 1986). Distance between broad-wing nests varies among regions with an average internest distance of 1,441 ± 331 m in New York (n = 11; Crocoll and Parker 1989), and in Wisconsin, a distance of 1,100 to 1,700 m (Rosenfield 1984). Nests in Pennsylvania appear more widely spaced (L. Goodrich, personal observation). Distance to openings and specific nest attributes vary considerably across its range.

Little is known about its habitat use on migration. Birds migrate in large flocks, and anecdotal observations suggest they seek out forests en route. During winter, broad-wings are seen using forest borders, coffee plantations, and both second-growth and primary forests. Observations in Veracruz, Mexico, show birds may roost several in one tree, where perch sites are limited or flock size is large (L. Goodrich, personal observation).

### *LIFE HISTORY AND ECOLOGY*

Broad-wings take a variety of food items with amphibians, insects, mammals, and juvenile birds the most common prey. Diet depends on local availability of prey. Overall, most prey captured are between 10 and 30 g body mass (Goodrich et al. 1996). Small mammals and amphibians are the most frequent prey and provide the greatest biomass in most studies (Errington and Breckenridge 1938, Fitch 1974, Matray 1974, Crocoll 1984). Their propensity for amphibians (mainly frogs and toads) may explain their association with water (Rusch and Doerr 1972, Mosher and Matray 1974, Crocoll 1984, Rosenfield et al. 1984). Continued declines in amphibian populations may affect this species in the future (Beebee and Griffiths 2005). A variety of invertebrates are taken by nesting birds (Goodrich et al. 1996).

Broad-wings are presumably monogamous, and at least one pair remained together longer than one year (Matray 1974). Nests are built by mid-May in Pennsylvania and surrounding states, and egg laying occurs in May (Goodrich et al. 1996). Hatching generally occurs in mid-June in the northeastern United States with fledging in July or early August (Goodrich et al. 1996). They have only one clutch per year but will replace a clutch if the first is destroyed (Burns 1911).

Most nest loss occurs in the egg stage and results from predation (Crocoll and Parker 1989). In western New York, predation accounted for half the loss of nests (Crocoll and Parker 1989). The great horned owl (*Bubo virginianus*) was responsible at most nests, including killing adults during the incubation phase of the nesting cycle. Other nest predators include the raccoon (*Procyon lotor*) and American crows (*Corvus brachyrhynchos*; Rosenfield 1984). In western New York, nest success was greater for a new nest than a rebuilt nest, greater for adult/adult pairs compared with adult/second-year bird pairs, greater in deciduous-mixed woodlands compared with conifer plantations, and greater for nests distant from woodland openings (Crocoll and Parker 1989).

Broad-wings usually do not breed until they are older than one year, but yearlings are reported breeding with adults on occasion (Burns 1911, Crocoll and Parker 1989). Adults probably attempt to breed every year (Crocoll and Parker 1989) but occasionally may not do so. In the only study in which adults were banded, one out of two pairs returned to nest within 400 m of the previous nest (Matray 1974).The broad-winged hawk is a complete migrant and leaves Pennsylvania by late September.

### *THREATS*

The increasing fragmentation of forests in Pennsylvania and the Northeast through human development may be increasing stress on nesting birds, although the extent of this effect is unclear. More than 50 percent of Pennsylvania forests are considered edge forest and would be unsuitable for nesting broad-wings (Goodrich et al. 2002). Nest predation and competition with other raptors can have a substantial effect on productivity in some areas (Crocoll 1984). Increasing populations of Cooper's hawks may limit nesting near forest edges.

Amphibians are an important component of the broad-wing diet. Global climate change and acidic deposition are some of the factors implicated in a widespread decline in amphibian populations, including species in the northeastern United States (Beebee and Griffiths 2005). The effects of this community-level change on broad-wing densities and productivity are unknown. They will feed on nestlings and small mammals; however, because a large proportion of their diet appears to be composed of amphibians, some effect on nest productivity is possible. Maintaining the health of all forest communities will benefit this sensitive pinnacle predator of the eastern deciduous forests.

Some threats for this long-distance migrant may occur outside of the state. Mortality on the first migration is probably quite high, given the extended flight and challenges faced en route. Preservation of forest patches along major flyways may be helpful for roosting. Starved or emaciated immature birds are found consistently in Panama during migration, suggesting that many birds may not survive their first migration (Senner and Fuller 1989). Deforestation trends in tropical America may increasingly limit available wintering habitat, but winter habitat-use studies are lacking.

Early in the century, shooting birds on migration and during breeding may have had significant effect on this species. Band recovery data suggest shooting on wintering range continues to affect this species (Robbins 1986). However, proportion of banded broad-wings recovered as shot in Latin America dropped from 100 percent in the 1950s to 71 percent in the 1970s (n = 38; Goodrich et al 1996). The tendency of this species to flock in large numbers on migration makes it particularly vulnerable to potential population effects from shooting during migration. One of the largest threats to this species may be the lack of knowledge about its health and abundance within the state and region.

### *CONSERVATION AND MANAGEMENT NEEDS*

Because the larger forest tracts of northern Pennsylvania counties may serve as a source population regionally, efforts should be made to limit fragmentation where possible. The Partners in Flight priorities for Pennsylvania recommend maintaining current populations of broad-wings. Because of the development patterns in some regions, such as the Poconos, populations may be unavoidably reduced, suggesting a need to expand other populations. Effects of fragmentation and disturbance on broad-wings should be considered in timber management plans, particularly on state lands. Short timber rotations of less than forty years may be inadequate to maintain this species as a breeder (Mitchell and Millsap 1990). Maintaining continuous forest habitat across the state should also limit interactions with potential predators, such as red-tailed hawks and great horned owls, and with nest predators, such as raccoons. Streams or wetlands need to be maintained in the vicinity of nesting territories where they may serve as important foraging sites (Keran 1978). Although shooting probably is not having a large effect on current populations, enhanced efforts in public education within and outside of the United States can only improve population sustainability in the future.

### *RESEARCH AND MONITORING NEEDS*

The immediate research and survey needs for the broad-winged hawk include determining their abundance and density during nesting season within different forest types across the state, determining their home range size and critical habitat components in different regions of the state and assessing their tolerance for forest fragmentation and human development near nests. Research on nest productivity, population viability, and turnover among different forest types could be useful. A GIS-based landscape study comparing land cover data and degree of fragmentation in relation to broad-wing nesting distribution and densities could be informative in determining their level of sensitivity to forest loss and assisting managers in targeting areas to conserve. In addition, knowledge of how additional forest fragmentation may affect their metapopulation and individual nest success will inform the need for further conservation attention.

The Partners in Flight prioritization process noted that Atlantic Coast populations of Pennsylvania broad-wings need conservation attention. Special priority should be given to survey broad-wings in this region and to determine abundance, distribution, and any conservation threats. An intensive study on productivity and foraging patterns would be useful in determining long-term health of this subpopulation in the state. Periodic monitoring of this population may be warranted to ensure long-term viability.

Research also is needed to design and validate population survey methods for this and other woodland raptors, particularly in light of possible regional declines. Nesting season surveys stratified by forest type and conducted across the state regularly, in combination with annual migration surveys, would provide an effective monitoring program. If the source populations for migration count sites can be defined, migration counts can be a low-cost option for state-level, long-term monitoring, and the need for nesting season surveys could be eliminated.

Research on the foraging patterns and diet of broad-wings nesting in Pennsylvania could be useful in determining whether prey number or distribution, such as amphibians, could be limiting populations in some regions. Identifying prey and quantifying abundance and relative importance from incubation period through fledging would elucidate preferences in Pennsylvania. A comparison of nests from different regions or forest communities also should occur as birds may cue on different prey regionally and different pressures may limit prey in some regions.

Because broad-wings are long-distance migrants, threats outside of Pennsylvania and outside North America also could limit Pennsylvania populations. Research is needed to better define the migration routes and wintering locations of broad-wings originating in Pennsylvania. Analyses of banding recoveries and satellite-marked birds could be useful in defining potential threats and would be useful in developing effective monitoring programs for Pennsylvania birds. If migration routes can be mapped, certain migration count sites may prove more useful for monitoring than others. Such mapping also can guide conservation action for migration roosting habitat, as well as for wintering areas.

Research on the importance of Pennsylvania as a source population for potential nesting birds in neighboring states, such as New Jersey, Delaware, and Ohio, could assist in understanding the extent of metapopulation interactions. Studies of dispersal distance and site fidelity for young broad-wings could be useful for future conservation. Following radio- and color-marked individuals for a full annual cycle could provide some important population information, and a study of marked nesting pairs monitored over many years in different regions of the state would provide some important insight into population stability.

*Author:* LAURIE J. GOODRICH, HAWK MOUNTAIN SANCTUARY ASSOCIATION

## Sora

Order: Gruiformes
Family: Rallidae
*Porzana carolina*

The sora is a medium-sized rail with a yellow bill and black mask on its face (fig. 5.108). It was selected as a Species of Greatest Conservation Need because it is a rare breeder within the state and is perceived to have declined in numbers. The sora is a migratory game bird in Pennsylvania. The Pennsylvania Biological Survey ranks them as Candidate-Rare. Soras are also listed as a Species of Greatest Conservation Need by six other Northeastern states. Global populations are considered Secure (G5, NatureServe 2009).

### *GEOGRAPHIC RANGE*

Soras breed locally in North America in suitable wetland habitat. Soras are migrants breeding across Canada and the northern United States and wintering in costal areas, the southern United States south to Central America and northern South America. Pennsylvania is on the southern limit of the breeding range (Melvin and Gibbs 1996, McWilliams and Brauning 2000). The sora is the most abundant and widely distributed rail in North America (Melvin and Gibbs 1996).

Fig. 5.108. The Sora, *Porzana carolina*. Image courtesy of Elaine R. Wilson, naturespicsonline.com.

### *DISTRIBUTION AND RELATIVE ABUNDANCE IN PENNSYLVANIA*

The sora is a rare and local nester in suitable habitat throughout most of Pennsylvania. It is confined to fairly extensive marshes. Similar to most wetland-associates, soras have declined in number. They were formerly considered abundant in suitable habitat throughout most of the state. For example, A. Poole (unpublished manuscript) reported they were abundant in marshes of the lower Delaware River. Harlow (1913) described the sora as common in suitable habitat in the northern half of the state but rare in southern sections. They were considered to be more common than the Virginia rail in the north-central part of the state and abundant in Centre County wetlands (Harlow 1912, Burleigh 1931). Todd (1940) considered the sora to be fairly common in western-tier counties from Erie south through Crawford, Mercer, and Lawrence. By the beginning of the first Breeding Bird Atlas (1983), the sora was an uncommon breeder. Several regional coordinators for the atlas reported that the sora was much more common before the atlas (Brauning 1992b). Unfortunately, we lack adequate data to quantify these declines. Breeding bird survey trend data are not available for Pennsylvania because of low numbers and the inadequacy of Breeding Bird Survey routes for monitoring many secretive wetland species.

During the first Pennsylvania Breeding Bird Atlas (1983–1989), soras were reported in eighty-eight blocks, about 2 percent of the total and some of those in only one year. Almost half of the reports were from the glaciated northwest and northeast counties. Sora breeding activity is rarely reported at Pennsylvania Society for Ornithology Special Area Project locations, despite the emphasis on wetlands. They are rarely reported in Pennsylvania Birds accounts (D. A. Gross, personal communication). There were ninety-one sora reports to the second Pennsylvania Breeding Bird Atlas. Of those, eleven were confirmed and forty-three probable. Their greatest probability of occurrence continues to be in the northwest corner of the state (fig. 5.109).

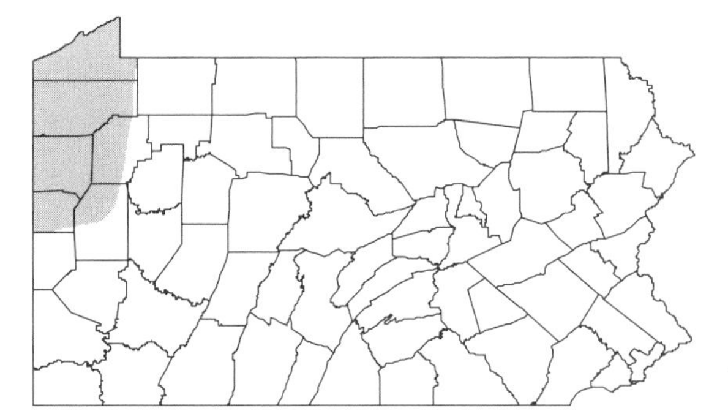

*Fig. 5.109.* Highest probability of occurrence of the Sora, *Porzana carolina*.

### *COMMUNITY TYPE/HABITAT USE*

Soras are found in shallow (40 cm) to intermediate-depth wetlands dominated by robust and fine-leaved emergents (e.g., cattails [*Typha* spp.], sedges [*Carex* spp., *Cyprus* spp.], bulrushes [*Scirpus* spp.], and bur reeds [*Sparganium* spp.]) interspersed with mud flats and open water with floating or submerged residual vegetation (Melvin and Gibbs 1996, McWilliams and Brauning 2000, Zimmerman et al. 2003). Soras occasionally use wet grassy fields and streamside areas (McWilliams and Brauning 2000). Soras do not appear to be highly area sensitive and have been recorded during the breeding season on wetlands <5 ha in size (Cashen and Brittingham 1998). However, it has been suggested that soras prefer more extensive marshes than Virginia rails. Soras are migratory birds and arrive in Pennsylvania as early as mid-March, with the majority arriving in mid- to late April (McWilliams and Brauning 2000). Soras feed primarily on seeds of wetland plants and secondarily on invertebrates (Melvin and Gibbs 1996). Soras are frequently found in the same habitat as Virginia rails (*Rallus limicola*) but are not thought to compete much due to a difference in feeding habits. Soras feed primarily on seeds and Virginia rails feed primarily on insects and other invertebrates (Horak 1970). Breeding soras leave Pennsylvania in late September or early October (McWilliams and Brauning 2000). In Pennsylvania, soras use similar habitats during migration as during the breeding season.

### *LIFE HISTORY AND ECOLOGY*

Soras nest in a variety of freshwater wetlands, bogs, and wet meadows, but their preferred habitat is shallow wetlands with emergent vegetation interspersed with open water and mud flats. Soras are monogamous, and pairs form soon after arrival on the breeding ground (Ehrlich et al. 1988, Melvin and Gibbs 1996). Both sexes build the nest, which is a loosely woven basket of emergent vegetation placed in clumps, with the base of the nest at or slightly below the water surface (Ehrlich et al. 1988, Melvin and Gibbs 1996). Soras may build many dummy nests close to the active nest. These serve as feeding and resting platforms (Melvin and Gibbs 1996). In Pennsylvania, nests with eggs have been found from the first week of May through mid-June (Brauning 1992b, McWilliams and Brauning 2000). Newly hatched chicks are fed entirely by their parents for the first couple of days. At two to three days post-hatching, the chicks begin to feed on their own but are still fed by their parents for two to three weeks. Juveniles become independent of their parents when they reach one month old (Melvin and Gibbs 1996). In Iowa, brood-rearing home ranges average 0.99 ha (Johnson and Dinsmore 1985). Soras raise one or possibly two broods per year. It is unclear whether second broods only occur if the first nest fails or whether soras ever raise two broods to independence (Melvin and Gibbs 1996, Zimmerman et al. 2003). Nest success, measured as the proportion of nests where at least one egg hatched, ranged from 0.58 to 0.83 (Melvin and Gibbs 1996). There is no information on age of first breeding, juvenile, or brood survival (Melvin and Gibbs 1996).

### *THREATS*

For soras, as for many other species associated with wetland habitat, the primary threat is loss and degradation of emergent wetland habitat. In Pennsylvania, more than 50 percent of historic wetlands have been lost, and many of the remaining areas are degraded (Goodrich et al. 2002). Loss of emergent wetlands in Pennsylvania exceeds both the regional and national average (Tiner 1990, Goodrich et al. 2002). From 1982 to 1992, 91 percent of wetland loss in the Northeast was due to development (Goodrich et al. 2002). Degradation of wetland habitat by runoff, pollution, and acidic deposition are problems, but there is no research looking specifically at how these pollutants directly or indirectly affect soras. Loss of emergent wetlands from conversion to lakes, ponds, and reservoirs has been a problem in the past and probably continues to be so (Tiner 1990, Goodrich et al. 2002). Effects of hunting on soras are presumed to be low, but because we lack adequate data on numbers harvested, this is really an unknown.

### *CONSERVATION AND MANAGEMENT NEEDS*

Soras are a game species legally hunted in thirty-one states and two Canadian provinces (Melvin and Gibbs

1996). In Pennsylvania, the daily limit is twenty-five birds and the season runs from September 1 through early November. The harvest in Pennsylvania is presumed to be negligible due to low hunter interest (J. Dunn, personal communication). However, data from the National Harvest Information Program (HIP) averaged over the years from 1994 to 2003 estimates that as many as 250 soras are harvested annually in Pennsylvania (P. I. Padding, Harvest Survey Section, U.S. Fish and Wildlife Service, personal communication).

A primary conservation and management goal is to minimize loss and degradation of currently available emergent wetland habitat and to increase the amount available through habitat restoration. Wetlands should be managed to provide shallow (<40 cm) to intermediate depths dominated by robust and fine-leaved emergents and interspersed with mud flats. Outside of Pennsylvania, preservation of emergent wetlands along the migration route and within the wintering range is also needed. We need better information on hunter effort and harvest. In addition, bag limits for soras should be reviewed to ensure annual harvest is within sustainable levels.

*MONITORING AND RESEARCH NEEDS*

A major issue with managing soras is the lack of adequate data on their distribution and abundance. Multispecies marsh bird survey protocols have been developed to help agencies design a standardized monitoring program (Conway 2004). These surveys are designed to monitor both the birds and the habitat. The Pennsylvania Game Commission and other public agencies with wetland habitat on their land should consider using this survey protocol.

In Pennsylvania, research needs include determining the abundance and distribution of soras in natural and restored wetlands across Pennsylvania. Because soras are not adequately monitored by the Breeding Bird Survey, we have little information on their abundance and distribution. In addition, we need to develop methodology to better estimate the numbers of soras harvested annually in Pennsylvania. Current estimates range from close to zero (J. Dunn, Pennsylvania Game Commission) to approximately 250 (P. I. Padding, U.S. Fish and Wildlife Service). We need to determine effects of habitat management on abundance and distribution and effects of habitat quality, predation, weather, and water-level fluctuations on reproductive success and breeding densities of soras (Melvin and Gibbs 1996). In addition, studies are needed to characterize site fidelity, annual survivorship, life span, and age of first breeding (Melvin and Gibbs 1996).

*Author:* MARGARET C. BRITTINGHAM, PENNSYLVANIA STATE UNIVERSITY

## Common Moorhen

Order: Gruiformes
Family: Rallidae
*Gallinula chloropus*

The common moorhen is one of the most aquatic and easily seen members of the rail family (fig. 5.110). It was selected as a Species of Maintenance Concern because of perceived declines in numbers of breeders and its dependence on large wetland complexes for breeding. It is a globally Secure species (G5), but the breeding population in Pennsylvania is listed as Vulnerable (S3B, NatureServe 2009). It is included on eight northeastern state lists of Species of Greatest Conservation Need and is listed as Threatened in Maine and Endangered in Connecticut.

*GEOGRAPHIC RANGE*

In eastern North America, common moorhens breed around the Great Lakes and in coastal areas from New England south into Mexico. Western populations are resident in Baja California and south through Mexico. The species also occurs throughout South America, Africa, and across the Pacific to Australia (Blake 1977). A disjunct subspecies occurs in Hawaii. Populations show considerable capacity to exploit wetland habitat as it becomes available. Inland populations are disjunct and reflect availability of large emergent wetland complexes, many of which have formed as a result of lake and pond construction by man.

*Fig. 5.110.* The Common Moorhen, *Gallinula chloropus.* Image courtesy of Alan D. Wilson, naturespicsonline.com.

### *DISTRIBUTION AND RELATIVE ABUNDANCE IN PENNSYLVANIA*

Pennsylvania's population appears to be peripheral to the Great Lakes population with disjunct colonies throughout the state associated with wetlands, which have developed along the margins of man-made lakes. Although considered locally common in suitable habitat (Leberman 1992c), most of the breeding records are from the Delaware River marshes in the southeastern corner of the state or the more extensive wetland complexes in the northwestern corner of the state (fig. 5.111). Elsewhere, the common moorhen's appearance is much less predictable. Results of the second Pennsylvania Breeding Bird Atlas indicate a decrease of 40 percent in distribution has occurred since the first atlas (Mulvihill and Brauning 2009). Before early 1900, when it was found nesting near Philadelphia (Miller 1946), there were no breeding records in the state. During the winter, the population migrates south, presumably to wetlands along the southern Atlantic and Gulf Coasts.

### *COMMUNITY TYPE/HABITAT USE*

Common moorhens in Pennsylvania prefer dense emergent marshes with adjacent deep water or interspersed channels of open water (Leberman 1992c). Cattail marshes are commonly used by moorhens for nesting, but a wide variety of other plants are often present, including phragmites and purple loosestrife (Brackney and Bookhout 1982, Kane 2001). Moorhens often forage among floating-leaved aquatics, but nests are more frequently placed in dense cattails or other emergent herbaceous vegetation (Bent 1926, Fredrickson 1971, Post and Seals 2000, Bannor and Kiviat 2002). Depth of water beneath nests averaged 40 cm in Ohio (Brackney and Bookhout 1982). Nests are typically built of plant materials from the immediate vicinity of the nest site (Helm et al. 1987). During the winter, a variety of emergent palustrine wetland vegetation types, ponds, streams, and lakes are used by birds in the southeastern United States.

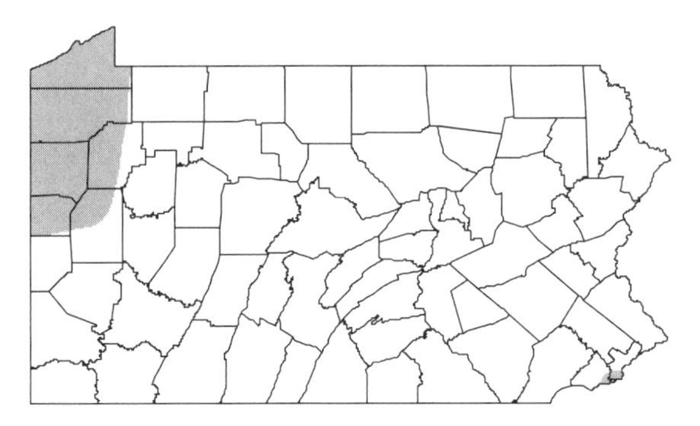

*Fig. 5.111.* Highest probability of occurrence of the Common Moorhen, *Gallinula chloropus*.

### *LIFE HISTORY AND ECOLOGY*

All populations of common moorhens in Pennsylvania are migratory, wintering in the Gulf Coast and returning to breed in April or May of each year (McWilliams and Brauning 2000). Breeding birds pair for the season and are highly territorial in the immediate vicinity of the nest. Pairs are vocal, but the context in which vocalizations are used is poorly understood (Bannor and Kiviat 2002).

Common moorhens forage on plant material and animal life while walking on floating plants or while swimming along the margins of emergent vegetation. Although their feet are not webbed or lobed, moorhens are capable swimmers and often dive for submerged plant material or to escape danger (Bannor and Kiviat 2002).

The peak of nesting occurs in the Northeast during the latter half of May. Nests are often placed over water, using dead reeds, cattails, or other firm substrates (e.g., logs, muskrat houses) to support the small structure (Fredrickson 1971). Trial nests and brooding platforms are commonly constructed (Helm et al. 1987). Nests are placed in emergent wetlands within a few meters of open water, often over water 40 to 50 cm in depth (Bannor and Kiviat 2002). Renesting occurs rapidly if the nest is depredated. Brood parasitism and dump nesting may occur (McRae 1996). Some pairs in Pennsylvania may be double brooded (Miller 1946).

Both parents share the three-week incubation chore, with the larger male doing most of the incubation at night when potential for heat loss may be greater (Bannor and Kiviat 2002). Chicks depend on the adults for three-and-a-half weeks, although they may leave the nest within twenty-four hours of hatching. Secondary satellite, or false ("sham"), nests may be used for brooding the young and for sleeping (Wood 1974). Juveniles may remain with their parents into the fall (Kibbe 1985a). Young common moorhens eat plant and insect material and fledge about forty days after hatching (Fredrickson 1971, McRae 1996, Post and Seals 2000). Survival rates and fidelity of young to natal areas are unstudied in our region.

### *THREATS*

The marshes in the vicinity of Tinicum were one of the former strongholds of the common moorhen

(Miller 1946). Breeding populations there apparently are much lower now than they were several decades ago. This area has been overrun with phragmites. Invasion of traditional nesting areas by phragmites and purple loosestrife is widely recognized as a concern (Bannor and Kiviat 2002); however, it has not been demonstrated that these plant species are less acceptable nesting habitat than more traditional cattail marshes (Kane 2001). Further research on this issue is desired. Human disturbance has also affected these marshes and all the reasons for the common moorhen's apparent decline have yet to be resolved. The major marsh complexes in northwestern Pennsylvania are less degraded; however, human disturbance and invasive plant species are issues of concern there also. Principal threats to the common moorhen appear to be invasion of emergent wetlands by phragmites and purple loosestrife and disturbance of nest sites by Canada geese, recreational boating, and fishing activities. The relative affect of these issues is unresolved at present.

*CONSERVATION AND MANAGEMENT NEEDS*

Common moorhens are federally regulated game birds. Although bag limits remain liberal in some regions (e.g., eighteen per day in California, fifteen per day in Florida, and ten per day in New Jersey), only three may be taken a day in Pennsylvania, where the harvest of rails, moorhens, and coots is considered "negligible" (Fergus 2004). Hunting pressure in Pennsylvania has always been light but may be more intense on the birds' wintering grounds. Unlike ducks and geese, common moorhens are seldom trapped and banded and harvest information is largely conjecture (Greij 1994).

The possible effect of hunting on this species is unknown at present. Greij (1994) suggested that hunting pressure on the wintering grounds may be a factor in the species decline; however, too little is known about the origin of the birds to draw conclusions about the effect it may have on breeding populations to the north. Because most major marsh systems are in public ownership, it is imperative that survey efforts be coordinated through the responsible agencies charged with managing the respective wetlands. Managers need to (1) identify principal breeding areas, (2) learn the relative population size in each area, and (3) determine whether disturbance and habitat change are affecting the species reproductive success.

*MONITORING AND RESEARCH NEEDS*

An inventory of major wetland complexes with a standardized survey approach similar to those presented by Crewe et al. (2005) and Conway (2005) to ascertain relative abundance and distribution is needed. Such surveys need to be continued on a regular basis to track changes in abundance and wetland usage, which may translate into impact assessments. At present the species tolerance of vegetation changes and disturbance levels are poorly understood.

Tape-playback surveys are recommended to obtain better data on the distribution and relative abundance of common moorhens (Conway 2005, Conway and Gibbs 2005, Conway and Nadeau 2006). A standardized approach needs to be selected and implemented in the major wetland to determine the current status of this (and other) wetland bird species. Survey tapes and a set protocol establishing length of survey, timing (diurnal or nocturnal), seasonal, and weather constraints need to be developed. Type of access (bank vs. boat) and duration of survey also need to be established. After a survey protocol is established, potential marsh systems need to be identified and prioritized for surveying. Surveys need to be continued regularly to track changes in abundance and wetland usage by community type, which may translate into impact assessments. At present, the species tolerance of vegetation changes and disturbance levels are poorly understood. The surveys, once implemented, will also serve as a basis for further investigations into factors, which may influence habitat choice, reproductive success, and population density. Additional study of the possible interaction of moorhen and Canada geese populations is warranted. As the Pennsylvania breeding goose population increases, opportunities to assess the effect of interspecific competition on other waterbirds may shed light on reasons for declines in some populations.

*Author:* DOUGLAS P. KIBBE, URS CORPORATION

## American Coot

Order: Gruiformes
Family: Rallidae
*Fulica americana*

The American coot is a hen-sized marsh bird in the rail family (fig. 5.112). It was selected as a Species of Greatest Conservation Need because it is a rare breeder in the state and is an obligate wetland species. State breeding populations are considered Vulnerable. Global populations are Secure (G5, NatureServe 2009).

*Fig. 5.112.* The American Coot, *Fulica americana*. Photo by Jacob W. Dingel III.

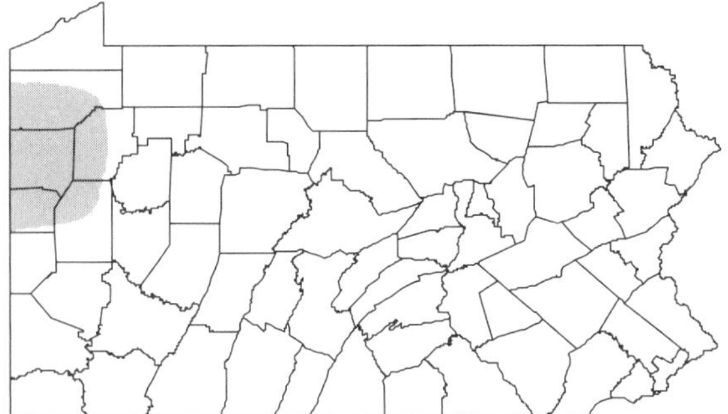

*Fig. 5.113.* Highest probability of occurrence of the American Coot, *Fulica americana*.

### *GEOGRAPHIC RANGE*

American coots can be found from Alaska and Canada to as far south as northern South America (Brisbin and Mowbray 2002). Migrant birds breed primarily in central Canada and the northern United States and winter in the eastern United States through South America and the West Indies. Residents can be found in temperate to tropical regions in the Pacific Northwest and the western United States south through Mexico and the West Indies. American coots include Pennsylvania in the southern edge of their breeding range and are most common during migration in the spring and especially the fall.

### *DISTRIBUTION AND RELATIVE ABUNDANCE IN PENNSYLVANIA*

The American coot is an uncommon to rare summer breeder and winter resident in Pennsylvania. Breeding birds have been documented during the summer on the western border of the state south of Lake Erie (fig. 5.113). Rare breeding residents occur at the Pymatuning Reservoir in Crawford and Mercer counties (Rappole 2002) and wetlands in Butler County. Regular breeding has been documented at Glen Morgan Lake in Berks County since 1955, and since 1960 in Centre, Erie, Lawrence, and Snyder counties (McWilliams and Brauning 2000). Birds in Crawford County nest mainly at Conneaut Marsh. Winter visitors prefer open lakes and slow rivers and are most common in the southeastern corner near the coast. In the winter, birds can be found swimming in large rafts of hundreds of birds on open water (Fergus 2004). During migration, American coots are common to abundant residents and prefer to feed in the open (McWilliams and Brauning 2000). Migrants are most abundant from late March to early April and during the last two weeks of September. Presque Isle has the greatest number of coots observed during migration (McWilliams and Brauning 2000). During 2005, the Mid-winter Waterfowl Survey (MWS) estimated 75 birds in Pennsylvania and 145,951 birds in the Atlantic Flyway (Serie and Raftovich 2005).

Nesting in Pennsylvania has been recorded since the late 1800s (Warren 1890) with regular breeding at Pymatuning Lake after flooding provided suitable habitat in 1932. By the end of the 1930s, coot numbers declined with the marsh habitat (McWilliams and Brauning 2000). Population trend data from Breeding Bird Survey routes are not available for Pennsylvania because of the low number of coots recorded on routes.

As of 1999, total numbers for the American coot were estimated to be around 3 million. In 2004, there were 400 ± 169 percent coots harvested in Pennsylvania and 100 ± 83 percent active coot hunters. Harvests for 2004 for the Atlantic Flyway were 13,100 ± 66 percent and 50,500 ± 66 percent nationally (U.S. Fish and Wildlife Service 2005), and most coots are shot incidentally by duck hunters (Alisauskas and Arnold 1994). Breeding Bird Survey trends indicate a slight increase from 1966 to 2003 in the United States, and a slight decrease in Canada during the same years. Since the 1970s, the total population numbers have been relatively stable yet fluctuate from year to year with precipitation (Sauer et al. 2004).

### *COMMUNITY TYPE/HABITAT USE*

American coots are found in a variety of freshwater wetlands. Coots require dense emergent aquatic vegetation on at least part of the shoreline during breeding, as well as standing water within the vegetation during nesting and natal development. Vegetation in breeding habitat commonly includes cattails (*Typha* spp.) and bulrush (*Scirpus acutus*) and sometimes reeds

(*Phragmites* spp.), sedges (*Carex* spp.), willows (*Salix* spp.), and grasses (*Poaceae*). Both permanent and seasonal wetlands can be used. The highest-breeding densities were observed at well-flooded, semipermanent wetlands with a mosaic of open-water and emergent vegetation. Any water body with sufficient vegetation and standing water within the vegetation can be used for breeding, including lakes, ponds, slow-moving rivers, prairie potholes, and swamps (Brisbin and Mowbray 2002).

Migrants exhibit broader habitat choices, including saline and brackish water, which breeding birds tend to avoid but winter residents prefer. Migrants also use coastal marine and estuarine habitats, as well as larger open-water lakes, which may not have emergent vegetation that is crucial for breeding. Winter residents use breeding and migratory habitats and more commonly choose coastal, brackish, and marine habitats and larger inland freshwater bodies. Terrestrial winter habitat includes agricultural areas, fields, and golf courses for grazing (Brisbin and Mowbray 2002).

*LIFE HISTORY AND ECOLOGY*

American coots prefer to nest in freshwater wetlands with a mosaic of open-water and emergent vegetation. They feed mainly on aquatic plants and algae by pecking, dabbling, and diving. Other food sources include terrestrial plants and grains; aquatic invertebrates, such as mollusks, crustaceans, and insects; and vertebrates, such as small fish and tadpoles.

Migration occurs in the spring from late February to mid-May. Most American coots in Pennsylvania are migrants mostly observed during March–May and August–October, but breeding populations exist. Pairs form after arriving on the breeding grounds. The pair establishes a territory, which is actively defended by both sexes, and begins nesting along the shore as soon as aquatic plants provide sufficient nesting substrate (Brauning 1992a). Coots are weakly philopatric to previous breeding areas (Alisauskas and Arnold 1994).

Coots generally have one or two clutches during the breeding season, but up to four clutches can be laid following clutch or brood loss. Productivity increases with age, especially between one and two years (Alisauskas and Arnold 1994). Average annual survivorship is 45 percent with the oldest-banded bird at least twenty-two years of age (Brisbin and Mowbray 2002). Fall migration occurs from late August to December, and some coots show winter site fidelity (Alisauskas and Arnold 1994).

*THREATS*

The primary threat to the American coot in Pennsylvania is loss and degradation of emergent wetland habitat. Because coots prefer semipermanent wetlands, such losses have had a greater effect on dabbling ducks, which prefer seasonal and temporary wetlands (Alisauskas and Arnold 1994, Brisbin and Mowbray 2004).

*CONSERVATION AND MANAGEMENT NEEDS*

American coots are federally protected game birds and are probably underharvested with the exception of a few states and are hunted at a sustainable level overall in North America (Brisbin and Mowbray 2002). During the 2004–2005 season in Pennsylvania, bag limits were set at fifteen coots a day with a maximum of thirty in possession. The primary conservation need is protection and enhancement of large wetlands that are used for breeders. Management should focus on conservation of current emergent wetlands and habitat restoration. Coots are known for their ability to pioneer new habitats and respond well to wetland restoration efforts in drained areas (Alisauskas and Arnold 1994). Flooding, mowing, burning, and herbicides can help to create mosaic patterns of open-water and emergent vegetation and have successfully increased coot numbers in North Dakota and Texas (Brisbin and Mowbray 2002). Habitat restoration projects should focus on areas with minimal human disturbances.

*MONITORING AND RESEARCH NEEDS*

There is a major lack of data on migratory patterns and environmental factors that influence the timing of migration. The Breeding Bird Survey is the only survey that provides an index of geographic patterns of abundance of breeding coots in North America, but this survey may be inadequate where coots are distributed in a few large breeding areas (Alisauskas and Arnold 1994). Because there is no data for Pennsylvania from the Breeding Bird Survey, a different monitoring protocol is needed. The United States Fish and Wildlife Service's annual Breeding Ground Surveys use a combination of aerial and ground surveys in the important midcontinent breeding area to estimate relative abundance (Brisbin and Mowbray 2002); however, the efficiency of this survey in estimating changes in coot abundance has not been addressed (Alisauskas and Arnold 1994).

Research needs include developing an effective population survey protocol. Monitoring should be done annually during the spring after migration (McWilliams and Brauning 2000). The Atlantic Flyway

Breeding Waterfowl Plot Survey, which surveys 1-km² plots in northeastern states (Virginia to New Hampshire), may be used. The Midwinter Waterfowl Survey, conducted annually in January, may be used to track trends in wintering coot populations. Coot harvest is currently being monitored through the federal Harvest Information Program (HIP), but refinements in the survey are needed to obtain more reliable estimates of harvest and hunter activity and potential sources of bias, including band reporting rates, failure to retrieve, and illegal kills should be considered (Serie and Raftovich 2005).

Monitoring of conservation efforts indicates that coots will use restored wetlands (Brisbin and Mowbray 2002). In Pennsylvania, the effect of specific treatments on increasing abundance has not been determined. Surveys before and after habitat restoration practices will provide an indication of the effectiveness of restorative efforts on improving coot habitat and determining which practices are best for increasing coot numbers.

An evaluation of the bias and precision of the Breeding Bird Survey and the Breeding Ground Survey in estimating coot populations is needed. More population data from accurate counts would allow distinction between breeding, migratory, winter, and permanent populations (Brisbin and Mowbray 2002). A geographic study of mitochondrial DNA would determine whether identifiable subpopulations occur and could also help to understand migratory patterns (Alisauskas and Arnold 1994, Brisbin and Mowbray 2002). Because coots are not secretive, they are ideal candidates for a radiotelemetry study to help understand migratory patterns, natal and breeding philopatry, winter-site fidelity, and population dynamics. Survival should be better understood by conducting an updated analysis of band recovery data, studying winter ecology and habitat-dependent mortality rates, and examining disease dynamics (Alisauskas and Arnold 1994).

*Author:* ELISSA OLIMPI, PENNSYLVANIA STATE UNIVERSITY

## Solitary Sandpiper

Order: Charadriiformes
Family: Scolopacidae
*Tringa solitaria*

The solitary sandpiper is a regular passage migrant in Pennsylvania (fig. 5.114). It was selected as a Species of Greatest Conservation Need as a representative

*Fig. 5.114.* The Solitary Sandpiper, *Tringa solitaria*. Photo by Jacob W. Dingel III.

of the group of shorebirds and plovers that migrate through the state, once in much greater numbers. Global abundance is estimated at 150,000 individuals, but the accuracy rating for this estimate is listed as poor (Morrison et al. 2006). The solitary sandpiper is considered a Species of High Concern because of declining population trends and potential threats on the breeding grounds (Brown et al. 2001, Morrison et al. 2006). Global populations are considered Secure (G5, NatureServe 2009).

### *GEOGRAPHIC RANGE*

The solitary sandpiper breeds in the boreal forest across Canada and Alaska. It breeds as far south as Minnesota, west-central Oregon, Nova Scotia, and New Brunswick (Moskoff 1995). It winters form northern Mexico and the extreme southern United States through Central America and into Northern South America (Moskoff 1995).

### *DISTRIBUTION AND RELATIVE ABUNDANCE IN PENNSYLVANIA*

The solitary sandpiper is a migratory species throughout Pennsylvania. Though uncommon, they are regular migrants traveling alone or in small groups of five to ten individuals (McWilliams and Brauning 2000). In the spring, solitary sandpipers arrive as early as the first week of April. During the normal period of migration (late April through late May), they have been recorded at a number of sites throughout the state. The North American Migration Count in May recorded a total of 443 solitary sandpipers in Pennsylvania in 2005 (Etter 2005). A high count of fifty-five individuals was recorded in Philadelphia in 1980 (Paxton et al. 1980). Stragglers from the spring migration may still be observed as late as the first week of June (McWilliams and Brauning 2000). In the fall, the earliest southbound migrants are seen during the first week of July. Juveniles will arrive by the second week of August. Peak migration is in August with the migration ending by the third week of October. Some birds have been observed as late as the first week of December. In the eastern part of the state, they are commonly recorded on the Great Monroe County Snipe Hunt (authors note) in early October. There are no documented records of solitary sandpipers nesting in Pennsylvania (McWilliams and Brauning 2000).

### *COMMUNITY TYPE/HABITAT USE*

In the breeding season, solitary sandpipers breed along the edges of freshwater lakes, ponds, and wetlands within coniferous forest (Moskoff 1995). In migration and the nonbreeding season, they use a wide range of wet areas. They are common in the grassy and muddy shorelines of marshes, woodland streams, pastures, and rivers. Solitary sandpipers are often found in isolated ditches and tiny temporary ponds where waders would not be expected (Hayman et al 1986). They may even be observed where water collects in parking lots, on lawns, and in ditches (McWilliams and Brauning 2000).

### *LIFE HISTORY AND ECOLOGY*

The solitary sandpiper is a long-distance migrant migrating from the boreal forests of Canada to wintering grounds in Central and South America. Its name refers to its relatively solitary habits (which may be observed in small flocks) during migration, which contrasts with the flocking behavior noted in other shorebirds (Moskoff 1995). It is adapted to life in shallow waters and temporary pools with its long legs and long probing bill. Solitary sandpipers are snatchers that tread on the mud, in vegetation, or in shallow water to forage for their prey (Palmer 1967). Using slow and deliberate movements, the solitary sandpiper will acquire food by pecking the ground. They will stir up the water, especially stagnant pools, by prodding the water with their feet. Food items often float to the surface and the bird will pick food from the surface. They eat small crustaceans, caterpillars, spiders, worms, aquatic animals, and other invertebrates.

The solitary sandpiper is a monogamous breeder. The nest is located in conifers between 1 m and 10 m off the ground (Street 1923). Solitary sandpipers have one brood a year. They are one of the first species on their breeding range to finish nesting. They begin their southbound migration by late June/July (Palmer 1967).

### *THREATS*

Although the solitary sandpiper can use a wide range of wet habitats during migration, it is still probably threatened by the loss of quality habitat available to use during migration. Threats include riparian corridor degradation, wetland loss, water pollution, and habitat fragmentation. Loss or alteration of wetlands along their migratory flyways likely has an effect on their populations.

### *CONSERVATION AND MANAGEMENT NEEDS*

In Pennsylvania, the solitary sandpiper is only a migratory visitor. However, there is growing evidence that the condition of migratory stopover sites has an effect on the condition of these birds on arrival at their breeding grounds. This in turn affects their breeding success. Identifying concentrated migratory stopovers and corridors are important to the conservation of this and other shorebird species. Several management activities could be undertaken to support migrating shorebirds. These actions may include identifying and protecting concentrated migratory stopover sites, identifying and protecting water habitats along migratory corridors, and support for comprehensive planning on public and private lands that emphasizes water resources and water quality, including encouraging wider buffers between development and water resources to protect water quality.

Breeding solitary sandpipers need wetlands with dense vegetation, reliable food sources, and reliable water availability. On the breeding grounds, solitary sandpipers have been affected by grazing and will opt not to breed in impacted areas (Popotnik and Giuliano 2000). Management strategies should address the need to control livestock grazing and allow for improved vegetative cover and structure.

### *MONITORING AND RESEARCH NEEDS*

Long-term monitoring through comprehensive programs, such as the International Shorebird Survey, e-bird surveys, and Migratory Bird Day surveys are good tools for monitoring current trends in this species and identifying areas of key international and regional importance (Morrison et al. 2000). In Pennsylvania, it is important to conduct field surveys to identify critical corridors and stopover sites for solitary sandpipers, as well as other migrant shorebirds. Research to determine factors affecting the quality of wetland stopover habitat would be useful for developing more specific management recommendations.

*Authors:* DARRYL SPEICHER, POCONO AVIAN RESEARCH CENTER; JACKIE SPEICHER, POCONO AVIAN RESEARCH CENTER

## Wilson's Snipe

Order: Charadriiformes
Family: Scolopacidae
*Gallinago delicate*

A cryptically colored brown shorebird recently separated from the common snipe (*Gallinago gallinago*) of Eurasia, the Wilson's snipe is often flushed from a marshland hiding spot before it is seen (fig. 5.115). The Wilson's snipe was selected as a Species of Maintenance Concern in Pennsylvania because it is a rare breeder in the state, has undergone perceived declines in both breeding and migrant populations, and is associated with declining wetland habitats. Breeding populations in Pennsylvania are considered Vulnerable (S3B), and global populations are Secure (G5, NatureServe 2009).

### *GEOGRAPHIC RANGE*

The Wilson's snipe breeds from sub-Arctic Alaska east to Labrador and south to California, Arizona, Nebraska, and northern portions of Ohio, West Virginia, western Maryland, Pennsylvania, and New Jersey. The species winters from the southern portion of its breeding range south to Columbia and Venezuela in South America (Tuck 1972, Mueller 2005). Wintering concentrations are greatest in wetland areas around the Gulf of Mexico.

*Fig. 5.115.* The Wilson's Snipe, *Gallinago delicate*. Image courtesy of Alan D. Wilson, naturespicsonline.com.

### *DISTRIBUTION AND RELATIVE ABUNDANCE IN PENNSYLVANIA*

The Wilson's snipe occurs in suitable wetland habitat throughout the state of Pennsylvania during the migration seasons but is rarely encountered during the breeding season. Summer records are mainly from the northern-tier counties, with a scattering in the Laurel Highlands and rarely elsewhere. These sightings sometime indicate breeding activity (e.g., flight displays), but the only regular breeding sites seem to be located in the northwestern portion of the state (Brauning 1992a, McWilliams and Brauning 2000; fig. 5.116).

Pennsylvania is near the southern edge of the Wilson's snipe's breeding range in eastern North America (Mueller 2005). The species occasionally lingers into winter in areas offering suitable foraging conditions. Predictions of global warming effects on breeding bird distributions suggest that this species is likely to be lost from the state's breeding avifauna (Matthews et al. 2004). Results of the second Pennsylvania Breeding Bird Atlas indicate a decrease of 10 percent in distribution has occurred since the first atlas (Mulvihill and Brauning 2009).

### *COMMUNITY TYPE/HABITAT USE*

The Wilson's snipe is a bird of northern boreal wetlands, which breeds only locally as far south as Pennsylvania. Snipes feed and breed in wetlands, including wet pastures, sphagnum bogs, sedge marshes, and brushy marshes. They are most commonly seen in wetlands with low vegetation and may frequent wet pastures, where the cropping of wetland vegetation by livestock suits their needs (Fogarty and Arnold 1977). They feed almost exclusively by probing for worms and other invertebrates in soft soil and mud. Wilson's snipes prefer open wetlands with short vegetation that allow them to view potential predators. Ground-water

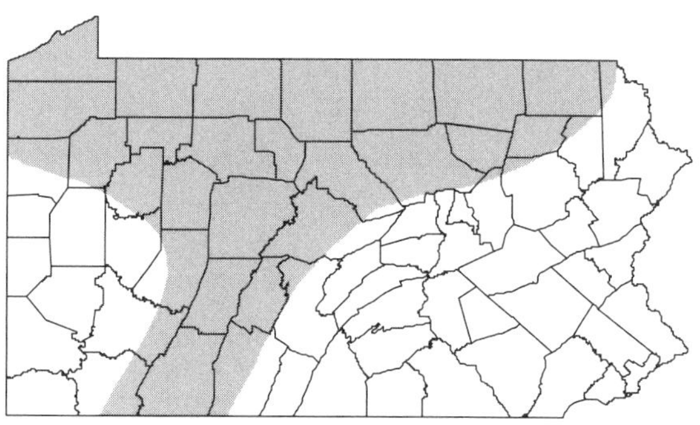

*Fig. 5.116.* Highest probability of occurrence of the Wilson's Snipe, *Gallinago delicate*.

saturation is essential because the species probes in moist, rich organic soil for most of its food. The birds rely on camouflage to avoid predation in the open wetlands they prefer. They generally avoid heavily wooded swamps and alder thickets, which are home to the closely related American woodcock (*Scolopax minor*).

During migration, Wilson's snipes may be found in a wider array of wetland situations, including tidal mud flats, flooded cropland, and other open wetlands, which offer less cover than seems to be sought during the breeding season. Lingering fall birds concentrate in areas offering good food densities (e.g., drainage from barnyards and spring outlets) until freezing conditions force them farther south.

### *LIFE HISTORY AND ECOLOGY*

Although a few Wilson's snipes may occasionally overwinter in southeastern Pennsylvania, most of the population migrates south and does not begin returning until mid-March. Birds seen in mid-April through May and into early June can be considered potential breeders, although northbound migrants may still be passing through at that time. Wilson's snipes have a distinctive aerial display called bleating, or "winnowing," during which the male flies to an elevation of several hundred feet and descends in a series of dramatic dives while making a vibrant winnowing sound, which is produced by air rushing through the bird's stiffly spread outer tail feathers. Aerial displays are infrequently given by migrating birds but commence as soon as the males reach their breeding areas. Females are also reported to participate in aerial displays (Mueller 2005), making counts of aerial birds unreliable as a survey parameter. Temperature may influence winnowing with greater activity at temperatures below 5°C (Tuck 1972). During the display and nesting period, which is poorly known for Pennsylvania, pairs may alight on posts and poles (and rarely wires) and scold intruders (Rand 1966, Laughlin and Kibbe 1985). The flight display may occur during the day or night, but the temporal frequency of flights is greatest during twilight hours of evening and morning. The influence of weather (e.g., wind, temperature) may be pronounced with little activity during wind, rain, or fog. Loud noises also are reported to stop displays. Flights apparently serve to establish territory among other males but are subsequently directed toward the female on the ground and at the nest. Winnowing diminishes substantially after pairs have started incubation, and late-season displaying birds are thought to be unpaired first-year males.

Wilson's snipes are seen in southbound migration as early as mid-July. Snipes often linger into late fall or early winter if there are open wetlands available. Successful overwintering in Pennsylvania is probably rare except possibly in the brackish marshes of the extreme southeastern portion of the state. The timing of migration by locally breeding birds is unknown due to the influx of migrants from more northern areas. The origin of lingering birds is also unknown.

The female builds a nest on or among wetland tussocks/hummocks. Renesting may occur within two weeks if the nest or young chicks are destroyed. Yearling birds may breed up to two months later than older pairs (Tuck 1972). Survival rates of juvenile snipes are unknown, and factors that contribute to their survival are numerous and highly variable (e.g., rainfall, drought, land use, vegetation condition, alternative prey population levels, disturbance, predator levels, pesticide poisoning, and grazing practices; Mueller 2005).

### *THREATS*

The Wilson's snipe has always been a rare breeding bird in most of the state as far as can be determined from available subjective evaluations. Breeding season occurrences are too few and breeding areas too remote for the species to even appear in the roadside-survey-based Breeding Bird Survey maps (Sauer et al. 2004). Because the demise of livestock farming in northern Pennsylvania, many wet meadows/pastures, which may have once been maintained in a low vegetative state are becoming too thick to attract Wilson's snipes. The effect of hunting on the species within the state and on its wintering grounds is unknown.

Factors that may influence wetland soil nutrient quality and biotic productivity may adversely affect Wilson's snipes. Acid mine drainage and loss of organic soil layers resulting from coal mining activities may contribute to the degradation of habitats, which may otherwise be used by the Wilson's snipe.

Climatic change may affect future Wilson's snipe populations in the state. Predictions for future population changes based on climatic change are for nearly complete extirpation of the species from the eastern United States (Matthews et al. 2004). However, this prediction is based on temperature criteria that have yet to be proved a significant factor in the species ecology and are unlikely to change substantially the vegetation of the wetlands currently occupied. This species is more likely to be effected by changes in rainfall/water tables than temperature. Given that the Wilson's snipe breeds at latitudes well south of Pennsylvania in the West, arguments for climatically induced breeding range changes should be cautiously evaluated.

The species is currently so uncommon that local area populations may be extirpated because of natural causes (e.g., mortality during migration). Seldom are more than one or two pairs found breeding in a given area in Pennsylvania. These habitats may go unoccupied for several years before being reoccupied by surplus birds from more populous breeding areas to the north where breeding densities in good habitat may exceed ten pairs per 100 ha.

*CONSERVATION AND MANAGEMENT NEEDS*

Wilson's snipes are migratory game birds protected by federal hunting regulations. Bag limits are liberal (eight per day), and the season long, although hunting pressure in the state is light. Hunters must obtain a state license and participate in the Migratory Bird Harvest Information Program (HIP) but need not purchase a duck stamp. Few hunters focus on snipe hunting, and most of the birds harvested are taken incidental to other hunting activities.

To achieve effective conservation management of the Wilson's snipe, we need to identify principal breeding areas within the state and take steps to manage these areas to the benefit of this species. In addition, we need to quantify and monitor the annual breeding population in the principal breeding areas within the state and develop and fund a monitoring program to identify population changes, threats to habitat, and if possible, contributing mortality factors, which affect the species reproductive success.

The large wetlands of the northwestern corner of Pennsylvania appear to be the current epicenter of breeding by the Wilson's snipe (Brauning 1992a). Preservation of these wetlands is considered critical to this species' continued survival as a breeding species in the state. Although the species is presumed to breed in a number of other wetlands in the state, the regularity with which these other areas are occupied appears to be much lower than the extensive wetlands of Crawford, Erie, Lawrence, and Mercer counties.

*MONITORING AND RESEARCH NEEDS*

Data from the Breeding Bird Atlas efforts should be used to determine primary breeding areas in the state. Monitoring programs (at least three surveys per breeding season) should be established in at least five areas to locate any supporting substantial populations (e.g., >3 pairs). Threats to habitat and breeding success should be evaluated at each site and steps taken to reduce these threats at each site, if feasible. Monitoring should be continued for not less than five years.

A high-priority research need is to determine the optimal period and survey techniques to locate Wilson's snipes on their breeding grounds. Data from the current Pennsylvania Breeding Bird Atlas may assist in locating areas for further study. Survey techniques (see Green 1985) should be applied to locate wetlands where substantial populations (greater than three pairs) currently breed. It is important to monitor annual fluctuations in the breeding populations at these wetlands and relate these fluctuations to causal factors. In areas where birds are relatively common, any factors that depress breeding or use of portions of the highly frequented wetlands can be determined.

Habitat-quality attributes sought by Wilson's snipes at the latitude of Pennsylvania are poorly understood at present. The species often breeds in wet pastures in New York and appears to tolerate at least limited grazing by livestock within these breeding areas. Tall marsh vegetation (e.g., cattails and *Phragmites*) may inhibit nesting. At present, purple loosestrife has invaded wet meadows throughout the East. This tall, introduced plant forms dense colonies and may inhibit usage by the Wilson's snipe, although research is lacking to support this supposition.

*Author:* DOUGLAS P. KIBBE, URS CORPORATION

## American Woodcock

Order: Charadriiformes
Family: Scolopacidae
*Scolopax minor*

The woodcock is a small, stocky upland game bird in the shorebird family (Scolopacidae; fig. 5.117). The American woodcock was selected as a Species of Greatest Conservation Need because of declining breeding and migrant populations. They are a Partners in Flight priority species and a United States shorebird conservation plan species of High Concern. They are included on the Species of Greatest Conservation Need list for all northeastern states. Global populations are considered Secure (G5, NatureServe 2009). The Woodcock is a game species in Pennsylvania with a thirty-day season from mid-October to mid-November. The Penn-

*Fig. 5.117.* The American Woodcock, *Scolopax minor*. Photo by Jacob W. Dingel III.

sylvania Game Commission, within guidelines set by the Fish and Wildlife Services (FWS), determines seasons and bag limits. Pennsylvania has 10,000–13,000 woodcock hunters (Kelley and Rau 2006), the most of any state in the Northeast.

### *GEOGRAPHIC RANGE*

The American woodcock occurs throughout eastern North America. They breed primarily in the northeastern United States and Canada, and winter in the southeastern United States. Breeding is documented as far north as 50°N, with Newfoundland the northeast limit and the northwest extent at the Manitoba-Saskatchewan border (Straw et al. 1994). Wintering range extends from eastern Texas into southeastern Kansas, then across southern portions of Missouri, Illinois, Indiana, Ohio, Pennsylvania, and New Jersey, south through the northern two-thirds of Florida.

### *DISTRIBUTION AND RELATIVE ABUNDANCE IN PENNSYLVANIA*

In Pennsylvania, breeding woodcock are found throughout the state (fig. 5.118). Woodcock singing ground surveys use the conspicuous courtship display of the male woodcock in early spring. Counts of these males are used as indices to populations and are used to monitor trends (Mendall and Aldous 1943). Based on singing ground survey counts from 1970 to 1988, estimated densities range from low (0–0.1 woodcock per route) in the Piedmont region of the state to relatively high (four to ten woodcock per route) in the extreme northwestern part of the state (Sauer and Bortner 1991). Migrating woodcocks may be encountered

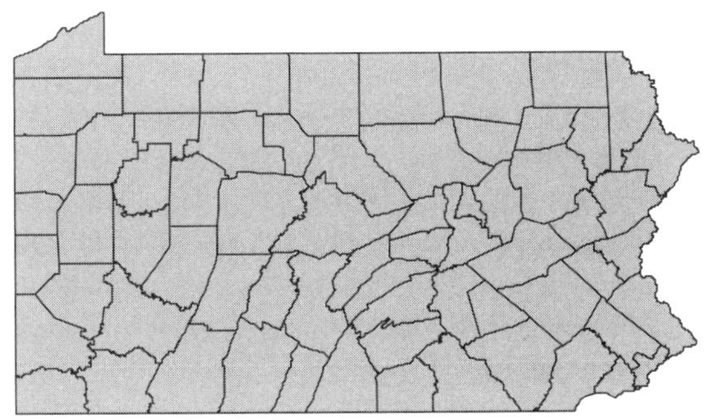

*Fig. 5.118.* Distribution of the American Woodcock, *Scolopax minor.*

throughout the state in spring and fall. Woodcocks winter in southern and southeastern counties during mild winters.

Breeding population indices for woodcock in Pennsylvania over the past thirty-six years show that the species is declining in abundance. From 1996 to 2006, there was no change in the number of woodcocks heard in the singing ground survey (Kelley and Rau 2006). However, from 1968 to 2006, woodcocks declined at an annual rate of –3.4 percent, which was nearly double the average decline for the eastern region as a whole. Continued declines in abundance since 1996 may warrant considering changes in the conservation status of woodcock populations.

### *COMMUNITY TYPE/HABITAT USE*

The woodcock can be described as an early successional habitat specialist. Their preferred habitat is often referred to as scrub-shrub or seedling-sapling, suggesting relatively low-growing, dense, woody cover. Abandoned agricultural lands and regenerating hardwood forests provide the bulk of woodcock habitat in Pennsylvania. Rarely are woodcock found in a mature forest with a dense canopy.

Breeding woodcock require a mix of early successional habitats, including small, scattered openings and dense stands of shrubs and young trees (Liscinsky 1972). Use of this mix of habitats varies with activity, time of day, and season. Openings varying in size from a fraction of an acre to several acres are used as singing grounds (courtship display) in the spring and by some woodcock for nighttime roosting during summer (Mendall and Aldous 1943, Sheldon 1971, Sepik and Derleth 1993, Dessecker and McAuley 2001). The suitability of a singing ground is generally determined by the quality of the adjacent habitat for nesting and brood rearing. Nocturnal roosts may be openings, including those used as singing grounds, or may be another forested site similar to diurnal cover. The woodcock nests in young to mixed-age stands but seem to prefer young hardwood stands (Dessecker and McAuley 2001).

Diurnal cover and brood cover are characterized by dense stands of early successional forest, which can be young hardwood trees or shrubs, on soils with an abundant supply of earthworms (Hudgins et al. 1985). Acceptable plant species composition varies widely at diurnal habitat sites. Woodcock are not restricted to specific plant assemblages as long as the habitat provides early successional structure (Dessecker and McAuley 2001), but several plant species groups are important indicators of potential woodcock habitat because they are typically early successional or have growth forms that provide proper habitat structure; stands of hawthorne (*Crataegus* spp.), alder (*Alnus* spp.), aspen (*Populus* spp.), and dogwood (*Cornus* spp.) are frequently indicators of good woodcock habitat (Straw et al. 1994).

Moist, fertile soils are required to support an abundance of earthworms and other soil invertebrates, the primary food sources for woodcocks. Woodcock are commonly found in damp thickets, riparian zones, brushy edges, and forest clearings. Woodcock use of coniferous stands is minimal, except during periods of drought when moist soils under the shade of dense conifers may offer a supply of earthworms (Sepik et al. 1983, Pennsylvania Game Commission, unpublished data). Burned areas or farmlands reverting to woodlands often provide favorable habitat (Brauning 1992a).

Diurnal migratory stopover habitat in Pennsylvania is similar to breeding habitat, but migrants may also use areas not inhabited by resident birds (Liscinsky 1972). Woodcocks seem to be slightly more flexible in their use of habitat during migration. While adequate breeding habitat for woodcocks is often emphasized, high-quality and well-distributed habitats are also crucial for migration. Opportunities for woodcocks to stop and "refuel" are crucial for surviving the perils of migration. At times, high densities of woodcock may be found in relatively small patches of suitable habitat during the spring and fall migration periods.

### *LIFE HISTORY AND ECOLOGY*

Woodcock are migratory, traveling almost exclusively at night (Mendall and Alduous 1943) and arriving in significant numbers in Pennsylvania starting in late February with peak arrival in March. Most birds continue to migrate north of Pennsylvania, with most migration completed by mid-April. Birds that remain in Pennsylvania after mid-April are considered resident-breeding birds. Woodcocks do not normally migrate in large flocks, but singly, in pairs or in loosely organized small flocks. Some woodcock may leave the breeding grounds in October for the southerly wintering grounds while others stay until December, especially when the weather is relatively mild. The peak of the southward migration is late October and early November but varies depending on the weather and ability to probe soils. Most resident birds leave Pennsylvania during November (Coon et al. 1976; Pennsylvania Game Commission, unpublished data). High densities of woodcocks encountered in late fall are often concentrations of migrating birds resting for the day.

Woodcocks leave obvious white droppings, often referred to as chalk or whitewash, as evidence of their use of a particular area. Possibly the most unusual feature of the woodcock is its bill. The long bill (females' noticeably longer than males') is used for probing in the soil for earthworms and other invertebrates.

Woodcocks are fairly secretive birds, rarely leaving the safety of dense cover during the day and therefore not often seen. The woodcock can be observed flying from diurnal cover to roosting cover just before dark. The best time to see woodcocks is during the spring mating season, when they perform their conspicuous aerial courtship display at dawn and dusk. In late winter through spring, males display in openings in or near dense early successional habitat (Mendall and Aldous 1943, Liscinsky 1972) in an attempt to attract a mate. Males will actively defend their singing area against intruding males. Mating takes place on or near the singing grounds. Males may display at numerous points along the migration route.

Most nesting occurs in March and April in Pennsylvania. Ground nests are simple depressions in the leaf litter. In central Pennsylvania, nests are often at the base of a tree or shrub (Coon et al. 1982). Chicks hatch ready to leave the nest. The female helps the chicks secure food until their beaks, which grow quickly, become more useful for soil probing (Gregg 1984). Chicks develop very rapidly and are able to fly in about two weeks. Survival of chicks is relatively high for a game bird, seemingly counteracting the relatively low annual reproductive output. By early summer, woodcock chicks are fully grown and spend their time feeding and loafing until fall migration.

*THREATS*

The most serious threat is habitat loss and degradation of habitat (Owen et al. 1977, Dwyer et al. 1983, Straw et al. 1994, Krementz and Jackson 1999). Woodcock habitat is in serious decline in Pennsylvania. From 1978 to 2002, total acreage in Pennsylvania forest land remained relatively stable, but the proportion in early successional stages (seedling, sapling, and nonstocked) declined from 21 percent to 11 percent (Alerich 1993, McWilliams et al. 2004). This was a fifty-year low in the proportion and area of young forests. While the simple maturing of forests is the primary factor in loss of key habitats, factors such as highway and urban development, intensification of agriculture, and slowing farm abandonment all contributed to decreasing quantity and quality of optimum woodcock habitat. Within old field habitats in particular, exotic shrubs are rapidly replacing native shrubs preferred by woodcocks (Department of Conservation and Natural Resources, unpublished data). The result in many cases has been more homogeneous habitats and lower plant species diversity. While woodcocks use landscapes dominated by exotic shrubs, primarily multiflora rose (*Rosa multiflora*), autumn olive (*Elaeagnus spp.*), and tartarian honeysuckle (*Lonicera tatarica*), the long-term effect of habitat changes caused by the invasion of exotic vegetation, if any, has yet to be determined.

The threat of contaminants on woodcock populations is a relatively unknown area. Acid deposition is a threat that can affect habitat; the effect on soil pH is a factor not only in forest regeneration but also on the supply of earthworms (Esher et al. 1993). Woodcocks accumulate pesticides in tissue (Clark and McLane 1974) and pesticides can affect the availability of preferred food. Scheuhammer et al. (1999) suggested that lead contamination was widespread in eastern Canada and speculated that it was likely problematic throughout regions in the United States where woodcocks are hunted.

Collision with man-made structures is a mortality factor for birds migrating nocturnally at low altitudes. Woodcocks have been found following collisions with tall office buildings, possibly attracted or disoriented by city lights. According to the Federal Communication Commission 2000 Antenna Structure Registry, the number of lighted towers is more than 74,000. Construction of towers is growing and will likely result in larger numbers of birds killed in collisions (Evans and Manville 2000). Still, in forty-seven studies summarized in 2002, only eight woodcocks were among the more than 500,000 birds killed by communication towers (Shire et al. 2000).

*CONSERVATION AND MANAGEMENT NEEDS*

The overriding need is to stop the woodcock population decline in Pennsylvania and to increase the population above current levels. Maintenance and expansion of old field habitat and rotational harvest of mature forests to increase and sustain the availability of zero- to ten-year-old forest is critical. It is particularly important to create and maintain early successional habitat on moist soils, particularly riparian zones. A concerted statewide effort, among public and private landowners to increase the quality and quantity of early successional habitat, is crucial. Aspen is a short-lived, low-value hardwood species that tends to grow very rapidly and form very dense stands when clear-cut. Aspen from one to ten years of age can provide high-quality habitat for the woodcock. Aspen is a declining species in Pennsylvania. Mature stands should be regenerated whenever possible to benefit woodcocks and maintain the forest type. Care should be taken to prevent the replacement of native shrubs by invasive, exotic species.

The United States Fish and Wildlife Service (FWS) has regulatory authority over woodcock and has both national (U.S. Department of Interior 1990) and northeast regional (U.S. Department of Interior 1996) woodcock management plans. Pennsylvania has no woodcock management plan. The FWS publishes a woodcock status report each year (Kelley and Rau 2006) that describes population trends based on their singing-ground surveys, trends in hunters and harvests based on the Harvest Information Program (Elden et al. 2002), and recruitment based on woodcock hunter wing returns. Under current regulations, hunting does not appear to lower annual survival rates (McAuley et al. 2005).

*MONITORING AND RESEARCH NEEDS*

Statewide monitoring is necessary to track woodcock population trends and is essential for regional population estimates as well as for Pennsylvania. This requires conducting singing-ground surveys and determining adequate sample sizes needed for reliable estimates of breeding populations. In addition, it is important to continue implementation of the wing-collection survey to determine recruitment and to ensure adequate samples for state estimates. The Harvest Information Program should be improved and continued because tracking numbers of hunters and their harvests is critical to understanding the role of hunting mortality on woodcock populations. All of these techniques need to be tested for effectiveness at the state level and enhanced if necessary. Additional surveys would enhance knowledge of wintering woodcock and their habitat. The continued monitoring of woodcock populations on special habitat management areas is important to demonstrate the response of birds to various treatments. Identification of habitat trends, both locally and statewide, is key to understanding the relationship between various habitat factors and the dynamics of woodcock populations. Sustained monitoring of Forest Inventory and Analysis data is essential because these data will provide the basis for many management actions.

Research on both population threats and habitat management are needed in Pennsylvania. Habitat research needs include developing a database of known woodcock habitats statewide; initiating a program to integrate woodcock habitat management techniques into timber management plans on public and private lands; determining land management incentives for (and how to provide them to) private landowners; and evaluating the effects of habitat management techniques on local woodcock populations. For woodcock populations, we need studies that result in better estimates of harvests and harvest rates and population trends that evaluate the effect of hunting and other sources of mortality on local woodcock populations in different regions of the state. In addition, we need to determine whether singing ground surveys accurately track population levels and can be used to monitor continental abundance (Sauer and Bortner 1991).

*Authors:* WILLIAM L. PALMER, PENNSYLVANIA GAME COMMISSION; MARK BANKER, RUFFED GROUSE SOCIETY

## Black-billed Cuckoo

Order: Cuculiformes
Family: Cuculidae
*Coccyzus erythropthalmus*

The black-billed cuckoo is a land bird that is similar in plumage, size, and behavior to its only congener in North America, the yellow-billed cuckoo (*Coccyzus americanas*) (fig. 5.119). The black-billed cuckoo was selected as a Species of Maintenance Concern because it has exhibited substantial population declines in Pennsylvania and throughout its range. Cuckoos are listed on Species of Greatest Conservation Need lists in nine

*Fig. 5.119.* The Black-billed Cuckoo, *Coccyzus erythropthalmus*. Photo by Jacob W. Dingel III.

northeastern states. Both the Pennsylvania breeding population, and the global population are ranked as Secure (SB5, G5, NatureServe 2009).

### *GEOGRAPHIC RANGE*

The breeding range of the black-billed cuckoo extends from northern Oklahoma and southern Missouri in the south, east through the Appalachian Mountains of western North Carolina, and north through central Alberta to Prince Edward Island and Nova Scotia (Hughes 2001). Within this range, densities are highest in the more northern latitudes and at high elevations in the southern latitudes (Hall 1983b, Robbins and Easterla 1992). The wintering range is less well known, but the species is known to occur (from November through April) from Colombia through western Venezuela and south to Peru and Bolivia (Hughes 2001).

### *DISTRIBUTION AND RELATIVE ABUNDANCE IN PENNSYLVANIA*

In Pennsylvania, the black-billed cuckoo occurs in all counties except those in the southeastern corner of the state (in the Piedmont and Coastal Plain physiographic regions; fig. 5.120). Densities are highest in the northwestern part of the state, particularly in the High Plateau physiographic region (Sauer et al. 2004). Across Pennsylvania, the black-billed cuckoo tends to occur in more heavily forested landscapes, and at higher elevations, than the yellow-billed cuckoo (McWilliams and Brauning 2000).

United States Geological Survey Breeding Bird Survey (BBS) data indicate significant annual decreases in black-billed cuckoo population densities range-wide of 1.6 percent from 1966 through 2003, with annual range-wide declines of 3.0 percent from 1980 to 2003

(Sauer et al. 2004). In Pennsylvania, black-billed cuckoos have declined by 3.6 percent annually from 1966 to 2003, with annual declines of 7.0 percent from 1980 to 2003 (Sauer et al. 2004; fig. 5.121). Black-billed cuckoo numbers tend to fluctuate with gypsy moth numbers often masking real declines. Declines throughout the entire Allegheny Plateau and Ridge and Valley physiographic regions (including areas outside of Pennsylvania) have been slightly lower, but still significant, with annual decreases of 5.9 percent and 5.7 percent in these two physiographic regions, respectively, since 1980 (Sauer et al. 2004).

### *COMMUNITY TYPE/HABITAT USE*

Primary breeding habitat includes mixed deciduous–coniferous forests and, in Pennsylvania, pine and hemlock woodlands. While yellow-billed cuckoo territories may occur within mature forest habitat (Hughes 1999), black-billed cuckoo territories are often located along forest edges and contain dense hedgerows and thickets (Hughes 2001). However, black-billed cuckoos are most abundant in Pennsylvania in areas of highest landscape-level forest cover (e.g., the northern-tier counties of the Allegheny Plateau, Ickes 1992a). Nests are often placed in groves of young deciduous trees, in thickets or tangles of vines, or along forest edges (Hughes 2001).

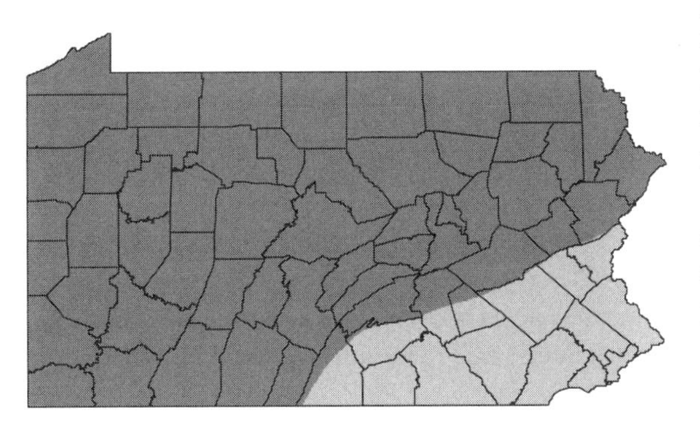

*Fig. 5.120* Primary (darker shading) and secondary (lighter shading) distribution of the of the Black-billed Cuckoo, *Coccyzus erythropthalmus*.

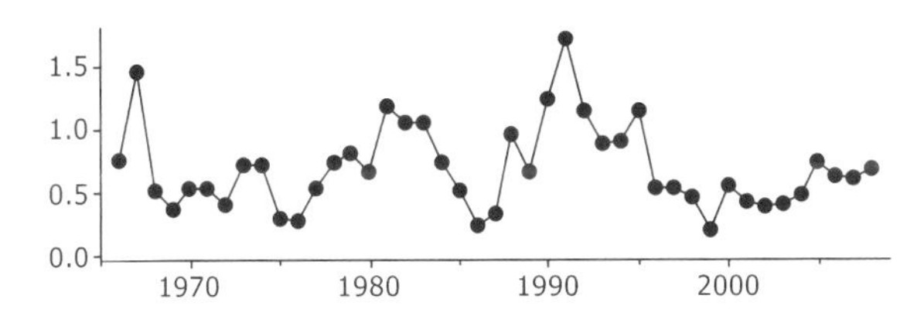


*Fig. 5.121.* Black-billed Cuckoo, *Coccyzus erythropthalmus*, population trends from the Breeding Bird Survey.

### *LIFE HISTORY AND ECOLOGY*

The black-billed cuckoo is a Neotropical-Nearctic migrant, wintering in northern and central South America and migrating to the United States and Canada to breed in the summers (Hughes 2001). Foraging is done mainly in the canopy, and primary food items taken include Lepidopteran larvae (including outbreaking species such as tent caterpillars (*Malacosoma* spp.) and gypsy moth larvae (*Lymantria dispar*; Bent 1940). Size of breeding territories is unknown for black-billed cuckoos (Hughes 2001).

Nesting occurs primarily from late May through August, with a peak in June and July (Peterjohn and Rice 1991, Sibley 1997). Nests are generally built along forest edges or in hedgerows or thickets and range in height from 0 to 13.5 m (mean of 1 to 2 m; Peck and James 1983), and both adults contribute to nest building. Although the yellow-billed cuckoo may be double brooded in some parts of its breeding range (Sutton 1967), the black-billed cuckoo is thought to be single brooded (Eastman 1991a). Black-billed cuckoos are also known to exhibit facultative intra- and interspecific brood parasitism during times of high food abundance (Nolan and Thompson 1975, Hughes 1997).

### *THREATS*

Little is known about the threats that negatively impact black-billed cuckoo population dynamics. Habitat fragmentation may play a role in some areas of the breeding range, as the species is not found in forest fragments smaller than 1.2 ha in Saskatchewan, 4.0 ha in New Jersey, and 4.5 ha in South Dakota (Galli et al. 1976, Martin 1981, Hughes 2001). Removal of understory vegetation and hedgerows and roadside shrubs may have led to decreases in abundance in areas of New Jersey and Illinois in the past (Graber and Graber 1963, Sibley 1997).

Use of pesticides on Lepidopteran pest species may also negatively impact black-billed cuckoos. Fat tissue of cuckoos collected in Florida in 1970 contained chlorinated hydrocarbon pesticide residues (Grocki and Johnston 1974), and large numbers of cuckoos were apparently killed by ingesting Lepidopteran larvae that had been sprayed with arsenical pesticides in orchards in Nova Scotia (Tufts 1986). The use of naturally occurring broad-spectrum pesticides to control specific Lepidopteran pest species (such as the aerial application of *Bacillus thuringiensis* to control gypsy moths) may also potentially impact black-billed cuckoo populations. While such effects have not been studied with cuckoos, *Bacillus thuringiensis* was found to decrease the number of young produced by red-eyed vireos (*Vireo olivaceus;* another migratory landbird species) as a result of a decrease in overall Lepidopteran abundance in West Virginia (Marshall et al. 2002).

*CONSERVATION AND MANAGEMENT NEEDS*

Because the cause of black-billed cuckoo population decline in Pennsylvania and elsewhere is unknown, research on demographic parameters and the factors limiting populations in various habitat and landscape types is necessary before conservation actions can be planned (see below).

*MONITORING AND RESEARCH NEEDS*

Because the factors that limit black-billed cuckoo populations are currently unknown, it is important to quantify productivity and survival in different habitat and landscape types and to identify those factors that are decreasing reproductive output or survival (including such potential factors as nest predation and food limitation). Such research and monitoring would include quantification of breeding habitat use and breeding territory size (in multiple habitat types and multiple landscapes) based on spot-mapping surveys and radiotelemetry; estimation of population productivity and adult survival rates in multiple habitat and landscape types; examination of nest success and season-long productivity in relation to food availability, habitat type, and landscape type; and assessment of effects of pesticides on food availability and on subsequent cuckoo productivity. Because of potential interannual and between-site variation in productivity and survival and in the factors limiting cuckoo populations, this research would need to be done over a period of several breeding seasons to obtain the minimum data necessary to formulate a conservation plan.

*Author:* ANGELA D. ANDERS, U.S. DEPARTMENT OF THE NAVY

## Barn Owl

Order: Strigiformes
Family: Tytonidae
*Tyto alba*

The barn owl is a medium-sized pale owl (fig. 5.122). It is a Species of Greatest Conservation Need in Pennsylvania, is listed as Vulnerable in both the breeding season and nonbreeding season because it is a rare and localized breeder, and has exhibited population declines. It is also listed as a Species of Greatest Conservation Need in eight northeastern states. This species has a globally Secure (G5) status (NatureServe 2009).

*Fig. 5.122.* The Barn Owl, *Tyto alba*. Photo courtesy of Hal Korber, PGC Photo.

*GEOGRAPHIC RANGE*

The barn owl is one of the most widely distributed land birds in the world, found on all continents except Antarctica. It is found in most of the lower forty-eight states, absent only from large parts of New England, and has small populations in Ontario and British Columbia. Populations in peripheral parts of its range are small. Pennsylvania is at the northern edge of this species' range. There have been almost no recent records from western New York state to the north, where there was a range contraction between the 1980s and early 2000s (McGowan and Corwin 2008).

*DISTRIBUTION AND RELATIVE ABUNDANCE IN PENNSYLVANIA*

The barn owl is a scarce and localized breeding bird in Pennsylvania, found predominantly in low altitude areas in the south and east of the state (fig. 5.123). Between 1983 and 1988, it was recorded in only 5 percent of breeding bird atlas blocks, while breeding was confirmed in less than half of those (Brauning 1992a). Away from the southeast a few scattered blocks were occupied in the Green and Washington counties in the southwest, Erie and Crawford counties in the northwest, and Bradford and Tioga counties on the state's border with New York. Although this is a secretive nocturnal species, the paucity of records during the fieldwork for the first Atlas of Breeding Birds in Pennsylvania suggests that this species is indeed a very scarce bird through most of the state.

Population declines have been reported in the Midwest and the Northeast of the United States (Rosenburg 1992), but population trends are difficult to ascertain as this species is rarely reported on Breeding Bird Survey (BBS) routes. Trends for the United States show a 1.3 percent per annum decrease between 1966 and 2004 (95% confidence limits –6.7 and +4.0), but this is estimated from a sample of just thirty-nine routes (Sauer et al. 2005). It is not possible to derive estimates of population change for Pennsylvania from the Breeding Bird Survey because of very small sample sizes. However, anecdotal evidence points to a decline since the first atlas, especially in the Southwest, where breeding was confirmed only once between 1993 and 2003. By the late 1990s and early 2000s, about ten nests were reported in any given year across Pennsylvania, a drastic reduction from reports of ten to fifteen years previous when there were as many as 150 nests (per D. Brauning). However, the Barn Owl Conservation Initiative of the Pennsylvania Game Commissions has successfully resulted in the reporting of more nests, confirming 49 active barn owl nests in 2008, bringing the total number of different nest sites to 102 since nest searches began in 2005 (D. Mummert and J. Zambo, personal communication).

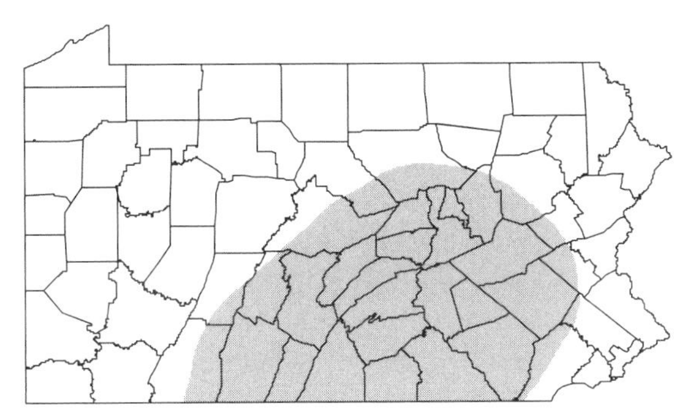

*Fig. 5.123.* Highest probability of occurrence of the Barn Owl, *Tyto alba*.

*COMMUNITY TYPE/HABITAT USE*

Low-altitude grasslands, including meadows, hay fields, and abandoned arable fields are the prime foraging habitat for barn owls (Rosenburg et al. 1992), but a wide range of open habitats are used elsewhere in the United States, including marshes and deserts. The distribution of this species is often limited by the availability of suitable nest sites. It is a cavity nester, using natural sites, such as holes in trees and crevices in rock faces, in addition to a wide range of human-created sites in abandoned buildings, grain silos, and especially barns, hence its common English name. Artificial nest boxes are readily occupied, especially if other nest sites are scarce. In a study in New Jersey, 50 percent of pairs nested in tree cavities and 31 percent in nest boxes (Colvin 1984). Home ranges of between 414 ha and 921 ha have been estimated in northeastern states (Rosenburg et al. 1992), but they do overlap if nest sites and prey are abundant (Smith et al. 1974). Winter habitats are similar to those occupied during the breeding season, although conifers may be used as winter roost sites (McWilliams and Brauning 2000).

*LIFE HISTORY AND ECOLOGY*

Although largely nocturnal, barn owls can sometimes be seen at dawn and dusk and will hunt in broad daylight if food is in short supply. Within eastern North America, the meadow vole (*Microtus pennsylvanicus*) is the preferred prey species, although shrews, mice, rats, and birds are also consumed, especially when vole populations are low (Colvin 1985, Solymár and McCracken 2002). Hunting is typically carried out on the wing by quartering low (1.5 m to 4.5 m) over the ground and locating prey with its excellent low-light vision and hearing (Bunn et al. 1982). Occasionally, barn owls hunt from perches, such as fence posts and trees. Although some individuals are residents, the barn owl is mainly a migratory bird in Pennsylvania (McWilliams and Brauning 2000). The winter quarters are not known but are likely to include the states to the south of Pennsylvania.

The breeding season is protracted, with the first eggs laid in March, while young may be found in the nest in late summer. A single brood is typical in most of temperate North America, but second broods are sometimes attempted (Marti 1992). Clutch size varies with condition of adults at the onset of breeding and with food supply, varying from one to thirteen eggs but typically between four and six, with a mean of 5.5 in a study in Maryland (Reese 1972). The young hatch asynchronously. The larger, older young outcompete their smaller siblings if food is scarce, thereby producing a food supply limitation on productivity. Dispersal during the late summer and fall may take young birds hundreds of kilometers from their natal range (Marti 1992). Breeding occurs at one year of age (Marti 1992). Mortality rates are high, especially for first-year birds, with an average life span reported to be less than two years in both North America and Europe.

Annual monitoring of nests in Ohio between 1988 and 2003 (eleven to forty-nine nests per year) revealed that an average 80 percent of nests were successful annually (range 70.6%–81.2%), producing an average of 4.4 fledged young (range of annual averages were 3.2–5.2) per successful nest (Ohio Department of Natural Resources 2004).

### *THREATS*

Changes in agricultural practices have been strongly implicated as a main cause of the decline in barn owl populations in both Europe and North America (Colvin 1984, Debruijn 1994, Tome and Valkama 2001). These changes include loss of grassland and marsh habitats for foraging, loss of nest sites due to barn conversion (Ramsden 1998) and removal of standing dead timber (Marti et al. 1979), and reduced food supply due to increased use of rodenticides. Poisoning from pesticides has been frequently confirmed as a cause of mortality (Blus 1996), but the population scale effects of this are not known (Newton et al. 1990). One of the major causes of mortality in many areas is collision with road traffic (Keran 1981, Massemin et al. 1998, Zorn 1998).

Between 1982 and 1997 more than 420,000 acres of cropland were lost in Pennsylvania to development, while more than 767,000 acres of pasture were lost. During this time, the rate of agricultural land loss increased from approximately 100 acres per day in 1982 to a current rate of more than 299 acres per day. These losses have been especially marked in southern Pennsylvania, where overall losses of agricultural land were estimated at as high as 37 percent in areas around Philadelphia between 1969 and 1992 (Goodrich et al. 2002). Further, the conversion of hay fields to row crops has led to a marked reduction in hay fields, which provide optimal habitat for meadow voles and other small mammals—the barn owl's key prey. Increased field size and mechanization of farming has also reduced the amount of marginal cover or buffer strips around field boundaries, resulting in a loss of noncultivated land within the farmed landscape. It is possible that the Conservation Reserve Enhancement Program, which has already resulted in the creation of almost 200,000 acres of grassland and noncultivated buffer strips in southern Pennsylvania, will provide an important foraging habitat for this species.

The enlargement of crops fields may have resulted in a loss of the isolated old trees that typically offer suitable nest cavities for this species. Availability of nest cavities may have been exacerbated by competition with other species, such as raccoons (*Procyon lotor*; Rosenburg 1992). Loss of human-made sites has also occurred as derelict barns and silos deteriorate, or collapse or are screened to prevent access by rock pigeons *Columbia livia* (Rosenburg et al. 1992).

A rapid increase in rural housing development coupled with increased car ownership has lead to a rapid increase in road traffic in some areas of Pennsylvania (Delaware Valley Regional Planning Commission 1997). Collisions with road traffic are known to be a major cause of mortality for this species. In addition, barn owls are susceptible to severe winter weather, especially deep snow cover as this makes prey inaccessible. Severe winters in the mid-1990s were thought to have extirpated this species form some parts of Pennsylvania (McWilliams and Brauning 2000).

### *CONSERVATION AND MANAGEMENT NEEDS*

The maintenance and enhancement of existing natural populations should be the conservation priority. Reintroduction programs might be considered in areas where this species is extirpated, but there is little evidence for the success of barn owl reintroductions elsewhere in the United States, Canada, or Great Britain (Solymár and McCracken 2002). The Moraine Preservation Fund has reintroduced small numbers of barn owls into southwestern Pennsylvania. These reintroduced birds are being monitored and their movements tracked using satellite telemetry. Existing reintroduction programs in Pennsylvania should be assessed to ensure that they are effective before they are either expanded or continued over a long time period.

The preservation of this species within the state depends on mediation of the changes in land management that have contributed to this species decline, and conservation efforts should be focused on improving habitat in areas in which this species is already established.

*MONITORING AND RESEARCH NEEDS*

The Pennsylvania Game Commission's Barn Owl Conservation Initiative has undertaken to establish an inventory of barn owl nest sites and monitor nesting success. This initiative will provide a much clearer picture of the status of this species within the state and will inform where targeted conservation measures, such as providing additional nest sites or hunting areas, might be most beneficial. Monitoring data from the Barn Owl Conservation Initiative should be analyzed annually so that changes in the species' abundance or nesting success are identified quickly and conservation action recommended. A population genetics study, incorporating both the birds in the established wild population, and the released individuals from the captive bred population should be considered. This could identify potential future gene flow between these populations of different provenance.

*Author:* ANDY WILSON, PENNSYLVANIA STATE UNIVERSITY

## Common Nighthawk

Order: Caprimulgiformes
Family: Caprimulgidae
*Chordeiles minor*

Most often observed in flight hawking insects at dawn or dusk, the common nighthawk is a medium-sized nightjar (fig. 5.124). The nighthawk was selected as a Species of Maintenance Concern because of perceived population declines. The nighthawk population in Pennsylvania is listed as Vulnerable to Apparently Secure (S3S4B). Global populations are Secure (G5, NatureServe 2009).

*GEOGRAPHIC RANGE*

A widespread species, the breeding distribution of the common nighthawk ranges from southern to central Canada south throughout most of the United States, with the exception of Southern California and Arizona, and into Mexico and Central America. The common nighthawk winters primarily in the lowlands of central or southern South America, although its winter distribution is poorly documented (Poulin et al. 1996).

*Fig. 5.124.* The Common Nighthawk, *Chordeiles minor*. Photo courtesy of Ronnie Maum.

*DISTRIBUTION AND RELATIVE ABUNDANCE IN PENNSYLVANIA*

Historically, a common and widespread species across the state, the common nighthawk has apparently exhibited a decline in abundance and a shift in breeding behavior over the past century (Brauning 1992e, McWilliams and Brauning 2000; fig. 5.125). Formerly, nighthawks used a variety of natural flat, open sites for nesting but appear to have shifted primarily to nesting on rooftops in cities and towns. As a result, the distribution of this species in Pennsylvania appears now to be tightly linked to humans. During the first Pennsylvania Breeding Bird Atlas (1983–1989), this species was detected in most counties, but overall, was located in only 15 percent of the state's atlas blocks and was only found in those blocks that contained towns and cities. Clusters of records, particularly those that include confirmations of breeding, corresponded closely with urban areas throughout the state. On the basis of long-term data (1966–2003) from the North American Breeding Bird Survey, common nighthawk populations have declined in the United States (–1.6% annual decline, $P < 0.0001$, 1,455 routes) and show a similar declining pattern in Pennsylvania (–7.7% annual decline, $P = 0.15$, 9 routes; Sauer et al. 2004; fig. 5.126). This apparent shift in breeding behavior in Pennsylvania, and the associated decline in population size, also appears to be occurring broadly at least through the northeastern United States (Brauning 1992e).

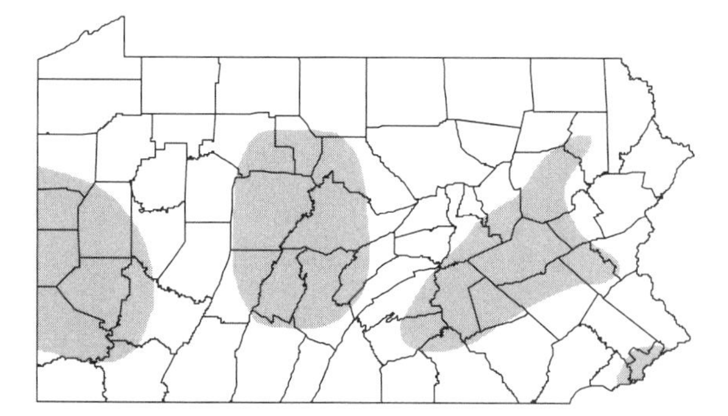

*Fig. 5.125.* Highest probability of occurrence of the Common Nighthawk, *Chordeiles minor.*

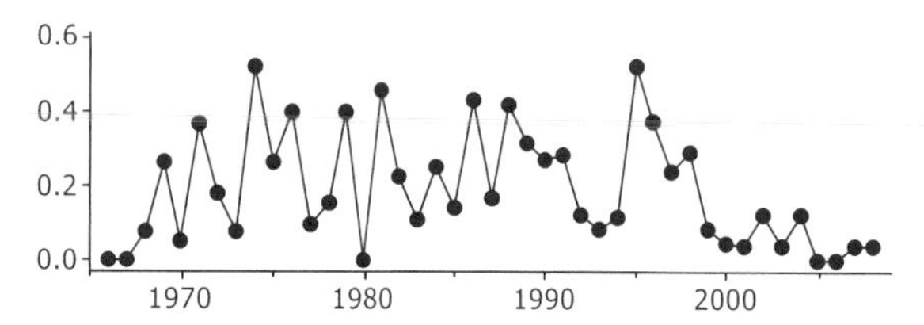


*Fig. 5.126.* Common Nighthawk, *Chordeiles minor*, population trends from the Breeding Bird Survey.

### COMMUNITY TYPE/HABITAT USE

Across their range, common nighthawks will breed in coastal sand dunes or beaches, logged or burned areas of forest, woodland clearings, sagebrush shrublands, prairies or farm fields, rock outcrops, and flat gravel rooftops in towns or cities (Poulin et al. 1996). Within Pennsylvania, nighthawks historically nested locally throughout the state on rock ledges, large boulders, or on other open microhabitats; however, in recent years, most individuals appear to have shifted to using gravel rooftops in urban areas (Brauning 1992e, McWilliams and Brauning 2000).

### LIFE HISTORY AND ECOLOGY

A long-distance complete migrant, the common nighthawk is typically present in Pennsylvania from early May through October (McWilliams and Brauning 2000). Primarily crepuscular in habit, this species most actively forages around dusk and dawn, spending the remainder of the time roosting nearly motionless on the ground, flat rooftops, or on limbs close to the ground with their bodies parallel to the branch. Strictly insectivorous, this species feeds on a variety of flying insects, particularly ants (Hymenoptera), beetles (Coleoptera), and true bugs (Hemiptera; Poulin et al. 1996). It forages while in flight catching prey by vision in its cavernous mouth at heights ranging from near ground level to >150 m.

Within Pennsylvania, nighthawk breeding activity occurs primarily between May and August, with most clutches initiated late May through June (McWilliams and Brauning 2000). Seasonally monogamous, this species does not build a nest; instead, the two eggs are laid directly on the bare ground or on gravel rooftops (Harrison 1975c). In most locations, common nighthawks appear to be single-brooded (Harrison 1975c), though two broods have apparently been observed in some locations (e.g., Weller 1958).

Rates of reproductive success and causes of nest failure are poorly known for this species. One hypothesized explanation for the shift of nighthawks from natural to rooftop nest sites is the probable reduction in predation risk on gravel roofs (Brauning 1992e). Recent shifts in urban areas from gravel to rubberized roofs, however, may help account for population declines in urban areas as rubberized roofs apparently do not provide suitable nesting substrate; the solid color of rubberized roofs provides no camouflage, the flat surface allows eggs to roll, and the dark rubber gets hotter than gravel in the sun, potentially damaging eggs (Marzilli 1989).

### THREATS

Causes of perceived population declines in the East are unclear but may be related to habitat development on the breeding grounds, shifts in methods of roof construction that have affected nest site availability and nesting success, the decline of food supplies as a result of pesticide applications, or problems on the wintering grounds (Poulin et al. 1996, McWilliams and Brauning 2000).

In the Northeast where nighthawks often depend on flat-topped, human-built structures for nesting, shifts from gravel-topped to rubberized roofs likely has negatively affected populations (Marzilli 1989, McWilliams and Brauning 2000). Increased road density and use also affects nighthawks as individuals often are killed by automobiles while roosting on roads or are killed while feeding low over roadways (Poulin et al. 1996). Activities that open up forested habitats, like clear-cutting or burning, may benefit this species by providing nesting sites, however, in some western locations nighthawks appear to avoid recent clear-cuts for breeding (Szaro and Balda 1986). In residential areas, increased densities of native (e.g., raccoon [*Procyon lotor*], Virginia opossum [*Didelphis virginianus*]), and introduced (e.g., domestic cat [*Felis catus*]) predators may affect both adult survival and nesting success, although detailed studies are needed (Poulin et al. 1996).

Pesticide use to battle gypsy moth infestations or to control mosquitoes in the eastern United States could negatively affect common nighthawks directly, or it could be affected indirectly by reducing availability of insect prey (Poulin et al. 1996). Effects of other pollutants, including industrial pollutants and acid rain, could also affect this species directly, or it could be affected indirectly by its prey. These potential factors remain untested, however.

### *CONSERVATION AND MANAGEMENT NEEDS*

Our current knowledge of common nighthawk abundance, distribution, population status, and threats is limited. This lack of knowledge prevents managers from knowing the status of the common nighthawk and threats to populations throughout Pennsylvania and, thus, seriously hampers any efforts to manage the species effectively. Because of this lack of information, the two most important initial conservation and management efforts needed for this species both relate to the gathering of information.

First, the author recommends that a statewide survey program be developed to monitor common nighthawk populations in Pennsylvania during both the breeding season and fall migration to better document breeding distribution and populations trends. Commonly used bird survey techniques (e.g., North American Breeding Bird Survey, Monitoring Avian Productivity and Survivorship [MAPS]) are not effective for monitoring this crepuscular species. Nighthawks, however, are relatively easy to monitor at dusk during the breeding season because they become active while it is still light, they call frequently in flight, and they are often associated with human structures (Brauning 1992e, British Columbia Ministry of Environment, Lands, and Parks 1998). Nighthawks are also relatively easily monitored during fall migration as they often travel in flocks (Poulin et al. 1996, Kinzey 2000). During the breeding season, standardized surveys of fixed locations in towns and cities would be useful to determine breeding abundance, distribution, and population trends.

Second, because most evidence suggests widespread declines, conservation efforts are apparently needed to maintain current population levels in the short term and provide for some recovery in the long term. Unfortunately, because the causes of the declines are not fully understood, appropriate management actions are not clear. As a result, the author recommends that the state supports research studying the breeding biology of nighthawks to evaluate how land use changes, road building, pesticide use, or changes in roof-building techniques affect nighthawk breeding success and survival. One possibility to consider is that current declines simply represent a return to natural, presettlement population levels for this species as a result of declines in artificial nesting habitats (i.e., gravel roofs). If this is the case, this could greatly alter perceived management needs for this species.

### *MONITORING AND RESEARCH NEEDS*

Current monitoring efforts could be improved for this species. Several eastern states have initiated volunteer-based, evening surveys in attempts to better monitor this species during the breeding season (Connecticut Department of Environmental Protection 2003, Audubon Society of New Hampshire 2005). These efforts generally entail volunteers visiting town parks, lighted athletic fields, shopping mall parking lots, or other open locations around dusk on clear evenings in July to look or listen for common nighthawks. These surveys, if conducted in a standardized manner, would be useful for determining local abundance and distribution and could provide data on population trends if conducted across years. During fall, common nighthawks migrate in flocks and are often visible in large numbers. common nighthawk "watches" have been performed at fixed locations in the state (e.g., near Pittsburgh; Kinzey 2000) and have demonstrated that it is possible to effectively count nighthawks during this migration. This technique could be used to better monitor broad population trends for this species across its range.

Evening surveys are needed during the summer breeding season across the state in a wide variety of locations and types of human settlements. Ideally, all major cities and towns could be sampled, however, even limited surveys at local sites would be useful to establish baseline data that could be used to begin evaluating local population trends. Further, an increase in the number and distribution of nighthawk watch sites in the state are needed during fall migration to better document migration patterns and broad population trends across time.

In addition to a large-scale, statewide monitoring program to better measure the abundance and distribution of common nighthawks research on the breeding biology of nighthawks is needed. At present, little is known about the breeding biology of this species, including rates of nesting success and major causes of nest failure. Given that this species currently nests primarily on rooftops in towns or in cities in Pennsylvania, detailed studies are needed to determine rates of nesting success and the characteristics of successful breeding sites. Comparison of nesting behavior and success on different types of rooftops, or in different types of human populations (e.g., rural towns versus cities), would be useful. In addition, it would be useful to test the effectiveness of various artificial gravel nesting pads placed on nongravel roofs. These pads have been used in some locations with apparent success (Marzilli 1989).

Much speculation has been made concerning the causes of population declines. Population-level studies are needed to scientifically examine other possible causes of common nighthawk declines by evaluating how pesticides, forest management, and other land use changes affect local populations. In addition, problems on the wintering grounds could be an important contributor to population declines detected in Pennsylvania (and could even be the major contributor). Knowledge of the importance of wintering ground events is critical to understanding population trends globally and in Pennsylvania.

*Author:* CHRISTOPHER B. GOGUEN, PENNSYLVANIA STATE UNIVERSITY

## Whip-poor-will

Order: Caprimulgiformes
Family: Caprimulgidae
*Caprimulgus vociferus*

More often heard than seen, the nocturnal whip-poor-will is an extremely cryptic medium-sized nightjar (fig. 5.127). It was selected as a Species of Maintenance Concern because of population declines throughout Pennsylvania and much of its range. The whip-poor-will is a Partners in Flight priority species and is included as a species of concern in most northeastern states. Breeding populations in Pennsylvania are Apparently Secure (S4B), and global populations are Secure (G5, NatureServe 2009).

### *GEOGRAPHIC RANGE*

The breeding distribution of the whip-poor-will is split into an eastern (one subspecies) and western (five subspecies) group. The eastern whip-poor-will inhabits southern Canada and the northern and central portions of the eastern United States, west to the edge of the Great Plains, and south to the northern edge of the Gulf Coast states. The western whip-poor-will reaches its northern limit in the southwestern United States and ranges into Mexico and Central America along the central highlands. Although much of the western group is nonmigratory, eastern whip-poor-wills migrate and overwinter, either in the extreme southeastern United States from Florida to Texas along the Gulf Coast, or south into eastern Mexico and northern Central America (Cink 2002).

*Fig. 5.127.* The Whip-poor-will, *Caprimulgus vociferus*. Photo by Jacob W. Dingel III.

### *DISTRIBUTION AND RELATIVE ABUNDANCE IN PENNSYLVANIA*

Currently, this species is locally present across much of the state but is uncommon in most locations. During the first Pennsylvania Breeding Bird Atlas (1983–1989), this species was detected in most counties but overall was located in only 17 percent of the state's atlas blocks (Santner 1992b). The species appears to be most common within the Ridge and Valley and the Poconos regions, while it is infrequently reported from the southeastern Piedmont of the state (McWilliams and Brauning 2000; fig. 5.128). In some isolated localities in Pennsylvania (i.e., on sites where strip-mined areas are regenerating forest), whip-poor-will abundance appears to be increasing (McWilliams and Brauning 2000). Although probably more abundant in Pennsylvania now than before nineteenth-century deforestation, most sources, including Breeding Bird Survey (BBS) data, suggest that whip-poor-will abundance has been declining for decades within the state (Genoways and Brenner 1985, McWilliams and Brauning 2000, Santner 1992b, Sauer et al. 2004; fig. 5.129).

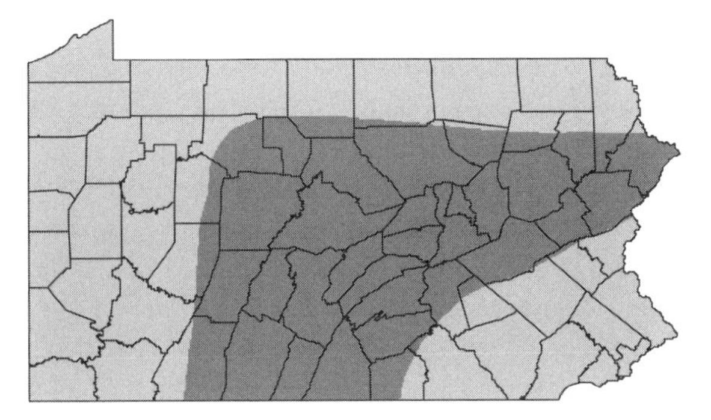

*Fig. 5.128.* Primary (darker shading) and secondary (lighter shading) distribution of the Whip-poor-will, *Caprimulgus vociferus*.

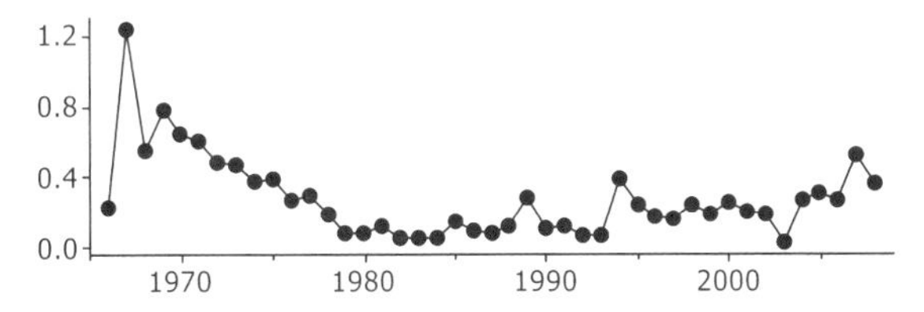


*Fig. 5.129.* Whip-poor-will, *Caprimulgus vociferus*, population trends from the Breeding Bird Survey.

### *COMMUNITY TYPE/HABITAT USE*

Across their range, whip-poor-wills breed in a variety of open woodlands, ranging from lowland moist deciduous forests to montane pine or pine-oak woodlands (Ehrlich et al. 1988). Regardless of habitat type used, the key elements required by this species appear to include shade, proximity to open areas for foraging, and fairly sparse ground cover (Eastman 1991b). As a result of these requirements, early to mid-successional and open, forested habitats appear to be preferred over contiguous, mature forest; however, no data are currently available on specific forest structure or size preferences (Cink 2002). Within Pennsylvania, whip-poor-wills breed in open woodlands, areas of secondary growth, and scrub barrens, particularly those with abandoned fields or cleared areas nearby (Genoways and Brenner 1985, McWilliams and Brauning 2000). In areas of contiguous mature forest, this species is restricted to areas near clear-cuts, burned, or otherwise open habitats (Santner 1992b).

### *LIFE HISTORY AND ECOLOGY*

A medium distance complete migrant, the whip-poor-will is present in Pennsylvania between mid-April and late October, although timing of fall migration is poorly documented because of the lack of calling after breeding (McWilliams and Brauning 2000). Crepuscular to nocturnal in habit, whip-poor-wills spend their days roosting nearly motionless, typically on low limbs with their bodies parallel to the branch but sometimes on the ground. Whip-poor-wills usually forage by sallying from perches in trees or from the ground, catching prey in their cavernous mouth. Strictly insectivorous, this species feeds particularly on moths and beetles but also regularly captures flying ants, flies, grasshoppers, and mosquitoes (Cink 2002).

Within Pennsylvania, whip-poor-will breeding activity seems to occur primarily between mid-May and July within open woodlands or areas of secondary growth near other open habitats (McWilliams and Brauning 2000). Seasonally monogamous, this species does not build a nest; instead, the two eggs are laid directly on the ground, often on leaf litter (Harrison 1975c). In most locations, at least a portion of the population raises two broods, although this may range from a majority of pairs in the South (e.g., 60% in Kansas) to a small percentage in northern populations (e.g., 20% in Ontario; Cink 2002).

Although well camouflaged, this ground-nesting species commonly suffers loss of eggs or nestlings to predation, particularly from mammalian predators but will readily renest following loss. Because of the cryptic nature of the species, reproductive success is poorly known; however, in a well-studied population in Kansas, 74 percent of twenty females reared at least one brood to independence (Cink 2002).

### *THREATS*

Causes of perceived population declines in the East are unclear, but increased nest losses due to forest fragmentation and increased predation, habitat loss to urbanization, maturation of eastern forests, declining food supplies due to pesticide applications, and problems on the wintering grounds have all been proposed (Genoways and Brenner 1985, Eastman 1991b, Santner 1992b).

Although in largely forested landscapes, some clearing of forests may benefit the whip-poor-will by creating the open foraging habitats it requires; excessive fragmentation created by agriculture or suburban sprawl may have several negative effects. Nesting success, for example, may be reduced because of introduced cats and dogs, as well as higher densities of native predators along edges (Santner 1992b). The increased density of roads may affect adult survival because foraging or roosting birds are struck by vehicles at night (Santner 1992b). Further, extreme fragmentation due to urbanization or agriculture may eliminate all breeding opportunities. Minimum forest size to sustain a whip-poor-will pair is unknown; however, small isolated woodlots in some agricultural settings do not appear to be used (Reese 1996).

Natural maturation of eastern forests has also been proposed as a possible explanation for declines (Genoways and Brenner 1985, Santner 1992b). Most of Pennsylvania's forests are relatively mature (80 to 120 years) and may be starting to exceed the age structure used by whip-poor-wills, because early to mid-successional stands appear to be preferred over mature forest. Yet 58 percent of Pennsylvania's forests are mature stands of large trees (U.S. Department of Agriculture 2004). Lack of forest regeneration due to overabundance of white-tailed deer (*Odocoileus virginianus*) has been implicated in declines of some bird species (Goodrich et al. 2002) and could similarly affect whip-poor-wills. Pesticides, acid rain, or other pollutants could also negatively affect the whip-poor-will, either directly by killing adults or indirectly by reducing the availability of its insect prey (Eastman 1991b); however, these potential factors remain untested.

### *CONSERVATION AND MANAGEMENT NEEDS*

Current knowledge of whip-poor-will abundance, distribution, population status, and threats is poor at best. This lack of knowledge prevents managers from knowing the status of the whip-poor-will and threats to populations throughout Pennsylvania and, thus, seriously hampers any efforts to manage the species effectively and prevents managers from knowing whether and where management is needed in the first place. Because of this lack of information, the two most important initial conservation efforts needed for this species both relate to information gathering.

First, the author recommends that a statewide whip-poor-will survey program be developed and performed annually to determine the distribution, areas of highest abundance, and population trends of whip-poor-wills in Pennsylvania. Surveys should be coordinated and methodology shared, with agencies or nongovernment organizations in other eastern states to maximize the information and conservation values of these efforts. A model survey protocol has been developed for the common poorwill (*Phalaenoptilus nuttallii*) in British Columbia (British Columbia Ministry of Environment, Land, and Parks 1998). These methodologies likely could be adapted for whip-poor-will monitoring in the eastern United States.

Second, because most evidence suggests widespread declines, conservation efforts are needed to maintain current population levels in the short term and provide for some recovery in the long term. Unfortunately, because the causes of the declines are not fully understood, appropriate management actions are not clear. As a result, the author recommends that the state supports research evaluating how human activities (e.g., forest fragmentation, road building, pesticide use) and how natural changes in Pennsylvania's forests (e.g., forest maturation, deer overabundance) affect whip-poor-will abundance, survival, and reproductive success.

### *MONITORING AND RESEARCH NEEDS*

Current monitoring efforts (i.e., Breeding Bird Survey) are inadequate for this species. Several eastern states have initiated volunteer-based surveys to attempt to better monitor this species (e.g., Audubon Society of New Hampshire 2004, Connecticut Department of Environmental Protection 2004, Massachusetts Division of Fisheries and Wildlife 2004). These efforts generally entail nocturnal surveys for whip-poor-will calls at fixed points along designated routes during the breeding season and probably are effective at determining whip-poor-will abundance and habitat associations at these local sites. Further, if repeated annually, they would provide population trend information. Similar surveys are needed in Pennsylvania as soon as possible. Ideally, large-scale surveys should be initiated to document patterns statewide. A modified Breeding Bird Survey methodology using volunteers along existing Breeding Bird Survey routes may be an effective and cost-efficient way to accomplish this need. However, even limited surveys on local sites would be useful to establish baseline data that could be used to begin evaluating local population trends.

Much speculation has been made concerning the causes of perceived population declines. Population-level studies are needed to study possible causes of whip-poor-will declines by evaluating how local populations are being impacted by human activities (e.g., fragmentation, pesticides), forest maturation, and deer overabundance. Current understanding of the habitat needs of the whip-poor-will is coarse at best. No data are currently available regarding the relationship between whip-poor-will abundance and forest type, structure, or size (Cink 2002), nor are data available about how habitat and landscape features affect breeding productivity. Intensive, population-level studies are needed to determine what features constitute high-quality whip-poor-will habitat in Pennsylvania.

Problems on the wintering grounds could be an important contributor to population declines detected in Pennsylvania. Knowledge of the importance of wintering ground events is critical to understanding population trends in Pennsylvania and globally.

*Author:* CHRISTOPHER B. GOGUEN, PENNSYLVANIA STATE UNIVERSITY

## Chimney Swift

Order: Apodiformes
Family: Apodidae
*Chaetura pelagica*

The chimney swift is a rapid-flying aerial insectivore often described as looking like a cigar with wings (fig. 5.130) It was selected as a Species of Maintenance Concern because of population declines both within Pennsylvania and range-wide (Sauer et al. 2008). It is a Partners in Flight IIA priority species. Global populations are Secure (G5, NatureServe 2009).

### *GEOGRAPHIC RANGE*

The chimney swift is common over the eastern half of North America. Its range extends from southern Canada to the Gulf of Mexico and from the Rocky Mountains and central Texas east to the Atlantic Coast. It is the only member of its family to nest east of the Rocky Mountains. It is primarily a trans-gulf migrant between the southern United States and Mexico. The chimney swift spends the winter months in the western portion of the Amazon Basin, with some populations wintering west of the Andes south to northern Chile (Collins and Farrand 1985, Baughman 2003, NatureServe 2005).

*Fig. 5.130.* The Chimney Swift, *Chaetura pelagic*. Photo courtesy of Robert S. W. Lin.

### *DISTRIBUTION AND RELATIVE ABUNDANCE IN PENNSYLVANIA*

The chimney swift is widely distributed across Pennsylvania with highest densities in suburban/urban settings (fig. 5.131). The abundance of chimney swifts across the state is directly tied to the availability of suitable nesting sites—the inside wall of a chimney or hollow tree. Geographically, chimney swifts are reported in greater densities in the piedmont region (southeast and western counties) with lower densities throughout the Ridge and Valley provinces and lowest densities in the north-central part of the state (McWilliams and Brauning 2000). The 2004 Pennsylvania Migration Count data (Etter 2004) indicate highest densities in areas surrounding the bigger cities, such as Philadelphia and Pittsburgh. Still common across most of the state, Breeding Bird Survey data reported a general decline in chimney swift abundance in Pennsylvania and over its range continent-wide (fig. 5.132; Sauer et al. 2008).

### *COMMUNITY TYPE/HABITAT USE*

The chimney swift is a species that has adapted well to human changes in the landscape. It is most often associated with human habitation in large cities,

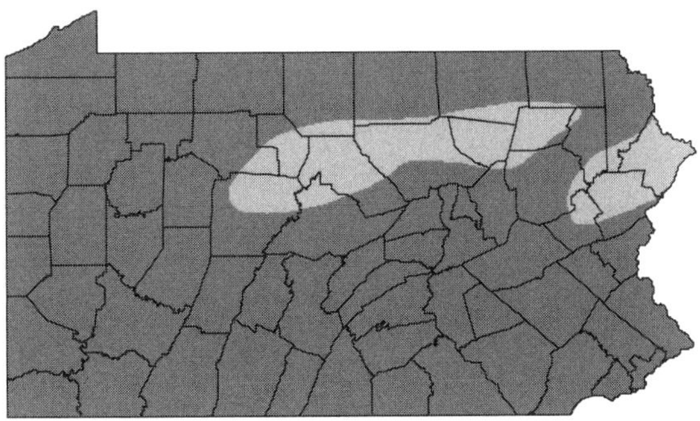

*Fig. 5.131.* Primary (darker shading) and secondary (lighter shading) distribution of the Chimney Swift, *Chaetura pelagic*.

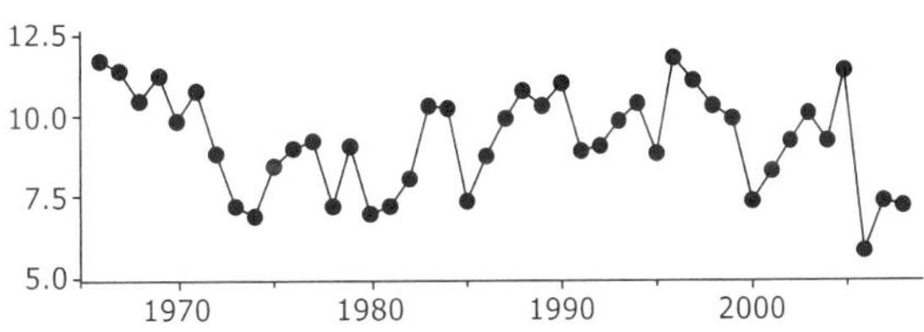


*Fig. 5.132.* Chimney Swift, *Chaetura pelagic*, population trends from the Breeding Bird Survey.

boroughs, villages, and rural areas. Human-made open chimneys mimic the preferred dark vertical shafts of large hollow trees used for nesting and roosting. Suitable chimneys are constructed from stone, firebrick, or any masonry that provides textured surface for grasping and nest building (chimneyswifts.org 2004, Kyle and Kyle 2004). Chimney swifts continue to use stands of old growth forest for nest sites when available (McWilliams and Brauning 2000). Distributions are higher in forest sites with abundant water resources (Wakeley and Roberts 1996).

### *LIFE HISTORY AND ECOLOGY*

The chimney swift is adapted to life in the air. The family name, Apodidae, meaning "without feet," indicates the reduced legs and feet of this species. With small limbs, it is difficult to walk on land or perch in trees. Food resources primarily consist of flying insects and on occasion spiders. During the breeding season, they are extremely vocal over nesting sights, emitting a series of rapid chirps that together sound like a metallic trilling (Baughman 2003, chimneyswifts.org 2004). They will fly in tandem with their partners alternating between rapid wing beats and gliding. At the conclusion of the breeding season, territorial behaviors are replaced by communal behaviors. Flocks and large aggregations normally occur at the onset of the fall migration (chimneyswifts.org 2004). Chimney swifts roost in chimneys or in hollow trees, clinging to the surface vertically. During the fall migration, they will communally roost with hundreds or thousands of individuals (Kyle and Kyle 2004).

Chimney swifts are monogamous and show strong mate fidelity, remaining with the same partner and returning to the same location for life (Dexter 1981). For nesting, they gather small twigs from the tree canopy (Shelly 1929, Kyle and Kyle 2004). They use their saliva to cement twigs to the wall of the chimney or hollow tree, forming a shallow cup. Chimney swifts have one brood a year between May and July but will double clutch if the first nesting fails early on (Allsop 2001, chimneyswifts.org 2004, NatureServe 2005).

### *THREATS*

Loss of nest sites and pesticide use are the two primary anthropogenic threats to chimney swift populations in Pennsylvania. Loss of nest sites is primarily due to loss of open chimneys. Many older structures have covered or screened chimneys to keep out unwanted nuisance species such as raccoons (NatureServe 2005). Some new structures are built without a chimney or with smooth metal chimney pipes. Metal chimney pipes should always be capped. These are unsuitable for swifts and any animal that enters this type of flue will be unable to climb out (Kyle and Kyle 2004). The second threat centers on the use of pesticides to control flying insects. Integrated pest management techniques (IPM) include inspection and insect monitoring, introduction of natural predators, and biological pesticides. It also includes use of chemical pesticides as needed. Use of pesticides and biological controls on farmlands and near human habitation affects the primary food source of these birds. Pesticides are also persistent in the environment after the original application. Other threats to chimney swift populations include natural predation. Predators, such as screech owls (Kyle and Kyle 2004), have been identified at nest and roost sites. Chimneys near low-lying branches generally have an increased chance of predation by snakes (Cink 1989). Reproductive failure has been linked to severe cold, wet conditions during the breeding season (NatureServe 2005).

### *CONSERVATION AND MANAGEMENT NEEDS*

Although not yet considered a Threatened Species in Pennsylvania, measures can be taken to maintain a healthy population of chimney swifts. Protecting high-density population areas and migratory roost sites can be initiated through community education. Recommendations to owners of historic structures to keep chimneys open and to avoid covering or screening would help keep nest and roost sites available. Local historical societies or commissions that are aware of the long association with chimney swifts could include use of open chimneys in development or renovation plans in historic districts.

### *MONITORING AND RESEARCH NEEDS*

A coordinated monitoring effort focused on urban species, including the chimney swift and common nighthawk (*Chordeiles minor*) is needed. Both species could be monitored at the same time. An effort should be made to locate large roost sites during the fall migration. This monitoring may indicate where declines are occurring and provide insight into the causes. Management priorities will become evident based on the data collected through intense monitoring. In addition, surveying old growth forests throughout the state for swifts nesting in natural habitats would be useful. These combined efforts will give us a better understanding of the overall status of chimney swifts in Pennsylvania.

The most pressing need is a robust monitoring program that emphasizes population density limits in suitable habitat (Rich et al. 2004). After determining population trends, important roost sites could be located and catalogued. Researchers could monitor for habitat change through changes in land use and management (changes in pesticide use, land use changes that affect the number of roost sites). In addition, a better understanding of pesticide use and its secondary affects to the food web would benefit all insect eaters.

*Authors:* DARRYL SPEICHER, POCONO AVIAN RESEARCH CENTER; JACKIE SPEICHER, POCONO AVIAN RESEARCH CENTER

## Red-headed Woodpecker

Order: Piciformes
Family: Picidae
*Melanerpes erythrocephalus*

The red-headed woodpecker is one of the showiest members of its family (fig. 5.133). It was listed as a Species of Maintenance Concern because of declining populations and a spotty distribution within the state. The red-headed woodpecker is listed as a Species with High Continental Concern and Low Regional Responsibility by Partners in Flight for the Allegheny Plateau Region. Global populations are considered Secure (G5, NatureServe 2009).

### *GEOGRAPHIC RANGE*

The red-headed woodpecker occurs across most of the United States east of the Rocky Mountains. Once one of the most common and conspicuous members of its family, the red-headed woodpecker has become increasingly uncommon in the last half-century. The species is still commonly encountered in the Midwest and locally in the southeastern United States but becomes increasingly rare in the Northeast, particularly in the New England states. Populations in northern states are highly localized and generally in decline.

*Fig. 5.133.* The Red-headed Woodpecker, *Melanerpes erythrocephalus*. Photo courtesy of Joe Kosack, PGC photo.

### *DISTRIBUTION AND RELATIVE ABUNDANCE IN PENNSYLVANIA*

During the first Pennsylvania Breeding Bird Atlas (1984–1989), they were found in nearly every county but were considered much reduced from previous decades. Today, in Pennsylvania, the red-headed woodpecker is an uncommon species found principally in the western and south-central portion of state (fig. 5.134). Results of the second Pennsylvania Breeding Bird Atlas indicate a decrease of 50 percent in distribution has occurred since the first atlas (Mulvihill and Brauning 2009). The Breeding Bird Survey route data indicate an annual rate of decline throughout the species range of approximately 2.5 percent per year since the 1960s, and populations have shown nonsignificant declines in Pennsylvania (Sauer et al. 2005; fig. 5.135). Pennsylvania is near the edge of the species current breeding range and hosts a relatively small, mostly migratory breeding population. Overwintering birds are rare and may originate from breeding populations farther north.

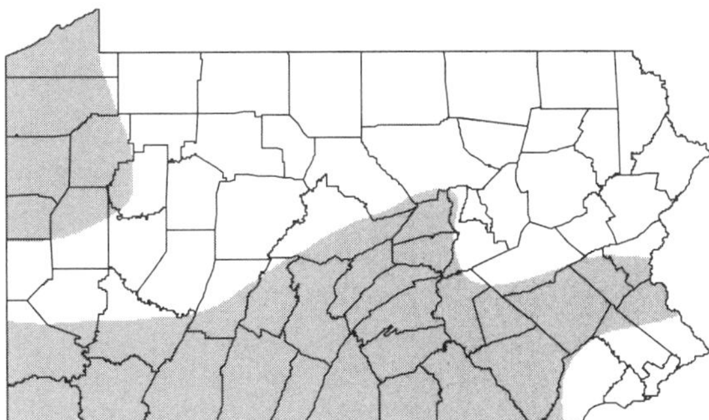

*Fig. 5.134.* Highest probability of occurrence of the Red-headed Woodpecker, *Melanerpes erythrocephalus*.

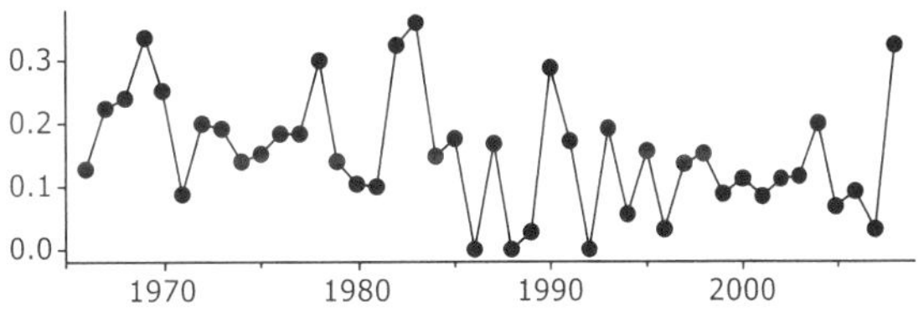


*Fig. 5.135.* Red-headed Woodpecker, *Melanerpes erythrocephalus*, population trends from the Breeding Bird Survey.

This species has exhibited great population swings in North America. Once such a common crop pest that it had a bounty on its head, red-headed woodpeckers have increased and decreased at least twice since North America was settled. Forest tree mortality, which increases dead standing snags preferred for nesting and roosting sites, seems to have been an important factor in past population increases (Smith et al. 2000). Within Pennsylvania, the red-headed woodpecker is the rarest breeding member of its family. Once considered common to abundant in many areas, it is at best an uncommon bird in prime habitats. Predictions based on possible climatic changes related to global warming call for increases in the red-headed woodpecker population throughout the northeastern United States, including Pennsylvania (Matthews et al. 2004).

### *COMMUNITY TYPE/HABITAT USE*

The red-headed woodpecker inhabits open deciduous forest with mast-producing oak, hickory or beech trees, and standing dead trees (snags) or dead limbs. Oak-savannah, forest edges, and woodlots bordered by agriculture and flooded forest, and swamps and beaver ponds with numerous snags are common concentration points (Conner et al. 1994, Smith et al. 2000). Nearby orchards and agricultural cropland may be an important component of this species' habitat. Forest edges in agricultural areas and other open areas with small woodlots on the Piedmont Province and the glaciated section of the Appalachian Plateau are prime habitats in Pennsylvania. Removal or clearing of fencerows adjacent to agricultural areas, burning of elm trees killed by Dutch elm disease, and heavy usage of dead trees for firewood has resulted in the loss of thousands of dead snags, which could have served as nesting or roosting sites for this species. Red-headed woodpeckers tolerate considerable disturbance and accept altered environments provided they have adequate food and nesting and roost sites.

### *LIFE HISTORY AND ECOLOGY*

Red-headed woodpeckers spend much of their time flycatching rather than scouring tree trunks and branches for prey. Presumably, it is their penchant for aerial sorties that lies behind their preference for open habitats. Dead snags in beaver ponds are commonly used by this species (Lochmiller 1979), which prefers dead snags for nesting, roosting, and foraging activities. In addition to flycatching, the species feeds on beech and acorn mast, which it often stores for later use. Foraging on the ground also occurs, as it does with northern flickers (*Colaptes auratus*). Early settlers considered them significant crop pests and frequently shot them as they foraged on fruit, berries, and corn. One hundred were reported shot from a single cherry tree by Audubon (1840).

This species normally migrates south of the state seeking areas of high mast production (Smith et al. 2000) although occasionally individuals (usually immatures) are noted at feeders during the winter. Individuals may remain in good mast years, relying on acorns stored during the fall (T. L. Master, personal communication). Migrants concentrate at locally abundant food supplies and may congregate in areas providing abundant food and roost sites (e.g., flooded dead timber areas). Most red-headed woodpeckers in Pennsylvania are migratory, returning in the spring to breed in late May.

The red-headed woodpecker is frequently found nesting along fencerows and on the border of forest fragments. Logged-over areas may be inhabited if they provide the numerous bare dead snags that this species selects for roosting and nesting sites. Even telephone and electric utility poles may be preferred nesting sites although the creosote-treated poles have been shown to cause reproductive failure (Rumsey 1970).

Nest cavities are always excavated in dead snags, usually in barkless stubs. Nest trees are often reused in subsequent years with the new nest excavated immediately below the previous one. Competition with other cavity nesters (e.g., European starlings, red-bellied woodpeckers) occurs (Ingold 1989), but red-headed woodpeckers are aggressive and appear capable of besting these competitors in the majority of cases (Smith et al. 2000) although competition with European starlings is commonly given as one reason for the species decline (e.g., see McWilliams and Brauning 2000). The male excavates a new gourd-shaped nest cavity in about two weeks (Smith et al. 2000). Compared with most members of the family, they are relatively late nesters, often taking over a site vacated by European starlings or other early cavity nesting species. Nesting at our latitude occurs from late April into July. Although southern birds often raise a second brood, in Pennsylvania, only a single clutch is generally laid (Schutsky 1992b).

### *THREATS*

Although the clearing of fencerows, harvesting of dead trees for firewood, and competition with European starlings (Schutsky 1992b, 2000) may have reduced available habitat somewhat, the specific causes behind their most recent decline remains largely conjecture. Further study to identify problems/threats/concerns that may adversely affect the species is required. Because red-headed woodpeckers typically use the same nest tree for several years, protection of all known nest sites is important. Removal of known or potential nesting trees is a primary threat to red-headed woodpeckers in Pennsylvania. Frequently dead or dying trees in suburban areas are removed for safety reasons before reaching the point of decay (i.e., when the bark has fallen off) where they are attractive to red-headed woodpeckers (D. Kibbe, personal observation). Because reproductive success in chemically treated utility poles is known to be adversely effected (Rumsey 1970), telephone poles near known colonies should be replaced with self-weathering steel poles. Colonies near major high-speed highways may be at risk from vehicular collisions.

*CONSERVATION AND MANAGEMENT NEEDS*

Historically, this species was common, but in recent decades, it has become increasingly uncommon. Population peaks have been linked to tree (American chestnut and American elm) mortality that provided abundant snags for nest and roost sites. Protection of suitable nesting trees and known nest locations are the only conservation steps currently identified.

To manage this species effectively, biologists need to identify key habitat components common to red-headed woodpecker habitat usage throughout Pennsylvania. Thinning of dense forest by removal of smaller trees and use of fire to create open savannah have been suggested (Smith et al. 2000) as possible habitat enhancement tools that warrant further investigation. Additional research into the type, spacing, and other resource components essential to establishment of a breeding site is needed. Goals for effective regional management of the species should be established and should focus on those areas currently harboring the majority of the state's population. This may be complicated because the nest site is a critical habitat component, and dead, bare snags have a finite existence after which they are no longer attractive to the birds as potential nest and roost sites. Consequently, long-term maintenance of a colony requires provision of a series of dead, decaying snags.

*MONITORING AND RESEARCH NEEDS*

Surveys need to be conducted on two levels. Statewide surveys should be initiated to identify sites for further long-term monitoring, and focused standardized searches are needed in historic strongholds of the species to establish the current baseline population level. There has been no qualitative assessment of the change in this species status. No surveys have been established at a local level to determine the breeding density or turnover in nest sites. Implementation of surveys of breeding areas need to be undertaken. Use of tape playbacks to locate all birds in study areas may be useful as would banding and color marking of breeding individuals and their young to track dispersal, site fidelity, and reproductive behavior.

Existing populations need to be located and quantified. Research needs to be initiated at existing colonies to clarify threats and important habitat components. Relationships between colony size and reproductive success need to be related to availability of dead snags, land use, agricultural practices, and human disturbance. Further research is needed to clarify the species nesting requirements. Although many factors common to occupied nest sites are well known, the critical features that contribute to nest-area selection are not known. Many sites with apparently excellent habitat are unoccupied. Additional research into the type and spacing of other resource components essential to establishment of a breeding site is necessary. At present, nesting density and the factors that may limit it are also unknown. In addition, little is known about many aspects of the life history of the species. Social structure of breeding colonies, behavior, and survival of juveniles are just a few of the poorly known aspects of the red-headed woodpecker's biology. Threats to existing colonies need to be identified and addressed. Fidelity of adult and juvenile birds to past nest sites is poorly documented and factors that may influence it are unknown.

*Author:* DOUGLAS P. KIBBE, URS CORPORATION

## Acadian Flycatcher

Order: Passeriformes
Family: Tyrannidae
*Empidonax virescens*

The Acadian flycatcher is a small eastern songbird in the genus *Empidonax* (fig. 5.136). It was selected as a Species of Greatest Conservation Need because it is an indicator of high-quality riparian forests and because

*Fig. 5.136.* The Acadian Flycatcher, *Empidonax virescens*. Photo by Jacob W. Dingel III.

of its close association with eastern hemlock (*Tsuga canadensis*). It is not listed as Threatened or Endangered by any state, and global populations are considered Secure (G5, NatureServe 2009). This species has been classified as a priority species in the Ohio Hill physiographical region by Partners in Flight (Rosenberg and Dettmers 2004).

### *GEOGRAPHIC RANGE*

The northern edge of the distribution of the Acadian flycatcher extends to southeastern Minnesota along southern edges of Wisconsin, Michigan, New York, and Ontario, extending into New England through Rhode Island, Connecticut, and Massachusetts. The range extends southward to Alabama, Mississippi, Louisiana (except coastal areas), and the northern peninsula of Florida. The western edge extends along eastern Iowa, southeastern Nebraska southward through eastern Kansas, Oklahoma, and Texas (Whitehead and Taylor 2002). The wintering grounds extend from the Caribbean slope of Nicaragua through Costa Rica and Panama into northern South America, but the flycatcher is most abundant from Panama southward (Whitehead and Taylor 2002).

### *DISTRIBUTION AND RELATIVE ABUNDANCE IN PENNSYLVANIA*

Within Pennsylvania, the Acadian flycatcher has a statewide distribution, recorded in every county except for Susquehanna in the Northeast (Brauning 1992b; fig. 5.137). The Acadian flycatcher is considered common to abundant in the southern portion of state, with the largest concentrations occurring in southwestern Pennsylvania, where they were detected in greater than 90 percent of the blocks during the first Breeding Bird Atlas. Detections decrease eastward to about 40 percent of the blocks in Delaware County. A larger decline occurs in the northward direction onto the Allegheny Plateau (18 percent of blocks). Overall, the flycatcher was detected in 36 percent of the blocks statewide (Brauning 1992b). The number of Acadian flycatchers detected on Breeding Bird Survey routes has remained constant in recent years (fig. 5.138; Breeding Bird Survey Data).

### *COMMUNITY TYPE/HABITAT USE*

The Acadian flycatcher uses a variety of habitats throughout its range, including moist mature deciduous forests, wooded ravines, floodplain forest, river swamps, hammocks and cypress bays, rhododendron thickets, and second growth and pine plantations (Whitehead and Taylor 2002, NatureServe 2004). Regardless of habitat type, there must be an open understory, a high dense canopy (Bushman and Therres 1988), and usually a nearby stream (Hamel et al. 1982, Becker et al. 2008). These flycatchers also appear to be area sensitive, requiring larger undisturbed tracts of forest (Blake and Karr 1984, Robbins et al. 1989, Peterjohn and Rice 1991). Smaller areas will be used if

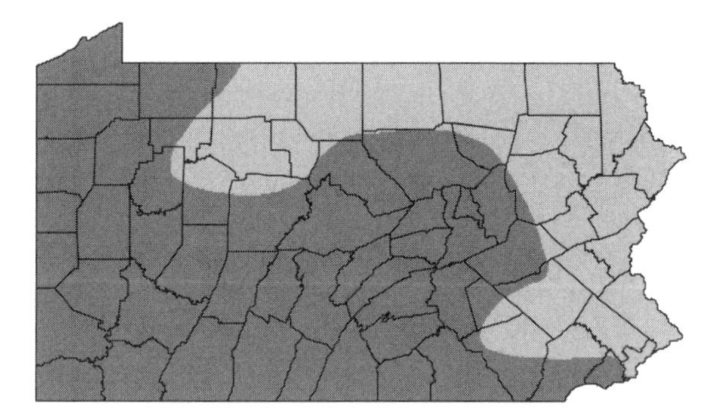

*Fig. 5.137.* Primary (darker shading) and secondary (lighter shading) distribution of the Acadian Flycatcher, *Empidonax virescens*.

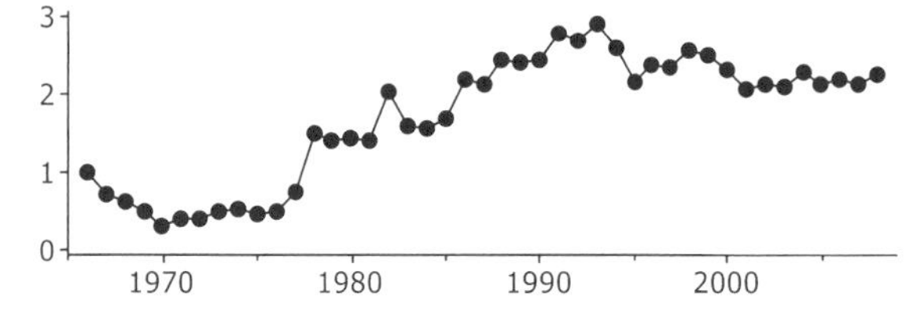


*Fig. 5.138.* Acadian Flycatcher, *Empidonax virescens*, population trends from the Breeding Bird Survey.

close to larger forested tracts (Bushman and Therres 1988). In Pennsylvania, the Acadian flycatcher uses two primary habitats (riparian deciduous forests and riparian eastern hemlock [*Tsuga canadensis*]) (Brauning 1992a). Habitat selection shifts from deciduous in the southwestern portion of the state to hemlock in the north; however, little is known about the interaction between these two habitats within transitional regions. In deciduous habitats, the flycatcher is often associated with American beech (*Fagus grandifolia*; Brauning 1992a) and witch hazel (*Hamamelis virginiana*; Becker et al. 2008).

### *LIFE HISTORY AND ECOLOGY*

The Acadian flycatcher forages within the understory up to the lower canopy (2–12 m), primarily on insects and their larvae but also on other arthropods such as spiders (Whitehead and Taylor 2002). This flycatcher is generalized and opportunistic in habitat use (Whitehead and Taylor 2002) but requires snags for foraging and exposed perches in the midstory (Hamel et al. 1982). From these snags, the flycatcher scans quietly and sallies forth to pluck prey items from the undersides of leaves. This species also occasionally will hawk insects in the air and glean insects from branches and boles (Whitehead and Taylor 2002).

Acadian flycatchers are migratory, initiating nesting in the mid-Atlantic region in late May and ending nesting in mid-August with the peak occurring from early June to early July (Bushman and Therres 1988). They migrate south in late August to early September (Whitehead and Taylor 2002).

Before the arrival of females, males establish a territory, typically averaging from –0.97 to 1.20 ha with a range from –0.53 to 1.70 ha (Mumford 1964, Walkinshaw 1966). Pair bonds are formed as soon as the females arrive and the female begins nest construction (Whitehead and Taylor 2002). The male defends the entire territory while the female defends the vicinity of the nest. Acadian flycatchers are usually single brooded; however, double brooding occurs, especially in the southern portion of its range (Whitehead and Taylor 2002, NatureServe 2004). Multiple nesting attempts will be made if initial nests fail because of depredation (Whitehead and Taylor 2002). Nests are usually found along woodland streams and placed between the understory and lower canopy (3–9 m) in the fork of a small horizontal branch with the nesting tree species varying by geographic region (Whitehead and Taylor 2002). In southern Pennsylvania, rich, wooded stream valleys with mature timber are used (Brauning 1992a) with a shift to riparian hemlock stands in the north (Morris et al 1984). Sheehan (2003) found significant preference for hemlock as a nesting substrate in the Delaware Water Gap National Recreation Area. Sites heavily infested with woolly adelgid (*Adelges tsugae*) exhibited considerably lower measures of productivity than lightly infested sites. Nesting success varies by location and year with estimates ranging from 10 percent to 25 percent (Mayfield Estimate, Wilson and Cooper 1998) to 43 percent (Mayfield Estimate, Becker et al. 2008) to 33–50 percent (percentage of nests fledged; Whitehead 1992). The primary sources of nest mortality are predation and cowbird parasitism (Wilson and Cooper 1998).

### *THREATS*

The primary threat to the Acadian flycatcher is the loss, fragmentation, and degradation of forested habitats to development. Acadian flycatchers, being area sensitive, rely on interior forest habitat associated with streams. Less than 20 percent of the southwestern portion of the state's forests is interior in nature and this region, along with the southeast, contains the highest density of roads in the state (Goodrich et al. 2002). Pennsylvania ranks second in the nation in the conversion of total acres of land to development (losing more than 1 million acres from 1992 to 1997, Samuels and Elliot 2000). Robbins (1980) predicts that a minimum of 30–50 ha are necessary to maintain viable flycatcher populations; however, the size could be much larger. Nott et al. (2001) suggests maintaining 500–900 ha of contiguous forest while Whitehead and Taylor (2002) suggest an ideal patch size of >10,000 ha. In addition, fragmentation increases parasitism and nest predation, which are the most important mortality sources in nesting success (Wilson and Cooper 1998).

The infestation of the hemlock woolly adelgid (HWA) has led to an overall decline in the quality of hemlock habitats in the state and, in some cases, complete loss of the tree. As of 2003, forty-two counties have confirmed HWA infestations, and this number will only increase because of the numerous dispersal vectors. With the decline of hemlocks, a shift to other riparian habitats can occur (Becker et al. 2008); however, the large-scale effects of this change on nest success and productivity are unknown. The continued deforestation of the Tropics will have a negative affect because Acadian flycatchers also rely on mature forests on their wintering grounds.

The Acadian flycatcher is a common host for the brown-headed cowbird (*Molothrus ater*); however, the parasitism rate is lower than for many other forest songbirds (Whitehead and Taylor 2002). Parasitism rate is usually related to the degree of forest fragmentation and positively correlated with percentage of forest cover, patch size, and percentage of forest interior (Whitehead and Taylor 2002). Parasitism varies annually and geographically, with rates from 0 percent to 50 percent across its range (Whitehead and Taylor 2002).

*CONSERVATION AND MANAGEMENT NEEDS*

In the southwestern portion of the state, large mature, riparian deciduous forest patches that remain should be identified, protected, and fragmentation reduced to provide suitable breeding habitat and maintain current population levels. In addition, a large-scale plan to protect hemlock habitat is necessary. The best plan of action toward this goal will be to support the Pennsylvania Department of Conservation and Natural Resources in its hemlock woolly adelgid management, to increase host resistance in eastern hemlock and continue trials of released predatory beetles. Finally, further management objectives should be outlined once the long-term reproductive effects of hemlock decline on Acadian flycatchers have been identified.

*MONITORING AND RESEARCH NEEDS*

The existing monitoring protocol from the Pennsylvania Department of Conservation and Natural Resources, as well as local volunteers, should allow for adequate monitoring of hemlock stands, while the Breeding Bird Survey and Breeding Bird Atlas should provide sufficient generalized statewide monitoring. In addition, researchers with experience monitoring hemlock health should be consulted to identify monitoring gaps or regions of special concern within hemlock habitats. The effectiveness of conservation actions can be determined through existing surveys. Besides regular statewide monitoring of the Acadian flycatcher and its habitat, a number of research needs exist. It is important to characterize the transition of flycatcher habitat selection from preferred riparian deciduous forests in the Southwest to hemlock habitats in the North and determine the variables that affect this habitat shift. In addition, researchers should evaluate the effects of hemlock decline on flycatcher ecology. While a few studies have evaluated the effects of hemlock decline on bird communities (Ross 2001, Tingley et al. 2002, Becker et al. 2008) long-term research is needed to determine shifts in habitat selection following hemlock decline and how these affect reproductive success. On the breeding grounds, research is needed to evaluate minimum-area requirements to determine better estimates for minimum viable population size and the effects of fragmentation (NatureServe 2004), determine the number of young produced per pair, annual survivorship, effects of cowbird parasitism, postbreeding dispersal, and natal dispersal using marked populations. Additional focus should be placed on evaluating source-sink landscape dynamics and understanding the scale at which patches are linked as metapopulations. The most effective method, as a result of low recapture rates of hatch-year individuals, would be DNA microsatellite markers (Whitehead and Taylor 2002).

*Author:* DOUGLAS BECKER, PENNSYLVANIA STATE UNIVERSITY

## Alder Flycatcher

Order: Passeriformes
Family: Tyrannidae
*Empidonax alnorum*

The alder flycatcher is a small flycatcher in the confusing genus *Empidonax* (fig. 5.139). It so closely resembles its look-alike congener, the willow flycatcher (*E. traillii*), that until 1973 they were considered a single species, Traill's flycatcher (American Ornithologists' Union 1973). As most published studies of Traill's flycatcher have been of willow populations, relatively little is known about many aspects of alder behavior and ecology. Alder flycatchers were selected as a Species of Greatest Conservation Need because of their limited range in Pennsylvania and their association with high-elevation shrub/scrub wetlands. Breeding populations in Pennsylvania are considered Vulnerable. Global populations are Secure (G5, NatureServe 2009).

*GEOGRAPHIC RANGE*

The alder flycatcher has the northernmost breeding range of any *Empidonax* flycatcher, extending across sub-Arctic Canada and Alaska, south to Ohio and Pennsylvania and through the Appalachians to North Carolina and Tennessee (Lowther 1999). It generally ranges north and east of the willow flycatcher, though the two species are widely sympatric, including in Pennsylvania (Lowther 1999). Alders migrate predominately east of the Great Plains to their winter-

*Fig. 5.139.* The Alder Flycatcher, *Empidonax alnorum*. Photo courtesy of Josiah LaCelle.

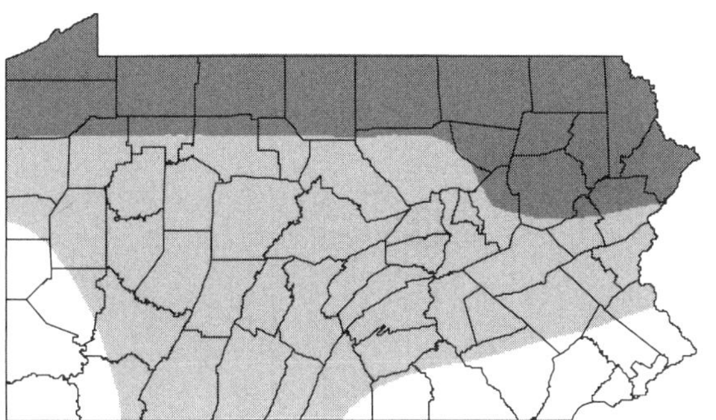

*Fig. 5.140.* Primary (darker shading) and secondary (lighter shading) distribution of the Alder Flycatcher, *Empidonax alnorum*.

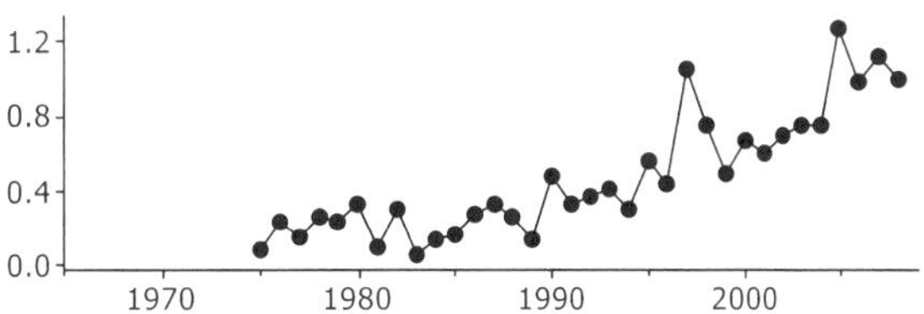


*Fig. 5.141.* Alder Flycatcher, *Empidonax alnorum*, population trends from the Breeding Bird Survey.

ing grounds in lowland areas of northern and central South America, south of the winter range of willow flycatcher (Stotz et al. 1996).

### *DISTRIBUTION AND RELATIVE ABUNDANCE IN PENNSYLVANIA*

In Pennsylvania, the alder flycatcher breeds primarily in the Glaciated Northeast and Northwest, and less commonly at higher elevations from the Allegheny High Plateau south through the Allegheny Mountains (McWilliams and Brauning 2000; fig. 5.140). Because of its specialized habitat requirements, the alder is local in distribution, having been recorded from only 7 percent of blocks (but two-thirds of the state's counties) in the first Pennsylvania Breeding Bird Atlas (Mulvihill 1992b). It probably occurs as a migrant in spring and fall across the entire state, but distinguishing it from the willow flycatcher in these seasons is difficult. Nowhere abundant within Pennsylvania, the alder flycatcher becomes increasingly common to the north of the state but is scarce and local to the south. Although ranked as Vulnerable, alders may be becoming more abundant in Pennsylvania; Breeding Bird Survey results suggest populations within the state have increased at a rate of 9.6 percent per year since 1980 ($P = 0.001$), although confidence in that estimate is not high (Sauer et al. 2004). Nationally, alders have increased since 1980 at a rate of 1.1 percent annually ($P = 0.04$; Sauer et al. 2004; fig. 5.141).

### *COMMUNITY TYPE/HABITAT USE*

Alder flycatchers breed in a variety of wet shrubby habitats, including brushy swamps, alder bogs, edges of beaver ponds, and wet meadows with woody vegetation. They occur less commonly in more upland habitats, such as overgrown fields or regenerating clear-cuts three to twenty years after harvest (Erskine 1984, Hobson and Schieck 1999). Elsewhere in the species' range, rights of way maintained in shrubs can support high densities of alder flycatchers (e.g., Marshall and Vandruff 2002), but whether they do so in Pennsylvania is not known (Yahner et al. 2002). Alders generally are found at higher elevations and in wetter, more wooded habitats than the willow flycatcher, although the two species can sometimes be found at the same site (Gorski 1970, Barlow and McGillivray 1983). Typical wetland breeding habitat consists of dense shrubby growth of willows (*Salix* spp.), alders (*Alnus* spp.), or dogwoods (*Cornus* spp.). More upland sites support shrubby viburnums (*Viburnum* spp.), *Spirea* spp., hawthorns (*Crataegus* spp.), elderberries (*Sambucus* spp.), roses (*Rosa* spp.), and briars (*Rubus* spp.; Mousley 1931, Mulvihill 1992b, Lowther 1999).

### *LIFE HISTORY AND ECOLOGY*

Alder flycatchers are summer residents only and migrate to their wintering grounds in the Neotropics.

They are among the last birds to arrive in the spring and the first to leave in the fall (Hussell 1991), rarely appearing in Pennsylvania before late May or staying beyond early September (McWilliams and Brauning 2000). On the breeding grounds, they establish and vigorously defend territories, which they advertise with their persistent, if undistinguished, song. Estimates of territory size vary considerably, from 0.2 to 3.0 ha. In some (but not all) areas of sympatry with the willow flycatcher, alders maintain interspecific territories but tend to be subordinate to the willows (Prescott 1987b, Lowther 1999). More often they segregate by habitat, alders preferring the taller, denser, and wetter areas. The diet of the alder flycatcher has not been quantified. Range-wide assessments of Traill's flycatcher diet (which included populations of both alder and willow) found the birds feed on a wide variety of arthropods, including bees and wasps (Hymenoptera), beetles (Coleoptera), flies and midges (Diptera), and butterflies and moths (Lepidoptera; Beal 1912, Bent 1942). Prey are caught primarily by aerial sallies or are gleaned from foliage.

Breeding usually begins well into June after females arrive on the breeding grounds. Female flycatchers select nest sites and build a somewhat sloppy nest relatively low (< 1 m) in a shrub. Pairs normally produce a single brood per season, though they are likely to rebuild and lay an additional clutch if their first attempt fails early. Alder flycatchers are sometimes parasitized by brown-headed cowbirds (*Molothrus ater*), although few data exist on species-specific parasitism rates. Alder populations in British Columbia, Ontario, and Quebec experienced brood parasitism rates of 10.5 percent, 15.0 percent, and 6.1 percent, respectively (Lowther 1999 and sources therein). The incidence of parasitism probably varies regionally and seasonally. Alders may respond to parasitism by abandoning the nest or burying the cowbird egg within the nest lining. Pairs that accept and raise cowbird chicks are likely to raise few or none of their own young, as is the case with willow flycatchers (Sedgwick and Iko 1999).

### THREATS

The principal threats to the sustainability of this species in Pennsylvania are the loss and conversion of its shrubby habitats. Between 1956 and 1979, Pennsylvania lost 6 percent of its vegetated wetlands; the most extensive losses have been in the Northeast, especially in the Pocono Plateau and Glaciated Northwest (Tiner 1990)—the areas of greatest alder flycatcher abundance. Early successional forests in the form of regenerating clear-cuts have decreased because of declines in timber harvests, shifts to uneven-aged management, and maturation of existing stands. Forest Inventory and Analysis data indicate that the area of Pennsylvania's forests in the sapling-seedling class has declined by more than 50 percent since 1950 (McWilliams et al. 2004). Chronic high deer densities can constrain or prevent forest regeneration and can limit the extent and height of new growth (Horsley et al. 2003) and therefore the suitability of such habitats for alder flycatchers. Also, in some areas (e.g., central New York), alders have been partly displaced by expanding populations of willow flycatchers (Stein 1963) but whether this pertains to Pennsylvania is unknown.

### CONSERVATION AND MANAGEMENT NEEDS

The increasing population trend of alder flycatchers within the state suggests no urgent need for conservation or management activities for the species. Over the longer term, it would be worthwhile to monitor trends in vegetated wetlands within the state through coordination with the National Wetland Inventory. Maintenance and restoration of shrubby wetlands could be encouraged through the development of an outreach program to educate private landowners, land managers, and wetland-oriented nongovernmental organizations (especially those involved with wetland restoration activities, e.g., Ducks Unlimited) about the importance of shrubby and forested wetlands (in addition to emergent wetlands) to nongame wildlife, including alder flycatchers.

### MONITORING AND RESEARCH NEEDS

Focused surveys using tape playback methodology would provide better estimates of distribution and abundance of both alder and willow flycatchers than are currently available through more general avian surveys, such as the United States Breeding Bird Survey or the Pennsylvania Breeding Bird Atlas. Protocols for such surveys could be readily developed from those already in use for the endangered southwestern subspecies of the willow flycatcher (Sogge et al. 1997). Monitoring the abundance and distribution of both species over time would reveal whether alders are being supplanted by willows in Pennsylvania, which has been reported elsewhere. Monitoring of demography at a subset of nesting sites would better elucidate any potential threats to alders, such as excessive brood parasitism by cowbirds, which, in turn, would determine whether future conservation actions need to be taken.

Effective management of the alder flycatcher in Pennsylvania would be facilitated through research in the following areas. Determine what features, at a variety of spatial scales (e.g., patch size, shape, connectivity, and landscape context), are correlated with source habitats for alder flycatchers so that managers can identify existing high-quality habitats (i.e., sources) for wetland protection, and incorporate habitat considerations into wetland restoration decisions and strategies. Changes to breeding habitats resulting from human activities or natural processes are likely to affect habitat quality for alder flycatchers, either positively (e.g., Norton and Hannon 1997) or negatively. Understanding how these changes affect productivity and population trends of alders will better enable their conservation in dynamic systems. Willows have supplanted alders in nearby areas of New York (Stein 1963), but whether such displacement occurs in Pennsylvania is unknown. A better understanding of differences in the microhabitat needs of the two species would facilitate the development of management strategies to sustain both species within the state.

*Author:* SCOTT H. STOLESON, PH.D., U.S. DEPARTMENT OF AGRICULTURE FOREST SERVICE

## Willow Flycatcher

Order: Passeriformes
Family: Tyrannidae
*Empidonax traillii*

One of several similar-appearing *Empidonax* flycatchers in Pennsylvania, it is practically impossible to differentiate the willow flycatcher visually from its former conspecific, the alder flycatcher (*Empidonax alnorum*; fig. 5.142). The willow flycatcher was selected as a Species of Greatest Conservation Need because it is a Partners in Flight Priority I (Continental importance) species. Both state and global populations are considered Secure (S5B, G5, NatureServe 2009).

### *GEOGRAPHIC RANGE*

The willow flycatcher is the most widely distributed *Empidonax* in North America (Sedgwick 2000), breeding throughout much of the United States and into southern portions of Canada (Kus and Sogge 2003) and wintering from central Mexico to northern South America (Lynn et al. 2003). Four or five subspecies are recognized, depending on the author; with *Empidonax traillii traillii* and *Empidonax traillii campestris* (which some synonymize with the former) breeding in Pennsylvania (Sedgwick 2000). The subspecies breeding in the Southwestern United States (southwestern willow flycatcher, *Empidonax traillii extimus*) is federally endangered (U.S. Fish and Wildlife Service 1995).

*Fig. 5.142.* The Willow Flycatcher, *Empidonax traillii*. Photo courtesy of Bill Moses.

### *DISTRIBUTION AND RELATIVE ABUNDANCE IN PENNSYLVANIA*

The history of the willow flycatcher in Pennsylvania is complicated by its former lumping with the alder flycatcher as Traill's flycatcher until 1973, when they were recognized as separate species (American Ornithologists' Union 1973). Early reports of the Traill's flycatcher in Pennsylvania list the bird only during spring and fall as a transient (Gentry 1877, Warren 1888). The first Traill's flycatcher nest was reported from Allegheny County in 1894 and was most likely that of a willow flycatcher (Mulvihill 1992c). By 1964, Traill's flycatcher bred locally throughout the state (Poole 1964), and this was primarily due to range expansion of the willow flycatcher (McWilliams and Brauning 2000). McWilliams and Brauning (2000) link this range expansion to the creation of suitable nesting habitat through extensive reversion of old farms and forest regrowth in the 1940s and 1950s. Range expansion has also been noted for this period in adjacent states (Mulvihill 1992c).

The species currently has a wide distribution within Pennsylvania (fig. 5.143). It was reported from every county, often as a confirmed or probable breeder, from the first Pennsylvania Breeding Bird Atlas (1983–1989) with the majority of reports coming from the western and southeastern portions of the state (Mulvihill 1992c). Data from Pennsylvania on the species' density as it occurs within different habitats in the state are unavailable. Density values vary widely according to sources reviewed in Sedgwick (2000); from 7.1 territories/km$^2$ in Michigan old growth woods and swamp forest to 111.1 territories/km$^2$ in Connecticut shrubby swamp and sedge hummock habitat. On average, 1.9 individuals were recorded on Breeding Bird Survey (BBS) routes in Pennsylvania from 1990 to 2003 (Sauer et al. 2005; fig. 5.144). Individual route means calculated from raw count data for this period varied considerably, however, from zero to sixteen individuals.

Statistical analysis provided by Sauer et al. (2005) for the Breeding Bird Survey in Pennsylvania supports the dramatic increase apparent in the mean/route trend information through the 1970s (1966–1979 trend +10.0%/year, P = 0.04, n = 44 routes), and further indicates that the population has been relatively stable since then (1980–2005 trend +0.6%/year, P = 0.43, n = 89 routes; fig. 5.144).

### COMMUNITY TYPE/HABITAT USE

Like the alder flycatcher, the willow flycatcher typically breeds in shrubby and wet habitats throughout its range, although the willow flycatcher appears to be more tolerant of drier habitats (Sedgwick 2000).

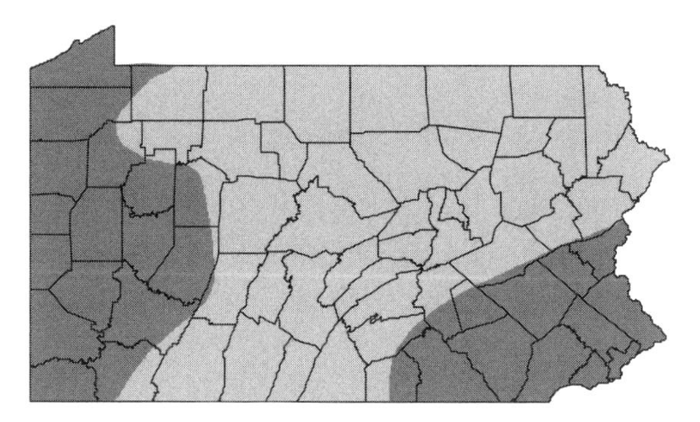

*Fig. 5.143.* Primary (darker shading) and secondary (lighter shading) distribution of the Willow Flycatcher, *Empidonax traillii.*

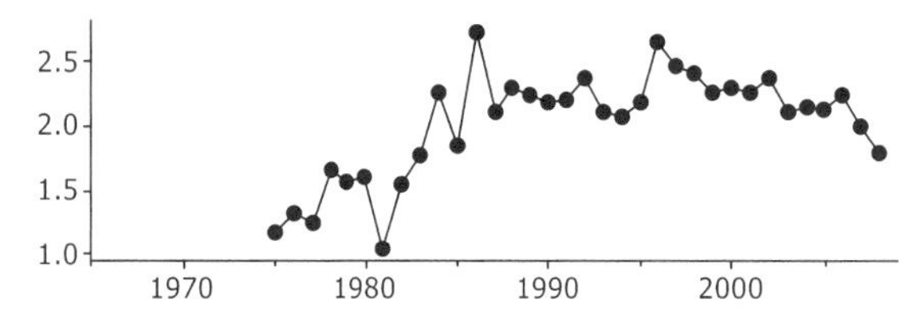


*Fig. 5.144.* Willow Flycatcher, *Empidonax traillii*, population trends from the Breeding Bird Survey.

In Pennsylvania, the species generally occurs below 1,800–2,000 feet, in wetlands such as shrub swamp, wet meadow, and brushy habitats along streams and the edges of ponds and marshes, and in uplands such as old field, shrubby pasture, and dry, brushy hillsides (Mulvihill 1992c, McWilliams and Brauning 2000). A tendency to use drier upland habitat is also noted for Ohio (Peterjohn and Rice 1991), although a recent survey in Alberta, Canada, found the species present in wetlands but absent from formerly occupied dry uplands (Kulba and McGillivray 2001). Habitat used in migration typically reflects breeding season habitat (McCabe 1991). Lynn et al. (2003) identify the presence of water, shrubs, patches or stringers of trees, and open areas as key wintering habitat components.

### LIFE HISTORY AND ECOLOGY

The willow flycatcher is a long-distance migrant and returns to the breeding grounds in late May and June (Sedgwick 2000). The male defends a breeding territory and may be visible on exposed perches from which it often sings (Sedgwick and Knopf 1992) while the female selects the nest site and builds the compact, open-cup nest low to the ground (usually 1 to 2 m) in a forked branch of a willow shrub (*Salix* spp.), or a variety of other shrubs and trees (Ehrlich et al. 1988, Sedgwick 2000). The nest is usually located on the periphery of the shrub or thicket and is often located near water (reviews in Sedgwick 2000). The species is almost wholly insectivorous and prey includes Hymenoptera, Coleoptera, Diptera, Lepidoptera, Hemiptera, and Odonata (Beal 1912, Prescott and Middleton 1988). Aerial hawking is the primary foraging mode, but the species will also hover-glean (Ehrlich et al. 1988). Breeding territory size estimates range from 0.1 ha to 1.8 ha (reviews in Sedgwick 2000).

The willow flycatcher has a short reproductive period (seventy to ninety days) following its return to the breeding grounds (Sedgwick 2000). Mammals, snakes, and birds are known nest predators, and parasitism by brown-headed cowbirds (*Molothrus ater*) may also affect nest success, but nest loss is usually followed by renesting attempts (Sedgwick 2000).

### THREATS

Loss of habitat, especially wetlands given the degree of their use, is likely the primary concern throughout the willow flycatchers' entire distribution. Pennsylvania's wetlands decreased by more than 50 percent in area from 1780 to 1980 (Dahl and Johnson 1991). Although this overall loss has slowed dramatically, substantial loss in the Poconos and the northwestern counties continues due to channelization, pond construction, and development (Tiner 1989, Goodrich et al. 2002). Potentially occupied upland habitat, such as early successional habitats and the mix of fields, hedgerows and woods characteristic of farmland areas, has been lost throughout the state, primarily due to urban and suburban sprawl and changes in farming practices (Goodrich et al. 2002). Although limited data on willow flycatcher ecology and distribution on the wintering grounds exist, habitat loss due to wetland draining, clearing of land for agriculture and livestock, logging, development, and erosion poses a considerable threat (Lynn et al. 2003).

Habitat degradation may also be a major threat. Possible degrading factors include exotic plants (e.g., invasion of wetlands in northwestern Pennsylvania by glossy buckthorn [*Rhamnus frangula*]; Possessky et al. 2000), livestock grazing of riparian habitat (Popotnik and Giuliano 2000), pollution on both the breeding and wintering grounds (Gard and Hooper 1995, Lynn et al. 2003), and changes in hydrology. Mortality due to collisions with towers (Shire et al. 2000) and window glass (Klem 1990) may also significantly affect populations.

*CONSERVATION AND MANAGEMENT NEEDS*

Because of continued, permanent losses of habitat and changes in habitat availability through natural and anthropogenic causes, populations should be maintained through preservation of high-quality habitats. Coordination of the various public agencies and private interests involved is vital to ensure the success of the conservation effort for such a widely distributed species. Habitats should be directly acquired when possible and the protection of habitat through landowner assistance programs such as the Conservation Reserve Enhancement Program (CREP) and educational outreach should be encouraged. Sources of habitat degradation specific to each site should be identified and mitigation measures devised and implemented if possible.

*MONITORING AND RESEARCH NEEDS*

Surveys and spatial data sources can be used to explore trends and availability of both wetland and upland habitats the willow flycatcher uses in Pennsylvania. An assessment of the condition and trends of land use and wetlands in the United States is produced every year and will eventually include state and substate data (U.S. Department of Agriculture 2002). The National Wetlands Inventory is mandated to provide status and trend reports to Congress at ten-year intervals (U.S. Fish and Wildlife Service 2002b). An updated land use classification for Pennsylvania based on remote-sensing data was recently produced, which provides information on the cover of woody wetlands and transitional habitats among others (Warner 2003). A new national land cover data set is now available that maps scrub-shrub wetlands and uplands, forest cover, commercial and residential development, and other useful data (U.S. Geological Survey's Multi-Resolution Land Characteristics Consortium, www.mrlc.gov). On the basis of these products, a five- to ten-year periodic assessment of potential habitat is recommended. Synthesis of this assessment with other potential sources of habitat information should provide a background for considering population trends and conservation actions. The primary population monitoring method should continue to be the annual Breeding Bird Survey because of its reliability and coverage. More intensive techniques that estimate density, such as distance sampling and territory mapping (see Bibby et al. 2000), could be employed in specific habitats or protected areas to address research needs or to predict the effects of conservation actions.

To best target conservation or management actions and more accurately forecast future population trends, several habitat-based research priorities are deemed important. The contribution of different habitats, especially water-associated versus dry habitats and stable versus transitional habitats to the state's willow flycatcher population, needs to be understood. A more fine-scale definition of Pennsylvania's suitable habitat, identification of regional habitat differences, and an assessment of temporal changes in habitat availability, such as the conversion of the state's emergent and forested wetlands to shrub-scrub wetlands (Tiner 1989), could be useful. If time and resources permit, additional effort should be allocated to studying the potential population effects of the habitat degradation and mortality factors identified earlier (e.g., invasive plants, tower and window kills, and pollution). Finally, habitat quality should be assessed by examining reproductive success in different habitats (sensu Van Horne 1983).

*Author:* JAMES SHEEHAN, MORGANTOWN, WEST VIRGINIA

## Yellow-throated Vireo

Order: Passeriformes
Family: Vireonidae
*Vireo flavifrons*

The yellow-throated vireo is the most brightly colored North American vireo in this family of predominantly drab birds (fig. 5.145). It was selected as a Species of Greatest Conservation Need as an indicator of forests with full canopies. It is listed by six other northeastern states. Breeding populations in Pennsylvania are Apparently Secure (S4B). Global populations are Secure (G5, NatureServe 2009). The yellow-throated vireo is reported as stable to increasing throughout its range and is not a federal or state-listed Threatened or Endangered Species (Zeller 1997).

### GEOGRAPHIC RANGE

The yellow-throated vireo is a bird of deciduous forests that breeds east of the Rockies and winters from southern Florida to Central and South America. The breeding range extends from southern Manitoba and the eastern Dakotas east across southern Canada and central Maine and south to Texas, the Gulf Coast, and central Florida (Scott 1987). This species winters from southern Florida to Central and South America.

### DISTRIBUTION AND RELATIVE ABUNDANCE IN PENNSYLVANIA

Yellow-throated vireos breed statewide in Pennsylvania, but the species' relative abundance across the state is somewhat uneven (fig. 5.146). In the High Plateau region of Pennsylvania, yellow-throated vireos are rare to absent breeders at elevations above 1,800 feet.

*Fig. 5.145.* The Yellow-throated Vireo, *Vireo flavifrons*. Photo courtesy of Bob Gress.

Yellow-throated vireos are common nesters in southwestern Pennsylvania and common to uncommon in the Glaciated Northeast, the southern Ridge and Valley, the northwest, and locally in the Piedmont. They are rare elsewhere in the state, particularly in the High Plateau. Except for the noted absence from higher elevations, the pattern of distribution for yellow-throated vireos in Pennsylvania does not appear to follow any known physical features (McWilliams and Brauning 2000). Population levels appear to be stable both in Pennsylvania and regionally (Sauer et al. 2008; fig. 5.147).

In Pennsylvania, migratory individuals are more often observed during the spring than in the fall. During the spring, yellow-throated vireos usually arrive in the southern portions of the state by the fourth week in April. Spring migration usually ends by the fourth week in May, with some individuals lingering until early June. During fall, yellow-throated vireos are rarely reported. Migration is under way by the fourth week in August and with all individuals departing by the first week in October (McWilliams and Brauning 2000).

### COMMUNITY TYPE/HABITAT USE

The yellow-throated vireo is a bird of deciduous forests, riparian woodland, tall floodplain forest, lowland swamp forest, mixed forest, orchards, and groves

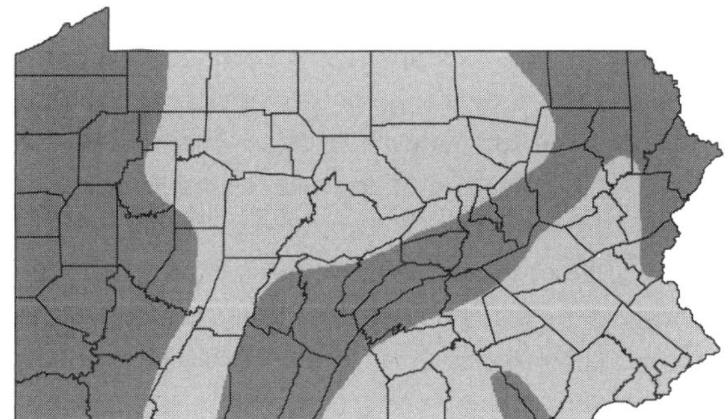

*Fig. 5.146.* Primary (darker shading) and secondary (lighter shading) distribution of the Yellow-throated Vireo, *Vireo flavifrons*.

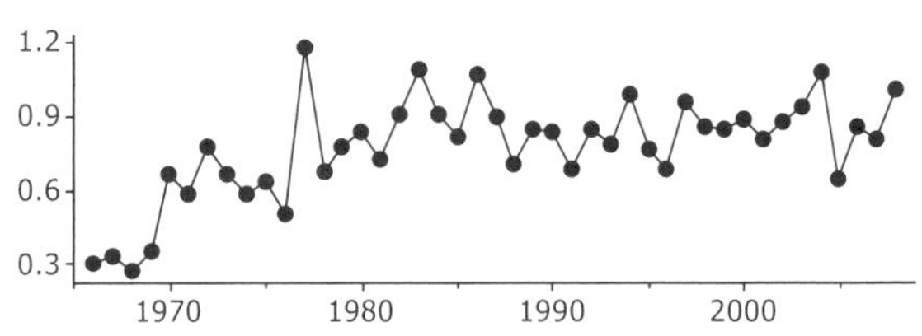


*Fig. 5.147.* Yellow-throated Vireo, *Vireo flavifrons*, population trends from the Breeding Bird Survey.

of shade trees. Yellow-throated vireos are not found in pure stands of conifers and generally do not breed in forest interior unless forest openings are present. The species spends much of its time in the upper canopy and seems to prefer habitats with an open understory during the breeding season (McWilliams and Brauning 2002). Zeller (1997) cites reports that the species requires a high, partially open canopy and may be intolerant of forest fragmentation; conversely Rodewald and James (1996) indicate yellow-throated vireos breed primarily in edge habitats.

Microhabitat of nest sites is not well documented. Forested habitats, including deciduous forest and woodland, riparian woodland, and floodplain forest, are key breeding habitat for this species. During migration, this vireo can be found in open woodlands, as well as more open habitats, including brushy woodland understories and edges (Zeller 1997, McWilliams and Brauning 2000).

*LIFE HISTORY AND ECOLOGY*

The diet of this species comprises mostly insects, but it may consume fruit and seeds during late summer and early fall (Chapin 1925). The yellow-throated vireo forages in the mid- to upper canopy, primarily on dead or dying interior limbs with little to no foliage. Males generally forage higher than females (Rodewald and James 1996).

In the mid-Atlantic region of the United States, yellow-throated vireos breed from mid-April to late July (Bushman and Therres 1988). Yellow-throated vireos are sensitive to human disturbance early in the nesting season. A female may desert a male or a pair may abandon a nest site if disturbed during pair formation or early in the nest-building stage. The species becomes more tolerant of disturbance once it is incubating (Rodewald and James 1996). Nests are generally located high in the tree canopy (1 to 18 m) in a horizontal twig fork (Baicich and Harrison 1997). McWilliams and Brauning (2000) provided egg dates of mid-May through early July in Pennsylvania. This species is sometimes parasitized by brown-headed cowbirds and has been documented successfully raising cowbird fledglings. If cowbird eggs are detected in the nest, a pair will sometimes build a second tier and renest over the parasitized clutch (Zeller 1997).

*THREATS*

Primary threats in Pennsylvania include forest loss and fragmentation, parasitism by brown-headed cowbirds, and widespread spraying to control forest insects, such as gypsy moths (Master 1992c). Little is known about the specific habitat requirements of the yellow-throated vireo and of the relationship between habitat condition and survival and reproductive success. Even less is known about the effects of habitat alteration or disturbance on this species. Therefore, it is difficult to identify more specific threats to this species.

*CONSERVATION AND MANAGEMENT NEEDS*

It is difficult to develop conservation and management needs for a species when knowledge of its biology and habitat requirements is limited. Therefore, research needs have been developed to identify conservation and management needs for yellow-throated vireo in Pennsylvania.

*MONITORING AND RESEARCH NEEDS*

Monitoring this species is needed to delineate the precise range in Pennsylvania. More intensive study is needed to identify specific habitats used in Pennsylvania and the condition of these habitats. Although additional information is needed on microhabitat requirements for this species, protection of large blocks of deciduous and riparian forest would undoubtedly benefit yellow-throated vireo, as well as other woodland species.

In Pennsylvania, the most important research needs are to identify natural history and specific habitat requirements and to identify the locations of key habitats in Pennsylvania. The species appears to range throughout Pennsylvania but exhibits differences in relative abundance in relation to physical/physiographic features. However, the relationships are poorly understood. An additional research need is to determine the effects of forest fragmentation, forest loss, and other habitat manipulations on survival and reproductive success.

*Author:* ROBERTA J. ZWIER, ENSR CONSULTING AND ENGINEERING

## Blue-headed Vireo

Order: Passeriformes
Family: Vireonidae
*Vireo solitarius*

The blue-headed vireo is a small, forest-dwelling songbird (fig. 5.148). It was selected as a Species of Maintenance Concern because of its close association with hemlocks and mixed-conifer forests. State and

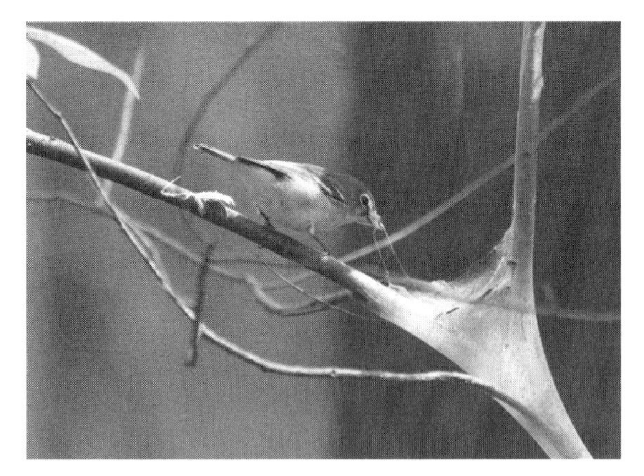

*Fig. 5.148.* The Blue-headed Vireo, *Vireo solitarius*. Photo by Jacob W. Dingel III.

global populations are considered Secure (S5B, G5, NatureServe 2009).

### *GEOGRAPHIC RANGE*

During the breeding season, the blue-headed vireo is widely distributed across much of the southern half of Canada and into the highlands of the eastern United States. Within Canada, the breeding range extends from the East Coast west as far as eastern British Columbia. In the United States, the breeding range includes northern Minnesota, Wisconsin, and Michigan, and in the East, the species ranges down the Appalachians through most of New England and New York, continuing south along the mountains to northern Georgia. During winter, the population is split between Mexico and the United States, with some individuals overwintering in the southeastern United States, from eastern North Carolina south through Florida and the southern halves of the other Gulf Coast states. The remainder of the population overwinters primarily in eastern and southern Mexico, mainly on the Gulf Coast mountains (James 1998).

### *DISTRIBUTION AND RELATIVE ABUNDANCE IN PENNSYLVANIA*

Historically, a common species throughout the mountains of Pennsylvania, the blue-headed vireo was detected in 30 percent of the state's atlas blocks in the first Breeding Bird Atlas (1983–1989; Brauning 1992f). The current distribution of this vireo corresponds closely with the most heavily forested areas of Pennsylvania (McWilliams and Brauning 2000; fig. 5.149), and this species appears to be increasing in abundance as a result of the ongoing recovery and regrowth of mature forests in the state (4.9% annual increase, P = 0.001, 66 routes; Sauer et al. 2004) (fig. 5.150). Blue-headed vireos are uncommon to fairly common breeders in the northern mountains of the Appalachian Plateau, ranging south through the Allegheny Mountains to the Maryland border. They are also uncommon to rare in the Ridge and Valley and Blue Ridge regions of the state. Locally, blue-headed vireos are more abundant in old growth forests containing pine and hemlock than in typical second growth woodlands (Haney 1999, McWilliams and Brauning 2000).

### *COMMUNITY TYPE/HABITAT USE*

A species of extensive forest, the blue-headed vireo breeds in a wide variety of mature coniferous or mixed deciduous–coniferous forests, particularly those with a high percentage of canopy closure and a well-developed woody understory. In the eastern United States, vireos are present in habitats ranging from conifer-dominated mountaintops or eastern hemlock (*Tsuga canadensis*) draws to mixed mesophytic forest, to pure hardwood forests (James 1998). Within Pennsylvania, the blue-headed vireo commonly uses mature and second-growth woodlands of all types during migration, but for breeding prefers mixed woodlands and conifer forests, and is often found in highest densities in hemlock-associated habitat types (Haney 1999, McWilliams and Brauning 2000, Swartzentruber and Master

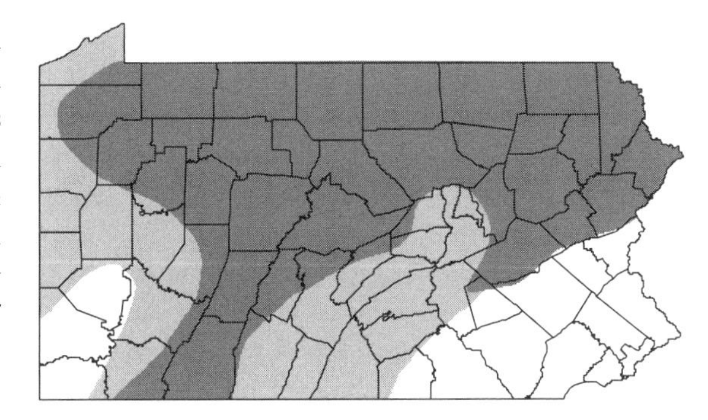

*Fig. 5.149.* Primary (darker shading) and secondary (lighter shading) distribution of the Blue-headed Vireo, *Vireo solitaries*.

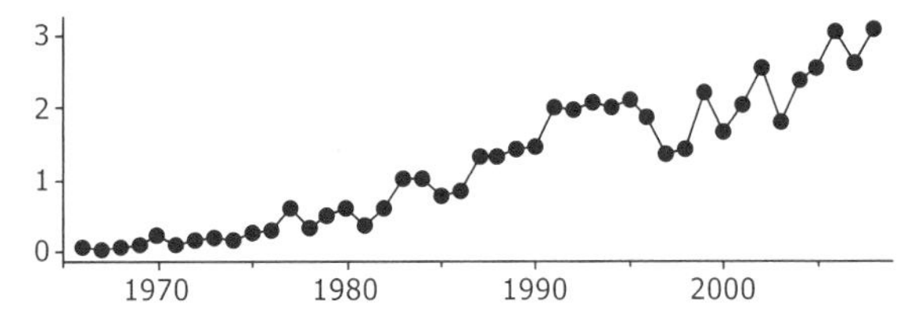


*Fig. 5.150.* Blue-headed Vireo, *Vireo solitaries*, population trends from the Breeding Bird Survey.

2003, Ross et al. 2004). This vireo breeds in deciduous woodlands above 600 m in northern counties (McWilliams and Brauning 2000).

*LIFE HISTORY AND ECOLOGY*

A medium distance complete migrant, the blue-headed vireo is present in Pennsylvania between mid-April and early November, although most nesting activity is restricted to May through July (McWilliams and Brauning 2000). During the breeding season, diet consists mostly of adult and larval insects, particularly lepidopterans and hemipterans. Although flexible in foraging site selection, this species tends to forage in the midcanopy of the forest more often than near the ground or in the upper canopy. Diet during migration or on the wintering grounds is poorly studied, but fruit may play an important role, particularly during winter in the Tropics (James 1998).

The blue-headed vireo breeds in a variety of mature conifer or mixed-conifer forest types that contain some, but not dense, understory shrubs or saplings (James 1998). Seasonally monogamous, this species builds a cuplike, hanging nest. Although poorly studied, double brooding may be common in southern populations. In northern populations, second broods are rare after an initial successful attempt; however, renesting following failure is common (James 1998).

Blue-headed vireos are common hosts of the brown-headed cowbird (*Molothrus ater*). Rates of parasitism vary widely among populations, but generally observed parasitism rates have been low (<15%) in populations in the northern United States and southern Canada, with higher rates (>40%) observed in some populations in the southern United States (James 1998, DeCecco et al. 2000). Although vireos will sometimes bury cowbird eggs into the nest lining when they are laid before clutch initiation, most eggs are accepted at great cost to the vireo. Blue-headed vireos rarely raise any of their own young in nests in which a cowbird hatches; for example, in the middle Appalachians, pairs fledged an average of 2.92 vireo young per unparasitized nest (n = 37) but only fledged 0.05 vireo young per singly parasitized nest (n = 20; DeCecco et al. 2000).

*THREATS*

The greatest threat to this species is the loss, fragmentation, and degradation of breeding habitat, particularly eastern hemlock habitats. Within Pennsylvania, loss of forest cover is primarily a problem limited to the more densely populated southeastern and southwestern regions. Fragmentation, however, is a problem throughout the state. Currently, the majority (58%) of forested habitat in Pennsylvania lies within 300 m of an improved road or edge (Goodrich et al. 2002). These edge habitats have often been found to be of lower quality for breeding songbirds than interior (core) forests due to higher densities of cowbirds or nest predators (e.g., Robinson et al. 1995). Given the vulnerability of the blue-headed vireo to cowbird parasitism, and its apparent preference for areas of extensive forest, increased fragmentation could lead to reduced reproductive success and abundance.

Perhaps the most immediate threat to the blue-headed vireo is the spread of the exotic hemlock woolly adelgid (*Adelges tsugae*). This insect pest infests hemlocks of all ages, causing reduction in new shoot production, thinning of foliage, lower branch dieback, and, eventually, mortality (Ward et al. 2004). At present, the hemlock woolly adelgid has spread throughout much of the southeastern half of Pennsylvania, in many areas eliminating entire hemlock stands. The loss of this habitat component could have a major effect on blue-headed vireo populations as the vireo is closely associated with hemlock in many eastern states (Yamasaki et al. 2000, Tingley et al. 2002, Ross et al. 2004). Within Pennsylvania, surveys indicate that blue-headed vireos occur more frequently in hemlock stands than in other surrounding habitats and occur more frequently in healthy hemlock ravines than in hemlock ravines that have been impacted by the hemlock woolly adelgid (Swartzentruber and Master 2003).

The high density of white-tailed deer (*Odocoileus virginianus*) in many parts of the state has greatly limited forest regeneration and reduced the density of understory and subcanopy forest layers (Goodrich et al. 2002). Because the blue-headed vireo often forages in midcanopy levels, these changes could negatively affect this species. Further, because hemlock is a preferred browse species, deer overabundance may also be limiting hemlock recruitment and recovery (Ward et al. 2004), exacerbating the threat of the hemlock woolly adelgid discussed above.

Finally, lighted towers and buildings can be an important mortality source, particularly for night-migrating songbirds. Although the blue-headed vireo is currently not a common victim of tower collisions, the growing numbers of new cellular and digital television towers built annually could become an important mortality source for migrating vireos (Goodrich et al. 2002).

*CONSERVATION AND MANAGEMENT NEEDS*

Pennsylvania currently supports a healthy and apparently increasing blue-headed vireo population. Primary conservation need for Pennsylvania is to maintain the current population size and a stable population trend within the state by maintaining abundant, high-quality breeding habitat. To accomplish this, the author recommends the following management efforts. Develop and support large-scale forest management plans that work to maintain current levels of forest coverage within the state and minimize fragmentation of remaining, large contiguous forest tracts. Continue to monitor the spread and effect of the hemlock woolly adelgid in the state, and support and encourage potential control or management efforts (e.g., biological control) where possible. Develop and support research to investigate how hemlock decline is affecting blue-headed vireo abundance, habitat use, and reproductive success. In areas already affected by the hemlock woolly adelgid, habitat restoration experiments should be conducted using alternative evergreen species to develop management techniques that provide habitat for vireos by restoring some of the ecological characteristics formerly provided by hemlock. Finally, implement management efforts that promote a well-developed woody shrub and sapling layer, and support research efforts that examine how activities that reduce these layers (e.g., deer overabundance) affect songbirds.

*MONITORING AND RESEARCH NEEDS*

Current monitoring efforts (i.e., Breeding Bird Survey, second Pennsylvania Breeding Bird Atlas) are adequate to monitor the general breeding population size and distribution and to measure the population trend for this species. Given the apparent importance of eastern hemlock habitats, the continued loss of this habitat type from the state could have major effects and could greatly change both the conservation needs for this species and the types of management actions required. At present, the blue-headed vireo population appears to be healthy and increasing in the state, and current research and conservation efforts should target hemlock decline and hemlock woolly adelgid management issues. Local survey efforts and studies, including nest searching and monitoring of blue-headed vireos and other songbird hemlock specialists, are needed in hemlock-dominated habitats across the state to monitor and study this threat. If hemlock continues to decline statewide, research and conservation efforts will need to shift to restoration of hemlock habitats, creation of alternative conifer habitats that provide breeding habitat for the vireo and other hemlock-nesting species, and identification and management of important alternative non-hemlock-breeding habitat types.

Research is needed to determine what constitutes high-quality breeding habitat (i.e., source habitat) for the blue-headed vireo, particularly regarding habitat type and forest fragmentation. It is important to determine the features associated with reproductive success for blue-headed vireos so that managers can identify high-quality habitats (i.e., sources) for protection and can initiate effective management efforts to improve the quality of habitats and landscapes for vireos. This research need may also link to hemlock decline; if hemlock habitats are largely lost in the state, knowledge of the quality of alternative habitats will be needed to effectively manage for this species. Activities that alter forest structure can potentially affect vireos, in both positive and negative ways, yet little is known about how these activities influence vireo habitat quality. Research focused on how forest harvest or management practices, natural forest maturation, and deer overabundance affect breeding habitat quality for vireos is recommended.

*Author:* CHRISTOPHER B. GOGUEN, PENNSYLVANIA STATE UNIVERSITY

## Bank Swallow

Order: Passeriformes
Family: Hirundinidae
*Riparia riparia*

The bank swallow is the smallest swallow found in Pennsylvania (fig. 5.151). This species has a globally Secure (G5) status, in part due to its extensive world range (NatureServe 2009). It is recognized as a Species of Greatest Conservation Need within Pennsylvania because population sizes and trends within the state are not adequately known, and breeding colonies are potentially vulnerable.

*GEOGRAPHIC RANGE*

A widely distributed species, the bank swallow breeds throughout most of North America, Europe, and Asia and winters in South America, Africa, and southern Asia (Garrison 1999). It is found through most of Canada, absent only from the Arctic, and is widely distributed in the United States, although it is localized south of the 40-degree line of latitude (Garrison 1999). Pennsylvania is therefore toward the southern edge of

*Fig. 5.151.* The Bank Swallow, *Riparia riparia*. Photo courtesy of Raymond Belhumeur.

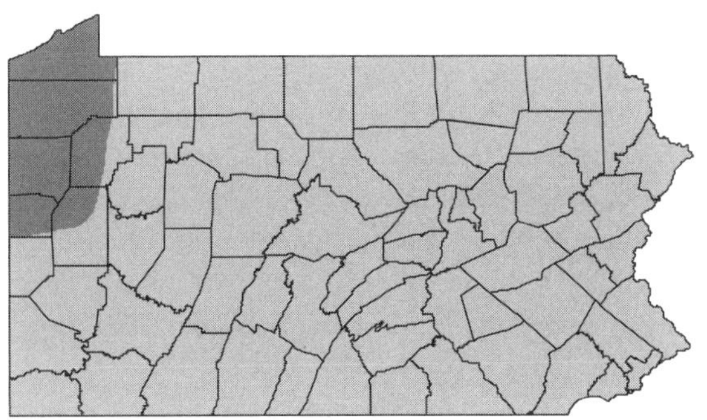

*Fig. 5.152.* Primary (darker shading) and secondary (lighter shading) distribution of the Bank Swallow, *Riparia riparia*.

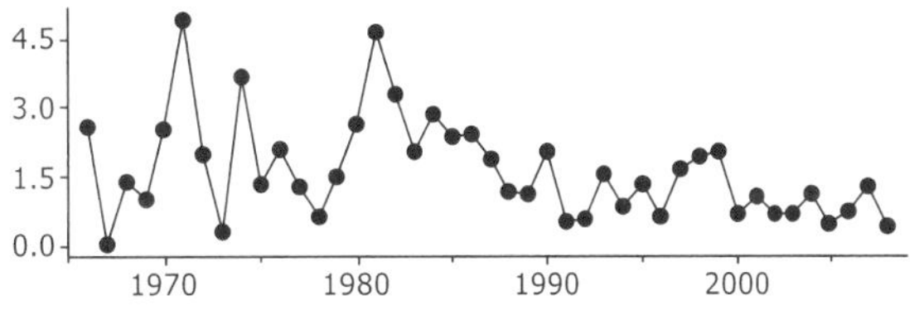


*Fig. 5.153.* Bank Swallow, *Riparia riparia*, population trends from the Breeding Bird Survey.

this species' extensive range and supports a tiny proportion of the North American or world populations.

### *DISTRIBUTION AND RELATIVE ABUNDANCE IN PENNSYLVANIA*

The bank swallow is a widespread but somewhat patchily distributed bird in Pennsylvania (fig. 5.152). During the *Atlas of Breeding Birds in Pennsylvania* (1983–1988), it was confirmed to breed in almost every county, but in only 5 percent of survey blocks (Brauning 1992a). The species is generally scarce in the south of the state. The largest concentration of colonies in Pennsylvania is along the escarpment of the Lake Erie shore and in the Glaciated Northwest, while large colonies are also found along the Delaware and Susquehanna rivers (McWilliams and Brauning 2000).

Because of the ephemeral nature of bank swallow colonies, this species is not adequately monitored by the Breeding Bird Survey. However, in the absence of other data, the Breeding Bird Survey is the only method by which it is possible to examine population trends. Trends for the United States show a nonsignificant 2.1 percent per annum increase between 1966 and 2002 with 95 percent confidence limits −1.1 and +5.3 (Breeding Bird Survey 2003). Breeding Bird Survey data for Pennsylvania are rather sparse, but since the 1980s, a decline in numbers has been noted (fig. 5.153). In New York state, where this species is more widespread, contraction and decline have been steady over the past three to four decades (McGowan and Corwin 2008).

### *COMMUNITY TYPE/HABITAT USE*

As the species' name implies, the bank swallow nests in banks and bluffs of rivers, streams, and lakes. The species excavates its own nest hole, usually in highly erodable substrates, especially sandy soils. Bank swallows are a colonial species, with between 20 and 200 pairs typical (Garrison 1999). A study in Pennsylvania and Vermont in the late 1950s and early 1960s revealed a mean colony size of ninety-five burrows (Spencer 1962).

The availability of a specific nesting habitat requirement limits populations of bank swallows. The principal nesting habitat is eroded banks along rivers, streams, and lakes. Banks chosen for nest sites are usually near vertical to deter predators. The most successful nests tend to be those higher up the bank (Garrison 1999). Sandy or sufficiently friable soils, in which nests are easily excavated, are favored (Garrison 1999), but the earth must be sufficiently solid to prevent burrows from collapsing, hence, a preference for compacted soils is shown (John 1991). In some areas, human-made banks in sand and gravel quarries have become an important additional nesting habitat, while the species will also use specially designed artificial next boxes and banks (LBP 2004). Of twenty-five colonies studied in Pennsylvania and Vermont during the late 1950s and early 1960s, thirteen were in sand or gravel pits and only one in a riverbank, the remainder being in other human-made banks (Spencer 1962); although this may not have been a representative sample of colonies within the state. The average length of nesting banks in this study was 59 m, with a range of 9 to 305 m (Spencer 1962).

Within sand quarries, bank swallows have been shown to avoid banks where the talus (sloping mass of soil or sand spoil) reaches to within 0.5–0.7 m of the overlaying sod (Ghent 2001). Spencer (1962) found that nest holes in Pennsylvania were in banks with an average height of 3.2 m (range 1.7 to 7.3) and averaged 85 cm from the top of the bank.

### *LIFE HISTORY AND ECOLOGY*

Bank swallows forage for insects during daylight hours over a wide variety of habitats. Because of the large volumes of emergent insects, wetlands are a favored, but by no means essential foraging habitat during the breeding season (Garrison 1999). During migration, wetlands often attract large feeding or staging flocks with trees overhanging water favored for communal roosts (Garrison 1999).

The bank swallow is one of the earliest Neotropical migrants to return north each spring, arriving on the Pennsylvanian breeding grounds from the end of March onward with the main arrival occurring in late April and early May (McWilliams and Brauning 2000). Bank swallows return to their South American winter quarters during late summer and early autumn, when staging birds can concentrate into flocks of hundreds if not thousands. Most have left the state by early October.

### *THREATS*

Flood and erosion control projects along river systems have resulted in a loss of suitable nesting banks for this species (Garrison 1998), although no information on the extent of this problem is available in Pennsylvania. In some parts of the state, this species may be heavily reliant on sand quarries for nest sites. These provide a very ephemeral habitat however and as such, any change in the extent of these operations is likely to have an effect on bank swallow numbers.

Studies have shown that intensive farming methods, and the increased application of pesticides in particular, have led to a decrease in the abundance of airborne insects in farmland (Benton et al. 2002). This could reduce food supply for aerial feeders such as the bank swallow, although proving a link is difficult due to the absence of long-term monitoring of airborne insect numbers.

Drought on the African wintering grounds has been shown to dramatically reduce the numbers of bank swallows returning to breed in Europe the following spring (Bryant and Jones 1995, Szep 1995). There is little evidence for such strong effects in the North American population, although it is interesting to note that numbers did fluctuate widely between years on Breeding Bird Survey routes during the 1970s. Furthermore, this species' habitat use and distribution in its South American winter quarters and on migration routes are little known (Garrison 1999), rendering it difficult to assess threats that may be caused by changes in land use in those areas.

### *CONSERVATION AND MANAGEMENT NEEDS*

Because sand and gravel mines provide important habitat for bank swallows, an educational program for mine owners would be useful for encouraging mining operations to be carried out with sensitivity to the needs of this species, wherever possible. Under such a program, mine workers could be encouraged to defer mining when birds are nesting.

Any engineering work in the state that might destroy an existing bank swallow colony or a bank that might be used in the future should be assessed and, if necessary, mitigated for. The enhancement of existing banks or creation of artificial new banks should be considered. Guidelines on the creation and maintenance of suitable nesting banks for this species would be useful and could benefit a range of other wildlife. Without maintenance to remove vegetation and prevent collapse, banks are likely to be abandoned by bank swallows after two or three years (Garrison 1999). A long-term commitment to maintaining artificial banks may therefore be necessary. Nest holes are evenly spaced within a bank surface (Spencer 1962); larger banks therefore have the potential to support larger colonies (Szep 1991) and should be afforded the highest conservation priority. If there is no alternative but to destroy a nesting bank, the work should be delayed until the end of August, by which time all of the fledgling bank swallows will have left the nest.

### *MONITORING AND RESEARCH NEEDS*

While the Breeding Bird Survey may be used to monitor bank swallow populations, the data from that scheme are rather sparse for this species and may be dominated by counts from just one or two areas, which are not representative of statewide population fluctuations. More robust population monitoring could be achieved by establishing an inventory of large colonies, which could then be surveyed at regular periods. The most efficient way of proceeding may be to use information gathered for the Pennsylvania Breeding Bird Atlas to guide the search effort. A monitoring program could be based on a network of volunteer counters.

Little is known about metapopulation dynamics of this species within Pennsylvania. Because it is a colonial species, and the colonies are ephemeral, it could be that a small number of large colonies are important to the long-term survival of populations over a wide area. A mark-recapture banding study within a specified study area, at both natural and man-made sites, would shed light on between-colony movements and nesting success. Such a study would need to be carried out for at least four years, and at a minimum of six colonies, preferably at various distances apart, ranging from 1 km to 20 km. A collaboration between local bird banders and a university or other research facility would be a useful way to establish a long-term banding program.

*Author:* ANDY WILSON, PENNSYLVANIA STATE UNIVERSITY

## Winter Wren

Order: Passeriformes
Family: Troglodytidae
*Troglodytes troglodytes*

The winter wren is Pennyslvania's smallest wren (fig. 5.154). Global populations are considered Secure (G5), and state populations are Apparently Secure (S4B, NatureServe 2009). It was selected as a Species of Greatest Conservation Need because of its close association with mature hemlocks, a currently threatened habitat type in Pennsylvania.

### *GEOGRAPHIC RANGE*

The winter wren is found throughout temperate areas of the Northern Hemisphere. Called simply "wren" in Eurasia (where it is the only wren species), it breeds from Iceland, the British Isles, Scandinavia, and Russia in the north, to northwestern Africa and east through portions of India, Tibet, China, Taiwan, Korea, and Japan (Cramp 1988).

*Fig. 5.154.* The Winter Wren, *Troglodytes troglodytes*. Photo courtesy of Glen Tepke.

In North America (summarized in Hejl et al. 2002), the winter wren breeds in southern Alaska, including the Aleutian and Pribilof Islands and across much of forested Canada. In the western United States, it breeds in forested habitats in Washington, Oregon, and Northern California and from Montana and Idaho south to scattered areas of Nevada and Arizona. In the east, its range extends from Canada into the upper midwestern states, New England, New York, Pennsylvania, and south through the Appalachian Mountains to Georgia. The greatest breeding density in North America occurs in boreal and northern Pacific rain forests where densities of forty to sixty pairs per 40 ha have been reported.

The winter range in western North America extends south to the lower Colorado River drainage in California and Arizona. In the East, they winter from their southern breeding range south to the Gulf Coast and northern Florida.

### *DISTRIBUTION AND RELATIVE ABUNDANCE IN PENNSYLVANIA*

Aside from small, scattered populations in the southern Appalachians, Pennsylvania is at the southern edge of the winter wren breeding range in eastern North America. The first Pennsylvania Breeding Bird Atlas (BBA) indicated this species to be a widespread breeder in the mountainous northern half of the state at high elevations (2,000 feet) on the Allegheny Plateau (fig. 5.155). Breeding also occurs in the Laurel Highlands in Somerset and Westmoreland counties and in favorable locations in the Ridge and Valley region, including Rothrock State Forest in Centre and Huntingdon counties, and the Tall Timbers Natural Area in Bald

Eagle State Forest in Snyder County. A few territorial individuals were recorded along the Kittatinny Ridge (Brauning 1992g). Preliminary results from the second Pennsylvania Breeding Bird Atlas suggest this distribution has not changed significantly since the 1980s, but the number of blocks reporting wrens has more than doubled (Mulvihill and Brauning 2009). True to its name, some birds winter in the state, primarily in major river valleys in the warmer southeast (McWilliams and Brauning 2000).

The relative abundance of the winter wren on the Allegheny plateau, including Pennsylvania, is lower than in any other physiographic region in North America that hosts a breeding population. It was recorded in only 15 percent of atlas blocks in its northern Pennsylvania breeding range and thus should be considered an uncommon breeder here (Brauning 1992g).

Despite its uncommon status, the population level of the winter wren in Pennsylvania since 1980 has apparently been stable and perhaps increasing (fig. 5.156). A positive trend is seen from Pennsylvania Breeding Bird Survey routes between 1980 and 2003, although that calculation is based on low numbers of recorded birds. winter wrens were recorded on twenty-three routes during that time frame with an annual average on those routes of 0.27 birds (Breeding Bird Survey 2006).

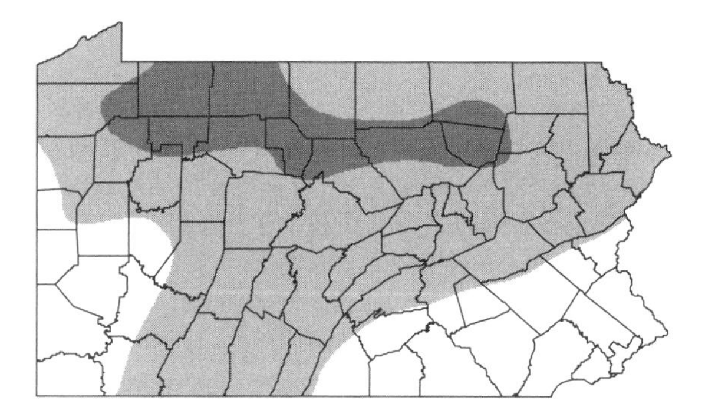

*Fig. 5.155.* Primary (darker shading) and secondary (lighter shading) distribution of the Winter Wren, *Troglodytes troglodytes*.

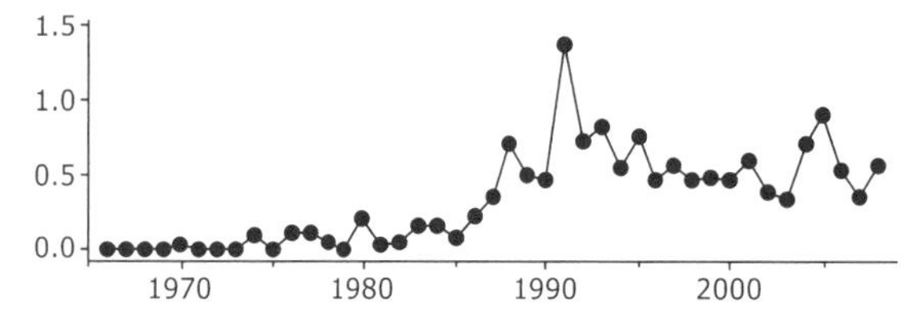


*Fig. 5.156.* Winter Wren, *Troglodytes troglodytes*, population trends from the Breeding Bird Survey.

### COMMUNITY TYPE/HABITAT USE

The breeding habitat of the winter wren in Pennsylvania is primarily northern coniferous forests, especially those featuring hemlocks. Territories are often near streams or other surface water, but some are on high steep slopes in cool, moist hemlock stands, usually north facing. Critical components of breeding territories are downed logs, woody debris, and standing dead trees (Todd 1940b, Hejl et al. 2002). With their dependence on mature hemlock habitat, the winter wren may indicate species for quality mature and old growth coniferous forests (Brauning 1992g, Haney 1999).

### LIFE HISTORY AND ECOLOGY

Within Pennsylvania, the winter wren occupies coniferous or mixed forests, usually with a hemlock component and substantial woody debris on the ground (Brauning 1992g). It feeds mostly on the ground, searching methodically among downed trees and woody debris and on exposed roots and bases of standing trees (Todd 1940b, Holmes and Robinson 1988). The diet consists of a variety of invertebrates but mostly insects, including beetles, spiders, caterpillars, and flies (Bent 1948, McLachlin 1983, Hejl et al. 2002). It is secretive during breeding and would seldom be detected except for its song.

In autumn, the winter wren withdraws from most of its summer range, but many birds winter at low elevation in the river valleys in the southern part of Pennsylvania. During spring and fall migration, it is found in brushy thickets and dense woodland undergrowth, often in areas away from typical summer and winter habitats (McWilliams and Brauning 2000).

Nests are built in cavities or crevices in tree bark or among rocks, on average about 1 m above ground level (Hejl et al. 2002). In Pennsylvania, nests with full egg sets have been found primarily between May 20 and June 4. There is no evidence to suggest production of two broods per season. No information is available on territory size in Pennsylvania, but in the western United States, territory size is generally 2–4 ha (Todd 1940b, Bent 1948, McLachlin 1983, Peck and James 1987, Van Horne 1995, Hejl et al. 2002).

### THREATS

The most substantial threat to the winter wren in Pennsylvania is the hemlock woolly adelgid (Ward et al. 2004). This pest has caused damage in southern and eastern parts of the state, mostly outside the summer range of the winter wren. However, the adelgid is spreading north, and it may eventually have a significant effect on northern hemlock forests and may possibly affect breeding populations of the winter wren and other hemlock-associated species in Pennsylvania. Adelgid infestations were found in eighteen counties in 1991 but that number had increased to thirty-eight by 2001 (Goodrich et al. 2002).

A more general threat is habitat degradation and fragmentation from logging and development. Various studies suggest the that fragmentation and some forms of timber management adversely affect winter wrens (Rosenberg and Raphael 1986, Fowler and Howe 1987, Lehmkuhl et al. 1991, Manuwal 1991, Darveau et al. 1995, McGarigal and McComb 1995, Hutto and Young 1999, Hejl et al. 2002). Much of the current winter wren habitat in Pennsylvania is on public lands and thus should be secure against development. Furthermore, this species breeds in areas unlikely to be logged—hemlock ravines, steep slopes, and fragments of old growth forest that are of high conservation priority. Habitat on private land is more likely to be disturbed.

### *CONSERVATION AND MANAGEMENT NEEDS*

To maintain the winter wren population in Pennsylvania, maturing and old growth coniferous forests, especially hemlock, must be protected and expanded. Pennsylvania has a limited amount of old growth coniferous forest. The total acreage of all kinds of old growth forests in Pennsylvania represents less than 1 percent of state forest lands and occurs mostly in patches of less than 500 acres (Goodrich et al. 2002). Stewards of public lands should be aware of the significance of maturing hemlock forests for the winter wren and other species with similar habitat affinities.

Detailed recommendations have been developed for managing winter wren habitat in western states (Altman 1999a, Hejl et al. 2002). These sources recommend the protection or creation of habitat blocks greater than 30 ha containing mature (80 to 200 years) and old growth (less than 200 years) forests on unmanaged or lightly managed lands. On managed lands, rotations should be greater than sixty years. Detailed recommendations are given for structure and composition to be retained in managed areas in terms of numbers of downed logs, amount of shrub cover, trunk diameter of standing trees, and preservation of minimum-width riparian buffer zones within harvest areas. However, substantial modification of the recommendations for western forests could be required because of the different suite of trees in eastern forests.

### *MONITORING AND RESEARCH NEEDS*

The winter wren is not well monitored by the Breeding Bird Survey. Territories are usually deep in forests and well off-road in rugged terrain. The second Pennsylvania Breeding Bird Atlas will allow a comparison with the breeding range delineated in the first Breeding Bird Atlas and permit detection of any changes. Beyond the completion of the second Breeding Bird Atlas, old growth areas containing populations of winter wrens and other associated species should be monitored at regular intervals.

For the winter wren in Pennsylvania, the most critically required research centers on the principal threat to its present stable status in the state, the hemlock woolly adelgid. Protocols for monitoring the spread of the adelgid have been developed (Evans 2002, Mayer et al. 2002). Other, less urgent research should address how forest management and fragmentation affects the winter wren within its typical Pennsylvania habitat. Following the completion of the second Breeding Bird Atlas in 2008, prime winter wren breeding areas in the state can be identified and monitored regularly.

*Author:* GREG GROVE, PENNSYLVANIA STATE UNIVERSITY

## Brown Thrasher

Order: Passeriformes
Family: Mimidae
*Toxostoma rufum*

The brown thrasher is Pennsylvania's only thrasher and one of three members of the family Mimidae found within the state (fig. 5.157). It was selected as a Pennsylvania Species of Greatest Conservation Need because of population declines and as a representative of the guild of species associated with old field and early successional habitats. Brown thrashers are not listed as Threatened or Endangered in any part of their range, but they are listed as a Species of Greatest Conservation Need on all northeastern state lists.

### *GEOGRAPHIC RANGE*

Brown thrashers breed throughout the eastern two-thirds of the United States and in southern Canada from Alberta eastward. They winter primarily from Maryland south to Florida and southwest to Arizona. A few will winter in southern Pennsylvania (Ickes 1992b, McWilliams and Brauning 2000).

*Fig. 5.157.* The Brown Thrasher, *Toxostoma rufum*. Photo by Jacob W. Dingel III.

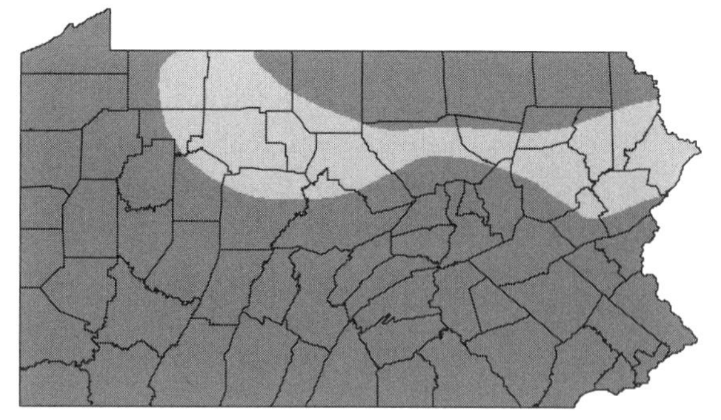

*Fig. 5.158.* Primary (darker shading) and secondary (lighter shading) distribution of the Brown Thrasher, *Toxostoma rufum*.

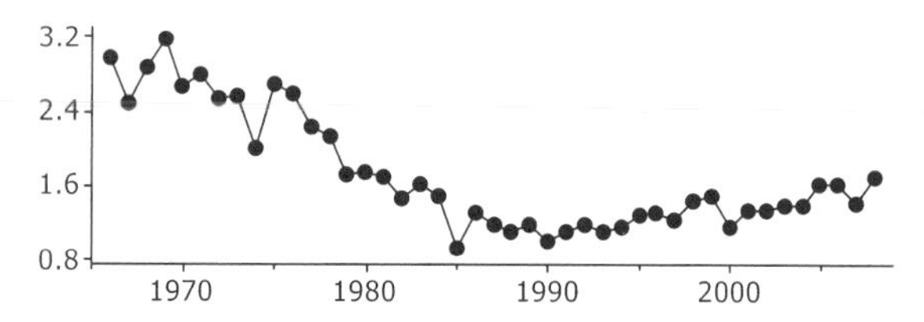


*Fig. 5.159.* Brown Thrasher, *Toxostoma rufum*, population trends from the Breeding Bird Survey.

### *DISTRIBUTION AND RELATIVE ABUNDANCE IN PENNSYLVANIA*

Historic records indicate that brown thrashers formerly had a much more restricted distribution than they do today (fig. 5.158). At the turn of the century, they were absent or rare in the north-central counties and rare or uncommon in northeastern Pennsylvania (referenced in Ickes 1992b). During the last century, their breeding range expanded as a result of the clearing of eastern forests, which created an abundance of suitable habitat (Cavitt and Haas 2000). Brown thrashers have had a widespread distribution across the state throughout the past century (Ickes 1992b). They are less widely distributed than the closely related catbird perhaps because they have more specialized habitat requirements, are more restricted to larger patches of habitat, and are less often found in suburban areas (see "Community Type/Habitat Use").

Similar to many other species associated with early successional habitat, the brown thrasher increased greatly in abundance during the early part of the century as forests were cleared and later as marginal farms were abandoned. In recent decades, as forests across the east have matured and old field habitat has become less common, this species has declined. Because our primary way of tracking changes in abundance is through breeding bird survey data, which did not begin until 1966, we are observing the decline from past population highs but have no information on how this relates to historic population sizes. Since 1966, the brown thrasher has been declining at a rate of −1.9 percent per year in Pennsylvania and −1.2 percent across its range (Sauer et al. 2005; fig. 5.159).

### *COMMUNITY TYPE/HABITAT USE*

Brown thrashers prefer brushy habitats, such as hedgerows, multiflora rose thickets, overgrown fields and pastures, power line corridors, regenerating clear-cuts, and forest edges (Ickes 1992b, Cavitt and Haas 2000, McWilliams and Brauning 2000). Naturally occurring barrens probably also provide suitable habitat for thrashers. The author's experience is that thrashers are primarily associated with large (>0.5 ha) overgrown fields that have open areas for walking and foraging, thick brushy areas for nesting, and an abundance of song perches. In New Jersey, thrashers are not found in woodlots <0.8 ha in size (Forman et al. 1976), and they only occasionally breed in suburban/urban settings (Bent 1948). Cade (1986) developed an HSI model for brown thrashers in forest habitat. They reached their maximum densities in shrub or mid-successional stages of forest, and the model included three variables. Suitability was highest when density of woody stems ≥1.0 m tall was 10,000–30,000/ha, canopy cover of trees was 10 percent to 30 percent, and ground covered by litter ≥1 cm deep was >80 percent.

### *LIFE HISTORY AND ECOLOGY*

The brown thrasher is a short-distance and partial migrant wintering primarily in the southern United States and arriving in Pennsylvania during the second to third week in April through early May (McWilliams and Brauning 2000). Brown thrashers migrate out of the state during September and October, with stragglers remaining throughout the first or second week of January (McWilliams and Brauning 2000). Thrashers are birds of early successional habitat, where they forage primarily on the ground using their sturdy bills to toss leaves and sticks and to rake the leaf litter in search of food (Cavitt and Haas 2000). Main food items include insects (primarily beetles), other arthropods, fruits, and nuts.

In Pennsylvania, the nesting season runs from May to July, and thrashers usually raise two broods (Ickes 1992b). They are vocal early in spring when they are setting up territories but are relatively quiet after that and are often difficult to see (McWilliams and Brauning 2000). Singing may increase again in mid- to late June before the second brood (D. A. Gross, personal communication). Breeding territories range from 0.5 to 8 ha, and all activities associated with breeding and feeding occur within the territory. Details on territory size are summarized in Cavitt and Haas 2000. Thrashers build a bulky cup nest, which is placed on the ground or in a vine, shrub, or small tree (Ehrlich et al. 1988). Thrashers are monogamous, and both adults build the nest, incubate the eggs, and feed the young.

*THREATS*

Current declines in brown thrashers appear to be tied to a loss of early successional habitats. These habitats, particularly old fields, overgrown farmsteads, abandoned orchards, and other areas that provide prime thrasher habitat, are not easily quantified by current inventory methods, making it difficult to track these habitats (Goodrich et al. 2002). However, it is clear that this habitat type has declined in extent as a result of cleaner farming practices, larger field sizes, larger farm sizes, and suburban development (Goodrich et al. 2002). Much of what remains is in small patches that may not be suitable for thrashers. Early successional forest also provides habitat for thrashers, but this habitat type has also declined as forests have matured. In 1955, 23 percent of the Pennsylvania forests were in seedling/sapling stands. Today, 10 percent of the forest is classified as seedling/sapling (U.S. Department of Agriculture 2004). This trend is occurring across the eastern United States (Askins 1993). Declines in thrashers may also be tied to fragmentation and degradation of natural barren habitats as has occurred in New Jersey (Kerlinger and Doremus 1981).

High rates of nest predation are a major source of mortality of eggs and young (Murphy and Fleischer 1986, Cavitt and Haas 2000). Brood parasitism does not appear to have a major effect on this species because of its large body size and propensity to eject cowbird eggs (Rothstein 1975, Haas and Haas 1998, Cavitt and Haas 2000). In agricultural areas, exposure to pesticides may be an issue (Cavitt and Haas 2000).

*CONSERVATION AND MANAGEMENT NEEDS*

The primary conservation goal of brown thrashers is to maintain population levels at or above historic levels. To reverse population declines, we need to develop management plans for naturally occurring early successional habitats such as barrens and for anthropogenic habitats such as transmission line corridors to provide high-quality habitats. In addition, managers should consider shifting management of small grassland patches (e.g., <6 ha) that have minimal value for high-priority grassland species to shrubland habitat (Kearney 2003). Private and public landowners should be encouraged to manage old field habitats to keep them from reverting to forests. Clustering groups of smaller fields in close proximity to one another is a way to increase the effective size of the early successional habitat patch. Conservation of shrubland habitat for brown thrashers will benefit a suite of early successional species that are currently facing population declines throughout the northeast (Dettmers 2003).

*MONITORING AND RESEARCH NEEDS*

Brown thrashers are currently adequately surveyed by the annual Breeding Bird Survey and Pennsylvania's second Breeding Bird Atlas. At a large scale, no additional surveys are needed. A primary research need is to understand how the abundance, distribution, and reproductive success of brown thrashers and other species associated with early successional–old field habitats vary within different natural and anthropogenic habitats. An initial study should examine abundance and reproductive success within different types of shrubland communities, including old fields, natural barrens, regenerating forests, and power line corridors. In addition, research should be conducted to determine the effect of different management practices on habitat use and reproductive success. If possible, this research should be coordinated with ongoing management practices occurring on public lands. Rights of way can also provide valuable habitat, but research is needed to determine the best way to manage these habitats for birds while meeting the needs of the utilities and minimizing the spread of invasive plant species.

*Author:* MARGARET C. BRITTINGHAM, PENNSYLVANIA STATE UNIVERSITY

## Black-throated Blue Warbler

Order: Passeriformes
Family: Parulidae
*Dendroica caerulescens*

The black-throated blue warbler is a small insectivorous warbler (fig. 5.160). It was selected as a Species of Greatest Conservation Need as an indicator of forest habitat with a well-developed understory. The black-throated blue warbler is considered globally Secure (G5) and Apparently Secure (S4B) at the state level (NatureServe 2009).

### *GEOGRAPHIC RANGE*

Black-throated blue warblers breed in the eastern United States and southeastern Canada. Their range extends from northern Minnesota, Wisconsin, and Michigan north to southern Canada from Ontario and east to the Maritimes. The breeding range extends to southern New England and south to northern Georgia in the high elevations of the Appalachian Mountains. This species winters primarily in the Greater Antilles, and sometimes in the Lesser Antilles, and along the Caribbean coast of Central America (Holmes 1994).

### *DISTRIBUTION AND RELATIVE ABUNDANCE IN PENNSYLVANIA*

In Pennsylvania, breeding bird densities are highest at higher elevations in the heavily forested areas of the Pocono Plateau, Allegheny High Plateau, and the Allegheny Mountain Section (Brauning 1992a; fig. 5.161). During the first Pennsylvania Breeding Bird Atlas (1983–1989), black-throated blue warblers were reported in 15 percent of all blocks (Brauning 1992a). Breeding Bird Survey data from 1966 to 2005 indicate significant population increases in the black-throated blue warbler in Pennsylvania (+3.4% change/year, n = 40 routes, P = 0.04, Sauer et al. 2006, fig. 5.162). In the United States, populations appear relatively stable, or slightly increasing (+0.8% change/year, n = 276 routes, P = 0.32; Sauer et al. 2006). However, both the state and national estimates are considered deficient because of the low overall abundance (less than one bird per route) of this species (Sauer et al. 2006).

*Fig. 5.160.* The Black-throated Blue Warbler, *Dendroica caerulescens*. Photo courtesy of Reinhard Geisler, Florida, www.reige.net/nature.

### *COMMUNITY TYPE/HABITAT USE*

Breeding birds require large, contiguous tracts of deciduous or mixed forests with dense understory of deciduous or broad-leaved evergreen shrubs and saplings. The composition of understory vegetation varies with location but often includes hobblebush (*Viburnum alnifolium*), mountain laurel (*Kalmia latifolia*), rhododendron (*Rhododendron* spp.), or saplings of sugar maple (*Acer saccharum*), beech (*Fagus grandifolia*), striped maple (*A. pennsylvanicum*), and conifers (Holmes 1994). In the southern part of its range, this species occurs mostly at elevations greater then 800 m, while farther

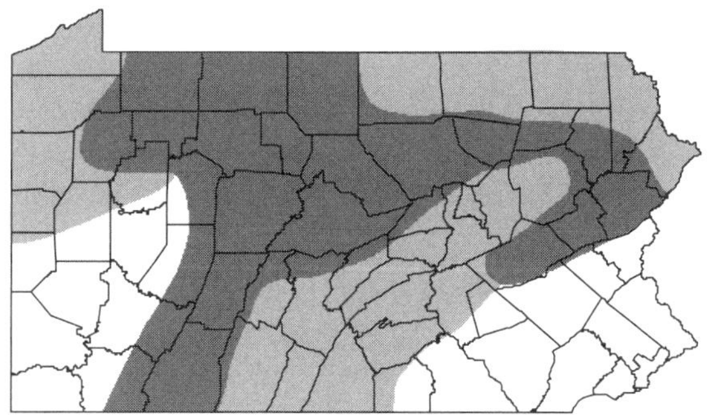

*Fig. 5.161.* Primary (darker shading) and secondary (lighter shading) distribution of the Black-throated Blue Warbler, *Dendroica caerulescens*.

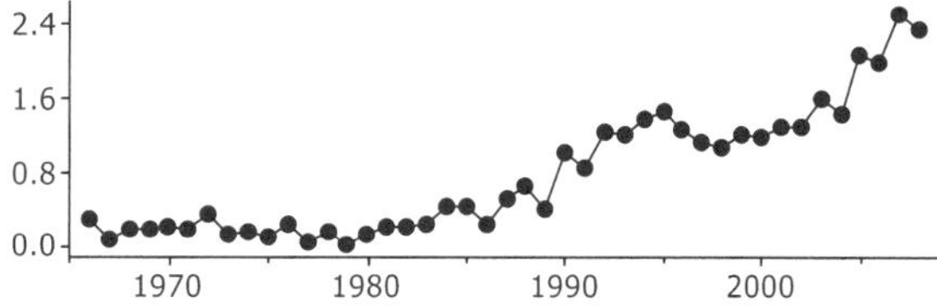


*Fig. 5.162.* Black-throated Blue Warbler, *Dendroica caerulescens*, population trends from the Breeding Bird Survey.

north, it is also found in lower elevations in hilly or mountainous landscapes (Holmes 1994). Selection of breeding habitat seems most closely related to nesting requirements rather then foraging needs (Holway 1991, Steele 1993). In Pennsylvania, black-throated blue warblers breed primarily in large tracts of northern hardwood forest with a dense understory of shrubs, such as mountain laurel at elevations above 1,000 feet (Brauning 1992a).

### *LIFE HISTORY AND ECOLOGY*

During the breeding season, the black-throated blue warbler feeds mainly on insects, including lepidoptera, diptera, coleoptera, and other arthropods (Holmes 1994). Birds forage from the forest floor to high in the forest canopy, although most time is spent foraging in the shrub and lower canopy strata, with males foraging higher than females (Holmes 1986). Birds capture most of their arthropod prey from foliage (Holmes 1986). Black-throated blue warblers are Neotropical migrants. They begin migrating to the breeding grounds in early April and arrive to breeding area in late April through May, with adult males arriving before females and first-year males (Holmes 1994). The breeding season lasts from May until August (Holmes 1994). Birds migrate back to wintering grounds from August to mid-October (Holmes 1994).

Males guard territories and are typically monogamous through the breeding season, but polygyny does occasionally occur (Holmes 1994). Nests are built primarily by females in the dense understory of deciduous and mixed forests, usually within 1–1.5 m of the ground (Holmes 1994). Multiple clutches represented both replacements for lost clutches, and second, or rarely third clutches after successful ones (Holmes et al. 2005). Mayfield nest mortality estimates range from 11 percent to 57 percent (Holmes et al. 2005). Black-throated blue warbler nests are infrequently parasitized by brown-headed cowbirds (*Molothrus ater*; Holmes et al. 2005).

### *THREATS*

Although this species' population is currently increasing in Pennsylvania, several potential threats exist, including forest habitat fragmentation, suppression of understory regeneration, invasive shrub species, and acid deposition. Black-throated blue warblers are area sensitive and avoid edge habitat (NatureServe 2004). Only 42 percent of forests in Pennsylvania are considered core forests. Of the remaining core forest, 70 percent is found in patches of 5,000 acres or less (Goodrich et al. 2002). Increased fragmentation could lead to reduced reproductive success and abundance of this species.

At present, forest regeneration in most of Pennsylvania is severely impaired because of deer overbrowsing, invasive species, and acid deposition (Goodrich et al. 2002, Rich et al. 2004). Already, many species that depend on dense understory vegetation are currently declining (Goodrich et al. 2002, Rich et al. 2004). Deer populations in the state are currently over statewide goals for habitat sustainability, resulting in reductions in understory, subcanopy, shrub, and herbaceous forest vegetation layers (Goodrich et al. 2002). The abundance of many understory birds declines as deer density increases (Goodrich et al. 2002), making deer a possible threat to the black-throated blue warbler habitat (although no studies have been specifically done on this species). Areas overbrowsed by deer are often reestablished with invasive grasses and fern that further impair regeneration (Goodrich et al. 2002). Pennsylvania receives the highest levels of acid deposition in the nation (Goodrich et al. 2002). Long-term soil acidification reduces the regeneration of acid sensitive vegetation such as sugar maple (*Acer saccharum*) and red oak (*Quercus rubra*; Sharpe 2002). The threat of repressed forest regeneration to black-throated blue warblers may be small in Pennsylvania as they prefer understory vegetation of mountain laurel and rhododendron. These species are acid tolerant and are not preferred forage for white-tailed deer (Forbes and Bechdel 1931, Johnson et al. 1995).

Nonnative understory plants could be reducing the quality of black-throated blue warbler habitat. Several bird species that nest in the shrub layer of mixed deciduous trees have higher nest predation rates when nesting in exotic shrubs (Schmidt and Whelan 1999 cited in NatureServe 2004), but these results may not be applicable to the black-throated blue warbler. This species also shows signs of nest location specificity to only a few species of native shrubs (NatureServe 2004), putting it at risk if exotic species outcompete the native shrubs.

In addition to suppressing the regeneration of acid-sensitive plants, acid deposition could be a threat to the black-throated blue warblers through reductions in available calcium and increases in the availability of toxic metals (Drent and Woldendorp 1989), although no studies have found strong support for this (Talliaferro et al. 2001).

*CONSERVATION AND MANAGEMENT NEEDS*

To ensure the maintenance of current populations of black-throated blue warblers, large tracts of forests with adequate understory vegetation need to be maintained. Management and conservation of other forest birds of concern, such as the scarlet tanager and black-throated green warbler, should also protect the preferred habitat of the black-throated blue warbler. Also, further consideration should be made whether to keep this species listed as a Pennsylvania Species of Concern.

*MONITORING AND RESEARCH NEEDS*

Current monitoring methods are adequate at this time. Black-throated blue warbler populations appear to be doing well, so no additional monitoring is needed. The Breeding Bird Survey and the second Pennsylvania Breeding Bird Atlas project should provide enough data to evaluate population conditions. To better understand current population trends and to better predict future risks to black-throated blue warblers, research is needed to investigate the effects of deer overabundance, acid deposition, and invasive species on their habitat. There is little evidence that these potential risks have any effect on them, especially because their preferred understory vegetation is acid tolerant and not preferred by deer. The author suspects these risks are minimal in the types of habitats occupied by black-throated blue warblers in Pennsylvania, but no specific research has been conducted on them.

*Author:* SARAH PABIAN, PENNSYLVANIA STATE UNIVERSITY

## Black-throated Green Warbler

Order: Passeriformes
Family: Parulidae
*Dendroica virens*

The black-throated green warbler is a small insectivorous wood warbler (fig. 5.163). It was selected as a Species of Greatest Conservation Need because of its association with hemlocks and mixed conifer forests. Black-throated green warbler populations are considered globally Secure (G5) and Secure (S5B) at the state level (NatureServe 2009).

*GEOGRAPHIC RANGE*

Black-throated green warblers breed from eastern British Columbia east to southern Labrador. The range

*Fig. 5.163.* The Black-throated Green Warbler, *Dendroica virens*. Photo courtesy of Kent Nickell.

dips south into the Great Lakes region and into the New England states and then extends south to northern Alabama and Georgia in the high-elevation areas of the Appalachian Mountains. There are also two disjunct breeding populations, one confined to the coastal plain of Virginia and the Carolinas (Morse 1993) and another located in northwestern Arkansas (Rodewald 1997). This species winters in southern Texas, Louisiana, and Florida south to Mexico, Central America, and in the West Indies (Morse 1993).

*DISTRIBUTION AND RELATIVE ABUNDANCE IN PENNSYLVANIA*

In Pennsylvania, black-throated green warblers are found at highest densities in the Allegheny High Plateau Section of the Appalachian Plateau. They are also found in lower densities in the Valley and Ridge Province (Brauning 1992a; fig. 5.164). During the first Pennsylvania Breeding Bird Atlas (1983–1989), black-throated green warblers were reported in 1,803 blocks or 37 percent of all blocks (Brauning 1992a). The Breeding Bird Survey data from 1966 to 2005 indicate a significant population increase in Pennsylvania (+2.6 % change/year, n = 67 routes, P = 0.03; Sauer et al. 2005; fig. 5.165).

*COMMUNITY TYPE/HABITAT USE*

Black-throated green warblers occupy a wide range of forested habitat types. In the northern portion of its range, they breed in boreal coniferous forests and transitional areas between coniferous and deciduous forests. In the southern portion of its range, they breed

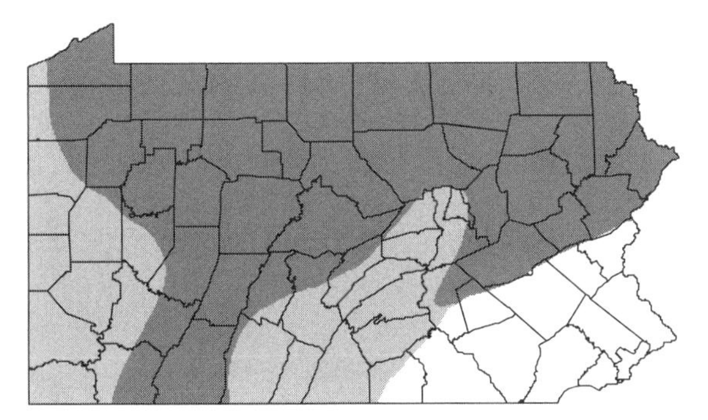

*Fig. 5.164.* Primary (darker shading) and secondary (lighter shading) distribution of the Black-throated Green Warbler, *Dendroica virens.*

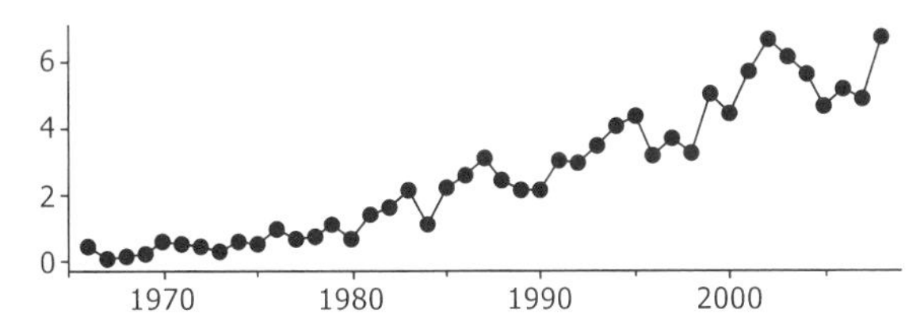


*Fig. 5.165.* Black-throated Green Warbler, *Dendroica virens,* population trends from the Breeding Bird Survey.

in mixed coniferous–deciduous forests or completely deciduous forests. This species may inhabit both young and mature forests (Morse 1993). This species requires a large feeding territory and therefore is generally not found in small woodlots (Brauning 1992a).

In Pennsylvania, the black-throated green warbler is found in highest concentrations in areas with large tracks of coniferous, deciduous, and mixed forests. The distribution is best explained by patch size and elevation (Brauning 1992a). Birds rarely nest in isolated forest patches or at elevations below 304 m. Historically, this species was reported in association with white pines and eastern hemlocks; however, in the high-elevation regions of the Appalachians, it is most common in pure northern hardwood forests (Brauning 1992a).

### LIFE HISTORY AND ECOLOGY

During the breeding season, the black-throated green warbler eats mostly insects, with caterpillars composing a large portion of the diet. Other insect prey include beetles, true bugs, gnats, hymenopterans, mites, moths, spiders, and plant lice. Most prey items are gleaned from leaves. Foraging behavior is more specialized in the north than in the south where there are fewer competing warbler species (Morse 1993). Foraging typically occurs in the middle and upper levels of vegetation (Dunn and Garrett 1997). Black-throated green warblers are Neotropical migrants. They begin migration to breeding grounds in March and arrive at the northern part of their breeding range by the end of May. Birds migrate back to winter ground from August to October (Morse 1993). In Pennsylvania, birds typically arrive between the fourth week of April to the fourth week of May and depart between the fourth week of July to the fourth week of September (McWilliams and Brauning 2000). Most of spring migration is trans-Gulf, but some birds migrate through Florida and even through the Rio Grande Valley of Texas (Dunn and Garrett 1997).

Males establish territories shortly after arrival to breeding grounds. Females build nests and lay one brood per season but may lay additional clutches if unsuccessful at previous attempts. In the northern part of its range, most nests are located in coniferous trees between 1 and 3 m, but occasionally between 15 and 20 m, from the ground (Morse 1993). In the central Appalachians, they will also build nests in deciduous trees. In deciduous forests, nests tend to be located between 6 and 24 m above the ground (Brauning 1992a). Black-throated green warblers accept brown-headed cowbird (*Molothrus ater*) eggs, but brood parasitism is mostly confined to individuals nesting near edges (Morse 1993).

### THREATS

Although this species' population is currently increasing in Pennsylvania, potential threats include forest habitat fragmentation, hemlock declines, and contaminants. Black-throated green warblers are area sensitive (Brauning 1992a, Morse 1993). Only 42 percent of forests in Pennsylvania are considered core forests. Of the remaining core forest, 70 percent is found in patches of 5,000 acres or less (Goodrich et al. 2002). This suggests that the large contiguous forests required by this bird are declining in Pennsylvania. Because this species is vulnerable to cowbird parasitism, and prefers areas with extensive forest, increased fragmentation could lead to reduced reproductive success and abundance.

Some of the primary habitat for black-throated green warblers is currently threatened by hemlock declines resulting from the hemlock woolly adelgid (*Adelges tsugae*), a nonnative insect that kills hemlock trees. The hemlock woolly adelgid was introduced in the 1950s and is steadily moving from southeast to northwest across Pennsylvania. In 2001, 1,595 acres of hemlock damage were reported (Goodrich et al. 2002). The resulting reduction of hemlock in Pennsylvania forests will have negative effects on conifer-associated species. The black-throated green warbler has been shown to have strong associations with hemlocks and therefore may be negatively affected by hemlock declines. The black-throated green warbler can breed in a variety of forest types, so hemlock declines may only present a moderate threat to this species. Populations of black-throated green warblers have shown marked decreases in association with the application of fenitrothian in spruce budworm control outside of Pennsylvania (Morse 1993). The effects of other pollutants, including acid rain, are currently unknown.

*Fig. 5.166.* The Blackburnian Warbler, *Dendroica fusca*. Photo courtesy of David McNicholas.

*CONSERVATION AND MANAGEMENT NEEDS*

Pennsylvania currently supports large and increasing numbers of black-throated green warblers. To maintain these populations, Pennsylvania needs to protect large tracts of forest, which can be accomplished by developing large-scale forest management plans to reduce forest fragmentation. Another conservation need is to continue monitoring the effects of the hemlock woolly adelgid and support any efforts to prevent its spread and to restore areas already destroyed.

*MONITORING AND RESEARCH NEEDS*

Current monitoring methods are adequate at this time. Black-throated green warbler populations are increasing in Pennsylvania, so no additional monitoring is needed. The Breeding Bird Survey and the current Pennsylvania Breeding Bird Atlas project should provide enough data to evaluate population conditions. Research needs include monitoring the spread and effects of the hemlock woolly adelgid on the black-throated green warbler. We do not know how dependent this species is on hemlock forests, even though it is not a hemlock specialist.

*Author:* SARAH PABIAN, PENNSYLVANIA STATE UNIVERSITY

## Blackburnian Warbler

Order: Passeriformes
Family: Parulidae
*Dendroica fusca*

The Blackburnian warbler is a small insectivorous warbler with a fiery orange throat (fig. 5.166). It was selected as a Species of Greatest Conservation Need because of its association with conifers and, in particular, old growth conifer forests. It serves as an indicator of conifer health and distribution throughout the state. The Blackburnian warbler is considered globally Secure (G5) and Apparently Secure (S4B) at the state level (NatureServe 2009).

*GEOGRAPHIC RANGE*

Blackburnian warblers breed from central Alberta, north through central Saskatchewan, and east to southern Quebec and the Maritimes. The breeding range continues south through New England and then along the Appalachian Mountains into Georgia. This warbler winters in south Central America and northern South America (Morse 1994).

*DISTRIBUTION AND RELATIVE ABUNDANCE IN PENNSYLVANIA*

In Pennsylvania, breeding is concentrated in the Appalachian Plateau Province (fig. 5.167). The range is delimited on the southeast by the Blue and South mountains, the southwest by Chestnut Ridge, and the northwest by the Allegheny River and French Creek (Brauning 1992a). Rosenberg (2004) estimated the Pennsylvania population at 160,000 Blackburnian warblers. During the first Pennsylvania Breeding Bird Atlas (1983–1989), Blackburnian warblers were reported in 917 blocks or 19 percent of all blocks (Brauning 1992b). Currently, populations of Blackburnian warblers in Pennsylvania seem to be relatively stable, or slightly increasing from 1966 to 2003

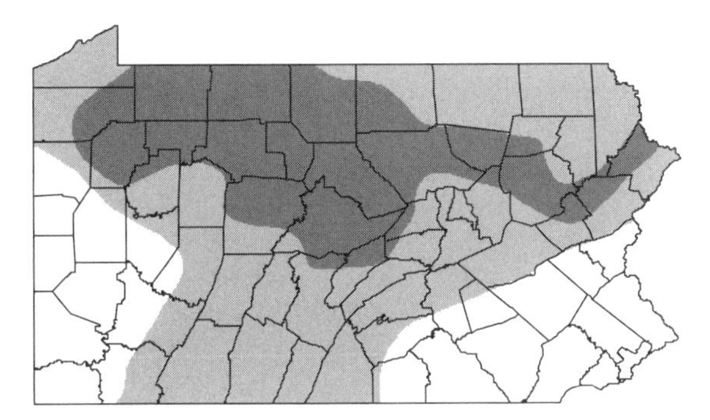

*Fig. 5.167.* Primary (darker shading) and secondary (lighter shading) distribution of the Blackburnian Warbler, *Dendroica fusca*.

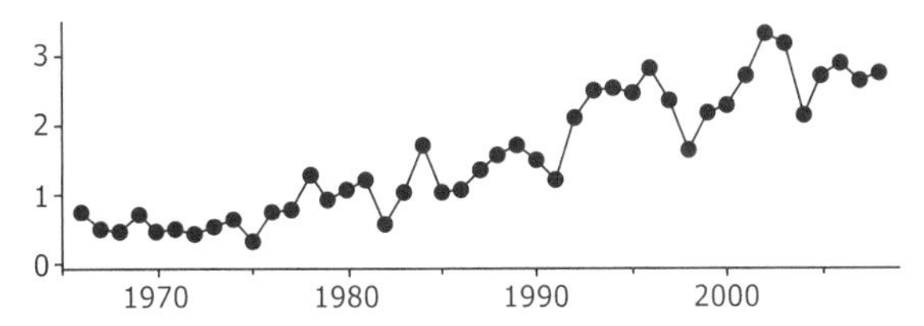


*Fig. 5.168.* Blackburnian Warbler, *Dendroica fusca*, population trends from the Breeding Bird Survey.

(+1.1 birds per route, n = 46, P = 0.39) (Sauer et al. 2004; fig. 5.168).

### *COMMUNITY TYPE/HABITAT USE*

Breeding Blackburnian warblers are typically associated with coniferous and mixed coniferous–deciduous forests but are often found in deciduous forest in the southern portion of its range (Morse 1994). Birds also show a preference for mature forests (Morse 1994, Haney 1999). In Pennsylvania, researchers found that Blackburnian warbler abundance was, on average, forty-five times greater in old growth hemlock-northern hardwood forests than in younger, successional forests (Haney and Schaadt 1996). An important habitat cue seems to be forest profile, with birds seldom nesting in forests without vegetation over eighteen meters with densely foliated crowns (Morse 1971, 1976, cited in NatureServe 2009). Nests are mostly placed in conifers, with preference for eastern hemlocks (Morse 1994). The density of nesting birds increases with increased percentage of coniferous trees (Morse 1994). In Pennsylvania, breeding birds were found primarily above 1,500 feet in areas with extensive forest cover, cooler summer temperatures, and above average rainfall (Brauning 1992a).

### *LIFE HISTORY AND ECOLOGY*

During the breeding season, Blackburnian warblers are primarily insectivorous, feeding mainly on lepidopteran larvae, spiders, beetles, and other arthropods (Morse 1994). Blackburnian warblers are separated from other *Dendroica* warbler ecologically by foraging at the highest treetop niche (Morse 1994). Males forage significantly higher then females, resulting from males foraging closer to song perches and females closer to nests (Holmes 1986). Spring migration to breeding grounds begins in March, with birds arriving at the northern part of range by mid-May (Morse 1994). Birds leave the breeding grounds from August to September and arrive at their wintering grounds in September and October (Morse 1994).

During the breeding season, males establish territories and hold them for most of the summer (Morse 1994). Territory sizes vary with habitat type, ranging from 1.1 ha in primarily deciduous forest, to 0.4 ha in spruce forest (Morse 1994). Pairs are formed once females arrive at the breeding grounds several days after the males (Morse 1994). Females build the nests in coniferous trees at a height generally greater then most other birds (Morse 1994). Females only have one brood but may lay more then one clutch if the nest is destroyed (Morse 1994).

### *THREATS*

Although Blackburnian warbler populations are currently stable in Pennsylvania, several potential threats exist. Blackburnian warblers are sensitive to the loss of mature native forests, resulting from accelerated forest harvest cycles. In general, timber harvest decreases or eliminates breeding populations (King and DeGraaf 2000). Current silvicultural practices in many areas have short harvest cycles and the levels of mechanization do not favor the development or maintenance of suitable breeding habitat (Doepker et al. 1992, cited in NatureServe 2009).

In Pennsylvania, Blackburnian warblers reach highest densities in mature hemlock–eastern hardwood forests (Haney and Schaadt 1996). The protection of these forests from harvest could be an important tool for conserving this species in the future. Blackburnian warbler habitat is currently threatened by hemlock declines resulting from the hemlock woolly adelgid (*Adelges tsugae*), a nonnative insect that kills hemlock trees. The hemlock woolly adelgid was introduced in the 1950s and is steadily moving from southeast to northwest across Pennsylvania. Blackburnian warblers depend on conifers, especially hemlocks, and are therefore susceptible to the negative effects of these declines. Chemical control of insect outbreaks may be detrimental to these birds in parts of their range. For example, the application of fenitrothian as a spruce budworm control was reported to decrease Blackburnian warbler numbers and probably caused mortalities (Morse 1994).

In addition, Blackburnian warblers may be affected by a loss of winter habitat. They are one of forty-five long-distance migratory birds most likely to suffer from alteration of its winter habitat (Petit et al. 1993, cited in NatureServe 2009). The loss of tropical broad-leaved forests in South America is occurring rapidly because of logging and the conversion of forest in agriculture (Diamond 1991, cited in NatureServe 2009).

*CONSERVATION AND MANAGEMENT NEEDS*

To ensure the maintenance of current populations of Blackburnian warblers, high-quality habitat needs to be maintained. Silviculture techniques need to be managed with long rotation cycles to ensure the maintenance of tall coniferous trees. The hemlock woolly adelgid, and resulting loss and degradation of Blackburnian warbler habitat, need to be managed.

*MONITORING AND RESEARCH NEEDS*

Current monitoring methods are adequate at this time. Blackburnian warbler populations seem to be stable, so no additional monitoring is needed. The Breeding Bird Survey and the Pennsylvania Breeding Bird Atlas project should provide enough data to evaluate population conditions. To better understand current population trends and to better predict future risks to the Blackburnian warbler, research is needed to investigate the effects of current silviculture practices and the hemlock woolly adelgid on their habitat. In addition, there is little information available on the reproduction of these birds. Information is lacking on nest success, hatchling period in nest, hatchling growth, and more (Morse 1994). This information is important in understanding and managing these birds. Because their nests are placed high in trees, it has been difficult to fill this gap in knowledge.

*Author:* SARAH PABIAN, PENNSYLVANIA STATE UNIVERSITY

## Prairie Warbler

Order: Passeriformes
Family: Parulidae
*Dendroica discolor*

The prairie warbler was selected as a Species of Greatest Conservation Need because of recent range-wide population declines (fig. 5.169). It is a Partners in Flight priority II species and is included in the state lists of Species of Greatest Conservation Need by most northeastern states. Breeding populations in Pennsylvania are Apparently Secure (S4B), and global populations are Secure (G5, NatureServe 2009).

*GEOGRAPHIC RANGE*

Prairie warblers occur throughout the eastern and south-central United States. The center of the prairie warbler winter distribution is the Caribbean, primarily in the West Indies (Dunn and Garrett 1997, Nolan et al. 1999). Prairie warblers also winter throughout Florida (Dunn and Garrett 1997). This species will rarely spend the winter in the United States north of Florida along coasts of Georgia, South Carolina, North Carolina.

*DISTRIBUTION AND RELATIVE ABUNDANCE IN PENNSYLVANIA*

This species is fairly common overall, and despite reported decreases, probability of extinction of prairie warblers is rated low (Reed 1992). Historically, prairie warblers were fairly uncommon species across much of Pennsylvania, occurring mainly in the southern half of Pennsylvania, breeding fairly localized even in these counties. Breeding birds were uncommon, and researchers knew it as a confirmed breeder only in Chester County or in the barrens of Fulton County where yellow pine is the prevailing forest tree (Leberman 1992d). Subsequently, breeding prairie warblers were found as far north as Indiana and Centre Counties, and expansion into the Poconos became evident (Leberman 1992d). Today, prairie warblers can be found throughout the state (fig. 5.170).

Prairie warblers were rare or absent from much of its present range before European settlement (Nolan 1978). As forests were opened and agricultural fields were abandoned, additional early successional habitat was created, allowing populations of prairie warblers to spread (NatureServe 2005). Population densities of this species apparently peaked between the late nineteenth- to the mid-twentieth century (Nolan 1978). This species tends to be declining in Pennsyl-

*Fig. 5.169.* The Prairie Warbler, *Dendroica discolor.* Photo courtesy of Reinhard Geisler, Florida, www.reige.net/nature.

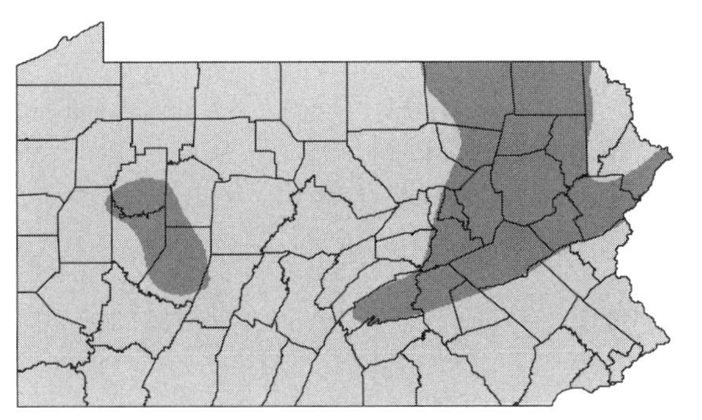

*Fig. 5.170.* Primary (darker shading) and secondary (lighter shading) distribution of the Prairie Warbler, *Dendroica discolor.*

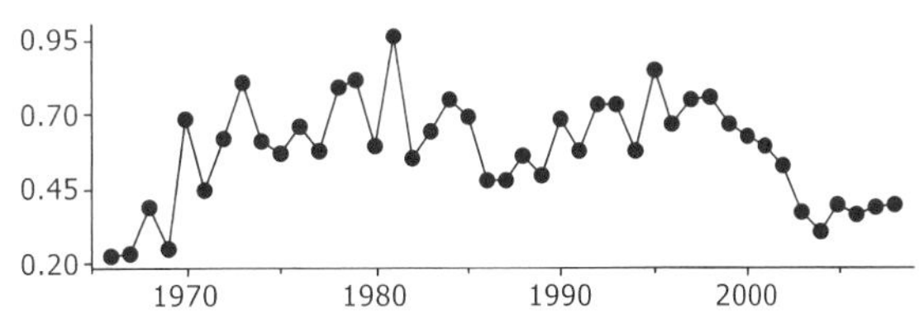


*Fig. 5.171.* Prairie Warbler, *Dendroica discolor*, population trends from the Breeding Bird Survey.

vania, but no significant population trend is apparent in Breeding Bird Survey data (Leberman 1992d, Sauer et al. 2005; fig. 5.171).

*COMMUNITY TYPE/HABITAT USE*

Prairie warblers are birds of scrubby habitats and require relatively open areas but with abundant small trees/shrubs that are used for perches and nesting areas. In Pennsylvania, prairie warblers prefer mostly open, brushy sites during the breeding season (Leberman 1992d). Prairie warblers will breed in scattered young pines and red cedars that invade old fields and pastures in the southeastern part of the state. This species occurs frequently in scattered pine and scrub oak barrens of the central and eastern part of Pennsylvania. Other habitats this species occupies during the breeding season include Christmas tree plantations, reclaimed strip mines, and old fields growing up into deciduous shrubs, hawthorns, crab apples, and other saplings (Leberman 1992d).

Prairie warblers are especially associated with scrub barrens in the Northeast, including Pennsylvania, where the species enters and leaves the habitat as it changes over time. It is a common species of the serpentine barrens of Lancaster County near the Maryland border (Morrin 1991) and the Long Pond and Moosic Mountain barrens of the Pocono Northeast (Gross 1999, 2001). In various sections of the Long Pond barrens, it was found in more than half of point counts conducted near Hypsie Gap Road and Grass Lake (Gross 1999). The dense shrub layer and conifer cover provided by pitch pine (*Pinus rigida*) are attractive to prairie warblers at any elevation. Scrub oak (*Quercus ilicifolia*), gray birch (*Betula populifolia*), sassafras (*Sassafras albidum*), and various ericaceous shrubs are dominant or conspicuous in these barrens. In New Jersey pine barrens, the period of suitability for prairie warblers was approximately five to twenty years after a fire with maximum density ten to twelve years after colonization of the site (Fables 1954 in Nolan 1978).

The prairie warbler belongs to a suite of bird species associated with early successional forests, thickets, and barrens that have declined in recent years (Rosenberg 2004, Sauer et al. 2005). Closely associated species of warblers include yellow warblers (*Dendroica petechia*), which prefer wetter habitats overall, and chestnut-sided warbler (*Dendroica pensylvanica*), which occurs at higher altitudes and latitudes, and on moister, cooler, more deciduous, north-facing slopes. Common yellowthroats (*Geothlypis trichas*) and yellow-breasted chat (*Icteria virens*) are closely associated with prairie warblers within the same habitat but prefer denser growth near ground (Nolan et al. 1999). Golden-winged warblers (*Vermivora chrysoptera*) share similar habitats, where the two species overlap, especially on high-ridge barrens habitat in the northeast. Blue-winged warbler (*Vermivora pinus*) shares fields and power line rights of way but invades several years after prairie warblers and prefers denser growth (Nolan et al. 1999). It also associates with pine warbler (*Dendroica pinus*) in pine barrens (Leberman 1992d, Gross 1999). Migrating prairie warblers tend to prefer habitats similar to their breeding habitats (Nolan et al. 1999). It is unknown whether this species frequents other natural communities during migration.

### *LIFE HISTORY AND ECOLOGY*

In Pennsylvania, the peak migration period is in the first or second week of May, but prairie warblers arrive regularly as early as the second week in April (McWilliams and Brauning 2000). They are much more readily observed in spring rather than in fall migration. In Luzerne and Columbia counties, pairs regularly nest as early as the third week of May (D. A. Gross, personal observation). Males generally arrived a few days before females to begin setting up territories. Territory size averages 0.5–3.5 ha (Nolan 1978). Brood parasitism by brown-headed cowbird (*Molothrus ater*) is high in certain areas, but this species tends to desert nests that have been parasitized by cowbirds. In Indiana, when individuals did not desert the nest, cowbird parasitism caused warbler deaths due to starvation, crowding of nest, and possibly suffocation, crushing, or chilling.

### *THREATS*

The main threats to this species are habitat fragmentation and conversion of habitat to development, succession of old field habitat to young forest, and suppression of fire in habitats such as pitch pine–scrub oak barrens. A net loss of suitable breeding habitat has occurred across the species range, as forests have matured and land converted to agriculture, residential, and industrial uses (NatureServe 2005). Available suitable habitat remains a threat to prairie warbler populations in Pennsylvania and throughout the breeding range. One threat is that succession continues to be halted in open fields by mowing, spraying, or agriculture, decreasing suitable habitat for breeding prairie warblers. At the other end of the scale, succession into later stages of forests from lack of fire will lead to dramatic decreases in habitat suitability for this species (Nolan et al. 1999, NatureServe 2005).

The lack of fire in pine/oak barrens in maintaining the ecological integrity of the habitat has been a possible factor in the decline of localized prairie warbler populations in the East. As more people move into areas where pine barrens dominate the landscape, fire becomes more obsolete in maintaining the structure of this ecosystem. Consequently, fire suppression of these habitats will lead to more mature pine/oak forests, which are unsuitable habitats for this species. There seems to be an optimum period of occupancy for this species after a burn or other major disturbance that probably differs by location (Nolan 1978).

Another major threat to the populations of this species lies in the wintering grounds of the Caribbean Islands (NatureServe 2005). Prairie warblers are especially vulnerable to declines in the nonbreeding season as most of the migratory populations occur in lowland scrub, dry forest, or wetland on just a cluster of Caribbean Islands (NatureServe 2005). A single or series of localized events, such as a strong hurricane, could affect a large proportion of a wintering population because of a much more localized wintering range compared with the breeding range. Any catastrophic event could cause decreases in the prairie warbler population in eastern North America (NatureServe 2005c).

### *CONSERVATION AND MANAGEMENT NEEDS*

The main needs for management of prairie warblers are in identifying priority early successional habitats in key areas where prairie warblers regularly occur. The protection and active management of large suitable habitats for prairie warblers across the breeding range of this species should be a goal. Another management need is to create a mosaic of sites across the landscape in different successional levels to provide continuous habitat for this species.

In many locations of prairie warbler breeding range, pine/oak barrens are the preferred habitat, especially barrens areas in early succession. Succession into forests is a main threat in many of these areas. It would be a priority to identify priority pine–oak barren habitats across species range in the middle Atlantic/Northeast and manage areas to produce suitable habitat for prairie warblers along with rare barrens species, including plants and invertebrates.

### *MONITORING AND RESEARCH NEEDS*

Prairie warblers occupy some areas of their breeding range only for a limited time, when temporary suitable habitat is available. Understanding the declines of this species through time means understanding the successional progression of habitats that this species depends on. It would be advantageous to conduct a thorough examination of the condition of prairie warbler habitat through various stages of succession and then manage these areas accordingly. Habitat attributes and changes in the habitat should be monitored at these locations to better understand the relationship between habitat structural changes and bird populations. This approach is particularly appropriate in locations with known high densities. After conducting surveys for prairie warblers across various areas of the breeding range, areas should be examined and monitored for nesting prairie warblers to describe which conditions are ideal for optimal prairie warbler habitat, which changing habitat conditions cause declines, whether any declines are occurring, and which management types should be conducted in various habitats to maintain optimal prairie warbler habitat.

A landscape analysis should be conducted to study habitat suitability over time across the species range (i.e., how much habitat is being lost to development and what present suitable habitats could be lost to successional changes in the future). To determine whether succession is a reasonable hypothesis for the species' decline, forest succession could be modeled over a period that Breeding Bird Survey routes have detected significant declines in prairie warblers (NatureServe 2005). Also, historical geographic distributions of the species, before European settlement through the present, should be reconstructed for summer and winter (NatureServe 2005).

Research is needed on this species' response to cutting, burning, mowing, or managing other types of suitable habitats and regimes (NatureServe 2005). In some areas, controlled burning may not be a viable management option, but other actions may imitate its effects. In addition, landscape-scale management studies are needed on postbreeding dispersal ecology, size of suitable habitat corridor for breeding, and effects of management on returning breeding adults (NatureServe 2005c). Research is also needed during migration and in the wintering range.

For Pennsylvania, research projects could be conducted on different suitable habitats, especially regenerating woodlands along power line rights of way, ridge-top regenerating shrublands, woods with clear-cuts, and pitch pine–scrub oak barrens to determine size requirements of early successional habitats, and time that is suitable for occupancy of habitat. Research on this species can be coupled with other species of similar habitat requirements that are conservation priorities (e.g., golden-winged warbler).

*Author:* FREDERICK C. SECHLER, JR., NEW YORK NATURAL HERITAGE PROGRAM, THE NATURE CONSERVANCY

## Kentucky Warbler

Order: Passeriformes
Family: Parulidae
*Oporornis formosus*

The Kentucky warbler is a medium-sized warbler with bright yellow underparts and a prominent black mask across the face (fig. 5.172). It is listed by Partners in Flight as a National Bird of Conservation Concern in the United States (Partners in Flight Watch List; Rich et al. 2004) and a Species of Greatest Conservation Need in Pennsylvania (Rosenberg 2004). It was selected as a Species of Maintenance Concern as an indicator of low-elevation, high-quality forest interior habitat. Breeding populations in Pennsylvania are still fairly widespread but apparently in decline.

### *GEOGRAPHIC RANGE*

The Kentucky warbler breeds entirely within the eastern United States from southeastern Texas, southern Louisiana, southern Mississippi, southern Alabama, and northwestern Florida, north to southeastern Nebraska, central Missouri, eastern Iowa, southwestern Wisconsin, central Indiana, northern Ohio, central Pennsylvania, central New Jersey, and (at least formerly) extreme southeastern New York and extreme southwestern Connecticut, west to eastern

*Fig. 5.172.* The Kentucky Warbler, *Oporornis formosus*. Photo courtesy of Glen Tepke.

Oklahoma, eastern Kansas, and southeastern Nebraska, and east to the Atlantic Coast.

The winter range extends from southeastern Mexico south through Central America (mainly along the Caribbean Slope) to portions of northern Colombia and Venezuela. Small numbers have also been recorded during winter on Bermuda and throughout the West Indies from Cuba east to the Virgin Islands. On rare occasions stray individuals have also been recorded wintering in the southernmost United States. Individuals occasionally recorded (various seasons) well north of usual breeding range and as far west as California (American Ornithologists' Union 1998, McDonald 1998).

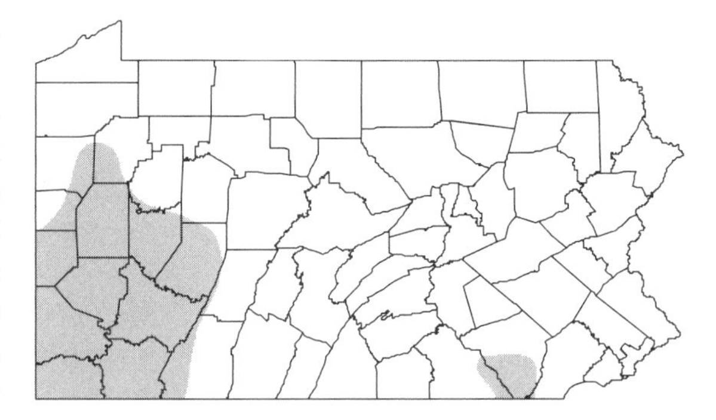

*Fig. 5.173.* Highest probability of occurrence of the Kentucky Warbler, *Oporornis formosus*.

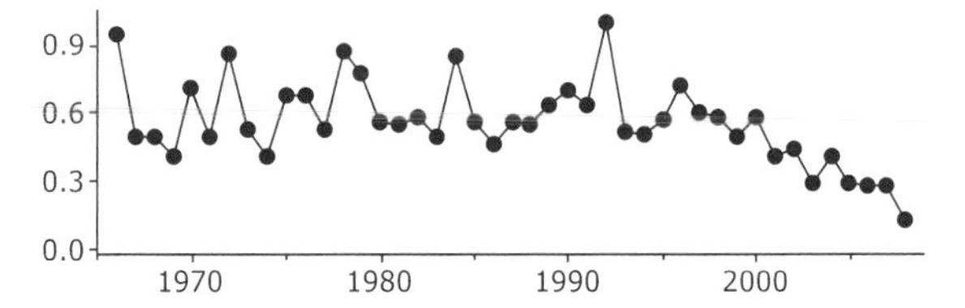


*Fig. 5.174.* Kentucky Warbler, *Oporornis formosus*, population trends from the Breeding Bird Survey.

### *DISTRIBUTION AND RELATIVE ABUNDANCE IN PENNSYLVANIA*

Pennsylvania lies at the northern edge of the Kentucky warbler's breeding range. Results from the first Pennsylvania Breeding Bird Atlas indicated that the species breeds mainly within the southern two-thirds of the state with a few isolated individuals or pairs scattered throughout the northernmost counties, especially in the northwest. The species' statewide breeding distribution is determined largely by elevation with most areas of low-elevation occupied and most areas of high-elevation unoccupied. The species was recorded most frequently in two areas: (1) the southwestern portion of the Appalachian Plateau physiographic province and (2) the Piedmont province (fig. 5.173; both areas are relatively low in elevation), while scattered populations occurred in many valleys within the Ridge and Valley provinces (Master 1992a).

During the first Pennsylvania Breeding Bird Atlas (1984–1989), the Kentucky warbler was only recorded in 19 percent of all atlas blocks within the state (Master 1992a). It was recorded at a high frequency throughout southwestern Pennsylvania (the southwestern portion of the Appalachian Plateau physiographic province) and frequently throughout the Piedmont province. During the second atlas (2004–2008) it was only recorded in 13 percent of all atlas blocks (28% decline) with the largest amount of decline observed in the piedmont (Mulvihill and Brauning 2009). The Kentucky warbler has been slowly declining as a breeder in Pennsylvania at a rate of 1–2 percent per year since 1966 and declining at a slightly slower rate throughout its breeding range during this period (Sauer 2005). The species has disappeared from Philadelphia County as a breeder since the mid-1990s (fig. 5.174). In 2004, Partners in Flight estimated the species' statewide population to be 14,000 individuals (Rosenberg 2004), which equals 1.3 percent of the species' estimated global population of 1.1 million (Rich et al. 2004).

### *COMMUNITY TYPE/HABITAT USE*

Kentucky warblers prefer lowland or bottomland second-growth deciduous forests with well-developed ground cover and shrubby, viney undergrowth as breeding areas. Breeding areas are also frequently located near streams (McDonald 1998), but swamps and other forested wetlands are generally avoided. Although lowland forests are preferred, the species may occasionally occupy mountain laurel thickets at higher elevations in Pennsylvania (McWilliams and Brauning 2000). In addition, woodlands 500 ha or larger appear to be necessary for successful breeding (Gibbs and Faaborg 1990), and forest size appears to be more important than the floristic or structural attributes of a forest in predicting the presence of the species during the breeding season (Lynch and Whigham 1984). Although the Kentucky warbler appears also to occur in areas of forest understory during the migration and winter periods more study is needed to fully understand the species' habitat requirements during these periods (McDonald 1998).

*LIFE HISTORY AND ECOLOGY*

The Kentucky warbler occurs in Pennsylvania as a breeder and transient. During the spring it usually arrives during the first week of May (occasionally during the last week of April) and migration continues through the fourth week of May (McWilliams and Brauning 2000). Kentucky warblers forage on the ground where they capture insects by rummaging through leaf litter or scratching with their feet. They also forage within forest understory vegetation where they often glean or hawk insects from the undersides of leaves.

Within its preferred breeding habitat the species can be difficult to see. As a result, it is most easily detected, at least during the breeding season, by its persistent song and by the fact that males often sing from exposed perches just above the forest understory. Kentucky warblers become difficult to detect after they stop singing at the end of the breeding season. As a result, their southbound movements are not as well known as their northbound movements. Banding records suggest, however, that dispersal or southbound migration begins in Pennsylvania by the third or fourth weeks of July. The species has usually left the state by mid-September, but stragglers have been recorded through the third week of October (McWilliams and Brauning 2000).

Nests are placed on or near the ground in dense forest understory where they are usually anchored to the base of a shrub or located within a cluster of branches. In Virginia, average territory size was 2.21 ha with a range of 1.21–3.75 ha (n = 493; McDonald 1998). The average date for a full clutch of eggs in Pennsylvania reported by E. L. Poole (unpublished manuscript) was June 3. Only one clutch is typically reared per season, although two have been reported (Whitcomb et al. 1981). Pairs will readily renest if their initial nesting attempt fails (McDonald 1998). Nests are frequently parasitized by the brown-headed cowbird (McDonald 1998). Cowbird parasitism of this species has been observed commonly in Pennsylvania since the 1880s (Friedmann 1929, Jacobs 1938).

*THREATS*

The principal threat to the Kentucky warbler throughout its breeding range is forest fragmentation. The Kentucky warbler appears to be an area-sensitive species, and forest fragmentation may create forest areas too small to attract the species during the breeding season. Forest fragmentation also allows larger numbers of potential nest predators, such as feral cats, skunks, and foxes, access to prime nesting areas. Another principal threat to breeding birds (Friedmann 1929, Jacobs 1938) is nest parasitism by brown-headed cowbirds, and forest fragmentation has also been linked to higher levels of cowbird parasitism in this species. In addition, the species' preferred breeding habitat (where nests are most successful) is second growth deciduous forest with shrubby, viney undergrowth. Any factors causing an overall loss of this type of habitat, including a lack of disturbance events that create this kind of habitat (Hunter et al. 2001), and development will also negatively affect this species. Another threat that may be operating on a more local level is loss of forest understory due to white-tailed deer overbrowsing. This problem may be most prevalent in areas where deer hunting is not occurring (e.g., urban areas and certain private lands). In addition, loss of wintering habitat and other threats affecting the species during migration may also be contributing to declines in species' breeding population in Pennsylvania.

*CONSERVATION AND MANAGEMENT NEEDS*

To protect the species in areas where it is declining, second-growth deciduous forests, particularly large tracts (500 ha or larger) that contain shrubby and viney undergrowth, must be protected from development and fragmentation, as well as from overbrowsing by white-tailed deer. In addition, many areas of suitable breeding habitat in Pennsylvania may now be aging past the stage of canopy closure preferred by this species (Hunter et al. 2001). In general, forest management techniques that would address this issue and otherwise improve existing habitat for Kentucky warblers are still in need of development.

*MONITORING AND RESEARCH NEEDS*

Kentucky warblers are not always well monitored by roadside surveys, such as the North American Breeding Bird Survey (BBS) because breeding territories are often located far from roads, well inside large tracts of forest. In addition, some important breeding areas in Pennsylvania are located in relatively urban areas (extreme southeastern Pennsylvania) where no Breeding Bird Survey routes exist. As a result, additional breeding surveys should be employed periodically (every fifteen years) in roadless and urban areas where Breeding Bird Survey routes may be lacking.

Long-term monitoring of breeding populations using point counts and spot-mapping techniques is needed to better determine minimum area requirements and specific habitat requirements for nest sites as they relate to breeding success. Additional research is also needed to determine minimal viable population sizes and to better assess the effects of forest fragmentation on breeding populations, especially in terms of its effect on predation and nest parasitism rates (McDonald 1998).

More information is also needed about how different forest management techniques may positively or negatively affect this species' breeding success within its preferred breeding habitat—second-growth deciduous forest. According to Partners in Flight (Rosenberg 2004) information on the effects of various silviculture practices on this species is particularly needed for breeding areas within the Allegheny Plateau and Ridge and Valley provinces.

*Author:* KEITH RUSSELL, AUDUBON PENNSYLVANIA

## Canada Warbler

Order: Passeriformes
Family: Parulidae
*Wilsonia canadensis*

The Canada warbler is a medium-sized, colorful, active wood-warbler (fig. 5.175). Globally, the Canada warbler is considered Secure (G5, NatureServe 2009). However, the United States Fish and Wildlife Service identified the species as a bird of Conservation Concern in 2002 (U.S. Fish and Wildlife Service 2002c), and the Canadian Wildlife Service has also identified the species as one of high regional Concern and Responsibility (Dunn et al. 1999). At the state level, the species is considered Apparently Secure (S4B, NatureServe 2009). Partners in Flight identified the Canada warbler as a species of high continental concern and regional Responsibility for the Allegheny Plateau physiographic region (Robertson and Rosenberg 2003). The Allegheny Plateau plays an important role in the species' survival within Pennsylvania because a significant portion of the state population is included within this physiographic region.

*Fig. 5.175.* The Canada Warbler, *Wilsonia canadensis*. Photo courtesy of Kent Nickell.

### *GEOGRAPHIC RANGE*

The breeding range of the Canada warbler includes the southern boreal region and much of southeastern Canada; northeastern United States, the Great Lakes region, and south in higher elevations along the Appalachians to northeast Georgia (Bent 1953, American Ornithologists' Union 1998). The species has a low overall density throughout its range. Core breeding range is in Canada; Ontario and Quebec alone hold 40 percent to 70 percent of the total population of Canada warblers (Rosenberg and Wells 2000). They winter in South America (American Ornithologists' Union 1998).

The fall migration route is generally in and west of the Appalachian Mountains, through coastal Texas to the Gulf Coast (Bent 1953, Rappole et al. 1979). Spring migration in North America is similar to the fall route except possibly more to the east (Clement and Gunn 1957).

### *DISTRIBUTION AND RELATIVE ABUNDANCE IN PENNSYLVANIA*

The Canada warbler is generally found at higher elevations (greater than 450 m) in areas with contiguous forest, cool summer temperatures, and above-average precipitation (Brauning 1992a). In northern Pennsylvania, this species is most frequently observed in the Pocono Plateau, Allegheny High Plateau and Glaciated Low Plateau sections (Brauning 1992a, Sauer et al. 2004; fig. 5.176). In southern Pennsylvania, the species' range includes the more prominent ridges: Chestnut Ridge, Laurel Hill, Negro Mountain, and Allegheny Mountain in the Allegheny Mountain Section; and Tussey, Broad, and Blue mountains in the Appalachian Mountain Section (Brauning 1992a, Sauer et al. 2004). Historical breeding populations were found in the southwestern part of the state (Armstrong, Butler, southern Indiana, and western Mercer counties; Todd 1940b).

North American Breeding Bird Survey (BBS) data indicate a region-wide decline of approximately 5.5 percent per year, one of the steepest declines in the northeast region (Rosenberg and Wells 2000). In Pennsylvania, the species has experienced a 1 percent decrease per year from 1993 to 2003, although numbers have shown fluctuations (Sauer et al. 2004; fig. 5.177). Significant declines of 40 percent have been noted since 1966 in the Adirondack Mountains and Northern Spruce Hardwoods physiographic regions; the latter is the region that includes the bulk of the total United States population (Franzreb and Rosenberg 1997).

### *COMMUNITY TYPE/HABITAT USE*

The Canada warbler typically breeds in forested wetlands and cool, moist, mixed coniferous–deciduous forests with a well-developed understory (Bent 1953, Beals 1960, Miller 1999, Weakland 2000). In Pennsylvania, the species favors hemlock (*Tsuga canadensis*) dominated ravines and wet sites in northern hardwood and mixed forest with a dense understory of mountain laurel (*Kalmia latifolia*) or rhododendron (*Rhododendron maximum*; Todd 1940a, Bent 1953, Brauning 1992a, McWilliams and Brauning 2000, Robertson and Rosenberg 2003). Alder (*Alnus spp.*) thickets and regenerating clear-cuts are also used by this species in the northern part of the state (S. H. Stoleson personal communication). Although tolerant of moderate disturbance, the species appears to be sensitive to forest fragmentation and habitat alteration (Freemark and Collins 1992, Miller 1999, Hobson and Bayne 2000, Cumming and Schmiegelow 2001). Canada warbler abundance is positively related to high foliage cover (between 0.3 and 6 m; Robbins et al. 1989, Miller 1999) and the species often uses low-growing vegetation for foraging and singing (Weakland 2000). Stand age is an important predictor of Canada warbler occurrences; the species is more typical of mature stands but may occur in younger stands (second-growth scrub) with lower canopy height and a greater number of large (>2.5 cm diameter) stems (Collins et al. 1982, Hannon 2000, Chace et al. 2002, Hagan and Meehan 2002).

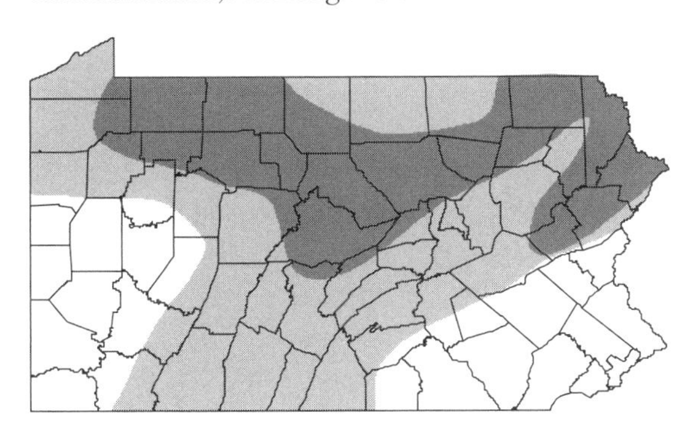

*Fig. 5.176.* Primary (darker shading) and secondary (lighter shading) distribution of the Canada Warbler, *Wilsonia canadensis*.

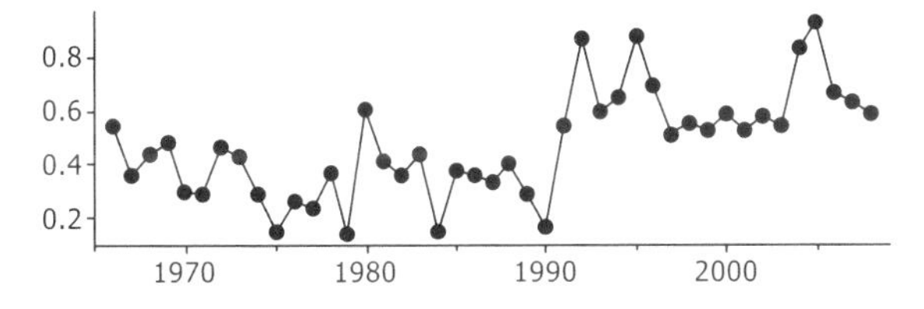


*Fig. 5.177.* Canada Warbler, *Wilsonia canadensis*, population trends from the Breeding Bird Survey.

### *LIFE HISTORY AND ECOLOGY*

Like many warbler species, the Canada warbler is a Neotropical migrant and largely insectivorous. Dense nest site cover appears to be an important habitat requirement for the Canada warbler (Kendeigh 1945). The species builds its well-concealed, bulky cup nest on or near the ground in thick shrub cover or in areas with dense ferns (Bent 1953, Harrison 1975a). In New York, paired males defended territories of 0.8 and 1.2 ha (Kendeigh 1945). Five nests were found along 46 m of stream in Vermont, and three pairs nested less than 30 m away from one another along a stream in West Virginia (Conway 1999).

The Canada warbler lays its eggs in Pennsylvania during late May to late June (Harlow 1918, Bent 1953). One brood is typically raised (Peterjohn and Rice 1991), although data from Pennsylvania's first Breeding Bird Atlas (PBBA) suggest the species may have two broods on occasion (Brauning 1992a). Typical fall migration period in Pennsylvania is from the second or third week of August to the first week of October. Most migrants are recorded from the fourth week of August to the second week of September (McWilliams and Brauning 2000). Their usual spring migration period in Pennsylvania is from the second week of May to the first week of June, with peak migration occurring during the third and fourth weeks of May (McWilliams and Brauning 2000).

### *THREATS*

Canada warbler populations may be affected by several factors in Pennsylvania, including loss of habitat, reduced productivity due to predation and brood parasitism, and white-tailed deer herbivory. Forested wetlands that are drained, filled, and developed contribute to the loss of primary breeding habitat for this species (Conway 1999). The westward migration of the hemlock woolly adelgid (*Adelges tsugae*) poses a threat to the hemlock trees found in habitat types used by this species (Pennsylvania Bureau of Forestry 2003). Impingement of urban development on heavily forested landscapes, such as development pressure in the Pocono region of Pennsylvania, is also a threat (Miller 1999). Forest succession may reduce habitat quality and availability of nest sites if, at a landscape scale, disturbance regimes have changed such that early successional forests with well-developed understories are not created. Predation by mammalian and avian predators and brown-headed cowbird (*Molothrus ater*) parasitism (Friedmann et al. 1977) may result in decreased productivity, especially in fragmented landscapes. Last, the Canada warbler has been shown to be sensitive to reduced understory vegetation by forest ungulates (DeGraaf et al. 1991). White-tailed deer herbivory influences the abundance of ground- and shrub-nesting bird species by reducing understory stem density.

### *CONSERVATION AND MANAGEMENT NEEDS*

Current habitats/community types used by the Canada warbler in Pennsylvania need to be protected to sustain productive breeding populations. Forested landscapes relatively undisturbed by human activity should be targeted for immediate protection. Large areas of quality habitat or smaller tracts surrounded by forested buffer zones to limit the encroachment of development should be protected. A minimum preserve size of 400 ha has been recommended for forest interior species such as the Canada warbler (Robbins et al. 1989); however, forested wetland patches as small as 6 ha may be used if buffered by contiguous upland forest (Miller 1999). White-tailed deer populations within preserves should be controlled.

Primary habitat for the Canada warbler in the state should be maintained and the secondary habitat should be improved by developing best management practices for land managers and foresters to implement in areas where the species has been recorded on a regular basis. In general, the species responds favorably to practices that increase density of understory vegetation within forests and unfavorably to those that reduce forest canopy or the amount of understory vegetation. Forest management practices that result in a well-developed understory while maintaining some forest canopy, for example, single-tree selection, should be used when feasible (James 1984). The Canada warbler population was enhanced by modest timber harvesting in Maine (Hagan et al. 1997), partial harvesting in the Adirondacks (Webb et al. 1977), and diameter-limiting harvesting in West Virginia (Weakland 2000). The conifer component of forest stands and the natural range of vertical and horizontal vegetative structure and composition should be retained, as well as a balance of forest-age classes. The retention of residual patches in clear-cuts provides some habitat for the species (Merrill et al. 1998).

### *MONITORING AND RESEARCH NEEDS*

Breeding Bird Survey data provide important information on population trends at a continental scale; however, limited coverage in some areas can make it difficult to use when characterizing regional population trends (Peterjohn et al. 1995). Species like the Canada warbler that prefer roadless, dense, wet, and inaccessible breeding habitat may not be adequately monitored by roadside surveys. A statewide, off-road, long-term monitoring program should be instituted to provide more localized information on abundance, distribution, and habitat use of the Canada warbler and other species of greatest conservation need. Monitoring sites could be established within appropriate Important Bird Areas and on public and private land where the species was recorded during the first and second Breeding Bird Atlas. Several regional and national monitoring programs already exist, including the Forest Bird Monitoring Program (Canadian Wildlife Service), the Vermont Forest Bird Monitoring Program (Vermont Institute of Natural Science), and Minnesota's Forest Bird Diversity Initiative (University of Minnesota). It would be useful to coordinate monitoring efforts among neighboring jurisdictions and bird initiatives.

In Pennsylvania, Canada warbler research is needed to further understand habitat requirements for the species and to determine causes of population declines in the state. For example, studies on specific habitat requirements, such as sensitivity to forest fragmentation and dependence on wetlands, should be conducted. The health of primary habitats in the state, for example, mixed forest with hemlock component, should be assessed and the amount, condition, and configuration of habitats on which this species depends should be monitored using landscape modeling.

Factors that are limiting populations should be studied through measurements of demographic parameters (adult and brood survival, nest success, productivity) using the Breeding Biology Research and

Monitoring Database (BBIRD), a national program that uses standardized methods for studies of nesting success and habitat requirements. The effect of white-tailed deer herbivory on habitat quality at nest sites and the effect of brood parasitism on productivity in Pennsylvania should be measured as well.

*Author:* RITA Y. HAWROT, PENNSYLVANIA NATURAL HERITAGE PROGRAM, WESTERN PENNSYLVANIA CONSERVANCY

## Yellow-breasted Chat

Order: Passeriformes
Family: Parulidae
*Icteria virens*

The chat was selected as a Species of Maintenance Concern because of long-term declines in Pennsylvania and throughout much of its range and as an indicator of thickets and early successional forest (fig. 5.178). It is a Partners in Flight Priority IIA (high regional concern) species. Global populations are Secure (G5) because of a large breeding range with local populations that are large in at least some areas (Hammerson et al. 1996, NatureServe 2009).

### *GEOGRAPHIC RANGE*

Yellow-breasted chats are migratory and breed from northern Mexico to southern Canada. Perhaps reflecting the distribution of suitable habitat, the breeding range of chats is moderately continuous through east-central United States and is patchy (and often with populations at lower local densities) west of the Rocky Mountains (Eckerle and Thompson 2001). Pennsylvania is on the northeast periphery of the breeding distribution. Chats were formerly common in New York, Connecticut, and Rhode Island, and their range may be contracting in the northeast (Eckerle and Thompson 2001, Sauer et al. 2004). Their migratory pattern is variable, with a minority of individuals wintering in the southeastern United States and most individuals wintering in Mexico or Central America (Eckerle and Thompson 2001).

Fig. 5.178. The Yellow-breasted Chat, *Icteria virens*. Photo courtesy of Reinhard Geisler, Florida, www.reige.net/nature.

### *DISTRIBUTION AND RELATIVE ABUNDANCE IN PENNSYLVANIA*

In Pennsylvania, chats breed primarily in the southern part of the state (McWilliams and Brauning 2000); they are most common in the southwest and locally common in the Piedmont and the Ridge and Valley regions (fig. 5.179). In the northern half of the state, they are sparsely and irregularly distributed. Chat densities on Breeding Bird Survey routes in Pennsylvania have declined for more than thirty years, at an average rate of 4.9 percent per year, and since 2000, the statewide average has been less than 0.5 detections per Breeding Bird Survey route (Sauer et al. 2004); fig. 5.180).

The distribution of chats in Pennsylvania has changed over the past 150 years, perhaps due to landscape changes. Before widespread logging in the late 1800s, chats may have been absent from the crest of the Alleghenies (Todd 1940a). After successional habitat was created by the forest clearing and, later, by abandonment of farmlands (Askins 2000), chats may have spread northward along river valleys and tributaries (Todd 1940a , Leberman 1992g), and anecdotal evidence suggests that chat populations declined in northern Pennsylvania as forests regenerated between 1900 and the late 1930s (Todd 1940a). By the time of the first Breeding Bird Atlas in Pennsylvania, chats were scarce in the northern part of the state (Leberman 1992g). Local declines may be continuing, as chats are currently absent from some parts of Pennsylvania that were occupied as recently as the 1980s (A. R. Schweinsberg, personal communication). A similar historical pattern has been seen for several other shrub-nesting birds in the eastern United States (Askins 2000). In Pennsylvania, chat populations seem to be declining statewide, which has led to their complete absence in much of northern and eastern Pennsylvania where they were previously uncommon but persistently present.

### *COMMUNITY TYPE/HABITAT USE*

Breeding habitat is typified by low, dense shrub vegetation without a closed canopy of trees (Leberman

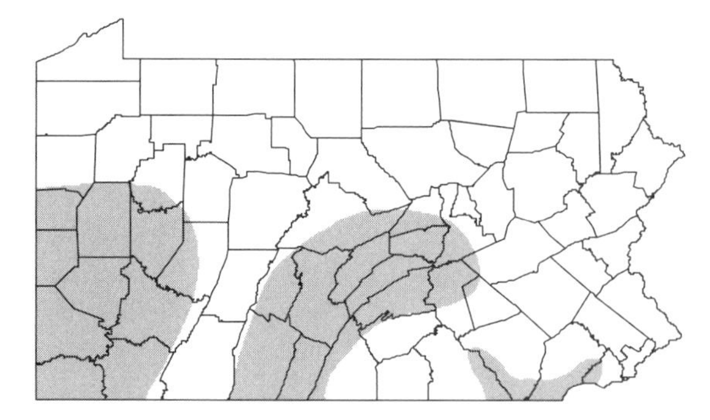

*Fig. 5.179.* Highest probability of occurrence of the Yellow-breasted Chat, *Icteria virens*.

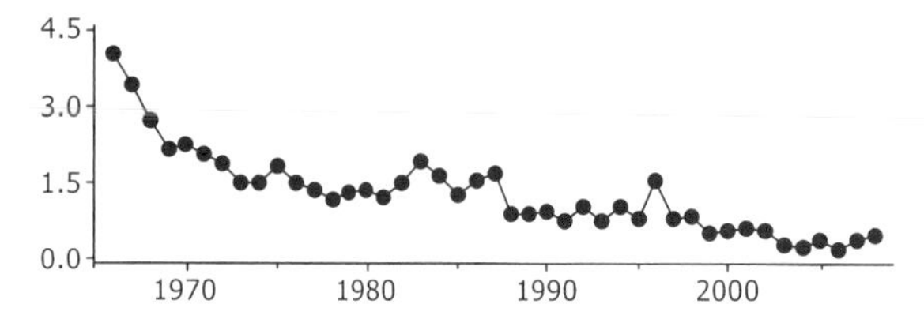


*Fig. 5.180.* Yellow-breasted Chat, *Icteria virens*, population trends from the Breeding Bird Survey.

1992g, Eckerle and Thompson 2001). Most commonly, this habitat is found in regenerating clear-cuts, forest edges, abandoned farmland, burned forest, and shrubby margins of ponds and streams. In Pennsylvania, breeding habitat is most permanently found along streamsides, power line cuts, thickets and hedgerows, and river bottoms (McWilliams and Brauning 2000). More transient patches of suitable shrub habitat are found in regenerating forests (particularly in even-aged stands resulting from clear-cuts), blueberry barrens, and recently abandoned pastures and farmlands. It is not clear whether different types of shrub habitat (e.g., early clear-cut regeneration versus blueberry barrens) are differentially suitable in their ability to support viable populations of chats, but larger patches are more likely to be occupied (Lehnen and Rodewald 2009).

In general, habitat availability for chats in Pennsylvania likely increased in the late nineteenth and early twentieth century and then decreased again in the mid-twentieth century, as widespread clear-cutting and later abandonment of farmland, initially created a great deal of shrub habitat, followed by a net loss of habitat as forest regeneration progressed and the rate of new timber harvest decreased (Todd 1940a, Leberman 1992g, Askins 2000). The amount of total forest cover in Pennsylvania has been relatively stable since the 1960s, but there is some evidence of a general decline in the amount of early successional habitat during that same time frame (Pennsylvania Department of Conservation and Natural Resources, unpublished data). A recent experimental study in Missouri showed that two timber harvest regimes (clear-cuts and group-selection cuts) led to increases in chat densities compared with unharvested control plots (Gram et al. 2003), but chats were absent from very small forest gaps created in a selective-logging treatment in Illinois (Robinson and Robinson 1999). Because data are lacking on the relative quality of different types of shrub habitat for chat breeding, no assessment is available for whether Pennsylvania offers high-quality breeding habitat.

Nonbreeding habitat is generally shrubby but has not been studied as well. Between the end of parental care and the start of their first fall migration, juvenile chats were found to disperse into home ranges in shrub habitats similar to those used by adults during the breeding season (Maxted 2001). Habitat use during migration is not well studied for adults or for juveniles. In winter, chats in Central America primarily use shrub areas, abandoned pastures, early successional forests, and a variety of disturbed older forests (summarized in Eckerle and Thompson 2001). Winter habitat use in the southeastern United States is not well quantified.

### *LIFE HISTORY AND ECOLOGY*

Chats display a social system that is territoriality in summer, with males defending territories of approximately 1 ha, occupied by a breeding pair and recently fledged offspring (Brewer 1955, Thompson and Nolan 1973). In winter, birds are solitary and somewhat territorial (Rappole and Warner 1980, Stiles and Skutch 1989). Diet is predominately insects, but important amounts of fruit are also eaten, both during breeding (Eckerle and Thompson 2001) and in winter (Stiles and Skutch 1989, Greenberg 1992).

Population dynamics are not well described. Recent radiotelemetry work has estimated that the probability of fledged young surviving to eight weeks is 40 percent (Maxted 2001). Survival of juveniles through their first year is, like with all Neotropical migrant songbirds, not known. Measuring adult survival across years is complicated by chats' apparently low breeding site fidelity (Thompson and Nolan 1973).

Their predominant mating system is social monogamy (with a low frequency of polygynous pairings), with extra-pair young found in approximately 33 percent of broods (Eckerle and Thompson 2001)—a value roughly comparable to many songbirds. They build open cup nests in low shrubs, often in large, dense patches of shrubbery (Burhans and Thompson 1999). Renesting after nest failure is probably common (Eckerle and Thompson 2001). Some pairs (8% of twenty-four nesting females in Thompson and Nolan 1973) rear two broods in a single season. Nest predation is the greatest cause of nest failure but does exhibit spatial variation in frequency (Woodward et al. 2001). Brood parasitism by cowbirds (primarily brown-headed cowbirds [*Molothrus ater*]) is common but geographically variable: frequency of parasitized nests in different populations ranges from 5 percent to 91 percent (Eckerle and Thompson 2001).

*THREATS*

Five main threats seem potentially important, and four are related to breeding. The first is habitat loss on the breeding grounds, which may be part of a long-term regional trend. Other unidentified factors are likely important, as populations are also declining in some areas that seem to have suitable habitat. Second is nest predation; this is not known to be an unusual problem in Pennsylvania but in general can have a strong effect on population viability in songbirds (Schmidt and Whelan 1999). Third is brood parasitism by brown-headed cowbirds. This may not be a severe threat in Pennsylvania relative to other parts of the chat's breeding range, as brown-headed cowbird densities are relatively low in Pennsylvania (Sauer et al. 2004). Fourth is spraying of *Bacillus thuringiensis* to control insect pests. This may be a threat along rivers (e.g., Susquehanna), where it could reduce caterpillar availability in river-bottom shrub habitat. Fifth is habitat loss on wintering grounds and along migratory routes. There is no evidence that this explains regionally specific population trends on breeding grounds, but it is not known whether particular breeding and wintering areas are demographically linked.

*CONSERVATION AND MANAGEMENT NEEDS*

Conservation priority for this species in Pennsylvania should recognize that chat populations are declining throughout the northeastern extreme of their breeding range; this may render local conservation efforts either especially important or especially futile. The eastern United States breeding distribution could be continuous enough that the Pennsylvania population is not genetically distinct or important, but this is fairly speculative, and the one study of genetic variation (Lovette et al. 2004) did not sample birds from Pennsylvania.

Conservation objectives are hard to define and implement, given the uncertainty about reasons for population declines. One possibility is to focus on protecting and maintaining good habitat in southern Pennsylvania where chats are still distributed fairly regularly. After researchers better understand the causes of breeding range contraction, we can reassess whether it is productive to aim for reestablishing populations in the northern part of Pennsylvania. Given the low breeding-site fidelity of chats and the rapid replacement of experimentally removed birds (Thompson 1977), chats may be unusually good at colonizing newly created or restored habitat. If habitat loss is the main cause of declines, then creation of new habitat may facilitate range expansion in the manner that appears to have happened in the late 1800s; such action might benefit a diverse suite of shrubland bird species (Askins 2000). However, two cautions are worth considering. First, the creation or maintenance of shrub habitat might have the drawback of fragmenting other habitat types. And second, if the recent declines in chat populations in Pennsylvania turn out to be simply a return to the natural distribution of chats before European settlement and widespread clearing of forests, it may be hard to justify active management for chat habitat in Pennsylvania.

*MONITORING AND RESEARCH NEEDS*

Monitoring should initially consist of point-count surveys (Ralph et al. 1995) in potentially suitable shrub habitat in early summer to establish detailed maps of presence or absence. Sites occupied by chats should be used for more intensive population studies, with studies focused on assessing reproductive success and site fidelity of individually marked birds. Given the successional nature of chat breeding habitat, the sites for these intensive studies should be shifted in adjustment to ongoing findings of presence/absence surveys. Survey design should allow for adequate comparison of population viability in different types of shrub habitat, with study sites of different habitats preferably stratified in relation to distance from edge of the species' breeding range.

Given that habitat availability has been suggested as an important issue, the priorities should be to identify and assess habitat use and habitat quality in Pennsylvania in three ways. First, presence/absence data (e.g., Annand and Thompson 1997) should be collected in shrub habitat across the state. A coarse assessment will be provided by the second Breeding Bird Atlas currently under way. More detailed data could be collected with point counts conducted early in the breeding season (Ralph et al. 1995).

Second, historical and current GIS data could be used to quantify change in breeding habitat availability at a variety of spatial scales—even across the chats' entire breeding range—to test whether those regions with long-term population declines (as revealed by Breeding Bird Survey data) are those with strongest loss of habitat over that time period.

Third, it is important to recognize that presence/absence data provide a good starting point for more intensive study but does not provide sufficient information for most management decisions because there is no information about the viability of populations in different regions or different habitats. Thus, the quality of breeding habitats, stability of populations, and causes of population declines in Pennsylvania need to be explored, which will require more intensive work over a longer time frame. A first step in this direction is the assessment of breeding productivity based on following the breeding success of individually marked birds over the entire breeding season (Anders and Marshall 2005). Other possible methods for assessing habitat quality are by measuring body condition of adults breeding in different habitats (Strong and Sherry 2000) or the degree of fluctuating asymmetry of nestlings or fledglings reared in different habitats (Lens et al. 2002).

Additional data that would be helpful include nestling growth rates, postfledging survival, and adult survival, though the latter will be made difficult by chats' reportedly low fidelity to particular breeding sites. Video cameras could be used to identify nest predators (sensu Thompson and Burhans 2003), though there is no a priori reason to expect that chats' nest predators would be different from those of other shrub-nesting songbirds; consequently, this may not be the most important threat for chats or target for management actions. The work described here is more expensive and time consuming than collecting simple presence-absence data, but much of it is crucial for a meaningful assessment of population viability.

At a broader (i.e., range-wide) spatial scale, it may be useful to explore the possibility of using geolocators to better link winter and breeding ranges (Stutchbury et al. 2009). This information could help explain the strong spatial variation in population trends, with some areas consistently increasing, others stable, and others consistently declining. Factors that limit the breeding range of chats are not known, and this may be an important topic, given evidence for recent range contraction in the northeastern United States.

*Author:* DONALD C. DEARBORN, BUCKNELL UNIVERSITY

## Grasshopper Sparrow

Order: Passeriformes
Family: Emberizidae
*Ammodramus savannarum*

The grasshopper sparrow is a small- to medium-sized, flat-headed, grassland sparrow (fig. 5.181). It was selected as a Species of Maintenance Concern because of range-wide declines in populations and as an indicator of large-scale grassland habitat. It is listed as a Species of Greatest Conservation Need in all northeastern states. Breeding populations in Pennsylvania are Apparently Secure (S4B), and global populations are Secure (G5, NatureServe 2009).

### *GEOGRAPHIC RANGE*

Grasshopper sparrows breed in widely scattered populations along the southern border of Canada and south across most of the United States. The main population is in the Great Plains from North Dakota south to Texas and east to Illinois (Vickery 1996). Significant populations breed in the eastern United States from north Georgia to Maine. Eastern populations winter primarily along the Gulf states and north to North Carolina (McWilliams and Brauning 2000).

### *DISTRIBUTION AND RELATIVE ABUNDANCE IN PENNSYLVANIA*

Grasshopper sparrows breed widely in the state in various grassland, pasture, field, and surface mine

*Fig. 5.181.* The Grasshopper Sparrow, *Ammodramus savannarum*. Photo courtesy of Joe Kosack.

habitats (McWilliams and Brauning 2000) (fig. 5.182). These sparrows are fairly common to locally common in the western Piedmont, Ridge and Valley, and southwest Appalachian Plateau physiographic province; they are uncommon in similar habitats of the northeast Glaciated Plateau (McWilliams and Brauning 2000). Rosenberg (2004), as part of the Partners in Flight (PIF) landbird conservation planning process, used Breeding Bird Survey (Sauer et al. 2004) data to estimate a rough, statewide population of grasshopper sparrows totaling 71,000 individuals. North American Breeding Bird Survey data indicate a significant population decline (estimated with moderate precision; Sauer et al. 2004) in North America (3.8% per year) and in Pennsylvania (6.1% per year) between 1966 and 2002 (fig. 5.183; Sauer et al. 2004).

In the eastern United States, and particularly in Pennsylvania, reclaimed bituminous coal fields are beneficial to grassland birds (Yahner and Rohrbaugh 1996). Widespread surface mining and subsequent reclamation in western Pennsylvania has resulted in an extensive patchwork of reclaimed sites among forests, woodlots, and agricultural fields. The acidic, nutrient-poor soils of reclaimed sites provide little potential for agricultural or timber production, and grasses and legumes tend to be the most successful and persistent vegetation types (Vogel 1981, W. G. Vogel, personal communication). These often undisturbed fields have a slow rate of ecological plant succession and are ideal for grasshopper sparrows, as well as are compatible for many other grassland associated species (Bajema et al. 2001).

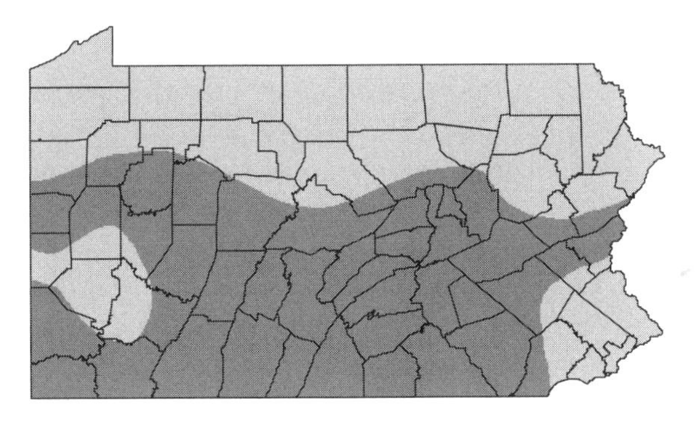

*Fig. 5.182.* Primary (darker shading) and secondary (lighter shading) distribution of the Grasshopper Sparrow, *Ammodramus savannarum*.

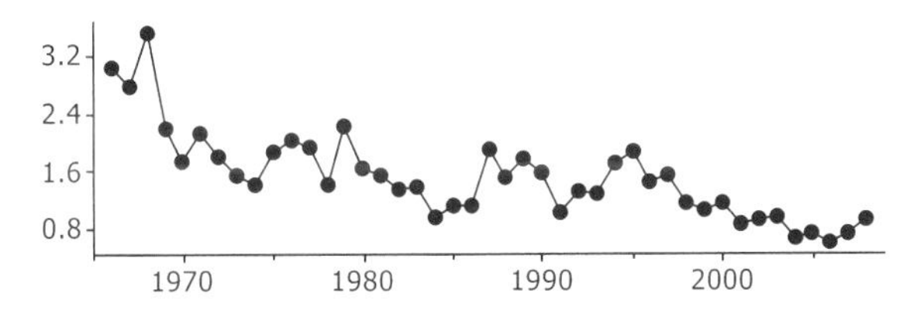


*Fig. 5.183.* Grasshopper Sparrow, *Ammodramus savannarum*, population trends from the Breeding Bird Survey.

For example, Mattice et al. (2005) estimated 35,373 ha of suitable reclaimed surface mine grassland habitat in their nine-county study area, the first estimate of its kind. Further, they estimated that 9,727 singing males (or, roughly, 20,000 individuals) were present in this nine-county area. Diefenbach et al. (2007) modified this abundance estimate (based on the proportion of singing males likely to be present but missed during surveys) to 46,845 singing males (or, roughly, 90,000 individuals) present in this nine-county area. A similar, but perhaps much more difficult, need is to estimate suitable habitat and sparrow abundance associated with hay fields, pastures, grassy fields, and other agriculturally related habitats. In Pennsylvania, grasshopper sparrows were the most numerous grassland obligate species found in fields enrolled in the Conservation Reserve Enhancement Program (CREP), but their estimated density was relatively low at 0.12 males per hectare (Wentworth et al. 2010).

### COMMUNITY TYPE/HABITAT USE

Grasshopper sparrows generally prefer moderately open grasslands with patchy bare ground (Vickery 1996); they select different components of vegetation, depending on the grassland ecosystem. In Pennsylvania, they inhabit meadows and grassy fields, including hay fields and pastures, but especially reclaimed surface mines and recently abandoned farm fields (McWilliams and Brauning 2000). In general, Pennsylvania offers two contrasting, broad habitat types key to maintaining this species in the commonwealth: (1) agricultural lands, such as hay fields, pastures, and fallow fields, and (2) grasslands associated with reclaimed surface mines. While the agricultural habitat types appear to be extensive (although declining) within Pennsylvania, several authors have concluded that most agricultural habitat essentially serves as a population sink (Bollinger et al. 1990, Kershner and Bollinger 1996, Rohrbaugh et al. 1999). Agricultural set-asides (e.g., Conservation Reserve Program lands) and prairie reserves provide reservoirs of grassland habitat that may help support remaining populations of some grassland bird species (Delisle and Savidge 1997, Koford 1999, Coppedge et al. 2001, Johnson and Igl 2001). Population trends of grasshopper sparrows in Pennsylvania were positively associated with the amount of farmland enrolled in the Conservation Reserve Enhancement Program, suggesting that this program is providing a direct benefit to grasshopper sparrows (Wilson 2009). The second broad habitat type, reclaimed surface mines, has inadvertently become the primary source of grassland bird habitat in Pennsylvania and other areas. Whitmore and Hall (1978) documented the presence of grassland birds on reclaimed surface mines twenty-five years ago, although the contribution of those populations was not recognized for many years. Recent studies have confirmed the existence of substantial grassland bird populations on reclaimed mines throughout the Midwest and Northeast, which indicate these habitats may be important for conserving many grassland species (Yahner and Rohrbaugh 1996, Bajema et al. 2001).

### *LIFE HISTORY AND ECOLOGY*

Boundaries of the typically <1-ha territory are delineated by conspicuous song perches, flight displays, and agonistic encounters (Vickery 1996). Mattice et al. (2005) investigated landscape and patch-level habitat characteristics for grasshopper sparrows on reclaimed surface mines and found that patches generally needed to be >30 ha in area for high probability of occupancy.

Pairs form on the breeding grounds immediately after the spring arrival of females (Vickery 1996). Grasshopper sparrows nest on the ground or in a clump of vegetation with the lip of the nest at ground level. Breeding generally begins in mid-May in Pennsylvania and can be protracted, depending on favorable weather; generally, pairs are able to produce two broods annually in Pennsylvania if nests are not predated (M. R. Marshall and D. R. Diefenbach, unpublished data). Grasshopper sparrows will renest within a breeding season if a nest fails because of predation. The ability to fledge multiple broods per season likely contributes greatly to season-long productivity (e.g., Holmes et al. 1992). Grasshopper sparrows are occasionally parasitized by brown-headed cowbirds (*Molothrus ater*); however, none of fourteen nests monitored by M. R. Marshall and D. R. Diefenbach in 2003 (unpublished data) on a reclaimed surface mine in Pennsylvania were parasitized. In general, little is known about the spatial and temporal patterns of nest success, causes of nest failure (i.e., nest predators), season-long productivity, or levels of cowbird parasitism for grasshopper sparrows in Pennsylvania.

Annual return rates of males to the same breeding site vary greatly, apparently being much lower in the Midwest and prairie regions than in the East. Vickery (1996) summarizes several studies, indicating 0 percent to 21 percent of banded birds return to midwestern sites and 31 percent to 50 percent returning to sites in the East. Marshall and Diefenbach (unpublished data) found return rates of 45 percent (twenty-one of forty-seven banded males) to a reclaimed surface mine in Pennsylvania. No data exist on return rates and site fidelity of females or juveniles. Unbiased annual survival estimates are also lacking.

### *THREATS*

The primary factors influencing the species decline in Pennsylvania include the changes in agricultural practices during the past fifty years that have made much agricultural habitat unsuitable for native grassland species (Warner 1994, Bolgiano 1999, 2000). Population declines are also due to the accelerating loss of agricultural areas and surface mine areas to urban sprawl (Vickery et al. 1999, P. D. Vickery, personal observation). Agricultural set-aside programs may increase the amount of grassland habitat in agricultural areas; however, these programs often only create small, highly fragmented habitats that may not be high-quality habitat for this species. Many of these programs are relatively new and more time is needed to assess their value to grassland birds.

Reclamation practices, such as planting trees, as well as natural succession, can lead to an excessive density of woody vegetation on reclaimed surface mines thus rendering them inappropriate for grassland species. Habitat protection for this species must include large tracts (>20 ha) of grassland habitat and, if necessary, continual habitat management to prevent vegetation successional changes. Improved reclamation practices may be a threat to this species as well because a greater percentage of planted trees may become established, again rendering these areas unsuitable for obligate grassland species.

### *CONSERVATION AND MANAGEMENT NEEDS*

A major conservation and management need is to identify the location and extent of suitable, high-quality reclaimed surface mine habitat for grasshopper sparrows across the nine-county bituminous coal area of Pennsylvania and coordinate protection and management of these locations. Mattice et al. (2005) estimated that roughly 35,000 ha of suitable habitat exist with the nine-county area of Pennsylvania coincident with the bituminous coal fields. This assessment should not be done without considering other priority grassland species.

Until a better understanding of population status and dynamics within these areas of Pennsylvania and a clearer understanding of Pennsylvania's regional/global responsibility for this species emerges, it seems reasonable to recommend maintaining the current estimated acreage of suitable habitat. Because land is currently being mined and subsequently reclaimed while reclaimed areas are simultaneously undergoing succession, these 35,000 ha will likely be secured within a shifting landscape mosaic where not all 35,000 ha need protection as long as the acreage exists within the landscape. Some large blocks, however (such as the Piney Tract, Clarion County), should be preserved and managed. Finally, it is imperative to work with the Pennsylvania Department of Environmental Protection on the reclamation process. Specific areas that have a high probability of becoming important grassland bird habitat (e.g., open patches >20 ha near other nonforested habitats) should not have trees planted as part of the reclamation process to increase the likelihood of providing quality grassland bird habitat. In agricultural landscapes, efforts should continue to increase enrollment of farmland in the Conservation Reserve Enhancement Program as this provides a direct benefit to grasshopper sparrows (Wilson 2009).

### *MONITORING AND RESEARCH NEEDS*

Until more information is gathered about density, demographic rates, and source habitats, it would be difficult to propose a fully specified monitoring strategy. However, Diefenbach et al. (2007) and Mattice et al. (2005) have developed efficient, statistically defensible methods of estimating grassland sparrow abundance over broad geographic areas that could be implemented as needed. This type of effort, while efficient and informative, is fairly time consuming and costly. It is therefore prudent to ensure that Breeding Bird Survey routes continue to be run throughout the state. These are the best data available for statewide population trends and continue to be valuable for most bird species, including the grasshopper sparrow. It may be valuable to augment the existing Breeding Bird Survey routes with additional routes in areas of Pennsylvania with grassland bird populations and relatively few active Breeding Bird Survey routes (such as the nine-county bituminous coal area). In this manner, the less expensive, volunteer-based Breeding Bird Survey data could be used for general monitoring and when a particular threshold (say, a decline of >5% per year) is reached, the more intensive (but more precise) monitoring methods of Mattice et al. (2005) and Diefenbach et al. (2007) could be triggered and implemented (Houser et al. 2006). The Pennsylvania Game Commission Conservation Reserve Enhancement Program monitoring surveys should be continued and used to monitor use of Conservation Reserve Enhancement Program areas by grasshopper sparrows and other grassland birds and to better estimate the effect of Conservation Reserve Enhancement Program on grassland bird populations (Wilson 2009).

Not all patches of reclaimed surface mine habitat are equal to grasshopper sparrows because of such factors as size of the patch, vegetation characteristics, and landscape context. Because of this, and because we do not advocate maintaining *all* reclaimed surface mines as grassland bird habitat, it is imperative to identify and prioritize the best-quality areas. Mattice et al. (2005; J. A. Mattice, unpublished data) have investigated and developed models explaining the relationship and relative importance of patch size, vegetation, and landscape context in terms of grasshopper sparrow patch occupancy and abundance. These models should be tested, verified, and then implemented widely throughout the nine-county area with the majority of reclaimed surface mine habitat to identify the most important high-quality patches of habitat available for conservation and management.

Little demographic data exist for the grasshopper sparrow in Pennsylvania in agricultural or surface mine areas. Estimates suggest high sparrow abundance for much of the surface mine habitat in western Pennsylvania; however, empirical data on nest success, productivity, and annual survival estimates are needed in these habitats to assess their quality.

A better understanding of vegetation succession on reclaimed surface mines as it relates to grassland bird populations, especially in light of improved reclamation practices, is needed to evaluate and prescribe management actions to maintain surface mines as high-quality habitat. For example, the authors (unpublished data) and Mattice et al. (2005; J. A. Mattice, unpublished data) found high densities of grassland sparrows in areas with more than thirty years postreclamation. Other reclaimed surface mines have undergone more rapid natural or anthropogenic (i.e., establishing tree cover through planting) succession resulting in woody vegetation densities too high for obligate grassland species. Populations of grasshopper sparrows among patches of reclaimed surface mine may be acting as a metapopulation with implications for conservation and management. Documenting movement patterns and exchange among patches are needed to understand these dynamics.

*Authors:* MATT R. MARSHALL, NATIONAL PARK SERVICE; DUANE R. DIEFENBACH, U.S. GEOLOGICAL SURVEY, PENNSYLVANIA COOPERATIVE FISH AND WILDLIFE RESEARCH UNIT

## Eastern Meadowlark

Order: Passeriformes
Family: Emberizidae
*Sturnella magna*

The eastern meadowlark is a bird of open country that was selected as a Species of Greatest Conservation Need because of widespread population declines in Pennsylvania and throughout its range (fig. 5.184). It is a Partners in Flight Priority IIA species and is included on birds of Conservation Concern lists throughout the Northeast. Global populations are considered Secure (G5, NatureServe 2009).

### *GEOGRAPHIC RANGE*

The eastern meadowlark breeds from Ontario, Nova Scotia, and Maine southward to Florida, Texas, New Mexico, Arizona, and into Central America and northern South America. They are migratory in the extreme northern part of their ranges and residents elsewhere (Lanyon 1995).

*Fig. 5.184.* The Eastern Meadowlark, *Sturnella magna*. Photo courtesy of Michael Brown.

### *DISTRIBUTION AND RELATIVE ABUNDANCE IN PENNSYLVANIA*

The eastern meadowlark is a common but declining breeder in parts of the state that are not heavily forested or urbanized (fig. 5.185). They are virtually absent from the forested north-central areas, rare in the fields bordering suburbanized eastern counties, and common in north and western agricultural fields and reclaimed strip mines. This is a common migrant in the northern portions of the state. It arrives in spring before the last week of February, especially in the Piedmont Plain and Coastal regions (McWilliams and Brauning 2000). In winter, the eastern meadowlark is a resident primarily in the southern half of the state. Flocks of eastern meadowlarks have also been recorded in Northampton and Bucks counties (in northwestern counties of the state). Corresponding with current trends toward reforestation and increased urban development in the state, they have been declining as a breeding species in Pennsylvania since the 1960s. According to Breeding Bird Survey results, eastern meadowlarks have been in decline over the past thirty years regionally and in

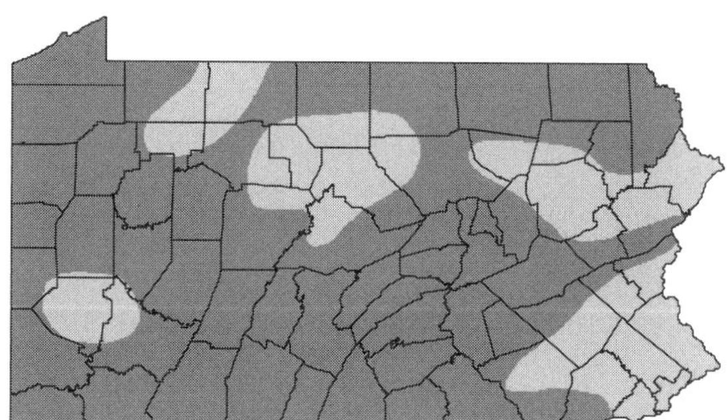

*Fig. 5.185.* Primary (darker shading) and secondary (lighter shading) distribution of the Eastern Meadowlark, *Sturnella magna*.

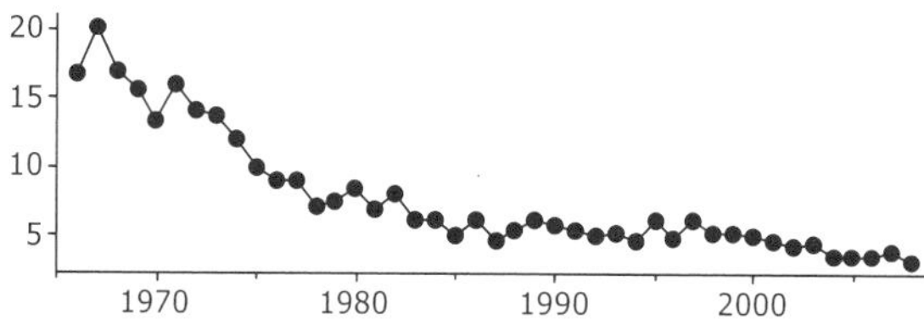


*Fig. 5.186.* Eastern Meadowlark, *Sturnella magna*, population trends from the Breeding Bird Survey.

Pennsylvania (fig. 5.186). Eastern meadowlarks are still reported in approximately 75 percent of the surveys, but at low quantities of one to two birds per route (McWilliams and Brauning 2000).

### *COMMUNITY TYPE/HABITAT USE*

One of the best-known birds of American farmlands, the eastern meadowlark delivers its cheerful song from the treetops, telephone poles, wires, and fenceposts. It is found in various open habitats, including agricultural fields, reclaimed strip mines, meadows, pastures, hay fields, and fallow land. The eastern meadowlark prefers to nest in open grasslands, and it has become localized in areas where fields are still maintained for livestock and hay production. It prefers relatively large fields (5 ha and larger) although it may sing from a perch in smaller grassy areas (Norment et al. 1999). Preferred fields have a variety of grass species, limited shrub layer, and a low understory. The eastern meadowlark maintains a wide distribution because it is capable of using smaller grassland sites, reclaimed strip mines, highway median strips, and interchanges that are periodically mowed.

### *LIFE HISTORY AND ECOLOGY*

The eastern meadowlark is a bird of open grasslands. In appropriate habitat, it is common to see meadowlarks perching on fenceposts and other high perches during the day. The diet of the eastern meadowlark consists primarily of insects and small invertebrates. It is also a granivore, often feeding on seeds that it forages from the ground (NatureServe 2005). Eastern meadowlarks remain in pairs or family groups through most of the year.

Northern populations of eastern meadowlarks are migratory. They return to their breeding grounds in early spring. The males often return in early April, approximately two weeks before the return of the females (NatureServe 2005).

The eastern meadowlark is often polygamous and more than one female may be found nesting in the territory of a single male. Females build a cup-shaped nest of grasses on the ground in a meadow depression. A second clutch may be attempted if conditions are suitable (i.e., food resources are adequate).

### *THREATS*

As with almost all obligate grassland species, the eastern meadowlark is on a nationwide decline. Declines are primarily due to loss of habitat resulting from the loss of native grasslands throughout their range. These losses are a result of increased human activity, such as conversion of fields to row crops, and residential and commercial development. Eastern meadowlarks are area sensitive, meaning they are more likely to nest in large fields than in small fields. Area requirements may vary regionally and with the surrounding landscape (Johnson et al. 2001). They will often occur in smaller fields and grassy areas, however not as breeders (Horn et al. 2000). Mowing of hayfields frequently destroys nests and nestlings. In pastures, grazing livestock trample nests, nestlings, and incubating females. Agricultural practices, including the application of pesticides to farm fields may affect nests, nestlings, and brooding females and indirectly affect their food supply. The eastern meadowlark is also a common cowbird host.

### *CONSERVATION AND MANAGEMENT NEEDS*

The primary conservation and management need for meadowlarks in Pennsylvania and throughout their breeding range is to improve the quality and quantity of nesting habitat. Meadowlarks frequently nest in hay fields where their nests are lost to mowing. Encouraging farmers to delay mowing and mow less frequently would improve nest success. Encouraging farmers to enroll their land in the Conservation Reserve Enhancement Program (CREP) and convert fields to a mix of grasses would be best. In Pennsylvania, meadowlarks are one of the most frequently reported grassland obligates using Conservation Reserve Enhancement Program fields (Wentworth et al. 2010), and their populations have responded positively with reduced rates of decline in landscapes with more fields enrolled in the Conservation Reserve Enhancement Program (Wilson 2009).

### *MONITORING AND RESEARCH NEEDS*

Large-scale monitoring through the Breeding Bird Atlas and Breeding Bird Survey effectively monitors current trends in this species. Additional targeted monitoring of agricultural lands as part of the Conservation Reserve Enhancement Program's monitoring program provides specific information on the meadowlarks' response to the Conservation Reserve Enhancement Program and should be continued (Wilson 2009).

A primary research need is to determine the effects of different management strategies on reproductive success of meadowlarks in agricultural habitats and in reclaimed strip mines. Use of reclaimed strip mines by meadowlarks should also be evaluated. Further studies on eastern meadowlarks may identify potential threats of pesticide use within its range (such as in suburban lawn care, farming practices, and other nonpoint sources as may be identified).

*Authors:* DARRYL SPEICHER, POCONO AVIAN RESEARCH CENTER; JACKIE SPEICHER, POCONO AVIAN RESEARCH CENTER

6

# The Mammals

## *Introduction*

Joseph F. Merritt
Michael A. Steele

Among the diverse assemblage of mammals residing in the mid-Atlantic states, sixty-four mammals are found within the borders of Pennsylvania. A few of these species are introduced. Nine other mammal species are assumed to have occurred in the state at sometime in the past. Five of these species (rice rat [*Oryzomys palustris*], wolf [*Canis lupis*], wolverine [*Gulo gulo*], moose [*Alces americanus*], and American bison [*Bison bison*]) are now extirpated from the state, and four others (American marten [*Martes americana*], fisher [*Martes pennanti*], Canadian lynx [*Lynx canadensis*], and cougar [*Puma concolor*]) are considered incidental or of uncertain occurrences sometime in the past. Similarly, the absence of bone debris of moose, American bison, and wolverine in cave deposits or Indian villages suggests that these species may have been only occasional transients in the state. Frequent recent sightings of cougars are classified as either unconfirmed sightings or individuals released from captivity, although we suggest that all reports should be taken seriously and that the potential for recolonization of the state by the species should not be discounted.

The sixty-one native species currently found in Pennsylvania reflect the diversity of habitats, climate, and topography across the state. In essence, Pennsylvania is a crossroads of diverse physiographic provinces where mammal assemblages common to the north intersect those from the midwestern United States, the south-central United States, and the eastern Coastal Plain (Merritt 1987, Steele and Kwiecinski 1994). Included among the extant Pennsylvania mammals are one marsupial (order Marsupialia), eleven insectivores (shrews and moles, order Soricomorpha), eleven bats (order Chiroptera), three rabbits (order Lagomorpha), twenty-three rodents (order Rodentia), thirteen carnivores (order Carnivora), and two members of the deer family (family Cervidae, order Artiodactyla).

More than one-third of these species—twenty-two—are formerly registered in the Pennsylvania's Wildlife Action Plan (WAP) and identified for conservation management by both the Pennsylvania Game Commission (PGC) and the Mammal Technical Committee of the

Pennsylvania Biological Survey. These species include five of Immediate Concern, six of High-Level Concern, one of Responsibility Concern, one classified as Pennsylvania Vulnerable, and the remaining nine of Maintenance Concern. In this book, we present species accounts on twenty-one of these species. The least weasel (*Mustela nivalis*), which is rare throughout its range, has received virtually no research in the state and is therefore not covered here.

## Threats to Mammals of Pennsylvania

As with most terrestrial vertebrates, the primary threats to mammals of Pennsylvania involve a range of factors: some are obvious, such as the expansion of the human population; others involve subtle but equally significant or long-term effects, and still others are only beginning to emerge. Here, we recognize six factors that we believe have the greatest effect on mammal species within the state. These are no doubt unique to Pennsylvania and may include most of the factors that influence mammal species throughout the northeastern United States. These include (1) *habitat loss and habitat fragmentation*, (2) more subtle impacts to the habitat collectively categorized as *habitat degradation* (i.e., loss of water quality), (3) *synergistic interactions* and (4) *ecological cascades* that result from subtle shifts in ecological stability, which lead to an indirect but significant effect on target species, (5) *new and emerging threats* that have only begun to influence species and whose effect will be only fully realized long into the future, and (6) problems *of natural rarity* that result from a species' specialized habits, the occurrence of a species on the periphery of its geographic range, or its naturally low population levels that are widely separated and in constant fluctuation.

The northern flying squirrel (*Glaucomys sabrinus*) represents a particularly interesting case study whose status and reasons for conservation priority are only recently emerging. Always considered a Species of Concern, it was long assumed that the northern flying squirrel was relatively abundant in suitable habitat unlike populations to the south in the Appalachian Mountains of Virginia, West Virginia, and North Carolina, where two subspecies have long been listed as federally Endangered. Intense survey efforts over the past eight years in Pennsylvania, however, demonstrate that populations of this species are relatively small, isolated, and perhaps now locally extinct from sites where it was previously known to occur. No doubt this decline has been due to habitat loss and fragmentation as a result of human encroachment, especially in the Pocono region, but also habitat degradation due to the loss of hemlocks as a result of invasive woolly adelgid aphids (forests are now subject to significant mortality from the invasive woolly adelgid [*Adelges tsugae*]). As discussed in the final chapter, this problem is further exacerbated by several other complicating factors, including the potential for parasite-mediated competition with the southern flying squirrel and what appears to be an emerging hybrid zone between the two flying squirrel species that is potentially mediated by climate change. Indeed, in the case of the northern flying squirrel now registered by the Pennsylvania Game Commission as Endangered in Pennsylvania, the factors contributing to decline are only beginning to emerge after nearly a decade of intensive investigation.

Other species have done remarkably well over the past two decades. Monitoring programs for two species, the river otter (*Lontra canadensis*) and fisher (*Martes pennanti*), for example, provide evidence that their recovery effort has been successful. Formerly extirpated from the state, these two species were reintroduced (see species accounts). On the basis of surveys of wildlife conservation officers and accidental trappings, the Pennsylvania Game Commission suggests that river otter populations are stable or increasing in approximately half the counties in the state and that density of the species has more than doubled since 1995. Similar data, as well as sightings, indicate that fisher populations have increased in more than a third of the state's counties and that their numbers statewide have increased as much as fourfold in the past five years.

Despite these successes, there is much yet to do, but conservation of mammals presents a variety of challenges somewhat different from those specific to the reptiles, amphibians, and birds. By far the most significant factor interfering with mammal conservation is the ability to detect the presence and absence of mammals and monitor their densities once their presence is established. Although secretive like many other vertebrates, detection probability of most mammal species, with the exception of larger mammals such as deer, elk, and bear, is far from 100 percent. This means simply that even when intensive survey techniques fail to produce evidence of a species, we cannot be sure they are absent from an area.

The accounts presented here provide our best estimate on the distribution of these species and outline the approaches we believe must be taken for future research management, and conservation. It is our hope that these accounts will help focus future efforts and inspire a new generation of biologists to study and protect our mammal fauna.

## IMMEDIATE CONCERN

### West Virginia Water Shrew

Order: Soricomorpha
Family: Soricidae
*Sorex palustris punctulatus*

The West Virginia water shrew can be distinguished from other species of Pennsylvania shrews by its large size (138 to 164 mm total length), long tail (63 to 72 mm), fringes of hair on both the tail and hind feet, and slightly webbed feet (Whitaker and Hamilton 1998; fig. 6.25). This species was ranked as a Species of Immediate Concern although it has a global rank of Secure (G5, NatureServe 2009), it is a Northeast Region Priority Species, and it is Threatened in Pennsylvania because of declining water quality and increasing fragmentation of populations due to road construction and development.

*GEOGRAPHIC RANGE*

The global distribution of the West Virginia water shrew appears restricted to the western portion of the Appalachian Plateau Province within the Allegheny Mountains Section extending southward through West Virginia into Tennessee along the Great Smoky Mountains (Hall 1981). *Sorex palustris* has a wide range extending from Alaska to New York and south into West Virginia and in the west to New Mexico.

*DISTRIBUTION AND RELATIVE ABUNDANCE IN PENNSYLVANIA*

In Pennsylvania, the West Virginia water shrew appears to be restricted to the Allegheny Plateau having been taken only in Somerset (Grimm and Roberts 1950), Fayette, and Westmoreland counties (fig. 6.1). Few individuals have been collected across its range, suggesting small population sizes (Kirkland and Schmidt 1982); however, potential riparian habitat for the subspecies is extensive. Relatively little work has been done concerning both abundance and population densities. Throughout the distribution of the West Virginia water shrew, its rarity results in either an S1 or an S2 status (NatureServe 2004) although the species as a whole is listed as G5T3; the T3, or taxonomic category,

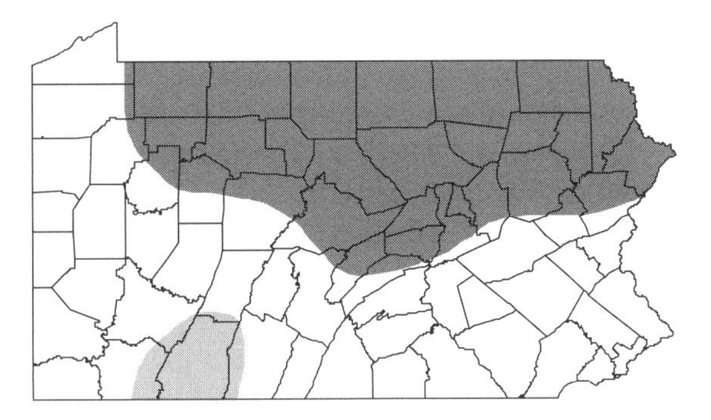

*Fig. 6.1.* Light shading = distribution of the West Virginia Water Shrew, *Sorex palustris punctulatus*.

is a result of a lack of clarity about the relationship between the northern water shrew (*Sorex palustris albibarbis*) and the southern water shrew (*S. p. punctulatus*). Further investigations along the perceived gap within the species distribution in Pennsylvania may lead to a future revision in taxonomy, if warranted.

*COMMUNITY TYPE/HABITAT USE*

As with the American water shrew, the West Virginia water shrew is found along moderate to sometimes swift-flowing streams that have undercut banks and a large amount of structure, such as rocks, logs, and crevices (Baker 1983). Researchers within Pennsylvania have found the West Virginia water shrew occurring in primarily mixed forests with canopy closures of up to 90 percent. The West Virginia water shrew is usually not found within steep gradient streams although Pagels et al. (1998) noted its occurrence along streams with gradients between 7 percent and 14 percent slope. Laerm et al. (1999) characterized the West Virginia water shrew's habitats south of Pennsylvania as occurring at elevations of between 2,900 and 5,000 feet in spruce-fir, northern hardwood, cove hardwood, and white pine-hemlock forests. Generally, there is a heavy cover of forbs and moderate structure to the understory.

Although high-quality habitat exists for this subspecies in undisturbed areas of the interior portion of the Allegheny Plateau, it appears that increased road building and logging coupled with increased development in surrounding habitats may affect the quality of habitat left for this shrew. Key habitats remain fairly intact because of their isolation, but these isolated populations may be affected in the near future as dispersal corridors decline. More research is needed to define the movements within and between known populations.

*LIFE HISTORY AND ECOLOGY*

As with the American water shrew, the West Virginia water shrew occurs most often in clear mountain streams at elevations between 1,500 and 2,000 feet in Pennsylvania. The streams are characteristically of high quality and moderate flow and usually have deeply undercut banks (Beneski and Stinson 1987). In Wyoming, Clark (1973) found that this species of shrew was most common when ground cover was 75 percent or higher. The author has found this to be fairly accurate for the American water shrew in northeastern Pennsylvania as well. Unlike the American water shrew, the West Virginia water shrew has been found most often in mixed coniferous/deciduous forests in southwestern Pennsylvania where hemlock is prevalent.

Although there may be a connection between water quality and the occurrence of water shrews, a study by Rozanski et al. (1996) in northern Pennsylvania was inconclusive because of low sample sizes. Any disturbance to the forest environment that promotes warming of streams, sediment loading, and dramatic changes in pH is likely to have significant negative effects on the habitat of the water shrew.

As with the American water shrew, West Virginia water shrews are capable divers and can remain submerged for more than forty-five seconds (Beneski and Stinson 1987). They remain active during the winter and are able to continue diving for prey because of the ability to reduce their metabolic demands (Boernke 1977). The diet of the West Virginia water shrew is comparable to the American water shrew's, which includes stonefly, caddisfly, and mayfly nymphs, as well as cranefly larva. Other important foodstuffs are most likely snails, slugs, earthworms, small fish, fish eggs, and salamanders (Whitaker and Hamilton 1998).

Although there is little information concerning the reproductive biology of water shrews in general, it is believed that they become reproductively active in late February and remain active until late summer (Kurta 1995). They produce two to three litters annually although the gestation period remains undefined (Whitaker and Hamilton 1998). It is also believed that, unlike other insectivores, they do not become sexually mature until their second season after birth.

*THREATS*

As with the American water shrew, several threats are thought to occur, chief among these are pollution of the streams in which they forage. This pollution may be acid mine drainage or increased degradation due to acid precipitation. It is unknown how these factors may affect the continued existence of this species in the long term. Very little research has been conducted concerning these threats.

Another possible threat is the isolation of local populations due to development and increased road building. As the tourist industry increases in southwestern Pennsylvania, infrastructure is improved to provide adequate access to areas that have, until now, remained relatively isolated from these threats. It is likely that this pressure may serve to "cut off" populations from sources of repopulation in times of natural disturbances, such as flooding or forest fire. Sedimentation of portions of streams may also serve to isolate populations by reducing the food supply available to the West Virginia water shrew and, hence, its survival and dispersal.

*CONSERVATION AND MANAGEMENT NEEDS*

The West Virginia water shrew is listed in NatureServe (2004) as "S1." It is designated as Threatened (Kirkland and Krim 1990) by both the Pennsylvania Biological Survey and Pennsylvania Game Commission and receives protection under the Pennsylvania's Threatened and Endangered Species legislation.

The most pressing research priority is to determine the extent of the distribution of the West Virginia water shrew in southwestern Pennsylvania. Along with this research would be the collecting of macro- and microhabitat variables that may begin to let researchers better define the optimal habitat for the subspecies. At present, a lack of information concerning the effects of differing water quality, acid rain levels, and sedimentation on the occurrence of the American water shrew precludes defining parameters for optimal habitat. Another research priority is to define existing optimal habitat and its relationship to future human impacts, such as development and infrastructure expansion.

Ecological studies pertaining to reproductive success, population and community structure and the effects of both physical and temporal changes to the habitat supporting the West Virginia water shrew should be initiated to better understand this species' requirements. Such studies should include: (1) developing a network of cooperators that will increase the likelihood of discovering new occurrences of the West Virginia water shrew by locating existing viable habitat; (2) continuing inventory and monitoring of known populations, including the collection of micro- and landscape-level habitat characteristics; (3) monitoring both chemical and physical properties of streams at existing West Virginia water shrew sites for changes that may affect the shrews continued success; (4) coordinating development of a stream condition database with the Department of Environmental Protection (DEP), the Pennsylvania Fish and Boat Commission (PFBC), the Pennsylvania Natural Heritage Program (PANHP), and other conservation organizations that may be based partially on the new invertebrate classification system currently being produced by PANHP personnel; (5) developing low-impact trapping regimes increasing trap survivability and thereby lessening the impacts to West Virginia water shrew populations. This, in turn, will lead to enhanced information concerning the ecology of the West Virginia water shrew; (6) creating maps based on the Geographic Information System, or GIS, for existing habitat as well as maps and layers based on predicted occurrence of viable habitat in association with the current Pennsylvania Gap Analysis Project (GAP) model (Myers et al. 2000); and (7) incorporating appropriate habitat in conservation planning issues at the state level to begin to develop habitat corridors encouraging dispersal of West Virginia water shrews.

*MONITORING AND RESEARCH NEEDS*

At present, there is no conservation management plan in effect for the West Virginia water shrew; minimal research has focused on this subspecies. Existing populations should be monitored at five-year intervals using livetraps to assess their health and to possibly ascertain population densities at the known locations.

Habitat data at existing sites should be collected and monitored for changes, both natural and human induced. Several issues arise when considering habitat conservation, chief among these are (1) identifying viable West Virginia water shrew habitat based on remote-sensed data as well as stream condition; (2) developing low-impact methods for the capture and release of American water shrews, thereby reducing trap mortality; and (3) assessing the current distribution of the American water shrew in terms of connectivity and landscape changes due to human alterations.

Research needs include understanding what viable habitat consists of, determining the distribution in Pennsylvania and what barriers exist that preclude dispersal, and creating a map of possible habitat that can be verified. Another research need is understanding population densities at the existing sites. The creation of livetraps that promote survival of shrews is necessary if researchers are to capture and mark water shrews for determining population densities, dispersal, home range, and other basic ecological parameters. Other research needs include (1) creating GIS-based maps of possible habitat in Pennsylvania based on existing information; (2) collecting data based on intensive habitat analysis targeted at producing usable habitat maps for predicting West Virginia water shrew habitat; (3) assessing and sampling all predicted habitats at least one time using Pennsylvania Game Commission protocols; and (4) incorporating information on population into conservation planning for the area involved.

*Author:* JAMES HART, PENNSYLVANIA NATURAL HERITAGE PROGRAM

## Indiana Bat

Order: Chiroptera
Family: Vespertilionidae
*Myotis sodalis*

The Indiana bat was first identified through type specimens obtained at Wyandotte Cave located in southern Indiana (Miller and Allen 1928), hence the names Indiana bat, Indiana *Myotis*, and Wyandotte Cave bat. The Indiana bat usually has a distinctly keeled calcar composed of cartilage support on the edge of the interfemoral membrane (fig. 6.2). The species was identified for Conservation Concern because it is listed as a federally Endangered Species. The recent occurrence (winter 2008–2009) of white-nose syndrome (WNS) in Pennsylvania hibernacula

*Fig. 6.2.* The Indiana Bat, *Myotis sodalis*. Photo courtesy of Cal Butchkoski.

will likely cause significant reductions in numbers of all species of hibernating bats, including *Myotis sodalis*. The global status of the Indiana bat is Imperiled (G2, NatureServe 2009).

### *GEOGRAPHIC RANGE*

The Indiana bat has been found throughout much of the eastern United States from Oklahoma, Iowa, and Wisconsin, east to Vermont and south to northwestern Florida. The winter habitat of Indiana bats closely follows regions of well-developed limestone caverns and abandoned mines within the species range. These underground shelters serve as hibernacula and are categorized in the U.S. Fish and Wildlife Service (USFWS) Indiana bat recovery plan draft (2007) as: Priority 1 (P1), hibernacula with a recorded population >10,000 Indiana bats with two subcategories of P1A for sites that have held >5,000 Indiana bats during one or more winter surveys within the past ten years, and P1B for sites that recorded <5,000 Indiana bats over the past ten years; Priority 2 (P2), hibernacula with a current or recorded population ≥1,000 Indiana bats; Priority 3 (P3), hibernacula with a current or recorded population of 50–1,000 Indiana bats; and Priority 4 (P4), hibernacula having current or observed populations of less than 50 bats.

P1 and P2 categories must also currently have suitable microclimates and not be ecological traps. The goal of this most recent hibernacula ranking (USFWS 2007) is to recognize a wider, more even distribution of important hibernacula across the species range and replaces the 1999 plan (USFWS 1999). Positive winter occurrence since 1995 spans nineteen states with twenty-three P1 hibernacula in Illinois ($n = 1$), Indiana ($n = 7$), Kentucky ($n = 5$), Missouri ($n = 6$), New York ($n = 2$), Tennessee ($n = 1$), and West Virginia ($n = 1$) and fifty-three P2 hibernacula located within those seven, plus another four states of Arkansas, Ohio, Pennsylvania, and Virginia. A total of 150 P3 are located in sixteen states and 213 P4 within twenty-three states (USFWS 2007). More than 90 percent of the Indiana bat population hibernates in the five states of Indiana (45.2%), Missouri (14.2%), Kentucky (13.6%), Illinois (9.7%), and New York (9.1%). The range-wide distribution of the Indiana bat has been divided into four recovery units based on taxonomic studies, banding returns, range-wide genetic variation, and ecoregions (USFWS 2007). The four units are (1) the Ozark-Central, (2) the Midwest, (3) the Appalachian Mountains, and (4) the Northeast. Pennsylvania is in both the Appalachian Mountains and northeast units. The draft plan (2007) reports 2005 population numbers for Indiana bats to be approximately 457,000, which is up from the 1997 estimate of 353,000 bats (USFWS 1999).

### *DISTRIBUTION AND RELATIVE ABUNDANCE IN PENNSYLVANIA*

The Indiana bat was officially listed as a federally Endangered Species on March 11, 1967 (32 FR 4001), under the Endangered Species Preservation Act of October 15, 1966 [80 Stat. 926; 16 U.S.C. 668aa(c)]. Following along with the federal listing, the species is also categorized as a Pennsylvania State Endangered Species. Pennsylvania has no recorded P1 hibernacula. The only currently active P2 site is an abandoned limestone mine in Blair County where in 1965 Hall (1979b) estimated 1,000 Indiana bats (fig. 6.3). More recently, 774 Indiana bats were counted during the biennial PGC survey on February 20, 2007 (Turner 2007). The other, a Centre County cave that was reported by Mohr (1932a, 1932b) to have 2,000 Indiana bats, no longer supports an Indiana bat population. P3 hibernacula with extant Indiana bat populations include two limestone caves (Mifflin and Centre counties), one abandoned railroad tunnel (Somerset County), and one abandoned limestone mine (Armstrong/Butler County). The remaining P4 hibernacula with Indiana bat populations are two abandoned anthracite coal mines in Luzerne County, one limestone cave in Centre County, one limestone cave in Huntingdon County, one limestone cave in Mifflin County, one limestone cave in Blair County, one abandoned limestone mine in Somerset County, one abandoned limestone mine in Armstrong County, one abandoned limestone mine in Beaver County, two abandoned limestone mines in Lawrence County, and one abandoned clay mine and one abandoned limestone mine in Fayette County. Museum records, literature, and PGC surveys identify thirty-one hibernacula in fifteen counties with extant overwintering populations at eighteen sites in ten counties. The USFWS estimates 1,038 wintering Indiana bats in Pennsylvania based on 2006/2007 hibernacula surveys conducted by the PGC (A. King, United States Fish and Wildlife Service, personal communication).

Genoways and Brenner (1985) estimated the total known past populations of Indiana bats at about 5,000 individuals in Pennsylvania. In the 1970s, John Hall of Albright College conducted surveys for the Indiana bat in Pennsylvania (Hall 1979a, 1979b) and reported only one hibernating colony of 150 individuals in the Blair

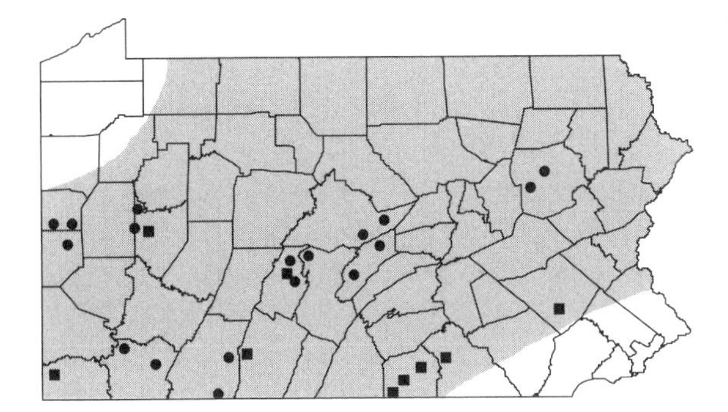

*Fig. 6.3.* Distribution of the Indiana Bat, *Myotis sodalis*. Circles = hibernacula, squares = maternity sites, shading = summer distribution.

County limestone mine and observed one individual in a Mifflin County cave. Since that time the mine hibernation site was afforded the protection of a bat-friendly gate to keep people from disturbing the hibernating population. The population appears to be responding with the most recent count of 774 individuals on February 20, 2007 (Turner 2007). Seventeen Indiana bat hibernacula also have been identified since Hall's report, fifteen of which also have been protected with bat-friendly gates. At one of these fifteen sites, gating by the Nature Conservancy of a limestone cave in Mifflin County, a historic location, resulted in the reestablishment of a small colony shortly after gate placement. Security needs to be provided to two hibernation sites.

Gannon and Blackburn (2002) report male Indiana bat use of a building and trees during the summer in the Allegheny National Forest in Elk and McKean counties. On August 6, 2007, Pennsylvania Game Commission Wildlife Diversity staff captured a male Indiana bat in Butler County near the Armstrong/Butler County hibernaculum (Butchkoski and Turner 2008). In addition, a wildlife consultant reported the capture of two male Indiana bats in the summer of 2007 and another two male Indiana bat captures in 2008 near the same location in Somerset County (Butchkoski and Turner 2008). None of these male captures were associated with a maternity roost.

In Pennsylvania, the first summer maternity site was documented and located within the attic of a decommissioned country church at Canoe Creek State Park, Blair County, in 1997 (Butchkoski and Hassinger 2002a). This was the first-known Indiana bat maternity roost within a building. Research and management has allowed this maternity colony to expand to nearby artificial roosts and an adjacent garage attic. This site has been purchased by the Department of Conservation and Natural Resources (DCNR) Canoe Creek State Park through funds provided by the Pennsylvania Wild Resource Conservation Fund. To find additional summer roosts, six female Indiana bats exiting the Blair County limestone mine in April 2005 were radio-tagged by PGC researchers, one of these was intensively followed with aircraft and ground support. Two of the six bats were found in farmland woodlots of Carroll County, Maryland, in two different roost trees, 135 and 148 km from the Hartman mine. None of the migrating radio-tagged bats remained at Canoe Creek, suggesting that most of the Indiana bats travel elsewhere for the summer with few staying at Canoe Creek (Butchkoski and Turner 2005a). Four of these animals were not recovered. Continued PGC migration telemetry using females as they emerge from hibernacula has resulted in the following summer maternity roost finds: one roost in Berks County in 2006 from a Luzerne County mine; one roost in Armstrong County in 2007 from an Armstrong/Butler County mine; and three roosts in Adams County and one in York County in 2008 from the previously mentioned Blair County limestone mine (Butchkoski and Turner 2006, 2007, 2008).

In 2007, a wildlife consultant conducted a similar female migration project from a Somerset County tunnel and tracked them to Bedford County roosts (Butchkoski and Turner 2008). A different wildlife consultant located a Greene County maternity site while conducting summer netting in 2007 (Butchkoski and Turner 2008). On January 25, 2008, a juvenile female banded at the Greene County, Pennsylvania, site on July 28, 2007, was found by West Virginia Department of Natural Resources (WVDNR) among 115 hibernating Indiana bats in a cave in Randolph County, West Virginia, 141 km distant (C. Stihler, personal communication). Although it appears that the downward trend has been halted in Pennsylvania, much work is still needed to identify and understand the summer habitats and migration patterns used by the species. In summary, post-1994 records for Indiana bats in Pennsylvania include male mist-net capture in four counties, nine summer maternity sites in seven counties, and eighteen hibernacula in eleven counties.

### COMMUNITY TYPE/HABITAT USE

*Summer habitat.* Most Indiana bats use trees as roosts in the summer. In general, it appears that the largest available trees with exfoliating bark and some daily exposure of sun are preferred maternity roosts.

Most primary roosts are well exposed to extensive solar radiation (Menzel et al. 2001). Choosing maternity roosts with high solar exposure increases the roost temperature, which may decrease the time of fetal development and juvenile growth (Callahan et al. 1997). Most roosts are within 1 km of water. The quality of habitat for roosting appears to be best where canopy closures are between 60 percent and 80 percent. Males are less selective and will use trees of almost any size as roosts, if they have loose bark or cavities to roost in or under (Kiser and Elliott 1996). Males are not restricted to maternity colonies and do not need the high temperatures (Callahan et al. 1997). Males therefore seek cooler roosts to conserve energy. Females in maternity colonies use multiple roosts having at least one primary roost. One to three primary roosts were used in Missouri (Callahan et al. 1997). Several secondary roosts occur in the vicinity of these primary roosts (Gardner et al. 1991, Callahan et al. 1997). The primary roosts were standing dead trees with solar exposure. Alternate roosts included live and dead trees located in more shaded locations. The use of alternate trees may be influenced by weather. Roost trees are ephemeral. They are only suitable for use before all bark falls off or the tree falls (Kurta et al. 1996, Callahan et al. 1997).

More recently (1997), Indiana bats have been found to form a maternity roost in a Blair County church attic where the bats seek surface roost temperatures averaging 36.5°C ± 1.3°C (Butchkoski and Hassinger 2002a). The PGC and DCNR manage the site as a summer bat roost, and the colony has expanded to other artificial structures. The attic maternity roost is located 2.3 km from a P2 hibernaculum. Other documented summer roosts include a garage beside the church; the bat condo, a large, artificial bat roost located 300 m from the church; and various bat boxes located between the church and condo, all of which are managed as bat roosts (Butchkoski and Turner 2005b). Within these roosts, Indiana bats are typically found mixed among little brown bats (*Myotis lucifugus*). The center of the Indiana bat maternity colony's foraging area is located ca. 1.3 km southwest of the managed roosts.

In the summer of 2004, five north/south transects, sampling 24.5 ha, were walked to collect data on standing dead or dying trees to evaluate potential Indiana bat roost trees within the foraging area (Butchkoski and Turner 2005a). This primary portion of the foraging area is approximately 580 ha in size. Four hundred and eight standing dead or dying trees more than 2 m in height and >10.2 cm diameter breast height (dbh) were tallied. Indiana bats are specific in roost tree requirements used for maternity colonies (Farmer et al. 2002). Necessary traits for a typical maternity roost tree include a diameter of 22 cm or greater (64 per hectare recommended), height of 3 m or more, exfoliating bark of at least 25 percent (Gardner et al. 1991), and a density of at least twelve such trees per hectare (Farmer et al. 2002). Our samples found fifteen snags, 0.61 per hectare, that were >22 cm dbh, >3 m high, and with 25 percent or more exfoliating bark, far below the recommended 12 per hectare. In addition, most primary roosts are well exposed to solar radiation (Menzel et al. 2001). Only two trees meeting the aforementioned criteria also received an estimated four or more hours of solar exposure, resulting in 0.08 trees per hectare. There are five potential reasons why a bat will change its roost (Kunz 1982): (1) to be closer to foraging grounds; (2) to disrupt parasite life cycles; (3) to minimize predation; (4) to find a more desirable microclimate; and (5) to develop/maintain awareness of suitable alternate roosts. This apparent lack of suitable natural roosts may explain the high success rate of artificial structures at the Blair County maternity site and the lack of roosting behavior within the foraging area.

Bats from this attic roost were similar to those studied elsewhere in that they concentrated foraging in wooded areas (LaVal et al. 1977, Brack 1983, Gardner et al. 1991, Murray and Kurta 2002). Within the foraging range of the building roost, there were large amounts of riparian and lakeside forests, as well as forested mountainsides. Despite availability of these wooded areas, Indiana bats restricted foraging to the largest island (ca. 1,300 ha) of upland slopes with slopes <10° (using a 100-m resolution slope model). Telemetry data were collected through five summers (2000 through 2004) for twelve Indiana bats and eight little brown bats at the Blair County maternity roost area. From telemetry points of primary foraging areas (where the bats spend most foraging time), primary foraging core areas were created using a fixed kernel home range of 50 percent utilization distribution (Hooge et al. 1999). It was found that primary foraging cores for little brown bats are on and adjacent to major bodies of water (rivers and lakes), whereas Indiana bat cores are located on intermittent streams and dry-forested hillsides (Butchkoski and Turner 2005a).

Work by the PGC in 2004 also documented Indiana bat reproduction in two bat boxes and the condominium, all located near (ca. 300 m) the original attic maternity roost. From June 22 through July 23, 2004, three lactating Indiana bats were captured at the Blair County maternity roosts, radio-tagged, and followed to gain more information about their foraging habitat. Two were captured exiting a large bat box and the third was captured by hand within the church attic. All three bats showed exclusive or significant use of two bat boxes and the condo during the 2004 reproductive season (Butchkoski and Turner 2005a).

In 2002, a timber stand analysis of the foraging area was conducted by PGC foresters (Butchkoski and Hassinger 2002c). The core foraging areas typically occurred in deciduous forests on gentle to moderately sloping southerly aspects (south, southeast, and southwest) and included hills and shallow drainages. The two main forest types were mixed oak on the upland sites and cove hardwood/northern hardwood mixtures in some of the drainages. This is similar to most forests in the Ridge and Valley provinces. The predominant timber-size classes in the foraging areas consisted of large poletimber and small sawtimber. Canopy closure was variable throughout the foraging areas and resulted in mixed understory conditions. Foraging areas contained both sparse and dense understory, depending on the number and size of canopy openings. Canopy openings were generally small and the result of both natural tree mortality, as well as semirecent timber cutting; possibly a diameter limit cut (a commercial high-grade) within the past ten years. In nearly every instance, a core foraging area included a drainage feature as well as an upland site. This predominantly mixed oak stand is typical of immature oak stands found throughout the Ridge and Valley Province. Total basal area (BA) is 27.7 $m^2$/ha (121 square feet per acre), with the number of trees at 914 stems per hectare (370 stems per acre). Stand density is high, with current stocking at 114 percent (considered overstocked in a silvicultural context). This is a poletimber stand with 62 percent of the basal area composed of poletimber-sized trees (10.2 cm to 27.9 cm dbh/4 inches to 11 inches dbh). Average diameter of the stand is 19.5 cm (7.7 inches) dbh. More than one-quarter (28%) of the BA consists of sawtimber-sized trees, and most of these are oaks. Overall species composition of all size classes based on the percentage of total basal area is red oak (16%), black oak (16%), white oak (10%), chestnut oak (9%), scarlet oak (8%), all oaks (59%), red maple (13%), black locust (9%), black cherry (7%), white ash (5%), hickory (2%), American elm (2%), Virginia pine (2%), and beech (1%). The overstory is in an overstocked condition, and the forest canopy is correspondingly tight. Plots indicate 10 percent overstory mortality, but the majority of this consists of dead pole-sized stems in the intermediate and suppressed crown classes that have died naturally because of competition for growing space from adjacent trees (i.e., overcrowding; Butchkoski and Hassinger 2002c).

Since the first find of a building maternity roost in Pennsylvania, maternity roosts have been found in a barn in southeast Iowa (Chenger 2003); a house in Dutchess County, New York (A. Hicks, personal communication); and a house in Morris County, New Jersey (M. Craddock, personal communication). All four of these sites also receive high solar exposure, which increases roost temperatures (as with maternity trees).

*Winter habitat.* True hibernators, Indiana bats enter hibernation sites in the autumn and survive on stored fat until spring. Hibernation sites have stringent requirements. Indiana bat hibernacula have noticeable airflow (Henshaw 1965). Tuttle and Kennedy (1999) hypothesized that Indiana bats prefer unique hibernacula with the lowest nonfreezing temperatures possible. In the species core range, midwinter cave temperatures of 2°C to 5°C were reported for Indiana bat roost sites (Hall 1962, Henshaw 1965, Henshaw and Folk 1966, Thomson 1982). By placing temperature data loggers within hibernacula, Tuttle and Kennedy (1999) recorded an overwinter range of –8.3°C to 13.1°C from fifteen important hibernacula in Kentucky (4), Illinois (1), Indiana (5), Missouri (3), Tennessee (1), and Virginia (1). An analysis of temperature and population trend for some of these caves revealed population increases in four of six caves where overwinter temperatures ranged from 3°C to 7.2°C and population declines in all four caves/mines where overwinter temperatures exceeded 8.1°C or were less than 0°C (Tuttle and Kennedy 1999). Warmer temperatures may increase metabolic rates in Indiana bats and cause premature fat depletion during the hibernation period (Richter et al. 1993). Stable midwinter temperatures of 1°C to 10°C may represent a thermal threshold for hibernacula occupancy by *M. sodalis* (Clawson 1984). A recent examination of long-term data suggests that a range of 3°–6°C may be ideal for the species (USFWS 1999). Only a small percentage of available hibernacula provide these unique temperatures. In Pennsylvania, extant winter roosts consist of eleven abandoned mines, six caves, and one abandoned railroad tunnel. Mine types are limestone (8), anthracite coal (2), and clay (1). All caves are limestone, and the railroad tunnel, dating back to the 1880s, was never completed with an opening at only one end. Documentation of Indiana bat use of the two coal mines is through swarm trapping because of the dangerous nature of the sites. During the years 2001–2002, data loggers were placed in the Blair County limestone mine to record every four hours from August 25, 2001, through July 15, 2002. The data loggers placed at three locations at the bottom of the mine, where most of the bats overwinter, recorded average temperatures and standard deviations of 4.8°C ± 0.34°C, 4.7°C ± 0.37°C, and 5.28°C ± 0.18°C (Butchkoski 2003). This is well within the ideal temperature range of 3°C to 6°C (USFWS 1999).

*LIFE HISTORY AND ECOLOGY*

Indiana bats primarily use trees in the summer throughout their range. However, use of man-made structures is becoming more evident (Belwood 2002, Butchkoski and Hassinger 2002a). Indiana bats generally forage in wooded areas (LaVal et al. 1977, Brack 1983, Gardner et al. 1991, Butchkoski and Hassinger 2002a, Murray and Kurta 2002). Summer habitat descriptions include riparian, bottomland or upland forests, old fields, and pastures. This species forages on insects, as do all of Pennsylvania's bats. The bat generates high-frequency sounds that exit the mouth (they fly with their mouths open) and have sensitive ears to hear and use the returning echoes to navigate, to hunt, and to intercept prey in the dark. The Indiana bat begins entering caves and mine tunnels to hibernate in mid-September; most are in hibernation by early November. While they are in hibernation, this species usually roosts in dense clusters of ca. 250 bats per square foot or more. The tendency to cluster densely during hibernation is distinctive for the species and explains their common names: social bat, companion bat, and cluster bat. Also, when Indiana bats are in these dense clusters, the pinkish cast to the fur and pinker lips contrast significantly with the less densely packed, darker clusters of little brown bats, hence their common name, pink bat. Indiana bats are also found mixed among hibernating clusters of little brown bats, which is the primary situation in Pennsylvania. Low statewide numbers may cause Indiana bats to use little browns as surrogate roosting partners. The lack of pure hibernation clusters makes it extremely difficult to identify Indiana bats in Pennsylvania hibernacula. The spring exit begins in mid-March, with most having left for summer habitats by mid-May. Before hibernating in late summer/fall and again during the spring emergence, bat activity around the entrances increases and is known as swarming. This behavior is not thoroughly understood. Mating occurs during the fall swarms, and swarming may also act as a behavior to familiarize juveniles with available habitats. The maximum migration distance from hibernacula to summer habitats is estimated at ca. 520 km (Gardner and Cook 2002).

Reproductive female Indiana bats usually form small maternity colonies under exfoliating bark of trees during the summer months (Whitaker and Hamilton 1998) and more rarely in buildings. Mating occurs in autumn, and females store the sperm through the winter. Fertilization occurs in the spring. Females arrive at the summer maternity site in mid-April to late May and give birth to a single young in mid-June to early July. The young are able to fly by mid-July to early August. Volant juvenile Indiana bats have been found mixed in groups of volant juvenile little brown bats within the Blair County building roost (Butchkoski 2003). It is not known whether females routinely rear pups their first reproductive season (the following year after birth) because of a lack of recapture data. A female Indiana bat at the Blair County maternity site was banded as a juvenile on July 24, 2001. She was recaptured twice, on July 22 and July 29, 2002, and was nonreproductive. She was then recaptured on August 12, 2003, and was again nonreproductive. On July 12, 2004, she was found to be lactating, and on August 29, 2005, she was in a postlactating condition. This bat did not reproduce until her third reproductive season (Pennsylvania Game Commission, Wildlife Diversity Section, unpublished banding data).

*THREATS*

Of the eighteen hibernation sites in Pennsylvania, only six are natural caves. Because of disturbance, alterations to natural caves, and surrounding habitats, cave use is minimal. Human entry to hibernation sites in winter causes bats to arouse and burn up valuable fat reserves—a finite energy supply required for survival through the hibernation period. Alteration of natural caves and surface landscape can alter interior temperatures, making the site less suitable. Loss of surface habitat also reduces foraging potential during prehibernation swarms, causing bats to enter the hibernation period with lower fat reserves. In Pennsylvania, most known Indiana bats have resorted to man-made structures for hibernation (mines and tunnels). Although this ability to adapt is encouraging, the long-term effects may be unsuitable. Underground tunnels were created by blasting and are not as stable as natural cave features. Tunnels are unstable and prone to collapses that can alter the interior habitats and potentially make them unsuitable in the future. Collapses over the years within the abandoned railroad tunnel in Somerset County may be contributing to warmer temperatures recorded during biennial PGC surveys. In the tunnel, Indiana bat roost locations regularly exceed 10°C during midwinter surveys. These temperatures are generally considered too warm and unsuitable for Indiana bat hibernation. Site fidelity may be the reason for continued use. Unfortunately, the site was not documented or studied before the existing collapses that probably modified the airflow within the tunnel. Two of the known hibernacula in Pennsylvania do not have adequate security to prevent disturbances during the hibernation period, requiring more intense efforts toward landowner and interagency cooperation. Throughout the Indiana bat's range, deforestation around hibernacula has decreased available habitat for prehibernation foraging (Clawson 1984).

Forest habitat must be preserved around the hibernacula to provide readily available foraging resources before entering hibernation. Kiser and Elliot (1996) found Indiana bats foraging within 2.5 km of hibernacula in October and encourage the retention of potential roost snags within 2.5 km of hibernacula. On the evening of September 18, 2006, PGC personnel radio-tagged a male Indiana bat trapped at the Armstrong/Butler County mine swarm (Butchkoski and Turner 2007). This bat foraged and day-roosted 14.5 km from the mine. During six days of nonconsecutive monitoring this male visited the hibernaculum once, returning to its foraging and roosting area by dawn after that visit. Development, such as urban sprawl and highways, contributes significantly to habitat degradation around hibernacula, causing longer commuting distances to foraging areas and day roosts.

One of the nine maternity roosts in Pennsylvania is in an attic and other nearby artificial roosts. As with the hibernation sites, Indiana bats appear to be adapting to man-made structures for summer roosts. This may be due to forest management practices that no longer allow for large trees in necessary quantity to provide a regular supply of natural roosts. Most primary roosts are under exfoliating bark (Menzel et al. 2001), such as dead snags, and are temporary, requiring a steady supply of suitable roosts. Most maternity roosts are found in larger trees. Callahan et al. (1997) found primary roost trees to average 58.4 ± 4.5 cm dbh. Most primary maternity roosts are also well exposed to solar radiation (Menzel et al. 2001). Most maternity sites for Pennsylvania's hibernating Indiana bats are unknown. This lost link poses a threat because any impacts to the sites would go unobserved. The primary foraging area for Pennsylvania's attic maternity site is on privately owned property. The future of the property is uncertain, posing a threat to both the foraging habitat and the maternity colony.

Indiana bat travel routes between roosts and foraging areas can take the bats over highways. A dead Indiana bat was found on U.S. Route 22 in Blair County at a known bat travel corridor and was likely killed by traffic (Butchkoski and Hassinger 2002b, Russell et al. 2008). At this location, the maternity roosts are on one side of the highway with foraging areas on the other. Lactating females have been found to cross the highway up to eight times per night to nurse young and roost (Butchkoski and Hassinger 2002a). To minimize conflicts, roosts near highways should be studied to evaluate this potential threat and to preserve or create safe travel corridors over or under highways. Initiated by a planned highway upgrade that would increase the deforestation of a travel corridor at the Indiana bat maternity area in Blair County from 20 m to ca. 55 m, a different colony where bats cross a 55-m mowed field was monitored to sample behavior (Butchkoski 2003, Russell et al. 2008). Little brown bats were used as surrogates to sample traveling behavior from roosts to foraging areas at a Huntingdon County bat box site. At the Blair County site, it was observed that traveling bats stay high in canopy cover when it is available, but the question arose about what behavior would be exhibited over a cleared area. At the Huntingdon County site, more than 1,300 little brown bats exited seven bat boxes mounted 8 m high on a building and across a 55-m cleared field. Two 8-m fiberglass poles were erected approximately 30 m apart in the middle of the mowed field. Mason's string was stretched between the two poles at heights of 1 m, 2 m, 4 m, 6 m, and 8 m. Two people held the poles, stretching the string tight. Another four people spread across the count area near the measuring strings, lying flat on the ground to minimize disturbance to approaching bats. Count sectors were divided among the counters to avoid counting a bat twice. As the bats crossed the mowed field, flying through the stretched strings, they were tallied by height. The flight heights of 453 and 1,346 bats were recorded on two summer evenings. The low count was due to a thunderstorm that stopped the count. On both evenings, at least 76 percent (79.47% and 76.59%) crossed <2 m in height above the ground. More than 95 percent (96.47% and 95.61%) crossed the 55-m field flying <4 m above the ground and would have been in traffic had this been a highway. These flight heights may be skewed upward somewhat. Surveyors indicated that as bats approached the measuring lines in the field they often gained height to "check out" the line above them, often flying over the higher line. After exiting the bat boxes, the animals used the side of the building as cover, flew into a treeline, and then exited the treeline to cross the field. Once the high cover was gone, the animals dropped low, presumably to avoid detection by avian predators. This behavior of seeking cover while traveling is consistent with Indiana bat findings in Michigan where traveling Indiana bats consistently used a treeline for cover while commuting (Murray and Kurta 2002). Actual distance flown from the day roost to a foraging area averaged 55 percent ±11 percent greater than straight-line distance; the extra distance flown varied from 0.2 to 3.4 km (Murray and Kurta 2004). Traveling bats seek out and use travel corridors that provide cover even when the straight-line distance (with no cover) is energetically more efficient. Understanding this behavior will allow managers to plan safe corridors for "commuting" bats.

In Pennsylvania, the rapidly increasing construction of wind turbines to generate electricity poses potential threats to the Indiana bat. In April 2003, eleven female Indiana bats exiting the limestone mine in Blair County were radio-tagged in an unsuccessful attempt to follow them to new summer habitats (Butchkoski and Mehring 2004). However, it was observed that during migration the bats foraged high on ridge tops where the temperatures were probably ten to twenty degrees warmer, resulting in greater insect activity for foraging. The bats also readily crossed ridges in their flight path. In April 2005, six Indiana bats were again radio-tagged at the same mine with two successfully found roosting in Carroll County, Maryland. One of these was intensively followed both on the ground and by air, again demonstrated the ability to readily cross ridges (Butchkoski and Turner 2005a). LaVal et al. (1977) and Brack (1983) reported Indiana bats foraging on hillsides and ridge tops during prehibernation swarms. Kiser and Elliot (1996) theorized that cooler fall temperatures limit insect abundance in lower elevations because of cold air drainage in hilly terrain while higher terrains have more favorable warm exposures. The aforementioned foraging areas describe ideal locations for wind farms. Kerns and Kerlinger (2004) estimated total bat mortalities of 47.53 bats per turbine for the Mountaineer Wind Turbine project in Tucker County, West Virginia, during 2003. Although only 13.9 percent of mortalities were *Myotis* species, the potential exists for Indiana bat conflicts if wind farms bisect the path of migration routes to or from hibernacula and preswarming habitats around hibernacula.

Unnatural predation has been observed at the mine hibernaculum in Blair County during the swarming period. Feral cats were found residing at the gated entrances during the swarming period, creating an unnatural predation situation. One Indiana bat was found among other uneaten carcasses at the Blair County site (Butchkoski 2003). This is likely another effect of urban encroachment on hibernacula. Natural predation by screech owls and black snakes is routinely observed at hibernation sites during bat swarming periods, and the pressure is minimal when compared with feral cats.

### *CONSERVATION AND MANAGEMENT NEEDS*

Primary needs are to identify and protect key habitats that include hibernacula, summer roosts, foraging habitats, and migration routes. The small size, mobility, and nocturnal habits of this species make it a difficult animal to study. Abandoned mine lands scheduled for reclamation should be searched for open portals. Any openings should be evaluated for bat use by trapping the spring/fall swarms or interior surveys. Protocols have been developed for these surveys and are available from the Pennsylvania Game Commission, Bureau of Wildlife Management, Wildlife Diversity Section, 2001 Elmerton Avenue, Harrisburg, Pennsylvania 17110-9797. Biennial surveys of the existing Indiana bat hibernacula should be conducted in accordance with the Indiana bat recovery plan (USFWS 2007). Contributing to this monitoring continually updates species population trends by region. Gating to eliminate human disturbance should be used to manage important hibernacula. Gated sites should be monitored during the swarming periods to identify vandalism and control unnatural predators, such as feral cats. To protect and manage known habitats, land purchases and landowner agreements should target important areas. The development and evaluation of forest management plans that incorporate Indiana bat requirements for foraging and roosting habitat is needed.

*MONITORING AND RESEARCH NEEDS*

Where possible, temperature monitoring of hibernacula with remote data loggers is needed, as was done at the Blair County mine, to document the winter roost environment, recognize changes, and evaluate roost conditions. This is especially important at sites where collapses or alteration of entrances could modify the interior airflow. Research is needed to design a basic plan for underground roosts that will provide suitable hibernation habitat for Indiana bats. Having such a plan would guide reclamation of abandoned mine lands to create new, more stable hibernacula in areas where existing mine structures used by the species are prone to collapse. Given the potential for forest management conflicts, there is a need to research the use of artificial summer roosts (bat houses) as a tool for managing summer colonies where a lack of natural roosts are found. The mobility and foraging behavior of this and other bat species make them potentially susceptible to the bioaccumulation of environmental contaminants; work is needed to identify contaminants and their effects on bat populations. With the increased construction of wind farms, there is a need to understand the bat mortality identified with these structures. Research is needed to understand wind turbine/bat interactions around roosts, foraging areas, and migration routes to develop bat deterrent/avoidance measures. Better marking and recapture techniques are needed to identify and understand survival and recruitment rates and to recognize the species' limiting factors.

Continued research is needed to understand behavior and habitat requirements of the species and to identify its abilities or inabilities to deal with environmental changes. Unknown maternity roost sites need to be located, described, and afforded protection. This can be accomplished by radio-tagging bats exiting hibernacula in April and following the migrating animals to summer habitats. Intensive mist-netting surveys combined with refined acoustic equipment also have the potential to locate summer habitats. Acoustic recording equipment, software, and call analysis research is needed to more reliably separate the echolocation calls of *Myotis* species. The development of reliable acoustic surveys would provide for more efficient detection of Indiana bats and identify areas for capture work. Biotelemetry could then be used to research habitat usage and identify landscape problems specific to the species. Travel corridors between summer roosts and foraging areas must be identified and studied to avoid conflicts with human activities. Once conflicts are identified, successful designs must be developed for use in planning, such as travel corridors over major highways. Migration patterns and travel behavior between winter and summer habitats remain largely unstudied. Biotelemetry again can be used to assess behavior and habitat usage during migration. Hibernacula have been researched for years; however, many questions remain. Landscape use during prehibernation swarms needs to be researched to define critical surface components for individual sites. This information can then be used to establish management buffers.

*Author:* CALVIN BUTCHKOWSKI, PENNSYLVANIA GAME COMMISSION

## Eastern Small-footed *Myotis*

Order: Chiroptera
Family: Vespertilionidae
*Myotis leibii*
(also *Responsibility Concern*)

One of the smallest bats in eastern North America, the eastern small-footed *Myotis*, has dull, dark brownish fur, often with a golden sheen, that contrasts with its blackish face and ears and its blackish brown wing and interfemoral membranes (fig. 6.4). This species is considered a Species of Concern because of numerous threats, such as disturbance of hibernacula sites, destruction of foraging areas, and the recent occurrence (winter 2008–2009) of white-nose syndrome in Pennsylvania. Its global rank is Vulnerable (G3, NatureServe 2009). Genoways (1985) also considered this species to be Threatened in Pennsylvania, as did Kirkland and Krim (1990). However, Kirkland and Krim noted that new hibernacula found by Dunn and Hall (1989) may justify upgrading the rank of *Myotis leibii* from Threatened to At Risk.

*Fig. 6.4.* The Eastern Small-footed Bat, *Myotis leibii*. Photo courtesy of Cal Butchkoski.

### GEOGRAPHIC RANGE

The eastern small-footed *Myotis* is a species of the upland forests of eastern North America. It occurs from southern Maine, southern Quebec, and southeastern Ontario southward along the Appalachian Mountains to northern Georgia and Alabama and westward as far as Arkansas and eastern Oklahoma (Best and Jennings 1997, Whitaker and Hamilton 1998). This bat was once referred to as *Myotis subulatus*, but Glass and Baker (1968) argued that the name *subulatus* probably applied originally to a different species, and they recommended *M. leibii* as the proper name for this species. The matter is further complicated because western populations of small-footed bats were once considered a subspecies of *M. leibii* but are now considered a distinct species, the western small-footed *Myotis* (*Myotis ciliolabrum*; van Zyll de Jong 1984). As a result, care is necessary when using the older literature for this species: some reports on this species refer to it as *M. subulatus*, and some reports referring to *M. leibii* are actually about *M. ciliolabrum*.

### DISTRIBUTION AND RELATIVE ABUNDANCE IN PENNSYLVANIA

The eastern small-footed *Myotis* was first reported from Pennsylvania in the early 1930s, when Charles Mohr found it in caves in the central and southwestern part of the state (Mohr 1932a, 1932b, 1936). Extensive surveys in the 1980s added new state records for this species, documenting it from twenty-one hibernacula in eight counties (Dunn and Hall 1989). Most of the known occurrences for *M. leibii* in Pennsylvania are from the center of the state in the Ridge and Valley Physiographic Province. However, recent unpublished surveys have extended the known state distribution into twenty counties, including counties in the northeastern corner of the state (fig. 6.5).

This species has traditionally been considered one of the rarest bats in eastern North America (Barbour and Davis 1969, Best and Jennings 1997). It appears never to have been common in Pennsylvania (Mohr 1932a, Hall 1985), and there are some indications that it may be declining. For example, Hall (1985) reported that it had disappeared from many of the Pennsylvania caves where it had been found in the 1930s and 1940s. However, Krutzsch (1966) and Martin et al. (1966) suggested that this species is more common than generally believed and that its solitary habits and choice of roosting and hibernating sites just make it difficult to detect. Much still needs to be learned about the status of this species in Pennsylvania.

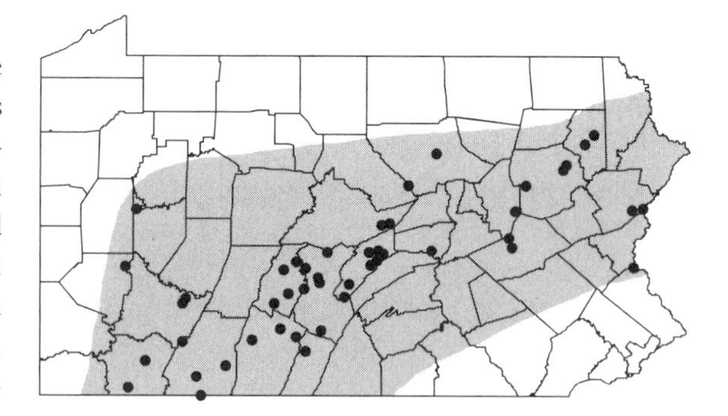

*Fig. 6.5.* Distribution of the Eastern Small-footed Bat, *Myotis leibii*. Dots = hibernacula, shading = summer distribution.

### COMMUNITY TYPE/HABITAT USE

*Myotis leibii* is a species of eastern forests and most commonly occurs in mountainous areas. The majority of observations are of hibernating individuals in mines, tunnels, and caves. It has been found in narrow cracks in rock walls, under rocks, and in crevices on cave floors (Tuttle 1964, Krutzsch 1966, Whitaker and Hamilton 1998). Dunn and Hall (1989) reported that most of the hibernating sites where they found this species in Pennsylvania were small caves <150 m in length. Little information is available on its summer habitat, although individuals have been found in summer in buildings, in caves and mines, under rocks, and in crevices in rock walls (Hitchcock 1955, Doutt et al. 1966, Krutzsch 1966, Best and Jennings 1997, Bogan 1999, Harvey et al. 1999). There are also reports of this species using bridge expansion joints as summer roosts in the southern part of its range (Erdle and Hobson 2001).

### LIFE HISTORY AND ECOLOGY

The eastern small-footed *Myotis* is a solitary species and is usually found roosting alone (Mohr 1936, Dunn and Hall 1989). It appears more tolerant of cold temperatures than most other eastern bats, and it hibernates comparatively closer to cave and mine entrances, where temperatures are cooler and humidity is lower (Martin et al. 1966, Fenton 1972). It enters hibernacula later in the fall than other species, and leaves earlier, in late winter or early spring (Mohr 1933, Mohr 1936).

This species emerges shortly after sunset to forage, flying slowly and erratically, usually within 1–3 m of the ground (Davis et al. 1965, Merritt 1987, Harvey et al. 1999). There are no published dietary analyses for *M. leibii*, but Harvey et al. (1999) stated that it consumes flies, mosquitoes, true bugs, beetles, ants, and other insects.

There is little information on reproduction in this species. Apparently, a single young is typical, with birth occurring in late spring or early summer (Barbour and Davis 1969, Merritt 1987, Harvey et al. 1999). The life span is unknown, but Hitchcock (1965) recaptured an individual that he had banded twelve years earlier. Hitchcock et al. (1984) found that *M. leibii* has a significantly lower survival rate from year to year than *Myotis lucifugus*, with survival for males significantly higher than for females.

### *THREATS*

An important threat to *M. leibii* is the destruction and disturbance of hibernacula. Sealing caves and mines may prevent bats from entering and can disrupt the temperature and airflow characteristics required by this species. Disturbance of hibernating bats by cave visitors during the winter months (November to March) may lead to increased mortality. Some eastern small-footed *Myotis* are known to roost in rock crevices and along rock walls and therefore may be affected by rock climbers. Destruction of foraging areas, particularly of forested areas with abundant rock outcrops that serve as summer roosts, may also be a concern; more information on summer habits is needed to assess this threat. A new and very serious threat is white-nose syndrome (WNS), an emerging disease of hibernating bats, including this species, with very high mortality of affected populations (Cohn 2008, Blehert et al. 2009).

### *CONSERVATION AND MANAGEMENT NEEDS*

Known hibernacula should be protected from complete closure and also from disturbance during winter (November to March). Caves and mines that are important hibernating sites for this species should be gated to restrict access by cave visitors. An outreach program to private landowners and public land managers should be developed to increase awareness of the need to protect hibernating bats and their hibernacula. Although little is known about the summer habitat of this species, it may be necessary to limit summer access to caves where roosting bats might be disturbed. In addition, rock climbing may need to be restricted in areas where this activity could affect roosting bats. Finally, forested areas with abundant rock outcrops used for summer roosting and foraging may need to be protected from logging and development.

### *MONITORING AND RESEARCH NEEDS*

Continued monitoring of eastern small-footed *Myotis* at hibernacula is necessary to better understand the current distribution and population size of the species. This effort should include techniques for detecting bats hibernating individually or in small numbers in cracks or crevices in the hibernation sites. Intensive summer monitoring, including telemetry, should be conducted to determine the summer roosting and maternity habits of this species. Monitoring programs should include netting and trapping at caves and in forested areas near rock outcrops and rock walls.

Detailed inventories need to continue to estimate current population levels and determine geographic distribution in the state. Identification of summer habitats and location of maternity roosts, preferably through telemetry, is particularly important. Research also needs to be conducted to determine the importance of rock crevices as summer roosting sites. Location of additional hibernacula for this species, including determination of preferred hibernating site characteristics, is also necessary. In addition, much still needs to be learned about the basic biology of this species. For example, there are no published data on its food habits, and little is known of its reproductive biology.

*Author:* HOWARD P. WHIDDEN, EAST STROUDSBURG UNIVERSITY

## Delmarva Fox Squirrel

Order: Rodentia
Family Sciuridae
*Sciurus niger cinereus*

The largest of the ten subspecies of fox squirrels and the largest tree squirrel in North America, the Delmarva fox squirrel is recognized by its uniformly steel gray color (fig. 6.6). It was selected as a Pennsylvania Species of Greatest Conservation Need because it is listed as a federally Endangered Species and was recently reintroduced to Pennsylvania but is now considered extirpated from the state. Its global rank is Secure (G5, NatureServe 2009).

### *GEOGRAPHIC RANGE*

Represented by ten subspecies, the fox squirrel's native range extends across the eastern two-thirds of the

*Fig. 6.6.* The Delmarva Fox Squirrel, *Sciurus niger cinereus*. Photo courtesy of Bill Swindaman.

United States from western New York to Florida in the east, and from northern Mexico and Texas to Manitoba, Canada, in the west (Koprowski 1994). The species has also spread along river valleys to the front range of the Rocky Mountains and has been introduced to other portions of the west (Edwards et al. 2003). Range of the Delmarva subspecies of the fox squirrel (*Sciurus niger cinereus*) formerly included the entire Delmarva Peninsula, northward into southern Pennsylvania and New Jersey (Rhoads 1903, Poole 1932, Taylor 1973, U.S. Fish and Wildlife Services 1993). Relic populations currently exist in Kent, Queen Anne's, Talbot, and Dorchester counties in Maryland. Reintroductions of the subspecies were conducted throughout its historic range between 1978 and 1992 in six counties in Maryland, two counties in Delaware, two counties in Virginia, and Chester County, Pennsylvania (U.S. Fish and Wildlife Service 1993).

### *DISTRIBUTION AND RELATIVE ABUNDANCE IN PENNSYLVANIA*

The historic range of the subspecies in Pennsylvania appeared to include the southeastern corner of the state in Chester, Dauphin, and Lancaster counties (Poole 1932, Derge and Steele 1999). Initial extirpation of the subspecies from Pennsylvania is estimated to have occurred by 1900 (Poole 1932, Derge and Steele 1999). Kirkland and Krim (1990) concluded that there was no firm evidence, other than historic natural history notes, that the subspecies ever occurred in Pennsylvania. Poole (1932), however, reported ten specimens from the state, and Derge and Steele (1999) confirmed two of these and found four additional museum specimens, thereby documenting the past presence of the subspecies in Pennsylvania.

Reintroductions of the subspecies to southern Chester County (Chadsford), conducted in 1987 to 1988, were considered unsuccessful because of heavy predation and dispersal from the reintroduction site (Dunn 1989). Tentative reports of sightings, six to eight years after the reintroductions, however, suggest that a small population may have established in Chester County (Derge and Steele 1999, M. A. Steele, personal observation). Among eleven reintroductions conducted in Maryland between 1978 and 1992, the only one that failed was the translocation at the Fairhill site, located just south of the Pennsylvania border, <30 km from the translocation in Chadsford, Pennsylvania (Therres and Willey 2002). Although additional surveys are needed to determine whether Delmarva fox squirrels still reside in this region, available records suggest that the species is currently extirpated from the state.

### *COMMUNITY TYPE/HABITAT USE*

Like most subspecies of *Sciurus niger*, the Delmarva fox squirrel prefers open stands of forests with little understory. Edwards et al. (2003) provide detailed information on habitat selection in various parts of the species' range. On the Delmarva Peninsula, habitat use has been studied most extensively by Taylor (1976) and Dueser et al. (1988). Delmarva fox squirrels are regularly found in smaller stands of older timber, often in association with agricultural fields (U.S. Fish and Wildlife Service 1993). Habitat suitability models (1988), based on the data collected by Taylor (1973, 1976) at both occupied and unoccupied sites, indicate that Delmarva fox squirrels are more likely to occur on sites with larger trees (>30 cm dbh), a lower percentage of shrubby ground cover, and a lower understory density. Although Delmarva fox squirrels are now often associated with coniferous and mixed coniferous/hardwood stands, it is quite possible that the forests available today differ significantly from the old growth hardwood stands in which the subspecies evolved (U.S. Fish and Wildlife Service 1993).

In Pennsylvania, potential habitat is abundant in the southeastern corner of the state where the subspecies once resided. The habitat at the Chadsford translocation site (Dunn 1989) was indeed suitable habitat with significant patches of mature forests and open understory, embedded in an extensive agricultural matrix. Such suitable habitat is likely abundant in this region, but without appropriate connections (dispersal corridors) to other stable populations, further translocations have been discouraged for the present time by the Delmarva Fox Squirrel Recovery Team. As a federally Endangered Species (World Wildlife Fund), the Delmarva fox squirrel is given top priority for protection. However, the subspecies' limited historic range in Pennsylvania, coupled with a significant spatial hiatus between current populations (in Maryland, Virginia, and Delaware) and the southern border of Pennsylvania, raise serious doubts regarding the states' current role in the recovery of this subspecies. Despite the potential for suitable habitat in portions of southern Pennsylvania, the current distribution of the subspecies in the Delmarva Peninsula appears to be too isolated from Pennsylvania to justify future translocations in the state. Efforts at this time should be directed at first verifying descendants of the 1987 to 1988 translocations to Chester County. The potential role of Pennsylvania in recovering this subspecies also depends on documentation of viable populations in northern Delaware and northeastern Maryland.

Historic records of Delmarva fox squirrels in Pennsylvania are currently limited to six specimens: two of the ten original specimens reported by Poole (1932) and four additional specimens located more recently by Derge and Steele (1999). Records on distribution and abundance of the subspecies are virtually nonexistent (Zegers 1985, Derge and Steele 1999).

### *LIFE HISTORY AND ECOLOGY*

The fox squirrel feeds heavily on seeds of pine, oak, beech, walnut, and hickory and consumes animal material when available (e.g., insects, bird eggs; Steele and Koprowski 2001). Like gray squirrels, fox squirrels scatter-hoard much of their food in individual, widely dispersed cache sites, and as a result of this activity, they may significantly influence seed dispersal and forest regeneration (Stapanian and Smith 1978, Stapanian and Smith 1984, Steele and Koprowski 2001). Fox squirrels also feed heavily on buds and flowers during the spring when energy demands are high and food availability is low. Fox squirrels are more cursorial than gray squirrels, usually traveling across open ground rather than through canopy.

The species is active year-round, but activity levels vary with season. Fox squirrels, like gray squirrels, exhibit overlapping home ranges and a dominance hierarchy. Home ranges vary with season and food supply but are usually <8 ha. Males are dominant over females, adults are dominant over subadults, and lactating females are exceptionally dominant over other individuals, often exhibiting territoriality around the nest tree (Brown and Batzli 1985, Steele and Koprowski 2001).

Fox squirrels breed primarily between December and January, and if food is available, they breed again in late spring and summer. Litter sizes range from one to seven, gestation is forty-four to forty-five days, and weaning occurs ten weeks after birth (Steele and Koprowski 2001). Juvenile mortality is particularly high, but drops significantly for adult squirrels (Hansen et al. 1986). Fox squirrels can live up to 12.5 years in the wild (Koprowski et al. 1988).

### *THREATS*

The Delmarva fox squirrel was first listed as an Endangered Species in 1967, at which time it was estimated that endemic populations accounted for <10 percent of the subspecies' historic range (Taylor 1976). The primary factor attributed to this decline was the loss of suitable forested habitat, due first to agricultural expansion, and more recently to human development (Taylor 1973, 1976). Hunting pressure was also considered a potential factor leading to the initial decline of this subspecies (U.S. Fish and Wildlife Service 1993). Secondary factors posing important threats to Delmarva fox squirrel populations are predation risks (Edwards et al. 2003), traffic mortality (U.S. Fish and Wildlife Service 1993), and competition from gray squirrels, which may limit the size and distribution of fox squirrel populations (Brown and Batztli 1985, Drake and Brenner 1995, Edwards et al. 1998). Predation may be a significant failure of some translocations (Dunn 1989).

### *CONSERVATION AND MANAGEMENT NEEDS*

Because this species is likely extirpated from the state, we identify no plan at this time for conservation and management or monitoring. The Delmarva Fox Squirrel Recovery Plan (U.S. Fish and Wildlife Service 1993) is the best available resource for identifying conservation objectives.

### *MONITORING AND RESEARCH NEEDS*

The most important priority is to determine, through surveys and reconnaissance livetrapping, whether descendants of the 1988 reintroduction still reside in Chester County. If such descendents occur there, immediate work should begin on determining the health and stability of the population.

Genetic comparisons between Delmarva fox squirrels (*S. n. cinereus*) and the eastern (*Sciurus niger vulpinus*) and western fox squirrels (*Sciurus niger rufiventer*), which still occur in Pennsylvania, are sorely needed. Such comparisons will help determine the relative distinctiveness of these three subspecies from each other and other subspecies.

Several public, private, and nonprofit stakeholders are involved in managing and recovering this species: U.S. Fish and Wildlife Service Office in University Park, Pennsylvania; Delmarva Fox Squirrel Recovery Team directed by the U.S. Fish and Wildlife Service; Chesapeake Field Office; Pennsylvania Game Commission and Department of Natural Resources; Virginia Museum of Natural History; and the Pennsylvania Mammal Technical Committee of the State Biological Survey.

*Author:* MICHAEL A. STEELE, WILKES UNIVERSITY

## Allegheny Woodrat

Order: Rodentia
Family: Cricetidae
*Neotoma magister*
(also *Responsibility Concern*)

The Allegheny woodrat (*Neotoma magister*) is a native North American rodent similar in size and build to the introduced Norway and black rats (genus *Rattus*) but distinct from them in a number of physical details and behavior and ecology (fig. 6.7). Until recently, the Allegheny woodrat was considered a subspecies of the eastern or Florida woodrat, *Neotoma floridana*, but morphological and mitochondrial DNA studies (Hayes and Harrison 1992, Hayes and Richmond 1993) show it to be differentiated enough to be considered a distinct species. The species has a global rank of Vulnerable; it is a Northeast Region Priority Species and is Threatened in Pennsylvania. Populations are declining because of factors that may include loss of hard-mast food supply, increased predation by generalist avian predators, fatal infection by raccoon roundworms, human-generated disturbances, and increasing forest fragmentation. The global status of the Allegheny woodrat is Vulnerable/Apparently Secure (G3/G4, NatureServe 2009).

Fig. 6.7. The Allegheny Woodrat, *Neotoma magister*. Photo courtesy of Will Evans.

### *GEOGRAPHIC RANGE*

The Allegheny woodrat inhabits rocky habitat of the Appalachian Mountains and along cliffs and river gorges extending eastward and westward from the Appalachians. Museum specimens collected around the turn of the twentieth century show that woodrats were found as far northeast as western Massachusetts, westward through Indiana, south to Alabama, northward through western North Carolina and Virginia, and down the Potomac River almost to the District of Columbia (Hall and Kelson 1959). The species' present range is more restricted, however; no current populations are known in New York or in New England, only isolated small populations are known in Ohio, Indiana, and New Jersey, and numbers are diminishing elsewhere as well (NatureServe 2004). An essential habitat feature appears to be rockiness, in the form of caves, cliffs, talus slopes, or ridge-top outcrops (Merritt 1987).

### *DISTRIBUTION AND RELATIVE ABUNDANCE IN PENNSYLVANIA*

In Pennsylvania, woodrat populations are concentrated along the rocky ridge tops of the Ridge and Valley Physiographic Province across the central and eastern counties from Maryland to New Jersey, on the Appalachian Plateau southeast of Pittsburgh, and along the gorges of tributaries to the West Branch of the Susquehanna in central Pennsylvania (fig. 6.8). Historical records, and surveys for old woodrat signs in

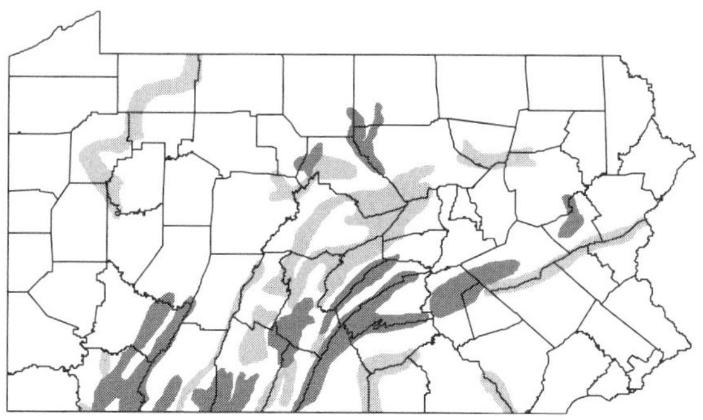

Fig. 6.8. Distribution of the Allegheny Woodrat, *Neotoma magister*. Dark shading = current geographic range, light shading = historic geographic range.

appropriate habitat, show that woodrats were formerly found in numerous sites along the eastern ridges of the Appalachians, in central and northwest Pennsylvania and southeast along the Susquehanna River but have disappeared from these areas in the past half-century (Hall 1985).

### *COMMUNITY TYPE/HABITAT USE*

The most essential feature of Allegheny woodrat habitat is rock—cliffs, ledges, outcrops, boulder fields, and caves—where woodrats build their nests and food caches. Heavily dissected rock with many crevices is preferred; junked cars and hillside rubbish heaps near rock piles may also be used. In Pennsylvania, the most appropriate rock types are sandstone and limestone, both of which are abundant and widespread as surface rock, still relatively undisturbed (Myers et al. 2000). These rock types are typically distributed in small patches interspersed with forest, however. As a consequence, woodrats are typically found in population groups of fewer than twenty individuals each centered on one rock patch. Such populations are too small to be self-sustaining over the long term in the face of random fluctuations in population size; instead, woodrats in Pennsylvania appear to conform to a classic metapopulation model in which individual population units wink out from time to time and are reestablished through interpatch dispersal, while the overall large population persists (McCullough 1996). Even with abundant habitat, such a population structure is vulnerable if connections among the population units are lost and empty habitat patches are not recolonized. Hassinger et al. (1996) found that a large surrounding buffer of intact forest may be important in population persistence.

Vegetation associated with woodrat habitat may be coniferous, deciduous, or mixed forest; the eclectic contents of food caches suggest that, although individual animals have foraging preferences, there are no obligate food plant species. One consideration for food resources is the presence of mast crops, as in some areas woodrats accumulate large nut caches. Castleberry et al. (2001) found that woodrats increased their home range size in years of poor mast production, so this factor deserves further attention.

### *LIFE HISTORY AND ECOLOGY*

Allegheny woodrats are asocial and tolerate one another's presence only briefly during the mating season (Kinsey 1977). An individual woodrat builds a sleeping nest of plant material in a rock crevice or on a cave ledge and may surround the nest with dry leaves and twigs, possibly as an alarm system. It emerges at dusk to forage for leaves, fruits, nuts, seeds, fungi, and twigs; through summer and fall, forage is collected into storage caches for winter use, as woodrats do not hibernate. From July onward, piles of wadded, drying leaves and other materials are stuffed into rock crevices and protected ledges. Nonfood items, including wasp nests, dry sticks, bones, snakeskins, and human debris, such as candy wrappers and shotgun shells, are also added to the cache; the significance of this pack-rat behavior is unknown. Another distinctive behavior is their tendency to defecate in latrine areas, usually flat rock surfaces protected by an overhang; latrines are apparently used by a number of woodrats over several years, as the pile may be several centimeters deep. Fresh droppings are shiny and black and indicate an active woodrat site (Poole 1940).

Predators of the Allegheny woodrat as revealed in scats and pellets include great horned owl, raccoon, coyote, weasel, and black rat snake (J. Wright, personal observation). Although the timber rattlesnake is common in many of the woodrat-inhabited rock areas, its status as predator is not clear; at least one close relative of the Allegheny woodrat, *Neotoma micropus* of Texas, has considerable immunity to rattlesnake venom (Poole 1940, Perez et al. 1978).

In Pennsylvania, female Allegheny woodrats may be found lactating as early as April and well into the autumn, so the breeding season does not appear to be strongly pulsed, and some females possibly produce two litters in a year (Merritt 1987). After a gestation period of about thirty-five days, one to four naked blind pups weighing about 14 g are born (Merritt 1987). Of eighteen litters reported in the literature, average litter size was 2.24 (Poole 1940, Kinsey 1977, Mengak 2002). Their eyes open at about two-and-one-half weeks (Poole 1936). Pups nurse for about a month and then may accompany their mother on short foraging trips. Female woodrats typically reproduce in their second year (Merritt 1987). The relatively small litter size and late age of first reproduction imply rather low reproductive potential for the Allegheny woodrat compared with other rodents.

### *THREATS*

No single factor has been identified to explain population declines, but several factors have been proposed, including (1) loss of hard mast food supply with the disappearance of American chestnut (*Castanea dentata*) and decline of oaks (*Quercus* species) from gypsy moth defoliation and other causes (Hall 1988, Wright and Kirkland 2000); (2) an increase in predation by generalist predators, such as great horned owls (Balcom and Yahner 1996); (3) fatal infection by raccoon roundworm (*Baylisascaris procyonis*) from raccoon scats gathered into food caches (McGowan 1993, LoGiudice 2003); and (4) human-generated disturbances and barriers that limit successful recolonization of empty patches in the habitat mosaic, leading to metapopulation collapse. Forest fragmentation within and near woodrat habitat may exacerbate most of these threats.

The Allegheny woodrat is listed as a Threatened Species in Pennsylvania. Its global G3G4 rating (NatureServe 2004) reflects that it is considered relatively secure in West Virginia and Kentucky, but population declines at the northern and eastern range margins have resulted in considerable range contraction in Pennsylvania.

*CONSERVATION AND MANAGEMENT NEEDS*

At least one woodrat management area, with a core of occupied rock patches surrounded by 2 km of intact forest buffer, should be designated in each of the three major regions of woodrat habitat in Pennsylvania (Appalachian Plateau, Ridge and Valley, and Susquehanna River West Branch), and a management plan developed for each that considers mast supply, contamination by raccoon roundworm, human encroachment, forest fragmentation, and monitoring of woodrat occurrence on each rock patch. Reintroduction of woodrats to favorable and protected sites, such as Hawk Mountain Sanctuary in eastern Pennsylvania, may be necessary for continued conservation.

*MONITORING AND RESEARCH NEEDS*

The Pennsylvania Game Commission has amassed a large database of sites in Pennsylvania surveyed for the presence of active or former woodrat occupation (Butchkoski 2003). This effort should continue, with a regular cycle to revisit sites for updated status, particularly at the range margin. Public interest in the Allegheny woodrat should be harnessed with information for landowners and land managers to identify woodrat presence and enhance woodrat habitat.

The most pressing research needs are (1) a statewide evaluation of the raccoon roundworm threat by assessing the prevalence of this parasite in raccoons and the exposure level to infection in woodrat habitat, (2) experimental analysis of the effect of mast forage levels on woodrat populations, (3) radiotelemetry studies of dispersal behavior of woodrats among habitat patches and use of rock patches for denning and foraging, (4) studies to determine effects of human encroachment (highways, urban areas, agriculture) and forest fragmentation on habitats, and (5) time-series data on extirpation and recolonization rates of individual rock patches. In addition, GIS-based analysis of existing data may reveal essential features that favor woodrat population persistence. The overall goal should be to understand basic woodrat biology well enough to project the probable fate of metapopulations under different management plans.

*Author:* JANET WRIGHT, DICKINSON COLLEGE

### HIGH-LEVEL CONCERN

## Least Shrew

Order: Soricomorpha
Family: Soricidae
*Cryptotis parva*

The least shrew (*Cryptotis parva*) is a diminutive insectivore resembling the northern short-tailed shrew, *Blarina brevicauda*, only smaller (fig. 6.9). Total length ranges from 64 mm to 86 mm with a tail length of 12–18 mm (Kurta 1995), and weights generally range between 4 g and 6.5 g (Merritt 1987, Kurta 1995, Whitaker and Hamilton 1998). The least shrew was selected as a Pennsylvania Species of Special Concern primarily because of factors associated with habitat modification and fragmentation and loss of grassland habitats due to intensive farming. It was once common across the state but is now limited to a few counties; as such, it

Fig. 6.9. The Least Shrew, *Cryptotis parva*. Photo courtesy of Bob Gress.

is now listed as Endangered in Pennsylvania. This is a Northeast Priority Species. Its global rank is Secure (G5, NatureServe 2009).

### *GEOGRAPHIC RANGE*

The range for *C. parva* in the United States extends from central New York to northern Florida and from the Atlantic Seaboard to northeastern Colorado and southward into Texas in the western United States (Hall 1981). Four subspecies occur in the United States with several others occurring in Central America.

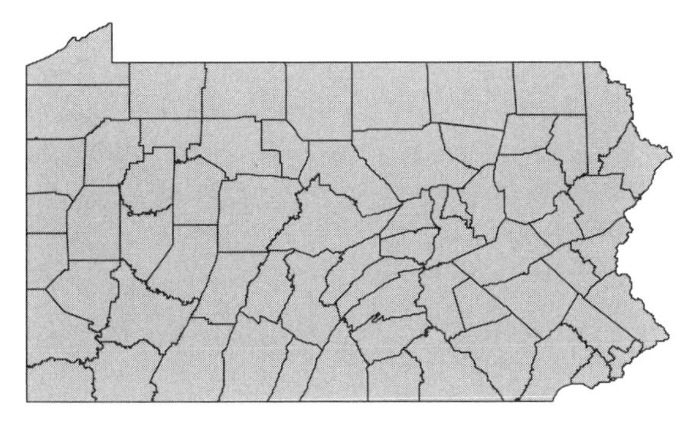

*Fig. 6.10.* Historic distribution of the Least Shrew, *Cryptotis parva*.

### *DISTRIBUTION AND RELATIVE ABUNDANCE IN PENNSYLVANIA*

Even though the least shrew occurred throughout much of Pennsylvania historically (Richmond and Rosland 1949, Grimm and Roberts 1950, Gifford and Whitebread 1951, Roslund 1951, Grimm and Whitebread 1952, Roberts and Early 1952), significant sampling by Pennsylvania Game Commission personnel and other researchers over the previous twenty years has not turned up more than three recently known localities in south-central Pennsylvania along with one isolated occurrence in Westmoreland County. This includes more than 42,000 trap nights of sampling effort at more than 150 sites within forty-five counties (Hart, Annual Reports, Pennsylvania Game Commission, Job Code 70007, Project Code 06710). In addition to sampling-type surveys, recent analysis of more than 3,000 barn owl (*Tyto alba*) pellets has only revealed one additional occurrence. During the Pennsylvania Mammal Surveys of the 1940s and early 1950s, the remains of as many as sixteen least shrews were recovered from pellets collected in Franklin County alone (Latham 1950). At present, the least shrew remains one of the least-encountered mammal species in Pennsylvania, with only eight specimens having been taken over the past forty years. It currently appears to be restricted to the Piedmont Province of the state (Kirkland and Hart 1999; fig. 6.10).

A trend toward range extension in the western portion of the global distribution of *C. parva* is attributable to habitat changes created by irrigation and climate change (Hafner and Shuster 1996); however, the least shrew appears to be declining in more interior areas along its eastern range because of loss of habitat (Kirkland and Hart 1999).

Although the species is considered globally Secure (G5), it is of Conservation Concern in several states, including Connecticut (S1), Iowa (S2), Maryland (S3), Minnesota (S3), New Mexico (S1), South Dakota (S3), and West Virginia (S2) (NatureServe.org data 2004). In several other states, it is listed as either Unknown (Delaware-SNR) or Extirpated (Wisconsin-SX). It remains unknown as to what research effort, if any, has been conducted in surrounding states although the least shrew is currently listed as a northeastern United States Species of Regional Concern (Therres 1999). This species remains a priority species in terms of research and monitoring efforts and is currently listed as Endangered in Pennsylvania (Genoways and Brenner 1985).

### *COMMUNITY TYPE/HABITAT USE*

The primary habitat association for least shrews in the northern United States consists of old fields, fallow fields, and ungrazed native grasslands. Although Ford et al. (2005) captured this species in several forested habitat types along the central and southern Appalachian Mountains, their survey sites were adjacent to old fields or successional habitats created from regenerating timber harvests. In Pennsylvania, populations of least shrews are typically associated with a water source of some form. This may range from intermittent swales to well-established small streams (J. Hart, unpublished data). Heavily managed agricultural fields and pastures do not appear to support populations of least shrews.

Habitat seemed to be widespread throughout Pennsylvania before the 1960s as evidenced by the occurrences of the least shrew during the Pennsylvania Game Commission Mammal Survey (Richmond and Rosland 1949, Grimm and Roberts 1950, Gifford and Whitebread 1951, Roberts and Early 1952). Since then, grassland habitat has been declining rapidly (Goodrich et al. 2000) because of natural succession and development. Currently, large blocks of suitable habitat are restricted to the southern-tier counties of Pennsylvania with most of this limited to the south-central and southeastern portion of the state along the Piedmont and Ridge and Valley physiographic regions. Although other grasslands are scattered throughout the state, they mainly occur as relatively small patches of fewer than 100 acres (Goodrich et al. 2000).

*LIFE HISTORY AND ECOLOGY*

The least shrew inhabits grasslands and old fields throughout most of its range, although along the coastal areas and in the south, it is common in marshes and adjacent nonforested habitats (Whitaker and Hamilton 1998). Although historic accounts for Pennsylvania have this shrew inhabiting stony, shaley habitats with sparse cover (Gifford and Whitebread 1951), present surveys have recorded the species in grasslands with moderately heavy vegetation, as well as old fields. Least shrews are active throughout the year and are diurnal, although most daily activity occurs during the evening hours. Taking advantage of microtine runways (Doutt et al. 1966) for most of its activities, it also constructs pencil-thin passageways within the vegetation for both foraging and dispersal (Merritt 1987, Kurta 1995, Whitaker and Hamilton 1998).

The diet of least shrews consists of adult insects, caterpillars, beetle larvae, earthworms, sowbugs, and crickets along with many other invertebrate items (Whitaker and Hamilton 1998). Observations of a captive individual taken from salt marshes in Accomack County, Virginia, showed that food hoarding occurs in this species (J. Hart and G. L. Kirkland, Jr. unpublished data). Although most species of shrews exhibit aggression toward one another, the least shrew has been known to exhibit colonial behavior. In one instance, as many as thirty-one adult individuals were found nesting together (McCarley 1959). The nests of least shrews consist of shredded grass, leaves, or other natural materials and are most often found under rocks and logs or sometimes under man-made debris (Doutt et al. 1966). Approximately 14 cm in diameter, nests generally have two openings leading to adjoining runways, one above the nest and one below (Whitaker 1974).

Little is known of the reproductive biology of least shrews except for anecdotal information derived from a few captive individuals. Hamilton (1944) reported that this species most likely breeds from March to November in the northern part of its range. In general, three litters averaging five altricial young are born after a gestation period of around twenty-one to twenty-three days (Merritt 1987, Whitaker and Hamilton 1998). The young are weaned after approximately three weeks; adult weight is reached by around one month (Conway 1958). Some evidence indicates that both female and male adults care for the young as pairs of adults have been observed interacting with the young in wild nests in several cases (Cooper 1960).

*THREATS*

The key threat to the continued existence of the least shrew in Pennsylvania is loss of grassland habitats and isolation of existing populations. Increased sprawl and use of intensive farming practices that do not provide the fallow fields that were once prevalent throughout Pennsylvania are the most evident direct threats to known habitats. Existing populations of least shrews are becoming increasingly isolated from one another and will, most likely, suffer localized extinction. The population located near East Berlin in York County has become more isolated because of the increase of residential development surrounding it. This is not an isolated incident, as more and more open space is lost because of sprawl and a rise in rural housing. Since 1996, only two sizable agricultural fields exist surrounding the East Berlin sites, as two other sizable fields were lost to large housing projects.

Primary factors influencing the decline of the least shrew in Pennsylvania include (1) loss of primary grassland and associated marginal habitats that provide habitat for least shrew populations and serve as dispersal corridors and (2) creation of isolated populations that suffer localized extinction events due to alteration of surrounding habitats such that these habitats become ineffectual as either core habitat or dispersal corridors. Current protection of the species does not necessarily extend to habitats.

*CONSERVATION AND MANAGEMENT NEEDS*

Currently listed as a State Endangered Species by the Pennsylvania Biological Survey, the least shrew is protected by the Pennsylvania Game and Wildlife Code. Of a more important note is the lack of protection for its basic habitats. Programs that support the conservation of both agricultural and native grasslands, such as the Conservation Reserve Enhancement Program (CREP) and the Pennsylvania Game Commission's Landowner Incentive Program (LIP), will also promote the conservation of the least shrew by providing habitats necessary for its survival.

Another program that may serve to enhance protection of the least shrew is the Important Mammal Areas Project, initiated by the Mammal Technical Committee of the Pennsylvania Biological Survey to describe important habitats across the state for the conservation of Pennsylvania mammal species. This program will delineate areas where the least shrew occurs and allow for planning corridors and connections between sites to expand habitats that support least shrews.

At present, no conservation management plan is in effect or planned for the least shrew as few researchers have been involved with the least shrew in the northeast region. The perceived ecological unimportance of the least shrew has limited conservation efforts directed at this species. The creation of the Important Mammal Areas Project may serve to highlight areas of importance for conservation of the least shrew. Further work is needed to delimit the extent of the currently known populations in both Adams and York counties. When these core areas become known they can be incorporated into the large Least Shrew Conservation Area that is currently described by the IMAP as within the Piedmont Section of Gettysburg and extending into York County.

*MONITORING AND RESEARCH NEEDS*

To maintain a viable monitoring program, it is imperative to determine where sustainable habitat occurs and can be protected or conserved. The Pennsylvania Gap Analysis Project (GAP; Myers et al. 2000) identified areas throughout Pennsylvania that appeared to have suitable habitat for least shrews, albeit at a relatively large scale. These areas need to be assessed to verify whether the habitat is suitable. After determining the appropriate habitat, other programs, such as CREP, need to be incorporated into any conservation planning for the least shrew. Combining these programs with a viable monitoring plan can lead to designing conservation areas that may ensure the least shrew's survival in Pennsylvania.

As more information is gathered about the least shrew and its required habitats, adaptive management plans that incorporate grassland nesting bird species as well can be created and implemented on both public lands and on private lands with the cooperation of landowners. Pilot studies that look at the long-term effects of these plans should be initiated at such sites as Middle Creek and within the confines of the Gettysburg Least Shrew Important Mammal Area.

The natural history of the least shrew is not well understood. Research needs to focus on determining home range, basic habitat needs, relationship to other soricid species, and what types of habitats may serve to sustain healthy least shrew populations. Shrews, by virtue of their inherently high metabolic needs, are poor dispersers and seem to be very restricted locally.

Several basic questions that need to be answered include the following: (1) At what size does a habitat become functionally important to least shrews? (2) Do habitats covered by other conservation programs, such as stream bank fencing, serve as dispersal habitats for least shrews to move between areas of historically superior habitat?

*Author:* JAMES HART, PENNSYLVANIA NATURAL HERITAGE PROGRAM

## Silver-haired Bat

Order: Chiroptera
Family: Vespertilionidae
*Lasionycteris noctivagans*

A medium-sized migrant bat, the silver-haired bat can be distinguished from all other bats in Pennsylvania by its color. Its dorsal pelage is dark, blackish to brown, with silvery/white tips imparting a frosted, or silver, appearance (fig. 6.11). It is a habitat specialist

*Fig. 6.11.* The Silver-haired Bat, *Lasionycteris noctivagans*. Photo courtesy of Cal Butchkoski.

vulnerable to habitat alteration and considered Rare and Protected by the Pennsylvania Game Commission (Kirkland and Krim 1990, Steele and Kwiecinski 1994, Pennsylvania Game Commission 2006). The silver-haired bat was selected as a Pennsylvania Species of Greatest Conservation Need because of the elimination of mature and old growth and mixed deciduous forests as a result of human activity, coupled with the loss of permanent ponds and streams located near forested habitats. It has a global rank of Secure (G5, NatureServe 2009).

### *GEOGRAPHIC RANGE*

*Lasionycteris noctivagans* occurs only in North America, from the Atlantic to the Pacific, from a line drawn across southern Canada roughly even with the south tip of James Bay and into southeastern Alaska east of the Wrangall Mountains, south to central California, northern Mexico, and east to Georgia (Hall 1981, Kunz 1982, Cryan 2003).

### *DISTRIBUTION AND RELATIVE ABUNDANCE IN PENNSYLVANIA*

The silver-haired bat has been found in coniferous and mixed forests in Pennsylvania (Merritt 1987; fig. 6.12). Records of occurrence within Pennsylvania and adjacent states indicate that this species occurs throughout the commonwealth, but records in Pennsylvania are few. Locally, in Pennsylvania, this bat occurs most often in forested areas along streams and near forest ponds.

### *COMMUNITY TYPE/HABITAT USE*

Across its range, the silver-haired bat is a species of coniferous and mixed forests and not found in open country. In Pennsylvania, this bat is most commonly captured during spring or fall migration season. Bird banders regularly capture this species in mist-nets set along bird migratory routes, especially in September. This bat is usually solitary, but at these times, more than one individual is captured, suggesting these bats travel in groups (Genoways and Brenner 1985). No summer roosts have been identified in Pennsylvania. During migration, this bat has been found in buildings, rock crevices, woodpiles, tunnels, caves, and mines (Genoways and Brenner 1985, Merritt 1987, Pennsylvania Game Commission 2003).

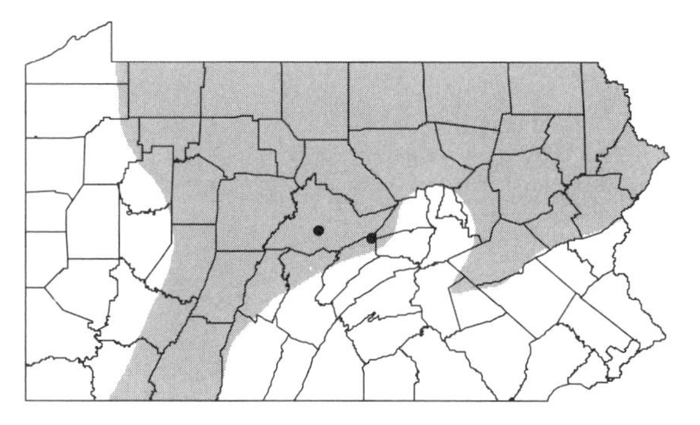

*Fig. 6.12.* Distribution of the Silver-haired Bat, *Lasionycteris noctivagans*. Circles = hibernacula, shading = summer distribution.

### *LIFE HISTORY AND ECOLOGY*

In Pennsylvania, the silver-haired bat can be found primarily in coniferous and mixed forests throughout the commonwealth (Genoways and Brenner 1985, Merritt 1987). Although usually solitary, groups of three or four individuals have been found, usually during migration. Throughout its range, the silver-haired bat usually roosts singly in dense foliage of trees, under bark, in hollow trees, in woodpecker holes, and in birds' nests. Summer maternity and solitary male and female roosts have been located in tree cavities and under bark in South Dakota and Canada (Mattson et al. 1966, Parsons et al. 1986, Barclay et al. 1988, Vonhof 1996). This bat winters in the Pacific Northwest, in disjointed areas of the Southwest, and in middle latitudes of the eastern United States south of Michigan and east of the Mississippi River (Cryan 2003). Although the migratory patterns of *L. noctivagans* have not been studied by banding, analyses of specimens in mammal collections throughout North America indicate that, with the progression of spring, populations of *L. noctivagans* from eastern parts of its range disperse east and north from wintering areas, whereas western populations disperse northward (Cryan 2003). For eastern populations, males remain in wintering areas while females migrate north and east (Cryan 2003). For western populations, females apparently migrate before males (Izor 1979). Females begin moving south into areas occupied by males during late summer and early autumn, when distribution of sexes overlap and ranges shift south (Cryan 2003). Silver-haired bats in Pennsylvania probably migrate south for winter and migrate north in spring. Hibernating bats have been found in Valley View Tunnel and Poe Paddy Rail Road Tunnel, Center County, in March (Pennsylvania Game Commission 2003).

*Lasionycteris noctivagans* forages after dusk, usually taking a direct course to water to feed over ponds, streams, and other bodies of water in or near coniferous or mixed deciduous forests. The silver-haired bat has an intermediate wing loading and aspect ratio and is relatively slow but maneuverable compared with other North American tree bats (Barclay 1985). These bats often feed in clearings, at least several meters aboveground, and pursue prey for relatively short distances (Barclay 1985). When foraging, silver-haired bats fly along erratic courses taking many twists and frequent short glides (Whitaker et al. 1977, Barclay 1985). *Lasionycteris noctivagans* uses broadband, multiharmonic search-approach calls with an initial frequency sweep and constant frequency tail (Barclay 1986). Such calls are suited to foraging in the open, allowing detection of nearby obstacles, and for pursuing prey detected at relatively close range (Barclay 1986). Reported dietary elements include a wide range of small insects, including moths, beetles, caddis flies, houseflies, and flying ants (Kunz 1982, Barclay 1985, 1986). Natural predators include owls, hawks, and striped skunks. Ectoparasites include mites, bat flies, and fleas, and endoparasites include nematodes, trematodes, and cestodes (Kunz 1982). A unique strain of rabies has been identified with this species, but the incidence is very low. Because silver-haired bats rarely come in contact with humans, individuals infected with silver-haired bat rabies variant provided a conundrum to public health workers. Laboratory experiments showed the rabies variant associated with silver-haired bats replicates more efficiently in epithelial cells and at lower temperatures than a coyote rabies variant (Morimoto et al. 1966). Superficial contact and failure to recognize risk of exposure with a bat or other infected animal may carry a greater risk of infection with the silver-haired bat rabies variant than similar contact with a different strain (Messenger et al. 2003). Rabies records from the Pennsylvania Department of Public Health (1992–2000) indicated that only 3.9 percent of bats suspected of having rabies actually tested positive, that most positives (41 of 48) were from *Eptesicus fuscus*, and all suspect bats of the other ten species combined (this group included *L. noctivagans*) had a positive rate of 0.57 percent (Olnhausen and Gannon 2004).

It is presumed that mating and reproductive cycling in *L. noctivagans* occurs as it does in other temperate-zone vespertilionid bats, with mating from late summer throughout winter, followed by sperm storage in females during the winter (Kunz 1982). Spermatogenesis in males peaks during the summer, whereas the peak in female ovulations occurs in late April and early May. Pregnant females roost singly or form maternity colonies in trees where they give birth, nurse, and wean young (Parsons et al. 1986). Gestation is fifty to sixty days with birth occurring in late June or early July (Kunz 1982). The most complete description of reproduction in this species indicates females give birth to one to two young (usually two) with an equal sex ratio (Kunz 1971). The body mass of two young at birth was 1.8 and 1.9 g. Young remain in their roost while mothers forage. The lactation period was estimated at thirty-six days, and young mature sexually their first summer.

### *THREATS*

A major threat to this species is the lack of available information on specific habitat needs and habitat uses in Pennsylvania. The primary environmental factor influencing this species is the loss of mature and old growth coniferous and mixed deciduous forests as a result of human activity. Also, a great threat to this species is the loss of permanent ponds/streams in and associated with forested habitats.

### *CONSERVATION AND MANAGEMENT NEEDS*

Protection of coniferous or mixed deciduous ecotones with ponds, lakes, and streams are essential to providing habitat for this species. We recommend that logging of all kinds cease and mature and old growth coniferous and mixed deciduous forested habitat with documented *L. noctivagans* activity (migrating and hibernating) be preserved. Once summer roosts have been identified, logging within five miles of roosting sites should cease.

### *MONITORING AND RESEARCH NEEDS*

Intensive monitoring is necessary (1) to determine precise habitats and use of space, particularly preferred migratory routes (by replicate sampling), summer roosting sites, and relationships with sympatric species (e.g., shared habitats between species will strengthen conservation of critical habitat) in Pennsylvania; (2) to identify roost trees, and (3) to monitor regularly demographic, reproductive, and roost-site characteristics. To determine abundance and habitat use, a monitoring program using mist-netting/harp traps and bat detectors would be most economical, despite the lack of understanding of detector probability (accuracy/precision cannot be determined). Summer-caught bats could be fitted with transmitters to locate roost sites. Marking with bands/pit tags and resighting/recapture would be most useful for roosting/hibernating bats and might prove beneficial for regularly monitored migration routes. Bat detector surveys are of use only if care is taken to ensure adequate spatial and temporal replication (O'Shea and Bogan 2003). Radiotelemetry studies coupled with multiyear demographic studies should be preformed, preferably at roosting and hibernation sites.

In Pennsylvania, the most pressing research needs are to understand roosting habitat, space use, and migratory pathways. This calls for regular statewide monitoring for *L. noctivagans* using mist-netting and bat detector surveys, particularly during the summer to identify roost trees and during spring and fall to define migratory pathways. Once summer roosts are identified, an intensive radiotelemetry study would help understand habitat use. A statewide mist-netting program will provide demographic data, and coupled with long-term radiotelemetry studies, will identify roost habitats and roost characteristics requiring conservation and management. It would also be valuable to Pennsylvania to know distances potentially traveled by migrating *L. noctivagans* (to answer where migrating bats come from and where they are going) by applying stable hydrogen isotope analyses from hair of captured bats (Cryan et al. 2004).

*Author:* GARY KWIECINSKI, UNIVERSITY OF SCRANTON

## Eastern Spotted Skunk

Order: Carnivora
Family: Mephitidae
*Spilogale putorius*

The eastern spotted skunk is smaller than the striped skunk (*Mephitis mephitis*; Crabb 1944, Hall and Kelson 1959, Walker 1964). The eastern spotted skunk exhibits a unique checkerboard pattern, which easily distinguishes it from other skunks (Merritt 1987; fig. 6.13). Recognized as a Species of High-Level Concern, the spotted skunk may be Extirpated from Pennsylvania, although its current status in the state is Undetermined. Globally it is Secure (G5, NatureServe 2009).

*Fig. 6.13.* The Eastern Spotted Skunk, *Spilogale putorius*. Photo courtesy of Bob Gress.

### *GEOGRAPHIC RANGE*

Eastern spotted skunks range up the southern Appalachians as far as south-central Pennsylvania (Bedford, Franklin, and Fulton counties only), across the southern United States from Florida to Texas, and northward through the central United States in the Great Plains west to the Continental Divide (Kirkland 1985, Kinlaw 1995, Patterson et al. 2005). Spotted skunks are not known to occur in portions of the Mississippi River Valley, the Great Lakes basin, or the Atlantic Coastal Plain.

### *DISTRIBUTION AND RELATIVE ABUNDANCE IN PENNSYLVANIA*

Pennsylvania is on the northeast periphery of their range, and they have been known to occur only in Bedford, Franklin, and Fulton counties (Kirkland 1985, Merritt 1987). Even here, populations are low and sightings are uncommon. No well-documented occurrences exist within the past two decades. The Pennsylvania Gap Analysis Project (GAP) has identified potential habitats for this species in Pennsylvania (Myers and Bishop 2000; fig. 6.14).

### *COMMUNITY TYPE/HABITAT USE*

In Pennsylvania, *Spilogale putorius* inhabits the dry oak (*Quercus*), Virginia pine (*Pinus virginianus*), and pitch pine (*Pinus rigida*) forested ridges and ravines that are typically rocky (Kirkland 1985, Merritt 1987). Elevations range from 700 to 3,000 feet (213 to 914 m). Dense cover is preferred to open areas, although spotted skunks can be numerous around human residences. Eastern spotted skunk occurrences have often been associated with agriculture. No evidence exists to suggest habitat segregation from striped skunks in sympatric portions of their range (Kirkland 1985).

### *LIFE HISTORY AND ECOLOGY*

Although eastern spotted skunks are still omnivores, they are more carnivorous than other skunks, feeding primarily on small mammals, especially in the

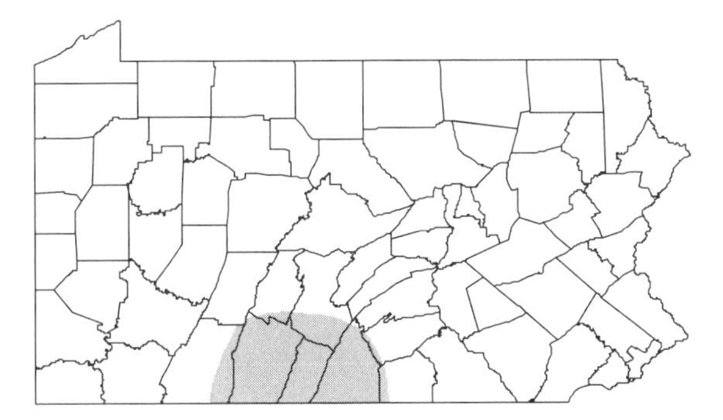

*Fig. 6.14.* Distribution of the Eastern Spotted Skunk, *Spilogale putorius*.

winter months. Carrion is also a common winter food item. In other seasons, invertebrates, fruits, amphibians, reptiles, birds' eggs, grain, and other items are consumed as available (Howard and Marsh 1982). *Spilogale putorius*, like other skunks, sprays its anal gland musk as a defense. First, however, they attempt to intimidate would-be attackers by raising the tail and hairs, arching the back, and stomping the feet. If they are unsuccessful, eastern spotted skunks do a "handstand" on the forelegs to appear even larger. Finally, for particularly tenacious attackers, they bend their body so the head and anal glands both point toward the offender and spray. At this point, the attacker quickly becomes the victim, often experiencing temporary blindness from the musk (Merritt 1987). Predators of the eastern spotted skunk include large raptors, foxes, coyotes, bobcats, domestic cats and dogs, and trappers (Kinlaw 1995). Occasionally, they are struck by vehicles.

Denning locations include rock cavities, hollow logs, woodpiles, deserted buildings, and woodchuck burrows. It is common for eastern spotted skunks to dig their own burrows. Usually solitary, they sometimes den communally in cold weather, sleeping for several days of inactivity (Merritt 1987). Spotted skunks are dexterous climbers that are known to den in tree cavities.

*Spilogale putorius* exhibits delayed implantation of fertilized ovum, like most mustelids. Breeding occurs from March through April. Following a fifty- to sixty-five-day gestation period (about thirty of these days are postimplantation; Howard and Marsh 1982), the litter of four to nine young is born in June, at a rate of only one litter per year in Pennsylvania (Merritt 1987). Young skunks are born blind and helpless, but within forty-six days, they can emit musk and are weaned around fifty-four days. By three months, they are adult sized (Howard and Marsh 1982).

*THREATS*

Eastern spotted skunks have experienced a precipitous range-wide population decline (Patterson et al. 2005), with no direct evidence linking this decline to any particular factor (Gompper and Hackett 2005). Such a large-scale decline could be particularly decimating to a peripheral population such as Pennsylvania's. Potential factors contributing to the decline include overharvest, habitat change, pesticide use, changes in agricultural practices to "clean farming," predation, road kill, or disease outbreaks. Eastern spotted skunks are susceptible to the rabies virus, giving them their other common name, "hydrophobia cat" (Merritt 1987). Other diseases that could affect the species include distemper, parvoviruses, or the mink enteritis virus. The range-wide population declines, however, have not followed the trend patterns typical of a rabies or distemper-related population crash (Gompper and Hackett 2005).

*CONSERVATION AND MANAGEMENT NEEDS*

Although *S. putorius* is listed as furbearer by the Pennsylvania Game Commission, it is still a legally trapped and hunted species with no closed season. However, the Pennsylvania Biological Survey has recommended it be listed as Endangered. Fortunately, few, if any, trappers and hunters pursue this species, and most of the take is accidental. No records of this species have been reported to the Pennsylvania Game Commission between 1985 and 2005 (M. Lovallo, personal communication).

*MONITORING AND RESEARCH NEEDS*

Inventory and monitoring are the most immediate conservation needs but not necessarily urgent. Populations across the range are in strong decline, and few states have monitoring programs (Gompper and Hackett 2005). No current populations are verified to exist in the state, and verification must be done before any subsequent conservation measures are taken. As extant populations are located, assessments should be done to study demographics, susceptibility to rabies, and other mortality factors, and management plans should be developed. If rabies were determined to be a primary threat to this population, then oral vaccine could be administered through a targeted program similar to the current oral rabies vaccination program (ORV) of the Pennsylvania Department of Agriculture (2006–present). If no populations were found, reintroductions could be considered. Monitoring within the historical range through livetrapping and sighting reports is the important first step to managing this population. Current rabies control efforts could be focused on this part of the state if monitoring indicated that rabies was a significant threat.

*Author:* J. MERLIN BENNER, WILDLIFE SPECIALISTS, LLC

## Northern Flying Squirrel

Order: Rodentia
Family: Sciuridae
*Glaucomys sabrinus*

A small, nocturnal, gliding mammal, the northern flying squirrel (*Glaucomys sabrinus*) is similar in appearance and size to its only congener the southern flying squirrel (*Glaucomys volans*) but is distinguished by its larger body size, grayish white belly hairs, a longer upper tooth row, and a shorter, more robust baculum (Wells-Gosling and Heaney 1984, Merritt 1987; fig. 6.15). Despite the federal Endangered listing of *Glaucomys sabrinus fuscus* (now de-listed) and *Glaucomys sabrinus coloratus* (Weigl 1977, Austin et al. 1990), *Glaucomys sabrinus macrotis* has not received similar consideration because of its relatively high abundance in New York state. This species was selected because intensive monitoring over the past decade has shown it to be far less common than suggested by its historic range because of loss of mature coniferous forests, declining health of the hemlock forests, and the possibility of parasite-mediated competition with the southern flying squirrel (*G. volans*). Hybridization between the two flying squirrel species may also contribute to loss of the genetic integrity of *G. sabrinus* as well. The status of the northern flying squirrel was recently elevated to Endangered in Pennsylvania. Its global rank is Secure (G5, NatureServe 2009).

Fig. 6.15. The Northern Flying Squirrel, *Glaucomys sabrinus*. Photo courtesy Ralph Palmer.

### *GEOGRAPHIC RANGE*

A denizen of boreal forests (conifer and mixed conifer/hardwood forests) of Canada and the northern United States, the northern flying squirrel is relatively common across this region. Its range extends southward through the Appalachian Mountains (in the east) and the Rockies and Sierra Nevada Range (in the west). Across these southern extensions of its range, the species' distribution is far more fragmented by isolated patches of suitable montane habitat (Wells-Gosling and Heaney 1984). In the southern Appalachian Mountains such isolation has resulted in the federal listing of two subspecies (*G. sabrinus fuscus* and *G. sabrinus coloratus*) as Endangered (Payne et al. 1989, Austin et al. 1990).

### *DISTRIBUTION AND RELATIVE ABUNDANCE IN PENNSYLVANIA*

In Pennsylvania, the range of the northern flying squirrel is reported to extend across the northern third of the state, southward into Maryland through a narrow stretch in Somerset and Fayette counties (Merritt 1987). Mahan et al. (1999), however, reported only twenty historic locations in the northern tier of the state (fig. 6.16). More recent records over the past twenty years indicate only one location in the western portion of the state and six in northeastern Pennsylvania (Mahan et al. 1999). These populations in the Northeast appear to comprise one larger metapopulation that extends throughout the Pocono region (Mahan et al. 1999). Significant surveying and monitoring of the species over the past ten years in Pennsylvania

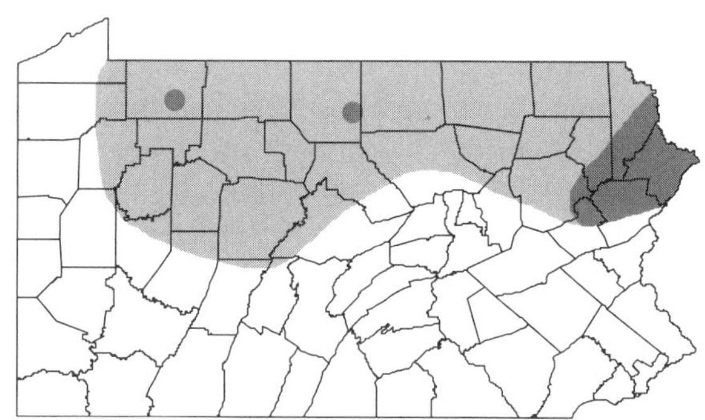

Fig. 6.16. Distribution of the Northern Flying Squirrel, *Glaucomys sabrinus*. Dark shading = current geographic range, light shading = historic geographic range.

strongly suggests that the species is declining in range and abundance and may be far more threatened than previously realized. Recent unpublished reports indicate that *G. sabrinus* is declining in other locations on the southern periphery of its range (i.e., Ontario and Michigan).

*COMMUNITY TYPE/HABITAT USE*

Across its range, including the Appalachians, *G. sabrinus* is reported from a wide range of forests, in addition to those containing a significant conifer component (Wells-Gosling and Heaney 1984). The first record of the species in Pennsylvania was reported from a stand of beech (*Fagus*), yellow birch (*Betula alleghaniensis*), and sugar maple (*Acer saccharum*; Doutt 1930). In New York, the species is also known to occur in oak (*Quercus*) forests (Wells-Gosling and Heaney 1984). Recent studies in Pennsylvania indicate heavy use of mixed stands, especially those with a significant hemlock (*Tsuga*) component. Where it occurs, red spruce (*Picea rubens*) also appears to be an important component of the habitat of *G. sabrinus* in Pennsylvania (Mahan et al. 1999, Merritt 1987). Although more commonly associated with pure stands of hardwoods, the southern flying squirrel is regularly found in association with *G. sabrinus* in the state.

*LIFE HISTORY AND ECOLOGY*

Similar to other squirrels, including its congener, the diet of *G. sabrinus* includes the nuts and seeds of numerous species, as well as fruits, catkins, staminate cones, and various animal material (Wells-Gosling and Heaney 1984, Merritt 1987, Steele and Koprowski 2001). However, *G. sabrinus* also appears to specialize on lichens and hypogeous fungi, both of which represent key dietary components during some parts of the year. The species is nocturnal, active year-round, and forms conspecific nest associations for energy conservation during colder months (Wells-Gosling and Heaney 1984, Merritt 1987). Both nest cavities and dreys (twig and leaf nests) are used for nesting. Because of their size, numerous avian and mammalian predators feed on flying squirrels. The species relies on gliding locomotion for predator escape and traversing much of their home range. Both species of *Glaucomys*, however, frequently forage on the ground where they also engage in scatter-hoarding behavior (Wells-Gosling and Heaney 1984, Merritt 1987).

Two breeding periods are recorded for the species, but in many parts of the range, individuals may only breed once per year (Wells-Gosling and Heaney 1984). Merritt (1987) cites late March and late August as the time of birth in Pennsylvania, but recent observations by the authors suggest that breeding may occur later in the spring in some locations. Gestation is thirty-seven to forty-two days (Wells-Gosling and Heaney 1984); four young are typically found in a litter, and weaning occurs within eight weeks of birth (Merritt 1987).

*THREATS*

The primary factors influencing the species decline in Pennsylvania include (1) the loss of older and old growth conifer and mixed stands as a result of increased human activity (e.g., development pressure in the Pocono region of Pennsylvania), (2) the declining health of the hemlock forests throughout the state, and (3) the potential competitive pressure exerted by southern flying squirrels (Weigl 1969, Weigl 1978). The southern flying squirrel appears to be a more aggressive competitor for resources such as nest cavities (Weigl 1978) and carries a nematode parasite (*Strongyloides robustus*) that may be debilitating or even lethal to the northern flying squirrel (Weigl 1969, Weigl et al. 1999, Krichbaum et al. 2010). Moreover, a fourth, and a potentially greater, concern for this species follows from the recent discovery that in parts of the Northeast (Pennsylvania and Ontario, Canada), the northern flying squirrel is now hybridizing with its congener, the southern flying squirrel (Garroway et al. 2009). Detailed molecular analyses (nuclear and mitochondrial DNA) document a broad hybrid zone where the two species are sympatric, as well as evidence of backcrossing, which verifies the viability of hybrids. Garroway et al. (2010) contends that this hybridization event is a recent phenomenon brought on by several warm winters that have allowed northward range expansion of *G. volans* and the subsequent increase in sympatry between the two species. In Pennsylvania, this trend may be further exacerbated by loss and fragmentation of habitat, which increases the extent and duration of habitat overlap between these two species. Such anthropogenic-induced hybridization is especially alarming given the Endangered status of *G. sabrinus* in Pennsylvania.

*CONSERVATION AND MANAGEMENT NEEDS*

Current sites of occupation, as well as any new locations where the species is discovered, should be designated for immediate protection. We recommend that wherever possible, logging of all kinds cease within a one-mile radius of locations where northern flying squirrels are found. Second, we recommend a more limited level of forest protection within a five-mile radius of known sites of current occupation. This limited protection should include efforts to maintain a matrix of 50 percent of the most mature stands of conifer and mixed conifer/hardwood. Older snags, especially those with nest cavities, should also be maintained within this secondary boundary of protection. Statewide efforts to delineate and protect hemlock forests in general, and older growth conifer and mixed stands in particular, regardless of the known presence of the species, are absolutely essential for the protection of *G. sabrinus* in Pennsylvania.

*MONITORING AND RESEARCH NEEDS*

Intensive monitoring is necessary to delineate the precise range of *G. sabrinus* in Pennsylvania, to track demographic and reproductive trends, and to document patterns of habitat and space use by *G. sabrinus* and its ecological interactions with *G. volans*. Statewide monitoring of >500 nest boxes (distributed statewide in both potential and occupied stands) was initiated in 2003 and should be continued for several years. Additional reconnaissance livetrapping is needed in sites where potential populations may occur but have not been documented in the past twenty years. More intensive study is recommended at benchmark sites where the status, behavior, and ecology of the species can be followed more closely. Immediate protection of forests, including the cessation of all logging, should be enforced at all locations of known occupancy.

In Pennsylvania, the three most pressing research needs, in addition to the regularly statewide monitoring of *G. sabrinus*, are (1) an intensive radiotelemetry study that is coupled with a multiyear demographic study at one or more suitable benchmark sites, (2) a detailed experimental and field investigation that explores the ecological relationship between *G. sabrinus* and *G. volans* and the potential impact of *G. volans* on its congener, and (3) a detailed genetic study to further delineate the extent of the hybrid zone in Pennsylvania.

*Author:* MICHAEL A. STEELE, WILKES UNIVERSITY; CAROLYN MAHAN, PENNSYLVANIA STATE UNIVERSITY; GREGORY TURNER, PENNSYLVANIA GAME COMMISSION.

## Eastern Fox Squirrel

Order: Rodentia
Family Sciuridae
*Sciurus niger vulpinus*

The eastern fox squirrel, one of the least understood of the ten subspecies of fox squirrel, is similar in appearance to the western fox squirrel (*Sciurus niger rufiventer*), but it has a white belly and less orange in its grizzled dorsal pelage (fig. 6.17). *Sciurus niger vulpinus* was identified for Conservation Concern in Pennsylvania and listed as a High-Level Concern because its status is poorly understood in the state. Although stable fox squirrel populations occur within the historic range of this subspecies, it is not known to what extent the original genetic integrity of this subspecies has been lost because of gene flow from populations of the western fox squirrel (*S. n. rufiventer*) in western Pennsylvania. The species' global status is Secure (G5, NatureServe 2009).

*GEOGRAPHIC RANGE*

The fox squirrel's native geographic range extends across the eastern two-thirds of the United States (see species account on Delmarva fox squirrel). The range of the eastern fox squirrel, one of ten subspecies, originally extended across much of western Virginia, eastern West Virginia, western and central Maryland northward into the south-central counties of Pennsylvania (Zegers 1985, Merritt 1987). Today, this range is highly fragmented with stable popula-

*Fig. 6.17.* The Eastern Fox Squirrel, *Sciurus niger vulpinus*. Photo courtesy of Nancy Szymanski.

tions isolated throughout the original range (Derge and Steele 1999).

### DISTRIBUTION AND RELATIVE ABUNDANCE IN PENNSYLVANIA

In Pennsylvania, surveys of land managers conducted in both 1980 and again in 1997 confirmed fox squirrel populations within the historic range of the eastern fox squirrel in eleven counties of south-central Pennsylvania (Bedford, Blair, Cumberland, Franklin, Dauphin, Huntingdon, Juniata, Mifflin, Perry, Somerset, and York; Derge and Steele 1999; fig. 6.18). Fox squirrels were reported in one county (Lebanon) in 1980 but not in 1997 and two counties (Adams and Fulton) in 1997 but not 1980. Five museum specimens taken several decades ago are reported from Berks, Centre, and Cumberland counties, thus suggesting that the current range has contracted in the north and east (Derge and Steele 1999).

### COMMUNITY TYPE/HABITAT

The general habitat preferences of eastern fox squirrels are similar to those of other subspecies of fox squirrels (see species account of Delmarva fox squirrel). However, compared with other subspecies in the southeastern United States, which are often associated with pine (*Pinus* spp.) or oak (*Quercus* spp.) stands, the eastern fox squirrel is typically found in hardwood forests (oak-hickory [*Carya* spp.]; Derge 1997). Surveys reported by Derge and Steele (1999) indicated that eastern fox squirrel populations were most often found in open stands (>76% of surveys) of oak-hickory with limited understory (>60%), often adjacent to agricultural crops (>67%). Although survey results suggest that relatively few populations of the eastern fox squirrel are found in riparian, floodplain forests (<24%), a stable population appears to persist along some of the tributaries (e.g., Condoguinet Creek) of the Susquehanna River in south-central Pennsylvania (M. A. Steele, unpublished data).

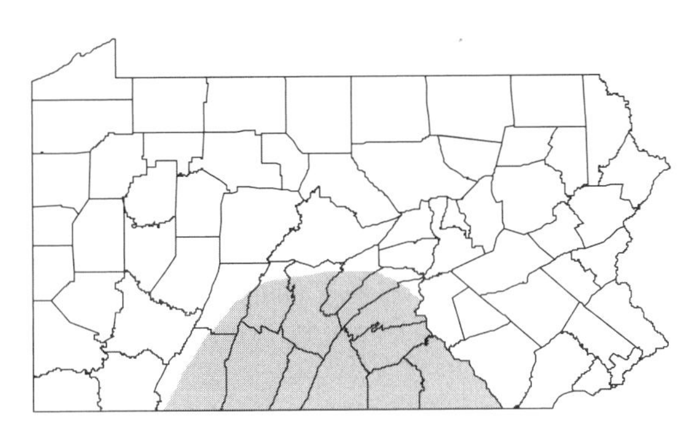

*Fig. 6.18.* Distribution of the Eastern Fox Squirrel, *Sciurus niger vulpinus*.

### LIFE HISTORY AND ECOLOGY

See account on the Delmarva fox squirrel.

### THREATS

In Pennsylvania—and perhaps other portions of the eastern fox squirrels' range—the most significant threat may be genetic exchange with the western fox squirrel (*S. n. rufiventer*). Common throughout the midwestern United States, the western fox squirrel historically resided in the western third of Pennsylvania where the subspecies is reported to be common and abundant in more than fifteen counties (Merritt 1987). More recently, it appears that the range of the western fox squirrel has now extended into three or more counties in the central portion of the state, and as many as six counties in the Northeast (Derge and Steele 1999). This range extension unfortunately appears not to be natural but rather due to anthropogenic factors. Over the past several decades, populations of western fox squirrels have been translocated by a number of organizations to central and eastern portions of the state. In some cases, these introductions may have resulted in significant genetic exchange between the two subspecies. Consequently, it is currently difficult to determine the extent to which fox squirrels occupying the historic range are truly eastern fox squirrels or if the genetic integrity of this subspecies has now been compromised or even lost.

A second important factor affecting the species within the range of the eastern fox squirrel is hunting pressure. Like gray squirrels, fox squirrels are hunted in Pennsylvania. Because populations of eastern fox squirrels may be smaller, more fragmented, and more isolated than those of the western subspecies, they may be highly vulnerable to hunting pressure and more susceptible to local extinctions. Thus, although such hunting pressure is not likely to influence populations of the western subspecies, the same is unlikely for the eastern fox squirrel within this range.

### CONSERVATION AND MANAGEMENT NEEDS

More detailed surveys and demographic studies are needed to determine the precise distribution of eastern fox squirrels in Pennsylvania. Particular efforts should be directed at understanding the distribution of populations and the metapopulation structure across those counties where the subspecies is still found. Wherever possible, hunting pressure should be discouraged where this subspecies is found. Such measures are not easily enforced, however, because of the popularity of gray squirrel hunting in the state and the difficulty in distinguishing these two species of tree squirrels.

*MONITORING AND RESEARCH NEEDS*

To date, only one study on habitat use (Derge 1997) and another unpublished demographic study (M. A. Steele, unpublished data) have been conducted on this subspecies. Hence, more information on distribution and population structure is clearly needed. Three or more benchmark sites should be identified for long-term monitoring.

The most immediate research priority for this subspecies is a thorough genetic study aimed at determining (1) genetic variability within the *S. n. vulpinus* subspecies, including comparisons with populations from Virginia, Maryland, and West Virginia; (2) the extent to which the western fox squirrel has spread across the state and influenced the genetic integrity of the eastern fox squirrel; and (3) comparisons of the eastern fox squirrel with that of both the western fox squirrel and the other fox squirrel subspecies found in the eastern United States (*S. n. cinereus* and *S. n. niger*).

*Authors:* MICHAEL A. STEELE, WILKES UNIVERSITY; DAVID A. ZEGERS, MILLERSVILLE UNIVERSITY

## Appalachian Cottontail

Order: Lagomorpha
Family: Leporidae
*Sylvilagus obscurus*
(also *Responsibility Concern*)

Chapman (1999, 690) described the Appalachian cottontail as "a medium-size rabbit with fine, silky fur. Its upperparts are pinkish buff to reddish buff and the back is overlaid with a distinct black wash, giving a penciled effect." The species has a distinct black spot between the ears (fig. 6.19). The Appalachian cottontail (*Sylvilagus obscures*) was selected as a Pennsylvania species of special concern primarily due to factors associated with habitat modification by human activities and habitat loss due to forest succession. The latter factors coupled with massive introductions of other species of cottontail rabbits in Pennsylvania influence the welfare of the species in the commonwealth. Global rank is G4, Apparently Secure (NatureServe 2009).

*Fig. 6.19.* The Appalachian Cottontail, *Sylvilagus obscurus*. Photo courtesy of Art Dragulis.

*GEOGRAPHIC RANGE*

Appalachian cottontails have a disjunct range from Alabama in the south to Pennsylvania (Chapman et al. 1992, Chapman 1999). Efforts to locate this species in New York have proved unsuccessful (S. Hicks, personal communication). The New England cottontail, with which the Appalachian cottontail was once conspecific (Chapman et al. 1992), is found in a small area in New York as far west as the Hudson River and in the other northeastern states (Chapman 1999). Nowhere in its range in the northern Appalachians can this cottontail be considered common or abundant (Kirkland 1986).

*DISTRIBUTION AND RELATIVE ABUNDANCE IN PENNSYLVANIA*

Published range maps have typically overestimated the known range of the Appalachian cottontail (Merritt 1987, Kurta 1995). Museum specimens document this rabbit's occurrence in modern times in just fifteen Pennsylvania counties (Harnishfeger 2004). Past range estimates have often included what was thought to be suitable habitats in a more extensive original range. Remains of animals found in caves and fissures suggest a wider range in the past (Fonda 2003). Chapman and Stauffer (1982) discuss the probable fragmentation of a formerly more widespread population. The detailed specimen record reported by Kirkland (1985) named fourteen Pennsylvania counties. Kirkland's narrative on this species range mentions the data collected in the six regional mammal surveys from 1948 to 1952; unfortunately, many counties listed in the regional surveys as having Appalachian cottontails were incorrectly listed on the basis of specimens now identified as eastern cottontails. Appalachian cottontails are not common in Pennsylvania and apparently never have been in modern times. A recent three-year statewide survey (Harnishfeger 2004) found only three populations in two contiguous counties, Clinton and Centre (fig. 6.20). One population in Centre County, on State Game Lands 176, also known as "The Barrens," appears to be relatively stable. This estimate is based on comparisons of capture success data in 1976 and 1977 (Holdermann 1978) with those from 2002 livetrapping (Harnishfeger 2004). A second population in northern Centre County located on private property is small with only a single animal captured. The third population occurs over a rather extensive area in northern Centre and adjacent Clinton County on the Sproul State Forest. The dramatic decline in the range of this species in Pennsylvania suggests that this species may be more threatened than previously believed.

### *COMMUNITY TYPE/HABITAT USE*

Published reports have often associated the Appalachian cottontail with conifer and heath habitats at high elevations as reported from the type locality in the Dolly Sods Scenic Area, West Virginia (Chapman 1999). The presence of blueberry and mountain laurel (*Kalmia latifolia*) is indicative, although use is also made of young (under ten years old) clear-cuts. This species supposedly eats a variety of grasses and other herbaceous plants (Spencer and Chapman 1986) and has been reported to be somewhat unique among cottontails in feeding extensively on conifer needles (Chapman 1999). Recent observations in Pennsylvania have found the rabbit in a wide array of habitats, most at higher elevations (Harnishfeger 2004). This observation is similar to Sole's (1999) who found this species in Kentucky in a wider variety of habitat types than previously reported, although at lower elevations. In Pennsylvania, flat ridge tops dominated by mountain laurel with interspersed grassy openings are occupied, as are small, recently planted pine plantations with significant grass and forb cover. Pine plantations occupied by Appalachian cottontails were planted following salvage harvest of trees toppled by a tornado in the mid-1980s. Treetops and other woody debris not removed during the harvest were windrowed by bulldozer. These windrows of decaying treetops and soil have developed significant cover of blackberry (*Rubus* spp.) that provides nearby concealment cover for cottontails using the young pine plantations. Areas extensively burned in the early 1990s also have grown up into thick blackberry cover. Bramble flats with substantial young saplings of maples (*Acer* spp.) and oaks (*Quercus* spp.) have very high densities of Appalachian cottontails. Stevens and Barry (2002) reported extensive use of areas with dense understory of blackberry (*Rubus alleganiensis*), as the dominant shrub species. They also suggested that the presence of high stem densities for concealment and escape cover were likely more important than the species of vegetation to Appalachian cottontails in Maryland. In Pennsylvania, use of grasses, forbs, and young blackberry stems was extensive as was debarking of young saplings. One population was limited to a relatively small area with cover provided by a dense stand of multiflora rose in a wooded valley partially inundated by beaver activity. Grasses, rushes, sedges, and other herbaceous plants were available in abundance where the water regime prevented shrub and tree growth. The geologically distinct barrens area had Appalachian cottontails in moderate densities in dense stands of small oak species, collectively called scrub oaks (Holdermann 1978, Harnishfeger 2004). One interesting observation made during the recent Pennsylvania study (Harnishfeger 2004) appears to be the avoidance of habitat with dense blueberry ground cover. This contrasts quite sharply with previous reports noting the use of areas with blueberry (Stevens and Barry 2002, Chapman 1999). An explanation may be the avoidance of areas with lowbush blueberry with preference for highbush blueberry. Stands with dense but upright stem growth, facilitating easy movement, appear to be preferred by Appalachian cottontails.

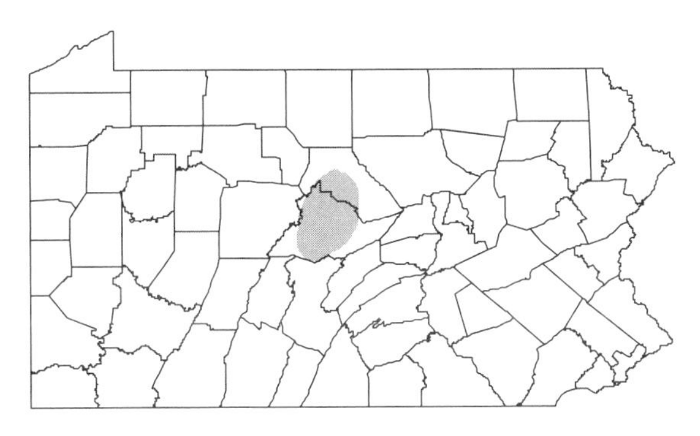

Fig. 6.20. Distribution of the Appalachian Cottontail, *Sylvilagus obscurus*.

### *LIFE HISTORY AND ECOLOGY*

No detailed study of the behavior of this species in Pennsylvania has been conducted. Chapman (1999) summarized knowledge of this species, stating that both solitary and social behavior have been reported and the basic behavior patterns appear similar to those of other rabbits. Solitary grooming and dusting were noted. Vocalizations were elicited by individuals in groups with some evidence that they form social hierarchies. Activity is primarily nocturnal with some movement occurring near dawn and dusk, particularly on overcast days. The home range of Appalachian cottontails has been studied in western Maryland by Stevens and Barry (2002). Median home range size of radio-collared cottontails was 5.4 ha with a range of 1.7 to 9.0 ha when all locations were included in the analysis. Home range size did not differ between adult males and females. Home ranges of eastern cottontails determined using radiotelemetry ranged from 1.0 to 2.8 ha (Chapman et al. 1982). Kurta (1995) reported a home range of 2.0 ha for the eastern cottontail. Eastern cottontail males generally have larger home ranges than females (Trent and Rongstad 1974). Livetrapping in a recent Pennsylvania survey found many animals retrapped in the same or a nearby trap, and movements of individuals beyond 100 m were not documented in the winter season (Harnishfeger 2004). Cottontail rabbits of all species are vulnerable to many predators. Survey work in Pennsylvania suggested that avian predation may be significant, and preferred habitat always had substantial upright stem density with protection from overhead (Harnishfeger 2004).

Breeding in Appalachian cottontails has not been studied in detail in Pennsylvania. Work in Centre County by Holdermann (1978) on a syntopic population of eastern and Appalachian cottontails suggested similar times of onset, durations, and rates of reproduction. These conclusions were based on similar age and sex ratios in livetrapped and hunter-killed rabbit samples. Cottontails are generally reported to build nests lined with dried grass and fur in shallow natural or dug holes or depressions. Merritt (1987) has suggested that nests of eastern cottontails are located in more open grassy habitats than the dense brush preferred by Appalachian cottontails. Nest coverings may well be different in these two distinct habitats. Cottontails have several litters of three to eight young per season. Mean litter size is five in eastern cottontails but reported to be the same (Merritt 1987) or somewhat less, three to four young, in Appalachian cottontails (Kurta 1995). Spring arrives somewhat later at the higher-elevation habitats of Appalachian cottontails, and this probably limits the number of litters to two or less. Gestation is believed to be similar to the twenty-eight days reported for the New England cottontail (Chapman 1975, Merritt 1987, Tefft and Chapman 1987). Newborn rabbits are naked, and the nest materials help conserve body heat. Nestling mortality can be substantial when heavy rains flood nest depressions. Growth of young rabbits is rapid with a reported lactation period of sixteen days (Tefft and Chapman 1983, Merritt 1987, Kurta 1995). It is likely that more southern populations of this species begin breeding as early as March and continue into September; in Pennsylvania, some populations may not begin mating until April or even early May. Kurta (1995, 1999) reported that one in five females reproduce in the same summer as their birth; the author suspects that precocial breeding in the summer of birth would be rare in Pennsylvania populations because of the later onset of breeding in the summer.

### *THREATS*

The substantial time interval between the regional statewide mammal surveys from 1948 to 1952 (Richmond and Roslund 1949, Grimm and Roberts 1950, Gifford and Whitebread 1951, Roslund 1951, Grimm and Whitebread 1952, Roberts and Early 1952) and the recent survey (Harnishfeger 2004) make identifying specific causes of decline difficult. Habitats formerly occupied by Appalachian cottontails have been developed for housing, sand and gravel mining, construction and flooding of the Allegheny Reservoir, and intensive agricultural use. Other likely, and more numerous, causes of habitat loss have been through forest succession and changes in forest composition. This species requires brushy cover, a habitat type not generally favored by private landowners or the forest industry because of its lack of economic return. Kirkland (1985) recognized two major factors of decline: (1) habitat alteration and (2) massive introductions of other species of cottontails in Pennsylvania. Private individuals, organized sportsman groups, and the Pennsylvania Game Commission introduced cottontail rabbits from outside Pennsylvania and relocated rabbits within the state. Merritts (1943) chronicled the history of known official rabbit introductions from Missouri and Kansas and relocations from 1916 through 1942 using Game Commission reports (Pennsylvania Game Commission 1942). She suggested that the effect of such introductions, 700,000 rabbits in just the last twelve years of her study, was minimal because of the 3 million to 4 million rabbits harvested by hunters during this period each year. Movement of large numbers of cottontail rabbits may be responsible for the displacement and decline of Appalachian cottontails in Pennsylvania. It has been suggested that eastern cottontails have a competitive advantage over Appalachian cottontails in more open habitats (Ruedas et al. 1989, Smith 1997). Holdermann (1978) did not find any evidence of such competitive advantage in his study in Centre County where both species of cottontail continue to coexist. Appalachian cottontails were found in dense scrub oak stands with no eastern cottontails present, but also used more open habitats where eastern cottontails were abundant.

### *CONSERVATION AND MANAGEMENT NEEDS*

*Sylvilagus obscurus* is considered a cottontail rabbit and thus a game species and, as such, is regulated in the same manner as the more abundant and widespread eastern cottontail. The current hunting season is split in time and occurs from late October to late November, closes during the statewide deer season, and reopens in January and extends into early February. The daily bag limit is four rabbits per day. The species was considered Status Undetermined by Kirkland (1985), At Risk by Kirkland and Krim (1990), and a Species of Northeastern Regional Concern by Therres (1999). The Mammal Technical Committee of the Pennsylvania Biological Survey has recommended that the Appalachian cottontail be considered Threatened in Pennsylvania on the basis of our current understanding of its distribution and abundance (Pennsylvania Biological Survey 2001).

As for all species, it would appear that suitable habitat is the key to continued survival of the Appalachian cottontail rabbit. The occurrence of the two largest populations on state-owned land is fortuitous and offers more opportunities than would be likely if private property were involved. State game lands and state forest offer ownership and management stability, access for continued research, and both agencies recognize and consider the value of wildlife in their management planning. The species appears to be doing well in these two locations under the current management as a game species. Hunting pressure appears to be light and is not believed to be significantly affecting the current population. Storm et al. (1993) reported that maintenance of aspen-scrub oak stands using clear-cutting and short rotations may be beneficial. Implementation of such an approach on the Sproul State Forest offers the potential to maintain a larger rabbit population longer than would occur naturally if forest succession were not interrupted. The suppression of fire, a natural force for retarding or reversing succession, may be undesirable for the continued existence of the Appalachian cottontail. The Sproul State Forest has a number of historically occupied sites that have permanent brushy cover due to high water table, poor soil, or other characteristics. The tornado and fire events of the 1980s and 1990s apparently allowed these refugia populations to expand rapidly and occupy the extensive early successional stages created. Maintenance of an interconnected pattern or mosaic of early successional habitats, either natural or anthropogenic, should be considered. Historically, the forests of Pennsylvania were more complex with a substantial shrub component. Appalachian cottontails would appear to benefit from a healthy forest with a diverse shrub component on the forest floor. The population on private land in Centre County could benefit from habitat management and perhaps predator control. Kirkland (1985) recommended that private landowners be encouraged to manage their lands to benefit this species. The proximity of this private holding to State Game Lands offers potential.

### *MONITORING AND RESEARCH NEEDS*

The significant decline in range of the Appalachian cottontail suggests that continued intensive monitoring of known populations is necessary. DNA isolation and screening technology from cells in fecal pellets offer the potential to expand the monitoring survey far beyond the few hundred areas visited during the livetrapping survey of historically occupied and other reported sites completed in 2004 (Harnishfeger 2004). Pellets are small and lightweight, and a sample can be collected and transported easily by untrained volunteers from remote areas difficult to trap or access, except by foot. It is important to find any additional occupied sites, no matter how small in extent or population size. DNA technology also offers methods to investigate intraspecific variation and relatedness among individuals and populations (Weir 1996). Low intraspecific variation and high relatedness within a population would suggest potential inbreeding problems and low relatedness between populations would indicate isolation and fragmentation of the more widespread historical populations.

It will be difficult to manage this species effectively in Pennsylvania until basic population characteristics can be determined. The highest priority is learning whether the few remaining populations are increasing, are decreasing, or are stabilizing. Limiting factors need to be identified next. We need to know how habitat-specific these animals are. Radiotelemetry can be used to assess habitat use, home range size, some behavioral traits, the rate of predation, and seasonal mortality. These rabbits are not difficult to trap in the winter; an adequate population size should be reasonably easy to obtain for such a study. Telemetry has been used successfully in studies of various aspects of the ecology of the Appalachian cottontail in nearby western Maryland and West Virginia (Stevens and Barry 2002). The response of rabbits to likely forestry practices would be of great interest and help in planning future management (Yoakum et al. 1980) to maintain or enhance local populations. Questions could also be answered from such work about the adaptability of this species should existing populations prove to be adequate, and the likelihood of successful reintroduction into formerly occupied but now vacant habitats. Also of interest is whether competition between Appalachian and eastern cottontails exists and, if so, whether it is a possible cause for decline and eventual loss of the Appalachian cottontail. The Barrens, in Centre County, would appear to be an excellent location to address this question, given the syntopy of these species.

*Author:* RALPH HARNISHFEGER, LOCK HAVEN UNIVERSITY

### RESPONSIBILITY CONCERN

## Northern *Myotis* (Northern Long-eared Bat)

Order: Chiroptera
Family: Vespertilionidae
*Myotis septentrionalis*

This is a medium-sized *Myotis* with brownish fur, long ears, and a long, pointed tragus. It is distinguished from other Pennsylvania *Myotis* by the greater length of its ears (14–18 mm from the notch) and the long (8–10 mm from the notch) and pointed tragus (fig. 6.21). The calcar is not keeled or only slightly keeled. The species is designated as one of Responsibility Concern because it is a habitat specialist and it is vulnerable to habitat destruction and disturbance of hibernacula and the loss of foraging areas that provide crucial roosting trees. It is considered Vulnerable at the state level; its global rank is Vulnerable.

*GEOGRAPHIC RANGE*

The northern *Myotis* is widely distributed in forested areas of North America. It occurs across southern Canada from Newfoundland to eastern British Columbia, extending south in the United States as far as northern Florida and west as far as Kansas and the Dakotas (Whitaker and Hamilton 1998, Caceres and Barclay 2000). Older literature may refer to this species as *Myotis keenii*, but van Zyll de Jong (1979) presented evidence that eastern long-eared bats represent a distinct species, *Myotis septentrionalis.*

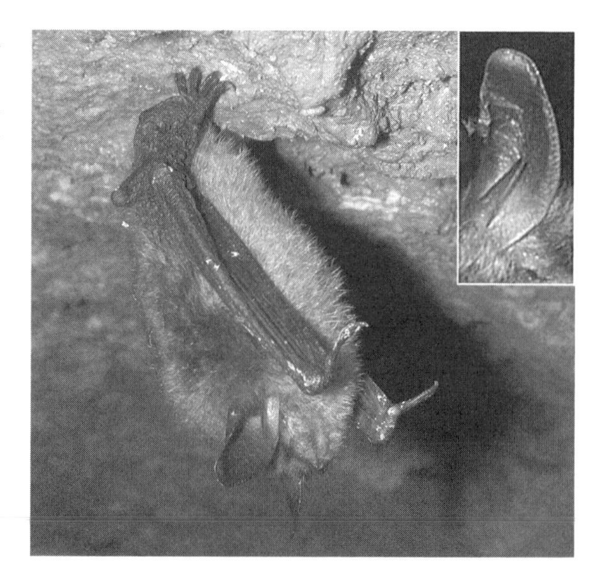

*Fig. 6.21.* The Northern Long-eared Bat, *Myotis septentrionalis.* Photo courtesy of Cal Butchkoski.

*DISTRIBUTION AND RELATIVE ABUNDANCE IN PENNSYLVANIA*

The northern *Myotis* is found throughout Pennsylvania but apparently never in large numbers, and its distribution is considered local and irregular (Hall 1985, Merritt 1987). Little is known of population trends in this species. In contrast to the related little brown and Indiana bats, this species has never been found in large groups or clusters (Hall 1985; fig. 6.22), and this makes it difficult to estimate population size. The northern bat is most commonly observed during the summer in Pennsylvania, and hibernating colonies appear relatively rarely and consist of few individuals (Mohr 1932, Merritt 1987). Dunn and Hall (1989) found this species hibernating at 34 out of 190 caves and mines they examined in Pennsylvania. The total number of northern *Myotis* at these thirty-four sites was 253, and the maximum number at a single site was ninety-three; twenty-six out of the thirty-four occupied sites contained ten or fewer individuals.

*COMMUNITY TYPE/HABITAT USE*

*Myotis septentrionalis* is a species of forested habitats across much of North America. Forested upland

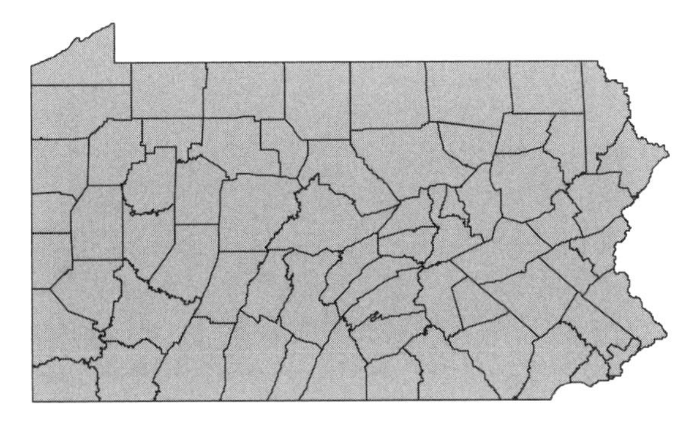

*Fig. 6.22.* Distribution of the Northern Long-eared Bat, *Myotis septentrionalis*.

areas appear to be key summer foraging habitat for this species (LaVal et al. 1977), although Kunz (1973) also found them common in riparian areas in Iowa. An important feature of summer habitat appears to be older and larger trees that contain cavities and have exfoliating bark (Foster and Kurta 1999). Caves, mines, and tunnels are key winter habitats for this species. Raesly and Gates (1987) found that this species chooses large hibernacula with large passages, sizable entrances, and regions with high-relative humidity.

### LIFE HISTORY AND ECOLOGY

During the summer, the northern *Myotis* roosts under exfoliating bark or in crevices and hollows within trees (Merritt 1987, Foster and Kurta 1999). A telemetry study by Foster and Kurta (1999) found northern *Myotis* roosting during the summer in maples and ashes in wetlands in southern Michigan. These bats frequently changed their roost trees over during a season. This species is also reported to roost behind shutters and in buildings (Whitaker and Hamilton 1998). During the summer, females apparently form maternity colonies separate from the male colonies (Clark et al. 1987).

In winter, northern bats hibernate in caves and in mines in or near their summer range, often hiding in tiny cracks or crevices within the rock walls (Whitaker and Hamilton 1998). Raesly and Gates (1987) assessed winter habitat selection by hibernating northern *Myotis* in caves in Maryland, Pennsylvania, and West Virginia. This species chose hibernacula that were large, with large passages, sizable entrances, and regions with high relative humidity. Within these caves, *M. septentrionalis* generally selected cooler areas, and most individuals hibernated in crevices along the side walls, usually in clusters.

A study in Missouri found that this species foraged exclusively among trees, usually in hillside and ridge forest rather than riparian or floodplain forest (LaVal et al. 1977). It may forage only 1–3 m above the ground, flying over and among the understory shrubs (LaVal et al. 1977). *Myotis septentrionalis* emits echolocation calls that are relatively low in intensity, and this apparently makes the calls all but inaudible to most moths (Fenton 1999). They frequently feed by gleaning, taking insects off the ground or vegetation and then carrying them to perches to eat them; this apparently allows this species to eat larger prey than related *Myotis* species (Fenton 1999). Analysis of the diet of the northern *Myotis* indicates that it feeds largely on moths, beetles, dipterans, and lacewings; spiders and orthopterans are sometimes important components of its diet (Griffith and Gates 1985, Whitaker and Hamilton 1998, Lee and McCracken 2004). A diet analysis of this species in Missouri found that it consumed more orthopterans and more large beetles than either *Myotis lucifugus* or *Myotis sodalis* (Lee and McCracken 2004).

Relatively little information is available on the reproductive habits of this species. A single young per year is typical, with birth occurring in late spring or early summer (Merritt 1987, Whitaker and Hamilton 1998). Nursery colonies may contain up to 100 females (Whitaker and Hamilton 1998). The northern *Myotis* has been recorded to live at least eighteen years (Hall et al. 1957).

### THREATS

An important threat to *M. septentrionalis* is destruction and disturbance of hibernacula, either through the closing of caves and mines or through disturbance by cave visitors during the winter months. Destruction of foraging areas, particularly of forested areas with abundant older trees that contain cavities and have exfoliating bark, may also be a threat. A new and serious threat is white-nose syndrome (WNS), an emerging disease of hibernating bats, including this species, with very high mortality of affected populations (Cohn 2008, Blehert et al. 2009).

### CONSERVATION AND MANAGEMENT NEEDS

Northern *Myotis* are currently listed as protected in Pennsylvania. Genoways (1985) listed this species as Vulnerable in Pennsylvania, and Kirkland and Krim (1990) considered it to be rare. Known hibernacula should be protected from complete closure and from disturbance during winter (November to March). Caves and mines that are important hibernating sites for this species should be gated to restrict access. An outreach program to private landowners and public land managers should be developed to increase awareness of the need to protect hibernating bats and their hibernacula. Forested areas known to be summer roosting sites or summer foraging areas should be given protection. This particularly includes forested areas with older trees that have exfoliating bark and contain cavities.

*MONITORING AND RESEARCH NEEDS*

Continued monitoring of the northern *Myotis* at hibernacula will be necessary to understand better the current distribution and population size of the species. This effort should include techniques to detect bats that hibernate individually or in small number in cracks or crevices in the hibernation sites. Intensive summer monitoring, including telemetry, should be conducted to determine the summer roosting habits and maternity roosting habits of this species. Monitoring programs should include mist-netting at caves and in forested areas.

More detailed inventorying needs to be performed to obtain better estimates of current population levels and current geographic distribution in the state. Identification of summer habitats and location of maternity colonies, preferably using telemetry, are particularly important. Research also needs to be conducted to determine the types of forests used for summer roosting sites. Location of additional hibernacula for this species, including determining preferred hibernating site characteristics, should also be performed. In addition, much still needs to be learned about the basic biology of this species. For example, little is known of its reproductive biology.

*Author:* HOWARD P. WHIDDEN, EAST STROUDSBURG UNIVERSITY

**PENNSYLVANIA VULNERABLE**

## Rock Vole

Order: Rodentia
Family: Cricetidae
*Microtus chrotorrhinus*

The rock vole is very similar to the more common meadow vole (*Microtus pennsylvanicus*) in most characteristics (fig. 6.23) but can be distinguished by its yellowish to saffron-colored snout (Merritt 1987, Kurta 1995) and the four reentrant folds on the side of the M2 molars (Kurta 1995). The rock vole (*Microtus chrotorrhinus*) was selected as a Pennsylvania Species of Greatest Conservation Need because of a noted reduction

*Fig. 6.23.* The Rock Vole, *Microtus chrotorrhinus*. Photo courtesy of Cal Butchkoski.

in optimal habitat attributable to various commercial and recreational development in the northeastern region of the commonwealth. Global rank is Apparently Secure (G4, NatureServe 2009).

*GEOGRAPHIC RANGE*

The distribution of the rock vole extends from Labrador, southern Quebec, and Ontario into the New England states south into Pennsylvania where it remains disjunct from the southern subspecies, *Microtus chrotorrhinus carolinensis* (Kirkland and Jannett 1982). The southern distribution for the species appears to be centered in the Appalachian Mountains south to western North Carolina. In the northern region of its distribution, two subspecies occur, *M. c. ravus* (occurring in Labrador), and the subspecies occurring in Pennsylvania, *M. c. chrotorrhinus*. Both *M. c. carolinensis* and *M. c. ravus* are noted to be of Conservation Concern according to Kirkland (1998).

*DISTRIBUTION AND RELATIVE ABUNDANCE IN PENNSYLVANIA*

Pennsylvania is at the southern limits of the subspecies' distribution. Although recently found only in the northeast portion of the state (fig. 6.24), the rock vole occupied a wider distribution historically as noted by Guilday et al. (1964, 1966).

The rock vole is found in moist, talus areas in the northeast portion of Pennsylvania. These habitats are generally at elevations of more than 762 m (2,500 feet) in typical northern hardwood forests. Recent specimens are known to exist from Wayne, Sullivan, Wyoming, and Lycoming counties. Historic specimens are also known from Luzerne County. Although Kirkland and Hart (1999) reported on several new records for the rock vole from Wayne, Wyoming, and Sullivan counties, there is no information on relative abundances as these were all from short-term surveys.

### *COMMUNITY TYPE/HABITAT USE*

Primary habitats consist mainly of talus and boulder strewn areas in high canopy areas within northern hardwood forests in Pennsylvania. Although this species has been found in red spruce forests to the north and south of Pennsylvania (Kirkland 1976, 1977), it has not been found in coniferous or mixed forests in Pennsylvania.

Current investigations in Sullivan and Lycoming counties suggest that it is most likely associated with low to moderate gradient streams in mature hardwood forests (Hart 2003). These streams are bordered by heavy forb growth, including many species of ferns along with large amounts of medium-sized boulders and talus. Moss, denoting moist conditions, is evident all along the sites that have produced specimens.

### *LIFE HISTORY AND ECOLOGY*

The rock vole's primary habitat consists of elevated forests of greater than 3,000 feet. Most authors recognize the conspicuous occurrence of rocks and talus as part of the primary habitat (Grimm and Whitebread 1952, Kirkland 1977, Kirkland and Knipe 1979).

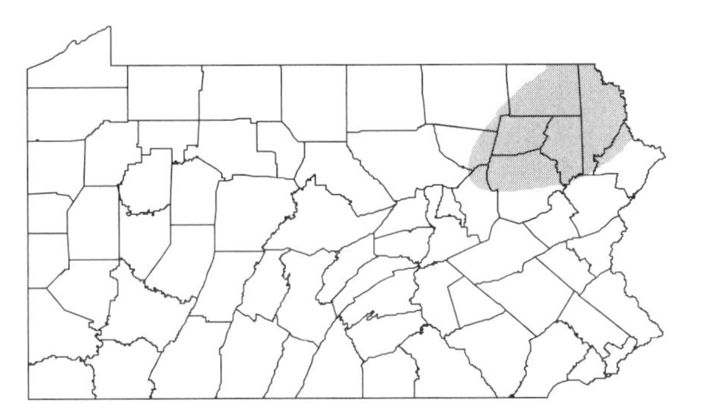

*Fig. 6.24.* Distribution of the Rock Vole, *Microtus chrotorrhinus*.

Although the rock vole is mainly associated with red spruce/northern hardwoods forests, Kirkland (1976) took several specimens from mine waste sites in northern New York as well as from recent clear-cuts in West Virginia (Kirkland 1974). Kirkland and Knipe (1979) noted that the majority of specimens taken in New York were from traplines associated with small streams running through the study site. This is in concordance with what has recently been found (C. Butchkoski, personal communication). In Pennsylvania, all specimens have been taken along streams in northern hardwoods forests characterized with large amounts of talus and rock on the substrate and heavy amounts of forb cover.

Microhabitat for the rock vole consists of heavy vegetation and rock cover with underlying tunnels and runways (Kirkland and Knipe 1979). It is believed that primary activity occurs belowground, which probably explains why this species remains underrepresented in museum collections (Kirkland and Jannett 1982). Although there has been some thought that the meadow vole excludes the rock vole when the two occur sympatrically, Kirkland and Knipe (1979) reported the two species from transition forests in West Virginia. Whitaker and Martin (1977) reported that bunchberry and unidentified green vegetation made up more than 73 percent of the volume of forty-seven stomachs analyzed.

Rock voles breed from early spring until late fall (Kirkland and Jannett 1982). Litter sizes range from three to seven, based on embryo counts. Higher litter sizes tend to occur at more northern latitudes. Females begin to breed when they attain approximately 140 mm total length and 30 g weight, which is comparable to other microtenes. Besides reproduction, there is very little information about their development.

### *THREATS*

Although it is impossible to point out any immediate threats due to the restricted nature of this species' occurrence and the paucity of information on its distribution in Pennsylvania, it is probable that growth in the recreational skiing industry will significantly affect habitats occupied by this species in the future. The northeast portion of Pennsylvania is being developed at a high rate, and the ski industry continues to expand into mountainous areas that may possibly be occupied by this species. Runoff along the streams draining these areas may negatively affect the microhabitat necessary for the rock voles' survival.

*CONSERVATION AND MANAGEMENT NEEDS*

Current research conducted by Hart and Butchkoski has shown that it is restricted to medium gradient streams where rock and talus are common (Hart 2003). To conserve habitat for this species, dispersal habits and habitat requirements must be better understood. Basic ecology is not well understood in terms of home range, dispersal distances, and the use of marginal habitats. Barriers to dispersal need to be further delineated to provide adequate core and buffer areas for this species. Livetrapping studies as well as some form of tracking studies should be initiated.

*MONITORING AND RESEARCH NEEDS*

Although to date, most surveys have been short term, livetrapping studies should be conducted to determine basic ecological information. Although other species of microtenes appear to have high natality, the reproductive biology of the rock vole should be researched to determine possible differences from other, more common microtenes. Adaptive management needs include ensuring that heavy logging or reduction in forest understory along streams does not affect currently known sites for the rock vole. Mark and recapture studies may provide insight into the basic ecology of this species, including reproduction, dispersal, and habitat requirements, as well as fundamental population information.

*Author:* JAMES HART, PENNSYLVANIA NATURAL HERITAGE PROGRAM

**MAINTENANCE CONCERN**

## American Water Shrew

Order: Soricomorpha
Family: Soricidae
*Sorex palustris albibarbis*

The American water shrew can be distinguished from other species of Pennsylvania shrews by its large size (138 to 164 mm total length), long tail (63 to 72 mm), and fringes of hair on the tail and hind feet (Whitaker and Hamilton 1998), as well as its slightly webbed hind feet (fig. 6.25). Differences between the American water shrew and the West Virginia water shrew are negligible but in some cases, the underparts of the American water shrew may appear lighter.

This species was selected for Maintenance Concern because of its specialized habitat requirements, low population densities, and patchy distribution. It is highly susceptible to aquatic pollution, sedimentation of streams, and fragmentation of local populations due to human development and increased road construction. It is considered Vulnerable at the state level. Its global rank is Secure (G5, NatureServe 2009).

*Fig. 6.25.* The American Water Shrew, *Sorex palustris albibarbis*. Photo courtesy of Charlie Eichelberger.

*GEOGRAPHIC RANGE*

The species *Sorex palustris* has a wide range extending from Alaska, across Canada to New York, and south into the southern Appalachians and in the west to New Mexico. The American water shrew's range extends from Maine southward through the New England states and New York into the northern-tier counties of Pennsylvania (Whitaker and Hamilton 1998).

*DISTRIBUTION AND RELATIVE ABUNDANCE IN PENNSYLVANIA*

The American water shrew occupies many streams in the northern-tier counties in Pennsylvania (fig. 6.26). Its range in the state extends from Monroe County in northeastern Pennsylvania southwestward into Mifflin County and northwestward along the border of Elk and Forest counties. It is likely that the only northern-tier county in which the American water shrew does not occur is Erie County. Pennsylvania lies at the southern margin of the American water shrew's range. There are questions concerning the gap that lies between the American water shrew's southern boundary and the northernmost margin of the West Virginia water shrew's distribution. Whether this gap actually occurs or is an artifact due to insufficient data is unknown.

Although there are numerous records, both historic and recent, for the American water shrew, its abundance is most likely based on habitat quality, not on habitat quantity. Kirkland and Schmidt (1982)

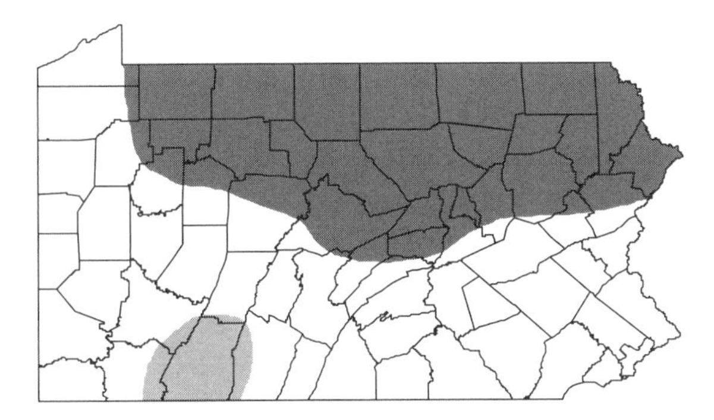

*Fig. 6.26.* Distribution of the American Water Shrew, *Sorex palustris albibarbis.*

noted low trapping success compared with other species of small mammals and suggested that this may be the result of low population densities. Significant sampling effort by the Pennsylvania Game Commission and personnel at the Shippensburg University Vertebrate Museum (Woleslagle 1994) has gathered considerable information concerning the distribution of this species.

### *COMMUNITY TYPE/HABITAT USE*

The primary habitat for the American water shrew is moderate to sometimes swift flowing streams that have undercut banks and a large amount of structure such as rocks, logs and crevices (Baker 1983). During surveys in northeastern Pennsylvania, Woleslagle (1994) captured American water shrew along marshes and within damp, floodplain forests. Deciduous forests appear to provide optimal cover as long as the understory is relatively heavy. This species is usually not found within steep gradient streams although Pagels et al. (1998) noted its occurrence along streams with gradients in slope ranging between 7 percent and 14 percent.

Within its range in Pennsylvania, the American water shrew's habitat occurs across the northern-tier counties in the mountainous portions of the Appalachian Plateau Province and along the northwestern portion of the Ridge and Valley Province in the Appalachian Mountains Section. It appears that elevation may be important, as most sites lie above 500 m. The predicted potential habitat as based on analysis of statewide habitat availability during the Pennsylvania Gap Analysis Project (GAP; Myers et al. 2000) indicated that habitats in the southwestern portion of the state are optimal for colonization by West Virginia water shrews, *S. p. punctulatus.*

### *LIFE HISTORY AND ECOLOGY*

The American water shrew occurs along clear mountain streams, most often at higher elevations. Streams are characteristically of high quality and moderate flow and usually have deeply undercut banks (Beneski and Stinson 1987). In Wyoming, Clark (1973) found that this shrew was most common when ground cover was 75 percent or higher. Similar habitat characteristics have been observed in Pennsylvania. Although generally occurring in hardwood forests, it has been found in mixed coniferous/deciduous forest at higher elevations (J. Hart, unpublished observations). A study by Rozanski et al. (1996) suggested a connection between water quality and the occurrence of water shrews, although this study was inconclusive because of low sample sizes. It is likely that any disturbance to the forest environment that promotes warming of streams, sediment loading, and dramatic changes in pH will negatively affect the water shrew's habitat. Water shrews are capable divers and can remain submerged for more than forty-five seconds (Beneski and Stinson 1987). They remain active during the winter and are able to continue diving for prey because of the ability to reduce their metabolic demands (Boernke 1977). There is also some suggestion that they are able to enter a daily torpor period. The diet of the American water shrew includes stonefly, caddisfly, and mayfly nymphs, as well as cranefly larva. Other important foodstuffs are snails, slugs, earthworms, small fish, fish eggs, and salamanders (Whitaker and Hamilton 1998).

Although there is little information concerning the reproductive biology of water shrews in general, it is believed that they become reproductively active in late February and remain active until late summer (Kurta 1995). They produce two to three litters of between three and ten young annually (Beneski and Stinson 1987), while Punzo (2004) noted litter sizes of five to eight for the species. Gestation period remains undefined (Whitaker and Hamilton 1998). It is believed that, unlike other insectivores, they do not become sexually mature until their second season after birth.

### *THREATS*

Likely threats to the persistence of the American water shrew in Pennsylvania include the effects of acid mine drainage, acid rain concentrations, and indiscriminate logging that degrades or destroys habitat along streams harboring the American water shrew. Sedimentation created during logging may disrupt the food chain to such an extant that water shrews are not able to persist long term within the logged areas. Stream channeling may also affect the survival of the water shrew although it is unknown at present whether this will be a key problem in the future.

Many areas in the Northeast have been developed at a high rate because of the demand for vacation homes and for tourism. This, in turn, will most likely affect both stream levels and quality as runoff increases. This problem will only increase, and needs to be investigated to determine the effect of increased human intrusion on the habitat of the American water shrew. Another possible threat may arise from the effect of deer overbrowsing of the understory, especially within the vicinity of stream habitat although this remains unstudied. The density of shrub cover along streams noted by most researchers who have collected water shrews may be important to maintaining microclimate variables, as well as a stable prey base.

### *CONSERVATION AND MANAGEMENT NEEDS*

At present, there is no conservation management plan in effect or planned for the American water shrew as few researchers have been involved with this species in Pennsylvania. The perceived economic unimportance of the American water shrew has limited conservation efforts directed at this species although Genoways and Brenner (1985) listed this species as warranting further protection because of the paucity of information concerning its ecology. Although this species would appear to be secure in Pennsylvania based on new information gathered during 2005 by the author, additional studies are needed to determine the link between this species and the water quality of streams in which it occurs.

Conservation and management efforts should focus on the following approaches: (1) developing a network of investigators that will increase the likelihood of discovering new occurrences of the American water shrew; (2) continuing inventory and monitoring of known populations, including the collection of micro- and landscape-level habitat characteristics; (3) monitoring both chemical and physical properties of streams at existing American water shrew sites for changes that may affect the shrews continued success; (4) developing a stream condition database with various state agencies and other conservation organizations that may be based partially on the new invertebrate classification system currently under development by the Pennsylvania Natural Heritage Program; (5) developing low-impact trapping techniques that reduce mortality of American water shrews; (6) creating GIS-based maps for existing habitat as well as maps and layers based on predicted occurrence of viable habitat in association with the current GAP model (Myers et al. 2000); and (7) incorporating appropriate habitat in conservation planning issues at the state level to begin to develop habitat corridors encouraging dispersal of American water shrews.

### *MONITORING AND RESEARCH NEEDS*

Long-term monitoring using livetrapping needs to be conducted to better understand the basic ecology of this species. Reproductive success, population and community structure, and the effects of both physical and temporal changes to the habitat supporting the American water shrew should be studied. Monitoring of the currently known populations should be conducted at five-year intervals to ensure that these populations remain active and successful. Conservation and protection of occupied habitat should include identifying "core" areas, as well as buffers that include a significant portion of necessary adjoining habitat.

The most pressing research priority is to determine the extent of the American water shrew's distribution along the northern tier of the state and along its southernmost boundary in Mifflin County. Such research should include the collection of macro- and microhabitat variables that may be used to better define the optimal habitat for the American water shrew. At present, a lack of information concerning the effects of differing water quality, acid rain levels, and sedimentation on the occurrence of the American water shrew precludes the development of more specific conservation plans.

Research activities should focus on (1) creating GIS-based maps of possible habitat in Pennsylvania based on existing information, (2) conducting intensive habitat analysis targeted at producing usable habitat maps for predicting American water shrew presence, (3) ground-truthing of predicted habitat, (4) sampling all predicted habitats at least one time using Pennsylvania Game Commission protocols, and (5) using information on existing populations to develop appropriate conservation plans for the future.

*Author:* JAMES HART, PENNSYLVANIA NATURAL HERITAGE PROGRAM

## Rock Shrew

Order: Soricomorpha
Family: Soricidae
*Sorex dispar*

The rock shrew is a medium-sized (103–136 mm) slate gray, long-tailed shrew that resembles the smoky shrew (*Sorex fumeus*) in size and appearance (Kirkland 1981; fig. 6.27). The tail of the rock shrew averages 55.5 mm, whereas the smoky shrew has a shorter tail, averaging around 42.5 mm (Kirkland and Van Deusen 1979). The rock shrew is considered globally Secure (NatureServe 2004). Within Pennsylvania, it is considered to be rare but does not receive protection as defined by current Pennsylvania law. A number of unknown factors concerning it basic ecology, population status, and taxonomy on the Pennsylvania/Maryland border justifies listing the rock shrew for Maintenance Concern. The global status of the rock shrew is Apparently Secure (G4, NatureServe 2009).

### GEOGRAPHIC RANGE

Two subspecies are currently recognized for the rock shrew and include *Sorex dispar dispar* and *Sorex dispar blitchi*. The distribution of the subspecies that occurs in Pennsylvania, *S. d. dispar*, extends from Maine through central Vermont and New Hampshire, into east central New York and western Massachusetts, and through much of Pennsylvania (Whitaker and Hamilton 1998). Although a boundary between *S. d. dispar* and *S. d. blitchi* occurs along the Pennsylvania/Maryland/West Virginia borders, it is unknown whether this is valid or just an artifact of sampling (Kirkland 1981).

### DISTRIBUTION AND RELATIVE ABUNDANCE IN PENNSYLVANIA

In Pennsylvania, the rock shrew appears to be absent from the Central Lowlands Province around Lake Erie, the Glaciated Pittsburgh Plateau Section of the Appalachians Plateau Province, and the Piedmont Province and Atlantic Coastal Plain (Kirkland and Hart 1999; fig. 6.28). The species appears to be more widespread than once thought, occupying much of the Appalachian Mountains and Allegheny Plateau Region although at present its population status in Pennsylvania remains unknown because of the difficulty of capturing shrews alive. In most cases in Pennsylvania, this species is more often caught when snaptraps are placed deep in crevices under rock and talus in deciduous forests. Although pitfalls seem more efficient for the capture of shrews, they are not good for this species due to the rocky nature of the species' preferred habitat.

### COMMUNITY TYPE/HABITAT USE

The primary habitats for the rock shrew are cool, moist forests with a talus-covered substrate (Kirkland 1981, Kirkland and Schmidt 1982). Laerm et al. (1999), in a study of soricid communities in the southern Appalachians, collected this species from sites within mesic oak-hickory communities with a shrub layer consisting of dense rosebay rhododendron (*Rhododendron maximus*) at elevations more than 1,000 m. Ford et al. (2005) found it to be prevalent in red spruce (*Picea rubens*), cove hardwood, and upland hardwood forests in both the central and southern Appalachians.

Although the majority of specimens taken have been from naturally occurring talus and rock, several specimens have been taken from artificially created talus such as that created around clear-cuts (Kirkland et al. 1976). This species has also been taken from small seeps where large amounts of talus were not a primary element of the habitat although loose gravel and some rock were present within the area of capture (Hart 2004). Specimens have also been taken from open-pit mine wastes in the Central Adirondack Mountains in

Fig. 6.27. The Rock Shrew, *Sorex dispar*. Photo courtesy of Roger Barbour.

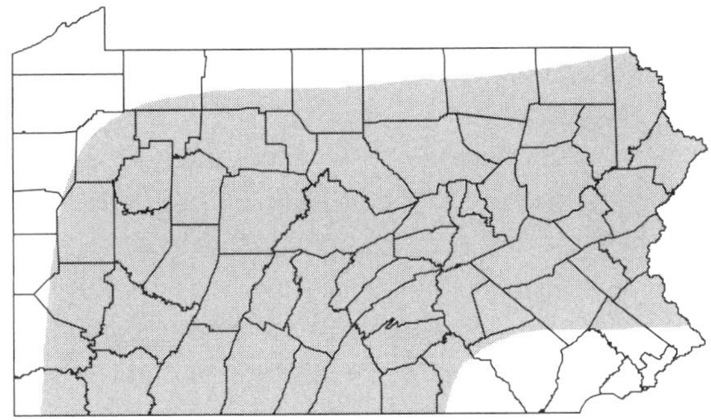

Fig. 6.28. Distribution of the Rock Shrew, *Sorex dispar*.

New York (Kirkland 1982). This species has also been taken in uncut red spruce stands in West Virginia (Kirkland 1974).

### *LIFE HISTORY AND ECOLOGY*

Little is understood concerning the ecology of the rock shrew. Kirkland (1981) reported that talus areas along mountain streams make up its primary habitat. Typically occupying areas deep within rocks, the species is rarely captured. Researchers have recently begun to set small traps deep within rocks and talus by pushing them farther under, using long sticks after setting the trap. Rock shrews have been found along seeps and floodplains in stream valleys primarily secluded from disturbance. A good example of this type of habitat for the rock shrew was located on property owned by a water company in Northumberland County, Pennsylvania (Hart 2004).

At present, there is not much information on the rock shrew's reproductive biology. Kirkland (1981) reports that reproductive activity may extend from early spring to late summer as pregnant or lactating females were taken between May and August.

### *THREATS*

Likely threats to the rock shrew include forest fragmentation, possible acid mine drainage, and the effects of acid precipitation on forest health and primary habitats. In addition, increased recreational development, such as ski resorts, may alter secluded stream valleys and rocky slopes and negatively affect the survival to this species. Documentation of threats, however, is scant because of the relatively few specimens available at present and the lack of any further ecological studies since Kirkland's work of the 1970s and 1980s (Kirkland 1974, Kirkland et al. 1976, Kirkland 1982, Kirkland and Schmidt 1982, Kirkland and Snoddy 1999).

### *CONSERVATION AND MANAGEMENT NEEDS*

Although managing for the conservation of rock shrews appears to be a daunting task, it most likely will be done within the scope of managing for larger, more recognizable species, such as the Allegheny woodrat and rock vole. Community analysis based on Kirkland and Snoddy's (1999) study that investigated the biogeography and diversity of insectivore communities in the northern Appalachian Mountains may also point to those habitats that are less restrictive to overall insectivore occurrence.

### *MONITORING AND RESEARCH NEEDS*

Monitoring and management will most likely be conducted during investigations of other species occupying the same habitat, such as carrying out small mammal surveys when conducting woodrat visual censuses. One method of monitoring would be to conduct long-term surveys of small mammal populations employing a variety of trapping techniques in a myriad of habitats and microhabitat types (Kirkland and Snoddy 1999).

Current research efforts should focus on describing the necessary micro- and macro-habitats for this species's survival. If, as suspected, the rock shrew inhabits many of the same habitats available to the woodrat, surveys should be conducted at sites where woodrats have been historically known to occur but are currently extirpated. This may lead to information that highlights habitat factors and threats common to both species. The extant of the rock shrew's distribution in Pennsylvania must be further understood through intensive shrew inventories of appropriate habitats. These habitats can be identified on the basis of the presence of rock and talus along streams within hardwood forests.

*Author:* JAMES HART, PENNSYLVANIA NATURAL HERITAGE PROGRAM

## Eastern Red Bat

Order: Chiroptera
Family: Vespertilionidae
*Lasiurus borealis*

The eastern red bat is a medium-sized vespertilionid with a distinctive red orange coat and whitish patches on shoulders and wrists (fig. 6.29). The tail membrane and parts of the underside of the wing are furred,

Fig. 6.29. The Eastern Red Bat, *Lasiurus borealis*. Photo courtesy of Charlie Eichelberger.

while the hairs on the torso tend to be lightly frosted at the tips. Sexual dimorphism is evident with females exhibiting lighter coloration and more frosting (Barbour and Davis 1969). The eastern red bat is recognized as a Species of Concern in the Pennsylvania because it is a habitat specialist and is susceptible to mortality resulting from collisions with wind turbines. It is globally ranked as G5, or Secure, and is known to be common throughout much of its range (NatureServe 2006). It is listed as a priority species for the Northeast states (State of Connecticut 2004).

### *GEOGRAPHIC RANGE*

*Lasiurus borealis* is distributed throughout the eastern United States, as well as the adjoining areas of southern Canada and the northeast tip of Mexico (Shump and Shump 1982). It is absent from the southern end of the Florida peninsula. The species' range extends west to the Rockies and is allopatric with its closely related congener, the western red bat (*Lasiurus blossevilli*), except for areas of southern New Mexico and western Texas where both species overlap.

### *DISTRIBUTION AND RELATIVE ABUNDANCE IN PENNSYLVANIA*

*Lasiurus borealis* is common and found throughout Pennsylvania (fig. 6.30). A survey of bat abundance and activity in south-central Pennsylvania by Hart et al. (1993) found that red bats were detected in 60 percent of all localities sampled and on 54 percent of all survey nights. They were more abundant than hoary bats but not as much as *Myotis* species.

Although this species is frequently encountered during surveys, no effort to evaluate population viability has been undertaken. Little is known about its population size or trends. In addition, given the cryptic and solitary nature of the bat, observations are primarily based on captures using mist-nets or on early evening emergence. Whitaker et al. (2002) examined the number of bats submitted to the Rabies Lab of the Indiana Department of Health and have suggested that the populations of many bat species, including the red bat, have declined since 1980. A similar study from bat submissions to the Arkansas Health Department Rabies Lab from 1983 to 1998 also suggests that there has been a significant decline in red bat numbers, and this may indicate a declining trend in population levels (Carter et al. 2003b).

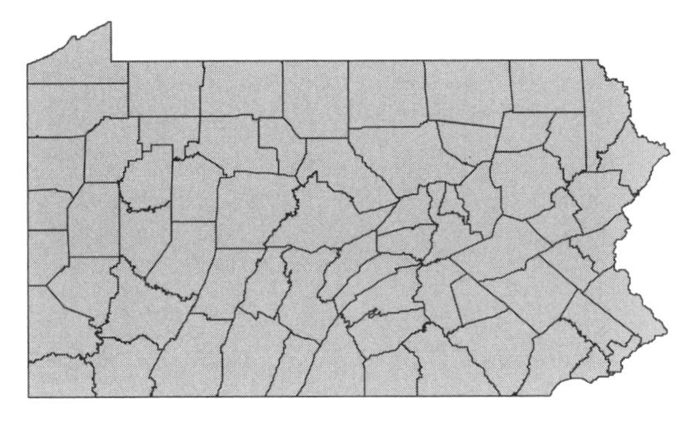

*Fig. 6.30.* Distribution of the Eastern Red Bat, *Lasiurus borealis.*

Historical anecdotes from more than 100 years ago describe large flocks of red bats migrating during the day in the Hudson Highlands of New York (Mearns 1898). Similar observations have been made along the East Coast in Cape Cod (Miller 1897) and in Washington, D.C. (Howell 1908). The absence of any such recent sightings may indicate that population levels have decreased in the twentieth century.

### *COMMUNITY TYPE/HABITAT USE*

*Lasiurus borealis* is known to roost in a variety of deciduous trees, often blending in with the dry foliage. In Iowa, they have been found along forest edges, fields, orchards, and hedgerows; the American elm is a favored roost tree (McClure 1942, Constantine 1966). Bats appear to choose roosts that are concealed on all sides except from the bottom, thus affording protection from predators (Barbour and Davis 1969). These roost sites also have few obstructions below them so that the bat can quickly drop out of the roost to avoid predators. Red bats are known to roost in caves (Myers 1960), in woodpecker cavities (Fassler 1975), on trunks of trees, in leaf litter, on Spanish moss (Constantine 1966), and even under sunflower leaves (Downes 1964). Red bats have been observed roosting in juniper and oak trees during late fall in Missouri and even overwintering in leaf litter at temperatures below 8°–10°C (Robbins et al. 2004).

Hardwoods appear to be the preferred trees for red bats in Georgia and South Carolina with sweetgum, blackgum, white oak, laurel-leaved oak, and water oaks as the most common (Menzel et al. 1998). The average diameter at breast height (37.75 cm) and the average height (24.85 m) of these roost trees were significantly higher than that of the entire sample plot. Menzel et al. (1998) also found that red bats roost only infrequently in intensively managed stands and did not use loblolly pines as often as expected. However, this is in contrast to a study by Miller (2003), who found red bats to be common in managed loblolly pine systems in Mississippi.

Red bats emerge early in the evening (Caire et al. 1988), often before dark. The foraging area of red bats average around 453 ha and can range from 113 to 925 ha, with radio-tracked bats observed to fly 1.2–7.4 km from the day roost (Carter 1998, Hutchinson and Lacki 1999). Red bats typically forage on moths (Acharya and Fenton 1992, Carter et al. 2003a, Whitaker 2004) and are also known to consume plant hoppers, leafhoppers, crickets, flies, mosquitoes, true bugs, beetles, cicadas, and other insects. They usually feed alone but have been known to congregate at lights (Hickey and Fenton 1990) and cornfields, foraging on emerging moths. Their foraging is primarily above the canopy, and they rarely venture into the trees to feed (Shump and Shump 1982). They often feed around streetlights and may be attracted by the echolocation calls of conspecifics (Balcombe and Fenton 1988, Obrist 1995, Fenton and Reddy 2003). Red bats typically catch prey in flight, though a diet of crickets suggests that it may glean insects from the ground as well. Blue jays are considered an important predator, especially of young bats, along with hawks, owls, opossums, and skunks (Shump and Shump 1982).

The Pennsylvania Gap Analysis Project (GAP) for red bat habitat in Pennsylvania suggests that the primary habitats for these bats are mixed forest, coniferous forest, woody scrubland, perennial herbaceous, and low-density urban areas (Myers and Bishop 2000). Secondary habitats include broadleaf forest, mixed forest, annual herbaceous, palustrine herb, palustrine woody, open water, and urban high-density areas. Riparian habitats were not evaluated in this analysis and the prevalence of low-density urban areas as a strong predictor in the model results in areas around cities being considered primary habitat. Other models of habitat use show similar correlations between red bat activity and urban indices (Gehrt and Chelsvig 2004). A more refined GAP analysis is needed for this species before the habitat conditions can be thoroughly evaluated.

Red bats mate in the fall between August and October, store sperm over winter, and after delayed fertilization, become pregnant in the spring (Shump and Shump 1982). Following an eighty- to ninety-day gestation period (Jackson 1961), they give birth to one to five (average 2.3) young in early summer (Constantine 1966). Young are born with their eyes closed, are hairless, and can weigh less than 0.5 g. They are carried from one roost to another by the mother and typically cling to a branch or leaf at night when the mother is foraging (Johnson 1932). The young are capable of flight within three to four weeks and are weaned by the time they are six weeks old (Barbour and Davis 1969).

Red bats have been seen congregating around the mouth of caves during the fall and sometimes even enter the caves. It is unknown why this occurs, though Barbour and Davis (1969) speculate that it may be related to mating because it coincides with the breeding season. Red bats appear to raise young as solitary individuals and not in maternity colonies. Although they do not form nursery colonies, Jones et al. (1967) found that summer captures in eastern Kansas were exclusively female and young, indicating differential habitat use among males and female.

*Lasiurus borealis* migrates south during the fall to the southeastern states (Cryan 2003) and typically absent from Pennsylvania during winter. Males and females are believed to migrate at different times and may have different summer and winter ranges as well (Grinnell 1918, Williams and Findley 1979). Migration appears to be the only time during the year when this typically solitary species forms large flocks. These large groups are also known to experience high mortality rates. Barbour and Davis (1969) describe thousands of red bats dying from collisions with the Empire State Building and television towers. More recently, high levels of mortality of migrating tree bats have been documented at wind power sites along ridge tops in West Virginia (Arnett 2005). Mormann et al. (2004) found bats in southern Missouri hibernating in trees when temperatures were above freezing, but they moved to leaf litter when temperatures fell below freezing. Moremann et al. (2004) suggest that these roosts provide a thermal advantage over trees in that they provide better insulation and are less susceptible to heat loss due to wind. Red bats are also known to arouse easily from hibernation and feed during warm days (Barbour and Davis 1969).

While red bats are not known to overwinter in Pennsylvania, a few have recently been documented in Lebanon County from November 2003 to March 2004, indicating that there might be a resident population during the winter. These bats were found on the roadside during a survey of bats killed along a ca. 5-mile stretch of highway 322 (C. Butchkoski, personal communication). Whether this is a chance occurrence or the bats are actually nonmigratory remains to be studied.

*THREATS*

The most apparent threat to red bat populations is the high mortality of tree bats at wind power sites (Johnson et al. 2003). During autumn 2004, studies at the Mountaineer wind power site in West Virginia and the Meyersdale site in Pennsylvania have found numerous dead bats (466 and 299, respectively) within a span of just six weeks (Williams 2004, Arnett 2005). Red bats accounted for roughly a quarter of all bats killed at both sites, with only the hoary bat (*L. cinereus*) having more fatalities (34 percent and 46 percent, respectively). Many of these bats may be migrants from the Pennsylvania region or from farther north, and given that the number of wind power sites in West Virginia and Pennsylvania is rapidly increasing, the long-term effect on population levels may be quite severe. It is not known why tree bats have such high mortality at these sites, nor do we have sufficient information about their migratory paths or behavior during migration to address these questions adequately; more research is necessary.

Carter et al. (2003b) and Mormann et al. (2004) describe the importance of leaf litter hibernation for red bats. Although it is not known how prevalent this behavior is among red bats, these authors suggest that this may put them at risk from wildfires and especially prescribed burns, which are often conducted in the winter. More research on the use of leaf litter in deciduous versus coniferous forests as hibernating sites is necessary to determine whether prescribed burns during winter are likely to have a detrimental effect on hibernating bats. Even though red bats are unlikely to use leaf litter hibernation in the state, bats from Pennsylvania that migrate south may be affected by prescribed fires in their wintering grounds, which may influence future population levels in the state.

*CONSERVATION AND MANAGEMENT NEEDS*

Given the high mortalities at wind power sites during the fall migratory season, these bats may face a severe population decline in coming years, especially as the abundance of wind farms increases and their placement continues to be in areas of high bat activity, such as ridge tops. Reliable pre- and post-construction monitoring of mortality rates at all major wind energy sites need to established. In addition, research needs to be undertaken to develop deterrents that would reduce bat activity around turbines, and to construct predictive models to determine sites of high bat activity, where turbines should not be located. If the effect of wind power on *L. borealis* mortality is as severe as the preliminary data suggest, then the time is right to consider creating a detailed conservation and management plan for this species in Pennsylvania.

Currently, little information is available on local and regional migration patterns of red bats. To better understand the effect of wind farm siting on migrating bats, we need more accurate information of flyways and habitats used during migration. This may be accomplished using region-wide telemetry of bats and mark-recapture studies to identify maternity sites and estimate home range.

*MONITORING AND RESEARCH NEEDS*

Additional research is needed to determine whether the presence of red bats in Lebanon County during the winter is a regular occurrence and whether they are present in other counties as well. Given that these bats were found dead along the roadside suggests that they are active in this area during the winter and may not be hibernating during warm periods.

The current GAP analysis of the red bat identifies urban and semiurban sites as primary habitats. While these bats are often encountered along the edge of an urban-forest transition, the predominance of this land-use type in the GAP analysis is disconcerting. The addition of more information about habitat use from the monitoring program and telemetry projects mentioned earlier should help to refine this model.

Although red bats appear to be abundant and common throughout the state, given the high levels of mortality associated with wind turbines, it is imperative that an in-depth, statewide population monitoring program be established to examine trends in abundance and activity levels.

*Author:* SHAHROUKH MISTRY,
WESTMINSTER COLLEGE

## Hoary Bat

Order: Chiroptera
Family: Vespertilionidae
*Lasiurus cinereus*

The hoary bat (*Lasiurus cinereus*) is the largest (20–35 g, 38–40 cm wingspan) bat found in eastern North America, and size alone is the single best characteristic for distinguishing the hoary bat from all other bats in Pennsylvania (fig. 6.31). Its yellowish brown to dark mahogany brown fur is frosted with silver over the entire body, giving the bat a pronounced "hoary" color, that is, white, gray, or grayish. The hoary bat was

*Fig. 6.31.* The Hoary Bat, *Lasiurus cinereus*. Photo courtesy of Cal Butchkoski.

identified as a Species of Concern because it is a habitat specialist and highly susceptible to habitat modifications, but little data are available for adequate assessment. This species has a global rank of Secure (G5), indicating the species is demonstrably widespread, abundant, and Secure (NatureServe 2005).

### *GEOGRAPHIC RANGE*

Both sexes of the hoary bat have been found throughout Pennsylvania, primarily between May and October. *Lasiurus cinereus* is the most widespread of American bats, ranging from near the limit of trees in Canada south to Argentina and Chile (Cryan 2003). They are also found in Hawaii, and there are records of presumed wayward migrant individuals from Southampton Island in northern Canada, Iceland, Bermuda, Hispaniola, and the Orkney Islands off Scotland (Shump and Shump 1982). They are common in the Great Plains and throughout the Pacific Northwest, and less abundant in the eastern United States and northern Rockies. Individuals summer in North America, winter in California, the southeastern United States, Mexico, and Guatemala, but a few individuals have been found in Michigan, New York, Connecticut, and Indiana during December and January (Shump and Shump 1982, Whitaker and Hamilton 1998, Cryan 2003). Spring migrants found in the southwestern United States seem to be hoary bats moving east from California, and this is the most reasonable explanation for lack of records of females from Pacific coastal regions during spring and summer (Cryan 2003). In summer, *L. cinereus* is found in many regions of the United States and southern Canada but is uncommon east of the Mississippi and south of the Ohio River (Cryan 2003). During summer, the sexes are segregated, with males primarily in mountainous western regions and females occurring in more eastern regions of North America, but the Great Plains are an area where both sexes commonly occur. During late summer, migration favors movements toward the Pacific Coast rather than movement toward the Atlantic Coast. Most eastern bats move toward the Southwest, rather than heading south across the Gulf of Mexico. A few records indicate hoary bats migrate south along the Atlantic Coast through northern Florida and then across the Gulf of Mexico to wintering grounds, but records are limited and exchange between continents seems unlikely (Cryan 2003).

### *DISTRIBUTION AND RELATIVE ABUNDANCE IN PENNSYLVANIA*

Bats of the genus *Lasiurus* are present in Pennsylvania only from early May until early October except for rare and unusual occurrences during winter months (Barbour and Davis 1969, Merritt 1987). *Lasiurus cinereus* abundance in the state is not known. Significant surveying and monitoring of this species has not been accomplished. No summer roosts have been identified in Pennsylvania. During the summer months, *L. cinereus* has been found at localities (fig. 6.32) that span the entire state (Pennsylvania Game Commission 2003). Monitoring bats using a remote ultrasonic detector technique Hart et al. (1993) found *L. cinereus* in eight (wooded stream, mixed stream, agriculture land, forest clear-cut, lake, forest, old field, pond habitats) of nine habitats studied in south-central Pennsylvania. It was not found in open stream habitat.

### *COMMUNITY TYPE/HABITAT USE*

*Lasiurus cinereus* is a solitary bat that roosts primarily among foliage of coniferous and deciduous trees near the forest edge. By day, hoary bats generally roost 3–5 m aboveground in trees where they are covered and invisible from above but visible from below. Unusual roosts include woodpecker holes, gray squir-

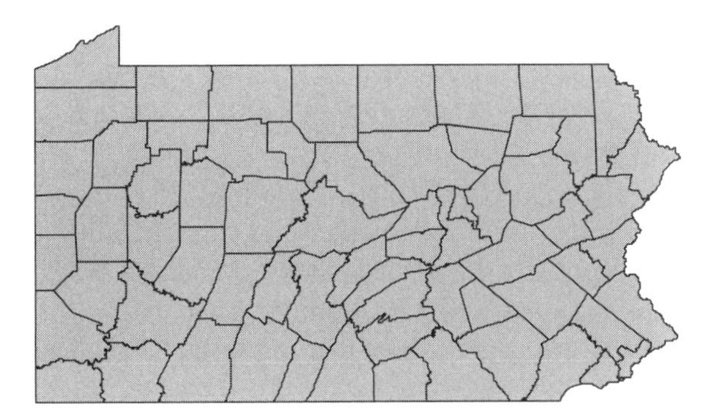

*Fig. 6.32.* Distribution of the Hoary Bat, *Lasiurus cinereus.*

rel nests, sides of buildings, and under a driftwood plank (Shump and Shump 1982). These bats are most commonly observed flying relatively high above the ground, near treetop level, and in open or edge situations (e.g., around street or building lights on beaches or north sides of a ridge), where it can fly in a straight line (Barclay 1985). *Lasiurus cinereus* appears to be a solitary forager that establishes feeding territories, and this use of habitat was most easily observed around lights (Barclay 1985).

### LIFE HISTORY AND ECOLOGY

*Lasiurus cinereus* is a solitary, tree-roosting species that has commonly been found in trees near clearing edges. These bats often fly in open areas, along treetops, and woodland streams and ponds, but are also often seen foraging around building and streetlights. Individuals foraging back and forth along 40- to 50-m routes would chase other bats that entered the area. While chasing other bats, territorial defense included emitting audible vocalizations and occasionally coming in contact with them (Barclay 1985). The sexes are separated during the summer.

*Lasiurus cinereus* emerges late in the evening to forage. These bats usually forage solitarily, but groups of females have been observed foraging together (Shump and Shump 1982). *Lasiurus cinereus* has one of the highest wing loadings and aspect ratios of any bat species in North America (Farney and Flaherty 1969) and is adapted for fast, unmaneuverable flight, as evidenced by their direct, swift (21.3 km per h) flight behavior (Hayward and Davis 1964). They employ single harmonic search-approach calls that are low (20–17 kHz) constant frequency signals designed for long-range target detection of prey in open-air situations (Barclay 1986). They feed consistently on large insects, including moths, beetles, grasshoppers, termites, dragonflies, and wasps (Shump and Shump 1982, Barclay 1985). The low, constant frequency design of their calls means that small prey is not detectable at close range and is difficult for these fast-flying bats to catch (Barclay 1986). The hoary bat is quite aggressive, as suggested by their establishment of foraging territories that they defend. When handled, it employs a threatening posture. It opens its mouth widely, spreads its wings rigidly, and emits a hissing or clicking noise. The only time the hoary bat has been seen in association with other bats (only Vespertilionids) was while foraging (Mumford 1969). Predators include hawks, owls, American kestrels, and a rat snake (Shump and Shump 1982). The only reported ectoparasites have been mites; endoparasites include helminthes and protozoa (Shump and Shump 1982, Whitaker and Hamilton 1998). *Lasiurus cinereus* contracts rabies and the incidence was reported as high (up to 25 percent), but this rate is questionable because in this species' "suspect bats" are those submitted for testing (Whitaker and Hamilton 1998). Rabies records from the Pennsylvania Department of Public Health (1992 to 2000) indicated that only 3.9 percent of bats suspected of having rabies actually tested positive and that most positives (41 of 48) were from a single species, and all suspect bats of the other ten species combined (this group included *L. cinereus*) had a positive rate of 0.57 percent (Olnhausen and Gannon 2004). Hoary bats make a pronounced southerly and westerly migration in North America in the fall and a northerly and eastern migration in the spring, and at these times are occasionally observed in migratory flocks. Females are residents of the boreal region during the summer, but in the east, they produce young as far south as Indiana (Whitaker and Hamilton 1998).

Copulation in *L. cinereus* probably occurs during autumn migration or on the wintering grounds. An overwinter delay in ovulation results in fertilization occurring in spring and parturition ranges from mid-May through July (Shump and Shump 1982, Cryan 2003). The usual number of a litter is two but ranges from one to four (Shump and Shump 1982). Newborn bats are covered with fine, silvery-gray hair, except on the venter, which is naked. The forearm lengths of newborn ranged from 16 to 20 mm (Munyer 1967). After birth, young cling to the mother by day but usually are left roosting (clinging to a twig, branch, or in a crevice) while the mother forages. The ears and eyes are closed at birth but open on day 3 and day 12, respectively. By day 33, purposeful flight occurs (Munyer 1967, Bogan 1972). In central Iowa, lactating hoary bats were caught from June 21 to July 22, volant young were taken beginning July 22, young males outnumbered females 5:1, and no females were captured in August (Kunz 1971).

*THREATS*

The major threats to these animals are the declining number and quality of suitable forest and aquatic habitats as a result of human activity and diseases. For example, woodlands are jeopardized in Pennsylvania by disease, such as the nonnative hemlock woolly adelgid (*Adelges tsugae*) that is causing mortality in eastern hemlock trees. Also, a great threat to this species is the loss of aquatic habitats associated with decreasing area of forested habitats. Another major threat is the lack of available information on specific habitat needs and habitat uses for this species in Pennsylvania.

*CONSERVATION AND MANAGEMENT NEEDS*

The primary conservation issue is the lack of knowledge and loss of habitat required for roosting in and migrating through Pennsylvania. Future efforts should be directed at increasing the database on summer habits and maintaining viable tracts of forested/aquatic habitats to sustain and increase summer roosting and spring and fall migrating populations of *L. cinereus* in Pennsylvania. Surveys are needed to determine the actual number of bats roosting and migrating in forest/aquatic habitats at benchmark sites.

*MONITORING AND RESEARCH NEEDS*

Efforts should be invested in a monitoring program that will survey forested regions throughout the state. Mist-netting is essential because sex and reproductive status cannot be determined from a bat detector. A regular monitoring program across the state should identify precise habitats and space use, particularly migratory and roosting areas and sympatric tolerances. Roost trees should be identified. All captured hoary bats should be assessed for age, reproductive status, parasites, and health. Marking with bands or pit tags and resighting/recapture would be most useful for roosting bats and might prove beneficial for regularly monitored migration routes. Bat detector surveys are of use only if care is taken to assure adequate spatial and temporal replication (O'Shea and Bogan 2003). Radiotelemetry studies that are coupled with multiyear demographic studies, preferably at suitable benchmark sites, will be useful for determining roosting sites and home ranges. This will increase the knowledge base for protecting and managing suitable roosting habitat for this species. Additional information is needed on the potential effects of forest decline on this species. Monitoring should be conducted at sites where *L. cinereus* has been found to occur to determine if their numbers are increasing, decreasing, or staying the same. Sites should be monitored for habitat quality on a short- and long-term basis to correlate with hoary bat occurrences. Immediate protection of forests, including the cessation of logging, should be enforced at all locations of known roosts and migratory routes.

In Pennsylvania, the most pressing research needs are to understand and accumulate data on roosting habitat, habitat space use, and migratory pathways. Regular statewide monitoring for *L. cinereus* should be a priority. This would best be accomplished by using mist-netting and bat detector surveys, particularly during the summer to identify roost trees, and during the spring and fall to determine preferred migratory pathways. Once summer roosts have been identified, an intensive radiotelemetry study would help elucidate habitat use. A statewide mist-netting program will provide demographic and parasitic data, and coupled with long-term radiotelemetry studies, will identify preferred habitat, roost sites, and roost site characteristics requiring conservation and management. It is possible to determine distances traveled by migrating hoary bats by applying stable hydrogen isotope analyses from hair of captured bats, to elucidate where hoary bats migrating through Pennsylvania come from and where they are going (Cryan et al. 2004).

*Author:* GARY KWIECINSKI, UNIVERSITY OF SCRANTON

## Fisher

Order: Carnivora
Family: Mustelidae
*Martes pennanti*

The fisher is a swift tree-climbing member of the weasel family (Mustelidae). Structurally, the fisher resembles a large weasel (fig. 6.33). The body is elongate and the limbs are short and powerful (Powell 1993). Fishers are sexually dimorphic. Adult males weigh 3.5–5.5 kg (7–12 pounds), and adult females weigh 2.0–2.5 kg (4–6 pounds). Fishers were extirpated from Pennsylvania by the late nineteenth century but reintroduced in northern Pennsylvania from 1994 to 1998. The fisher was selected as a Pennsylvania Species of Greatest Conservation Need because of the need

*Fig. 6.33.* The Fisher, *Martes pennanti*. Photo courtesy of Josh More.

to monitor populations, careful evaluation of timber harvesting activities in their habitat, and concerns regarding unregulated trapping in the commonwealth. The species' global rank is Secure (G5, NatureServe 2009).

### *GEOGRAPHIC RANGE*

Historical records indicate that fishers occurred throughout forested regions of northern North America, with their distribution extending southward along the Appalachian Mountains and adjacent Appalachian Plateau into North Carolina and Tennessee, the Sierra Nevada Mountains in California, and the Rocky Mountains possibly into Utah (Hagmeier 1956, Gibilisco 1994). Unregulated trapping and extensive timbering in the late 1800s and early 1900s caused fisher populations to become extirpated in Pennsylvania and throughout much of their southern and midwestern distribution (Powell 1993). Remnant populations persisted in northern New England, the northern Great Lakes region, portions of the northern Rockies, and Pacific Northwest (Powell 1993). Improvements in harvest and timber management have enabled fishers to become reestablished in portions of the eastern and upper midwestern United States through natural recolonization and implementation of reintroduction projects. Fishers in the Pacific states show greater habitat specialization and are closely associated with low- to mid-elevation forest with coniferous and older-growth components. Recovery in the Pacific states has been limited compared with eastern and upper midwestern populations; the fisher is considered extirpated in Washington and persists only as small, isolated populations in Oregon and California (Gibilisco 1994, Aubry and Lewis 2003).

### *DISTRIBUTION AND RELATIVE ABUNDANCE IN PENNSYLVANIA*

The fisher was reintroduced in northern Pennsylvania from 1994 to 1998 (fig. 6.34). This effort was made possible, in part, because of improvements in forest management practices, which have resulted in recovery of mature forest in many areas of the state. Reliable evidence of the occurrence of fishers, including incidental capture, recovery of dead or injured fishers, or sightings by reintroduction program personnel, has been gathered throughout many counties in the northern tier of the state, including Cambria, Cameron, Centre, Clarion, Clinton, Columbia, Forest, Elk, Erie, Luzerne, Lycoming, McKean, Potter, Sullivan, Tioga, Warren, and Wyoming (Serfass et al. 2001). Similar types of evidence demonstrate the occurrence of fishers in portions of southwestern Pennsylvania. The occurrence of fishers in this region is presumably from the northward expansion of a population reintroduced in West Virginia in 1969 (Pack and Cromer 1981). There is potential for the fisher to make a significant recovery in Pennsylvania.

Reestablished fisher populations in Pennsylvania have symbolic and ecological importance. The Pennsylvania Fisher Reintroduction Project is the most ambitious effort ever initiated to restore extirpated fisher populations (Proulx et al. 2004). This multiorganizational effort symbolizes improvements that have taken place in wildlife and forest resource management and in citizens' attitudes toward predator conservation. The fisher's charismatic nature and dependence on mature forest habitat make the species ideally suited to highlight the importance of scientifically based forest conservation in Pennsylvania and elsewhere. Reestablished populations in Pennsylvania also play a significant ecological role because they increase the likelihood of genetic interchange among populations north of Pennsylvania and the smaller, reintroduced population to the south (primarily in West Virginia and western Maryland; Proulx et al. 2004).

### *COMMUNITY TYPE/HABITAT USE*

There is no estimate for total population size, but the status of fishers has improved in many areas of North America since the early 1900s (Proulx et al. 2004). Generally, the fisher is considered widespread and secure (G5; NatureServe 2009). However, large-scale conversion of forest to agriculture and other land uses in several midwestern states (e.g., Illinois, Indiana, and Ohio) makes the reestablishment of fishers unlikely in these

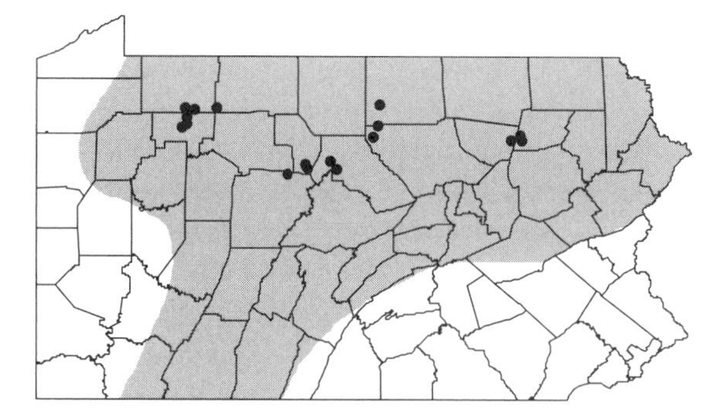

*Fig. 6.34.* Distribution of the Fisher, *Martes pennanti*. Shading = current distribution, Circles = release sites.

areas. Also, the fisher is doing poorly in several western states, including Washington, Oregon, and California, where populations are relatively small and isolated.

Post-release monitoring of reintroduced fishers demonstrated persistence and reproduction in Pennsylvania. Generating indices of abundance will require longer-term standardized monitoring efforts. Although appropriate to consider the status of the fisher improved and recovering, it would be premature to suggest a reliable estimate of abundance or to conclude that its long-term fate is secure. The fisher is listed as a state Species of Special Concern and characterized by NatureServe (2009) as Imperiled/Apparently Secure in Pennsylvania (S2S4).

Fishers once were widespread in forested regions in Pennsylvania (Genoways and Brenner 1985), but viable populations were eliminated by 1900. Fishers trapped in Clinton and Lancaster counties in 1901 and 1921 (Hagmeier 1956), respectively, were the only verified occurrences in Pennsylvania in the 1900s before a reintroduction project in 1994 (Serfass et al. 2001). A questionnaire survey designed to update the status of fishers in Pennsylvania was distributed among wildlife conservation officers in 1994. Results indicated that no viable fisher population occurred in the state (Serfass et al. 2001). From 1994 to 1998, 190 fishers were reintroduced among six sites in north-central and northwestern Pennsylvania. These sites were Fish Dam Wild Area, Quehanna Wild Area, Pine Creek Valley, Sullivan and Wyoming counties, Allegheny National Forest (ANF), and Kettle Creek Valley. Monitoring efforts conducted through 2000 demonstrated persistence and reproduction among reintroduced fishers throughout the northern tier of the state (Serfass et al. 2001, Peters 2002). The Pennsylvania Game Commission (PGC) monitors fishers through information derived from their wildlife conservation officers and reports that reintroduced fisher populations are expanded in northern regions of the state (Lovallo and Hardisky 2008).

Large tracts of contiguous forest are the primary descriptor of fisher habitat (Kelly 1977, Allen 1983, Douglas and Strickland 1987, Powell 1993). Coniferous forest has been considered an essential habitat component; however, fisher populations also occur in deciduous and mixed forested habitats (Kelly 1977). Fisher populations have recolonized or have been reintroduced successfully in mid-successional second-growth mixed deciduous woodlands (Pack and Cromer 1981, Roy 1991). Fishers generally avoid habitat with <50 percent canopy closure, but fishers can traverse and forage in nonforested and transitional habitat types (Arthur et al. 1989). Nonetheless, forest management strategies that include extensive clear-cutting will adversely affect fisher populations. Clear-cutting and similar forms of even-aged forest management practices fragment contiguous forest, thereby contributing to the reduction of critical components of the fisher's habitat, including large standing or downed trees, snags that are used as denning and resting sites, and dense, coarse woody debris (Allen 1983).

All known maternal dens of fishers occur in standing trees (Paragi 1990, Powell 1993, Gilbert et al. 1997). Powell (1993) reported that median diameter at breast height of den trees was 45 cm and median den entrance height was 6.3 m. Parturition and rearing of young occur in March and April, and females are known to use one to three dens per litter, so large standing live trees and snags can be considered a key seasonal component of fisher habitat. Fishers use a variety of resting sites, depending on food availability and weather. Resting sites are usually associated with dead and down woody material or coarse woody debris (Gilbert et al. 1997), which provides important shelter for thermoregulation (Powell 1993).

Fishers occupy a relatively large home range, are territorial, and, thus, occur at low population densities and require large, forested areas of appropriate habitat to maintain viable populations. Consequently, the size and distribution of forested habitats suitable for sustaining fishers should be considered at the landscape level. The landscape of Pennsylvania comprises more than17 million acres of publicly and privately owned forests (DCNR, Bureau of Forestry 2007). Because of the potential for various forms of development on privately owned forests, public lands should be regarded as offering the best potential for long-term maintenance of habitats suitable for fishers. Public lands in Pennsylvania are managed under various jurisdictions, including State Parks and Forests (Pennsylvania Department of Natural Resources), State Game Lands (PGC), and the ANF (U.S. Department of Agriculture) and include a substantial portion of the state's forested landscape (e.g., approximately 2.1 million acres of forest are managed as State Forests [DCNR, Bureau of Forestry 2007]), with the largest concentrations in northern Pennsylvania. The interconnectivity of public lands in northern Pennsylvania contributes to a generally contiguous forested landscape and should be regarded as composing primary habitat for fishers. Also, public lands at five primary areas in northern Pennsylvania were the focus of a successful program to reintroduce fishers. There is substantial evidence that the fisher population reintroduced in West Virginia has expanded through western Maryland into southwestern Pennsylvania. State forest and State park lands in portions of Fayette and Somerset counties in southwest Pennsylvania represent important areas that will retain permanent forest habitat for fishers. Forested ridges in portions of the Appalachian Plateau and Valley and Ridge Provinces provide habitat and important dispersal corridors for fishers.

Key habitat on public land in Pennsylvania is relatively intact. At this time, Pennsylvania generally provides high-quality habitat within the core of the fisher's historic northern distribution. Forest management practices that emphasize clear-cutting, diameter-limit cutting (i.e., high grading), or salvage harvesting reduce habitat quality by increasing fragmentation and by removing large trees and coarse woody debris, which are important for providing maternal dens and thermal cover, respectively. Public land should be considered critical for sustaining fisher populations because the habitat will not be compromised by development in the foreseeable future. Also, there are various examples of progressive forest management activities occurring on public land in Pennsylvania. For example, the Pennsylvania State Forest Resource Management Plan (DCNR, Bureau of Forestry 2007) stresses the importance of ecosystem management throughout the 2.1-million-acre state forest system. Many aspects of the plan, including maintenance and expansion of older-growth conditions (e.g., natural and wild areas), protection of riparian forests, and protection and enhancement of forested corridors all contribute positively to the conservation of fishers. Similarly, the ANF has implemented a plan to foster the development of an older-growth corridor system connecting existing wilderness areas (Nelson et al. 1997). Any comparable forest conservation activities providing for the development of mature and older-growth forest conditions that include coniferous and mixed components should be encouraged. Many ridges in portions of the Appalachian Plateau and Valley and Ridge Provinces are predominantly forested. However, the long-term potential for maintaining forest connectivity along these ridges is uncertain because substantial portions are privately owned.

### *LIFE HISTORY AND ECOLOGY*

Fisher populations exist at low density (one fisher per 2.6–11.7 km$^2$ [1.0–4.5 mi$^2$]). Habitat, prey base, and season influence local population density (Arthur et al. 1989). Outcomes from various telemetry studies demonstrated that home range size for males varied between 19 and 79 km$^2$ (7.3–30.5 mi$^2$; average = 38 km$^2$ [14.7 mi$^2$]) and for females between 4 km$^2$ and 32 km$^2$ (1.5–12.4 mi$^2$; average = 15 km$^2$ [5.8 mi$^2$]; Powell 1993). Fishers have been reported to disperse relatively short distances. In Maine, average natal dispersal of males and females was 10.8 km and 11.2 km (6.7 and 7.0 mi), respectively (Arthur et al. 1993). However, Aubry et al. (2004) provided evidence of male-biased dispersal and female philopatry in Oregon. Low population density, large home range, relatively short natal dispersal distance (especially by females), and a general preference for mature-forest habitat are factors that render fisher populations susceptible to extirpation, hinder natural reestablishment of extirpated populations, and limit interchange among isolated populations.

Female fishers are capable of breeding at one year of age, but typically become reproductively active in their second year (Eadie and Hamilton 1958, Douglas and Strickland 1987). Similarly, males generally do not achieve sexual maturity until their second year (Douglas and Strickland 1987). Fishers have obligate delayed implantation, generally giving birth and mating in late March and early April, respectively. Females achieve estrus two to eight days after parturition and breed during this time. Total gestation is 327 to 360 days, but the embryo implants in February so active pregnancy is a brief period (thirty to sixty days) in late winter (Powell 1993). Parturition may occur slightly earlier in southern latitudes (Leonard 1986). Typical litters are composed of two to three kits (range = 1–6; Powell 1993). Young are weaned when five to eight weeks old and generally disperse at nine to twelve months old (Arthur et al. 1993, Frost and Krohn 2004).

The fisher is a solitary, opportunistic predator. Prey abundance and susceptibility to capture largely influence the composition of the fisher's diet (Powell 1993). The diet varies regionally, but studies demonstrate that small mammals, including mice, voles, shrews, squirrels, snowshoe hare (*Lepus americanus*), and porcupine (*Erethizon dorsatum*), occur most commonly (deVos 1952, Coulter 1966, Clem 1977, Raine 1987, Powell 1993). Fishers also frequently scavenge carcasses of ungulates such as deer, elk (*Cervus elaphus*), and moose (*Alces americanus*) (Coulter 1966, Kelly 1977). Sources of ungulate carrion include natural mortality, road kills, deer parts discarded by hunters, or trap bait. The importance of vegetation in the diet varies seasonally and among regions (Arthur et al. 1989, Zielinski et al. 1999).

### *THREATS*

The fisher's distribution was reduced considerably in the 1800s because of extensive timber harvesting accompanied by unregulated trapping (Powell 1993). Presently, the immediacy of these threats is low in Pennsylvania, but several considerations warrant careful evaluation of trapping as a threat to recovery in Pennsylvania. Fishers exhibit low fecundity, short dispersal distance (especially by females), and are easily trapped (Powell 1993). Substantial data exist demonstrating that overtrapping was a primary initial cause of decline in fisher populations. Dixon (1925) and Scheffer (1992) showed that harvest-related mortality and associated injury, demographic modification, and isolation of fisher populations caused considerable declines before extensive timber harvest reduced habitat. These effects may be sufficient to hinder recovery and establishment of extirpated forest carnivore populations. Currently, key parameters characterizing variation in demography and population density among forest types, or models, predicting potential distribution have not been established for recovering populations in Pennsylvania. These data are necessary before implementing any management strategy that influences population structure and distribution (see "Monitoring and Research Needs"). Other factors that have been considered to be threats to fishers include fragmentation of forest habitat and mortality from collisions with vehicles (Aubry and Lewis 2003). Predation and disease do not appear to be significant threats to fisher populations.

### *CONSERVATION AND MANAGEMENT NEEDS*

A management plan is currently being developed by the PGC's Furbearer and Farmland Wildlife Sections. Demographic and spatial data on fishers in Pennsylvania are needed before developing and implementing specific management plans. A standardized monitoring protocol (see "Monitoring and Research Needs") is critical for assessing the long-term status of the recovering fisher population and should be a conservation priority (Zielinski and Stauffer 1996, Serfass et al. 2001, Peters 2002). Monitoring efforts should be designed to enhance understanding of the influence of forest age and composition, forest management practices, and various human perturbations on the distribution and density of fisher populations. Fishers are easily trapped and, therefore, vulnerable to being caught in traps set for legal furbearers (Powell 1993, Aubry and Lewis 2003). Incidental captures should be monitored closely to assess the effect on recovering populations; these events also provide an opportunity to gain additional information about the persistence and distribution of fishers in the state (Serfass et al. 2001). Compilations of sightings and other forms of miscellaneous information should be an ongoing activity among agencies responsible for the conservation of fishers and associated forest habitat. This information should be available to managing agencies, members of the Pennsylvania Mammal Technical Committee, and others with a professional interest in the conservation of fishers. An important aspect of the Pennsylvania Fisher Reintroduction Project was an initiative to educate and inform Pennsylvania's citizens about the ecological role of fishers in forested environments. Continued efforts to inform the public about the fisher's natural history and conservation in Pennsylvania, and to foster ongoing support for related forest conservation issues should be conservation priorities. Promoting sustainable, conservation-based management of privately owned forests also should be regarded as essential for maintaining viable fisher populations. Programs such as the Forest Stewardship Program (USDA, Forest Service 2009), which promote sustainable, ecologically sound use of private forest tracts, should be encouraged to benefit fishers and other forest wildlife.

### *MONITORING AND RESEARCH NEEDS*

The PGC documents the occurrence of fishers through various types of information provided by wildlife conservation officers, ranging from anecdotal (e.g., sightings reported by the public) to verified (e.g., identifying the carcass of a fisher killed along a highway; Lovallo and Hardisky 2008) but has not implemented geographically dispersed field-based assessments. Several techniques have been used to detect and monitor forest carnivore populations, including snow tracking and bait stations (Zielinski and Kucera 1995). Use of trained detection dogs to locate scat is emerging as an effective tool for noninvasive monitoring of species at risk. A critical element in monitoring low-density species such as the fisher is a thoroughly conceived design that includes considerations on data requirements and detection probability. An estimate of detection probability (i.e., from repeated visits to sampling units) can be used to determine necessary allocation of survey effort to reliably infer that fishers are absent in the sample area. Monitoring protocols should be accompanied by research on the ecology of fisher populations in Pennsylvania. An assessment of habitat use and associated demographic monitoring (i.e., habitat-specific differential fitness) must be conducted before it is possible to project the potential size and distribution that fisher populations can achieve in the state. Ongoing compilation and associated GIS-based mapping of point locations where fishers are documented to occur should be an ongoing activity of the PGC and should be available for review and interpretation by representatives from other agencies and organizations whose missions are related to the conservation of wildlife and forested landscapes in Pennsylvania (e.g., Pennsylvania Department of Conservation and Natural Resources [DCNR], Pennsylvania Mammal Technical Committee, and The Nature Conservancy).

The Pennsylvania Fisher Reintroduction Project was conceived, developed, and implemented by wildlife conservationists working through the School of Forest Resources at Pennsylvania State University, with subsequent involvement by the Department of Biology at Frostburg State University. Primary funding and other support for the reintroduction project were provided by the DCNR's Wild Resource Conservation Fund and Bureau of Forestry, the PGC, and the United States Department of Agriculture ANF. PGC is the primary agency responsible for wildlife management in the state. Currently, the fisher is listed as a state species of special concern and, as such, is technically under the jurisdiction of PGC's Wildlife Diversity Section, which is responsible for the conservation of nongame wildlife and species of special concern. However, although currently protected from legal harvest, the fisher also is classified as a furbearing animal. Consequently, if the fisher's status improves in the state, a trapping season could be initiated with conservation responsibilities formally shifting to the PGC's Furbearer and Farmland Wildlife Section. Decisions related to changes in the conservation status of the fisher should be made thorough objective and impartial deliberations conducted among representatives of the PGC's Wildlife Diversity and Furbearer and Farmland Wildlife Sections, DCNR, ANF, and university researchers responsible for coordinating the reintroduction effort. These deliberations should be facilitated through the Pennsylvania Mammal Technical Committee.

Monitoring spatial and demographic attributes of recovering fisher populations should be regarded as a primary research priority for the species in Pennsylvania. An objective field-based survey protocol that provides appropriate considerations of sample size, research design, and statistical power (i.e., the probability of detecting a change in the population) and calibrated to account for differing forest conditions throughout the state should be developed, implemented, and subjected to ongoing refinement (Dixon 1981, Kendall et al. 1992, Zielinski and Stauffer 1996). An important aspect of initial research should be to generate information on the current and potential distribution of fishers in Pennsylvania, population size, age and sex structure, fecundity, and mortality factors throughout representative portions of the forested landscape currently occupied by fishers in Pennsylvania.

Information on the ecology of reintroduced fisher populations in Pennsylvania would benefit the development and refinement of management and conservation strategies. Investigations addressing coarse-level habitat selection, microhabitat conditions associated with resting and denning sites, dispersal potential and success, with emphasis on constraints imposed by existing habitat conditions, and feeding habits should be implemented to complement Level-1 research needs.

*Author:* THOMAS SERFASS AND MATTHEW DZIALAK; FROSTBURG STATE UNIVERSITY, FROSTBURG

## North American River Otter

Order: Carnivora
Family: Mustelidae
*Lontra canadensis*

The North American river otter is an obligate wetland species with many unique attributes to facilitate a semiaquatic existence. The streamlined, muscular body with short, stocky legs, and dorsoventrally flattened tail, is superbly evolved for exploiting aquatic environments. The short legs, combined with fully webbed, pentadactyl paws, enhance the ability of river otters to make tight, rapid turns while pursuing aquatic prey (fig. 6.35). The river otter currently was chosen as a Species of Maintenance Concern because it was formally extirpated from the state but recently successfully reintroduced and now closely monitored. It is listed as Vulnerable in the state; its global rank is Secure.

### *GEOGRAPHIC RANGE*

Before European colonization, North American river otters occupied a wide variety of aquatic habitats in most major drainages in the continental United States, most of Canada, and Alaska (Polechla 1990). Unfortunately, unregulated fur trapping and various types of degradations to aquatic habitats initiated serious declines in many populations during the 1800s. By 1977, river otters were thought to occupy <75 percent of the historic range (Polechla 1990, Lariviere and Walton 1998, Melquist et al. 2003). Since the mid-1970s, twenty-one states have implemented reintroduction projects, which have contributed to the recovery of extirpated populations in many drainages (Melquist et al. 2003). River otters now are thought to have achieved at least some level of recovery in each of the continental United States, Alaska, and Canadian provinces, except for Prince Edward Island. Given their confinement to specific portions of the landscape (aquatic habitats) and relatively low populations densities (Melquist et al. 2003), river otters can generally best be described as uncommon in most areas of North America. The largest populations exist in coastal marshes associated with eastern and Gulf Coast states, particularly in Louisiana (Nilsson 1980, Melquist et al. 2003). Substantial populations also exist along portions of the Pacific Coast, from northern California through Alaska. The largest interior (noncoastal) populations occur in regions with an abundance of natural lakes and wetlands, such as states surrounding the Great Lakes and in portions of New England (Nilsson 1980, Melquist et al. 2003).

### *DISTRIBUTION AND RELATIVE ABUNDANCE IN PENNSYLVANIA*

As in other parts of the river otters' range in North America, intensive harvest and degradation to aquatic and riparian habitats were responsible for extirpation of populations in Pennsylvania (Rhoads 1903). By 1952, when the Pennsylvania Game Commission (PGC) prohibited legal trapping of river otters, populations had become limited to the Pocono Mountains region of northeastern Pennsylvania (Eveland 1978). In 1982, the Pennsylvania River Otter Reintroduction Project (PRORP) was initiated with the intent of restoring extirpated populations (Serfass et al. 1986, Serfass 1991, 1994). PRORP released 151 river otters in seven discrete drainage systems throughout the western and central portions of the state and has been successful in restoring extirpated populations (Serfass et al. 1999; Hubbard and Serfass 2005; fig. 6.36). Nonetheless, diverse aquatic habitats in northeastern Pennsylvania continue to support the state's largest population.

### *COMMUNITY TYPE/HABITAT USE*

Although North American river otters now occupy all major geographic regions of the state, northeastern Pennsylvania continues to retain the largest population. This population appears to be gradually expanding westward along upper reaches of the Susquehanna River drainage and southward along the Delaware River (Serfass et al. 1999). River otters reintroduced to

*Fig. 6.35.* The North American River Otter, *Lontra canadensis*. Image courtesy of Alan D. Wilson, naturespicsonline.com.

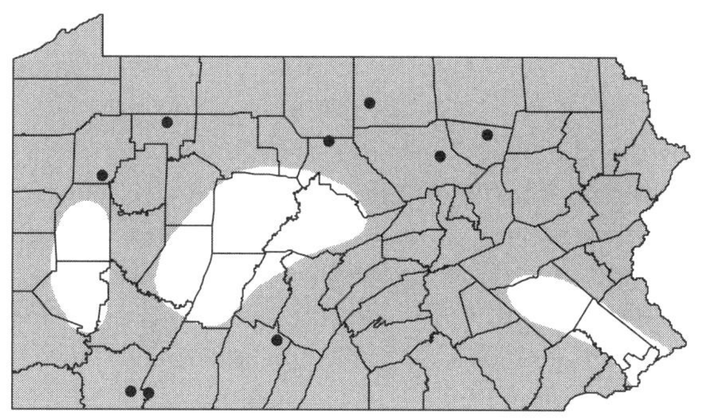

*Fig. 6.36.* Distribution of the North American River Otter, *Lontra canadensis*. Shading = current distribution, circles = release sites.

riverine habitats in north-central Pennsylvania from 1982 to 1986 have established populations and have expanded into connecting riverine systems, including the West Branch of the Susquehanna River (Serfass and Rymon 1985, Serfass et al. 1993, Serfass et al. 1999). Similar trends seem to be occurring among reintroduced populations in western and south-central Pennsylvania (Hubbard and Serfass 2005, Just 2007). Dispersal of individuals from the Chesapeake Bay in Maryland, and, possibly, from reintroduction sites in north-central Pennsylvania, appear to have resulted in repopulation of portions of the lower Susquehanna River, with river otters having been reported near the mouth of the Juniata River since the early 1990s (J. M. Benner, personal communication). Dispersal of river otters reintroduced in Ohio and New York are suspected to have contributed to the occurrence of individuals in northwestern Pennsylvania (Serfass et al. 1999). Similarly, river otters reintroduced in portions of western Maryland and West Virginia may be contributing to the expansion of populations in southwestern Pennsylvania (Serfass et al. 1999, Hubbard 2006). The PGC monitors accidental captures of river otters by trappers (mostly, but not exclusively, beaver trappers) through reports from wildlife conservation officers (WCOs). Since 1996, trappers pursuing other furbearers have annually caught greater than twenty-five river otters (fifty-seven were estimated to have been trapped during the 2007 to 2008 trapping season), mostly by beaver trappers in northeastern Pennsylvania (Lovallo and Hardisky 2008).

Through the successful reintroduction project and expansion of the remnant population from northeastern Pennsylvania and populations (both native and reintroduced) from adjacent states, the status of river otters has improved considerably during the past twenty years (Serfass et al. 1999, Hubbard 2006). However, many suitable habitats have yet to be pioneered or fully occupied by expansion of native or reintroduced populations. Unfortunately, degraded habitat conditions (e.g., riverine habitats polluted by acid mine drainage) limit or preclude reestablishment of river otters in portions of their historic range (Serfass et al. 1999, Hubbard 2006)

Portions of the landscape most suitable for North American river otters in Pennsylvania comprise a diversity and abundance of interconnected aquatic habitats, with suitable aquatic prey (e.g., fish and crayfish) and a relatively undisturbed riparian border (Genoways and Brenner 1985, Serfass et al. 1986, Polechla 1990, Melquist et al. 2003). Den and resting sites usually occur in riparian areas. Beaver dens (lodges and bank dens), undercut banks with extensive tree root systems, rock formations, and log jams with accumulated brush piles are known to serve as denning areas (Serfass and Rymon 1985, Swimley et al. 1998). Small steep gradient streams typically do not represent regularly used portions of the home range. However, portions of smaller streams may be occupied where impounded by beaver dams, which increase the volume of water, availability of prey, and number of potential den sites (Swimley et al. 1999). Suitable aquatic habitats must be relatively free of pollutants, such as heavy metals, pesticides, and effluents from industry (Melquist et al. 2003).

### *LIFE HISTORY AND ECOLOGY*

The North American river otter is an opportunistic carnivore, preying on a variety of aquatic and semi-aquatic organisms. Fish are the predominate prey item in most regions of North America, however, crustaceans, amphibians, insects, and birds may be of seasonal importance (Liers 1951, Melquist et al. 2003). For example, crayfish are a common prey item of river otters in Pennsylvania during late spring, summer, and early fall (Serfass et al. 1990). Typically, the slowest-moving and most abundant fish species are preyed on most frequently. Consequently, "forage fish," such as suckers (Catostomidae) and minnows (Cyprinidae), tend to be much more common in the river otter's diet than swift, agile game fish (e.g., trout [Salmonidae]; Serfass et al. 1990).

River otters engage in travel and other activities (e.g., foraging and scent marking) most frequently during nocturnal and crepuscular periods (Melquist and Hornocker 1983, Stevens 2005). The extent of travel is influenced by gender, region, season, and habitat conditions. Some of the largest home ranges (about 200 $km^2$ for males and 70 $km^2$ for females) have been reported from northern Alberta, Canada, where the abundance and quality of aquatic habitats are limiting, forcing river otters to travel long distances to obtain appropriate resources (Reid et al. 1994). Home ranges for river otters occupying riverine habitats in Idaho ranged from 8 to 78 km of the shoreline (Melquist and Hornocker 1983). In general, males occupy larger home ranges than females (Hornocker et al. 1983, Melquist and Hornocker 1983, Reid et al. 1994, Melquist et al. 2003). The difference is most pronounced during the breeding season when males tend to substantially increase movements while searching for mates (Melquist and Hornocker 1983, Spinola 2003). River otters are able to maintain low population densities (about one otter per 3.58 km of riverine habitat in Idaho; Melquist and Hornocker 1983), apparently without the need for overt (aggressive) displays of territoriality (Melquist et al. 2003). Home ranges of adjacent individuals sometimes overlap to varying degrees based on gender and season (Melquist and Hornocker 1983, Spinola 2003). The specific mechanisms by which river otters maintain low population densities are poorly understood. However, individuals appear able to practice mutual avoidance through olfactory communication, which likely is facilitated by scent marking at latrine sites (areas where river otters frequently deposit scats and secretions from anal glands; Hornocker et al. 1983, Melquist and Hornocker 1983).

The basic social group of river otters consists of a female and her offspring; however, lone river otters will interact with other individuals and family groups (Melquist and Hornocker 1983, Reid et al. 1994, Spinola 2003). Juveniles typically remain with the mother for about one year, dispersing from the natal area at twelve to fourteen months of age (Hornocker et al. 1983, Melquist et al. 2003). North American river otters are polygynous, with both males and females typically reaching sexual maturity at two years of age (Liers 1951, Hamilton and Eadie 1964, Lariviere and Walton 1998). Mating occurs between December and April, although this varies by latitude (Liers 1951, Hamilton and Eadie 1964). Hamilton and Eadie (1964) determined that river otters in New York bred between late March and early June. After a prolonged period of delayed implantation, two to three young are born the next March or April (Hamilton and Eadie 1964). River otters mate immediately after parturition (Liers 1951, Melquist et al. 2003).

### *THREATS*

The serious declines suffered by many river otter populations in North America have been somewhat offset by improved conservation practices (Melquist et al. 2003). Nonetheless, some of the original causative factors contributing to population declines persist. The specific types and relative contributions of the various factors undoubtedly differed among regions. In general, however, the decline of most populations can be in some way attributed to combinations of unregulated harvest, various categories of water pollutants, and disturbances to riparian habitats (Eveland 1978, Nilsson 1980, Melquist et al. 2003). Fortunately, improvements in water quality, resulting from implementation of clean water legislation (e.g., Federal Clean Water Act), better protection of wetlands and riparian habitats, and more progressive attitudes toward furbearer and predator conservation, have contributed to recovery or potential recovery of river otter populations in many regions of North America. A similar scenario exists in Pennsylvania, where improved habitat conditions combined with implementation of a reintroduction project have contributed to a substantial increase in the distribution of river otters (Serfass et al. 1999).

Ultimately, river otter populations are limited by the distribution of suitable aquatic habitats. Consequently, any factors that reduce the quantity or degrade the quality of aquatic environments will adversely affect populations. In many areas, the recovery of river otters remains limited by water pollution, loss of wetlands, and alteration of riparian habitats. In general, there is a poor understanding of the levels at which these types of perturbations impede the reestablishment or maintenance of river otter populations. Acid mine drainage, which is prevalent in many drainages in central and western Pennsylvania, prevents the reestablishment of river otters by eliminating aquatic prey. However, the individual and interactive effects that other aquatic disturbances (e.g., accumulation of pollutants in fish tissue, alteration of riparian habitats, acid rain, agricultural and other nonpoint source pollutants, and construction of dams) have on river otter populations are less obvious and poorly understood. Nonetheless, these types of disturbances should be regarded as potentially limiting to river otter populations.

The combination of low reproductive rates, low population densities, and dependence on aquatic habitats, which represents a relatively small portion of the landscape, contribute to river otters' susceptibility to overharvest. Although legally protected in Pennsylvania, river otters continue to be killed accidentally each year by trappers pursuing legal furbearers, particularly beavers (*Castor canadensis*; Serfass et al. 1999). These incidental captures do not appear to have a severe effect on established (core) river otter populations but could be impeding expansion of remnant and reintroduced populations (Serfass et al. 1999, Serfass 2001).

The extent and effect of many potential threats to river otters often are poorly quantified or understood. Fish culture is a substantial industry in Pennsylvania, and conflicts arise when piscivorous wildlife, such as river otters, is attracted to hatcheries by the easily accessible source of food. Hatchery owners are known to kill river otters, which cause depredations, but the actual numbers are difficult to discern. Another potential concern relates to negative publicity that could be generated from depredations at hatcheries, which may exacerbate the unfounded assumptions of some anglers that river otters are harmful to fish (particularly game fish) populations in natural aquatic environments. Some anglers undoubtedly kill river otters because of the misguided notion that such actions protect game fish populations. Most higher-order streams in Pennsylvania are adjacent to highways and collisions with vehicles contribute to the death of an undetermined number of river otters every year (Serfass et al. 1999). River otters are susceptible to a variety of diseases, and rabies has been reported from a river otter in Pennsylvania (Serfass et al. 1995). Predation is unlikely to cause significant mortality to river otters (Melquist et al. 2003).

*CONSERVATION AND MANAGEMENT NEEDS*

The Pennsylvania River Otter Reintroduction Project has been among the most successful and highly publicized wildlife projects implemented in the commonwealth (Serfass 1991, 1994). Support for the reintroduction project is widespread among diverse segments of the public (Serfass 1994). In many ways, the reintroduction project has resulted in the river otter becoming an important symbol for wildlife conservation in Pennsylvania. The popularity of this species therefore has the potential to serve as an important flagship species (a charismatic species used to attract attention to important conservation issues; see Simberloff 1998) for conserving aquatic resources. Conservation agencies should be encouraged to recognize and further promote the river otter as a flagship species. An educational program designed to inform the public about the ecological role and value of river otters in aquatic ecosystems should be an ongoing component of conservation efforts for this species.

The remnant river otter population in northeastern Pennsylvania exists within a matrix of private property interspersed with public lands. In contrast, public lands predominate in most areas where river otters were reintroduced. Unfortunately, the long-term integrity of habitats on private lands is often uncertain because of the potential for development. Any programs designed to minimize negative consequences of development on aquatic and riparian systems should be promoted. The Forest Stewardship Program is an example of an incentive program that promotes sustainable use of timber resources and provides guidelines designed to protect adjacent riparian and aquatic habitats. This program can potentially benefit river otters if it is implemented on a large scale. River otters will benefit on both public and private lands with the conscientious enforcement of the federal Clean Water Act and laws designed to minimize loss of wetlands.

Public lands should be regarded as essential for maintaining viable populations of river otters in many areas of the commonwealth. The Pennsylvania State Forest Resource Management Plan (Department of Conservation and Natural Resources, Bureau of Forestry 2007) and the ANF Land and Resource Management Plan (ANF 1986) cite watershed protection as an important conservation goal. Any public land management programs that promote protection of aquatic habitats throughout entire drainages should be encouraged.

The vulnerability of river otters to incidental capture by recreational trappers pursuing legal furbearers should be regarded as a Conservation Concern. River otters are particularly susceptible to being caught and killed in traps set for beavers. The frequency and impact of incidental captures should be carefully monitored throughout the river otter's range. The PGC provides recommendations to trappers on procedures thought to reduce the likelihood of accidentally capturing river otters. However, accidental captures remain a problem, and continuing effort should be put forth to identify trapping procedures that will reduce the likelihood of capturing river otters during seasons for beavers and other furbearers.

Developing strategies to deter depredation by river otters at fish hatcheries should be regarded as another conservation priority. The killing of river otters is the most direct threat at hatcheries. However, the possibility of negative publicity resulting from depredations by river otters at hatcheries rearing game fish is another potential concern. Specifically, discussions among hatchery personnel and local residents about depredations may foster the misconception that river otters prey primarily on game fish in natural aquatic systems. These misconceptions could contribute to unwarranted concerns among anglers that river otters adversely affect game fish populations and, subsequently, the development of animosity toward the species. Investigations are needed to assess factors related to the occurrence of river otters at hatcheries and to develop strategies to deter or minimize depredations.

Ultimately, ensuring the perpetuation of self-sustaining river otter populations depends on maintaining large areas of relatively undisturbed aquatic systems and associated riparian habitats. Applied research to understand individual and cumulative effects of various habitat disturbances are needed to identify and direct the implementation of specific conservation measures that will benefit river otters and other obligate wetland species. However, in many cases, the effect of various habitat disturbances on aquatic and riparian wildlife is obvious and reasonably well understood. Consequently, the need for research to understand fully an environmental perturbation should not preclude implementing conservation programs to address specific problems. North American river otters will benefit by encouraging the implementation or enhancement of programs that (1) reduce emissions causing acid rain, (2) implement stream-bank fencing projects to protect riparian and aquatic habitats in areas where livestock are grazed, (3) enhance existing regulations designed to protect or limit the loss of wetlands, (4) further regulate mining activities that cause acid mine drainage, (5) implement strategies to mitigate the effects of existing acid mine drainage, and (6) enhance policies and enforcement activities to control all forms of point and nonpoint sources of water pollution.

*MONITORING AND RESEARCH NEEDS*

Application and integration of existing approaches for monitoring river otter populations should be implemented by agencies responsible for managing the state's natural resources. Typically, questionnaire surveys of wildlife conservation officers (WCOs) of the PGC have been used to assess the occurrence of river otters within a county or WCO district (usually a county comprises two to three WCO districts; Eveland 1978). However, this approach substantially inflates the actual distribution of river otters across the landscape (see Hubbard and Serfass 2005, Hubbard 2006). More recently, riparian surveys to detect otter latrine sites (Swimley et al. 1998, Carpenter 2001, Hubbard 2006) and use of remote cameras (Stevens 2005, Stevens and Serfass 2008) have been conducted to determine whether river otters persist at reintroduction sites in Pennsylvania. These field assessments were designed to evaluate and refine existing techniques and define new approaches for detecting evidence of river otters. Practical outcomes applicable for enhancing the likelihood of detecting evidence of river otters during riparian surveys included identification of riparian features typically associated with latrine sites (e.g., rock formations, backwaters, and conifer cover; Swimley et al. 1998) and demonstrating that marking intensity at latrines was highest during the spring and fall (Carpenter 2001, Mills 2004, Stevens 2005, Stevens and Serfass 2008).

Questionnaire surveys of WCOs have been useful for assessing the distribution of river otters at a coarse scale. Unfortunately, questionnaire surveys offer less value in assessing population trends at finer levels. Conversely, riparian surveys yield site-specific information about the presence of river otters but are too time and labor intensive to quickly and efficiently obtain information at a larger scale. Hubbard and Serfass (2005) developed and recommend an integrated approach, which uses a GIS-based grid system (composed of 1 $km^2$ cells) developed in conjunction with the Pennsylvania GAP Project (PAGAP 2004) for mapping the statewide distribution of river otters. Information for completing the grid can be obtained from a variety of sources, ranging from interviews with WCOs to formal riparian surveys. This level of evaluation yields a more realistic assessment of the manner in which river otters are distributed and applied to identify expansions or contractions of populations over time. Also, each grid cell represents a sample unit and, therefore, facilitates quantification of associations between landscape-level habitat features and the distribution of river otters. Guidelines for applying the technology and methodology are well established for implementing this monitoring technique. Natural resource management agencies should be encouraged to implement and refine this technique for monitoring river otter populations.

Compilation of sightings and other forms of miscellaneous information pertaining to the presence of river otters (e.g., accidental captures by trappers and road kills) should be ongoing among natural resource agencies in the state. These data should be incorporated into the GIS-based approach recommended by Hubbard and Serfass (2005) for monitoring populations and shared among agencies, the Pennsylvania Mammal Technical Committee, and other professionals interested in conserving river otters.

Efforts to restore river otter populations through reintroduction efforts were conceived and conducted by university researchers (Serfass 1991, Serfass 1994, Serfass et al. 2003, Hubbard and Serfass 2005), with primary funding and logistical support provided by the

Pennsylvania Department of Conservation and Natural Resources' (DCNR) Wild Resource Conservation Program, the PGC, and the United States Department of Agriculture ANF. The PGC is the primary agency responsible for managing birds and mammals in the state. Currently, the river otter is protected from harvest, and responsibility for its conservation is under the PGC's Wildlife Diversity Section. Although the river otter also is classified as a furbearer, its current status protects it from legal harvest. Any future changes in the conservation status of river otters should be based on objective consultations among university researchers and others responsible for the restoration of river otter populations and appropriate representatives of the PGC, DCNR, and ANF. Consultations should be organized and mediated through the Pennsylvania Mammal Technical Committee.

A variety of research projects have been conducted to assess a few specific biological and ecological aspects of the North American river otter, many pertaining to food habits and reproductive biology (Melquist et al. 2003). During the past twenty years, most research and conservation attention has been devoted to river otter reintroduction projects, yielding useful information on the movement and spacing patterns of reintroduced individuals (Serfass and Rymon 1985, Johnson and Berkley 1999). Unfortunately, few comprehensive biological or ecological studies have been conducted on native river otter populations, including the population in northeastern Pennsylvania. Also, many of the investigations have been of regional significance and, therefore, not necessarily applicable for developing general conservation strategies for river otters throughout their range.

In Pennsylvania, a need exists to develop, implement, and evaluate approaches for clearly and accurately defining the status, distribution, densities, and limiting factors for both remnant and reintroduced river otter populations. Methods currently in place for monitoring river otters are of little value for accurately depicting and interpreting trends or patterns in the distribution of river otters. Research should focus on evaluating the efficacy and efficiency of existing protocols for monitoring river otters at various scales (from fine [site specific] to coarse [landscape] levels; see Hubbard and Serfass 2005). Ultimately, the intent of the research should be to integrate information from various levels to more accurately depict the size and distribution of the state's river otter population. The remnant river otter population has received less research attention than reintroduced populations and should be an important focus of future investigations.

The individual and interactive effects of environmental disturbances, including accumulation of toxic substances (e.g., polychlorinated biphenyls [PCBs] and chlorinated pesticides) in fish and other aquatic prey, alteration of riparian habitats, acid rain, and construction of dams, tolerable to river otter populations are poorly understood. The potential limiting influence of these types of perturbations should be conducted in conjunction with studies to determine the distribution of river otters in the state. Various human activities have direct impacts on individual river otters and may have implications for hindering the expansion of populations. For example, river otters are killed accidentally each year by trappers seeking legal furbearers, by collisions with vehicles, and for causing depredations at fish-rearing facilities. The extent and impact of these direct causes of mortality should be carefully and objectively evaluated to determine their effects on populations.

*Author:* THOMAS SERFASS AND DEVON ROTHSCHILD, FROSTBURG STATE UNIVERSITY, FROSTBURG

## Southern Bog Lemming

Order: Rodentia
Family: Cricetidae
*Synaptomys cooperi*

The southern bog lemming is a small terrestrial vole (adults 115–145 mm total length) that is dark brown above with a light gray underside (fig. 6.37). The species is generally identifiable by its very short tail (15–24 mm) and broad, grooved upper incisors. It was selected as a Pennsylvania Species of Special Conservation Concern because it is a habitat specialist with widely fragmented, small populations. The species is particularly sensitive to habitat modifications resulting from intensive farming practices, the destruction of wetlands, and the elimination of native grasslands. It has a rank of Apparently Secure in Pennsylvania; its global rank is Secure (G5, NatureServe 2009).

### *GEOGRAPHIC RANGE*

Southern bog lemmings are widely distributed in the eastern United States and adjoining southeastern Canada, reaching their western limits in southwestern Kansas and northeastern Arkansas and their southern limits in the mountains of eastern Tennessee and west-

*Fig. 6.37.* The Southern Bog Lemming, *Synaptomys cooperi*. Photo courtesy of Wayne van Devender.

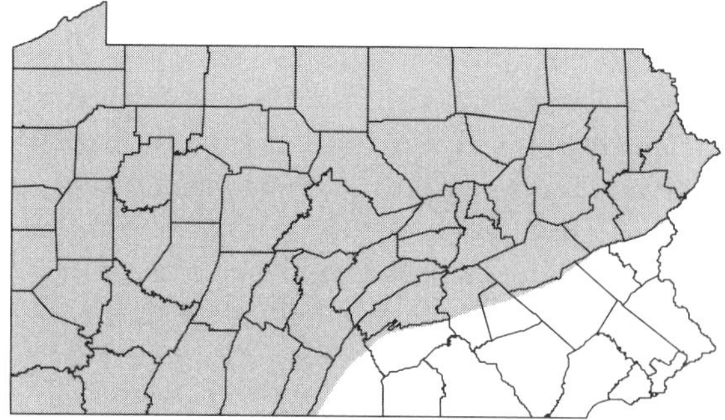

*Fig. 6.38.* Distribution of the Southern Bog Lemming, *Synaptomys cooperi*.

ern North Carolina. However, within this range, their occurrence is extremely patchy, and they are seldom found in significant numbers.

The fossil record indicates that southern bog lemmings once ranged much farther southward than at present, with late Pleistocene remains of this species having been found in Mexico and Texas (Linzey 1983 and references therein). Until approximately 11,000 years ago, the current range of *Synaptomys cooperi* was occupied by the northern bog lemming. (Formerly classified in the same genus, this species is now designated *Mictomys borealis*.)

### *DISTRIBUTION AND RELATIVE ABUNDANCE IN PENNSYLVANIA*

Southern bog lemmings are generally thought to range throughout Pennsylvania (fig. 6.38). However, the distribution of actual specimens collected suggests that this species is absent or nearly absent from the southeast quadrant of the state (see Kirkland and Hart 1999). However, this may result from a lack of survey efforts in these counties. It is unclear whether Pennsylvania is occupied by one or two subspecies of *S. cooperi*. According to Wetzel (1955), only *Synaptomys cooperi cooperi* is found in the state. However, the distribution map in Merritt (1987) indicates that *Synaptomys cooperi stonei* occurs in the southeastern counties (the region currently lacking specimen records).

### *COMMUNITY TYPE/HABITAT USE*

Despite its name, the southern bog lemming occurs in a wide variety of habitats, including old field communities, mixed deciduous/coniferous woodlands, spruce-fir forests, and margins of freshwater wetlands (Connor 1959, Linzey 1983, Merritt 1987). Small patches of habitat or microhabitat, such as clearings within woodlands or areas with slight differences in cover characteristics, may support small localized, populations. This pattern has resulted in misapplication of the word "colonial" to southern bog lemming populations, and these local concentrations are unlikely to be colonial in a social sense (Linzey 1983). Early Pennsylvania investigators commented that southern bog lemmings and meadow voles live in different portions of the same habitat (Richmond and Rosland 1949) and that the only factor in common among southern bog lemming habitats is that they are submarginal for meadow voles (Doutt et al. 1973). Later research in Virginia demonstrated that, where the two species co-occur, the distribution of southern bog lemmings is influenced by interactions with the meadow vole, with southern bog lemmings being a poorer competitor for living space. Only when meadow vole populations are low do southern bog lemmings secure better-quality habitat (Linzey 1984, Linzey and Cranford 1984).

### *LIFE HISTORY AND ECOLOGY*

Southern bog lemmings feed almost entirely on green vegetation, primarily grasses and sedges but also on mosses, fruit, fungi, bark, and roots (Linzey 1983 and references therein). That their foods tend to have low digestibility typically results in production of green fecal pellets. However, when southern bog lemmings gain access to better habitats from which they have been excluded by meadow voles, their access to higher-quality foods results in dark fecal pellets that resemble those of meadow voles (Linzey 1984).

Like most small rodents, *S. cooperi* is most active after sunset and at dawn, with constant, lower levels of activity between these peaks. Although southern bog lemmings are active throughout the year, there are seasonal variations in daily activity patterns, with diurnal activity declining in summer. In grassy habitats, southern bog lemmings make runways, which they use in their nightly travels. Their presence can be detected by neatly clipped piles of grass cuttings and characteristic green droppings. Where they live in high-elevation moist forests, they do not make runways and are more difficult to locate (Linzey 1983). Home ranges of individuals are quite small, ranging from 0.04 to 0.32 ha (Linzey 1983 and references therein).

Southern bog lemmings in the eastern portion of the range seem to enter livetraps reluctantly and, once captured, can be difficult to recapture. However, they will readily visit "dropping boards" placed in runways, which have been exploited to monitor their presence and track their movements (Linzey 1981, 1984). Pitfall traps may also reveal this species' presence in areas where they have not been detected by other trapping methods (Rose 1981, 2005). Individuals are docile and rarely attempt to bite when they are handled. However, they can be difficult to maintain in captivity and rarely, if ever, have reproduced under these circumstances.

Population densities tend to be low in the eastern portions of the range, where southern bog lemmings and meadow voles compete for living space (Linzey 1984, Linzey and Cranford 1984). Local temporary concentrations (e.g., up to 24 per ha reported by Linzey 1981) cannot be used as estimates of area-wide densities (Linzey 1983). Published densities reported for the species range from approximately 1.6 per ha (four per acre) in New Jersey (Connor 1959) to 106 per ha in Illinois (Beasley and Getz 1986). In general, they tend to be more abundant in the midwestern prairie states, where they apparently coexist more successfully with prairie voles (*Microtus ochrogaster*; Rose and Spevak 1978). Even there, however, studies in Kansas in the mid-1980s appear to have been based on an unusual "outbreak" (Beasley and Getz 1986, Danielson and Gaines 1987a, 1987b, Danielson and Swihart 1987). The term "cycle" has been used to describe southern bog lemming population variations (Gaines et al. 1977, Beasley and Getz 1986) but never in the eastern portion of the range.

Variation in southern bog lemming numbers has been noted in Pennsylvania (Grimm and Roberts 1950) and confirmed during a six-year study of small mammal populations at Yellow Creek State Park in the western portion of the state (A. V. Linzey and M. H. Kesner, unpublished data). Variation over both time and space in suitable habitat was noted during livetrapping (same effort on every grid in every season) on three 6-ha grids between spring 1998 and fall 2003. Numbers of *S. cooperi* on the "highest density" grid varied from zero (spring 2000, spring and fall 2002, spring and fall 2003) to six (=1 per ha, spring 1998). All three sites were consistent in having more individuals from 1998 to 1999 than in other years. However, over the six years of the study (108,000 total trap nights in 18 ha of habitat), only thirty-nine different animals were handled (0.00036 per trap night).

The status of *S. cooperi* is somewhat enigmatic. Globally, the species is considered Secure (NatureServe Ranking G5). However, information on population trends in Pennsylvania or in the region is lacking. Although there is a general perception of decline, there are no specific data to confirm this perception. Because they have apparently always been somewhat sparse (and often caught incidentally during research on other species), the difference between sparse and sparser may be difficult to detect. If they are less abundant in eastern North America than formerly, one could speculate that human-caused habitat changes have created conditions more favorable for meadow voles (deforestation, elimination of native grasslands, roadways that provide dispersal routes to habitat patches). The Pennsylvania state status (NatureServe Ranking S4) recognizes that the species is uncommon, and the Pennsylvania Biological Survey designation is I (Restricted Distribution). Surrounding states that have ranked this species consider it Uncommon (New York), Vulnerable (Maryland), or Imperiled (New Jersey and West Virginia).

Southern bog lemmings may breed throughout the year, although frequency of breeding in the eastern part of the range is very low between November and February. The young, usually numbering three or four, are born after a gestation period of twenty-three to twenty-six days. Neonates are mostly pink but have light gray pigmentation on their backs. In addition to whiskers on the sides of the snout, they have a sprinkling of hairs on the head and back. At birth, their eyes are closed and their ears are folded down against the head. Ears unfold by the second day, but eyes do not open until ten to twelve days old. The body is well-furred by seven days, and they look like smaller versions of adults by two weeks. Weaning is completed by three weeks, at which time they have nearly reached adult size. Adult pelage is attained by five to six weeks (Linzey 1983 and references therein).

*THREATS*

Given the paucity of specific information, any assessment of threats is speculative. Construction of roads through forest blocks likely provides dispersal routes for meadow voles, allowing them to colonize isolated patches of early successional habitats. Farming practices that create sharp edges between fields and woodlands would eliminate southern bog lemming habitat. Destruction of wetlands and elimination of natural grasslands would further degrade their potential living space.

*CONSERVATION AND MANAGEMENT NEEDS*

Other than general directives to restore natural grasslands and maintain large forested areas, information is insufficient to propose specific objectives at this time. There is currently no protection for this species. Monitoring is needed to determine population trends in this species.

*MONITORING AND RESEARCH NEEDS*

Although bog lemmings are likely to still be widely distributed in the state, there are two immediate needs. One is to determine whether the species occurs in the southeastern region. The other is population monitoring. However, use of classic survey methods for monitoring is likely to be unproductive and a poor return for time invested. Alternative monitoring tools that should be evaluated are use of pitfall traps or "dropping boards." Both of these methods have been used successfully to detect the presence of this species in Virginia (Rose 1981, Linzey 1984, Linzey and Cranford 1984, Rose 2005).

*Author:* ALICIA V. LINZEY, INDIANA UNIVERSITY OF PENNSYLVANIA

## Snowshoe Hare

Order: Lagomorpha
Family: Leporidae
*Lepus americanus*

The snowshoe hare is a lagomorph named for its disproportionately large hind feet (11–14 cm), which, with dense fur and stiff hairs, form "snowshoes" well adapted for locomotion in deep, powdery snow (fig. 6.39). It is also called "varying hare" because it has a pure white pelage in winter, except for black eyelids and ear tips, which changes to a black-peppered rusty brown or grayish summer pelage. This species was selected as one of Maintenance Concern because of its

*Fig. 6.39.* The Snowshoe Hare, *Lepus americanus*. Photo courtesy of Paul Shay.

sensitivity to habitat alterations and its importance for gene flow between states to the north and south. Habitat loss has resulted from a combination of factors, including human development, poor forest regeneration due to acid deposition and browsing by white-tailed deer (*Odocoileus virginianus*), climate change, and impact from invasive species. It is considered Vulnerable to Apparently Secure in Pennsylvania. Its global rank is Secure (G5, NatureServe 2009).

*GEOGRAPHIC RANGE*

Snowshoe hares are found only in North America. Snowshoe hare populations inhabit forest habitats from Alaska to Newfoundland (where they have been introduced) and south into the higher elevations of the mountains of the eastern and western United States. Within the conterminous United States, hares are found from Washington to California along the higher elevations of the Klamath, Cascade, and Sierra Nevada ranges and into northern New Mexico, along the northern-tier states, and at higher elevations southward along the Appalachian Mountains to North Carolina.

*DISTRIBUTION AND RELATIVE ABUNDANCE IN PENNSYLVANIA*

On the basis of harvest data from the Pennsylvania Game Commission, the range of snowshoe hares in Pennsylvania includes counties of the northern tier of the state and extends south to Maryland in counties that encompass the Laurel Highlands of southwestern Pennsylvania (fig. 6.40). In 2004, a survey of 223 randomly selected sites north of Interstate 80 and west of the Allegheny National Forest found fifty sites in which snowshoe hare signs (tracks or visual identification) or

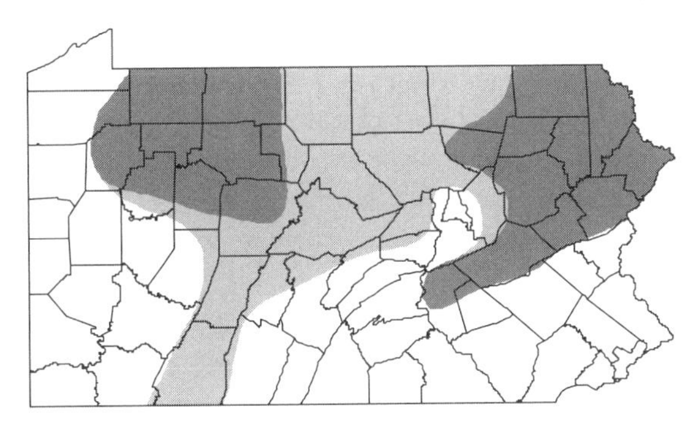

*Fig. 6.40.* Primary (darker shading) and secondary (lighter shading) distribution of the Snowshoe Hare, *Lepus americanus*.

pellets, identified using DNA analytical techniques, were detected (Diefenbach et al. 2005; fig. 6.40). The largest concentrations of sites where hares were detected were distributed similarly to the harvest data, which were Warren, McKean, Forest, and Elk counties in the west and the Poconos in the east; however, occupied sites were clustered and not distributed evenly across the state. Pennsylvania serves as a connector between hare populations of New York with those in Maryland and West Virginia. Pennsylvania's role in north-south gene flow for this species could be important.

### *COMMUNITY TYPE/HABITAT USE*

Hares are found in areas with dense vegetation that provides food, thermal cover, and protection from predators. Dense understory vegetation provides protection from predators, and clear-cuts will be avoided until woody vegetation reaches a height >2 m (Wolfe et al. 1982, Litvaitis et al. 1985). Ten- to thirty-year-old regenerating conifer stands provide excellent hiding cover for hares in Maine (Monthey 1986), but hares were more abundant in thirty-year-old stands than in twenty-year-old stands in Labrador (Newbury and Simon 2005). Because of the longer growing season and faster-growing tree species in Pennsylvania, suitable habitat for hares is likely to exist five to fifteen years after clear-cutting in northern hardwoods (Brown 1984) and mixed oak forests (Storm et al. 2003). Edge habitats between mature forest and regenerating clear-cuts have potential to be high-quality habitats because both cover and food are in proximity (Meslow and Keith 1968, Litvaitis et al. 1985, Forsey and Baggs 2001); however, these edge habitats may have greater predator densities (Forsey and Baggs 2001).

In Pennsylvania, hares are associated with areas of dense vegetation, particularly regenerating clear-cuts (Brown 1984, Scott and Yahner 1989). Scott and Yahner (1989) noted that this habitat may be marginal for hares, but Brown (1984) noted that areas with mountain laurel (*Kalmia latifolia*) or eastern hemlock (*Tsuga canadensis*) were used less than five- to fifteen-year-old clear-cuts when all three habitat types were in proximity. Home ranges are generally 3–6 ha and typically <10 ha. Brown (1984) recommended maintaining 16 percent of a forest management unit (ca. 900 ha) in the optimal age class for hares using a sixty-year rotation period to maintain suitable habitat. Extensive browsing by white-tailed deer may have important negative consequences for snowshoe hares (Glazer 1959, Brown 1984, Scott and Yahner 1989) because of their effect on forest regeneration.

### *LIFE HISTORY AND ECOLOGY*

An eight- to eleven-year cycle of abundance exists in northern snowshoe hare populations (Keith 1963), followed by a similar, but lagged, cycle in predator abundance. Although many hypotheses for these abundance cycles have been proposed, a large-scale, long-term study by Krebs et al. (1995) indicated that population cycles of snowshoe hares in the boreal forest are best explained as a result of a three trophic-level interaction among food supplies, predators, and prey. However, in southern parts of its range, where ecosystems are more stable and diverse, hare populations are weakly cyclic, irruptive, or stable. Population densities in noncyclic populations tended to be greater than cyclic populations at numerical lows (Murray 2000). In Pennsylvania, hare populations likely are noncyclic (Fergus 1976) and have been reported as noncyclic in other mid-Atlantic and northeastern states (Murray 2000).

Snowshoe hares are herbivores that, during summer, feed on grasses, wild berries, wildflowers, clover, horsetails, and new growth of trees and shrubs. In winter, their diet consists of twigs, bark, and evergreen needles (Hodges 1999). In Pennsylvania, snowshoe hares browse *Rubus*, striped maple (*Acer pennsylvanicum*), and yellow birch (*Betula alleghaniensis*) more than expected, and browse American beech (*Fagus grandifolia*), pin cherry (*Prunus pennsylvanica*), black cherry (*Prunus serotina*), red maple (*Acer rubrum*), and sugar maple (*Acer saccharum*) less than expected in northern hardwood forests (Brown 1984, Scott and Yahner 1989). However, areas with a high density of *Rubus* are avoided, supposedly because of the lack of overhead cover (Scott and Yahner 1989).

Beyond distribution, food habits, and habitat use,

little is known about snowshoe hares in Pennsylvania. Predation is likely the greatest source of mortality and probably involves most avian and mammalian predators that occur in the state. The large number of potential predators may explain why hare populations exhibit little evidence of population cycling in Pennsylvania. Hares tend to be solitary, except during the breeding season, and because aggressive behavior is most prevalent at peaks of population cycle, it is unlikely that aggressive behavior is common in Pennsylvania. A number of parasites and diseases have been reported in hares (summarized in Murray 2003), but outbreaks of disease or parasite infestations are most often associated with populations that experience abundance cycles.

Snowshoe hares are seasonal breeders, strongly controlled by photoperiod and have up to four litters per year (March–September). The average litter size is two to four young and the young are precocial and leave the birth site within twenty-four hours. Litters are larger in western populations, but females produce fewer litters; thus, productivity is similar across the range of the species (Murray 2000). Hares do not exhibit sex-biased dispersal, kin relationships do not affect spacing behavior, nor is spacing behavior likely to affect population dynamics (Burton and Krebs 2003).

### *THREATS*

The primary threat influencing the existence of snowshoe hares in Pennsylvania is loss of early successional habitat and corresponding lack of suitable habitat connectivity (e.g., for dispersal). The amount of early successional forest habitat has been decreasing for decades (Alerich 1993) in Pennsylvania because timber harvest rates have not kept pace with succession of forest vegetation. As a result, the remaining early successional habitats are becoming fragmented. Fortunately, this type of fragmentation does not have to be permanent and can be addressed through proper forest planning and the application of appropriate silvicultural techniques.

Permanent fragmentation of habitat caused by human development is another problem for snowshoe hares. For example, counties in the Poconos are now considered part of the New York City metropolitan area because of the number of commuters who reside in these counties, and these areas are experiencing some of the largest population growth and housing development rates in Pennsylvania. These changes are resulting in permanent fragmentation of forested habitats.

Poor forest regeneration caused by acid deposition (Sharpe and Drohan 1999) and browsing by white-tailed deer (Scott and Yahner 1989) likely is having an adverse effect on snowshoe hares. Glazer (1959) noted that introduced hares were less likely to become established in areas where white-tailed deer reduced available browse. In addition, poor silvicultural practices can adversely affect habitat quality for hares. Properly applied silvicultural practices ensure forest regeneration with high stem densities (e.g., clear-cuts or shelterwood cuts), but these techniques often are not applied on private lands. Instead, exploitative logging oftentimes results in too much shade such that tree regeneration is limited or outcompeted by invasive species (e.g., ferns). Further, regenerating tree species are usually shade tolerant and may not be preferred browse species.

Finally, climatic changes and invasive species could have significant effects on the viability of snowshoe hares in Pennsylvania. Global warming has already reduced winter severity in North America (e.g., currently lakes are ice covered for about two weeks less each winter; Magnuson et al. 2000). With milder winters will come less snow cover, which puts snowshoe hares at greater risk of predation because of their white pelage in winter. Some subspecies in the southern range have evolved to forgo pelage coloration change (Murray 2003), but it is unlikely other subspecies would be able to adapt fast enough in response to global warming. The spread of the hemlock woolly adelgid (*Adelges tsugae*), an Asian insect that causes mortality in northern hemlocks, could result in the total loss of the primary species that provides conifer cover in Pennsylvania. Especially in the Poconos, loss of hemlocks could result in degradation of existing habitats used by snowshoe hares.

### *CONSERVATION AND MANAGEMENT NEEDS*

At the state level, little can be done directly to address the threats of global warming because of the problem's national and international nature. No known means to prevent the spread of the hemlock woolly adelgid exists, and hemlock mortality, at this time, is expected to be near 100 percent. Given that viable snowshoe hare populations exist in Pennsylvania in areas with little conifer cover (Brown 1984, Scott and Yahner 1989), loss of hemlock stands could be mitigated readily by forest management practices, creating suitable early successional habitat. Global warming may result in restricting the range of hares in Pennsylvania to colder regions (i.e., higher elevations) and areas with greater snowfall (e.g., northwestern Pennsylvania). However, snowshoe hare populations along the Allegheny Mountain Range may become further isolated.

Certain conservation and management activities could help ensure the viability of the snowshoe hare in Pennsylvania. More programs are needed that engage private landowners in sharing information about the benefits of practicing sustainable forestry. Private landowners control the majority of forested habitats in Pennsylvania, and forestry practices on private lands will define the distribution and condition of early successional habitats in much of Pennsylvania. Forest landowners need to be educated about the benefits of sustainable forestry practices and the effects of forest management decisions on wildlife species at broader spatial and temporal scales. Furthermore, tax and other incentives for private landowners would encourage proper silvicultural practices on their lands. Central to this idea is to encourage landowners to develop forest management plans that reflect their intention to sustain forest resources, as well as adhere to the prescriptions in the management plan. Incentives would likely be more effective than a regulatory approach because of the myriad of factors effecting tree regeneration and the multitude of forestry practices that may be employed.

Integrating the forest management activities of the major public (Department of Conservation of Natural Resources, Bureau of Forestry, Pennsylvania Game Commission, and Allegheny National Forest) and private landowners (i.e., industrial forest landowners) to understand and better predict the spatiotemporal dynamics of early successional habitats in Pennsylvania also would be beneficial. The spatiotemporal dynamics of early successional habitats, at the scale relevant to the dispersal characteristics of snowshoe hares, will be critical to the viability of this species. Coordination of forest management activities within and among public and private landowners could ensure that early successional habitat is created in a manner that enhances snowshoe hare populations and other species dependent on this type of habitat.

Snowshoe hares are currently a hunted game species in Pennsylvania, although the hunting season is limited to one week following Christmas, a daily bag limit of one hare and a possession limit of two hares. The number of hunters pursuing hares has declined from >70,000 in 1980 to <6,000 since 1995. Hunter success rates (harvest per hunter) have remained constant over the past twenty-five years. Hunting is unlikely to be having an adverse effect on the snowshoe hare population and provides the only long-term data on relative abundance and distribution of hares.

### *MONITORING AND RESEARCH NEEDS*

Developing a protocol for monitoring snowshoe hares will be difficult and expensive because the species is cryptic and difficult to enumerate. The most accurate large-scale means of detecting presence of hares is genetic testing of fecal pellets (Diefenbach et al. 2005), which is still time consuming and expensive. Consequently, the Pennsylvania Game Commission's annual Game Take Survey is probably the most cost-effective means of monitoring snowshoe hare distribution and relative abundance in Pennsylvania. The Game Take Survey annually obtains harvest and hunter effort information from approximately 10,000 license buyers, but because so few hunters pursue snowshoe hares, sample sizes are small. Moreover, data are collected at the county or management unit level and will not provide fine-scale information on distribution and data must be aggregated over multiple years.

No immediate research or survey priorities have been identified. Current data (1997 to 2003 harvest data) suggest the distribution of hares is similar to what was known in the early 1980s (Brown 1984). Snowshoe hares, by nature of their habitat requirements, are patchily distributed and relatively low density in Pennsylvania. Ongoing research (D. Diefenbach, unpublished data) will provide a more detailed assessment of the distribution of snowshoe hares in the northern-tier counties, which contain the core populations in Pennsylvania. However, results from research on hare dispersal (e.g., distances and factors influencing dispersal) could be applied to habitat management activities to create the spatial and temporal distribution of early successional habitats that ensure the viability of local hare populations. Furthermore, if the Pennsylvania population is critical as a genetic connection between New York and Maryland–West Virginia, then knowledge about habitat conditions and status of hares in the Laurel Highlands will be critical for maintaining the connectivity and viability of Pennsylvania's population.

*Author:* DUANE DIEFENBACH, UNITED STATES GEOLOGICAL SURVEY AND PENNSYLVANIA STATE UNIVERSITY

7

# Critical and Emerging Issues in the Conservation of Terrestrial Vertebrates

Michael A. Steele
DeeAnn M. Reeder
Timothy J. Maret
Margaret C. Brittingham

## Introduction

Most of the issues contributing to species decline are part of an age-old assemblage of anthropogenic-related factors that have had a cumulative effect on vertebrates and other species over the past few centuries or more. Although the scientific and resource management communities are still trying to fully understand how these issues influence species decline, modern technology, greater public awareness of the value of biodiversity, and increased funding for research and management have all contributed significantly toward solutions to these historic conservation issues. Unfortunately, as progress is made on some of these long-standing problems, new ones are continually emerging—some result from recent and developing technologies, some because of rapid changes in the earth's climate, and some for which we don't even yet understand the proximate mechanisms. In this chapter, we review examples of a few critical ongoing and emerging issues related to the conservation of vertebrates in the eastern United States. Our goal here is not to provide a comprehensive overview but instead to focus on a few particularly vexing issues, some of which have only recently emerged within the region.

## Wind Energy and Wildlife

Growing concerns over the environmental impact of carbon emissions resulting from using fossil fuels have led to increasing emphasis and investment in "cleaner" alternative energy sources. At the center of this movement, wind power, especially that produced from commercial-level turbines (>1.0 megawatt), is perceived as a hopeful solution and has become one of the fastest-growing areas in the energy industry. However, an increasing number of studies have now documented that wind turbines are responsible for high levels of bird and bat fatalities (Kunz et al. 2007a, 2007b, Arnett et al. 2008) and may especially affect bat populations throughout the eastern United States. The result is a growing controversy that has those committed to wind power at odds with conservation biologists, who maintain that the benefits accrued from wind energy in many locations may not outweigh the costs to our wildlife. This controversy though is not so easily resolved, as both the benefits of a cleaner power source and the costs to our wildlife require detailed study and analysis. The following is a brief synopsis of what we currently know about this problem and how it is being addressed.

Large numbers of bat, songbird, and raptor fatalities are reported from wind turbines in the United States. However, deciphering and understanding the relative effect on wildlife populations is difficult because of the many factors that influence assessment of such fatalities, including (1) the types of species most affected, (2) the accuracy of mortality estimates, (3) the causes of mortality, (4) the actual effect on bat and bird populations, and (5) the considerable variation in mortality across locations, types of turbines, and regions of the country (Kunz et al. 2007b, National Research Council 2007, Arnett et al. 2008).

Raptor mortality appears to be heaviest at wind farms in the western United States (National Research Council 2007), whereas songbirds and bats appear to be at greatest risk in the East, where wind farms are usually constructed on forested ridges (Kunz et al. 2007a). In Pennsylvania, for example, the majority of the seventy-five wind-energy facilities are located along ridge tops (Capouillez and Limbrandi-Mumma 2008). In a review of songbird mortality at wind facilities across the United States (excluding California), Erickson et al. (2001, cited in Kunz et al. 2007a) found that mortality of songbirds accounted for 78 percent of the carcasses recovered. Similar reviews of bat mortality in the eastern United States (reviewed by Kunz et al. 2007b) report mortality rates between 15.3 bats per megawatt (MW) per year, to 53 bats/MW/year. All such estimates, however, are likely to underestimate actual mortality rates because of biases in human detection (Arnett 2006) and the loss of carcasses to scavengers (Kunz et al. 2007b and references therein). The effect on bird and bat populations also varies with behavior, habitat, and season. Among the bats, mortality is most common for migratory, leaf, and tree-roosting species, whereas for birds, it is the nocturnal, migratory species that are most commonly lost by turbines (National Research Council 2007).

Certainly, many more detailed studies are needed to assess the effect of wind turbines on bat and bird populations, yet the research to date clearly indicates a potential threat to wildlife. Kunz et al. (2007b) and others (Kunz et al. 2007a, National Research Council 2007) detailed methodology that is required to understand both the effect and options for mitigation. They call for cooperation between the conservation community and the wind-energy industry and careful quantitative research conducted before construction, after forests are cleared and construction is complete, and after turbines are fully operational. As detailed by Kunz et al. (2007b), such risk assessment requires systematic monitoring methods (e.g., capture surveys, night vision surveys, radar or acoustic monitoring, infrared imaging, or radiotelemetry), both costly and time consuming but absolutely necessary if we are to understand the effect of turbines on wildlife. On the basis of the concerns raised to date, several scientific organizations, including the National Research Council, U.S. Fish and Wildlife Service, North American Society of Bat Research, American Society of Mammalogists, the Wildlife Society, and the Pennsylvania Biological Survey, have released formal reports, resolutions, or policy statements that independently call for careful monitoring studies and mitigation treatments to reduce the threat.

In the northeastern United States, many states are already taking appropriate action to deal with this conservation challenge. In Pennsylvania, for example, where seventy-five wind farms were either proposed or in operation by 2008, the Pennsylvania Game Commission (PGC) coordinated the Pennsylvania Game Commission–Wind Energy Cooperative Agreement (Capouillez and Limbrandi-Mumma 2008), a voluntary arrangement between the Pennsylvania Game Commission and twenty of the twenty-four wind-energy developers operating in the state known as the Pennsylvania Wind and Wildlife Cooperative (PAWWC). The agreement calls for both pre- and postconstruction monitoring of both bat and avian populations, as well as a risk assessment provided by the Pennsylvania Game Commission that is used to determine the site selection of wind facilities and the design and intensity of survey protocols. By 2008, preconstruction monitoring efforts coordinated by the PAWWC resulted in the discovery of the first maternal colony of the silver-haired bat (*Lasiurus noctivagans*) and the second largest maternity colony of the Indiana bat (*Myotis sodalis*); these and other preconstruction results also have guided site selection of wind facilities across the state (Capouillez and Limbrandi-Mumma 2008).

## Amphibian Declines

In the late 1970s, biologists began to note the decline and disappearance of amphibians from habitats around the world (Vitt et al. 1990). Over the next two decades, the accumulating evidence of a severe global decline was irrefutable (Houlahan et al. 2000, Stuart et al. 2004). The scope of the decline is sobering. In 2008, the International Union for the Conservation of Nature (IUCN) estimated that at least 32 percent of the world's amphibian species were threatened or extinct, as many as 159 species were already extinct, and at least 42 percent of all amphibian species were declining in abundance (IUCN et al. 2008). Although the severest declines and majority of extinctions have occurred in the Tropics, amphibian declines have occurred throughout the world.

Although the causes are not completely understood, there is not a single universal cause to the many declining amphibian populations around the world. Collins and Storfer (2003) identified six leading hypotheses for the global decline in amphibian populations: (1) invasive species, (2) overexploitation, (3) land use change (i.e., habitat loss and degradation), (4) global change (e.g., increased ultraviolet radiation and global warming), (5) pesticides and other toxic chemicals (i.e., pollution), and (6) emerging infectious diseases. Habitat loss and degradation are by far the greatest present threat to amphibian populations. The IUCN estimates that nearly 4,000 species are affected. Pollution is considered to be the second greatest threat, affecting around 1,200 species. Although disease affects less than 500 species, its effects on those species can be devastating (IUCN et al. 2008). Several diseases are capable of causing widespread mortality in amphibian populations; however, the pathogenic chytrid fungus (*Batrachochytrium dendrobatidis*) has rapidly become the most threatening. By 2004, it had infected at least ninety-three species around the world (Speare and Berger 2004), often leading to massive declines or even extinction. One trademark of the chytrid fungus is that it often rapidly eliminates species from otherwise pristine habitats.

The status of amphibian populations in Pennsylvania is not as critical as in many parts of the world. The only known amphibian species extirpated from the state, the tiger salamander (*Ambystoma tigrinum*), occurred in the extreme southeast corner of the state and disappeared because of habitat loss. Five species of amphibians are presently listed as Threatened or Endangered in Pennsylvania; all five of these species are peripheral species whose ranges barely extend into the state. Many species of amphibians have declined in abundance in recent years, primarily due to habitat loss and degradation. However, most species have not yet reached levels at which their survival is threatened. This does not mean that all is well with amphibians in Pennsylvania. First, if the slow, gradual decline in many species is not eventually halted and reversed, abundant species will eventually become rare or extirpated. Second, several species in Pennsylvania have undergone substantial decline in recent years.

Populations of eastern hellbenders (*Cryptobranchus alleganiensis alleganiensis*) have declined substantially over the past several decades in many streams in Pennsylvania (A. Hulse, personal communication). The causes of this decline are fairly well understood and include stream siltation, acid mine drainage, water pollution, and habitat alteration due to construction of dams. Chytrid fungus may have played a role in the decline, as hellbenders in a stream in northern Pennsylvania have tested positive for *B. dendrobatidis*. However, the fungus is not fatal to all amphibian species, and it is not presently known what effects it has on hellbenders.

More problematic than the decline of the hellbender is the rapid disappearance of several anuran species within the tree frog (Hylidae) and true frog (Ranidae) families. Within the tree frogs, chorus frogs (*Pseudacris* spp.) and northern cricket frogs (*Acris crepitans*) have declined dramatically in recent years, as has the northern leopard frog (*Lithobates pipiens*) within the true frogs. Although declines in some areas are due to habitat loss and degradation and pollution, frogs also have disappeared from apparently pristine habitats. Adding to the mystery, closely related frogs within these same families are thriving. The cause of these declines must be identified and remedied. Because the chytrid fungus has already been identified in the state, a first step should be to test remaining populations, as well as closely related species from habitats where these species have recently disappeared, for the presence of the fungus. If the chytrid fungus is found to be present, then the susceptibility of these species to the fungus should be determined. Species, and even populations within the same species, appear to differ in their susceptibility to the fungus, which may explain why only some species are declining. If the chytrid fungus is found not to be the cause, other causal agents should be investigated. Quick action is necessary; otherwise, these species may soon disappear from the commonwealth.

## Complex and Unpredictable Consequences of Climate Change: A Case Study

Although many factors contributing to the imperiled status of our wildlife are obvious (e.g., habitat loss, white-nose syndrome), others are more subtle (e.g., gradual changes in vegetation due to climate change), while still others are quite complicated as a result of numerous intersecting factors that may have completely unexpected consequences. Here we describe one such situation that illustrates how several factors interact to produce rather significant and grave consequences for one of our mammals of concern: the northern flying squirrel (*Glaucomys sabrinus*).

This species—elevated to Endangered status in Pennsylvania in 2007—has been the subject of considerable investigation in the state for more than a decade. Early records indicated relatively stable populations in the state perhaps well into the 1960s (Merritt 1987). Intensive trapping efforts and nest box surveys over the past several years, however, reveal that current populations are small, spatially fragmented, and concentrated primarily in the Pocono Plateau, where they are subject to increasing expansion of the human population (Steele et al. 2004). Such studies also indicate that in Pennsylvania, unlike in most other portions of the species' geographic range, wherever northern flying squirrels are found so are southern flying squirrels (*Glaucomys volans*), a major competitor of *G. sabrinus*.

These two species generally show nonoverlapping geographic ranges. The northern species occurs in boreal, coniferous forests across Canada, the northern United States, and montane regions of the southern Appalachian, Rocky, and Cascade mountains, whereas the southern flying squirrel is typically found in oak-hickory or pine-oak associations in the eastern third of North America and as far south as Honduras.

Despite this dichotomy in habitat and geographic ranges, overlapping (sympatric) populations of the two species are reported from the southern Appalachians, Ontario, and now Pennsylvania (Steele et al. 2004). This sympatry, however, may be a recent occurrence, as rapid changes to the squirrels' habitat and environment may have brought the two together. In Ontario, for example, Bowman et al. (2005) recently documented the northward movement of southern flying squirrel populations, and a corresponding northward retraction in that of the northern flying squirrel, which they attribute to warming shifts in climate. Support from this hypothesis follows from the observation that in a single year of unusually low temperatures, the range of the southern flying squirrel appeared to contract more than 250 km (Bowman et al. 2005). In contrast, it appears that in Pennsylvania, habitat loss and habitat degradation are the key factors bringing the species together (Steele et al. 2004). Loss of mature conifer forests, especially those that include red spruce, appears to have resulted in considerable fragmentation and isolation of northern flying squirrel populations, with relic populations now isolated predominantly in hemlock stands. Such stands typically border riparian zones and are surrounded by hardwood forests where *G. volans* is often common. Moreover, the two species also tend to converge on hemlock patches for nest sites, and have even been recorded from the same nest structures during winter. But further concern follows from the condition of the state's hemlock forests, now subject to significant mortality from the invasive woolly adelgid (*Adelges tsugae*), which holds the potential for eliminating much of the squirrel's last remaining habitat in this region of the Appalachian Mountains (Orwig and Foster 1998).

The increasing association with southern flying squirrels raises yet other issues that may further contribute to the demise of *G. sabrinus* in Pennsylvania. Aside from competition for resources (e.g., nest structures), the southern flying squirrel is hypothesized to harbor a nematode parasite that is detrimental—possibly even lethal—to the northern flying squirrel (Weigl 1969). Indeed, this parasite-mediated competition between the two species may explain the nearly nonoverlapping ranges in these species and, as well, may have helped accelerate the formation of a reproductive barrier as these two species diverged (Price et al. 1988). Although strong experimental evidence for parasite-mediated competition is still lacking, recent preliminary parasite surveys in the Northeast, show that *Strongyloides robustus* is present in both species of flying squirrels in sympatric populations in Pennsylvania but is apparently absent from *G. sabrinus* in populations to the north in the Adirondack Mountains of New York where *G. volans* has always been absent, possibly until just recently (Krichbaum et al. 2010). Certainly, more information is needed on this interaction.

Finally, the close association between the two flying squirrels raises another situation only recently considered possible (Steele et al. 2004): hybridization between the two species. On the basis of extensive sampling and molecular analyses (nuclear and mitochondrial DNA), Garroway et al. (2009) demonstrate the presence of a hybrid zone in Pennsylvania and southern Ontario. Their results, which show evidence of backcrossing, but not extensive introgression, suggest that this is a recent phenomenon due possibly to both contemporary climate change and habitat degradation. Although detailed study is further needed to delineate this hybrid zone, this example illustrates how anthropogenic factors can influence interspecific hybridization and no doubt have important evolutionary and conservation consequences.

## Invasive Species: A Vertebrate Example

As outlined in chapter 1, the spread of invasive species, from pathogens to vertebrates, poses a significant problem for native biodiversity (Pimentel et al. 2000). The lists of invasive species that now plague the Northeast, however, are too extensive and varied to discuss here. Several, such as the woolly adelgid, pose significant conservation threats to vertebrate habitat and are mentioned throughout this volume. Here, we discuss one example—an invasive vertebrate—the feral hog (*Sus scrofa*) that potentially can inflict serious harm to ecosystems of the Northeast (Sweeney et al. 2003).

Also known as wild hogs, wild boars, and feral pigs, the feral hog is the same species as the domesticated pig but wild. This feral strain was introduced to the United States from Asia and Europe initially by Spanish explorers and is now present in approximately twenty-five states, sixteen in which the species has recently colonized or is now deemed a significant problem in forested ecosystems (Sweeney et al. 2003, World Conservation Union 2007, Campbell and Long 2009). Recent increases in feral pig populations are attributed to the escape of domestic pigs followed by interbreeding with feral pigs, escape of feral pigs from shooting preserves, and the deliberate release of feral pigs by hunters (Sweeney et al. 2003). As confirmed by the U.S. Department of Agriculture Animal and Plant Health Inspection Service and reported by the Western Pennsylvania Conservancy (www.wpconline.org/assets/feral-swine.pdf), breeding populations in the state are confirmed in at least five western counties and suspected in as many as five others as far east as Wyoming County. Similar problems with this invasive pig are reported in neighboring states (Ohio, Maryland, and New York), which suggest a potential problem for the mid-Atlantic.

The effect of feral hogs includes significant destruction of wildlife and their habitats; damage to crops, including those managed by the forestry product industry; and the spread of disease to wildlife, domestic animals, and humans (Graves 1984, Sweitzer and Van Vuren 2002, Sweeney et al. 2003, Witmer et al. 2003, Hartin et al. 2007). Feral pigs exhibit a broad opportunistic diet of most plant and animal material they encounter (Graves 1984, Sweeney et al. 2003). Moreover, rooting and digging behavior by the species results in serious damage to a broad range of terrestrial habitats and negatively influences forest regeneration, while wallowing behavior in streams and wetlands degrades water quality (Sweeney et al. 2003, Campbell and Long 2009). The feral pig is also known to consume a substantial biomass of small vertebrates (e.g., mammals, amphibians, and reptiles) when such species occur in high numbers because of episodes of explosive breeding or migration (Jolley 2007, Wilcox and Van Vuren 2009). And, perhaps of greatest concern, feral hogs are important disease vectors; they are susceptible to as many as twenty viral and ten bacterial diseases, many of which pose a significant threat to both wildlife species and domestic livestock (Witmer et al. 2003).

In Pennsylvania, the Department of Agriculture (PDA) was the sole agency dealing with feral swine until about 2005 when they developed a Feral Swine Task Force, which included a number of organizations such as the Governor's Invasive Species Council and eventually the Pennsylvania Game Commission (www.wpconline.org/assets/feral-swine.pdf). By May 2006, the Pennsylvania Game Commission had removed the statewide ban on feral swine and now allows the systematic and controlled removal of this invasive species (www.pgc.state.pa.us/pgc/lib/pgc/wildlife/feral_swine/executive_letter.pdf). However, management and eradication of feral swine is not easily accomplished and in most situations, requires a comprehensive integrated strategy of various removal techniques, public education, and long-term monitoring (Campbell and Long 2009).

## White-Nose syndrome

Bats have evolved a variety of physiological and behavioral mechanisms to survive and thrive in hypervariable environments, such as the highly seasonal temperate environment of Pennsylvania. In light of this, the emergence of what has been dubbed "white-nose syndrome" and its ensuing mortality and spread has caught wildlife biologists and pathologists by surprise. White-nose syndrome, now recognized as a disease in bats, was first noted in 2006 in New York state. It is named for the white fungus that grows on the muzzle and on the ears and wing membranes of hibernating bats (fig. 7.1). In its worst manifestation, white-nose syndrome has resulted in mortality rates that exceed 95 percent, most likely from starvation (United States Fish and Wildlife Service 2009). It is also associated with depleted fat reserves by midwinter, a potentially reduced ability to arouse from deep torpor, altered hibernation arousal patterns, changes in physiology, damage to wing membranes, and atypical behavior (e.g., "staging" near cave and mine entrances and emergence from the hibernacula) during the winter.

Since its first discovery (and as of the spring 2009 emergence from hibernation), white-nose syndrome has been identified in Canada, New York, Vermont, New Jersey, New Hampshire, Massachusetts, Connecticut, Pennsylvania, Virginia, and West Virginia (fig. 7.2). It has spread faster than expected and is predicted to continue to spread. White-nose syndrome was first noted in Pennsylvania in late December 2008, and by April 2009, at least ten affected hibernacula had been identified, including sites in Lackawanna, Luzerne, and Mifflin counties. White-nose syndrome affected hibernacula, including coal mines, limestone caves, and iron mines.

The characterization of the fungus associated with white-nose syndrome affected bats by Blehert et al. (2009) strongly suggests that this fungus (now identified as the newly described *Geomyces* spp.) may in fact be an emerging fungal pathogen. Experiments are under way to determine whether it is the causal agent and to examine the mechanisms by which this fungus is either directly or indirectly leading to bat death. Multiple laboratories as well as state and federal biologists are diligently collaborating to develop and test hypotheses related to white-nose syndrome (see Reeder and Turner 2008, United States Fish and Wildlife Service 2009) and to discuss mitigation and conservation strategies. Within Pennsylvania, white-nose syndrome work has included participation in regional efforts to monitor and conduct surveillance for white-nose syndrome, as well as partaking in studies examining the effects of white-nose syndrome on hibernation patterns and physiology and on summer survivorship and reproductive success.

From the emergence of this disease in 2006 through the winter of 2008-2009, it is estimated that more than 1 million bats have died from white-nose syndrome throughout the affected region. Thus far, white-nose syndrome has been identified in six species of cave-dwelling bats, including little brown *Myotis* (*Myotis lucifugus*), northern long-eared *Myotis* (*Myotis septentrionalis*), tri-colored bats (*Perimyotis subflavus* [also known as eastern pipistrelles]), big brown bats (*Eptesicus fuscus*), the threatened small-footed *Myotis* (*Myotis leibii*), and the endangered Indiana bat (*Myotis sodalis*). The number of affected sites and counties in Pennsylvania is likely an underestimate as many hibernacula are unknown or not able to be surveyed. In addition, to stop the potential anthropogenic spread of white-nose syndrome in Pennsylvania, some previously scheduled hibernacula surveys were suspended in the winter of 2008-2009, including sites within the known affected

Fig. 7.1. White-nose syndrome. Photo courtesy of Cal Butchkoski.

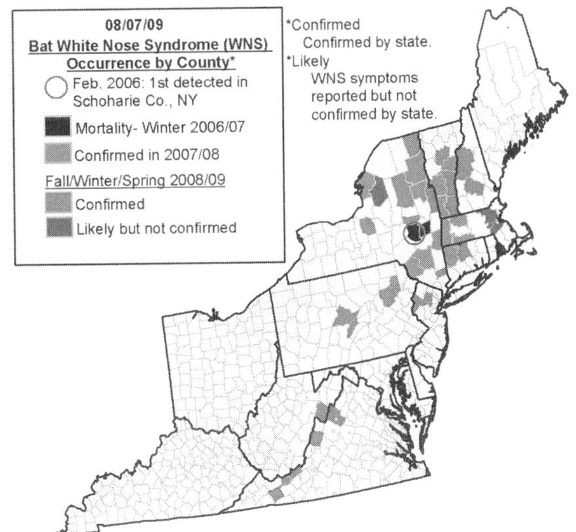


Fig. 7.2. Documented distribution of White-nose syndrome as of June 6, 2009. Map courtesy of Cal Butchkoski.

counties. Strict equipment and personal decontamination protocols have also been established (see the state white-nose syndrome updates and protocols available at www.pgc.state.pa.us).

Given the estimated millions of hibernating bats found throughout the affected region and in the projected path of the disease, white-nose syndrome represents a significant and unprecedented problem with likely dire consequences not only in Pennsylvania but also in continental North America as well. If the cold-loving *Geomyces* fungus is the disease-causing agent, other vertebrates that use daily or seasonal torpor may also be at risk.

## The Complexity of Forest Fragmentation and Its Effect on Avian Communities

Habitat fragmentation is one of the most pervasive and problematic threats to wildlife in Pennsylvania, the Northeast, and worldwide. It occurs when large contiguous blocks of habitat are broken up into smaller patches of habitat by other land uses or when blocks of habitat are penetrated by roads, transmission lines, or other corridors. Although habitat fragmentation can occur in any habitat type and does affect a vast array of species, we know the most about how fragmentation of contiguous blocks of forest habitat affects forest-dwelling Neotropical migrants (Faaborg et al. 1995, Walters 1998, Brittingham and Goodrich 2010).

Fragmentation results in both a quantitative and qualitative loss of habitat and can affect birds in a number of ways. As forests are converted to non-forest habitat, there is a direct loss of habitat for forest-dwelling birds. Perhaps more problematic is the resulting increase in forest edge habitat and decrease in forest interior (forest habitat away from an edge or opening). This change in amount of edge to interior is particularly detrimental to a group of birds known as area-sensitive or forest-interior songbirds. This group includes many of our migrant songbirds like warblers, thrushes, and tanagers that breed in Pennsylvania forests and winter in the Central and South America. For a variety of reasons, nest predators that feed on eggs and nestlings, and the brown-headed cowbird (*Molothrus ater*), an obligate brood parasite that lays its eggs in the nests of other species, tend to be more abundant close to edges and openings than within the forest interior. As a result, songbirds nesting near edges and openings are much less likely to successfully raise young than individuals that are able to nest away from edges and openings. As fragmentation increases, there is less forest interior and more forest edge. As a consequence more individuals nest near edges where nest success is low. Fewer young are produced, and eventually populations decline and disappear from the area.

When an area is fragmented, we see changes, or shifts, in the wildlife community as some species become more abundant while others decline. Species that tend to do well and increase in abundance are species that are habitat generalists, use a mix of habitat types, are tolerant of disturbance, and can easily coexist with people. This group includes birds, such as crows (*Corvus* spp.), blue jays (*Cyanocitta cristata*), and brown-headed cowbirds, and familiar backyard birds, such as chickadees (*Poecile* spp.) and cardinals (*Cardinalis cardinalis*). Species that tend to decline in number and eventually disappear from the area include habitat specialists, area-sensitive or forest-interior songbirds, and species such as the broad-winged hawk (*Buteo platypterus*) and northern goshawk (*Accipiter gentilis*) that are intolerant of disturbance. Many of our species of greatest conservation need are included in the group of species that decline in number as forests are fragmented. This is of great concern as we see an increase in abundance of wildlife species that are common throughout the state and a decline in abundance of rare species and species with regional and national significance.

The effects of fragmentation are complicated. They vary with the size of the opening (larger openings produce greater fragmentation effects), the type of opening (permanent openings are more detrimental than temporary openings), and the surrounding landscape. Fragmentation effects are most evident in landscapes that are around 45 percent to 55 percent forested (Thompson et al. 2002). With less forest, the entire landscape tends to be fragmented, and nest success is low throughout. In areas with >90 percent forest cover, the number of generalist nest predators and cowbirds are generally rare in the region, so they are initially less available to use the openings and edges.

In the past, agriculture was an important cause of fragmentation as forests were cleared for farms. Today, fragmentation is primarily due to suburban sprawl and the creation of permanent edges by roads and utility rights of way. Energy exploration and development also fragment habitats. Earlier in this chapter, we discussed the direct effects of wind-energy development on birds and bats, but there are also indirect effects through habitat fragmentation. In addition to habitat loss and fragmentation resulting from openings for the turbines, there are road networks and transmission line corridors that cause extensive fragmentation as they may extend for long distances, often traversing through extensive blocks of forest habitat. A new challenge to forest integrity is the accelerating pace of natural gas exploration and development in Pennsylvania. Much of the new drilling activity is targeted at natural gas found in the Marcellus Shale Formation. As with wind energy, fragmentation from the infrastructure of roads, and, in this case, pipelines, can be particularly detrimental. This is of particular concern because the Marcellus Shale Formation covers much of the Allegheny Plateau region, which encompasses the largest block of contiguous forest and is the stronghold for many forest-dwelling Neotropical migrants. It is also the location of much of the public lands that have played an important role in providing habitat for area-sensitive forest species.

From a regional and global perspective, Pennsylvania plays an important role in maintaining populations of area-sensitive songbirds and other forest habitat specialists. As we look to the future, we need to minimize future habitat fragmentation and maintain our remaining core forests and large blocks of contiguous forests to retain viable and abundant populations of the diversity of forest birds currently breeding within Pennsylvania forests.

## The Ever-Growing Challenge of Road Mortality

A nearly century-old threat that continues to grow throughout the Northeast and will continue to affect vertebrate populations long into the future is road mortality. As discussed throughout this volume, road mortality is a significant threat to many vertebrates of greatest conservation need (Glista and DeVault 2008). Road mortality, however, is not random; it varies considerably with road type, traffic volume, and habitat type and often results in hot spots of mortality for certain species or species assemblages (Gibbs and Shriver 2002, Glista and DeVault 2008, Langen et al. 2009). Such clusters of mortality are invariably associated with wetlands. In New York state, for example, spatial patterns of mortality of reptiles and amphibians were highly clustered, remained constant over time, and were most often associated with roads within 100 m of a wetland, especially when the road dissected a wetland (Langen et al. 2009). By both fragmenting habitats and serving as a major source of mortality, roads can significantly contribute to local extinction, and when this occurs near biodiversity hot spots, the effects can be catastrophic.

Road mortality is inevitable, and it will most likely increase as more roads are developed and traffic volume increases. Although it cannot be eliminated, future efforts must be directed at studying and managing this growing problem so that we optimize road networks, traffic patterns, and road designs to accommodate movement and dispersal of species and minimize mortality.

## Looking to the Future with a Lesson from the Past

Finally, we close the volume with a brief reference to a significant long-standing environmental concern that has contributed significantly to loss and degradation of habitat, one that often poses a significant threat to vertebrate populations throughout the Northeast: acid mine drainage. Historically, more than 50 percent of the wetlands that occurred in Pennsylvania before human settlement have been lost, an estimate that far exceeds both the national (14%) and regional (27%) averages (Tiner 1990). However, significant wetland degradation has also occurred as a result of acid mine drainage due to poor mining practices in both the northeastern and western portions of the state. Acid mine drainage, referenced repeatedly throughout this volume, has a significant long-term effect on water quality as a result of high levels of dissolved iron and sulfate and other pollutants, sedimentation, high levels of acidity, and physical disturbance to benthic substrates (see review by Bruns 2005). These effects significantly reduce macroinvertebrate richness, a primary component of the food webs in these systems (Bruns 2005) and, in turn, leave affected wetland systems virtually uninhabitable for many vertebrates.

Efforts to heal scarred lands from past mining practices have met with some success, but the repair will take many years and acid mine drainage problems will be around for decades before such lands are truly reclaimed. Indeed, such poignant lessons from the past must serve to guide future decision making.

We face numerous challenges in the conservation of biodiversity, and there is much to be done. Yet there are many reasons to be hopeful. Today, we understand many of the problems of rarity and species decline far better than we did a few decades ago. With more focused and comprehensive education, increased funding for research and conservation, growing public and government support, and the benefit of new technologies, we will continue to make the changes necessary to conserve Pennsylvania's biodiversity for generations to come.

# Appendix

A COMPOSITE LIST OF ALL VERTEBRATE SPECIES OF CONCERN FROM THE MID-Atlantic and northeastern states provided courtesy of Northeast Wildlife Diversity Technical Committee of the Northeastern Association of Fish and Wildlife Agencies. This appendix is a compilation of Species of Greatest Conservation Needs (SGCN) from the Northeast States' Wildlife Action Plans (WAP). Each state used its own criteria and categories for developing its respective SGCN lists for their action plans. Each state WAP should be consulted to explain the process and criteria used to identify SGCN and to explain what each category means to that state's conservation efforts. It should be cited as:

Whitlock, A. L., and L. L. Carpenter. 2007. A comprehensive list of Species of Greatest Conservation Need from the Northeast States' Wildlife Action Plans. Unpublished report for the Northeast Wildlife Diversity Technical Committee of the Northeastern Association of Fish and Wildlife Agencies.

Northeast States' Wildlife Action Plans (NES WAPs)
Comprehensive List of Species of Greatest Conservation Need (SGCN)—November 2007

## Key to Appendix Table

I) Consists of vertebrates by taxonomic groupings and lists species alphabetically by scientific name.

II) Includes seventeen columns (the states are arranged in geographic order, North to South). The first four columns include the following:

  a. Scientific Name: Multiple names (different names by different state for same species) are listed with "/" between. Example: *Glyptemys / Clemmys muhlenbergi*.

  b. Common Name: Multiple names are listed with "/" between each.

    i. * References those species identified by Northeast Endangered Species and Wildlife Diversity Technical Committee (Glenn D. Therres, Chairman) as Wildlife Species of Regional Conservation Concern in the Northeastern United States, published in *Northeast Wildlife*, vol. 54, 1999, pages 93–100.

    ii. ** Those species that further warrant federal Endangered or Threatened Species listing considerations, including prelisting status reviews.

c. U.S. federal regulatory status: E = Endangered, T = Threatened, C = Candidate.
d. Within the state columns, state *regulatory* status is given before the "/". Example: Maine E/2 = State Endangered/Priority 2 on SGCN List. State status ranks that do not carry regulatory authority were not included. Species are highlighted if they are believed currently present in the state; this includes migratory birds but does not include occasional or vagrant species or those with historical records.

Specific state notation follows:

ME: Maine:
- State Status: E = Endangered, T = Threatened (includes changes January 2007)
- State Plan: 1 = Very High Priority, 2 = High Priority; [Not included in Northeast Compilation: 3 = Moderate Priority; 4 = Low Priority].

NH: New Hampshire:
- State Status: E = Endangered, T = Threatened
- State Plan: X = Species on SGCN list; no priority ranking given

VT: Vermont
- State Status: E = Endangered, T = Threatened
- State Plan: HP = High Priority, MP = Medium Priority

MA: Massachusetts
- State Status: E = Endangered, T = Threatened, SC = Special Concern
- State Plan: X = Species on SGCN list; no priority ranking given

RI: Rhode Island
- State Status: E = Endangered
- State Plan: X = Species on SGCN list; no priority ranking given

CT: Connecticut
- State Status: E = Endangered, T = Threatened, SC = Special Concern
- State Plan: X = Species on SGCN list; no priority ranking given

NY: New York
- State Status: E = Endangered, T = Threatened
- State Plan: X = Species on SGCN list; no priority ranking given

NJ: New Jersey
- State Status: E = Endangered, T = Threatened, SC = Special Concern
- State Plan: X = Species on SGCN list; no priority ranking given

PA: Pennsylvania
- State Status: E = Endangered, T = Threatened
- State Plan: IC = Immediate Concern, HC = High-Level Concern, PV = Pennsylvania Vulnerable, RS = Responsibility Species, MC = Maintenance Concern

DE: Delaware
- State Status: E = Endangered, T = Threatened
- State Plan: 1 = Most in need, 2 = In need, not as urgent

MD: Maryland
- State Status: E = Endangered, T = Threatened
- State Plan: X = Species on SGCN list; no priority ranking given

DC: District of Columbia
- State Status: DC does not have State status ranking
- State Plan: X = Species on SGCN list; no priority ranking given

VA: Virginia
- State Status: E = Endangered, T = Threatened
- State Plan: Tiers I–IV

WV: West Virginia
- State Status: WVA does not have State status ranking
- State Plan: 1 = Priority 1, 2 = Priority 2

Taxonomic questions needing clarification for yellow-bellied sapsucker and southern fox squirrel.

| Mammals (n = 87)<br>Scientific Name | Common Name | US | ME | NH | VT | MA | RI | CT | NY | NJ | PA | DE | MD | DC | VA | WV |
|---|---|---|---|---|---|---|---|---|---|---|---|---|---|---|---|---|
| Alces alces | Moose | | | | | X | | | | | | | | | | |
| Balaenoptera borealis | Sei Whale | E | E/1 | | | E/X | X | | E/X | X | | 1 | E/X | | | |
| Balaenoptera musculus | Blue Whale | E | | | | E/X | X | | E/X | X | | 1 | E/X | | | |
| Balaenoptera physalus | Finback Whale | E | E/1 | | | E/X | X | | E/X | X | | 1 | E/X | | | |
| Balaenoptera acutorostrata | Northern Minke Whale | | | | | | X | | | | | | | | | |
| Canis latrans | Coyote | | | | | | | | | | | 2 | | | | |
| Canis lupus | Gray Wolf | E | 2 | X | MP | | | | E/X | | | | | | | |
| Clethrionomys gapperi | Southern Red-backed Vole | | | | | | | X | | | | | | | | |
| Condylura cristata | Star-Nosed Mole | | | | | | | | | | | | | | | 2 |
| Condylura cristata parva | Southeastern Star-nosed Mole | | | | | | | | | | | | X | | | |
| Corynorhinus rafinesquii | Rafinesque's Big-eared Bat | | | | | | | | | | | | X | | | 1 |
| Corynorhinus rafinesquii macrotis | Eastern Big-eared Bat | | | | | | | | | | | | | | E/I | |
| Corynorhinus townsendii virginianus | Virginia Big-eared Bat | E | | | | | | | | | | | | | E/II | 1 |
| Cryptotis parva | Least Shrew* | | | | | | | E/X | X | | E/HC | 2 | X | | | 2 |
| Didelphis virginiana | Virginia Opossum | | | | | | | | | | | | | X | | |
| Eptesicus fuscus | Big Brown Bat | | | | MP | | | | | | | | | | | |
| Erethizon dorsatum | North American Porcupine | | | | | | X | | | | | | I/X | | | |
| Eubalaena glacialis | Northern Right Whale | E | E/1 | | | E/X | X | | E/X | E/X | | 1 | E/X | | | |
| Glaucomys sabrinus | Northern Flying Squirrel | | | | MP | | | X | | | HC | | X | | | |
| Glaucomys sabrinus coloratus | Carolina Northern Flying Squirrel | E | | | | | | | | | | | | | E/I | |
| Glaucomys sabrinus fuscus | Virginia / West Virginia Northern Flying Squirrel | E | | | | | | | | | | | | | E/I | 1 |
| Glaucomys volans | Southern Flying Squirrel | | | | MP | | | | | | | | | X | | |
| Lasionycteris noctivagans | Silver-haired Bat** | | | X | HP | X | X | SC/X | X | X | HC | 2 | X | | | 2 |
| Lasiurus borealis | Eastern Red Bat* | | | X | HP | X | X | SC/X | X | X | MC | 2 | X | X | | 2 |
| Lasiurus cinereus | Hoary Bat* | | | X | HP | X | X | SC/X | X | X | MC | 2 | X | | | 2 |
| Lepus americanus | Snowshoe Hare | | | | | | | | | | MC | | X | | E/I | |
| Lontra canadensis | Northern River Otter | | | | MP | | | | X | X | MC | | | X | | 2 |
| Lynx canadensis | Lynx* | T | 2 | E/X | E/HP | | | | X | | | | | | | |
| Lynx rufus | Bobcat | | | X | MP | X | X | X | | E/X | | | I/X | | | |
| Martes americana | Pine Marten | | | T/X | E/HP | | | | X | | | | X | | | |

(continued)

**Mammals (n = 87)** *(continued)*

| Scientific Name | Common Name | US | ME | NH | VT | MA | RI | CT | NY | NJ | PA | DE | MD | DC | VA | WV |
|---|---|---|---|---|---|---|---|---|---|---|---|---|---|---|---|---|
| Martes pennanti | Fisher | | | | | | | | | | MC | | | | II | 2 |
| Megaptera novaeangliae | Humpback Whale | E | E/1 | | | E/X | X | | E/X | E/X | | 1 | E/X | | | |
| Microtus breweri | Beach Vole | | | | | X | | | | | | | | | | |
| Microtus chrotorrhinus carolinensis | Southern Rock Vole** | | | | HP | | | | | | PV | | E/X | | E/II | 1 |
| Microtus ochrogaster | Prairie Vole | | | | | | | | | | | | | | | 2 |
| Microtus pennsylvanicus provectus | Block Island Meadow Vole | | | | | | X | | | | | | | | | |
| Microtus pennsylvanicus shattucki | Penobscot Meadow Vole | | 1 | | | | | | | | | | | | | |
| Microtus pinetorum | Woodland Vole | | | | HP | | | X | | | | | | | | |
| Mustela erminea | Ermine Short-tailed Weasel | | | | | | | X | | | | | | | | |
| Mustela frenata | Long-tailed Weasel | | | | MP | | | X | | | | | | | | |
| Mustela nivalis | Least Weasel | | | | | | | | X | | MC | | I/X | | IV | |
| Mustela vison | Mink | | | | MP | | | X | | | | | | X | | |
| Myotis austroriparius | Southeastern Myotis** | | | | | | | | | | | | X | | IV | |
| Myotis grisescens | Gray Myotis | E | | | | | | | | | | | | | E/II | |
| Myotis leibii | Eastern Small-footed Bat** | | 2 | E/X | T/HP | SC/X | X | SC/X | SC/X | SC/X | T/IC-RS | 1 | I/X | X | III | 1 |
| Myotis lucifugus | Little Brown Myotis | | | | MP | | X | X | | | | | | | | |
| Myotis septentrionalis | Northern Long-eared Bat | | | X | MP | | X | X | | | RS | 2 | | | | |
| Myotis sodalis | Indiana Bat | E | | X | E/HP | E/X | X | E/X | E/X | E/X | E/IC | | E/X | | E/I | 1 |
| Napaeozapus insignis | Woodland Jumping Mouse | | | | | | | X | | | | | | | | |
| Neotoma magister | Allegheny Woodrat** | | | | | | | | E/X | E/X | T/IC-RS | | E/X | X | IV | 1 |
| Nycticelus humeralis | Evening Bat | | | | | | | | | | | 2 | | | | 2 |
| Ochrotomys nuttalli | Golden Mouse | | | | | | | | | | | | | | | 2 |
| Ondatra zibethicus | Muskrat | | | | MP | | | X | | | | | | | | |
| Oryzomys palustris | Marsh Rice Rat | | | | | | | | | X | | | | | | |
| Parascalops breweri | Hairy-tailed Mole | | | | MP | | | X | | | | | | | | |
| Peromyscus gossypinus | Cotton Mouse | | | | | | | | | | | | | | IV | |
| Peromyscus leucopus easti | Pungo White-footed Mouse | | | | | | | | | | | | | | III | |
| Peromyscus maniculatus | Deer Mouse | | | | | | | X | | | | | | | | |
| Phoca vitulina | Harbor Seal | | | | | | X | X | | | | | | | | |
| Phocoena phocoena | Harbor Porpoise** | | | | | X | | SC/X | SC/X | | | 1 | X | | | |

| | | | | | | | | | | | | | | | | |
|---|---|---|---|---|---|---|---|---|---|---|---|---|---|---|---|---|
| Physeter catodon | Sperm Whale | | E/1 | | | E/X | X | | E/X | X | | 1 | E/X | | | |
| Pipistrellus subflavus | Eastern Pipistrelle | | | X | HP | | X | X | | | | | | | | |
| Puma / Felis concolor | Eastern Cougar / Mountain Lion / Puma | E | | | E/MP | | | | E/X | | | 2 | | | | |
| Reithrodontomys humulis | Eastern Harvest Mouse | | | | | | | | | | | | X | | | 1 |
| Sciurus niger vulpinus[a] | Eastern Fox Squirrel[a] | | | | | | | | | | | | | | | |
| Sciurus n. niger[a] | Southern Fox Squirrel[a] | | | | | | | | | | HC | | | | III | |
| Sciurus niger cinereus | Delmarva Fox Squirrel | E | | | | | | | | | IC | E/1 | E/X | | E/II | |
| Sorex cinereus | Cinereus / Masked Shrew | | | | MP | | | | | | | | | | | |
| Sorex dispar | Long-tailed or Rock Shrew | | | | HP | SC/X | | | | X | MC | | I/X | | IV | 2 |
| Sorex fontinalis | Maryland Shrew | | | | | | | | | | | 2 | | | | |
| Sorex fumeus | Smoky Shrew | | | | MP | | X | | | | | | I/X | | | |
| Sorex hoyi winnemana | Pygmy Shrew | | | | HP | | | | | | | | X | | | 2 |
| Sorex longirostris | Southeastern Shrew | | | | | | | | | | | | X | | | |
| Sorex longirostris fisheri | Dismal Swamp / Southeastern Shrew | | | | | | | | | | | | | | T/IV | |
| Sorex palustris[a] | American Water Shrew | | | | | SC/X | X | X | | | | | | | II | |
| Sorex palustris albibarbis[a] | Northern Water Shrew | | | | | | | | | | MC | | | | | |
| Sorex palustris punctulatus | Southern / West Virginia Water Shrew** | | | | HP | | | | | | T/IC | | E/X | | | 1 |
| Spilogale putorius | Eastern Spotted Skunk | | | | | | | | | | HC | | X | | IV | 1 |
| Sylvilagus floridanus | Eastern Cottontail | | | | | | | | | | | | | X | | |
| Sylvilagus obscurus | Appalachian Cottontail* | | | | | | | | | | HC-RS | | | | IV | 2 |
| Sylvilagus palustris | Marsh Rabbit | | | | | | | | | | | | | | IV | |
| Sylvilagus transitionalis | New England Cottontail** | C | E/1 | SC/X | HP | X | X | X | SC/X | | | | I/X | | | |
| Synaptomys borealis | Northern Bog Lemming** | | T/2 | | HP | | | | | | | | | | | |
| Synaptomys cooperi | Southern Bog Lemming | | | | HP | SC/X | X | SC/X | | X | MC | | X | X | IV | 2 |
| Tamias striatus | Eastern Chipmunk | | | | | | | | | | | | | X | | |
| Urocyon cinereoargenteus | Common Gray Fox | | | | MP | | | | | | | | | X | | |
| Ursus americanus | Black Bear | | | BGP-X | MP | X | X | X | | | | | | | | |
| Zapus hudsonius | Meadow Jumping Mouse | | | | | | | X | | | | | | | | 2 |
| [a]Species requires further taxonomic clarification. | | | 11 | 14 | 33 | 20 | 24 | 27 | 21 | 18 | 22 | 18 | 34 | 11 | 24 | 25 |

**Birds (n = 267)**

| Scientific Name | Common Name | US | ME | NH | VT | MA | RI | CT | NY | NJ | PA | DE | MD | DC | VA | WV |
|---|---|---|---|---|---|---|---|---|---|---|---|---|---|---|---|---|
| Accipiter cooperii | Cooper's Hawk | | | T/X | MP | | | X | SC/X | T/X | | E/1 | | | | 2 |
| Accipiter gentilis | Northern Goshawk | | | X | MP | | X | X | SC/X | E/X | PV | | E/X | | | 1 |
| Accipiter striatus | Sharp-shinned Hawk | | | | | SC/X | X | E/X | SC/X | SC/X | MC | 1 | X | | | 2 |
| Actitis macularia | Spotted Sandpiper | | | | | | X | X | | SC/X | | 1 | | | | 2 |
| Aegolius acadicus | Northern Saw-whet Owl | | | | | | | SC/X | | | | | X | | II | 1 |
| Aimophila aestivalis | Bachman's Sparrow | | | | | | | | | | | | X | | T/I | 1 |
| Aix sponsa | Wood Duck | | | | | | | | | X | | | | X | | |
| Alca torda | Razorbill | | E/2 | | | | | | X | X | | | | | | |
| Ammodramus caudacutus | Saltmarsh Sharp-tailed Sparrow* | | 1 | X | | X | X | SC/X | X | X | | 1 | X | | II | |
| Ammodramus henslowii | Henslow's Sparrow* | | | | E/MP | E/X | | | T/X | E/X | HC-RS | E/1 | T/X | | T/I | 1 |
| Ammodramus maritimus | Seaside Sparrow | | | X | | X | X | SC/X | SC/X | X | | 1 | X | | IV | |
| Ammodramus nelsoni | Nelson's Sharp-tailed Sparrow | | 2 | X | | | | | | X | | | | | III | |
| Ammodramus savannarum | Grasshopper Sparrow | | E/2 | T/X | T/HP | T/X | X | E/X | X | T/X | MC | 2 | X | X | IV | 2 |
| Anas acuta | Northern Pintail | | | | | | | | X | X | | | | | | |
| Anas clypeata | Northern Shoveler | | | | | | | | | | | 2 | | | | |
| Anas crecca | Green-winged Teal | | | | | | X | | | | PV | | | | | 1 |
| Anas discors | Blue-winged Teal | | | | MP | | X | T/X | X | | | | | | | |
| Anas platyrhynchos | Mallard | | | | | | | | | | | 2 | | | | |
| Anas rubripes | American Black Duck | | 2 | X | HP | X | X | X | X | X | MC | 1 | X | X | II | 2 |
| Anas strepera | Gadwall | | | | | | X | | | | | | | | | |
| Anthus rubescens | American Pipit | | E/2 | X | | | | | | | | | | | | |
| Aquila chrysaetos | Golden Eagle* | | E/2 | E/X | | | | | E/X | | PV | | X | | | |
| Archliochus columbris | Ruby-throated Hummingbird | | | | | | | X | | | | | | | | |
| Ardea / Casmerodius alba | Great Egret | | 2 | | | | X | T/X | X | X | E/PV | 2 | X | | | |
| Ardea herodias | Great Blue Heron | | 2 | X | MP | | X | X | | SC/X | MC | 2 | X | | | 2 |
| Arenaria interpres | Ruddy Turnstone | | 2 | | | X | X | X | X | X | | 1 | X | | | |
| Asio flammeus | Short-eared Owl* | | T/1 | | MP | E/X | X | T/X | E/X | E/X | E/IC | E/1 | E/X | | | 1 |
| Asio otus | Long-eared Owl* | | 2 | | MP | SC/X | X | E/X | X | T/X | HC | 1 | X | | | 1 |
| Aythya affinis | Lesser Scaup | | | | | | | X | X | X | | 2 | | | | |
| Aythya americana | Redhead | | | | | | | | | | | 2 | | | III | |

| | | | | | | | | | | | | | | | | |
|---|---|---|---|---|---|---|---|---|---|---|---|---|---|---|---|---|
| Aythya marila | Greater Scaup | | 2 | | | | | X | X | X | | 2 | | | IV | |
| Aythya valisineria | Canvasback | | | | | | | X | | X | | 2 | X | | | |
| Bartramia longicauda | Upland Sandpiper* | | T/1 | E/X | E/HP | E/X | E/X | E/X | T/X | E/X | T/IC | E/1 | E/X | | T/I | 1 |
| Bonasa umbellus | Ruffed Grouse | | | X | MP | X | X | X | X | X | | 2 | | | | |
| Botaurus lentiginosus | American Bittern* | | 2 | X | HP | E/X | E/X | E/X | SC/X | E/X | E/HC | 2 | I/X | X | II | 1 |
| Branta bernicla | Atlantic Brant | | | | | | | | X | X | | 2 | X | | III | |
| Branta canadensis | Canada Goose | | | | | | | | | X | | 1 | | | | |
| Bubo virginianus | Great Horned Owl | | | | | | | X | | | | | | X | | |
| Bubulcus ibis | Cattle Egret | | 2 | | | | X | | X | SC/X | | 2 | | | | |
| Bucephala albeola | Bufflehead | | | | | | | | | X | | 2 | | | | |
| Bucephala clangula | Common Goldeneye | | | | | | | | | | | | | | | |
| Bucephala islandica | Barrow's Goldeneye | | T/2 | | | | | | | | | | | | | |
| Buteo lagopus | Rough-legged Hawk | | | | | | | X | | | | | | | | |
| Buteo lineatus | Red-shouldered Hawk | | | X | MP | | | X | SC/X | E/X | MC | 2 | X | X | | |
| Buteo platypterus | Broad-winged Hawk | | | | | X | X | X | | SC/X | MC | 1 | X | X | | |
| Butorides virescens | Green Heron | | | | | X | | X | | X | | | | | IV | |
| Calidris alba | Sanderling | | 2 | | | X | X | X | X | SC/X | | 1 | X | | | |
| Calidris alpina | Dunlin | | | | | | X | | X | | | 2 | X | | IV | |
| Calidris canutus | Red Knot* | C | 2 | | | X | X | | X | T/X | | 1 | X | | IV | |
| Calidris fuscicollis | White-rumped Sandpiper | | | | | | X | | | | | 2 | | | | |
| Calidris maritima | Purple Sandpiper | | 2 | X | | | X | | X | X | | 2 | X | | IV | |
| Calidris melanotos | Pectoral Sandpiper | | | | | | X | | | | | | | | | |
| Calidris minutilla | Least Sandpiper | | | | | | X | | | | | | | | | |
| Calidris pusilla | Semipalmated Sandpiper | | 2 | X | | | X | X | X | X | | 2 | X | | | |
| Calonectris diomedea | Cory's Shearwater | | | | | | | | X | | | | | | | |
| Caprimulgus carolinensis | Chuck-will's-widow | | | | | | | | | X | | | X | | IV | 1 |
| Caprimulgus vociferus | Whip-poor-will* | | 2 | X | HP | X | X | SC/X | SC/X | X | MC | 2 | X | | IV | 1 |
| Carpodacus purpureus | Purple Finch | | 2 | X | | | X | X | | X | | | | | | |
| Carduelis pinus | Pine Siskin | | | | | | | | | | PV | | | | | 1 |
| Catharus bicknelli | Bicknell's Thrush** | | 1 | X | HP | | | | SC/X | | | 1 | X | | IV | |
| Catharus fuscescens | Veery | | 2 | X | MP | | | X | | SC/X | | 2 | X | | | |
| Catharus guttatus | Hermit Thrush | | | | | | X | X | | | | | X | | | |

(continued)

**Birds (n = 267)** *(continued)*

| Scientific Name | Common Name | US | ME | NH | VT | MA | RI | CT | NY | NJ | PA | DE | MD | DC | VA | WV |
|---|---|---|---|---|---|---|---|---|---|---|---|---|---|---|---|---|
| Catharus minimus | Gray-cheeked Thrush | | | | | | | X | | SC/X | | | | | | |
| Catharus ustulatus | Swainson's Thrush | | | | | | | X | | | PV | | X | | | 1 |
| Catoptrophorus semipalmatus | Willet | | 2 | X | | | X | X | X | X | | 2 | X | | | |
| Cepphus grylle | Black Guillemot | | | X | | | | | | | | | | | | |
| Certhia americana | Brown Creeper | | | | | | X | X | | | | E/1 | X | X | IV | 2 |
| Ceryle alcyon | Belted Kingfisher | | | | | | X | X | | | | | | | | |
| Chaetura pelagica | Chimney Swift | | 2 | | MP | | X | X | | X | MC | 2 | | X | IV | |
| Charadrius melodus | Piping Plover | T | E/1 | E/X | | T/X | X | T/X | E/X | X | IC | E/1 | E/X | | T/I | |
| Charadrius semipalmatus | Semipalmated Plover | | | | | | X | | | | | | | | | |
| Charadrius wilsonia | Wilson's Plover | | | | | | | | | | | 2 | E/X | | E/I | |
| Chlidonias niger | Black Tern* | | E/1 | | E/HP | | | | E/X | SC/X | E/HC | 2 | X | | | |
| Chondestes grammacus | Lark Sparrow | | | | | | | | | | | | | | | 1 |
| Chordeiles minor | Common Nighthawk | | 2 | T/X | HP | | X | E/X | SC/X | SC/X | MC | 1 | X | | | 2 |
| Circus cyaneus | Northern Harrier* | | | E/X | HP | T/X | E/X | E/X | T/X | E/X | HC | E/1 | X | | III | 1 |
| Cistothorus platensis | Sedge Wren* | | E/1 | E/X | E/HP | E/X | | E/X | T/X | E/X | E/IC | E/1 | E/X | | III | 1 |
| Cistothorus palustris | Marsh Wren | | 2 | | | | X | X | | X | HC | 2 | X | X | IV | 1 |
| Clangula hyemalis | Long-tailed Duck / Old Squaw | | | | | X | | X | X | X | | 2 | | | | |
| Coccyzus americanus | Yellow-billed Cuckoo | | | | | | X | X | | X | | | | | IV | |
| Coccyzus erythropthalmus | Black-billed Cuckoo | | 2 | | MP | | X | X | X | SC/X | MC | 2 | X | | | 2 |
| Colaptes auratus | Northern Flicker | | 2 | | | | X | X | | X | | 2 | | | | |
| Colinus virginianus | Northen Bobwhite | | | | | X | X | X | X | X | IC | 2 | X | X | IV | 2 |
| Contopus cooperi | Olive-sided Flycatcher | | 2 | | MP | | | X | X | | IC | | E/X | | | 1 |
| Contopus virens | Eastern Wood-pewee | | | | | | | X | | X | | | | | IV | 1 |
| Coragyps atratus | Black Vulture | | | | | | | | | | | 2 | | | | 2 |
| Corvus corax | Common Raven | | | | | | | SC/X | | | | | X | | | |
| Coturnicops noveboracensis | Yellow Rail | | 2 | | | | | | X | | | 2 | | | IV | |
| Cygnus columbianus | Tundra Swan | | | | | | | | | | MC-RS* | 2 | | | | |
| Dendroica caerulescens | Black-throated Blue Warbler | | 2 | | MP | | E/X | X | X | SC/X | MC | | X | | | |
| Dendroica castanea | Bay-breasted Warbler | | 2 | X | MP | | | X | X | | | | | | | |
| Dendroica cerulea | Cerulean Warbler* | | | X | MP | | X | X | SC/X | SC/X | HC-RS | E/1 | X | X | II | 1 |

| | | | | | | | | | | | | | | | | |
|---|---|---|---|---|---|---|---|---|---|---|---|---|---|---|---|---|
| Dendroica coronata | Yellow-rumped Warbler | | | | | | X | X | | | | | | | | 2 |
| Dendroica discolor | Prairie Warbler | | 2 | | MP | X | X | X | X | X | MC | 1 | X | | IV | 1 |
| Dendroica dominica | Yellow-throated Warbler | | | | | | | | | X | | 2 | | | | |
| Dendroica fusca | Blackburnian Warbler | | 2 | | | | X | X | | X | MC | | T/X | | | 2 |
| Dendroica kirtlandii | Kirtland's Warbler | | | | | | | | | | | | | | IV | |
| Dendroica magnolia | Magnolia Warbler | | | | | | | X | | | | | X | | | |
| Dendroica palmarum | Palm Warbler | | | X | | | | | | | | | | | | |
| Dendroica pensylvanica | Chestnut-sided Warbler | | 2 | | MP | | X | X | | | | 2 | X | | | |
| Dendroica petechia | Yellow Warbler | | | | | | X | | | | | | | | IV | |
| Dendroica pinus | Pine Warbler | | | | | | | | | X | | | | | | |
| Dendroica striata | Blackpoll Warbler | | | | MP | SC/X | | | | | E/PV | | | | | |
| Dendroica tigrina | Cape May Warbler | | 2 | | | | | X | X | | | | | | | |
| Dendroica virens | Black-throated Green Warbler | | 2 | | | | X | X | | SC/X | MC | | X | | | |
| Dendroica virens waynei | Wayne's Black-throated Green Warbler | | | | | | | | | | | | X | | I | |
| Dolichonyx oryzivorus | Bobolink | | 2 | | MP | | X | SC/X | X | T/X | MC | 2 | X | X | | 2 |
| Dryocopus pileatus | Pileated Woodpecker | | | | | | X | X | | | | | X | | | |
| Dumetella carolinensis | Gray Catbird | | | | | | X | X | | X | | | | | IV | |
| Egretta caerulea | Little Blue Heron | | 2 | | | | X | SC/X | X | SC/X | | 2 | X | | II | |
| Egretta thula | Snowy Egret | | 2 | | | X | X | T/X | X | SC/X | | 2 | X | | | |
| Egretta tricolor | Tricolored Heron | | 2 | | | | | | X | SC/X | | 2 | X | | III | |
| Empidonax alnorum | Alder Flycatcher | | | | | | | SC/X | | | MC | | I/X | | | 2 |
| Empidonax flaviventris | Yellow-bellied Flycatcher | | | | | | | | | | E/PV | | | | | 1 |
| Empidonax minimus | Least Flycatcher | | | | | | X | X | | SC/X | | 2 | X | | | |
| Empidonax trailii | Willow Flycatcher | | 2 | | | X | X | X | X | X | MC | 2 | X | | IV | |
| Empidonax virescens | Acadian Flycatcher | | | | | | X | X | | X | MC | | X | X | | 1 |
| Eremophila alpestris | Horned Lark | | 2 | X | | | X | E/X | SC/X | SC/X | | | | | | 2 |
| Euphagus carolinus | Rusty Blackbird | | 2 | X | MP | | | | X | | | | | | IV | |
| Falcipennis canadensis | Spruce Grouse | | | X | E/HP | | | | E/X | | | | | | | |
| Falco peregrinus | Peregrine Falcon | | E/1 | E/X | HP | E/X | X | E/X | E/X | E/X | E/HC | 2 | I/X | | T/I | 1 |
| Falco sparverius | American Kestrel | | | | MP | X | X | T/X | | SC/X | | | | | | |
| Fratercula arctica | Atlantic Puffin | | T/2 | | | | | | | | | | | | | |
| Fulica americana | American Coot | | 2 | | | | | | | | MC | 2 | | | | 1 |

*(continued)*

**Birds (n = 267)** *(continued)*

| Scientific Name | Common Name | US | ME | NH | VT | MA | RI | CT | NY | NJ | PA | DE | MD | DC | VA | WV |
|---|---|---|---|---|---|---|---|---|---|---|---|---|---|---|---|---|
| Gallinago delicata | Wilson's Snipe | | | | | | X | | | | MC | | X | X | | 1 |
| Gallinula chloropus | Common Moorhen | | T/2 | X | | SC/X | X | E/X | | SC | MC | | I/X | | | 1 |
| Gavia immer | Common Loon | | 2 | T/X | HP | SC/X | | SC/X | SC/X | | | | X | | | |
| Gavia stellata | Red-throated Loon | | | | | | | X | X | X | | 2 | X | | | |
| Geothlypis trichas | Common Yellowthroat | | | | | | X | | | | | | | | | |
| Grus canadensis | Sandhill Crane | | 2 | | | | | | | | | | | | | |
| Haematopus palliatus | American Oystercatcher | | 1 | | | X | X | SC/X | X | SC/X | | E/1 | X | | II | |
| Haliaeetus leucocephalus | Bald Eagle | T | T/2 | E/X | E/HP | E/X | X | E/X | T/X | E/X | T/HC | E/1 | T/X | X | T/II | 1 |
| Helmitheros vermivorus | Worm Eating Warbler | | | | | | X | X | X | SC/X | RS | 2 | X | X | IV | 1 |
| Himantopus mexicanus | Black-necked Stilt | | | | | | | | | | | 2 | | | | |
| Hirundo rustica | Barn Swallow | | 2 | | | | | | | | | | | | | |
| Histrionicus histrionicus | Harlequin Duck** | | T/2 | | | X | X | | X | X | | | X | | | |
| Hylocichla mustelina | Wood Thrush | | 2 | X | MP | X | X | X | X | X | RS | 1 | X | X | IV | 1 |
| Icteria virens | Yellow-breasted Chat | | | | | | X | E/X | SC/X | SC/X | MC | 2 | | | IV | |
| Icterus galbula | Baltimore Oriole | | 2 | | | | X | X | | X | | 2 | | | | |
| Icterus spurius | Orchard Oriole | | | | | | X | X | | | | | | | | |
| Ixobrychus exilis | Least Bittern | | E/2 | X | HP | E/X | X | T/X | T/X | SC/X | E/PV | 2 | I/X | X | III | 1 |
| Junco hyemalis | Dark-eyed Junco | | | | | | | X | | | | | X | | | |
| Lanius ludovicianus | Loggerhead Shrike** | | 2 | | E | | | | E/X | E/X | E/IC | E/1 | E/X | | T/I | 1 |
| Larus argentatus | Herring Gull | | | | | | X | | | | | | | | | |
| Larus atricilla | Laughing Gull | | | | | X | | | X | | | | X | | | |
| Larus marinus | Great Black-backed Gull | | | | | | X | | | | | 2 | | | | |
| Larus minutus | Little Gull | | | | | | | | X | | | 2 | | | | |
| Larus philadelphia | Bonaparte's Gull | | 2 | | | | | | X | | | | | | | |
| Larus thayeri | Thayer's Gull | | | | | | | | X | | | | | | | |
| Laterallus jamaicensis | Black Rail | | | | | | | E/X | E/X | T/X | | E/1 | I/X | | I | |
| Limnodromus griseus | Short-billed Dowitcher | | | | | X | X | | X | | | 2 | X | | IV | |
| Limnothlypis swainsonii | Swainson's Warbler | | | | | | | | | X | | E/1 | E/X | | II | 2 |
| Limosa fedoa | Marbled Godwit | | | | | | | | X | X | | 2 | | | IV | |
| Limosa haemastica | Hudsonian Godwit | | | | | | | | X | X | | 2 | | | IV | |

| | | | | | | | | | | | | | | | | |
|---|---|---|---|---|---|---|---|---|---|---|---|---|---|---|---|---|
| Lophodytes cucullantus | Hooded Merganser | | | | | | | X | | | | 2 | | | | 1 |
| Loxia curvirostra | Red Crossbill | | 2 | | | | | | | | PV | | | | I | |
| Megascops asio | Eastern Screech-owl | | 2 | | | | | X | | X | | | | | | |
| Melanerpes erythrocephalus | Red-headed Woodpecker | | | | | | | E/X | SC/X | T/X | MC | E/1 | X | | | 2 |
| Melanitta fusca | White-winged Scoter | | | | | | | X | X | X | | 2 | | | | |
| Melanitta nigra | Black Scoter | | | | | | | X | X | X | | 2 | | | | |
| Melanitta perspicillata | Surf Scoter | | | | | | | X | X | X | | 2 | | | | |
| Melospiza georgiana nigrescens | Coastal Plain Swamp Sparrow | | | | | | | | | | | | I/X | | ? | |
| Mergus merganser | Common Merganser | | | | | | | X | | | | | | | | |
| Mniotilta varia | Black-and-White Warbler | | 2 | | | | X | X | | X | | 2 | X | | IV | |
| Morus bassanus | Northern Gannet | | | | | | | | | X | | | X | | | |
| Myiarchus crinitus | Great Crested Flycatcher | | 2 | | | | X | X | | X | | 2 | | | | |
| Numenius borealis | Eskimo Curlew | | | | | X | | | | | | | | | | |
| Numenius phaeopus | Whimbrel | | 2 | | | X | X | | X | SC/X | | 1 | X | | IV | |
| Nyctanassa violacea | Yellow-crowned Night-heron | | | | | | X | SC/X | X | T/X | E/PV | E/1 | X | | II | 1 |
| Nyctea scandiaca | Snowy Owl | | | | | | | X | | | | | | | | |
| Nycticorax nycticorax | Black-crowned Night-heron | | T2 | | MP | X | X | X | X | T/X | E/PV | E/1 | X | X | III | 1 |
| Oceanodroma leucorhoa | Leach's Storm-petrel | | | | | E/X | | | | | | | | | | |
| Oporornis formosus | Kentucky Warbler | | | | | | | | X | SC/X | MC | 2 | X | X | IV | 1 |
| Oporornis philadelphia | Mourning Warbler | | | | | SC/X | | | | | | | E/X | | | |
| Oxyura jamaicensis | Ruddy Duck | | 2 | | | | | | X | | MC | | X | | | |
| Pandion haliaetus | Osprey | | | T/X | MP | | X | X | SC/X | T/X | T/PV | 1 | | | | 2 |
| Parula americana | Northern Parula | | 2 | | | T/X | X | SC/X | | SC/X | | E/1 | X | | IV | |
| Passerculus sandwichensis | Savannah Sparrow | | | | | | X | SC/X | | T/X | | 2 | X | | | |
| Passerculus sandwichensis princeps | Ipswich Sparrow | | | | | | | SC/X | | | | | | | | |
| Passerina cyanea | Indigo Bunting | | | | | | X | X | | X | | | | | | |
| Pelecanus erythrorhynchos | American White Pelican | | | | | | | | | | | 2 | | | | |
| Pelecanus occidentalis | Brown Pelican | | | | | | | | | | | 2 | X | | | |
| Perisoreus canadensis | Gray Jay | | | | MP | | | | | | | | | | | |
| Petrochelidon pyrrhonota | Cliff Swallow | | | | | | X | X | | SC/X | | 2 | | | | 2 |
| Phalacrocorax auritus | Double-crested Cormorant | | | | | | X | | | | | 2 | | | | |
| Phalacrocorax carbo | Great Cormorant | | T/2 | | | | | X | | | | 2 | | | | |

(continued)

**Birds (n = 267)** *(continued)*

| Scientific Name | Common Name | US | ME | NH | VT | MA | RI | CT | NY | NJ | PA | DE | MD | DC | VA | WV |
|---|---|---|---|---|---|---|---|---|---|---|---|---|---|---|---|---|
| Phalaropus lobatus | Red Necked Phalarope | | 2 | | | | | | X | | | 2 | | | | |
| Phalaropus tricolor | Wilson's Phalarope | | | | | | | | | X | | 2 | | | | |
| Phasianus colchicus | Ring-necked Pheasant | | | | | | X | | | | | | | | | |
| Pheuctitus ludovicianus | Rose-breasted Grosbeak | | 2 | | | | X | X | | X | | | | | IV | |
| Picoides arcticus | Black-backed Woodpecker | | | | MP | | | | | | | | | | | |
| Picoides borealis | Red-cockaded Woodpecker | E | | | | | | | | | | | X | | E/I | |
| Picoides dorsalis-see below | American Three-toed Woodpecker | | 2 | T/X | | | | | X | | | | | | | |
| Picoides villosus | Hairy Woodpecker | | | | | | X | | | | | | X | | | |
| Pipilo erythrophthalmus | Eastern / Rufous-sided Towhee | | 2 | X | HP | X | X | X | | X | | 2 | X | X | IV | |
| Piranga olivacea | Scarlet Tanager | | 2 | | | | X | X | X | X | RS | 2 | X | X | IV | |
| Piranga rubra | Summer Tanager | | | | | | | | | X | HC | | X | | | |
| Plegadis falcinellus | Glossy Ibis | | 2 | | | | X | SC/X | X | SC/X | | 2 | X | | III | |
| Pluvialis dominica | American Golden-plover | | | | | | | | X | X | | 2 | | | | |
| Pluvialis squatarola | Black-bellied Plover | | | | | | X | | X | | | 2 | X | | IV | |
| Podiceps auritus | Horned Grebe | | | | | | | X | X | X | | 2 | X | | IV | |
| Podiceps grisegena | Red-necked Grebe | | | | | | | X | | | | | | | | |
| Podilymbus podiceps | Pied-billed Grebe* | | | E/X | HP | E/X | E/X | E/X | T/X | E/X | MC | E/1 | X | | | 2 |
| Polioptila caerulea | Blue-gray Gnatcatcher | | 2 | | | | | X | | | | | | | | |
| Pooecetes gramineus | Vesper Sparrow | | 2 | X | HP | T/X | | E/X | SC/X | E/X | | 2 | X | | | 2 |
| Porzana carolina | Sora Rail | | | | MP | X | X | X | | SC/X | MC | 2 | | X | | 1 |
| Progne subis | Purple Martin | | 2 | E/X | HP | | X | T/X | | | | | | | | |
| Protonotaria citrea | Prothonotary Warbler | | | | | | X | | X | X | HC | 2 | X | X | IV | 2 |
| Puffinus gravis | Greater Shearwater | | 2 | | | | | | X | X | | 2 | | | | |
| Puffinus Iherminieri | Audubon's Shearwater | | | | | | | | | X | | 1 | | | | |
| Puffinus puffinus | Manx Shearwater | | | | | | | | | X | | | | | | |
| Quiscalus major | Boat-tailed Grackle | | | | | | | | | | | | X | | | |
| Rallus elegans | King Rail | | | | | T/X | X | E/X | T/X | SC/X | E/PV | 2 | X | | II | 1 |
| Rallus limicola | Virginia Rail | | | | | | | X | | SC/X | HC | | | X | IV | 1 |
| Rallus longirostris | Clapper Rail | | | | | | X | X | | X | | | | | IV | |
| Regulus satrapa | Golden-crowned Kinglet | | | | | | X | X | | | | | X | | | |

| | | | | | | | | | | | | | | | | |
|---|---|---|---|---|---|---|---|---|---|---|---|---|---|---|---|---|
| Rhodostethia rosea | Ross' Gull | | | | | | | | | | | 2 | | | | |
| Riparia riparia | Bank Swallow | | | | | | X | X | | | MC | 2 | X | | | 2 |
| Rynchops niger | Black Skimmer | | | | | | X | X | SC/X | E/X | | E/1 | E/X | | II | |
| Scolopax minor | American Woodcock | | 2 | X | MP | X | X | X | X | X | MC | 1 | X | X | IV | 1 |
| Seiurus aurocapillus | Ovenbird | | | | | | X | X | | | | | X | X | IV | |
| Seiurus motacilla | Louisiana Waterthrush* | | 2 | | | X | X | X | X | X | RS | 2 | X | X | IV | 1 |
| Seiurus noveboracensis | Northern Waterthrush | | | | | | X | X | | | | | X | | | 2 |
| Setophaga ruticilla | American Redstart | | | | | | | X | | | | 1 | X | | | |
| Sitta canadensis | Red-breasted Nuthatch | | | | | | X | X | | | | | X | | | |
| Sitta pusilla | Brown-headed Nuthatch | | | | | | | | | | | 2 | X | | IV | |
| Somateria mollissima | Common Eider | | 2 | | | X | | | X | X | | 1 | | | | |
| Sphyrapicus varius[a] | Yellow-bellied Sapsucker | | 2 | | | | | | | | | | | | | |
| Sphyrapicus varius appalachiensis[a] | Appalachian Yellow-bellied Sapsucker | | | | | | | | | X | | | X | | I | 1 |
| Spiza americana | Dickcissel | | | | | | | | X | X | E/HC | | X | | | 2 |
| Spizella pusilla | Field Sparrow | | 2 | | MP | X | X | X | | X | | 2 | X | X | IV | 1 |
| Stelgidopteryx serripennis | Northern Rough-winged Swallow | | | | | | | X | | | | | | | IV | |
| Sterna anaethetus | Bridled Tern | | | | | | | | | X | | 2 | | | | |
| Sterna antillarum | Least Tern* | | E/1 | E/X | | SC/X | X | T/X | T/X | E/X | | E/1 | T/X | | II | |
| Sterna caspia | Caspian Tern | | | | | | | | X | SC/X | | | | | | |
| Sterna dougallii | Roseate Tern | E | E/1 | E/X | | E/X | X | E/X | E/X | E/X | | 1 | X | | E/IV | |
| Sterna forsteri | Forster's Tern | | | | | | | | X | X | | E/1 | X | | IV | |
| Sterna hirundo | Common Tern* | | 2 | E/X | E/HP | SC/X | X | SC/X | T/X | SC/X | E/PV | E/1 | X | | III | |
| Sterna maxima | Royal Tern | | | | | | | | | X | | | E/X | | II | |
| Sterna nilotica | Gull-Billed Tern | | | | | | | | X | SC/X | | 2 | E/X | | T/I | |
| Sterna paradisaea | Arctic Tern | | T/2 | T/X | | SC/X | | | | | | 2 | | | | |
| Strix varia | Barred Owl | | 2 | | | | X | X | | T/X | | 2 | X | | | |
| Sterna sandvicensis | Sandwich Tern | | | | | | | | | | | | X | | | |
| Sturnella magna | Eastern Meadowlark | | 2 | X | MP | X | X | SC/X | X | SC/X | MC | | X | X | IV | |
| Thryomanes bewickii altus | Appalachian Bewick's Wren** | | | | | | | | | | | | E/X | | E/I | 1 |
| Toxostoma rufum | Brown Thrasher | | 2 | | MP | X | X | SC/X | X | SC/X | MC | 2 | X | X | IV | |
| Tringa flavipes | Lesser Yellowlegs | | | | MP | | X | | | | | | | | | |
| Tringa melanoleuca | Greater Yellowlegs | | 2 | | | | X | | X | X | | 2 | X | | | |

*(continued)*

**Birds (n = 267)** *(continued)*

| Scientific Name | Common Name | US | ME | NH | VT | MA | RI | CT | NY | NJ | PA | DE | MD | DC | VA | WV |
|---|---|---|---|---|---|---|---|---|---|---|---|---|---|---|---|---|
| Tringa solitaria | Solitary Sandpiper | | | | | | X | | | | MC | 2 | X | | | |
| Troglodytes troglodytes | Winter Wren | | | | | | X | X | | SC/X | MC | | X | | | |
| Troglodytes troglodytes pullus | Appalachian Winter Wren | | | | | | | | | | | | | | II | |
| Tryngites subruficollis | Buff-breasted Sandpiper | | | | | | | | X | | | 2 | | | | |
| Tyrannus tyrannus | Eastern Kingbird | | 2 | | | | X | X | | X | | 2 | | | IV | |
| Tyto alba | Barn Owl | | | | MP | SC/X | E/X | E/X | X | SC/X | MC | 2 | X | | III | 1 |
| Uria aalge | Common Murre | | 2 | | | | | | | | | | | | | |
| Vermivora chrysoptera | Golden-winged Warbler* | | | X | HP | E/X | | E/X | SC/X | SC/X | HC-RS | 2 | X | | I | 1 |
| Vermivora peregrina | Tennessee Warbler | | | | | | | | X | | | | | | | |
| Vermivora pinus | Blue-winged Warbler | | 1 | | MP | X | X | X | X | X | RS | 1 | I/X | | | 1 |
| Vermivora ruficapilla | Nashville Warbler | | | | | | X | | | SC | | | X | | | 1 |
| Vireo flavifrons | Yellow-throated Vireo | | 2 | | | | X | X | | X | MC | 2 | X | X | IV | |
| Vireo gilvus | Warbling Vireo | | | | | | | X | | | | 2 | | | | |
| Vireo griseus | White-eyed Vireo | | | | | | | X | | | | | | X | | |
| Vireo olivaceus | Red-eyed Vireo | | | | | | | | | | | | X | | | |
| Vireo solitarius | Blue-headed Vireo | | | | | | X | X | | SC/X | MC | | X | | | |
| Wilsonia citrina | Hooded Warbler | | | | | | X | X | | X | | E/1 | X | X | | |
| Wilsonia canadensis | Canada Warbler* | | 2 | X | HP | X | X | X | X | SC/X | MC | 2 | X | | IV | |
| Zonotrichia albicollis | White Throated Sparrow | | | | | X | | | | | | | | | | |
| [a]Species requires further taxonomic clarification. | | | 102 | 52 | 58 | 63 | 129 | 148 | 120 | 151 | 79 | 146 | 141 | 35 | 96 | 74 |

**Reptiles (n = 65)**

| Scientific Name | Common Name | US | ME | NH | VT | MA | RI | CT | NY | NJ | PA | DE | MD | DC | VA | WV |
|---|---|---|---|---|---|---|---|---|---|---|---|---|---|---|---|---|
| Agkistrodon contortrix | Copperhead | | | | | E/X | | X | X | SC/X | MC | 2 | | X | | |
| Apalone mutica mutica | Midland Smooth Softshell | | | | | | | | | | | | | | | 1 |
| Apalone spinifera | Eastern Spiny Softshell | | | | T/HP | | | | | | | | I/X | | IV | |
| Caretta caretta | Atlantic Loggerhead Turtle | T | T/1 | | | T/X | X | T/X | T/X | X | | E/1 | T/X | | T/I | |
| Carphophis amoenus | Worm Snake | | | | | T/X | | | X | | | | | X | | 2 |
| Cemophora coccinea | Eastern Scarletsnake | | | | | | | | | | | 1 | X | X | IV | |

| | | | | | | | | | | | | | | | | |
|---|---|---|---|---|---|---|---|---|---|---|---|---|---|---|---|---|
| Chelonia mydas | Atlantic Green Turtle | T | | | | T/X | X | T/X | T/X | X | | E/1 | T/X | | | |
| Chelydra s. serptina | Common Snapping Turtle | | | | | | | | X | | | | | | | |
| Chrysemys p. picta | Eastern Painted Turtle | | | | | | | | | | | | | X | | |
| Clemmys guttata | Spotted Turtle* | | T/2 | SC/ X | E/HP | X | X | X | SC/X | SC/X | IC-RS | 1 | X | X | III | 1 |
| Clonophis kirtlandii | Kirtland's Snake | | | | | | | | | | E/IC | | | | | |
| Cnemidophorus/Aspidoscelis s. sexlineatus | Eastern Six-lined Racerunner | | | | | | | | | | | | | | | 1 |
| Coluber c. constrictor | Northern Black Racer | | E/2 | SC/X | T/HP | X | | X | X | | | | | X | | |
| Crotalus horridus | Timber/Canebrake Rattlesnake** | | 2 | E/X | E/HP | E/X | | E/X | T/X | E/X | IC-RS | | X | X | T/ IV | 1 |
| Deirochelys reticularia | Chicken Turtle | | | | | | | | | | | | | | E/ I | |
| Dermochelys coriacea | Atlantic Leatherback Turtle | E | 1 | | | E/X | X | E/X | E/X | E/X | | E/1 | E/X | | | |
| Diadophis punctatus edwardsii | Northern Ringneck Snake | | | | | | | | | | | | | X | | |
| Elaphe g. guttata | Corn Snake | | | | | | | | | E/X | | E/1 | X | X | | 1 |
| Elaphe obsoleta | Eastern/Black Rat Snake | | | | T/HP | E/X | X | | X | | | | | | | |
| Emydoidea (Emys) blandingii | Blanding's Turtle** | | E/1 | SC/X | | T/X | | | T/X | | IC | | | | | |
| Eretmochelys i. imbricata | Atlantic Hawksbill Turtle | E | | | | E/X | X | | E/X | X | | 2 | E/X | | | |
| Eumeces a. anthracinus | Northern Coal Skink* | | | | | | | | X | | HC-RS | | E/X | | | 2 |
| Eumeces fasciatus | Five-lined Skink | | | | E/HP | | | T/X | X | | | | | X | | |
| Eumeces laticeps | Broadhead Skink* | | | | | | | | | | HC | 1 | X | | | 2 |
| Farancia abacura | Mudsnake | | | | | | | | | | | | | | IV | |
| Farancia erytrogramma | Rainbow Snake | | | | | | | | | | | | X | | IV | |
| Glyptemys/Clemmys insculpta | Wood Turtle** | | 2 | SC/X | HP | SC/X | X | SC/X | SC/X | T/X | IC-RS | | X | X | T/ I | 1 |
| Glyptemys/Clemmys muhlenbergii | Bog Turtle | T | | | | E/X | | E/X | E/X | E/X | E/IC-RS | E/1 | T/X | X | E/ I | |
| Graptemys geographica | Northern Map Turtle | | | | | | | | X | | MC | | (E)/X | | IV | 2 |
| Graptemys pseudogeographica | False Map Turtle | | | | | | | | | | | | | | | 1 |
| Heterodon platirhinos | Eastern Hognose Snake* | | | T/X | | X | X | SC/X | SC/X | | MC | 2 | X | X | IV | 2 |
| Kinosternon subrubrum | Eastern Mud Turtle | | | | | | | | E/X | | | | | X | | |
| Lampropeltis g. getula | Eastern Kingsnake | | | | | | | | | SC/X | | 2 | | | | 2 |
| Lampropeltis getula nigra | Eastern Black Kingsnake | | | | | | | | | | | | | | III | |
| Lampropeltis t. triangulum | Eastern Milk Snake | | | | | | | | | | | 1 | | | | |
| Lampropeltis t. triangulum × elapsoides | Coastal Plain Milk Snake | | | | | | | | | SC/X | | | | | | |
| Lepidochelys kempii | Kemp's (Atlantic) Ridley | E | 1 | | | E/X | X | E/X | E/X | E/X | | E/1 | E/X | | | |
| Liochlorophis vernalis | Smooth Green Snake | | | SC/X | MP | | | X | X | | MC | | | | III | |

*(continued)*

**Reptiles (n = 65)** *(continued)*

| Scientific Name | Common Name | US | ME | NH | VT | MA | RI | CT | NY | NJ | PA | DE | MD | DC | VA | WV |
|---|---|---|---|---|---|---|---|---|---|---|---|---|---|---|---|---|
| Malaclemys t. terrapin | Northern Diamondback Terrapin** | | | | | T/X | E/X | X | X | SC/X | | 1 | X | | II | |
| Nerodia erythrogaster | Redbelly Water Snake | | | | | | | | | | | 1 | X | | | |
| Nerodia s. sipedon | Northern Water Snake | | | | MP | | | | | | | | | | | |
| Opheodrys aestivus | Rough Green Snake | | | | | | | | | | E/PV | 2 | | X | | 2 |
| Ophisaurus attenuatus | Eastern Slender Glass Lizard | | | | | | | | | | | | | | IV | |
| Ophisaurus ventralis | Eastern Glass Lizard | | | | | | | | | | | | | | T/ II | |
| Pituophis m. melanoleucus | Northern Pine Snake* | | | | | | | | | T/X | | | X | | I | |
| Pseudemys concinna | River Cooter | | | | | | | | | | | | | | | 2 |
| Pseudemys rubriventris | Redbelly / Red-bellied Cooter / Turtle* | | | | | E/X | | | | | T/HC | 2 | X | X | | 2 |
| Regina r. rigida | Glossy Crayfish Snake | | | | | | | | | | | | | | III | |
| Regina septemvittata | Queen Snake* | | | | | | | | E/X | E/X | HC | 2 | X | X | IV | |
| Sceloporus undulatus | Fence Lizard | | | | | | | | T/X | | MC | | | X | | |
| Scinecella lateralis | Ground Skink | | | | | | | | | | | 2 | | | | 2 |
| Sistrurus c. catenatus | Eastern Massasauga** | C | | | | | | | E/X | | E/IC | | | | | |
| Sternotherus minor peltifer | Striped-necked Musk Turtle | | | | | | | | | | | | | | IV | |
| Sternotherus odoratus | Common Musk Turtle / Stinkpot | | | | MP | | | | X | | | | | X | | |
| Storeria dekayi | Brown Snake | | | | MP | | | | | | | | | X | | |
| Storeria occipitomaculata | Redbelly Snake | | | | | | | | | | | 2 | | | | |
| Tantilla coronata | Southeastern Crowned Snake | | | | | | | | | | | | | | IV | |
| Terrapene c. carolina | Eastern Box Turtle* | | E/1 | X | | SC/X | X | SC/X | SC/X | SC/X | MC | 1 | X | X | III | |
| Thamnophis brachystoma | Short-headed Garter Snake | | | | | | | | X | | HC-RS | | | | | |
| Thamnophis s. sauritus | Eastern Ribbon Snake* | | | X | HP | X | X | SC/X | X | | HC | 2 | E/X | X | IV | 2 |
| Thamnophis s. sirtalis | Eastern Garter Snake | | | | | | | | | | | | | X | | |
| Trachemys s. scripta | Yellow-bellied Slider | | | | | | | | | | | | | | IV | |
| Trachemys scripta troostii | Cumberland Slider | | | | | | | | | | | | | | III | |
| Virginia v. valeriae | Smooth / Eastern Earthsnake | | | | | | | | | | PV | 2 | | | | 2 |
| Virginia valeriae pulchra | Mountain Earthsnake* | | | | | | | | | | HC-RS | | E/X | | II | 1 |
| | | | 9 | 9 | 12 | 18 | 12 | 15 | 28 | 16 | 21 | 23 | 25 | 22 | 27 | 20 |

| Amphibians (n = 73) Scientific Name | Common Name | US | ME | NH | VT | MA | RI | CT | NY | NJ | PA | DE | MD | DC | VA | WV |
|---|---|---|---|---|---|---|---|---|---|---|---|---|---|---|---|---|
| Acris c. crepitans | Eastern Cricket Frog | | | | | | | | | | | | | | | 2 |
| Acris crepitans | Northern Cricket Frog | | | | | | | | E/X | | HC | | | X | | |
| Acris crepitans blanchardi | Blanchard's Cricket Frog | | | | | | | | | | | | | | | 2 |
| Ambystoma barbouri | Streamside Salamander | | | | | | | | | | | | | | | 1 |
| Ambystoma jeffersonianum | Jefferson Salamander* | | | SC/X | HP | SC/X | | SC/X | SC/X | SC/X | RS | | X | | IV | 2 |
| Ambystoma laterale | Blue-spotted Salamander* | | 2 | X | MP | SC/X | | SC/X | SC/X | E/X | | | | | | |
| Ambystoma mabeei | Mabee's Salamander | | | | | | | | | | | | | | T/ II | |
| Ambystoma maculatum | Spotted Salamander | | | | MP | | | X | | SC | | 2 | | X | | |
| Ambystoma opacum | Marbled Salamander | | | E/X | | T/X | X | X | SC/X | SC/X | MC | | | X | | |
| Ambystoma talpoideum | Mole Salamander | | | | | | | | | | | | | | II | |
| Ambystoma texanum | Smallmouth Salamander | | | | | | | | | | | | | | | 1 |
| Ambystoma tigrinum | Eastern / Tiger Salamander* | | | | | | | | E/X | E/X | | E/1 | E/X | | E/ II | |
| Aneides aeneus | Green Salamander* | | | | | | | | | | T/IC | | E/X | | II | 1 |
| Bufo (woodhousii) fowleri | Fowler's Toad | | | X | HP | | X | X | X | SC/X | MC | | | X | | |
| Bufo americanus | American Toad | | | | | | | | | | | | | X | | |
| Bufo quercicus | Oak Toad | | | | | | | | | | | | | | II | |
| Cryptobranchus alleganiensis | Eastern Hellbender** | | | | | | | | SC/X | | IC-RS | | E/X | | II | 1 |
| Desmognathus fuscus | Northern Dusky Salamander | | | | | | X | X | | | | | | X | | |
| Desmognathus marmoratus | Shovel-nosed Salamander | | | | | | | | | | | | | | III | |
| Desmognathus monticola | Seal Salamander | | | | | | | | | | | | X | | | |
| Desmognathus ochrophaeus | Mountain Dusky Salamander | | | | | | | | | | | | X | | | |
| Desmognathus orestes | Blue Ridge Dusky Salamander | | | | | | | | | | | | | | IV | |
| Desmognathus quadramaculatus | Black-bellied Salamander | | | | | | | | | | | | | | | 2 |
| Desmognathus welteri | Black Mountain Salamander | | | | | | | | | | | | | | | 2 |
| Desmognathus wrighti | Pygmy Salamander | | | | | | | | | | | | | | III | |
| Eurycea bislineata | Northern Two-lined Salamander | | | | | | | | | | | | | X | | |
| Eurycea longicauda | Longtail Salamander* | | | | | | | | X | T/X | | 2 | X | | | |
| Eurycea lucifuga | Cave Salamander | | | | | | | | | | | | | | | 2 |
| Eurycea wilderae | Blue Ridge Two-lined Salamander | | | | | | | | | | | | | | III | |
| Gastrophryne carolinensis | Eastern Narrow-mouthed Toad | | | | | | | | | | | | E/X | | | |

(continued)

**Amphibians (n = 73)** *(continued)*

| Scientific Name | Common Name | US | ME | NH | VT | MA | RI | CT | NY | NJ | PA | DE | MD | DC | VA | WV |
|---|---|---|---|---|---|---|---|---|---|---|---|---|---|---|---|---|
| Gyrinophilus p. porphyriticus | Northern Spring Salamander | | | | | X | X | T/X | | SC/X | | | | | | |
| Gyrinophilus subterraneus | West Virginia Spring Salamander | | | | | | | | | | | | | | | 1 |
| Hemidactylium scutatum | Four-toed Salamander | | | | MP | SC/X | X | | X | | MC | 2 | | | | |
| Hyla andersonii | Pine Barrens Treefrog | | | | | | | | | T/X | | | | | | |
| Hyla chrysocelis | Cope's Gray Treefrog | | | | | | | | | E/X | | 2 | | | | |
| Hyla gratiosa | Barking Treefrog | | | | | | | | | | | E/1 | E/X | | T/ II | |
| Hyla versicolor | Gray Tree Frog | | | | | | | X | | | | | | | | |
| Necturus maculosus | Common Mudpuppy | | | | HP | | | | X | | | | X | | III | |
| Necturus punctatus | Dwarf Waterdog | | | | | | | | | | | | | | III | |
| Notophthalmus v. viridescens | Eastern / Red-spotted Newt | | | | | | X | X | | | | | | X | | |
| Plethodon cinereus | Northern Red-backed Salamander | | | | | | | | | | | | | X | | |
| Plethodon dorsalis | Southern Zigzag Salamander | | | | | | | | | | | | | | II | |
| Plethodon glutinosus | Northern Slimy Salamander | | | | | | | T/X | | | | | | | | |
| Plethodon hubrichti | Peaks of Otter Salamander | | | | | | | | | | | | | | II | |
| Plethodon kentucki | Cumberland Plateau Salamander | | | | | | | | | | | | | | IV | |
| Plethodon nettingi | Cheat Mountain Salamander | T | | | | | | | | | | | | | | 1 |
| Plethodon punctatus | Cow Knob Salamander | | | | | | | | | | | | | | II | 1 |
| Plethodon shenandoah | Shenandoah Salamander | E | | | | | | | | | | | | | E/ I | |
| Plethodon virginia | Shenandoah Mountain Salamander | | | | | | | | | | | | | | III | 1 |
| Plethodon wehrlei | Wehrle's Salamander | | | | | | | | | | | | I/X | | | |
| Plethodon welleri | Weller's Salamander | | | | | | | | | | | | | | II | |
| Plethodon yonahlossee | Yonahlossee Salamander | | | | | | | | | | | | | | IV | |
| Pseudacris brachyphona | Mountain Chorus Frog* | | | | | | | | | | IC-RS | | T/X | | II | |
| Pseudacris c. crucifer | Northern Spring Peeper | | | | | | | | | | | | | X | | |
| Pseudacris nigrita | (Striped) Southern Chorus Frog | | | | | | | | | | MC | | | | IV | |
| Pseudacris ocularis | Little Grass Frog | | | | | | | | | | | | | | IV | |
| Pseudacris triserata feriarum | Upland / Southeastern Chorus Frog | | | | | | | | | | PV | | | X | IV | 1 |
| Pseudacris triseriata | Western (Striped) Chorus Frog | | | | E/HP | | | | X | | PV | | | | | |
| Pseudacris triseriata kalmi | New Jersey Chorus Frog* | | | | | | | | | | E/HC | | X | | | |
| Pseudotriton m. montanus | Eastern Mud Salamander* | | | | | | | | | T/X | E | 2 | X | X | IV | |

| | | | | | | | | | | | | | | | | |
|---|---|---|---|---|---|---|---|---|---|---|---|---|---|---|---|---|
| Pseudotriton montanus diastictus | Midland Mud Salamander | | | | | | | | | | | | | | | 1 |
| Pseudotriton ruber | Northern Red Salamander | | | | | | | | X | | | | X | X | | 2 |
| Rana catesbeiana | Bullfrog | | | | | | | | | | | | | X | | |
| Rana palustris | Pickerel Frog | | | | | | | | | | | | | X | | |
| Rana pipiens | Northern Leopard Frog* | | | SC/X | | X | X | SC/X | | | MC | | | | | 2 |
| Rana septentrionalis | Mink Frog | | | X | | | | | | | | | | | | |
| Rana sphenocephala | Southern / Coastal Plain Leopard Frog | | | | | | | | SC/X | | E/PV | | | | | |
| Rana sylvatica | Wood Frog | | | | | | X | X | | | | | | X | | |
| Rana virgatipes | Carpenter Frog* | | | | | | | | | SC/X | | 2 | I/X | | III | |
| Scaphiopus holbrookii | Eastern Spadefoot Toad* | | | | | T/X | X | E/X | X | | E/HC | 2 | X | | IV | 1 |
| Siren intermedia | Lesser Siren | | | | | | | | | | | | | | III | |
| Siren lacertina | Greater Siren | | | | | | | | | | | | | | IV | |
| Stereochilus marginatus | Many-lined Salamander | | | | | | | | | | | | | | IV | |
| | | | 1 | 6 | 7 | 7 | 9 | 13 | 14 | 12 | 16 | 9 | 17 | 16 | 32 | 18 |

**Fish (n = 299)**

| Scientific Name | Common Name | US | ME | NH | VT | MA | RI | CT | NY | NJ | PA | DE | MD | DC | VA | WV |
|---|---|---|---|---|---|---|---|---|---|---|---|---|---|---|---|---|
| Acantharcus pomotis | Mud Sunfish* | | | | | | | | T/X | | | 1 | I/X | | IV | |
| Acipenser brevirostrum | Shortnose Sturgeon | E | 1 | E/X | | E/X | X | E/X | E/X | E/X | E/IC-RS | 1 | E/X | X | E/I | |
| Acipenser fulvescens | Lake Sturgeon** | | | | E/HP | | | | T/X | | E/IC | | | | | 1 |
| Acipenser oxyrhynchus | Atlantic Sturgeon* | | 1 | X | | E/X | X | T/X | T/X | SC/X | E/IC-RS | E/1 | X | X | II | |
| Alopias superciliosus | Bigeye Thresher Shark | | | | | | | | X | | | | | | | |
| Alopias vulpinus | Thresher Shark | | | | | | | | X | | | | | | | |
| Alosa aestivalis | Blueback Herring | | | X | MP | X | X | X | X | | | | | X | | |
| Alosa chrysochloris | Skipjack Herring | | | | | | | | | | T/MC | | | | | |
| Alosa mediocris | Hickory Shad | | | | | | | X | | X | PV | 2 | I/X | X | | |
| Alosa pseudoharengus | Alewife | | | X | | X | X | X | X | | MC | | | X | IV | |
| Alosa sapidissima | American Shad | | 2 | X | MP | X | X | X | X | | | | I/X | X | IV | |
| Ambloplites cavifrons | Roanoke Bass | | | | | | | | | | | | | | II | |
| Amblyraja radiata | Thorny Skate | | | | | | | | X | | | | | | | |

(continued)

**Fish (n = 299)** *(continued)*

| Scientific Name | Common Name | US | ME | NH | VT | MA | RI | CT | NY | NJ | PA | DE | MD | DC | VA | WV |
|---|---|---|---|---|---|---|---|---|---|---|---|---|---|---|---|---|
| Ameiurus brunneus | Snail Bullhead | | | | | | | | | | | | | | IV | |
| Ameiurus catus | White Catfish | | | | | | | | | | MC | | X | | | |
| Amerius melas | Black Bullhead | | | | | | | | | | E/PV | | | | | 1 |
| Amia calva | Bowfin | | | | | | | | | | MC | | X | X | | |
| Ammocrypta clara | Western Sand Darter | | | | | | | | | | | | | | T/II | |
| Ammocrypta/Etheostoma pellucida | Eastern Sand Darter** | | | | T/HP | | | | T/X | | E/IC-RS | | | | | 2 |
| Ammodytes americanus | American Sand Lance | | | | | | X | X | | | | | | | | |
| Anchoa mitchilli | Bay Anchovy | | | | | | | X | X | | | | | | | |
| Anguilla rostrata | American Eel | | 1 | X | HP | X | X | X | X | | MC | | | X | IV | 2 |
| Apeltes quadracus | Fourspine Stickleback | | | | | | | X | X | | | 2 | | | | |
| Aphredoderus sayanus gibbosus | Western Pirate Perch | | | | | | | | X | | | | | | | |
| Aphroderus sayanus | Pirate Perch | | | | | | | | | X | | | | | | |
| Aplodinotus grunniens | Freshwater Drum | | | | | | | | | | | | | | IV | |
| Brevoortia tyrannus | Menhaden | | | | | | X | X | X | | | | | | | |
| Campostoma anomalum | Central Stoneroller | | | | | | | | | | | | | X | | |
| Carcharhinus obscurus | Dusky Shark | | | | | | | | X | | | 2 | | | | |
| Carcharhinus plumbeus | Sandbar Shark | | | | | | | X | X | | | | | | | |
| Carcharius taurus | Sand Tiger | | | | | | | | X | | | | | | | |
| Carcharodon carcharias | White Shark | | | | | | | | X | | | 2 | | | | |
| Carpoides carpio | River Carpsucker | | | | | | | | | | PV | | | | | 2 |
| Carpiodes cyprinus | Quillback | | | | HP | | | | | | | | | | | |
| Carpiodes velifer | Highfin Carpsucker | | | | | | | | | | PV | | | | | 1 |
| Catostomus catostomus | Longnose Sucker | | 2 | | | SC/X | | SC/X | | | E/IC-RS | | EX/X | | | |
| Catostomus commersoni | White Sucker | | | | | X | | X | | | | | | | | |
| Centrarchus macropterus | Flier | | | | | | | | | | | | T/X | | | |
| Cetorhinus maximus | Basking Shark | | | | | | | | X | | | 2 | | | | |
| Chologaster cornuta | Swampfish | | | | | | | | | | | | | | IV | |
| Clinostomus elongatus | Redside Dace | | | | | | | | | | | | X | | | 2 |
| Clinostomus funduloides | Rosyside Dace | | | | | | | | | | | | X | | | |
| Clupea harengus | Atlantic Herring | | | | | | | X | | | | | | | | |

| | | | | | | | | | | | | | | | | |
|---|---|---|---|---|---|---|---|---|---|---|---|---|---|---|---|---|
| Coregonus artedi | Cisco / Lake Herring | | | | MP | | | | | | E/PV | | | | | |
| Coregonus clupeaformis | Lake Whitefish | | 1 | X | MP | | | | | | | | | | | |
| Coregonus hoyi | Bloater | | | | | | | | X | | | | | | | |
| Coregonus kiyi | Kiyi | | | | | | | | X | | | | | | | |
| Coregonus reighardi | Shortnose Cisco | | | | | | | | X | | | | | | | |
| Coregonus zenithicus | Shortjawed Cisco | | | | | | | | X | | | | | | | |
| Cottus baileyi | Black Sculpin | | | | | | | | | | | | | | IV | |
| Cottus bairdi | Mottled Sculpin | | | | MP | | | | | | | | X | | | |
| Cottus caeruleomentum | Blue Ridge Sculpin | | | | | | | | | | | 1 | | | | |
| Cottus carolinae | Banded Sculpin | | | | | | | | | | | | | | | 2 |
| Cottus carolinae kanawhae | Kanawha Sculpin | | | | | | | | | | | | | | | 1 |
| Cottus cognatus | Slimy Sculpin | | | X | | X | | X | | X | | | | | IV | 2 |
| Cottus girardi | Potomac Sculpin | | | | | | | | | | | | | | | 2 |
| Cottus ricei | Spoonhead Sculpin* | | | | | | | | E/X | | | | | | | |
| Cottus sp1 | Bluestone Sculpin | | | | | | | | | | | | | | III | 1 |
| Cottus sp4 | Clinch Sculpin | | | | | | | | | | | | | | III | |
| Cottus sp5 | Holston Sculpin | | | | | | | | | | | | | | III | |
| Cottus sp7 | Checkered Sculpin | | | | | | | | | | HC-RS | | X | | | |
| Couesius plumbeus | Lake Chub | | | | | E/X | | | | | | | | | | |
| Crystallaria asprella | Crystal Darter | | | | | | | | | | | | | | | 1 |
| Culaea inconstans | Brook Stickleback | | | | | | | | | | MC | | | | | |
| Cycleptus elongatus | Blue Sucker | | | | | | | | | | | | | | | 1 |
| Cyclopterus lumpus | Lumpfish | | | | | | | X | | | | | | | | |
| Cynoscion regalis | Weakfish | | | | | | X | X | | | | | | | | |
| Cyprinella analostana | Satinfin Shiner | | | | | | | | | | | | | | | 1 |
| Cyprinella labrosa | Thicklip Chub | | | | | | | | | | | | | | IV | |
| Cyprinella whipplei | Steelcolor Shiner | | | | | | | | | | | | | | T/III | |
| Cyprinodon variegatus | Sheepshead Minnow | | | | | | | X | | | | | | | | |
| Dasyatis centroura | Roughtail Stingray | | | | | | | X | X | | | | | | | |
| Dipterus laevis | Barndoor Skate | | | | | | | | X | | | | | | | |
| Enneacanthus chaetodon | Blackbanded Sunfish* | | | | | | | | | X | | 2 | T/X | | E/I | |
| Enneacanthus gloriosus | Bluespotted Sunfish | | | | | | | | | | | | X | | | |

(continued)

**Fish (n = 299)** *(continued)*

| Scientific Name | Common Name | US | ME | NH | VT | MA | RI | CT | NY | NJ | PA | DE | MD | DC | VA | WV |
|---|---|---|---|---|---|---|---|---|---|---|---|---|---|---|---|---|
| Enneacanthus obesus | Banded Sunfish* | | | X | | X | X | SC/X | T/X | X | E/HC | 2 | X | | IV | |
| Ericymba buccata | Silverjaw Minnow | | | | | | | | | | | | X | X | | |
| Erimonax monachus | Spotfin Chub | T | | | | | | | | | | | | | T/I | |
| Erimystax cahni | Slender Chub | T | | | | | | | | | | | | | T/I | |
| Erimystax dissimilis | Streamline Chub | | | | | | | | SC/X | | RS | | | | IV | |
| Erimystax insignis | Blotched Chub | | | | | | | | | | | | | | IV | |
| Erimystax x-punctatus | Gravel Chub** | | | | | | | | T/X | | E/IC | | | | | 2 |
| Erimyzon sucetta | Lake Chubsucker* | | | | | | | | T/X | | | | | | IV | |
| Erimyzon oblongus | Creek Chubsucker | | | | | X | X | X | | | | | | | | 1 |
| Esox americanus americanus | Redfin Pickerel | | E/1 | X | MP | | | X | | | | | | | | |
| Esox americanus vermiculatus | Grass Pickerel | | | | | | | | | | | | | | | 1 |
| Esox masquinongy | Muskellunge | | | | HP | | | | | | | | | | | |
| Esox niger | Chain Pickerel | | | | | | | X | | | | | | | | |
| Etheostoma acuticeps | Sharphead Darter | | | | | | | | | | | | | | E/I | |
| Etheostoma blennioides | Greenside Darter | | | | | | | | | | | | X | X | | |
| Etheostoma caeruleum | Rainbow Darter | | | | | | | | | | | | | | IV | |
| Etheostoma camurum | Bluebreast Darter** | | | | | | | | E/X | | T/HC | | | | III | 2 |
| Etheostoma chlorobranchium | Greenfin Darter | | | | | | | | | | | | | | T/II | |
| Etheostoma cinereum | Ashy Darter | | | | | | | | | | | | | | I | |
| Etheostoma collis | Carolina Darter | | | | | | | | | | | | | | T/II | |
| Etheostoma exile | Iowa Darter * | | | | | | | | X | | E/PV | | | | | |
| Etheostoma fusiforme | Swamp Darter | | T/1 | X | | X | | X | T/X | | | | I/X | | | |
| Etheostoma kanawhae | Kanawha Darter | | | | | | | | | | | | | | III | |
| Etheostoma longimanum | Longfin Darter | | | | | | | | | | | | | | | 1 |
| Etheostoma maculatum | Spotted Darter** | | | | | | | | T/X | | T/IC-RS | | | | | 1 |
| Etheostoma nigrum | Johnny Darter | | | | | | | | | | | | X | | | |
| Etheostoma olmstedi | Tessellated Darter | | | X | | X | | | | | | | | | | 1 |
| Etheostoma osburni | Candy Darter* | | | | | | | | | | | | | | II | 1 |
| Etheostoma percnurum | Duskytail Darter | E | | | | | | | | | | | | | E/I | |
| Etheostoma podostemone | Riverweed Darter | | | | | | | | | | | | | | IV | |

| | | | | | | | | | | | | | | | | |
|---|---|---|---|---|---|---|---|---|---|---|---|---|---|---|---|---|
| Etheostoma sellare | Maryland Darter | E | | | | | | | | | | | E/X | | | |
| Etheostoma stigmaeum | Speckled Darter | | | | | | | | | | | | | | IV | |
| Etheostoma swannanoa | Swannanoa Darter | | | | | | | | | | | | | | IV | |
| Etheostoma tippecanoe | Tippecanoe Darter** | | | | | | | | | | T/HC | | | | T/III | 2 |
| Etheostoma variatum | Variegate Darter | | | | | | | | | | | | | | E/II | |
| Etheostoma vitreum | Glassy Darter | | | | | | | | | | | 2 | T/X | | | |
| Etheostoma vulneratum | Wounded Darter | | | | | | | | | | | | | | III | |
| Etheostoma zonale | Banded Darter | | | | | | | | | | | | | | IV | |
| Exoglossum laurae | Tonguetied Minnow | | | | | | | | | | RS | | | | | 2 |
| Exoglossum maxillingua | Cutlips Minnow | | | | | | | X | | X | | | | | | |
| Fundulus catenatus | Northern Studfish | | | | | | | | | | | | | | IV | |
| Fundulus diaphanus | Banded Killifish | | | | | | | | | | | | | | | 1 |
| Fundulus heterolitus | Mummichog | | | | | | | X | X | | | | | | | |
| Fundulus lineolatus | Lined Topminnow | | | | | | | | | | | | | | IV | |
| Fundulus luciae | Spotfin Killfish | | | | | | X | X | X | | | | X | | | |
| Fundulus majalis | Striped Killfish | | | | | | | | X | | | | | | | |
| Fundulus rathbuni | Speckled Killifish | | | | | | | | | | | | | | IV | |
| Galeocerdo cuvier | Tiger Shark | | | | | | | | X | | | | | | | |
| Gasterosteus aculeatus | Threespine Stickleback | | | | | T/X | | | X | | E/PV | | | | | |
| Hemitripterus americanus | Sea Raven | | | | | | X | X | | | | | | | | |
| Hiodon alosoides | Goldeye | | | | | | | | | | T/PV | | | | | 1 |
| Hiodon tergisus | Mooneye* | | | | MP | | | | T/X | | T/PV | | | | | 2 |
| Hippocampus eresctus | Lined Seahorse | | | | | | X | X | X | | | | | | | |
| Hybognathus hankinsoni | Brassy Minnow | | | | HP | | | | | | | | | | | |
| Hybognathus regius | Eastern Silvery Minnow | | | | | SC/X | | | | | | | | | | 1 |
| Hybopsis hypsinotus | Highback Chub | | | | | | | | | | | | | | IV | |
| Hypentelium nigricans | Northern Hog Sucker | | | | | | | | | X | | | X | | | |
| Hypentelium roanokense | Roanoke Hog Sucker | | | | | | | | | | | | | | IV | |
| Ictiobus cyprinellus | Bigmouth Buffalo | | | | | | | | | | PV | | | | | 1 |
| Ictalurus natalis | Yellow Bullhead | | | | | | | | | | | 1 | | | | |
| Ichthyomyzon bdellium | Ohio Lamprey* | | | | | | | | X | | RS | | | | III | 2 |
| Ichthyomyzon fossor | Northern Brook Lamprey* | | | | E/HP | | | | | | E/HC | | | | | 1 |

(continued)

**Fish (n = 299)** *(continued)*

| Scientific Name | Common Name | US | ME | NH | VT | MA | RI | CT | NY | NJ | PA | DE | MD | DC | VA | WV |
|---|---|---|---|---|---|---|---|---|---|---|---|---|---|---|---|---|
| Ichthyomyzon greeleyi | Mountain Brook Lamprey* | | | | | | | | X | | T/RS | | | | III | 1 |
| Ichthyomyzon unicuspis | Silver Lamprey* | | | | MP | | | | | | | | | | | 2 |
| Ictiobus bubalus | Smallmouth Buffalo | | | | | | | | | | T/MC | | | | | |
| Iciobus niger | Black Buffalo | | | | | | | | | | PV | | | | | 1 |
| Isurus oxyrinchus | Shortfin Mako Shark | | | | | | | | X | | | | | | | |
| Isurus paucus | Longfin Mako Shark | | | | | | | | X | | | | | | | |
| Labidesthes sicculus | Brook Silverside | | | | | | | | | | MC | | | | IV | |
| Lampetra aepyptera | Least Brook Lamprey | | | | | | | | | | MC | 2 | X | | IV | 2 |
| Lampetra appendix | American Brook Lamprey* | | | X | T/HP | T/X | X | E/X | | X | MC | 2 | T/X | | IV | 2 |
| Lamna nasus | Porbeagle Shark | | | | | | | | X | | | | | | | |
| Lepisosteus oculatus | Spotted Gar* | | | | | | | | | | E/PV | | | | | |
| Lepisosteus osseus | Longnose Gar | | | | | | | | | X | MC | | X | | | |
| Lepomis auritus | Redbreast Sunfish | | | | MP | | X | X | | | | | | | | |
| Lepomis gibbosus | Pumpkinseed | | | | | | | X | | | | | | | | |
| Lepomis gulosus | Warmouth | | | | | | | | | | E/PV | | X | X | | 2 |
| Lepomis humilis | Orangespotted Sunfish | | | | | | | | | | | | | | | 2 |
| Lepomis megalotis | Longear Sunfish | | | | | | | | T/X | | E/PV | | | | | |
| Leucoraja erinacea | Little Skate | | | | | | | X | X | | | | | | | |
| Leucoraja garmani | Rosette Skate | | | | | | | | X | | | | | | | |
| Leucoraja ocellata | Winter Skate | | | | | | | X | X | | | | | | | |
| Lophius americanus | American Goosefish | | | | | | X | | | | | | | | | |
| Lota lota | Burbot | | 2 | X | | SC/X | | E/X | | | E/PV-MC | | | | | |
| Luxilus chrysocephalus | Striped Shiner | | | | | | | | | | | | I/X | | | |
| Luxilus cornutus | Common Shiner | | | | | X | X | X | | | | | | | | 2 |
| Lythrurus ardens | Blueside Shiner | | | | | | | | | | | | | | | 1 |
| Lythrurus lirus | Mountain Shiner | | | | | | | | | | | | | | IV | |
| Lythrurus umbratilis | Redfin Shiner | | | | | | | | SC/X | | E/PV | | | | | 2 |
| Macrhybopsis hyostoma | Speckled Chub | | | | | | | | | | | | | | | 1 |
| Macrhybopsis storeriana | Silver Chub* | | | | | | | | E/X | | E/PV | | | | | 2 |
| Macrozoarces americanus | Ocean Pout | | | | | | | X | | | | | | | | |

| | | | | | | | | | | | | | | | |
|---|---|---|---|---|---|---|---|---|---|---|---|---|---|---|---|
| Malacoraja senta | Smooth Skate | | | | | | | | X | | | | | | | |
| Manta birostris | Manta | | | | | | | | X | | | | | | | |
| Margariscus margarita | Pearl Dace | | | | | | | | | | | | T/X | | IV | 1 |
| Menidia beryllina | Inland Silverside | | | | | | | | X | | | | | | | |
| Menidia menidia | Atlantic Silverside | | | | | | | X | X | | | | | | | |
| Menticirrhus saxatilis | Northern Kingfish | | | | | | X | | | | | | | | | |
| Merluccius bilinearis | Silver Hake | | | | | | | X | | | | | | | | |
| Microgadus tomcod | Atlantic Tomcod | | | | | | X | X | X | | | | | | | |
| Micropterus dolomieu | Smallmouth Bass | | | | | | | X | | | | | | | | |
| Micropterus salmoides | Largemouth Bass | | | | | | | X | | | | | | | | |
| Minytrema melanops | Spotted Sucker | | | | | | | | | | T/PV | | | | | |
| Morone saxatilis | Striped Bass | | 1 | | | | | X | | | | | | | | |
| Moxostoma anisurum | Silver Redhorse | | | | HP | | | | | | | | | | | |
| Moxostoma ariommum | Bigeye Jump Rock | | | | | | | | | | | | | | III | |
| Moxostoma carinatum | River Redhorse* | | | | | | | | X | | MC | | | | III | 2 |
| Moxostoma duquesnei | Black Redhorse | | | | | | | | SC/X | | | | | | | |
| Moxostoma macrolepidotum | Shorthead Redhorse | | | | MP | | | | | | | 1 | | | | |
| Moxostoma robustum | Smallfin Redhorse | | | | | | | | | | | | | | IV | |
| Moxostoma valenciennesi | Greater Redhorse | | | | HP | | | | | | | | | | | |
| Mustelus canis | Smooth Dogfish | | | | | | X | X | | | | | | | | |
| Myoxocephalus octodecemspinosus | Longhorn Sculpin | | | | | | | X | | | | | | | | |
| Myoxocephalus thompsoni | Deepwater Sculpin* | | | | | | | | E/X | | | | | | | |
| Nocomis biguttatus | Hornyhead Chub | | | | | | | | | | PV | | | | | |
| Nocomis leptocephalus | Bluehead Chub | | | | | | | | | | | | | | | 2 |
| Nocomis platyrhynchus | Bigmouth Chub | | | | | | | | | | | | | | | 2 |
| Notemigonus crysoleucas | Golden Shiner | | | | | | | X | | | | | | | | |
| Notropis alborus | Whitemouth Shiner | | | | | | | | | | | | | | T/IV | |
| Notropis amblops | Bigeye Chub | | | | | | | | X | | | | | | | |
| Notropis amoenus | Comely Shiner | | | | | | | | X | X | | 2 | T/X | | | 2 |
| Notropis anogenus | Pugnose Shiner | | | | | | | | E/X | | | | | | | |
| Notropis ariommus | Popeye Shiner | | | | | | | | | | | | | | II | 1 |
| Notropis atherinoides | Emerald Shiner | | | | | | | | | | | | | | T/III | |

*(continued)*

**Fish (n = 299)** *(continued)*

| Scientific Name | Common Name | US | ME | NH | VT | MA | RI | CT | NY | NJ | PA | DE | MD | DC | VA | WV |
|---|---|---|---|---|---|---|---|---|---|---|---|---|---|---|---|---|
| Notropis bifrenatus | Bridle Shiner* | | | X | HP | SC/X | X | X | | X | E/HC | 1 | E/X | | I | |
| Notropis blennius | River Shiner | | | | | | | | | | PV | | | | | 1 |
| Notropis boops | Bigeye Shiner | | | | | | | | | | | | | | | 1 |
| Notropis buchanani | Ghost Shiner | | | | | | | | | | E/PV | | | | | 2 |
| Notropis chalybaeus | Ironcolor Shiner | | | | | | | | SC/X | SC/X | E/HC | 1 | E/X | | IV | |
| Notropis chiliticus | Redlip Shiner | | | | | | | | | | | | | | IV | |
| Notropis dorsalis | Bigmouth Shiner | | | | | | | | | | T/PV | | | | | |
| Notropis heterodon | Blackchin Shiner | | | | HP | | | | X | | E/PV | | | | | |
| Notropis heterolepis | Blacknose Shiner | | | | HP | | | | | | | | | | | |
| Notropis hudsonius | Spottail Shiner | | | | | | X | | | | | | | | | |
| Notropis procne | Swallowtail Shiner | | | | | | | | X | | | | | | | 1 |
| Notropis scabriceps | New River Shiner | | | | | | | | | | | | | | IV | 1 |
| Notropis semperasper | Roughhead Shiner | | | | | | | | | | | | | | II | |
| Notropis sp. A | Sawfin Shiner | | | | | | | | | | | | | | IV | |
| Notropis spectrunculus | Mirror Shiner | | | | | | | | | | | | | | IV | |
| Notropis stramineus | Sand Shiner | | | | | | | | | | | | | | IV | |
| Noturus eleutherus | Mountain Madtom | | | | | | | | | | E/HC | | | | IV | 2 |
| Noturus flavipinnis | Yellowfin Madtom | T | | | | | | | | | | | | | T/I | |
| Noturus flavus | Stonecat | | | | E/HP | | | | | | | | E/X | | IV | |
| Noturus gilberti | Orangefin Madtom | | | | | | | | | | | | | | T/II | |
| Noturus gyrinus | Tadpole Madtom | | | | | | | | | | E/PV | | | | | |
| Noturus insignis | Spotted / Margined Madtom | | | | | | | | | X | | 2 | | | II | |
| Noturus miurus | Brindled Madtom | | | | | | | | | | E/PV | | | | | |
| Noturus stigmosus | Northern Madtom | | | | | | | | | | E/HC-RS | | | | | 1 |
| Opsanus tau | Oyster Toadfish | | | | | | X | X | X | | | | | | | |
| Osmerus mordax | Rainbow Smelt | | 2 | X | | | X | T/X | X | X | HC | | | | | |
| Pararhinichthys bowersi | Cheat Minnow | | | | | | | | | | IC-RS | | EX/X | | | 1 |
| Paralichthys oblongus | Fourspot Flounder | | | | | | | X | | | | | | | | |
| Peprilus triacanthus | Butterfish | | | | | | | X | | | | | | | | |
| Perca flavescens | Yellow Perch | | | | | | | X | | | | | | | | |

| | | | | | | | | | | | | | | | | |
|---|---|---|---|---|---|---|---|---|---|---|---|---|---|---|---|---|
| Percina aurantiaca | Tangerine Darter | | | | | | | | | | | | | | IV | |
| Percina burtoni | Blotchside Logperch | | | | | | | | | | | | | | II | |
| Percina caprodes / nebulosa | Logperch / Chesapeake Logperch | | | | | | | | | | IC-RS | | T/X | | IV | |
| Percina copelandi | Channel Darter* | | | | E/HP | | | | | | T/PV | | | | III | 2 |
| Percina crassa | Piedmont Darter | | | | | | | | | | | | | | IV | |
| Percina evides | Gilt Darter** | | | | | | | | E/X | | T/HC | | | | IV | 2 |
| Percina / Etheostoma gymnocephala | Appalachia Darter | | | | | | | | | | | | | | IV | 1 |
| Percina macrocephala | Longhead Darter* | | | | | | | | T/X | | T/RS | | | | T/II | 2 |
| Percina maculata | Blackside Darter | | | | | | | | | | | | | | IV | |
| Percina notogramma | Stripeback Darter | | | | | | | | | | | | E/X | | | 1 |
| Percina oxyrhynchus | Sharpnose Darter* | | | | | | | | | | | | | | IV | |
| Percina peltata | Shield Darter | | | | | | | | | X | | 1 | X | | | 1 |
| Percina phoxocephala | Slenderhead Darter | | | | | | | | | | | | | | | 1 |
| Percina rex | Roanoke Logperch | E | | | | | | | | | | | | | E/I | |
| Percina sciera | Dusky Darter | | | | | | | | | | | | | | IV | 2 |
| Percina shumardi | River Darter | | | | | | | | | | | | | | | 1 |
| Percopsis omiscomaycus | Trout-perch | | | | | | | | | | | | EX/X | | IV | |
| Petromyzon marinus | Sea Lamprey | | | X | MP | | | X | | | | | | | | |
| Phenacobius crassilabrum | Fatlips Minnow | | | | | | | | | | | | | | III | |
| Phenacobius mirabilis | Sucker Mouth Minnow | | | | | | | | | | | | | | IV | |
| Phenacobius teretulus | Kanawha Minnow | | | | | | | | | | | | | | III | 1 |
| Phenacobius uranops | Stargazing Minnow | | | | | | | | | | | | | | IV | |
| Phoxinus cumberlandensis | Blackside Dace | | | | | | | | | | | | | | III | |
| Phoxinus eos | Northern Redbelly Dace | | | X | | E/X | | | | | | | | | | |
| Phoxinus erythrogaster | Southern Redbelly Dace | | | | | | | | | | T/PV | | | | | 2 |
| Phoxinus neogaeus | Finescale Dace | | | X | | | | | | | | | | | | |
| Phoxinus oreas | Mountain Redbelly Dace | | | | | | | | | | | | | | | 2 |
| Phoxinus sp1 | A Dace | | | | | | | | | | | | | | II | |
| Phoxinus tennesseensis | Tennessee Dace | | | | | | | | | | | | | | E/I | |
| Pimephales vigilax | Bullhead Minnow | | | | | | | | | | | | | | IV | 1 |
| Polyodon spathula | Paddlefish | | | | | | | | E/X | | HC | | | | T/II | 1 |
| Pomoxis nigromaculatus | Black Crappie | | | | | | | X | | | | | | | | |

*(continued)*

**Fish (n = 299)** *(continued)*

| Scientific Name | Common Name | US | ME | NH | VT | MA | RI | CT | NY | NJ | PA | DE | MD | DC | VA | WV |
|---|---|---|---|---|---|---|---|---|---|---|---|---|---|---|---|---|
| Prionace glauca | Blue Shark | | | | | | | | X | | | | | | | |
| Prionotus carolinus | Northern Searobin | | | | | | | X | | | | | | | | |
| Prionotus evolans | Striped Searobin | | | | | | | X | | | | | | | | |
| Pristis pectinata | Small Toothed Sawfish | | | | | | | | | | | 1 | | | | |
| Prosopium cylindraceum | Round Whitefish* | | 2 | X | MP | | | | E/X | | | | | | | |
| Pseudopleuronectes americanus | Winter Flounder | | | | | | X | X | X | | | | | | | |
| Pungitius pungitius | Ninespine Stickleback | | | | | | | | X | | | | | | | |
| Raja eglanteria | Clearnose Skate | | | | | | | X | X | | | | | | | |
| Rhinichthys atratulus | Blacknose Dace | | | | | X | X | X | | | | | | | | |
| Rhinichthys cataractae | Longnose Dace | | | | | X | X | X | | | | | | | | |
| Rhinoptera bonasus | Cownose Ray | | | | | | | | X | | | | | | | |
| Salmo salar | Atlantic Salmon | E | 1,2 | X | HP-MP | X | X | X | X | | | | | | | |
| Salvelinus aureolus | Sunapee Trout / Arctic Char | | 1 | E/X | HP | | | | | | | | | | | |
| Salmo trutta | Brown Trout (wild) | | | | | | | X | | | | | | | | |
| Salvelinus fontinalis | Brook Trout | | 2 | X | MP | X | X | X | X | X | | | X | | | 1 |
| Salvelinus namaycush | Lake Trout | | 1 | X | MP | | | | | | | | | | | |
| Scaphirhynchus platorynchus | Shovelnose Sturgeon | | | | | | | | | | | | | | | 1 |
| Scomber scombrus | Atlantic Mackerel | | | | | | | X | | | | | | | | |
| Scophthalmus Aquosos | Windowpane Flounder | | | | | | X | X | | | | | | | | |
| Semotilus atromaculatus | Creek Chub | | | | | X | | | | | | | | | | |
| Semotilus corporalis | Fallfish | | | | | X | | X | | | | | | | | |
| Squalus acanthias | Spiny Dogfish | | | | | | | X | | | | | | | | |
| Squatina dumeril | Atlantic Angel Shark | | | | | | | | | | | 2 | | | | |
| Sphoeroides maculatus | Northern Puffer | | | | | | | X | X | | | | | | | |
| Sphyrna lewini | Scalloped Hammerhead Shark | | | | | | | | X | | | | | | | |
| Sphyrna tiburo | Bonnethead Shark | | | | | | | | X | | | | | | | |
| Sphyrna zygaena | Smooth Hammerhead Shark | | | | | | | | X | | | | | | | |
| Stizostedion canadense | Sauger | | | | HP | | | | X | | | | | | IV | |
| Syngnathus fuscus | Pipefish | | | | | | | X | X | | | | | | | |
| Tautoga onitis | Tautog | | | | | | | X | X | | | | | | | |

| | | | | | | | | | | | | | | | | |
|---|---|---|---|---|---|---|---|---|---|---|---|---|---|---|---|---|
| Tautogolabrus adspersus | Cunner | | | | | | X | X | X | | | | | | | |
| Thoburnia hamiltoni | Rustyside Sucker | | | | | | | | | | | | | | III | |
| Thoburnia rhothoeca | Torrent Sucker | | | | | | | | | | | | | | | 2 |
| Torpedo nobiliana | Atlantic Torpedo Ray | | | | | | | | X | | | | | | | |
| Trinectes maculatus | Hogchoker | | | | | | X | X | | | | | | | | |
| Umbra limi | Central Mud Minnow | | | | | | | | | | MC | | | | | 1 |
| Umbra pygmaea | Eastern Mud Minnow | | | | | | | | | | MC | | | | | |
| Urophycis chuss | Red Hake | | | | | | | X | | | | | | | | |
| | | | 16 | 24 | 33 | 27 | 33 | 74 | 90 | 18 | 67 | 23 | 40 | 12 | 96 | 76 |

# Literature Cited


Acharya, L., and M. B. Fenton. 1992. Echolocation behavior of vespertilionid bats (*Lasiurus cinereus* and *Lasiurus borealis*) attacking airborne targets including arctiid moths. Canadian Journal of Zoology 70:1292–1298.

Adair, P. 1982. The short-eared owl (*Asio accipitrinus*, Pallus) and the kestrel (*Falctinnunculus*, L.) in the vole plague districts. Annals of Scottish Natural History 1982:219–231.

Adolph, S. C., and W. P. Porter. 1996. Growth, seasonality, and lizard life histories: age and size at maturity. Oikos 77:267–278.

Akre, T. S. B., and T. F. Robinson. 2003. *Opheodrys vernalis* (smooth greensnake). Herpetological Review 34:389.

Aldrich, J. W. 1946. The United States races of the bobwhite. Auk 63:493–508.

Aldridge, R. D., M. J. Dreslik, C. A. Phillips, B. C. Jellen, M. Allender, and J. M. Cox. 2005. Reproductive biology of the massasauga (*Sistrurus catenatus*) from south central Illinois (Abstract). Program of the Biology of Rattlesnakes Symposium, Loma Linda University, California, USA.

Alerich, C. L. 1993. Forest statistics for Pennsylvania—1978 and 1989. Resource Bulletin NE-126. U.S. Department of Agriculture Forest Service, Northeastern Forest Experiment Station, Radnor, Pennsylvania, USA.

Alisauskas, R. T., and T. W. Arnold. 1994. American coot. Pages 127–143 *in* T. C. Tacha and C. E. Braun, editors. Migratory shore and upland game bird management in North America. International Association of Fish and Wildlife Agencies, Washington, D.C., USA.

Allan, B. F., R. B. Langerhans, W. A. Ryberg, W. J. Landesman, N. W. Griffin, R. S. Katz, B. J. Oberle, M. R. Schutzenhofer, K. N. Smyth, A. de St Maurice, L. Clark, K. R. Crooks, D. E. Hernandez, R. G. McLean, R. S. Ostfeld, and J. M. Chase. 2009. Ecological correlates of risk and incidence of West Nile virus in the United States. Oecologia 158:699–708.

Allegheny National Forest. 1986. Land and resource management plan—Allegheny National Forest. U.S. Department of Agriculture Forest Service, Warren, Pennsylvania, USA.

Allegheny National Forest. 2004. Species at risk (mammals)—process summary. Species Viability Evaluation Team, Allegheny National Forest, Pennsylvania, USA.

Allen, A. W. 1983. Habitat suitability index models: fisher. U.S. Fish and Wildlife Service, Habitat Evaluation Procedures Group, Fort Collins, Colorado, USA.

Allen, R. P. 1958. A progress report on the wading bird survey. National Audubon Society, Tavernier, Florida, USA.

Allombert, S., S. Stockton, and J. L. Martin. 2005. A natural experiment on the impact of overabundant deer on forest invertebrates. Conservation Biology 19:1917–1929.

Altman, B. 1999a. Conservation strategy for landbirds in coniferous forests of western Oregon and Washington. Version 1.0. American Bird Conservancy and Oregon-Washington Partners in Flight. <www.gorge.net/natres/pif.html>.

Altman, B. 1999b. Nest success and habitat relationships of the olive-sided flycatcher in managed forests of northwestern Oregon. Unpublished report submitted to U.S. Fish and Wildlife Service, Portland, Oregon, USA.

Altman, B., and R. Sallabanks. 2000. Olive-sided flycatcher (*Contopus cooperi*). *In* A. Poole and F. Gill, editors. The birds of North America, 502. The Birds of North America, Philadelphia, Pennsylvania, USA.

American Birds. 2000. One-hundredth Christmas count: National Audubon Society 1999–2000. National Audubon Society, New York, New York, USA.

American Ornithologists' Union. 1973. Thirty-second supplement to the American Ornithologists' Union check-list of North American birds. Auk 90:411–419.

American Ornithologists' Union. 1983. Checklist of North American birds. Sixth edition. American Ornithologists' Union, Washington, D.C., USA.

American Ornithologists' Union. 1998. Checklist of North American birds. Seventh edition. American Ornithologists' Union, Washington, D.C., USA.

Amico, T. M., R. M. Schutsky, and J. E. Witmer. 1984. Annotated checklist. Pages 125–174 *in* H. B. Morrin, editor. A guide to the birds of Lancaster County, Pennsylvania. Lancaster County Bird Club, Lancaster, Pennsylvania, USA.

Anders, A. D., D. C. Dearborn, J. Faaborg, and F. R. Thompson III. 1997. Juvenile survival in a population of Neotropical migrant birds. Conservation Biology 11:698–707.

Anders, A. D., J. Faaborg, and F. R. Thompson III. 1998. Postfledging dispersal, habitat use, and home-range size of juvenile wood thrush. Auk 115:349–358.

Anders, A. D., and M. R. Marshall. 2005. Increasing the accuracy of productivity and survival estimates in assessing landbird population status. Conservation Biology 19:66–74.

Andrew, J. M., and J. A. Mosher. 1982. Bald eagle nest site selection and nesting habitat in Maryland. Journal of Wildlife Management 42(2):383–390.

Angilletta, M. J., Jr. 2001a. Variation in metabolic rate between populations of a geographically widespread lizard. Physiological and Biochemical Zoology 74:11–21.

Angilletta, M. J., Jr. 2001b. Thermal and physiological constraints on energy assimilation in a widespread lizard (*Sceloporus undulatus*). Ecology 82:3004–3056.

Angilletta, M. J., Jr., P. H. Niewiarowski, A. E. Dunham, A. D. Leache, and P. Warren. 2004. Bergmann's clines in ectotherms: illustrating a life-history perspective with sceloporine lizards. American Naturalist 164:E168–E183.

Annand, E. M., and F. R. Thompson III. 1997. Forest bird response to regeneration practices in central hardwood forests. Journal of Wildlife Management 61:159–171.

Anonymous. 1934. Kill waterdogs and watersnakes. Pennsylvania Angler 3(9):10.

Anonymous. 1935. Just to remind you—its time to get after the watersnake. Pennsylvania Angler 4(5):11.

Anonymous. 1936. Kills 108 snakes in contest. Pennsylvania Angler 5(12):11.

Anonymous. 1982. Seasons and bag limits 1982–1983. Pennsylvania Game News 53(6):40-42.

Anonymous. 1999. The Nature Conservancy species management abstract: northern goshawk (*Accipiter gentilis*). Nature Conservancy, Arlington, VA.

Anonymous. 2004. The state of the forest: a snapshot of Pennsylvania's updated forest inventory. Pennsylvania Department of Conservation and Natural Resources Bureau of Forestry, Harrisburg, Pennsylvania, USA.

Anton, T. G., D. Mauger, C. A. Phillips, M. J. Dreslik, J. E. Petzing, A. R. Kuhns, and J. M. Mui. 2003. *Clonophis kirtlandii* (Kirtland's snake): aggregating behavior and site fidelity. Herpetological Review 34:248–249.

Apfelbaum, S., and P. Seelbach. 1983. Nest tree, habitat selection, and productivity of seven North American raptor species based on the Cornell University nest record card program. Journal of Raptor Research 17:97–113.

Aresco, M. J. 2005. The effect of sex-specific terrestrial movements and roads on the sex ratio of freshwater turtles. Biological Conservation 123:37–44.

Armstrong, E., and D. Euler. 1983. Habitat usage of two woodland *Buteo* species in central Ontario. Canadian Field-Naturalist 97:200–207.

Arndt, R. G., and W. A. Potter. 1973. A population of the map turtle, *Graptemys geographica*, in the Delaware River, Pennsylvania. Journal of Herpetology 7:375–377.

Arnett, E. 2005. Relationships between bats and wind turbines in Pennsylvania and West Virginia: an assessment of fatality search protocols, patterns of fatality, and behavioral interactions with wind turbines. A final report prepared for the Bats and Wind Energy Cooperative.

Arnett, E. B. 2006. A preliminary evaluation on the use of dogs to recover bat fatalities at wind energy facilities. Wildlife Society Bulletin 34:1444–1445.

Arnett, E. B., K. Brown, W. P. Erickson, J. Fielder, T. H. Henry, G. D. Johnson, J. Kerns, R. Kolford, C. P. Nicholson, T. O'Connell, M. Piorkowski, and R. Tankersly. 2008. Patterns of fatality of bats at wind energy facilities in North America. Journal of Wildlife Management 72:61–78.

Arthur, S. M., W. B. Krohn, and J. A. Gilbert. 1989. Habitat use and diet of fishers. Journal of Wildlife Management 53:680–688.

Arthur, S. M., T. F. Paragi, and W. B. Krohn. 1993. Dispersal of juvenile fishers in Maine. Journal of Wildlife Management 57:868–874.

Artman, V. L., and J. F. Downhower. 2003. Wood thrush (*Hylocichla mustelina*) nesting ecology in relation to prescribed burning of mixed-oak forest in Ohio. Auk 120:874–882.

Arvisais, M., J. C. Bourdeois, E. Levesque, C. Daigle, D. Masse, and J. Jutras. 2002. Home range and movements of a wood turtle (*Clemmys insculpta*) population at the northern limit of its range. Canadian Journal of Zoology 80:402–408.

Askins, R. A. 1993. Population trends in grassland, shrubland, and forest birds in eastern North America. Current Ornithology 11:1–34.

Askins, R. A. 2000. Restoring North America's birds: lessons from landscape ecology. Second edition. Yale University Press, New Haven, Connecticut, USA.

Askins, R. A. 2001. Sustaining biological diversity in early successional communities: the challenge of managing unpopular habitats. Wildlife Society Bulletin 29: 407–412.

Asplund, K. K. 1963. Ecological factors in the distribution of *Thamnophis brachystoma* (Cope). Herpetologica 19:128–132.

Atkinson, D. A. 1901. The reptiles of Allegheny County, Pennsylvania. Annals of the Carnegie Museum 1:145–157.

Atkinson, D. A., and M. G. Netting. 1927. The distribution and habits of the Massasauga. Bulletin of the Antivenin Institute of America 1:40–44.

Atkisson, C. S. 1996. Red crossbill. *In* A. Poole and F. Gill, editors. The birds of North America, 256. Academy of Natural Sciences, Philadelphia, Pennsylvania, USA, and American Ornithologists' Union, Washington, D.C., USA.

Aubry, K., S. Wisely, C. Raley, and S. Buskirk. 2004. Zoogeography, spacing patterns, and dispersal in fishers: insights gained from combining field and genetic data. Pages 201–220 *in* D. J. Harrison, A. K. Fuller, and G. Proulx, editors. Martens and fishers (*Martes*) in human-altered environments: an international perspective. Springer Science and Business Media, New York, New York, USA.

Aubry, K. B., and J. C. Lewis. 2003. Extirpation and reintroduction of fishers (*Martes pennanti*) in Oregon: implications for their conservation in the Pacific states. Biological Conservation 114:79–90.

Audubon, J. J. 1840. The birds of the Americas. Volume 4. J. J. Audubon, New York, New York, USA.

Audubon Society of New Hampshire. 2004. Audubon Society of New Hampshire research: whip-poor-will monitoring in 2003. <www.nhaudubon.org/research/03whippoorwill.htm>.

Audubon Society of New Hampshire. 2005. Audubon Society of New Hampshire current research: rare species monitoring in 2002. <www.nhaudubon.org/research/02rare.htm>.

Austin, K., M. Fies, J. Jacobs, N. Murdock, C. Stihler, and P. Weigl. 1990. Appalachian northern flying squirrels (*Glaucomys sabrinus fuscus* and *G. s. coloratus*) recovery plan. U.S. Fish and Wildlife Service, Newton Corner, Massachusetts, USA.

Austin, O. L., editor. 1968. Life histories of North American cardinals, grosbeaks, buntings, towhees, finches, sparrows, and their allies. U.S. National Museum Bulletin 237.

Badzinski, S. S. 2003. Influence of tundra swans on aquatic vegetation and staging waterfowl at Long Point, Ontario. Dissertation, University of Western Ontario, London, Ontario, Canada.

Baicich, P. J., and C. J. O. Harrison. 1997. A guide to the nests, eggs, and nestlings of North American birds. Second edition. Academic Press, Boston, Massachusetts, USA.

Bajema, R. A., T. L. DeVault, P. E. Scott, and S. L. Lima. 2001. Reclaimed coal mine grasslands and their significance for Henslow's sparrows in the American Midwest. Auk 118:422–431.

Baker, R. H. 1983. Michigan mammals. Michigan State University Press, East Lansing, USA.

Bakker, V. J., and D. H. Van Vuren. 2004. Gap-crossing decisions by the red squirrel, a forest-dependent small mammal. Conservation Biology 18:689–697.

Balcom, B. J., and R. H. Yahner. 1996. Microhabitat and landscape characteristics associated with the threatened Allegheny woodrat. Conservation Biology 10:515–525.

Balcombe, J., and M. B. Fenton. 1988. Eavesdropping by bats: the influence of echolocation call design and foraging strategies. Ethology 79:158–166.

Ballinger, R. E., and K. S. Watts. 1995. Path to extinction: impact of vegetational change on lizard populations on Arapaho Prairie in the Nebraska sandhills. American Midland Naturalist 134:413–417.

Balph, D. F., and M. H. Balph. 1979. Behavioral flexibility of pine siskins in mixed species foraging groups. Condor 81:211–212.

Banfield, A. W. F. 1947. The mammals of Canada. University of Toronto Press, Toronto, Canada.

Bannor, B. K., and E. Kiviat. 2002. Common moorhen (*Gallinula chloropus*). *In* A. Poole and F. Gill, editors. The birds of North America, 685. Birds of North America, Philadelphia, Pennsylvania, USA.

Barber, N. A., and R. J. Marquis. 2009. Spatial variation in top-down direct and indirect effects on white oak (*Quercus alba* L.). American Midland Naturalist 162:169–179.

Barbour, R. W., and W. H. Davis. 1969. Bats of America. University of Kentucky Press, Lexington, USA.

Barclay, J. H. 1988. Peregrine restoration in the eastern United States. Pages 549–558 *in* T. J. Cade, J. H. Enderson, C. G. Thelander, and C. M. White, editors. Peregrine falcon populations; their management and recovery. The Peregrine Fund, Boise, Idaho, USA.

Barclay, J. H. 1995. Patterns of dispersal and survival of eastern peregrine falcons derived from banding data. Published privately by Biosystems Analysis.

Barclay, J. H., and T. J. Cade. 1983. Restoration of the peregrine falcon in the eastern United States. Bird Conservation 1:3–57.

Barclay, R. M. R. 1985. Long- versus short-range foraging strategies of hoary (*Lasiurus cinereus*) and silver-haired (*Lasionycteris noctivagans*) bats and the consequences for prey detections. Canadian Journal of Zoology 63:2507–2515.

Barclay, R. M. R. 1986. The echolocation calls of hoary (*Lasiurus cinereus*) and silver-haired (*Lasionycteris noctivagans*) bats as adaptations for long- versus short-range foraging strategies and the consequences for prey selection. Canadian Journal of Zoology 64:2700–2705.

Barclay, R. M. R., P. A. Faure, and D. R. Farr. 1988. Roosting behavior and roost selection by migrating silver-haired bats (*Lasionycteris noctivagans*). Journal of Mammalogy 69:821–825.

Barker, M. W., and M. J. Caduto. 1984. Geographic distribution: *Bufo woodhousii fowerli*. Herpetological Review 15:51–52.

Barlow, J. C., and W. B. McGillivray. 1983. Foraging and habitat relationships of the sibling species willow flycatcher (*Empidonax traillii*) and alder flycatcher (*E. alnorum*) in southern Ontario. Canadian Journal of Zoology 61:1510–1516.

Bart, J., and S. Earnst. 2002. Double sampling to estimate density and population trends in birds. Auk 119:36–45.

Barton, A. J., and J. W. Price, Sr. 1955. Our knowledge of the bog turtle, *Clemmys muhlenbergi*, surveyed and augmented. Copeia 1955:159–165.

Barzilay, S. 1980. Aggressive behavior in the wood turtle, *Clemmys insculpta*. Journal of Herpetology 14:89–91.

Basili, G. D., and S. A. Temple. 1999. Winter ecology, behavior, and conservation needs of dickcissels in Venezuela. Studies in Avian Biology 19:289–299.

Basore, N. S., L. B. Best, and J. B. Wooley. 1986. Bird nesting in Iowa no-tillage and tilled cropland. Journal of Wildlife Management 50:19–28.

Bassett, R. L., D. A. Boyce, Jr., M. H. Reiser, R. T. Graham, and R. T. Reynolds. 1994. Influence of site quality and stand density on goshawk habitat in southwestern forests. Studies in Avian Biology 16:41–45.

Baughman, M., editor. 2003. Reference atlas to the birds of North America. National Geographic Society, Washington, D.C., USA.

Bavetz, M. 1993. Geographic variation, status, and distribution of Kirtland's snake (*Clonophis kirtlandii* Kennicott) in Illinois. Thesis, Southern Illinois University at Carbondale, Carbondale, USA.

Bayer, R. 1984. Foraging ground displays of great blue herons at Yaquina Estuary, Oregon. Colonial Waterbirds 7:45–54.

Beal, F. E. L. 1912. Food of our more important flycatchers. U.S. Department of Agriculture Biological Survey Bulletin 44.

Beals, E. 1960. Forest bird communities in the Apostle Islands of Wisconsin. Wilson Bulletin 72:156–181.

Beasley, L. E., and L. L. Getz. 1986. Comparison of demography of sympatric populations of *Microtus ochrogaster* and *Synaptomys cooperi*. Acta Theriologica 31:385–400.

Becker, D. A., M. C. Brittingham, and C. B. Goguen. 2008. Effects of hemlock wooly adelgid on breeding birds at Fort Indiantown Gap, Pennsylvania. Northeastern Naturalist 15:227–240.

Bednarz, J. C, D. Klem, Jr., L. J. Goodrich, and S. E. Senner. 1990. Migration counts of raptors at Hawk Mountain, Pennsylvania, as indicators of population trends, 1934–1986. Auk 107:96–109.

Beebee, T., and R. Griffiths. 2005. The amphibian decline crisis: a watershed for conservation biology? Biological Conservation 125(3):271–285.

Beer, D. 2007. Dispatch on Philly's first bald eagle nest. Presented at the Delaware Valley Ornithological Club Meeting, April 5, 2007. Academy of Natural Sciences, Philadelphia.

Beissinger, S. R., J. R. Walters, D. G. Catanzaro, K. G. Smith, J. B. Dunning, S. M. Haig, B. R. Noon, and B. M. Stith. 2006. Modeling Approaches in Avian Conservation and the Role of Field Biologists. American Ornithologists' Union, Washington, D.C., USA.

Bell, R, K. 1984. Dickcissel nest found in Clarion County, Pennsylvania. Redstart 57:68–69.

Bellocq, M. I., J. F. Bendell, and B. L. Cadogan. 1992. Effects of *Bacillus thuringiensis* on *Sorex cinereus* (masked shrew) populations, diet, and prey selection in a jack pine plantation in northern Ontario. Canadian Journal of Zoology 70:505–510.

Bellrose, F. C. 1976. The ducks, geese, and swans of North America. Stackpole Books, Harrisburg, Pennsylvania, USA.

Bellrose, F. C. 1980. Ducks, geese, and swans of North America. Third edition. Stackpole Books, Harrisburg, Pennsylvania, USA.

Belwood, J. J. 2002. Endangered bats in suburbia: observations and concerns for the future. Pages 193–198 *in* A. Kurta and J. Kennedy, editors. The Indiana bat: biology and management of an endangered species. Bat Conservation International, Austin, Texas, USA.

Beneski, J. T., and D. W. Stinson. 1987. *Sorex palustris*. Mammalian Species 296:1–6.

Benkman, C. W. 1987. Crossbill foraging behavior, bill structure, and patterns of food profitability. Wilson Bulletin 99:351–368.

Benkman, C. W. 1988. Flock size, food dispersion, and the

feeding behavior of crossbills. Behavioural Ecology and Sociobiology 23:167–175.

Benkman, C. W. 1990. Foraging rates and the timing of crossbill reproduction. Auk 107:376–86.

Benkman, C. W. 1993a. Adaptation to single resources and the evolution of crossbill (*Loxia*) diversity. Ecological Monographs 63:305–325.

Benkman, C. W. 1993b. Logging, conifers, and the conservation of crossbills. Conservation Biology 7:473–479.

Bennetts, R. E., and R. L. Hutto. 1985. Attraction of social fringillids to mineral salts: an experimental study. Journal of Field Ornithology 56:187–189.

Bent, A. C. 1926. Life histories of North American marsh birds. U.S. National Museum Bulletin 135, Washington, D.C., USA.

Bent, A. C. 1937. Life histories of North American birds of prey. Pt. 1. U.S. Natural Museum Bulletin 167, Washington, D.C., USA.

Bent, A. C. 1938. Life histories of North American birds of prey. Pt. 2. U.S. National Museum Bulletin 170, Washington, D.C., USA.

Bent, A. C. 1940. Life histories of North American cuckoos, goatsuckers, hummingbirds, and their allies. U.S. National Museum Bulletin 176, Washington, D.C., USA.

Bent, A. C. 1942. Life histories of North American flycatchers, larks, swallows, and their allies. U.S. National Museum Bulletin 179, Washington, D.C., USA.

Bent, A. C. 1948. Life histories of North American nuthatches, wrens, thrashers, and their allies. U.S. National Museum Bulletin 195, Washington, D.C., USA.

Bent, A. C. 1949. Life histories of North American thrushes, kinglets, and their allies. U.S. National Museum Bulletin 196, Washington, D.C., USA.

Bent, A. C. 1950. Life histories of North American wagtails, shrikes, vireos, and their allies. U.S. National Museum Bulletin 197, Washington, D.C., USA.

Bent, A. C. 1953. Life histories of North American wood warblers. U.S. National Museum Bulletin 203, Washington, D.C., USA.

Bent, A. C. 1958. Life histories of North American blackbirds, orioles, tanagers, and their allies. U.S. National Museum Bulletin 211, Washington, D.C., USA.

Bent, A. C. 1961. Life histories of North American birds of prey. Pt. 1. Dover Publications, New York, New York, USA.

Bent, A. C. 1963. Life histories of North American gulls and terns. Dover Publications, New York, New York, USA.

Bentley, E. L. 1994. Use of a landscape level approach to determine the habitat requirements of the yellow-crowned night-heron, *Nycticorax violacea*, in the lower Chesapeake Bay. Thesis, College of William and Mary, Williamsburg, Virginia, USA.

Benton, T., D. M. Bryant, L. Cole, and H. Q. P. Crick. 2002. Linking agricultural practice to insect and bird populations: a historical study over three decades. Journal of Applied Ecology 39:673–687.

Benzinger, J. 1994. Hemlock decline and breeding birds. I. Hemlock ecology. Records of New Jersey Birds 20:2–12.

Berger, D. D., C. E. Sindelar, Jr., and K. E. Gambel. 1969. The status of breeding peregrines in the eastern United States. Pages 165–173 *in* J. J. Hickey, editor. Peregrine falcon populations: their biology and decline. University of Wisconsin Press, Madison, USA.

Berger, L., R. Speare, P. Daszak, D. E. Green, A. A. Cunningham, C. L. Goggin, R. Slocombe, M. A. Ragan, A. D. Hyatt, K. R. McDonald, H. B. Hines, K. R. Lips, G. Marantelli, and H. Parkes. 1998. Chytridiomycosis causes amphibian mortality associated with population declines in the rain forests of Australia and Central America. Proceedings of the National Academy of Sciences of the United States of America 95:9031–9036.

Bergstrom, P. W. 1991. Incubation temperatures of Wilson's plovers and killdeers. Condor 91:634–641.

Berthold, P., W. van den Bossche, and W. Fiedler. 2001. Detection of a new important staging and wintering area of the white stork *Ciconia ciconia* by satellite tracking. Ibis 143:450–455.

Bertin, R. I. 1977. Breeding habitat of the wood thrush and veery. Condor 79:303–311.

Best, L. B., H. Campa III, K. E. Kemp, R. J. Robel, M. R. Ryan, J. A. Savidge, H. P. Weeks, Jr., and S. R. Winterstein. 1997. Bird abundance and nesting in CRP fields and cropland in the Midwest: a regional approach. Wildlife Society Bulletin 25:864–877.

Best, T. L., and J. B. Jennings. 1997. *Myotis leibii*. Mammalian Species 547:1–6.

Bibby, C. J., N. D. Burgess, and D. A. Hill. 1992. Bird census techniques. Academic Press, San Diego, California, USA.

Bibby, C. J., N. D. Burgess, D. A. Hill, and S. H. Mustoe. 2000. Bird census techniques. Second edition. Academic Press, London, United Kingdom.

Bier, C. W. 1985. Geographic distribution: *Aneides aeneus*. Herpetological Review 16:60.

Biggins, R. G. 1987. Notice of status review—Blue Ridge Mountain population of the green salamander (*Aneides aeneus* [Cope and Packard]). U.S. Fish and Wildlife Service, USA.

Bildstein, K., J. P. Smith, E. R. Inzunza, and R. R. Veit, editors. 2008. State of North America's Birds of Prey. Series in Ornithology 3. Nuttall Ornithological Club and American Ornithologist's Union. Cambridge, Massachusetts and Washington, D.C., USA.

Bildstein, K. L. 1987. Behavioral ecology of red-tailed hawks (*Buteo jamaicensis*), rough legged hawks (*Buteo lagopus*), northern harriers (*Circus cyaneus*), and American kestrels (*Falco sparverius*) in south central Ohio. Ohio Biology Survey Biological Notes 18.

Bildstein, K. L., and J. B. Gollop. 1988. Northern harrier. Pages 251–303 *in* R. S. Palmer, editor. Handbook of North American birds. Volume 4. Yale University Press, New Haven, Connecticut, USA.

Bildstein, K. L., and K. Meyer. 2000. Sharp-shinned hawk (*Accipiter striatus*). *In* A. Poole and F. Gill, editors. The birds of North America, 782. The Birds of North America, Philadelphia, Pennsylvania, USA.
Bird Studies Canada. 2005. Canadian Migration Monitoring Network: bird population indices. <www.bsc-eoc.org>.
Birney, E. C., W. E. Grant, and D. D. Baird. 1976. Importance of vegetative cover to cycles of *Microtus* populations. Ecology 57:1043–1051.
Bishop, S. C. 1941. The salamanders of New York. New York State Museum Bulletin 324.
Bishop, S. C. 1947. Handbook of salamanders. Comstock, Ithaca, New York, USA.
Black, J. H., and A. N. Bragg. 1968. New addition to the herpetofauna of Montana. Herpetologica 24:247.
Blahnick, J. F., and P. A. Cochran. 1994. *Opheodrys vernalis* (smooth green snake). USA: Wisconsin. Herpetological Review 25:77.
Blake, E. R. 1977. Manual of Neotropical birds. Volume 1. University of Chicago Press, Chicago, Illinois, USA.
Blake, J. G., and J. R. Karr. 1984. Species composition of bird communities and the conservation benefit of large versus small forests. Biological Conservation 30: 173–187.
Blanchard, F. N. 1923. The snakes of the genus *Virginia*. Papers of the Michigan Academy of Science, Arts, and Letters 3:343–365.
Blanchard, H. 1964. Weight of a large fisher. Journal of Mammalogy 45:487–488.
Blancher, P., and J. Wells. 2005. The boreal forest region: North America's bird nursery. Canadian Boreal Initiative and Boreal Songbird Initiative, Ottawa, Ontario.
Blaustein, A. R., L. K. Belden, D. H. Olson, D. M. Green, T. L. Root, and J. M. Kiesecker. 2001. Amphibian breeding and climate change. Conservation Biology 15:1804–1809.
Blehert, D. S., A. C. Hicks, M. Behr, C. U. Meteyer, B. M. Berlowski-Zier, E. L. Buckles, J. T. H. Coleman, et al. 2009. Bat white-nose syndrome: an emerging fungal pathogen? Science 323:227.
Blem, C. R., and L. B. Blem. 1985. Notes on Virginia (Reptilia: Colubridae) in Virginia. Brimleyana 11:87–95.
Bloom, P. H. 1994. The biology and current status of the Long-eared Owl in coastal southern California. Bulletin of the Southern California Academy of Sciences 93:1–12.
Blumstein, D. T., E. Fernandez-Juricic, P. A. Zollner, and S. C. Garity. 2005. Inter-specific variation in avian responses to human disturbance. Journal of Applied Ecology 42:943–953.
Blumton, A. K. 1989. Factors affecting loggerhead shrike mortality in Virginia. Thesis, Virginia Polytechnical Institute and State University, Blacksburg, USA.
Blus, L. J. 1996. Effects of pesticides on owls in North America. Journal of Raptor Research 30:198–206.
Boal, C. W., D. E. Anderson, and P. L. Kennedy. 2003. Home range and residency status of northern goshawks breeding in Minnesota. Condor 1005:811–16.
Boal, C. W., D. E. Anderson, and P. L. Kennedy. 2006. Foraging and nesting habitat of breeding male northern goshawks in the Laurentian mixed forest province, Minnesota. Journal of Wildlife Management 69:1516–1527.
Bock, C. E., and L. W. Lepthien. 1976. Synchronous eruptions of boreal seed-eating birds. American Naturalist 110:559–579.
Bodie, J. R. 2001. Stream and riparian management for freshwater turtles. Journal of Environmental Management 62:443–455.
Boernke, W. E. 1977. A comparison of arginase maximum velocities from several poikilotherms and homeotherms. Comparative Biochemical Physiology 56: 113–116.
Bogan, M. A. 1972. Observation on parturition and development in the hoary bat, *Lasiurus cinereus*. Journal of Mammalogy 53:611–614.
Bogan, M. A. 1999. Eastern small footed myotis, *Myotis leibii*. Pages 93–94 *in* D. E. Wilson and S. Ruff, editors. The Smithsonian book of North American mammals. Smithsonian Institution Press, Washington D.C., USA.
Bogart, J. P., and M. W. Klemens. 1997. Hybrids and genetic interactions of mole salamanders (*Ambystoma jeffersonianum* and *A. laterale*) (Amphibia: Caudata) in New York and New England. American Museum Noviatates 3218. American Museum of Natural History, New York, USA.
Bohlen, H. D. 1989. The birds of Illinois. Indiana University Press, Bloomington, USA.
Bohlen, P. J., S. Scheu, C. M. Hale, M. A. McLean, S. Migge, P. M. Groffman, and D. Parkinson. 2004. Non-native invasive earthworms as agents of change in northern temperate forests. Frontiers in Ecology and the Environment 2:427–435.
Bolgiano, N. 1999. The story of the ring-necked pheasant in Pennsylvania. Pennsylvania Birds 13:2–10.
Bolgiano, N. C. 1997. Pennsylvania Christmas Bird Count: counts of sharp-shinned and Cooper's hawks. Pennsylvania Birds 11:134–137.
Bolgiano, N. C. 2000. A history of northern bobwhites in Pennsylvania. Pennsylvania Birds 14:58–68.
Bolgiano, N. C. 2004. Changes in boreal bird irruptions in eastern North America relative to the 1970s spruce budworm infestation. American Birds 58:26–33.
Bolgiano, N. C. 2005. Was the rise and fall of eastern sharp-shinned hawk migration counts linked to the 1970s spruce budworm infestation? Hawk Migration Studies 31:9–14.
Bollinger, E. K., P. B. Bollinger, and T. A. Gavin. 1990. Effects of hay-cropping on eastern populations of the Bobolink. Wildlife Society Bulletin 18:142–150.
Bollinger, E. K., and J. D. Maddox. 2000. A double-brooded dickcissel. Prairie Naturalist 32:253–255.

Bortner, J. B. 1985. Bioenergetics of wintering tundra swans in the Mattamuskeet region of North Carolina. Thesis, University of Maryland, College Park, USA.

Bosakowski, T. 1982. Roost selection and behavior of the long-eared owl (*Asio otus*) wintering in New Jersey. Raptor Research 18:137–142.

Bosakowski, T. 1986. Short-eared owl winter roosting strategies. American Birds 40:237–40.

Bosakowski, T., R. Kane, and D. G. Smith. 1989. Decline of the long-eared owl in New Jersey. Wilson Bulletin 101:481–485.

Bothner, R. C. 1963. A hibernaculum of the short-headed garter snake, *Thamnophis brachystoma* Cope. Copeia 1963:572–573.

Bothner, R. C. 1976. *Thamnophis brachystoma*. Catalog of American Amphibians and Reptiles 190.1–190.2.

Bothner, R. C. 1986. A survey of New York populations of the short-headed garter snake, *Thamnophis brachystoma* (Cope) (Reptilia: Colubridae). Unpublished report for the New York State Department of Environmental Conservation Endangered Species Unit, Albany.

Bothner, R., and T. C. Moore. 1964. A collection of *Haldea valeriae pulchra* from western Pennsylvania, with notes on some litters of their young. Copeia 1964:709–710.

Bottitta, G. E. 1997. Piping plovers produce two broods. Wilson Bulletin 109:337–339.

Boulinier, T., J. D. Nichols, J. D. Hines, J. R. Sauer, C. H. Flather, and K. H. Pollock. 1998. Higher temporal variability of forest breeding bird communities in fragmented landscapes. Proceedings of the National Academy of Sciences 95:7497–7501.

Bowlin, M. S., W. W. Cochran, and M. C. Wikelski. 2005. Biotelemetry of New World thrushes during migration: physiology, energetics, and orientation in the wild. Integrative and Comparative Biology 45: 295–304.

Bowman, J., G. L. Holloway, J. R. Malcolm, K. R. Middel, and P. J. Wilson. 2005. Northern range boundary dynamics of southern flying squirrels: evidence for an energy bottleneck. Canadian Journal of Zoology 83:1486–1494.

Brack, V., Jr. 1983. The non-hibernating ecology of bats in Indiana with emphasis on the endangered Indiana bat, *Myotis sodalis*. Dissertation, Purdue University, West Lafayette, Indiana, USA.

Brackney, A. W., and T. A. Bookhout. 1982. Population ecology of common gallinules in southwestern Lake Erie marshes. Ohio Journal of Science 82:229–237.

Bragg, A. N. 1965. Gnomes of the night. University of Pennsylvania Press, Philadelphia, USA.

Brambilla, M., D. Rubolini, and F. Guidali. 2004. Rock climbing and raven (*Corvus corax*) occurrence depress breeding success of cliff-nesting peregrines (*Falco peregrinus*). Ardeola 51(2):425–430.

Brandes, D. 1998. Spring golden eagle passage through the northeast U.S.—evidence for a geographically concentrated flight? Hawk Migration Association of North America Hawk Migration Studies 23(2):38–42.

Brandes, D., T. Katzner, T. Miller, M. Lanzone, K. Bildstein, and D. Ombalski. 2007. A terrain based dynamic model for simulating raptor migration through the Appalachians. 2007 Joint meeting of Raptor Research Foundation and Hawk Migration Association of North America.

Brandes, D., and D. Ombalski. 2004. Modeling raptor migration pathways using a fluid flow analogy. Journal of Raptor Research 38(3):195–207.

Branson, B. A., and E. C. Baker. 1947. An ecological study of the queen snake, *Regina septemvittata* (Say) in Kentucky. Tulane Studies in Zoology and Botany 18:153–171.

Braun, J., and G. R. Brooks, Jr. 1987. Box turtles, *Terrapene carolina*, as potential agents for seed dispersal. American Midland Naturalist 117:312–318.

Brauning, D. W. 1988. Peregrine falcon food habits. Cassinia 62:63–64.

Brauning, D. W., editor. 1992a. Atlas of breeding birds in Pennsylvania. University of Pittsburgh Press, Pittsburgh, Pennsylvania, USA.

Brauning, D. W. 1992b. Recent history and current status of nesting bald eagles, *Haliaeetus leucocephalus*, in Pennsylvania. Pennsylvania Birds 6:2–5.

Brauning, D. W. 1992c. King rail, *Rallus elegan*. Pages 122–123 *in* D. W. Brauning, editor. Atlas of breeding birds in Pennsylvania. University of Pittsburgh Press, Pittsburgh, Pennsylvania, USA.

Brauning, D. W. 1992d. Swainson's thrush. Pages 268–269 *in* D. W. Brauning, editor. Atlas of breeding birds in Pennsylvania. University of Pittsburgh Press, Pittsburgh, Pennsylvania, USA.

Brauning, D. W. 1992e. Common nighthawk. Pages 168–169 *in* D. W. Brauning, editor. Atlas of breeding birds in Pennsylvania. University of Pittsburgh Press, Pittsburgh, Pennsylvania, USA.

Brauning, D. W. 1992f. Solitary vireo. Pages 290–291 *in* D. W. Brauning, editor. Atlas of breeding birds in Pennsylvania. University of Pittsburgh Press, Pittsburgh, Pennsylvania, USA.

Brauning, D. W. 1992g. Winter wren. Pages 254–255 *in* D. W. Brauning, editor. Atlas of Breeding Birds in Pennsylvania. University of Pittsburgh Press, Pittsburgh, Pennsylvania, USA.

Brauning, D. W. 1998a. Final project report: wetland nesting bird population surveys in Pennsylvania. Unpublished report. Pennsylvania Game Commission, Harrisburg, USA.

Brauning, D. W. 1998b. Peregrine falcon research and management. Annual report to the Pennsylvania Game Commission, Harrisburg, USA.

Brauning, D. W. 2002a. Bald eagle breeding and wintering

surveys. Annual report. Pennsylvania Game Commission. <www.pgc.state.pa.us//pgc/lib/pgc/reports/2002>.

Brauning, D. W. 2002b. Bald eagles in the 21st century. Pennsylvania Game News 73:29–31.

Brauning, D. W. 2003. Bald eagle breeding and wintering surveys. Annual report. Pennsylvania Game Commission. <www.pgc.state.pa.us//pgc/lib/pgc/reports/2003>.

Brauning, D. W. 2004. Falcon wanderings: daily movements of five juvenile peregrine falcons. Pennsylvania Birds 18:168–171.

Brauning, D. W. 2005. Summary of the season—June through July 2005. Pennsylvania Birds 19:164–165.

Brauning, D. W. 2008. Management and biology of the peregrine falcon (*Falco peregrinus*) in Pennsylvania, ten year plan (2008–2017). Draft. Pennsylvania Game Commission, Harrisburg, USA.

Brauning, D. W., M. C. Brittingham, D. A. Gross, R. C. Leberman, T. L. Master, and R. S. Mulvihill. 1994. Pennsylvania breeding birds of special concern: a listing rationale and status update. Journal of the Pennsylvania Academy of Science 68:3–28.

Brauning, D. W., and C. Dooley. 1991. Recent history and current status of nesting peregrine falcons in Pennsylvania. Pennsylvania Birds 5:59–61.

Brauning, D. W., and J. Hassinger. 2000. Pennsylvania recovery and management plan for the bald eagle. Pennsylvania Game Commission, Harrisburg, USA.

Brauning, D. W., and J. Hassinger. 2001. Osprey management plan. Unpublished draft.

Brauning, D. W., S. Hoffman, and L. Mangel. 2002. Conneaut Marsh important bird areas wetland surveys. Unpublished annual report 72302-02. Pennsylvania Game Commission, Harrisburg, USA.

Brauning, D. W., and D. Siefken. 2003. Protected wildlife research/management. Project annual job report. Bureau of Wildlife Management, Pennsylvania Game Commission, Harrisburg, USA.

Brauning, D. W., and D. Siefken. 2004. Protected wildlife research/management. Project annual job report. Bureau of Wildlife Management, Pennsylvania Game Commission, Harrisburg, USA.

Brauning, D. W., and D. Siefken. 2005a. Loggerhead shrike research/management: loggerhead shrike nesting survey/habitat enhancement, Adams and Franklin counties. Project annual job report. Bureau of Wildlife Management, Pennsylvania Game Commission, Harrisburg, USA.

Brauning, D. W., and D. Siefken. 2005b. Osprey nesting surveys. Annual report 71701. Pennsylvania Game Commission, Harrisburg, USA.

Breeding Bird Survey. 2003. Breeding bird survey home page. <www.mbr.pwrc.usgs.gov/bbs>. Accessed 2006 June 19.

Brennan, L. A. 1999. Northern bobwhite (*Colinus virginianus*). *In* A. Poole and F. Gill, editors. The birds of North America, 397. Birds of North America, Philadelphia, Pennsylvania, USA.

Brenner, F. J. 1985. Aquatic and terrestrial habitats in Pennsylvania. Pages 7–17 *in* H. H. Genoways and F. J. Brenner, editors. Species of special concern in Pennsylvania. Special Publication of the Carnegie Museum of Natural History 11. Pittsburgh, Pennsylvania, USA.

Brewer, R. 1955. Size of home range in eight bird species in a southern Illinois swamp-thicket. Wilson Bulletin 67:140–141.

Bridges, C. M. 2000. Long-term effects of pesticide exposure at various stages of the southern leopard frog (*Rana sphenocephala*). Archives of Environmental Contamination and Toxicology 39:91–96.

Bridges, C. M. 2002. Tadpoles balance foraging and predator avoidance: effects of predation, pond drying, and hunger. Journal of Herpetology 36:627–634.

Brisbin, I. L., Jr., and T. B. Mowbray. 2002. American coot (*Fulica americana*). *In* A. Poole and F. Gill, editors. The birds of North America, 697. Birds of North America, Philadelphia, Pennsylvania, USA.

British Columbia Ministry of Environment, Land, and Parks. 1998. Inventory methods for nighthawk and poorwill. Standards for Components of British Columbia's Biodiversity, 9. Vancouver, British Columbia, Canada.

Britson, C. A., and R. E. Kissell, Jr. 1996. Effects of food type on developmental characteristics of an ephemeral pond-breeding anuran, *Pseudacris triseriata feriarum*. Herpetologica 52:374–382.

Brittingham, M. C., and L. J. Goodrich. 2010. Habitat fragmentation: a threat to Pennsylvania's forest birds. Pages 204–216 *in* S. K. Majumdar, T. L. Master, M. C. Brittingham, R. M. Ross, R. S. Mulvihill, and J. E. Huffman, editors. Avian ecology and conservation: a Pennsylvania focus with national implications. Pennsylvania Academy of Science, Easton, Pennsylvania, USA.

Brodeur, S., R. DeCarie, D. M. Bird, and M. Fuller. 1996. Complete migration cycle of golden eagles breeding in northern Quebec. Condor 98:293–299.

Brodman, R., S. Cortwright, and A. Resetar. 2002. Historical changes of reptiles and amphibians of Northwest Indiana fish and wildlife properties. America Midland Naturalist 147:135–144.

Brooks, R. J., C. M. Shilton, G. P. Brown, and N. W. S. Quinn. 1992. Body size, age distribution, and reproduction in a northern population of wood turtles (*Clemmys insculpta*). Canadian Journal of Zoology 70:462–469.

Brooks, R. T. 2004. Weather related effects on woodland vernal pond hydrology and hydroperiod. Wetlands 24:104–114.

Brown, B. W., and G. O. Batzli. 1985. Field manipulations of fox and gray squirrel populations: how important is interspecific competition? Canadian Journal of Zoology 63:2134–2140.

Brown, D. F. 1984. Snowshoe hare populations, habitat, and management in northern hardwood forest regeneration areas. Thesis, Pennsylvania State University, University Park, USA.

Brown, E. E. 1979. Some snake food records from the Carolinas. Brimleyana 1:113–124.

Brown, J. D., and J. M. Sleeman. 2002. Morbidity and mortality of reptiles admitted to the Wildlife Center of Virginia, 1991 to 2000. Journal of Wildlife Diseases 38:699–705.

Brown, L., and D. Amadon. 1968. Eagles, hawks, and falcons of the world. McGraw-Hill Book Company, New York, New York, USA.

Brown, M., and J. J. Dinsmore. 1986. Implications of marsh size and isolation for marsh bird management. Journal of Wildlife Management 50:392–397.

Brown, S., C. Hickey, B. Harrington, and R. Gill, editors. 2001. United States shorebird conservation plan. Second edition. Manomet Center for Conservation Sciences, Manomet, Massachusetts, USA.

Brown, W. P., and R. R. Roth. 2004. Juvenile survival and recruitment of wood thrushes, *Hylocichla mustelina*, in a forest fragment. Journal of Avian Biology 35:316–326.

Brown, W. S. 1991. Female reproductive ecology in a northern population of the timber rattlesnake, *Crotalus horridus*. Herpetologica 47:101–115.

Brown, W. S. 1993. Biology, status, and management of the timber rattlesnake (*Crotalus horridus*): a guide for conservation. Society for the Study of Amphibians and Reptiles Herpetological Circular No. 22:1–78.

Brown, W. S. 1995. Heterosexual groups and mating season in a northern population of timber rattlesnakes, *Crotalus horridus*. Herpetological Natural History 3:127–133.

Brown, W. S., M. Kery, and J. E. Hines. 2005. Long-term ecology of *Crotalus horridus*: dens, survival, and longevity. Program of the Biology of Rattlesnakes Symposium, Loma Linda University, California. (Abstract.)

Brown, W. S., and F. M. MacLean. 1983. Conspecific scent-trailing by newborn timber rattlesnakes, *Crotalus horridus*. Herpetologica 39:430–436.

Brua, R. B. 1999. Ruddy duck nesting success: do nest characteristics deter nest predation? Condor 101:867–870.

Brua, R. B. 2002. Ruddy duck (*Oxyura jamaicensis*). *In* A. Poole, editor. The birds of North America online. Cornell Lab of Ornithology, Ithaca, New York, USA. Retrieved from the Birds of North America Online. <http://bna.birds.cornell.edu/species/696>.

Brucker, E. F. 1992. Conducting a census of heron colonies in Pennsylvania. Unpublished paper, available from Pennsylvania Game Commission, Harrisburg, USA.

Bruno, J. F., and B. J. Cardinale. 2008. Cascading effects of predator richness. Frontiers in Ecology and the Environment 6:539–546.

Bruns, D. A. 2005. Macro-invertebrate response to land cover, habitat, and water chemistry in a mining-impacted river ecosystem: A GIS watershed analysis. Aquatic Sciences 67:403–423.

Brunton, D. F., and W. J. Crins. 1975. Status and habitat preference of the yellow-bellied flycatcher in Algonquin Park, Ontario. Ontario Field Biology 29:25–28.

Brush, T. 1991. Effects of competition and predation on prothonotary warblers and house wrens nesting in eastern Iowa. Journal of Iowa Academy of Sciences 101:28–30.

Bryan, G. G., and L. B. Best. 1991. Bird abundance and species richness in grassed waterways in Iowa rowcrop fields. American Midland Naturalist 126:90–102.

Bryant, D. M., and G. Jones. 1995. Morphological-changes in a population of sand martins *Riparia riparia* associated with fluctuations in population size. Bird Study 42:57–65.

Buckelew, A. R., Jr., and G. A. Hall. 1994. The West Virginia breeding bird atlas. University of Pittsburgh Press, Pittsburgh, Pennsylvania, USA.

Buckwalter, M. 1988. Short-eared owls in Clarion County. Pennsylvania Birds 2:55–56.

Buech, R. R., L. G. Hanson, and M. D. Nelson. 1997. Identification of wood turtle nesting areas for protection and management. Pages 383–391 *in* J. Van Abbema, editor. Proceedings: conservation, restoration, and management of tortoises and turtles—an international conference. Turtle and Tortoise Society, New York, USA.

Buhlmann, K. A., and J. W. Gibbons. 2001. Terrestrial habitat use by aquatic turtles from a seasonally fluctuating wetland: implications for wetland conservation boundaries. Chelonian Conservation Biology 4:115–127.

Bull, J. 1974. Birds of New York State. Natural History Press, Garden City, New York, USA.

Bull, J. J., J. M. Legler, and R. C. Vogt. 1985. Non-temperature dependent sex determination in two suborders of turtles. Copeia 1985:784–786.

Bull, J. J., and R. C. Vogt. 1979. Temperature-dependent sex determination in turtles. Science 206:1186–1188.

Bunn, D. S., A. B. Warburton, and R. D. S. Wilson. 1982. The barn owl. Buteo Books, Vermillion, South Dakota, USA.

Burger, A. E., and S. A. Shaffer. 2008. Application of tracking and data-logging technology in research and conservation of seabirds. Auk 125:253–264.

Burger, J. 1978. The pattern and mechanism of nesting in mixed-species heronries. Pages 45–58 *in* A. Sprunt IV, J. C. Ogden, and S. Winckler, editors. Wading birds. National Audubon Society research report 7, National Audubon Society, New York, New York, USA.

Burger, J. 1991. Foraging behavior and the effect of human disturbance on the piping plover (*Charadrius melodus*). Journal of Coastal Research 7:39–52.

Burger, J. 1994. The effect of human disturbance on foraging behavior and habitat use in the piping plover (*Charadrius melodus*). Estuaries 3:695–701.

Burhans, D. E., and F. R. Thompson III. 1999. Habitat patch size and nesting success of yellow breasted chats. Wilson Bulletin 111:210–215.

Burke, V. J., and J. W. Gibbons. 1995. Terrestrial buffer zones and wetland conservation: a case study of freshwater turtles in a Carolina bay. Conservation Biology 9:1365–1369.

Burke, V. J., J. E. Lovich, and J. W. Gibbons. 2000. Conservation of freshwater turtles. Pages 156–179 *in* M. Klemens, editor. Turtle conservation. Smithsonian Institution Press, Washington, D.C., USA.

Burleigh, T. D. 1931. Notes on the breeding birds of State College, Centre County, Pennsylvania. Wilson Bulletin 73:37–54.

Burns, F. L. 1911. A monograph of the broad-winged hawk (*Buteo platypterus*). Wilson Bulletin 23(3 and 4):1–320.

Burns, J. T. 1982. Nests, territories, and reproduction of sedge wrens (*Cistothorus platensis*). Wilson Bulletin 94:338–349.

Burt, W. 2001. Rare and elusive birds of North America. Universal Publishing, New York, New York, USA.

Burton, C., and C. J. Krebs. 2003. Influence of relatedness on snowshoe hare spacing behavior. Journal of Mammalogy 84:1100–1111.

Bury, R. B. 1979. Review of the ecology and conservation of the bog turtle, *Clemmys muhlenbergii*. U.S. Fish and Wildlife Service. Special Scientific Report. Wildlife 219:1–9.

Bush, W. L. 1989. Black tern (*Chlidonias niger*) nesting platform and habitat study in Crawford and Erie counties, Pennsylvania. Unpublished report to Pennsylvania Wild Resources Conservation Fund.

Bushar, L. M., H. K. Reinert, and L. Gelbert. 1998. Genetic variation and gene flow within and between local populations of the timber rattlesnake, *Crotalus horridus*. Copeia 1998:411–422.

Bushar, L. M., H. K. Reinert, and A. H. Savitsky. 2005. Isolation and reduced genetic variation in the timber rattlesnake, *Crotalus horridus*, of the New Jersey Pine Barrens. Program of the Biology of Rattlesnakes Symposium, Loma Linda University, California. (Abstract.)

Bushman, E. S., and G. D. Therres. 1988. Habitat management guidelines for forest interior breeding birds of coastal Maryland. Maryland Department of Natural Resources, Wildlife Technical Publication 88–1.

Butchkoski, C. 2003a. Eastern woodrat research/management. Project 6718 annual report, Pennsylvania Game Commission Bureau of Wildlife Management, Harrisburg, USA. <www.pgc.state.pa.us/pgc/cwp/view.asp?A=495&Q=161129#71801>. Accessed 1 Jan 2005.

Butchkoski, C. 2003b. Indiana bat (*Myotis sodalis*) investigations at Canoe Creek, Blair County Pennsylvania. Annual job report. Pennsylvania Game Commission, Harrisburg, USA.

Butchkoski, C., and J. D. Hassinger. 2002a. Ecology of a building maternity site. Pages 130–142 *in* A. Kurta and J. Kennedy, editors. The Indiana bat: biology and management of an endangered species. Bat Conservation International, Austin, Texas, USA.

Butchkoski, C., and J. D. Hassinger. 2002b. Impacts of a heavily traveled highway, U.S. Route 22, intersecting a major travel corridor for bats. Annual job report. Pennsylvania Game Commission, Harrisburg, USA.

Butchkoski, C., and J. D. Hassinger. 2002c. Timber stand analysis of Indiana bat core foraging sites. Annual job report. Pennsylvania Game Commission, Harrisburg, USA.

Butchkoski, C., and A. Mehring. 2004. Indiana bat (*Myotis sodalis*) investigations at Canoe Creek, Blair County Pennsylvania. Annual job report. Pennsylvania Game Commission, Harrisburg, USA.

Butchkoski, C., and G. Turner. 2005a. Indiana bat (*Myotis sodalis*) investigations at Canoe Creek, Blair County Pennsylvania. Annual job report. Pennsylvania Game Commission, Harrisburg, USA.

Butchkoski, C., and G. Turner. 2005b. Indiana bat hibernacula surveys. Annual job report. Pennsylvania Game Commission, Harrisburg, USA.

Butchkoski, C. M., and G. Turner. 2006. Indiana bat (*Myotis sodalis*) summer roost investigations. Annual job report. Pennsylvania Game Commission, Harrisburg, USA.

Butchkoski, C. M., and G. Turner. 2007. Indiana bat (Myotis sodalis) summer roost investigations. Annual job report. Pennsylvania Game Commission, Harrisburg, USA.

Butchkoski, C. M., and G. Turner. 2008. Indiana bat (*Myotis sodalis*) summer roost investigations. Annual job report. Pennsylvania Game Commission, Harrisburg, USA.

Butler, R. W. 1989. Breeding ecology and population trends of the great blue heron (*Ardea herodias fannini*) in the Strait of Georgia. Pages 112–117 *in* K. Vermeer and R. W. Butler, editors. The ecology and status of marine and shoreline birds in the Strait of Georgia, British Columbia. Canadian Wildlife Service Special Publication, Ottawa, Canada.

Butler, R. W. 1992. Great blue heron. *In* A. Poole, P. Stettenheim, and F. Gill, editors. The birds of North America, 25. Academy of Natural Sciences, Philadelphia, Pennsylvania, USA, and American Ornithologists' Union, Washington, D.C., USA.

Byrd, M. A. 1978. Dispersal and movements of six North American ciconiiforms. Pages 161–185 *in* A. Sprunt IV, J. C. Ogden, and S. Winckler, editors. Wading birds. Research report 7, National Audubon Society, New York, New York, USA.

Byrd, M. A., G. D. Therres, and W. N. Weimeyer. 1990. Chesapeake Bay Region Bald Eagle Recovery Plan, first revision. U.S. Fish and Wildlife Service. Region Five. Newton Corner, Massachusetts, USA.

Caceres, M. C., and R. M. R. Barclay. 2000. *Myotis septentrionalis*. Mammalian Species 634:1–4.

Cade, B. S. 1986. Habitat suitability index models: brown thrasher. U.S. Department of Agriculture, U.S. Forest and Wildlife Service Biological Report 82, Washington, D.C., USA.

Cade, T. J., J. H. Enderson, C. G. Thelander, and C. M. White, editors. 1988. Peregrine falcon populations; their management and recovery. Peregrine Fund, Boise, Idaho, USA.

Caire, W., R. M. Hardisty, and K. E. Lacy. 1988. Capture heights and times of *Lasiurus borealis* (Chiroptera: Vespertilionidae) in southeastern Oklahoma. Proceedings of the Oklahoma Academy of Science 68:51–53.

Caldwell, J. P. 1986. Selection of egg deposition sites: a seasonal shift in the southern leopard frog, *Rana sphenocephala*. Copeia 1986:249–253.

Callahan, E. V., R. D. Drobney, and R. L. Clawson. 1997. Selection of summer roosting sites by Indiana Bats (*Myotis sodalis*) in Missouri. Journal of Mammalogy 78:818–825.

Callicott, J. B., L. B. Crowder, and K. Mumford. 1999. Current normative concepts in conservation. Conservation Biology 13:22–35.

Campbell, T. A., and D. B. Long. 2009. Feral swine damage and damage management in forested ecosystems. Forest Ecology and Management 257:2319–2326.

Canadian Peregrine Foundation. 2000. Canadian Peregrine Foundation Web site. <www.peregrine-foundation.ca/programs/trackem/track.html>.

Canterbury, R. A, N. J. Kotesovec, Jr., and B. Catuzza. 1995a. A preliminary study of the effects of brown-headed cowbird parasitism on the reproductive success of blue-winged warblers in northeastern Ohio. Ohio Cardinal 18:124–125.

Canterbury, R. A., N. J. Kotesovec, Jr., B. Catuzza, and B. M. Walton. 1995b. Effects of brown headed cowbird (*Molothrus ater*) parasitism on habitat selection and reproductive success in blue-winged warblers (*Vermivora pinus*) in northeastern Ohio. Unpublished report. West Virginia Division of Natural Resources, Elkins, West Virginia, USA.

Canterbury, R. A., and T. K. Pauley. 1994. Time of mating and egg deposition of West Virginia populations of the salamander *Aneides aeneus*. Journal of Herpetology 28:431–434.

Capouillez, W., and T. Limbrandi-Mumma. 2008. Pennsylvania Game Commission wind energy voluntary cooperation agreement. First annual report. Pennsylvania Game Commission, Harrisburg, USA.

Carey, C., A. P. Pessier, and A. D. Peace. 2003. Pathogens, infectious disease, and immune defenses. Pages 127–136 *in* R. D. Semlitsch, editor. Amphibian conservation. Smithsonian, Washington, D.C., USA.

Carpenter, C. C. 1952. Comparative ecology of the common garter snake, the ribbon snake, and Butler's garter snake in mixed populations. Ecological Monographs 22:235–258.

Carpenter, C. C., and J. C. Gillingham. 1990. Ritualized behavior in *Agkistrodon* and allied genera. Pages 523–531 *in* H. K. Gloyd and R. Conant, editors. Snakes of the *Agkistrodon* complex: a monographic review. Contributions to Herpetology 6. Society for the Study of Amphibians and Reptiles, Athens, Ohio, USA.

Carpenter, C. P. 2001. Scat marking and the use of latrine sites by river otters along Tionesta Creek, northwestern Pennsylvania. Thesis, Frostburg State University, Frostburg, Maryland, USA.

Carr, L. W., and L. Fahrig. 2001. Effect of road traffic on two amphibian species of differing vagility. Conservation Biology 15:1071–1078.

Carter, J. D. 1904. Summer birds of Pocono Lake, Monroe County, Pennsylvania. Cassinia 8:29–35.

Carter, M., G. Fenwick, C. Hunter, D. Pashley, D. Petit, J. Price, and J. Trapp. 1996. WatchList 1996—for the future. National Audubon Society Field Notes 50:238–240.

Carter, T. C. 1998. The foraging ecology of three species of bats at the Savannah River Site, South Carolina. Thesis, University of Georgia, Athens, USA.

Carter, T. C., M. A. Menzel, S. F. Owen, J. W. Edwards, J. M. Menzel, and W. M. Ford. 2003a. Food habits of seven species of bats in the Allegheny plateau and ridge and valley of West Virginia. Northeastern Naturalist 10:83–88.

Carter, T. C., M. A. Menzel, and D. A. Saugey. 2003b. Population trends of solitary foliage-roosting bats. *In* T. J. O'Shea and M. A. Bogan, editors. Monitoring trends in bat populations of the United States and territories: problems and prospects. U.S. Geological Survey, Biological Resources Discipline, Information and Technology Report, USGS/BRD/ITR-2003-0003.

Cashen, S. T., and M. C. Brittingham. 1998. Avian use of restored wetlands in Pennsylvania. Final report to Pennsylvania Game Commission, Harrisburg, USA.

Casper, G. S. 1996. Geographic distribution. *Opheodrys vernalis*. Herpetological Review 27:214.

Castleberry, S. B., W. M. Ford, P. B. Wood, and N. L. Castleberry. 2001. Movements of Allegheny woodrats in relation to timber harvesting. Journal of Wildlife Management 65:148–156.

Caughley, G., and A. Gunn. 1996. Conservation biology in theory and practice. Blackwell Science, Cambridge, Massachusetts, USA.

Cavitt, J. F., and C. A. Haas. 2000. Brown thrasher (*Toxostoma rufum*). *In* A. Poole and F. Gill, editors. The birds of North America, 557. Birds of North America, Philadelphia, Pennsylvania, USA.

Chace, J. F., J. E. Gillis, and S. D. Faccio. 2002. Habitat selection of a declining warbler population: the Canada warbler in northern Vermont. Abstracts from the third North American Ornithological Conference, New Orleans, Louisiana, USA.

Chalfoun, A. D., M. J. Ratmaswamy, and F. R. Thompson III. 2002b. Songbird nest predators in forest-pasture edge and forest interior in a fragmented landscape. Ecological Applications 12:858–867.

Chalfoun, A. D., F. R. Thompson III, and M. J. Ratmaswamy. 2002a. Nest predators and fragmentation: a review and meta-analysis. Conservation Biology 16:306–318.

Chapin, E. A. 1925. Food habits of vireos: a family of insectivorous birds. U.S. Department of Agriculture Bulletin 1355.

Chapin, F. S., III, E. S. Zavaleta, V. T. Eviner, R. L. Naylor, P. M. Vitousek, H. L. Reynolds, D. U. Hooper, S. Lavorel, O. E. Sala, S. E. Hobbie, M. C. Mack, and S. Diaz. 2000. Consequences of changing biodiversity. Nature 405:234–242.

Chapman, J. A. 1975. *Sylvilagus transitionalis*. Mammalian Species 55:1–4.

Chapman, J. A. 1999. Appalachian cottontail, *Sylvilagus obscurus*. Pages 690–691 *in* D. E. Wilson and S. Ruff, editors. The Smithsonian book of North American mammals. Smithsonian Institution, Washington, D.C., USA.

Chapman, J. A., K. L. Cramer, N. J. Dippenaat, and T. J. Robinson. 1992. Systematics and biogeography of the New England Cottontail, *Sylvilagus transitionalis* (Bangs, 1895), with a description of a new species from the Appalachian Mountains. Proceedings of the Biological Society of Washington 105:841–866.

Chapman, J. A., J. G. Hockman, and W. R. Edwards. 1982. Cottontails. Pages 83–123 *in* J. A. Chapman, and G. A. Feldhamer, editors. Wild mammals of North America: biology, management, and economics. John Hopkins University Press, Baltimore, Maryland, USA.

Chapman, J. A., and J. R. Stauffer, Jr. 1982. The status and distribution of the New England Cottontail. Pages 973–983 *in* K. Myers and C. D. MacInnes, editors. Proceedings of the World Lagomorph Conference, University of Guelph, Guelph, Canada.

Chazal, A. C., and P. H. Niewiarowski. 1998. Responses of mole salamanders to clearcutting: using field experiments in forest management. Ecological Applications 8:1133–1143.

Chen, J., J. F. Franklin, and T. A. Spies. 1995. Growing-season microclimatic gradients from clearcut edges into old-growth douglas-fir forests. Ecological Applications 5:74–86.

Chenger, J. 2003. Iowa army ammunition plant 2003 Indiana bat investigations. Unpublished report to Iowa Army Ammunitions Plant, Middletown, Iowa. Bat Conservation and Management, Carlisle, Pennsylvania, USA.

Choate, J. R., E. D. Fleharty, and R. J. Little. 1974. Status of the spotted skunk, *Spilogale putorius*, in Kansas. Transatlantic Kansas Academy of Science 76:226–233.

Christian, K. A. 1982. Changes in the food niche during postmetamorphic ontogeny of the frog *Pseudacris triseriata*. Copeia 1982:73–80.

Christy, B. H. 1926. Bob-white in Pennsylvania. Cardinal 1(7):7–18.

Church, K. E., J. R. Sauer, and S. Droege. 1993. Population trends in quails of North America. Pages 44–54 *in* K. E. Church and T. V. Dailey, editors. Quail III: National Quail Symposium. Kansas Department of Wildlife and Parks, Pratt, USA.

Cink, C. L. 1989. Snake predation on chimney swift nestlings. Journal of Field Ornithology 61:288–289.

Cink, C. L. 2002. Whip-poor-will. *In* A. Poole and F. Gill, editors. The birds of North America, 620. The Birds of North America, Philadelphia, Pennsylvania, USA.

Clark, B. S., J. B. Bowles, and B. K. Clark. 1987. Summer occurrence of the Indiana bat, Keen's myotis, evening bat, silver-haired bat and eastern pipistrelle in Iowa. Proceedings of the Iowa Academy of Science 94:89–93.

Clark, D. R., Jr., and M. A. R. McLane. 1974. Chlorinated hydrocarbon and mercury residues in woodcock in the United States, 1970–1971. Pesticides Monitoring Journal 8:15–22.

Clark, K. E., L. J. Niles, and W. Stansley. 1998. Environmental contaminants associated with reproductive failure in Bald Eagle (*Haliaeetus leucocephalus*) eggs in New Jersey. Bulletin of Environmental Contaminants and Toxicology 671:247–254.

Clark, R. J. 1975. A field study of the short-eared owl (*Asio flammeus*) in North America. Wildlife Monographs 47:1–67.

Clark, R. J., D. Euler, and E. Armstrong. 1983. Habitat associations of breeding birds in cottage and natural areas in central Ontario. Wilson Bulletin 95:77–96.

Clark, R. W. 2002. Diet of the timber rattlesnake, *Crotalus horridus*. Journal of Herpetology 36:494–499.

Clark, T. W. 1973. Distribution and reproduction of shrews in Grand Teton National Park, Wyoming. Northwest Science 47:128–131.

Clark, W. 1985. The migrating sharp-shinned hawk at Cape May Point: banding and recovery results. Pages 137–148 *in* M. Harwood, editor. Proceedings of the North American Hawk Migration Conference IV, Rochester, New York, March 1983. Hawk Migration Association of North America.

Clausen, R. T. 1938. Notes on *Eumeces anthracinus* in central New York. Copeia 1938:3–7.

Clawson, R. L. 1984. Recovery efforts for the endangered Indiana bat (*Myotis sodalis*) and gray bat (*Myotis grisescens*). Pages 301–307 *in* W. C. Comb, editor. Proceedings of Workshop on Management of Nongame Species and Ecological Communities. Agricultural Experiment Station, University of Kentucky, USA.

Clegg, S. M., J. F. Kelly, M. Kimura, and Thomas B. Smith. 2003. Combining genetic markers and stable isotopes to reveal population connectivity and migration patterns in a Neotropical migrant, Wilson's warbler (*Wilsonia pusilla*). Molecular Ecology 12:819–830.

Clem, M. K. 1977. Interspecific relationship of fishers and martens in Ontario during winter. Pages 165–182 *in* R. L. Phillips and C. Jonkel, editors. Proceedings of the 1975 predator symposium. Montana Forest and Conservation Experiment Station, University of Montana, Missoula, USA.

Clement, R., A. Harris, and J. Davis. 1993. Finches and sparrows: an identification guide. Princeton University Press, Princeton, New Jersey, USA.

Clement, R. C., and W. W. H. Gunn. 1957. Canada warbler. Pages 238–338 *in* L. Griscom and A. Sprunt, Jr., editors. The warblers of America. Devin-Adair, New York, New York, USA.

Cliburn, J. W., and A. B. Porter. 1987. Vertical stratification of the salamanders *Aneides aeneus* and *Plethodon glutinosus* (Caudata: Plethodontidae). Journal of Alabama Academy of Science 58:18–22.

Cohen, M. 2004a. Pennsylvania important bird areas 27: Marsh Creek Wetlands—"The Muck." Pennsylvania Audubon Society, Harrisburg, USA.

Cohen, M. 2004b. Pennsylvania important bird areas 56: Conejohela Flats. Pennsylvania Audubon Society, Harrisburg, USA.

Cohen, M., and J. Johnson. 2004. Pennsylvania important bird areas 73: John Heinz National Wildlife Refuge at Tinicum. Pennsylvania Audubon Society, Harrisburg, USA.

Cohn, J. P. 1999. Tracking wildlife. BioScience 49:12–17.

Cohn, J. P. 2008. White-nose syndrome threatens bats. BioScience 58:1098.

Coker, D. R., and J. L. Confer. 1990. Brown-headed cowbird parasitism on golden-winged and blue-winged warblers. Wilson Bulletin 102:550–552.

Coleman, J. L., and D. Bird. 1999. Habitat selection by sharp-shinned hawks (*Accipiter striatus*) (Abstract). Annual meeting of the Raptor Research Foundation, La Paz, Baja, California, USA.

Collins, C. T., J. Farrand, Jr., editors. 1985. "Chimney swift," The Audubon Society Master Guide to Birding. Volume 2. Gulls to dippers. Alfred A. Knopf, New York, New York, USA.

Collins, J. P., and A. Storfer. 2003. Global amphibian declines: sorting the hypotheses. Diversity and Distributions 9:89–98.

Collins, J. T. 1974. Amphibians and reptiles in Kansas. University of Kansas Museum of Natural History Public Education Series One, Lawrence, USA.

Collins, S. L., F. C. James, and P. G. Risser. 1982. Habitat relationships of wood warblers (Parulidae) in northern central Minnesota. Oikos 39:50–58.

Colvin, B. A. 1984. Barn owl foraging and secondary poisoning hazard from rodenticide use on farms. Dissertation, Bowling Green State University, Bowling Green, Ohio, USA.

Colvin, B. A. 1985. Common barn-owl population decline in Ohio and the relationship to agricultural trends. Journal of Field Ornithology 56:224–235.

Combs, K. P., and S. M. Melvin. 1989. Population dynamics, habitat use, and management of short-eared owls on Nantucket Island, Massachusetts. Unpublished progress report. Massachusetts Division of Fish and Wildlife, Natural Heritage and Endangered Species Program, Boston, Massachusetts, USA.

Committee on the Status of Endangered Wildlife in Canada. 2004. Committee on the Status of Endangered Wildlife in Canada assessment and status report on the red crossbill *percna* subspecies *Loxia curvirostra percna* in Canada. Committee on the Status of Endangered Wildlife in Canada, Ottawa, Ontario, Canada.

Compton, B. W., J. M. Rhymer, and M. McCollough. 2002. Habitat selection by wood turtles (*Clemmys insculpta*): an application of paired logistic regression. Ecology 83:833–843.

Conant, R. 1938. On the seasonal occurrence of reptiles in Lucas County, Ohio. Herpetologica 1:137–144.

Conant, R. 1943. Studies on North American water snakes. Pt. 1. *Natrix kirtlandii* (Kennicott). American Midland Naturalist 29:313–341.

Conant, R. 1950. On the taxonomic status of *Thamnophis butleri* (Cope). Bulletin of the Chicago Academy of Science 9:71–77.

Conant, R. 1951. The red-bellied terrapin, *Pseudemys rubriventris* (Le Conte) in Pennsylvania. Annals of the Carnegie Museum 32:281–290.

Conant, R. 1960. The queen snake, *Natrix septemvittata*, in the interior highlands of Arkansas and Missouri, with comments upon similar disjunct distributions. Proceedings of the Academy of Natural Sciences of Philadelphia 112:25–40.

Conant, R. 1978. Distributional patterns of North American snakes: some examples of the effects of Pleistocene glaciation and subsequent climatic changes. Bulletin of the Maryland Herpetological Society 14:241–259.

Conant, R. 1990. An annotated checklist of the breeding birds of Dutch Mountain, Pennsylvania. Cassinia 63:61–71.

Conant, R., and J. T. Collins. 1991. A field guide to reptiles and amphibians: eastern and central North America. Houghton Mifflin, Boston, Massachusetts, USA.

Conant, R., and J. T. Collins. 1998. A field guide to reptiles and amphibians: eastern and central North America. Houghton Mifflin, Boston, Massachusetts, USA.

Confer, J. L. 1992. Golden-winged warbler. *In* A. Poole, P. Stettenheim, and F. Gill, editors. The birds of North America, 20. Academy of Natural Sciences, Philadelphia, Pennsylvania, USA, and American Ornithologists' Union, Washington, D.C., USA.

Confer, J. L. 2006. Secondary contact and introgression of golden-winged warblers (*Vermivora chrysoptera*): documenting the mechanism. Auk 123:958–961.

Confer, J. L., and K. Knapp. 1977. Hybridization and interaction between blue-winged and golden-winged warblers. Kingbird 27:181–190.

Confer, J. L., and K. Knapp. 1981. Golden-winged warblers and blue-winged warblers: the relative success of a habitat specialist and a generalist. Auk 98:108–114.

Confer, J. L., J. L. Larkin, and P. E. Allen. 2003. Effects of vegetation, interspecific competition, and brood parasitism on golden-winged warbler (*Vermivora chrysoptera*) nesting success. Auk 120:138–144.

Congdon, J. D., A. E. Dunham, and R.C. van Loben Sels. 1993. Delayed sexual maturity and demographics of Blanding's turtles: implications for conservation and management of long-lived organisms. Conservation Biology 7:826–833.

Congdon, J. D., A. E. Dunham, and R. C. van Loben Sels. 1994. Demographics of common snapping turtles: implications for conservation and management of long-lived organisms. American Zoologist 34:397–408.

Congdon, J. D., R. D. Nagle, O. M. Kinney, M. F. Osentoski, R. C. van Loben Sels, H. W. Avery, and D. W. Tinkle. 2000. Nesting ecology and embryo mortality: implications for hatching success and demography of Blanding's turtles, *Emydoidea blandingii*. Chelonian Conservation and Biology 3:569–579.

Congdon, J. D., D. W. Tinkle, G. L. Breitenbach, and R. C. Van Loben Sels. 1983. Nesting ecology and hatching success in the turtle *Emydoidea blandingii*. Herpetologica 39:417–429.

Congdon, J. D., and R. C. Van Loben Sels. 1991. Growth and body size in Blanding's turtles (*Emydoidea blandingii*): relationships to reproduction. Canadian Journal of Zoology 69:239–245.

Connecticut Department of Environmental Protection. 2003. Night hawk surveys. <http://dep.state.ct.us/cgnhs/nddb/nhawk.htm>. Accessed 5 May 2005.

Connecticut Department of Environmental Protection. 2004. Volunteer for bird surveys. <http://dep.state.ct.us/cgnhs/nddb/volun.htm>. Accessed 8 Dec 2004.

Conner, R. N., S. D. Jones, G. D. Jones. 1994. Snag condition and woodpecker foraging ecology in a bottomland hardwood forest. Wilson Bulletin 106:242–257.

Connor, P. F. 1959. The bog lemming *Synaptomys cooperi* in southern New Jersey. Publications of the Museum, Michigan State University, Biological Series 1:161–248.

Conroy, M. J., M. W. Miller, and J. E. Hines. 2002. Identification and synthetic modeling of factors affecting American black duck populations. Wildlife Monographs 150.

Constantine, D. G. 1966. Ecological observation of lasiurine bats in Iowa. Journal of Mammalogy 47:34–41.

Convention on International Trade in Endangered Species [CITES]. 2007. CITES species database. CITES, Geneva, Switzerland. <www.cites.org>. Accessed 9 Oct 2008.

Convention on International Trade in Endangered Species [CITES]. 2009. CITES species database. CITES, Geneva, Switzerland. <www.cites.org>. Accessed 25 Jan 2009.

Conway, C. H. 1958. Maintenance, reproduction, and growth of the least shrew in captivity. Journal of Mammalogy 39:507–512.

Conway, C. J. 1995. Virginia rail (*Rallus limicola*). *In* A. Poole, editor. The birds of North America online. Cornell Lab of Ornithology, Ithaca, New York, USA. Retrieved from the Birds of North America online <http//bna.birds.cornell.edu.bnaproxy.birds.cornell.edu/bna/species/173 doi:10.2173/bna.173>.

Conway, C. J. 1999. Canada warbler (*Wilsonia canadensis*). *In* A. Poole and F. Gill, editors. The birds of North America, 421. Birds of North America, Philadelphia, Pennsylvania, USA.

Conway, C. J. 2004. Standardized North American Marsh Bird Monitoring Protocols. U.S. Geological Survey. Arizona Cooperative Fish and Wildlife Research Unit, 104 Biological Sciences, East University of Arizona, Tucson, USA.

Conway, C. J. 2005. Standardized North American Marsh Bird Monitoring Protocols. Wildlife research report 2005–04. U.S. Geological Survey. Arizona Cooperative Fish and Wildlife Research Unit, Tucson, USA.

Conway, C. J., and J. P. Gibbs. 2005. Effectiveness of call-broadcast surveys for monitoring marsh birds. Auk 122:26–35.

Conway, C. J., and C. P. Nadeau. 2006. Development and field testing of survey methods for a continental marsh bird monitoring program in North America. Wildlife Research report 2005-11. U.S. Geological Survey. Arizona Cooperative Fish and Wildlife Research Unit, Tucson, Arizona, USA.

Cook, F. R. 1964. Communal egg laying in the smooth green snake. Herpetologica 20:206.

Cook-Haley, B. S., and K. F. Millenbah. 2002. Impacts of vegetative manipulations on common tern nest success at Lime Island, Michigan. Journal of Field Ornithology 73(2):174–179.

Coon, D. R., B. K. Williams, J. S. Lindzey, and J. L. George. 1982. Examination of woodcock nest sites in central Pennsylvania. Pages 55–62 *in* T. Dwyer and G. L. Storm, editors. Woodcock ecology and management. U.S. Fish and Wildlife Service. Wildlife Research Report 14. Washington, D.C. USA.

Coon, D. R., P. D. Caldwell, and G. L. Storm. 1976. Some characteristics of a fall migration of female woodcock. Journal of Wildlife Management 40:91–95.

Cooper, J. E. 1960. Notes on a specimen of the least shrew. Maryland Naturalist 29:21–22.

Cooper, J. L. 1999. Special animal abstract for *Buteo lineatus* (red-shouldered hawk). Michigan Natural Features Inventory, Lansing, USA.

Cooper, R. J., K. M. Dodge, P. J. Martinat, S. B. Donahue, and R. C. Whitmore. 1990. Effect of diflubenzuron application on eastern deciduous forest birds. Journal of Wildlife Management 54:486–493.

Cooper, W. E., Jr., and N. Burns. 1987. Social significance of ventrolateral coloration in the fence lizard, *Sceloporus undulatus*. Animal Behaviour 35:526–532.

Cooper, W. E., Jr., and L. J. Vitt. 1986. Interspecific odour discrimination by a lizard (*Eumeces laticeps*). Animal Behaviour 34:367–376.

Cope, F. R., Jr. 1901. Observations on the summer birds

of parts of Clinton and Potter counties, Pennsylvania. Cassinia 5:8–21.

Cope, T. M. 1936. Observations of the vertebrate ecology of some Pennsylvania virgin forests. Thesis, Cornell University, Ithaca, New York, USA.

Coppedge, B. R., D. M. Engle, R. E. Masters, and M. S. Gregory. 2001. Avian response to landscape change in fragmented southern great plains grasslands. Ecological Applications 1:47–59.

Corn, P. S., and J. C. Fogleman. 1984. Extinction of montane populations of the northern leopard frog (*Rana pipiens*) in Colorado. Journal of Herpetology 18:147–152.

Cornell Laboratory of Ornithology. 1997. Birds in forested landscapes. Cornell Laboratory of Ornithology, Ithaca, New York, USA.

Corser, J. D. 2001. Decline of disjunct green salamander (*Aneides aeneus*) populations in the southern Appalachians. Biological Conservation 97:119–126.

Costanza, R., R. d'Arge, R. de Groot, S. Farber, M. Grasso, B. Hannon, K. Limburg, S. Naeem, R. V. O'Neill, J. Paruelo, R. G. Raskin, P. Sutton, and M. van den Belt. 1997. The value of the world's ecosystem services and natural capital. Nature 387:253–260.

Cottam, C. F., and F. M. Uhler. 1945. Birds in relation to fishes. U.S. Fish and Wildlife Service Leaflet 272.

Coupe, B. 2002. Pheromones, search patterns, and old haunts: how do male timber rattlesnakes (*Crotalus horridus*) locate mates? Pages 139–148 *in* G. W. Schuett, M. Hoggren, M. E. Douglas, and H. W. Greene, editors. Biology of the vipers. Eagle Mountain Publishing. Eagle Mountain, Utah, USA.

Coulter, M. W. 1966. Ecology and management of fishers in Maine. Dissertation, Syracuse University, Syracuse, New York, USA.

Cox, G. W. 1993. Conservation ecology. William C. Brown, New York, New York, USA.

Crabb, W. D. 1944. Growth, development, and seasonal weights of spotted skunks. Journal of Mammalogy 25:213–221.

Craighead, J. J., and F. C. Craighead, Jr. 1956. Hawks, owls, and wildlife. Stackpole Company, Harrisburg, Pennsylvania, USA, and the Wildlife Management Institute, Washington, D.C., USA.

Cramp, S., editor. 1988. The birds of the western Palearctic. Volume 5. Tyrant flycatchers to thrushes. Oxford University Press, Oxford, United Kingdom.

Crawford, R. L. 1981. Bird casualties at a Leon County, Florida TV tower: a 25-year migration study. Bulletin of the Tall Timbers Research Station 22.

Crewe, T., S. Timmermans, and K. Jones. 2005. The Marsh Monitoring Program annual report, 1995–2003: annual indices and trends in bird abundance and amphibian occurrence in the Great Lakes basin. Bird Studies Canada, Port Rowan, Ontario.

Crocoll, S. T. 1984. Breeding biology of broad-winged and red-shouldered hawks in western New York. Thesis, State University College, Fredonia, New York.

Crocoll, S. T. 1994. Red-shouldered hawk. *In* A. Poole and F. Gill, editors. The birds of North America, 107. Philadelphia Academy of Natural Sciences, Philadelphia, Pennsylvania, USA, and American Orinthologists' Union, Washington, D.C., USA.

Crocoll, S. T., and J. W. Parker. 1989. The breeding biology of broad-winged and red-shouldered hawks in western New York. Journal of Raptor Research 23:125–139.

Crooks, K. R. 2002. Relative sensitivities of mammalian carnivores to habitat fragmentation. Conservation Biology 16:488–502.

Crooks, K. R., and M. E. Soulé. 1999. Mesopredator release and avifaunal extinctions in a fragmented system. Nature 400:563–566.

Crossley, G. J. 1999. A guide to critical bird habitat in Pennsylvania. Pennsylvania Audubon Society. Harrisburg, Pennsylvania, USA.

Crother, B. I. 1992. Genetic characters, species concepts, and conservation biology. Conservation Biology 6:314.

Crouch, S., C. Paquette, and D. Vilas. 2002. Relocation of a large black crowned night-heron colony in southern California. Waterbirds 25:474–478.

Crouse, D. T., L. B. Crowder, and H. Caswell. 1987. A stage-based population model for loggerhead sea turtles and implications for conservation. Ecology 68:1412–1423.

Crozier, G. E., and D. E. Gawlik. 2003. The use of decoys as a research tool for attracting wading birds. Journal of Field Ornithology 74:53–58.

Cryan, P. M. 2003a. Migration in North American tree bats. Journal of Mammalogy 84:579–593.

Cryan, P. M. 2003b. Seasonal distribution of migratory tree bats (*Lasiurus* and *Lasionycteris*) in North America. Journal of Mammalogy 84:579–593.

Cryan, P., M. A. Bogan, R. O. Rye, G. P. Landis, and C. L. Kester. 2004. Stable hydrogen isotope analysis of bat hair as evidence of seasonal molt and long-distance migration. Journal of Mammalogy 85:995–1001.

Cumming, S. G., and F. K. A. Schmiegelow. 2001. Effects of habitat abundance and configuration, and the forest matrix, on distributional patterns of boreal birds. Online publication of the Sustainable Forest Management Network, University of Alberta, Edmonton, Alberta, Canada. <www.ualberta.ca/sfm>.

Cupp, P. V., Jr. 1971. Fall courtship of the green salamander, *Aneides aeneus*. Herpetologica 27:308–310.

Cupp, P. V., Jr. 1980. Territoriality in the green salamander, *Aneides aeneus*. Copeia 1980:463–468.

Curson, J., D. Quinn, and D. Beadle. 1994. Warblers of the Americas: an identification guide. Houghton-Mifflin, Boston, Massachusetts, USA.

Cushman, S. A., K. S. McKelvey, and M. K. Schwartz. 2008. Use of empirically derived source-destination models to map regional conservation corridors. Conservation Biology 23:368–376.

Cuthbert, F. J., B. Scholtens, L. C. Wemmer, R. McLain. 1999. Gizzard contents of piping plover chicks in northern Michigan. Wilson Bulletin 111:121–123.

Cuthbert, F. J., J. L. D. Smith, C. L. Jolls, B. Scholtens, and L. Wemmer. 1998. Conservation of biodiversity in coastal systems of Michigan. Unpublished report. U.S. Forest Service Experiment Station, St. Paul, Minnesota, USA.

Cuthbert, F. J., L. R. Wires, and K. Timmerman. 2003. Status assessment and conservation recommendations for the common tern (*Sterna hirundo*) in the Great Lakes Region. U.S. Department of the Interior, Fish and Wildlife Service, Fort Snelling, Minnesota, USA.

Cuthbert, N. L. 1954. A nesting study of the black tern in Michigan. Auk 71:36–63.

Czech, B., and P. R. Krausman. 1997. Distribution and causation of species endangerment in the United States. Science 277:1116–1117.

Dabrowski, A., R. Fraser, J. L. Confer, and I. J. Lovette. 2005. Geographic variability in mitochondrial introgression among hybridizing populations of golden-winged (*Vermivora chrysoptera*) and blue-winged (*V. pinus*) warblers. Conservation Genetics 6:843–853.

Dahl, T. E. 1900. Wetlands losses in the United States 1790s to 1980s. U.S. Department of the Interior, Fish and Wildlife Service, Washington, D.C., USA.

Dahl, T. E., and C. E. Johnson. 1991. Status and trends of wetlands in the conterminous United States, mid-1970s to mid-1980s. U.S. Department of the Interior, Fish and Wildlife Service, Washington, D.C., USA.

Daily, G. C., and P. A. Matson. 2008. Ecosystem services: from theory to implementation. Proceedings of the National Academy of Sciences of the United States of America 105:9455–9456.

Danielson, B. J., and M. S. Gaines. 1987a. Spatial patterns in two syntopic species of microtines: *Microtus ochrogaster* and *Synaptomys cooperi*. Journal of Mammalogy 68:313–322.

Danielson, B. J., and M. S. Gaines. 1987b. The influences of conspecific and heterospecific residents on colonization. Ecology 68:1778–1784.

Danielson, B. J., and R. K. Swihart. 1987. Home range dynamics and activity patterns of *Microtus ochrogaster* and *Synaptomys cooperi* in syntopy. Journal of Mammalogy 68:160–165.

Darveau, M., P. Beauchesne, L. Belanger, J. Huot, and P. LaRue. 1995. Riparian forest strips as habitat for breeding birds in boreal forest. Journal Wildlife Management 59:67–78.

Darwin, C. 1859. Origin of species. John Wiley, London.

Davidson, C. 2004. Declining downwind: amphibian population declines in California and historical pesticide use. Ecological Applications 14:1892–1902.

Davis, A. 2004. *Aneides aeneus* and *Plethodon glutinosus*: nesting observations. Herpetological Review 35:51–52.

Davis, A. F., J. A. Lundgren, B. Barton, J. R. Belfonti, J. L. Farber, J. R. Kunsman, and A. M. Wilkinson. 1995. A natural areas inventory of Wyoming County, Pennsylvania. Pennsylvania Science Office of the Nature Conservancy, Middletown, Pennsylvania, USA.

Davis, M. B. 1993. Old growth in the east: a survey. Cenezoic Society, Richmond, Vermont, USA.

Davis, M. B. 1996. Eastern old growth forests: prospects for rediscovery and recovery. Island Press, Washington, D.C., USA.

Davis, W. E., Jr. 1992. Are accipiter populations in winter affected by bird feeders? Bird Observer 20:253–257.

Davis, W. E., Jr. 1993. Black-crowned night-heron. *In* A. Poole and F. Gill, editors. The birds of North America, 74. Academy of Natural Sciences, Philadelphia, and American Ornithologists' Union, Washington, D.C., USA.

Davis, W. H., M. D. Hassell, and C. L. Rippy. 1965. *Myotis leibii leibii* in Kentucky. Journal of Mammalogy 46:683–684.

Dawson, W. R. 1997. Pine siskin (*Carduelis pinus*). *In* A. Poole and F. Gill, editors. The birds of North America, 280. Academy of Natural Sciences, Philadelphia, Pennsylvania, USA, and American Ornithologists Union, Washington, D.C., USA.

Debruijn, O. 1994. Population ecology and conservation of the barn owl *Tyto alba* in farmland habitats in Liemers and Achterhoek (The Netherlands). Ardea 82:1–109.

DeCalesta, D. S. 1994. Effect of white-tailed deer on songbirds within managed forests in Pennsylvania. Journal of Wildlife Management 58:711–718.

DeCecco, J. A., M. R. Marshall, A. B. Williams, G. A. Gale, and R. J. Cooper. 2000. Comparative seasonal fecundity of four Neotropical migrants in middle Appalachia. Condor 102:653–663.

Dechant, J. A., M. F. Dinkins, D. H. Johnson, L. D. Igle, C. M. Goldade, B. D. Parkin, and B. R. Euliss. 2002. Effects of management practices on grassland birds: upland sandpiper. Northern Prairie Wildlife Research Center, Jamestown, North Dakota, USA.

Dechant, J. A., M. L. Sondreal, D. H. Johnson, L. D. Igl, C. M. Goldade, M. P. Nenneman, and B. R. Euliss. 2003. Effects of management practices on grassland birds: northern harrier. Northern Prairie Wildlife Research Center, Jamestown, North Dakota. Northern Prairie Wildlife Research Center home page. <www.npwrc.usgs.gov/resource/literatr/grasbird/noha.htm (version 12DEC2003)>.

Dechant, J. A., M. L. Sondreal, D. H. Johnson, L. D. Igl, C. M. Goldade, M. P. Nenneman, A. L. Zimmerman, and B. R. Euliss. 1998a (revised 2003). Effects of management practices on grassland birds: loggerhead shrike. Northern Prairie Wildlife Research Center, Jamestown, North Dakota, USA.

Dechant, J. A., M. L. Sondreal, D. H. Johnson, L. D. Igl, C. M. Goldade, B. D. Parkin, and B. R. Euliss. 1998b (revised 2003). Effects of management practices on

grassland birds: sedge wren. Northern Prairie Wildlife Research Center, Jamestown, North Dakota, USA.
DeGraaf, R. M., W. M. Healy, and R. T. Brooks. 1991. Effects of thinning and deer browsing on breeding birds in New England oak woodlands. Forest Ecology Management 41:179–191.
DeGraaf, R. M., and J. H. Rappole. 1995. Neotropical migratory birds: natural history, distribution, and population change. Cornell University Press, Ithaca, New York, USA.
DeGraaf, R. M., and D. D. Rudis. 1983. Amphibians and reptiles of New England: habitats and natural history. University of Massachusetts Press, Amherst, USA.
DeGraaf, R. M., and D. D. Rudis. 1986. New England wildlife: habitat, natural history, and distribution. Northeastern Forest Experiment Station General Technical Report NE-108, USDA Forest Service, Amherst, Massachusetts, USA.
Delannoy, C. A., and A. Cruz. 1988. Breeding biology of the Puerto Rican sharp-shinned hawk (*Accipiter striatus venator*). Auk 105:649–662.
Delaware Valley Regional Planning Commission. 1997. Traffic: Delaware Valley Regional Planning Commission, highway traffic trends in the Delaware Valley Region 1960 to 1995. Delaware Valley Regional Planning Commission, Philadelphia, Pennsylvania, USA.
Delisle, J. M., and J. A. Savidge. 1997. Avian use and vegetation characteristics of conservation reserve program fields. Journal of Wildlife Management 61:318–325.
Dellinger, R. L., P. B. Wood, and P. D. Keyser. 2007. Occurrence and nest survival of four thrush species on a managed central Appalachian forest. Forest Ecology and Management 243:248–258.
DeMaynadier, P. G., and M. L. Hunter, Jr. 1999. Forest canopy closure and juvenile emigration by pool-breeding amphibians in Maine. Journal of Wildlife Management 63:441–450.
DePari, J. A., M. H. Linck, and T. E. Graham. 1987. Clutch size of the Blanding's turtle, *Emydoidea blandingi*, in Massachusetts. Canadian Field-Naturalist 101:440–442.
Department of Conservation and Natural Resources. 1992. Presque Isle State Park Resource Management Plan, Department of Conservation and Natural Resources, Harrisburg, Pennsylvania, USA.
Department of Conservation and Natural Resources. 2005. <www.dcnr.state.pa.us/forestry/pndi/vertebrates.asp>.
Department of Conservation and Natural Resources—Bureau of Forestry. 2007. State Forest Management Plan. Department of Conservation and Natural Resources—Bureau of Forestry, Harrisburg, Pennsylvania, USA. <www.birdpop.org/maps.htm>.
Derge, K. L. 1997. Habitat use by sympatric eastern fox squirrels (*Sciurus niger vulpinus*) and gray squirrels (*Sciurus carolinensis*) at forest farmland interfaces of the Valley and Ridge Province, Pennsylvania. Thesis, Pennsylvania State University, University Park, USA.
Derge, K. L., and M. A. Steele. 1999. Distribution and status of the fox squirrel (*Sciurus niger*) in Pennsylvania. Journal of the Pennsylvania Academy of Science 73:43–50.
DeSante, D. F., K. M. Burton, P. Velez, and D. Froehlich. 2001. MAPS manual: 2006 protocol. The Institute for Bird Populations, Point Reyes Station, California, USA. Institute for Bird Populations Web site version 2002. <www.birdpop.org/maps.htm>.
Desrochers, A., and S. J. Hannon. 1997. Gap crossing decisions by forest songbirds during the post-fledging period. Conservation Biology 11:1204–1210.
Dessecker, D. R., and D. G. McAuley. 2001. Importance of early successional habitat to ruffed grouse and American woodcock. Wildlife Society Bulletin 29:456–465.
Dettmers, R. 2003. Status and conservation of shrubland birds in the northeastern U.S. Forest Ecology and Management 185:81–93.
Dettmers, R., and J. Bart. 1999. A GIS modeling method applied to predicting forest songbird habitat. Ecological Applications 9:152–163.
Detwiler, D. 2008. Habitat use, foraging behavior, and competitive interactions of black-crowned night-herons (*Nycticorax nycticorax*) on the Susquehanna River, at Wade Island in Harrisburg, Pennsylvania. Masters Thesis, East Stroudsburg University of Pennsylvania, Pennsylvania, USA.
DeVos, A. 1952. Ecology and management of fisher and marten in Ontario. Technical Bulletin of the Ontario Department of Lands and Forests, Ontario, Canada.
DeVos, T., and B. S. Mueller. 1993. Reproductive ecology of northern bobwhite in north Florida. Pages 83–90 *in* K. E. Church and T. V. Dailey, editors. Quail III: National Quail Symposium. Kansas Department of Wildlife and Parks, Pratt, Kansas, USA.
Dexter, R. W. 1981. Chimney swifts reuse ten-year-old nest. North American Bird Bander 6:136 137.
Diamond, A. W. 1991. Assessment of the risks from tropical deforestation to Canadian songbirds. Transactions of the North American Wildlife and Natural Resources Conference 56:177–194.
Diamond, S. A., G. S. Peterson, J. E. Teitge, and G. R. Ankley. 2002. Assessment of the risk of solar ultraviolet radiation to amphibians. III. Prediction of impacts in selected Northern Midwestern wetlands. Environmental Science and Technology 36:2866–2874.
Diana, S. G., and V. R. Beasley. 1998. Amphibian toxicology. Pages 266–277 *in* M. J. Lannoo, editor. Status and conservation of midwestern amphibians. University of Iowa Press, Iowa City, USA.
Dickerman, R. W. 1987. The "old northeastern" subspecies of red crossbill. American Birds 41:188–194.
Dickson, J. G., F. R. Thompson III, R. N. Conner, and K. E. Franzeb. 1995. Silviculture in central and southeastern oak-pine forests. Pages 245–266 *in* T. E. Martin

and D. M. Finch, editors. Ecology and management of Neotropical migratory birds: a synthesis and review of critical issues. Oxford University Press, New York, New York, USA.

Diefenbach, D. R., M. R. Marshall, J. A. Mattice, and D. W. Brauning. 2007. Incorporating availability for detection in estimates of bird abundance. Auk 124:96–106.

Diefenbach, D. R., S. Rathbun, and J. K. Vreeland. 2005. Distribution and coarse-scale habitat association of snowshoe hares in Pennsylvania. Final report submitted to Pennsylvania Game Commission, State Wildlife Grants, Harrisburg, USA.

DiGiovanni, M., and E. D. Brodie, Jr. 1981. Efficacy of skin glands in protecting the salamander *Ambystoma opacum* from repeated attacks by the shrew *Blarina brevicauda*. Herpetologica 37:234–237.

Dimmick, R. W., M. J. Gudlin, and D. F. McKenzie. 2002. The northern bobwhite conservation initiative. Miscellaneous publication of the Southeastern Association of Fish and Wildlife Agencies, South Carolina. <www.qu.org/seqsq/nbci/nbci.cfm>.

Ditmars, R. L. 1907. The reptile book: a comprehensive, popularized work on the structure and habits of the turtles, tortoises, crocodilians, lizards and snakes which inhabit the United States and Northern Mexico. Doubleday, Page and Company, New York, New York, USA.

Dixon, J. 1925. A closed season needed for fisher, marten, and wolverine in California. California Fish and Game 11:23–25.

Dixon, K. R. 1981. Data requirements for determining the status of furbearer populations. Proceedings of the Worldwide Furbearer Conference 2:1360–1373.

Dobson, A., D. Lodge, J. Alder, G. S. Cumming, J. Keymer, J. McGlade, H. Mooney, J. A. Rusak, O. Sala, V. Wolters, D. Wall, R. Winfree, and M. A. Xenopoulos. 2006. Habitat loss, trophic collapse, and the decline of ecosystem services. Ecology 87:1915–1924.

Dodd, M. G., and T. M. Murphy. 1995. Accuracy and precision in techniques for counting great blue heron nests. Journal of Wildlife Management 59:667–673.

Doepker, R. V., R. D. Earle, and J. J. Ozoga. 1992. Characteristics of Blackburnian warbler, *Dendroica fusca*, breeding habitat in Upper Michigan. Canadian Field-Naturalist 106:366–371.

Dole, J. W. 1965. Summer movements of adult leopard frogs, *Rana pipiens* (Schreber), in northern Michigan. Ecology 46:236–255.

Donnelly, M. A., and M. L. Crump. 1998. Potential effects of climate change on two Neotropical amphibian assemblages. Climate Change 39:541–561.

Donovan, T. M., F. R. Thompson III, J. Faaborg, and J. R. Probst. 1995. Reproductive success of migratory birds in habitat sources and sinks. Conservation Biology 9:1380–1395.

Douglas, C. W., and M. A. Strickland. 1987. Fisher. Pages 510–529 *in* J. A. Novak, J. A. Baker, M. W. Obbard, and B. Malloch, editors. Wild furbearer management and conservation in North America. Ontario Ministry of Natural Resources, Toronto, Canada.

Douglas, M. E., and B. L. Monroe. 1981. A comparative study of topographical orientation in *Ambystoma* (Amphibia: Caudata). Copeia 1981:460–463.

Doutt, J. K. 1930. *Glaucomys sabrinus* in Pennsylvania. Journal of Mammalogy 11:239–240.

Doutt, J. K., C. A. Heppenstall, and J. E. Guilday. 1966. Mammals of Pennsylvania. Pennsylvania Game Commission, Harrisburg, USA.

Doutt, J. K., C. A. Heppenstall, and J. E. Guilday. 1973. Mammals of Pennsylvania. Third edition. Pennsylvania Game Commission, Harrisburg, USA.

Downes, W. L. 1964. Unusual roosting behavior in red bats. Journal of Mammalogy 45:143–144.

Dragoo, J. W., and R. L. Honeycutt. 1997. Systematics of mustelids-like carnivores. Journal of Mammalogy 78:426–443.

Drake, J. C., and F. J. Brenner. 1995. Comparison of habitat preferences of gray and fox squirrels in northwestern Pennsylvania. Journal of the Pennsylvania Academy of Science 69:73–76.

Drent, P. J., and J. W. Woldendorp. 1989. Acid rain and eggshells. Nature 339:431.

Driftwood Wildlife Association. 2004. Chimneyswifts.org Home Page. <www.chimneyswifts.org>.

Driscoll, M. J. L., and T. M. Donovan. 2004. Landscape context moderates edge effects: nesting success of wood thrushes in central New York. Conservation Biology 18:1330–1338.

Driscoll, M. J. L., T. Donovan, R. Mickey, A. Howard, and K. K. Fleming. 2005. Determinants of wood thrush nest success: a multi-scale model selection approach. Journal of Wildlife Management 69:699–709.

Driver, E. C. 1936. Observations of *Scaphiopus holbrooki* (Harlan). Copeia 1:67.

Dueser, R. D., J. L. Dooley, Jr., and G. J. Taylor. 1988. Habitat structure, forest composition, and landscape dimensions as components of habitat suitability for the Delmarva fox squirrel. Pages 417–421 *in* Symposium for management of amphibians, reptiles, and small mammals in North America. Flagstaff, Arizona, USA.

Duffy, J. E., B. J. Cardinale, K. E. France, P. B. McIntyre, E. Thebault, and M. Loreau. 2007. The functional role of biodiversity in ecosystems: incorporating trophic complexity. Ecology Letters 10:522–538.

Duguay, J. P., P. B. Wood, and J. V. Nichols. 2001. Songbird abundance and avian nest survival rates in forests fragmented by different silvicultural treatments. Conservation Biology 15:1405–1415.

Duncan, C. D. 1996. Changes in the winter abundance of sharp-shinned hawks in New England. Journal of Field Ornithology 62:254–262.

Duncan, J. R., and P. A. Duncan. 1997. Increase in distribution records of owl species in Manitoba based on a volunteer nocturnal survey using boreal owl (*Aegolius funereus*) and great gray owl (*Strix nebulosa*) playback. Pages 519–524 *in* Biology and conservation of owls of the Northern Hemisphere. Second International Symposium held 5 February 1997, U.S. Department of Agriculture Forest Service General Technical Report NC-190. Winnipeg, Manitoba, Canada.

Dundee, H. A., and D. A. Rossman. 1989. The amphibians and reptiles of Louisiana. Louisiana State University Press, Baton Rouge, USA.

Dunn, E. H., and D. J. Agro. 1995. Black tern (*Chlidonias niger*). *In* A. Poole and F. Gill, editors. The birds of North America, 147. Academy of Natural Sciences, Philadelphia, Pennsylvania, USA, and American Ornithologists' Union, Washington, D.C., USA.

Dunn, E. H., and P. Blancher. 2004. Managing for the "best of the rest": stewardship species. Bird Conservation August:14–15.

Dunn, E. H., D. J. T. Hussell, and D. A. Welsh. 1999. Priority-setting tool applied to Canada's landbirds based on concern and responsibility for species. Conservation Biology 13:1404–1415.

Dunn, E. H., and D. Tessaglia. 1994. Predation on birds at feeders in winter. Journal of Field Ornithology 65:8–16.

Dunn, J. L., and K. L. Garrett. 1997. A field guide to warblers of North America. Houghton Mifflin, Boston, Massachusetts, USA.

Dunn, J. P. 1989. Translocation of Delmarva fox squirrels to Chester County, Pennsylvania. Pennsylvania Game Commission, Harrisburg, USA.

Dunn, J. P., and J. S. Hall. 1989. Status of cave-dwelling bats in Pennsylvania. Journal of the Pennsylvania Academy of Science 63:166–172.

Dusi, J. L. 1985. Use of sounds and decoys to attract herons to a colony site. Colonial Waterbirds 8:178–180.

Dwight, J., Jr. 1892. Summer birds of the crest of the Pennsylvania Alleghenies. Auk 9:129–141.

Dwyer, T. J., D. G. McAuley, and E. L. Derleth. 1983. Woodcock singing-ground counts and habitat changes in the northeastern United States. Journal of Wildlife Management 47:772–779.

Eadie, W. R., and W. J. Hamilton, Jr. 1958. Reproduction in the fisher in New York. New York Fish and Game Journal 5:77–83.

Eastern Population Tundra Swan Committee. 2007. A management plan for the eastern population of Tundra Swans. Unpublished report. Atlantic, Mississippi, Central, and Pacific Flyway Councils.

Eastman, J. 1991a. Black-billed cuckoo. Pages 232–233 *in* R. Brewer, G. A. McPeek, and R. J. Adams, Jr., editors. The atlas of breeding birds of Michigan. Michigan State University Press, East Lansing, USA.

Eastman, J. 1991b. Whip-poor-will. Pages 252–253 *in* R. Brewer, G. A. McPeek, and R. J. Adams, Jr., editors. The atlas of breeding birds of Michigan. Michigan State University Press, East Lansing, USA.

Ebbers, B. C. 1989. Relationships between red-shouldered hawk reproduction and the environment in northern Michigan. Unpublished report to Michigan Department of Natural Resources.

Eckerle, K. P., and C. F. Thompson. 2001. Yellow-breasted chat (*Icteria virens*). *In* A. Poole, and F. Gill, editors. The birds of North America, 575. Academy of Natural Sciences, Philadelphia, Pennsylvania, USA, and American Ornithologists' Union, Washington, D.C., USA.

Edgren, R. A. 1955. The natural history of the hog-nosed snakes, genus *Heterodon*: a review. Herpetologica 11:105–117.

Edwards, J. W., M. Ford, and D. C. Guynn, Jr. 2003. Fox and gray squirrels. Pages 247–267 *in* G. A. Feldhamer, B. C. Thompson, J. A. Chapman, editors. Wild mammals of North America. Johns Hopkins University Press, Baltimore, Maryland, USA.

Edwards, J. W., D. G. Heckel, and D. C. Guynn, Jr. 1998. Niche overlap in sympatric populations of fox and gray squirrels. Journal of Wildlife Management 62:354–363.

Ehrlich, P. R., D. S. Dobkin, and D. Wheye. 1988. The birders handbook: a field guide to the natural history of North American birds. Simon and Schuster, New York, New York, USA.

Elden, R. C., W. V. Bevill, P. I. Padding, J. E. Frampton, and D. L. Shroufe. 2002. Pages 7–16 *in* J. M. Ver Steeg and R. C. Elden, compilers. Harvest Information Program: evaluation and recommendations. International Association of Fish and Wildlife Agencies, Migratory Shore and Upland Game Bird Working Group, Ad Hoc Committee on Harvest Information Program, Washington, D.C., USA.

Elkins, K. C. 1994. Swainson's thrush, *Catharus ustulatus*. Pages 220–221 *in* Atlas of breeding birds in New Hampshire. Audubon Society of New Hampshire by Arcadia, Dover, USA.

Elliot, J. E., R. W. Butler, R. J. Norstrom, and P. E. Whitehead. 1989. Environmental contaminants and reproductive success in great blue herons, *Ardea herodias*, in British Columbia, 1986–87. Environmental Pollution 59:91–114.

Elliot, J. E., and P. Martin. 1994. Chlorinated hydrocarbons and shell thinning in eggs of *Accipiter* hawks in Ontario, 1986–1989. Environmental Pollution 86:189–200.

Elliot, J. E., and L. Shutt. 1993. Monitoring organochlorines in blood of sharp-shinned hawks (*Accipiter striatus*) migrating through the Great Lakes. Environmental Toxicology and Chemistry 12:241–250.

Emery, A. R., A. H. Berst, and K. Kodaira. 1972. Under-ice observations of wintering sites of leopard frogs. Copeia 1972:123–126.

Erdle, S. Y., and C. S. Hobson. 2001. Current status and conservation strategy for the eastern small-footed myotis (*Myotis leibii*). Virginia Department of Conservation and Recreation, Division of Natural Heritage, Natural Heritage Technical Report 00-19, Richmond, USA.

Erdman, T. C., D. F. Brinker, J. P. Jacobs, J. Wilde, and T. O. Meyer. 1998. Productivity, population trend and status of northern goshawks, *Accipiter gentilis atricapillus,* in northeastern Wisconsin. Canadian Field-Naturalist 112:17–27.

Ernst, C. H. 1970. Home range of the spotted turtle, *Clemmys guttata*. Copeia 1970:473–474.

Ernst, C. H. 1976. Ecology of the spotted turtles *Clemmys guttata* (Reptilia, Testudines, Testudinidae), in southeastern Pennsylvania. Journal of Herpetology 10:25–33.

Ernst, C. H. 1977. Biological notes on the bog turtle, *Clemmys muhlenbergii*. Herpetologica 33:241–246.

Ernst, C. H. 1985. Bog turtle. Pages 270–273 *in* H. H. Genoways and F. J. Brenner, editors. Species of special concern in Pennsylvania. Carnegie Museum of Natural History Special Publication 11. Pittsburgh, Pennsylvania, USA.

Ernst, C. H. 2001a. An overview of the North American turtle genus *Clemmys* Ritgen, 1828. Chelonian Conservation and Biology 4:211–216.

Ernst, C. H. 2001b. Some ecological parameters of the wood turtle, *Clemmys insculpta* in southeastern Pennsylvania. Chelonian Conservation and Biology 4:94–99.

Ernst, C. H., and R. W. Barbour. 1972. Turtles of the United States. University Press of Kentucky, Lexington, USA.

Ernst, C. H., and R. W. Barbour. 1989. Snakes of eastern North America. George Mason University Press, Fairfax, Virginia, USA.

Ernst, C. H., and E. M. Ernst. 2003. Snakes of the United States and Canada. Smithsonian Books, Washington, D.C., USA.

Ernst, C. H., J. E. Lovich, and R. W. Barbour. 1994. Turtles of the United States and Canada. Smithsonian Institution Press, Washington, D.C., USA.

Ernst, C. H., R. T. Zappalorti, and J. E. Lovich. 1989. Overwintering sites and thermal relations of hibernating bog turtles, *Clemmys muhlenbergii*. Copeia 1989:761–764.

Errington, P. L., and N. J. Breckenridge. 1938. Food habits of buteo hawks in north central United States. Wilson Bulletin 50:113–121.

Erskine, A. J. 1977. Birds in boreal Canada: communities, densities, and adaptations. Canadian Wildlife Service Report Series 41.

Erskine, A. J. 1984. A preliminary catalog of bird census plot studies in Canada. Pt. 5. Canadian Wildlife Service, Progress Notes 144.

Erskine, A. J. 1992. Atlas of breeding birds of the Maritime Provinces. Nimbus Publishing and Nova Scotia Museum, Halifax, Nova Scotia, Canada.

Erwin, R. M., J. G. Haig, D. B. Stotts, and J. S. Hatfield. 1996. Reproductive success, growth and survival of black-crowned night-heron and snowy egret chicks in coastal Virginia. Auk 113:119–130.

Eschtruth, A. K., and J. J. Battles. 2009. Acceleration of exotic plant invasion in a forested ecosystem by a generalist herbivore. Conservation Biology 23:388–399.

Esher, R. J., R. L. Baker, S. J. Ursic, and L. Miller. 1993. Responses of invertebrates to experimental acidification of the forest floor under southern pines. Pages 75–83 *in* J. R. Longcore and G. F. Sepik, editors. Proceedings of the eighth American woodcock symposium. U.S. Fish and Wildlife Service Biological Report 16.

Etter, B. 2004. Observations from the 2004 Pennsylvania Migration Count. Pennsylvania Birds 18:79–108.

Etter, B. 2005. Observations from the 2005 Pennsylvania migration count. Pennsylvania Birds 19:84–112.

Euliss, N. H., Jr., R. L. Jarvis, and D. S. Gilmer. 1991. Feeding ecology of waterfowl wintering on evaporation ponds in California. Condor 93:582–590.

Evans, M. L., B. J. M. Stutchbury, and B. E. Woolfenden. 2008. Off-territory forays and genetic mating system of the wood thrush (*Hylocichla mustelina*). Auk 125:67–75.

Evans, R. 2002. An ecosystem unraveling? Pages 23–33 *in* B. Onken, R. Reardon, and J. Lashomb, editors. Proceedings, Hemlock woolly adelgid in the Eastern U.S. symposium, 5–7 February 2002. Rutgers University, New Brunswick, New Jersey, USA.

Evans, W. R., and A. M. Manville II, editors. 2000. Avian mortality at communication towers. Transcripts of Proceedings of the Workshop on Avian Mortality at Communication Towers, 11 August 1999, Cornell University, Ithaca, New York, USA.

Eveland, T. 1978. The status, distribution, and identification of suitable habitat of river otters in Pennsylvania. Thesis, East Stroudsburg University, East Stroudsburg, Pennsylvania, USA.

Evers, D., N. Burgess, L. Champoux, B. Hoskins, A. Major, W. Goodale, R. Taylor, R. Poppenga, and T. Daigle. 2005. Patterns and interpretation of mercury exposure in freshwater avian communities in northeastern North America. Ecotoxicology 14:193–221.

Evers, D. C. 1992. A guide to Michigan's endangered wildlife. University of Michigan Press, Ann Arbor, USA.

Ewert, M. A. 1985. Embryology of turtles. Pages 329–491 *in* C. Gans, F. Billett, and P. F. A. Maderson, editors. Biology of the reptilia. John Wiley and Sons, New York, New York, USA.

Ewing, H. E. 1943. Continued fertility in female box turtles following mating. Copeia 1943:112–114.

Faaborg, J., M. Brittingham, T. Donovan, and J. Blake. 1995. Habitat fragmentation in the temperate zone. Pages 357–380 *in* T. E. Martin and D. M. Finch, editors. Ecology and management of Neotropical migratory birds. Oxford University Press, New York, New York, USA.

Fables, D. 1954. Breeding-bird census of a pine barrens and cedar bog. Audubon Field Notes 8:374–375.

Faccio, S. D. 2003. Postbreeding emigration and habitat use by Jefferson and spotted salamanders in Vermont. Journal of Herpetology 37:479–489.

Falcon Trak Program. 2000. Web page of the Dominion peregrine falcon satellite telemetry project. <www.dom.com/about/environment/falcon/>.

Fallon, S. M. 2007. Genetic data and the listing of species under the U.S. Endangered Species Act. Conservation Biology 21:1186–1195.

Fanok, S., T. Gagnolet, and N. Johnson. 2008. The Upper Delaware Connectivity Project. The Nature Conservancy, Pennsylvania Chapter, Harrisburg, USA.

Farmer, A. H., B. S. Cade, and D. F. Stauffer. 2002. Evaluation of a habitat suitability index model. Pages 172–179 *in* A. Kurta and J. Kennedy, editors. The Indiana bat: biology and management of an endangered species. Bat Conservation International, Austin, Texas, USA.

Farmer, C. J., L. J. Goodrich, E. Ruelas, and J. Smith. 2008. Conservation status of North American raptors. Pages 303–420 *in* K. L. Bildstein, J. P. Smith, E. Ruelas Inzunza, and R. R. Veit, eds. State of North America's birds of prey. Series in Ornithology No. 3. Nuttall Ornithological Club. Cambridge, MA U.S.A., and American Ornithologists' Union, Washington, D.C., U.S.A.

Farney, J., and E. D. Fleharty. 1969. Aspect ratio, loading, wing span, and membrane areas of bats. Journal of Mammalogy 50:362–367.

Farnsworth, G. L., K. H. Pollock, J. D. Nichols, T. R. Simmons, J. E. Hines, and J. R. Sauer. 2002. A removal model for estimating detection probabilities from point-count surveys. Auk 119:414–425.

Farnsworth, G. L., and T. R. Simons. 2000. Observations of wood thrush nest predators in a large contiguous forest. Wilson Bulletin 112:82–87.

Farrell, R. F., and T. E. Graham. 1991. Ecological notes on the turtle *Clemmys insculpta* in northwestern New Jersey. Journal of Herpetology 25:1–9.

Fasola, M. 1984. Activity rhythm and feeding success of nesting night herons, *Nycticorax nycticorax*. Ardea 72:217–222.

Fassler, D. J. 1975. Red bat hibernating in a woodpecker hole. American Midland Naturalist 93:254.

Fauth, P. T. 2001. Wood thrush populations are not all sinks in the agricultural midwestern United States. Conservation Biology 15:523–527.

Federal Aviation Administration. 2000. Obstruction marketing and lighting. Advisory Circular AC 70/7450–1K. Air Traffic Airspace Management, March 2000.

Fenton, M. B. 1972. Distribution and overwintering of *Myotis leibii* and *Eptesicus fuscus* (Chiroptera: Vespertilionidae) in Ontario. Life Sciences Occasional Papers, Royal Ontario Museum 21:1–8.

Fenton, M. B. 1999. Northern long-eared myotis—*Myotis septentrionalis*. Page 96 *in* D. E. Wilson and S. Ruff, editors. Smithsonian book of North American mammals. Smithsonian Institution Press, Washington, D.C., USA.

Fenton, M. B., and E. Reddy. 2003. Exploiting vulnerable prey: moths and red bats (*Lasiurus borealis*; Vespertilionidae). Canadian Journal of Zoology 81:1553–1560.

Fergus, C. 1976. The varying hare. Pennsylvania Game News 47:2–6.

Fergus, C. 2004. Wildlife notes: rails, moorhen and coot. Pennsylvania Game Commission. Harrisburg, Pennsylvania, USA. <www.pgc.state.pa.us/pgc/cwp/view.asp?a=458&q=150447&tx=0>. Accessed 4 May 2006.

Ferguson, R. M. 1968. The timber resources of Pennsylvania. U.S. Forest Service Resource Bulletin NE-8.

Ferland, C. L., and S. M. Haig. 2002. 2001 International Piping Plover Census. U.S. Geological Survey, Forest and Range Ecosystem Science Center, Corvallis, Oregon, USA.

Fialkovich, M. 2001. Summary of the season—April through June 2001. Pennsylvania Birds 15:89–141.

Fialkovich, M. 2002. Summary of the season—April through June 2002. Pennsylvania Birds. 16:91–144.

Ficken, M. S., and R. W. Ficken. 1968a. Courtship of blue-winged warblers, golden winged warblers, and their hybrids. Wilson Bulletin 80:442–451.

Ficken, M. S., and R. W. Ficken. 1968b. Territorial relationships of blue-winged warblers, golden-winged warblers, and their hybrids. Wilson Bulletin 80:442–451.

Fike, J. 1999. Terrestrial and palustrine plant communities of Pennsylvania. The Nature Conservancy, the Western Pennsylvania Conservancy, and Pennsylvania Department of Conservation and Natural Resources, Harrisburg, USA.

Finck, E. J. 1984. Male dickcissel behavior in primary and secondary habitats. Wilson Bulletin 96:672–680.

Fingerhood, E. D. 1992a. Red crossbill. Pages 437–438 in D. W. Brauning, editor. Atlas of breeding birds in Pennsylvania. University of Pittsburgh Press, Pittsburgh, Pennsylvania, USA.

Fingerhood, E. D. 1992b. Common tern. Pages 432–33 *in* D. W. Brauning, editor. Atlas of breeding birds in Pennsylvania. University of Pittsburgh Press, Pittsburgh, Pennsylvania, USA.

Fingerhood, E. D. 1992c. Ruddy duck. Pages 430–431 *in* D. W. Brauning, editor. Atlas of breeding birds in Pennsylvania. University of Pittsburgh Press, Pittsburgh, Pennsylvania, USA.

Finke, D. L., and R. F. Denno. 2004. Predator diversity dampens trophic cascades. Nature 429:407–410.

Fitch, H. S. 1960. Autecology of the copperhead. University of Kansas Public Museum of Natural History 13(4):85–288.

Fitch, H. S. 1970. Reproductive cycles in lizards and snakes. University of Kansas Museum of Natural History Miscellaneous Publications 52:1–247.

Fitch, H. S. 1974. Observations on the food and nesting of the broad-winged hawk (*Buteo platypterus*) in northeastern Kansas. Condor 76:331–360.

Fitch, H. S. 1985. Variation in clutch and litter size in New World reptiles. University of Kansas Museum of Natural History Miscellaneous Publications 76:1–76.

Fitch, H. S. 1999. A Kansas snake community: composition and changes over 50 years. Krieger, Malabar, Florida, USA.

Fitch, H. S. 2002. An exceptionally large natural assemblage of female copperheads (*Agkistrodon contortrix*). Herpetological Review 33:94–95.

Fitch, H. S., and H. W. Shirer. 1971. A radiotelemetric study of spatial relationships in some common snakes. Copeia 1971:118–128.

Fitchel, C. 1985. Olive-sided flycatcher, *Contopus borealis*. Pages 170–171 *in* S. B. Laughlin and D. P. Kibbe, editors. The atlas of breeding birds of Vermont. Published for Vermont Institute of Natural Sciences by University Press of New England, Vermont, USA.

Fitzner, R. E., L. J. Blus, C. J. Henny, and D. W. Carlile. 1988. Organochlorine residues in great blue herons from the northwestern United States. Colonial Waterbirds 11:293–300.

Fitzpatrick, J. W. 1980. Wintering of North American tyrant flycatchers in the Neotropics. Pages 67–78 *in* A. Keast and E. S. Morton, editors. Migrant birds in the Neotropics: ecology, behavior, distribution, and conservation. Smithsonian Institute Press, Washington, D.C., USA.

Flaspohler, D. J. 1996. Nesting success of prothonotary warblers in the upper Mississippi River bottomlands. Wilson Bulletin 108:457–466.

Flather, C. H., M. S. Knowles, and I. A. Kendall. 1998. Threatened and endangered species geography. BioScience 48:365–376.

Fleming, W. J., D. R. Clar, Jr., and C. J. Henny. 1983. Organochlorine pesticides and PCBs: a continuing problem for the 1980s. Transactions of North American Wildlife and Natural Resource Conference 48:186–99.

Fogarty, M. J., and K. A. Arnold. 1977. Common snipe. Pages 189–209 *in* G. C. Sanderson, editor. Management of migratory shore and upland game birds in North America. University of Nebraska Press, Lincoln, USA.

Fonda, S. 2003. The Hollidaysburg Fissure Bone Deposit. *In* J. H. Way and G. M. Fleeger, editors. Geology on the edge: selected sites from Bedford, Blair, Cambria, and Somerset Counties, Pennsylvania: guidebook, 68th Annual Field Conference of Pennsylvania Geologists, Altoona, USA.

Fontenot, L. W., G. P. Noblet, and S. G. Platt. 1996. A survey of herpetofauna inhabiting polychlorinated biphenyl contaminated and reference watersheds in Pickens County, South Carolina. Journal of the Elisha Mitchell Scientific Society 112:20–30.

Forbes, E. B., and S. I. Bechdel. 1931. Mountain laurel and rhododendron as foods for the white tailed deer. Ecology 12:323–333.

Ford, N. B. 1982. Courtship behavior of the queen snake, *Regina septemvittata*. Herpetological Review 13:72.

Ford, T. B., D. E. Winslow, D. R. Whitehead, and M. A. Koukol. 2001. Reproductive success of forest-dependent songbirds near an agricultural corridor in south-central Indiana. Auk 118:864–873.

Ford, W. M., T. S. McCay, M. A. Menzel, W. D. Webster, C. H. Greenberg, J. F. Pagels, and J. F. Merritt. 2005. Influence of elevation and forest type on community assemblage and species distribution of shrews in the central and southern Appalachian Mountains. Pages 303–315 *in* J. F. Merritt, S. Churchfield, R. Hutterer, and B. I. Sheftel, editors. Advances in the biology of shrews II. Special Publication, International Society of Shrew Biologists, Number 01:1–468.

Forest Wetlands Task Force (D. B. Brown, compiler). 1993. Best management practices for silvicultural practices in Pennsylvania's forested wetlands, a pocket guide for foresters, loggers, and other forest land managers. School of Forest Resources, Pennsylvania State University, University Park, USA.

Forester, D. C., S. Knoedler, and R. Sanders. 2003. Life history and status of the mountain chorus frog (*Pseudacris brachyphona*) in Maryland. Maryland Naturalist 46:1–15.

Forman, R. T. T., A. E. Gall, and C. F. Leck. 1976. Forest size and avian diversity in New Jersey woodlots with some land use implications. Oecologia 26:1–8.

Forsey, E. S., and E. M. Baggs. 2001. Winter activity of mammals in riparian zones and adjacent forests prior to and following clearcutting at Copper Lake, Newfoundland, Canada. Forest Ecology and Management 145:163–171.

Foster, R. W., and A. Kurta. 1999. Roosting ecology of the northern bat (*Myotis septentrionalis*) and comparisons with the endangered Indiana bat (*Myotis sodalis*). Journal of Mammalogy 80:659–672.

Fowells, H. A. 1965. Silvics of forest trees in the United States, 271. Agricultural handbook, U.S. Department of Agriculture, Washington, D.C., USA.

Fowler, N. E., and R. W. Howe. 1987. Birds of remnant riparian forests in northeastern Wisconsin. Western Birds 18(1):77–83.

Francis, C. M., and M. S. W. Bradstreet. 1997. Monitoring boreal forest owls in Ontario using tape playback surveys with volunteers. Pages 175–184 *in* Biology and conservation of Owls of the Northern Hemisphere. Second International Symposium held 5–9 February 1997, at Winnipeg, Manitoba, Canada, U.S. Department of Agriculture Forest Service General Technical Report NC-190.

Francis, C. M., J. R. Sauer, and J. R. Serie. 1998. Effect of restrictive harvest regulations on survival and recovery rates of American black ducks. Journal of Wildlife Management 62:1544–1557.

Franzreb, K. E., and K. V. Rosenberg. 1997. Are forest songbirds declining? Status assessment from the southern Appalachians and Northeastern forests. Transactions of the 62nd North American Wildlife and Natural Resources Conference.

Frederick, P. C., N. Dwyer, S. Fitzgerald, and R. E. Bennetts. 1990. Relative abundance and habitat preferences of least bitterns (*Ixobrychus exilis*) in the Everglades. Florida Field Naturalist 18:1–20.

Fredrickson, L. H. 1971. Common gallinule breeding biology and development. Auk 88:914–919.
Fredrickson, T. 1996. Forest transitions in the central Appalachian Mountains. *In* 1996 Central Appalachian Ecological Integrity Conference, Massanetta Springs, Virginia, USA.
Freemark, K., and B. Collins. 1992. Landscape ecology of birds breeding in temperature forest fragments. Pages 443–454 *in* J. M. Hagan and D. W. Johnston, editors. Ecology and conservation of Neotropical migrant landbirds. Smithsonian Institution Press, Washington, D.C., USA.
Fricke, R. L. 1930. Unusual observations for western Pennsylvania. Auk 47:572–573.
Friedman, H. 1963. Host relations of the parasitic cowbirds. U.S. National Museum Bulletin 233:1–276.
Friedmann, H. 1929. The cowbirds. Charles C. Thomas Publishing, Baltimore, Maryland, USA.
Friedmann, H., L. F. Kiff, and S. J. Rothstein. 1977. A further contribution of knowledge of the host relations of the parasitic cowbirds. Smithsonian Contributions to Zoology 235:1–75.
Friesen, L., M. D. Cadman, and R. J. MavKay. 1999. Nesting success of Neotropical migrant songbirds in a highly fragmented landscape. Conservation Biology 13: 338–346.
Friesen, L. E., P. F. Eagles, and R. J. Mackay. 1995. Effects of residential development on forest-dwelling Neotropical migrant songbirds. Conservation Biology 9:1408–1414.
Fritts, T. H. 1968. Intrabrood variation in *Opheodrys vernalis* (Harlan). Herpetologica 24:79–82.
Frost, H. C., and W. B. Krohn. 1994. Capture, care, and handling of fishers (*Martes pennanti*). Maine Agriculture and Forest Experimental Station, Technical Bulletin Number 157, University of Maine, Orono, USA.
Frost, H. C., and W. Krohn. 2004. Postnatal growth and development in fishers. Pages 253–263 *in* D. J. Harrison, A. K. Fuller, and G. Proulx, editors. Martens and fishers (*Martes*) in human-altered environments: an international perspective. Springer Science and Business Media, New York, New York, USA.
Fuselier, L., and D. Edds. Habitat partitioning among three sympatric species of map turtles, genus *Graptemys*. Journal of Herpetology 28:154–158.
Futuyma, D. J. 1998. Wherefore and whither the naturalist? American Naturalist 151:1–6.
Gabbe, A. P., S. K. Robinson, and J. D. Brawn. 2002. Tree-species preferences for foraging insectivorous birds: implications for floodplain forest regeneration. Conservation Biology 16:462–470.
Gaines, M. S., R. K. Rose, and L. R. McClenaghan, Jr. 1977. The demography of *Synaptomys cooperi* populations in eastern Kansas. Canadian Journal of Zoology 55:1584–1594.
Galli, A. E., C. F. Leck, and R. T. T. Forman. 1976. Avian distribution patterns in forests islands of different sizes in central New Jersey. Auk 93:356–364.
Galligan, J. H., and W. A. Dunson. 1979. Biology and status of timber rattlesnake (*Crotalus horridus*) populations in Pennsylvania. Biological Conservation 15:13–58.
Gancz, A.Y., I. K. Barker, R. Lindsay, A. Dibernardo, K. McKeever, and B. Hunter. 2004. West Nile virus outbreak in North American owls, Ontario, 2002. Emerging Infectious Diseases 10(12). <www.cdc.gov/ncidod/EID/vol10no12/04–0167.htm>.
Gannon, M. R., and T. E. Blackburn. 2002. Telemetry study of bats in Pennsylvania with emphasis on the Indiana bat (*Myotis sodalis*). Final report to the U.S. Department of Agriculture Forest Service, Warren, Pennsylvania, USA.
Garber, S. D., and J. Burger. 1995. A 20-year study documenting the relationship between turtle decline and human recreation. Ecological Applications 5:1151–1162.
Gard, N. W., and M. J. Hooper. 1995. An assessment of potential hazards of pesticides and environmental contaminants. Pages 294–310 *in* T. E. Martin and D. M. Finch, editors. Ecology and management of Neotropical migratory birds: a synthesis and review of critical issues. Oxford University Press, New York, New York, USA.
Gardner, J. E., and E. A. Cook. 2002. Seasonal and geographic distribution and quantification of potential summer habitat. Pages 9–20 *in* A. Kurta and J. Kennedy, editors. The Indiana bat: biology and management of an endangered species. Bat Conservation International, Austin, Texas, USA.
Gardner, J. E., J. D. Gardner, and J. E. Hofmann. 1991. Summer roost selection and roosting behavior of *Myotis sodalis* (Indiana bat) in Illinois. Final report. Illinois Natural History Survey, Illinois Department of Conservation, Champaign, USA.
Garman, H. 1892. A synopsis of the reptiles and amphibians of Illinois. Bulletin of the Illinois State Laboratory of Natural History 3:215–390.
Garrison, B. A. 1998. Bank swallow (*Riparia riparia*). *In* The Riparian Bird Conservation Plan: a strategy for reversing the decline of riparian-associated birds in California. California Partners in Flight. <www.prbo.org/calpif/htmldocs/riparian_v-2.html>.
Garrison, B. A. 1999. Bank swallow. *In* A. Poole, P. Stettenheim, and F. Gill, editors. The birds of North America, 414. Academy of Natural Sciences, Philadelphia, Pennsylvania, USA, and American Ornithologists' Union, Washington, D.C., USA.
Garroway, C. J., J. Bowman, T. J. Cascaden, G. L. Holloway, C. G. Mahan, J. R. Malcolm, M. A. Steele, G. Turner, and P. J. Wilson. 2009. Climate change induced hybridization in flying squirrels. Global Change Biology, doi:10.1111/j.1365–2486.2009.01948.x.
Gauthreaux, S. A., Jr., and C. G. Belser. 2003. Radar ornithology and biological conservation. Auk 120:266–277.

Gauthreaux, S. A., C. Belser, and D. Blaricom. 2003. Using a network of WSR-88D weather surveillance radars to define patterns of bird migration at large spatial scales. Pages 335–345 *in* P. Berthold, E. Gwinner, and E. Sonnenschein, editors. Avian Migration. Springer-Verlag. New York, USA.

Gehrt, S. D., and J. E. Chelsvig. 2004. Species-specific patterns of bat activity in an urban landscape. Ecological Applications 14:625–635.

Genoways, H. H. 1985. Mammals. Pages 355–423 *in* H. H. Genoways and F. J. Brenner, editors. Species of Special Concern in Pennsylvania. Carnegie Museum of Natural History, Special Publication 11:1–430. Pittsburgh, Pennsylvania, USA.

Genoways, H. H., and F. J. Benner, editors. 1985. Species of Special Concern in Pennsylvania. Carnegie Museum of Natural History. Special Publication 11. Pittsburgh, Pennsylvania, USA.

Gentry, T. G. 1876. Life histories of the birds of eastern Pennsylvania. Volume 1. Published privately, Philadelphia, Pennsylvania, USA.

Gentry, T. G. 1877. Life-histories of the birds of eastern Pennsylvania. Volume 2. The Naturalists' Agency, Salem, Massachusetts, USA.

George, G. A. 2004. Resource partitioning and habitat use among a guild of resident and migratory riparian passerines in Costa Rica. Thesis, East Stroudsburg University, East Stroudsburg, Pennsylvania, USA.

Ghazoul, J. 2007. Challenges to the uptake of the ecosystem service rationale for conservation. Conservation Biology 21:1651–1652.

Ghent, A. W. 2001. Importance of a low talus in location of bank swallow (*Riparia riparia*) colonies. American Midland Naturalist 146:447–449.

Gibbons, J. W. 2003. Terrestrial habitat: a vital component for herpetofauna of isolated wetlands. Wetlands 23:630–635.

Gibbons, J. W., and M. E. Dorcas. 2004. North American watersnakes: a natural history. University of Oklahoma Press, Norman, USA.

Gibbs, H. L., K. A. Prior, and C. Parent. 1998. Characterization of DNA microsatellite loci from a threatened snake, the eastern massasauga rattlesnake (*Sistrurus catenatus*) and their use in population studies. Journal of Heredity 89:169–173.

Gibbs, H. L., K. A. Prior, and P. J. Weatherhead. 1994. Genetic analysis of threatened snake species using RAPD markers. Molecular Ecology 3:329–337.

Gibbs, H. L., K. A. Prior, P. J. Weatherhead, and G. Johnson. 1997. Genetic structure of the threatened Eastern Massasauga rattlesnake, *Sistrurus catenatus catenatus*: evidence from microsatellite DNA makers. Molecular Ecology 6:1123–1132.

Gibbs, J. P. 1998. Amphibian movements in response to forest edges, roads, and streambeds in southern New England. Journal of Wildlife Management 62:584–589.

Gibbs, J. P., and J. Faaborg. 1990. Estimating the viability of ovenbird and Kentucky warbler populations in forest fragments. Conservation Biology 4:193–196.

Gibbs, J. P., J. R. Longcore, D. G. McAuley, and J. K. Ringelman. 1991. Use of wetland habitats by selected nongame water birds in Maine. Fish and Wildlife Resources Report 9, U.S. Fish and Wildlife Services.

Gibbs, J. P., and S. M. Melvin. 1992a. American bittern. Pages 51–69 *in* K. J. Schneider and D. M. Pence, editors. Migratory nongame birds of management concern in the Northeast. U.S. Fish and Wildlife Service, Newton Corner, Massachusetts, USA.

Gibbs, J. P., and S. M. Melvin. 1992b. Least bittern, *Ixobrychus exilis*. Pages 71–88 *in* K. J. Schneider and D. M. Pence, editors. Migratory nongame birds of management concern in the Northeast. U.S. Fish and Wildlife Service, Newton Corner, Massachusetts, USA.

Gibbs, J. P., and S. M. Melvin. 1992c. Sedge wren, *Cistothorus platensis*. Pages 191–209 *in* K. J. Schneider and D. M. Pence, editors. Migratory nongame birds of management concern in the Northeast. U.S. Department of the Interior, Fish and Wildlife Service, Newton Corner, Massachusetts, USA.

Gibbs, J. P., S. M. Melvin, and F. A. Reid. 1992a. American bittern (*Botaurus lentiginesus*). *In* A. Poole and F. Gill, editors. The birds of North America, 18. Academy of Natural Sciences, Philadelphia, Pennsylvania, USA, and Auk, Washington, D.C., USA.

Gibbs, J. P., F. A. Reid, and S. M. Melvin. 1992b. Least bittern (*Ixobrychus exilis*). *In* A. Poole, P. Stettenheim, and F. Gill, editors. The birds of North America, 17. Academy of Natural Sciences, Philadelphia, Pennsylvania, USA, and American Ornithologists' Union, Washington, D.C., USA.

Gibbs, J. P., and W. G. Shriver. 2002. Estimating the effects of road mortality on turtle populations. Conservation Biology 16:1647–1652.

Gibbs, J. P., S. Woodward, M. L. Hunter, and A. E. Hutchinson. 1988. Comparison of techniques for censusing great blue heron nests. Journal of Field Ornithology 59:130–134.

Gibilisco, C. J. 1994. Distributional dynamics of American martens and fishers in North America. Pages 59–71 *in* S. W. Buskirk, A. S. Harestad, M. G. Raphael, and R. A. Powell, editors. Martens, sables, and fishers: biology and conservation. Cornell University Press, Ithaca, New York, USA.

Gifford, C. L., and R. Whitebread. 1951. Mammal survey of south central Pennsylvania. Pennsylvania Game Commission, Harrisburg, USA.

Gilbert, J. H., J. L. Wright, D. J. Lauten, and J. R. Probst. 1997. Den and rest-site characteristics of American marten and fisher in northern Wisconsin. Pages 135–145 *in* G. Proulx, H. N. Bryant, and P. M. Woodard, editors. *Martes*: taxonomy, ecology, techniques, and management. The Provincial Museum of Alberta, Alberta, Canada.

Gill, F. B. 1980. Historical aspects of hybridization between blue-winged and golden-winged warblers. Auk 97:1–18.

Gill, F. B. 1985. Birds. Pages 299–351 *in* H. H. Genoways and F. J. Brenner, editors. Species of special concern in Pennsylvania. Carnegie Museum of Natural History Special Publication 11, Pittsburgh, Pennsylvania, USA.

Gill, F. B. 1992. Blue-winged warbler (*Vermivora pinus*). Pages 298–299 *in* D. W. Brauning, editor. Atlas of breeding birds in Pennsylvania. University of Pittsburgh Press, Pittsburgh, Pennsylvania, USA.

Gill, F. B. 1997. Local cytonuclear extinction of the golden-winged warbler. Evolution 51:519 525.

Gill, F. B. 2004. Blue-winged warblers (*Vermivora pinus*) versus golden-winged warblers (*V. chrysoptera*). Auk 121:1014–1018.

Gill, F. B., R. A. Canterbury, and J. L. Confer. 2001. Blue-winged warbler (*Vermivora pinus*). *In* A. Poole and F. Gill, editors. The birds of North America, 584. Birds of North America, Philadelphia, Pennsylvania, USA.

Glass, B. P., and R. J. Baker. 1968. The status of the name *Myotis subulatus* Say. Proceedings of the Biological Society of Washington 81:257–260.

Glazer, R. B. 1959. An evaluation of a snowshoe hare restocking program in Centre County, Pennsylvania. Thesis, Pennsylvania State University, University Park, USA.

Glista, D. J., and T. L. DeVault. 2008. Road mortality of terrestrial vertebrates in Indiana. Proceedings of the Indiana Academy of Science 117:55–62.

Gloyd, H. K., and R. Conant. 1990. Snakes of the *Agkistrodon* complex: a monographic review. Contributions to Herpetology Number 6. Society for the Study of Amphibians and Reptiles, Athens, Ohio, USA.

Goin, O. B., and C. J. Goin. 1951. Notes on the natural history of the lizard, *Eumeces laticeps*, in northern Florida. Quarterly Journal of the Florida Academy of Sciences 14:29–33.

Goldsmith, S. K. 1984. Aspects of the natural history of the rough green snake, *Opheodrys aestivus* (Colubridae). Southwestern Naturalist 29:445–452.

Gompper, M. E., and H. M. Hackett. 2005. The long-term, range-wide decline of a once common carnivore: the eastern spotted skunk (*Spilogale putorius*). Animal Conservation 8:195–201.

Goodrich, L. 1992a. Northern harrier. Pages 94–95 *in* D. W. Brauning, editor. Atlas of breeding birds of Pennsylvania. University of Pittsburgh Press, Pittsburgh, Pennsylvania, USA.

Goodrich, L. 1992b. The sharp-shinned hawk. Pages 96–97 *in* D. W. Brauning, editor. Atlas of breeding birds of Pennsylvania. University of Pittsburgh Press, Pittsburgh, Pennsylvania, USA.

Goodrich, L. J., S. T. Crocoll, and S. E. Senner. 1996. Broad-winged hawk. *In* A. Poole and F. Gill, editors. The birds of North America, 210. American Ornithologist's Union and the Philadelphia Academy of Natural Sciences, Philadelphia, Pennsylvania, USA.

Goodrich, L. M. J., M. B. Brittingham, J. A. Bishop, P. Barber. 2002. Wildlife habitat in Pennsylvania: past, present, and future. Report to state agencies. Department of Conservation and Natural Resources, Harrisburg, Pennsylvania, USA. <www.dcnr.state.pa.us/wlhabitat/toc.aspx>.

Goodrich, L. J., and J. Smith. 2008. Raptor migration in North America. *In* K. L. Bildstein, J. Smith, and E. Ruelas, editors. The state of North America's birds of prey. Hawk Mountain Sanctuary, Orwigsburg, Pennsylvania, USA.

Goodrich, W. 1989. Tioga County. Pennsylvania Birds 3:116.

Goodwin, R. E. 1960. A study of the ecology of the black tern *Chlidonias niger surinamensis* (Gmelin). Dissertation, Cornell University, Ithaca, New York, USA.

Gordon, D. M., and R. D. MacCulloch. 1980. An investigation of the ecology of the map turtle, *Graptemys geographica* (Le Sueur), in the northern part of its range. Canadian Journal of Zoology 58:2210–2219.

Gordon, R. E. 1952. A contribution to the life history and ecology of the plethodontid salamander *Aneides aeneus* (Cope and Packard). American Midland Naturalist 47:666–701.

Gordon, R. E. 1959. Homing in the green salamander, *Aneides aeneus*. Bulletin of the Ecological Society of America 40:114.

Gordon, R. E. 1961. The movement of displaced green salamanders. Ecology 41:200–202.

Gordon, R. E. 1967. *Aneides aeneus* (Cope and Packard). Catalogue of amphibians and reptiles. American Society of Icthyologists and Herpetologists 30.1–30.2.

Gorski, L. G. 1970. Banding the two song forms of Trail's flycatcher. Journal of Field Ornithology 41:204–206.

Graber, J. W., R. R. Graber, and E. L. Kirk. 1983. Illinois birds: wood-warblers. Biological Notes 118. Illinois Natural History Survey, Champaign, USA.

Graber, R. R., and J. W. Graber. 1963. A comparative study of bird populations in Illinois, 1906–1909 and 1956–1958. Illinois Natural History Survey Bulletin 28:383–528.

Graham, K., B. Collier, M. Bradstreet, and B. Collins. 1996. Great blue heron (*Ardea herodias*) populations in Ontario: data from and insights on the use of volunteers. Colonial Waterbirds 19:39–44.

Graham, R. T., R. T. Reynolds, M. H. Reiser, R. L. Bassett, and D. A. Boyce. 1994. Sustaining forest habitat for the northern goshawk: a question of scale. Studies in Avian Biology 16:12–17.

Graham, T. E. 1995. Habitat use and population of the spotted turtle, *Clemmys guttata*, a species of concern in Massachusetts. Chelonian Conservation and Biology 1:207–214.

Graham, T. E., and T. S. Doyle. 1977. Growth and population characteristics of Blanding's turtle, *Emydoidea blandingi*, in Massachusetts. Herpetologica 33:410–414.

Graham, T. E., and R. W. Guimond. 1995. Aquatic oxygen consumption by wintering red-bellied turtles. Journal of Herpetology 29:471–474.

Gram, W., P. A. Porneluzi, R. L. Clawson, J. Faaborg, and S. Richter. 2003. Effects of experimental forest management on density and nesting success of bird species in Missouri Ozark forests. Conservation Biology 17:1324–1337.

Graves, H. B. 1984. Behavior and ecology of wild feral swine (*Sus scrofa*). Journal of Animal Science 58: 482–492.

Green, D. M. 1997. Amphibians in decline: Canadian studies of a global problem. Society for the Study of Amphibians and Reptiles, St. Louis, Missouri, USA.

Green, N. B. 1963. The eastern spadefoot toad, *Scaphiopus holbrookii* Harlan, in West Virginia. Proceedings of the West Virginia Academy of Science 35:15–19.

Green, N. B., and T. K. Pauley. 1987. Amphibians and reptiles in West Virginia. University of Pittsburgh Press, Pittsburgh, Pennsylvania, USA.

Green, R. E. 1985. Estimating the abundance of breeding snipe. Bird Study 32:141–149.

Greenberg, R. 1987. Seasonal foraging specialization in the worm-eating warbler. Condor 89:158–168.

Greenberg, R. 1989. Neophobia, aversion to open space, and ecological plasticity in song and swamp sparrows. Canadian Journal of Zoology 67:1194–1199.

Greenberg, R. 1992. Forest migrants in non-forest habitats on the Yucatan Peninsula. Pages 273–286 *in* J. M. Hagan III and D. W. Johnston, editors. Ecology and conservation of Neotropical migrant landbirds. Smithsonian Institution Press, Washington, D.C., USA.

Gregg, I. D. 2002. Migration and wintering ecology of eastern population tundra swans in Pennsylvania. Annual job report 51901. Pennsylvania Game Commission, Harrisburg, USA.

Gregg, I. D. 2004. Migration and wintering ecology of eastern population tundra swans in Pennsylvania. Annual job report 51901. Pennsylvania Game Commission, Harrisburg, USA.

Gregg, I. D., J. P. Dunn, and K. J. Jacobs. 2003. Waterfowl population monitoring. Pennsylvania Game Commission Bureau of Wildlife Management Research Division Annual Job Report 51004. Pennsylvania Game Commission, Harrisburg, USA.

Gregg, L. E. 1984. Population ecology of woodcock in Wisconsin. Technical Bulletin 144, Department of Natural Resources, Madison, Wisconsin, USA.

Gregory, P. T. 1975. Aggregations of gravid snakes in Manitoba, Canada. Copeia 1975:185–186.

Greij, E. D. 1994. Common moorhen. Pages 144–157 *in* T. C. Tacha and C. E. Braun, editors. Migratory shore and upland game bird management in North America. International Association of Fish Wildlife Agencies, Washington, D.C., USA.

Griffith, L. A., and J. E. Gates. 1985. Food habits of cave-dwelling bats in the central Appalachians. Journal of Mammalogy 66:451–460.

Grimm, J. W., and R. H. Yahner. 1986. Status and management of selected species of avifauna in Pennsylvania with emphasis on raptors. Unpublished 1985 final report to Pennsylvania Game Commission, Harrisburg, USA.

Grimm, W. C., and H. A. Roberts. 1950. Mammal survey of southwestern Pennsylvania. Final report, Pittman-Robertson Project 24-R. Pennsylvania Game Commission, Harrisburg, USA.

Grimm, W. C. 1952. Birds of the Pymatuning Region. Pennsylvania Game Commission, Harrisburg, USA.

Grimm, W. C., and R. Whitebread. 1952. Mammal survey of northeastern Pennsylvania. Final report, Pittman-Robertson Project 42-R, Pennsylvania Game Commission, Harrisburg, USA.

Grinnell, H. W. 1918. A synopsis of the bats of California. University of California Publications in Zoology 17:223–404.

Griscom, L. 1937. A monographic study of the red crossbill. Proceedings of the Boston Society of Natural History 41:77–210.

Grobman, A. B. 1984. Scutellation variation in *Opheodrys aestivus*. Bulletin of the Florida State Museum. Biological Sciences 29:153–170.

Grobman, A. B. 1989. Clutch size and female length in *Opheodrys vernalis*. Herpetological Review 20:84.

Grocki, D. R. J., and D. W. Johnston. 1974. Chlorinated hydrocarbon pesticides in North American cuckoos. Auk 91:186–188.

Groskin, H. 1947. Duck hawks breeding in the business center of Philadelphia, Pennsylvania. Auk 64:312–313.

Groskin, H. 1952. Observations of duck hawks nesting on man-made structures. Auk 69:246–253.

Gross, A. O. 1956. The recent appearance of the dickcissel (*Spiza americana*) in eastern North America. Auk 73:66–70.

Gross, A. O. 1921. The dickcissel (*Spiza americana*) of the Illinois prairie. Auk 38:1–26.

Gross, D. 2004. Bald eagle breeding and wintering surveys. Annual report. Pennsylvania Game Commission, Harrisburg, USA.

Gross, D. 2005. Bald eagle breeding and wintering surveys. Annual report. Pennsylvania Game Commission, Harrisburg, USA.

Gross, D. 2006. Bald eagle breeding and wintering surveys. Annual report. Pennsylvania Game Commission, Harrisburg, USA.

Gross, D. 2008. Bald eagle breeding and wintering surveys. Annual report. Pennsylvania Game Commission. <www.pgc.state.pa.us/pgc/lib/pgc/reports/2008>.

Gross, D. A. 1991. Yellow-bellied flycatcher nesting in Pennsylvania with a review of its history, distribution, ecology, behavior, and conservation problems. Pennsylvania Birds 5:107–113.

Gross, D. A. 1992a. Louisiana waterthrush. Pages 344–345

*in* D. W. Brauning, editor. Atlas of breeding birds in Pennsylvania. University of Pittsburgh Press, Pittsburgh, Pennsylvania, USA.

Gross, D. A. 1992b. Olive-sided flycatcher, *Contopus borealis*. Pages 194–195 *in* D. W. Brauning, editor. Atlas of breeding birds in Pennsylvania. University of Pittsburgh Press, Pittsburgh, Pennsylvania, USA.

Gross, D. A. 1992c. Pine siskin *Carduelis pinus*. Pages 416–417 *in* D. W. Brauning, editor. Atlas of breeding birds in Pennsylvania. University of Pittsburgh Press, Pittsburgh, Pennsylvania, USA.

Gross, D.A. 1992d. Yellow-bellied flycatcher. Pages 198–199 *in* D. W. Brauning, editor. Atlas of breeding birds in Pennsylvania. University of Pittsburgh Press, Pittsburgh, Pennsylvania, USA.

Gross, D. A. 1993. An introductory overview of how Pennsylvania old growth forests benefit birds. Unpublished paper.

Gross, D. A. 1994. Discovery of a blackpoll warbler (*Dendroica striata*) nest: a first for Pennsylvania—Wyoming County. Pennsylvania Birds 8:128–132.

Gross, D. A. 1998. Birds: review of status in Pennsylvania. Pages 137–170 *in* J. D. Hassinger, R. J. Hill, G. L. Storm, and R. H. Yahner, editors. Inventory and monitoring of biotic resources in Pennsylvania. Pennsylvania Biological Survey, University Park, USA.

Gross, D. A. 1999. A summary of findings from breeding bird surveys including point counts: wings of the Americas Pocono Bioreserve Project. Report to the Nature Conservancy Pennsylvania Office by Ecology III, Berwick, USA.

Gross, D. A. 2000. Pennsylvania breeding bird survey of Northern saw-whet owl (*Aegolius acadicus*): a candidate—undetermined Species. Project Toot Route. Pennsylvania Game Commission Contract Number SP2341509901, Harrisburg, USA.

Gross, D. A. 2001. Report of breeding bird surveys: wings of the Americas Pennsylvania Ecuador Linkage Project. Report to the Nature Conservancy Pennsylvania Office by Ecology III, Berwick, USA.

Gross, D. A. 2002a. The status, distribution, and conservation of the yellow-bellied flycatcher. Final report for the Wild Resource Conservation Fund 380119. Harrisburg, Pennsylvania, USA.

Gross, D. A. 2002b. The status, distribution, and conservation of the yellow-bellied flycatcher. Final report for WS 023 01 4150 01 for the Pennsylvania Game Commission, Harrisburg, USA.

Gross, D. A. 2003. Avian population and habitat assessment project, Pennsylvania important bird area 48, State Game Lands 57, Wyoming, Luzerne, and Sullivan counties. Pennsylvania Audubon. Ecology III, Berwick, USA.

Gross, D. A., and P. E. Lowther. 2001. Yellow-bellied flycatcher (*Empidonax flaviventris*). *In* A. Poole and F. Gill, editors. The birds of North America, 566. The Birds of North America, Philadelphia, Pennsylvania, USA.

Groth, J. G. 1988. Resolution of cryptic species of Appalachian red crossbills. Condor 90:745 760.

Groth, J. G. 1993a. Evolutionary differentiation in morphology, vocalizations, and allozymes among nomadic sibling species in the North American red crossbill (*Loxia curvirostra*) complex. University of California Publication of Zoology 127:1–143.

Groth, J. G. 1993b. Call matching and positive assortive mating in red crossbills. Auk 110:398–401.

Groth, J. G. 1996. Crossbills audio-visual guide. American Museum of Natural History. <http://research.amnh.org/ornithology/crossbills/contents.html>.

Groves, J. D. 1985a. The New Jersey chorus frog. Pages 261–263 *in* H. H. Genoways and F. J. Brenner, editors. Species of special concern in Pennsylvania. Carnegie Museum of Natural History Special Publication 11, Pittsburgh, Pennsylvania, USA.

Groves, J. D. 1985b. Species account for the southern leopard frog. Pages 263–265 *in* H. H. Genoways and F. J. Brenner, editors. Species of special concern in Pennsylvania. Carnegie Museum of Natural History Special Publications 11, Pittsburgh, Pennsylvania, USA.

Grubb, T. C., Jr., and C. L. Bronson. 2001. On cognitive conservation biology: why chickadees leave patch of woodland. Journal of Avian Biology 32:372–376.

Grubb, T. C., Jr., and R. Yosef. 1994. Resource dependence and territory size in loggerhead shrikes (*Lanius ludovicianus*). Auk 111:465–469.

Guilday, J. E. 1985. The physiographic provinces of Pennsylvania. Pages 19–29 *in* H. H. Genoways and F. J. Brenner, editors. Species of Special Concern in Pennsylvania. Carnegie Museum of Natural History Special Publication 11, Pittsburgh, Pennsylvania, USA.

Guilday, J. E., H. W. Hamilton, and A. D. McCrady. 1966. The bone breccia of bootlegger sink, York County, Pennsylvania. Annals of Carnegie Museum 38:145–163.

Guilday, J. E., P. S. Martin, and A. D. McCrady. 1964. New Paris Number 4: a late Pleistocene cave deposit in Bedford County, Pennsylvania. Bulletin of the National Speleological Society, 26:121–194.

Guthery, F. S. 2000. On bobwhites. Texas A&M University Press, College Station, Texas, USA.

Gutzke, W. H. N., and G. C. Packard. 1987. The influence of temperature on eggs and hatchlings of Blanding's turtles, *Emydoidea blandingi*. Journal of Herpetology 21:161–163.

Haas, C. A., and K. H. Haas. 1998. Brood parasitism by brown-headed cowbirds on brown thrashers: frequency and rates of rejection. Condor 100:535–540.

Haas, F., B. Haas, and J. K. Ginaven. 1985a. Species accounts—birds. Pages 301–351 *in* H. H. Genoways and F. J. Brenner, editors. Species of special concern in Pennsylvania. Carnegie Museum of Natural History Special Publication 11, Pittsburgh, Pennsylvania, USA.

Haas, F., B. Haas, and J. Ginaven. 1985b. Northern bald eagle. Pages 301–303 *in* H. Genoways and F. Brenner, editors. Species of special concern in Pennsylvania. Carnegie Museum of Natural History Special Publication 11, Pittsburgh, Pennsylvania, USA.

Haas, F. C., and B. M. Haas. 1992. Rare and unusual bird reports. Pennsylvania Birds 6:128–129.

Haas, F. C., and B. M. Haas. 2005. Annotated list of the birds of Pennsylvania. Second edition. Pennsylvania Biological Survey, Ornithological Technical Committee.

Haffner, C. D., F. J. Cuthbert, and T. W. Arnold. 2009. Space use by Great Lakes piping plovers during the breeding season. Journal of Field Ornithology 80:270–279.

Hafner, D. J., and C. J. Shuster. 1996. Historical biogeography of western peripheral isolates of the least shrew, *Cryptotis parva*. Journal of Mammalogy 77:536–545.

Hagan, J. M., and A. L. Meehan. 2002. The effectiveness of stand-level and landscape level variables for explaining bird occurrence in an industrial forest. Forestry Science 48:231–242.

Hagan, J. M., P. S. McKinley, A. L. Meehan, and S. L. Grove. 1997. Diversity and abundance of landbirds in a northeastern industrial forest. Journal of Wildlife Management 61:718–735.

Hagan, J. M., and J. R. Walters. 1990. Foraging behavior, reproductive success, and colonial nesting in ospreys. Auk 107:506–521.

Hagen, Y. 1969. Norwegian studies on the reproduction on birds of prey and owls in relation to micro-rodent population fluctuations. Fauna 22:73–126.

Hagmeier, E. M. 1956. Distribution of marten and fisher in North America. Canadian Field-Naturalist 70:149–168.

Hahn, T. P. 1995. Integration of photoperiodic and food cures to time changes in reproductive physiology by an opportunistic breeder, the red crossbill, *Loxia curvirostra* (Aves: Carduelinae). Journal of Experimental Zoology 272:213–226.

Haig, S. M. 1992. Piping plover. *In* A. Poole, P. Stettenheim, and F. Gill, editors. The birds of North America, 2. American Ornithologists' Union, Philadelphia, Pennsylvania, USA.

Haig, S. M., and E. Elliott-Smith. 2004. The piping plover. *In* A. Poole, P. Stettenheim, and F. Gill, editors. The birds of North America, 2. American Ornithologists' Union, Philadelphia, Pennsylvania, USA.

Haig, S. M., and L. W. Oring. 1985. The distribution and status of the piping plover throughout the annual cycle. Journal of Field Ornithology 56:334–345.

Haig, S. M., and L. W. Oring. 1988. Genetic differentiation of piping plovers across North America. Auk 105: 260–267.

Haig, S. M., C. L. Ferland, D. Amirault, F. Cuthbert, J. Dingledine, P. Goossen, A. Hecht, and N. McPhillips. 2005. The importance of complete species censuses and evidence for regional declines in piping plovers. Journal of Wildlife Management 69:160–173.

Hall, E. R. 1981. Mammals of North America. Second edition. John Wiley and Sons, New York, New York, USA.

Hall, E. R., and K. R. Kelson. 1959. The mammals of North America. Volume 2. Ronald Press, New York, New York, USA.

Hall, G. 1983a. Appalachian region. American Birds 37:989.

Hall, G. 1983b. West Virginia birds: distribution and ecology. Carnegie Museum of Natural History Special Publication 7. Carnegie Museum of Natural History, Pittsburgh, Pennsylvania, USA.

Hall, G. 1984. Population decline of Neotropical migrants in an Appalachian forest. American Birds 38:14–18.

Hall, J. S. 1962. A life history and taxonomic study of the Indiana bat, *Myotis sodalis*. Gallery Publication 12. Reading Public Museum, Reading, Pennsylvania, USA.

Hall, J. S. 1979a. Status of the endangered Indiana bat, *Myotis sodalis*, in Pennsylvania. Unpublished report to the Pennsylvania Game Commission, Harrisburg, USA.

Hall, J. S. 1979b. Status of the endangered Indiana bat, *Myotis sodalis*, in Pennsylvania. Report 2. Unpublished report to the Pennsylvania Game Commission, Harrisburg, USA.

Hall, J. S. 1985. Eastern woodrat, *Neotoma floridana* Ord. Pages 362–365 *in* H. H. Genoways and F. H. Brenner, editors. Species of special concern in Pennsylvania. Carnegie Museum of Natural History Special Publication 11, Pittsburgh, Pennsylvania, USA.

Hall, J. S. 1985. Keen's little brown bat. Pages 365–367 *in* H. H. Genoways and F. J. Brenner, editors. Species of Special Concern in Pennsylvania. Carnegie Museum of Natural History, Special Publication 11, Pittsburgh, Pennsylvania, USA.

Hall, J. S. 1985. Small-footed bat. Pages 360–362 *in* H. H. Genoways and F. J. Brenner, editors. Species of Special Concern in Pennsylvania. Carnegie Museum of Natural History, Special Publication 11, Pittsburgh, Pennsylvania, USA.

Hall, J. S. 1988. Survey of the woodrat in Pennsylvania. Final report, Contracts 878903 and 878919, Pennsylvania Game Commission, Harrisburg, USA.

Hall, J. S., R. J. Cloutier, and D. R. Griffin. 1957. Longevity records and notes on tooth wear of bats. Journal of Mammalogy 38:407–409.

Hall, R. J., P. F. P. Henry, and C. M. Bunck. 1999. Fifty-year trends in a box turtle population in Maryland. Biological Conservation 88:165–172.

Hamel, P. B. 2000. Cerulean warbler (*Dendroica cerulea*). *In* A. Poole and F. Gill, editors. The birds of North America, 511. Birds of North America, Philadelphia, Pennsylvania, USA.

Hamel, P. B., D. K. Dawson, and P. D. Keyser. 2004. How we can learn more about the cerulean warbler (*Dendroica cerulea*). Auk 121:7–14.

Hamel, P. B., H. E. LeGrand, Jr., M. R. Lennartz, and S. A. Gauthreaux, Jr. 1982. Bird habitat relationships on southeastern forest lands. U.S. Forest Service General Technical Report SE-22, Washington, D.C., USA.

Hames, R. S., K. V. Rosenberg, J. D. Lowe, S. E. Barker, and A. A. Dhondt. 2002. Adverse effects of acid rain on the distribution of the wood thrush *Hylocichla mustelina* in North America. Proceedings of the National Academy of Sciences 99:11235–11240.

Hamilton, W. J., Jr. 1944. The biology of the little short-tailed shrew, *Cryptotis parva*. Journal of Mammalogy 25:1–7.

Hamilton, W. J., Jr., and W. R. Eadie. 1964. Reproduction in the otter, *Lutra canadensis*. Journal of Mammalogy 45:242–252.

Hammerson, G., F. Dirrigl, Jr., and C. F. Thompson. 1996. NatureServe Explorer home page: *Icteria virens* comprehensive report. <www.natureserve.org/explorer/>. Accessed 2004 Dec 22.

Hammerstrom, F. 1969. A harrier population study. Pages 367–383 *in* J. J. Hickey, editor. Peregrine falcon populations, their biology, and decline. University of Wisconsin Press, Madison, USA.

Hancock, J., and H. Elliot. 1978. The herons of the world. Harper and Row, New York, New York, USA.

Hancock, J., and J. A. Kushlan. 1984. The herons handbook. Harper and Row, New York, New York, USA.

Haney, J. C. 1999. Hierarchical comparisons of breeding birds in old-growth conifer hardwood forest on the Appalachian Plateau. Wilson Bulletin 111:89–99.

Haney, J. C. 1999. Old growth management plan for the Commonwealth of Pennsylvania, a special report to Pennsylvania Game Commission. The Wilderness Society, unpublished report.

Haney, J. C., and J. Lydic. 1999. Avifauna and vegetation structure in an old-growth oak-pine forest on the Cumberland Plateau, Tennessee. Natural Areas Journal 19:199–210.

Haney, J. C., and C. P. Schaadt. 1994a. Old-growth beech-hemlock forest I. Number 77. Breeding bird census in resident bird counts. Supplement to Journal of Field Ornithology 88–89.

Haney, J. C., and C. P. Schaadt. 1994b. Old-growth beech-hemlock forest II, Number 78. Breeding bird census in resident bird counts. Supplement to Journal of Field Ornithology 89–90.

Haney, J. C., and C. P. Schaadt. 1996. Functional roles of eastern old growth in promoting forest bird diversity. Chapter 6 *in* M. B. Davis, editor. Eastern old growth forests: prospects for rediscovery and recovery. Island Press, Washington, D.C., USA.

Hanners, L. A., and Patton, S. R. 1998. Worm-eating warbler (*Helmitheros vermivorum*). *In* A. Poole and F. Gill, editors. The birds of North America, 367. Birds of North America, Philadelphia, Pennsylvania, USA.

Hannon, S. 2000. Avian response to stand and landscape structure in the boreal mixed wood forest in Alberta. Project Report 2000–37 for Sustainable Forest Management Network, University of Alberta, Edmonton, Canada.

Hansen, L. P., C. M. Nixon, and S. P. Havera. 1986. Recapture rates and length of residence in an unexploited fox squirrel population. American Midland Naturalist 115:209–215.

Hardey, J, H., Q. P. Crick, C. V. Wernham, H. Riley, B. E. Etheridge, and D. B. A. Thompson. 2006. Raptors: a field guide to survey and monitoring. Stationary Office, Edinburgh, Scotland.

Harding, J. H. 1991. A twenty year wood turtle study in Michigan: implications for conservation. Pages 31–35 *in* K. R. Beamen, R. Kent, F. Caporaso, S. McKeown, and M. D. Graff, editors. Proceedings of the First International Symposium on Turtles and Tortoises: Conservation and Captive Husbandry. Chapman University, Orange, California, USA.

Harding, J. H. 1997. Amphibians and reptiles of the Great Lakes region. University of Michigan Press, Ann Arbor, USA.

Harding, J. H., and T. J. Bloomer. 1979. The wood turtle, *Clemmys insculpta*: a natural history. Bulletin of the New York Herpetological Society 15:9–26.

Harlow, R. C. 1912. The breeding birds of Southern Centre County, Pennsylvania. Auk 29:465–478.

Harlow, R. C. 1913. The breeding birds of Pennsylvania. Thesis, Pennsylvania State University, State College, USA.

Harlow, R. C. 1918. Notes on the breeding birds of Pennsylvania and New Jersey. Auk 39:399–410.

Harmeson, J. P. 1974. Breeding ecology of the dickcissel. Auk 91:348–359.

Harnishfeger, R. L. 2004. Appalachian cottontail rabbit distribution in Pennsylvania. Final report with Management Recommendations, Contract ME 231037. Pennsylvania Game Commission, Harrisburg, USA.

Harris, M. P. 1974. A field guide to the birds of the Galápagos. Collins, London, Great Britain.

Harrison, C. 1975a. A field guide to the nests, eggs and nestlings of North American birds. Collins, Cleveland, Ohio, USA.

Harrison, H. H. 1975b. A field guide to birds' nests of 285 species found breeding in the United States east of the Mississippi River. Houghton Mifflin, Boston, Massachusetts, USA.

Harrison, H. H. 1975c. A field guide to western bird's nests. Houghton Mifflin Company, Boston, Massachusetts, USA.

Hart, J. A. 2003. Surveys of terrestrial mammals species of special concern. Job Code 70007, Project Code 06710. Pennsylvania Game Commission annual reports, Harrisburg, USA.

Hart, J. A. 2004. Surveys of terrestrial mammals species of special concern. Job Code 70007, Project Code 06710. Pennsylvania Game Commission annual reports, Harrisburg, USA.

Hart, J. A., G. L. Kirkland, Jr., and S. C. Grossman. 1993. Relative abundance and habitat use by tree bats, *Lasi-*

*urus* species, in south-central Pennsylvania. Canadian Field-Naturalist 107:208–212.

Hartin, R. E., M. R. Ryan, and T. A. Campbell. 2007. Distribution and disease prevalence of feral hogs in Missouri. Human-Wildlife Conflict 1:186–191.

Hartman, F. E., and J. P. Dunn. 1991. Pennsylvania Waterfowl Management Plan. Unpublished report. Pennsylvania Game Commission, Harrisburg, USA.

Harvey, M. J., J. S. Altenbach, and T. L. Best. 1999. Bats of the United States. Arkansas Game and Fish Commission, Little Rock, USA.

Hassinger, J., C. Butchkoski, and D. Diefenbach. 1996. Fragmentation effects on the occupancy of forested Allegheny woodrat (*Neotoma magister*) colony areas. Paper presented to Allegheny Woodrat Recovery Group Meeting. Ferrum College, Ferrum, Virginia, USA.

Haxton, T., and M. Berrill. 1999. Habitat selectivity of *Clemmys guttata* in central Ontario. Canadian Journal of Zoology 77:593–599.

Haxton, T., and M. Berrill. 2001. Seasonal activity of spotted turtles (*Clemmys guttata*) at the northern limit of their range. Journal of Herpetology 35:606–614.

Hayes, J. P., and R. G. Harrison. 1992. Variation in mitochondrial DNA and the biogeographic history of woodrats (*Neotoma*) of the eastern United States. Systematic Biology 41:331–344.

Hayes, J. P., and M. E. Richmond. 1993. Clinal variation and morphology of woodrats (*Neotoma*) of the eastern United States. Journal of Mammalogy 74:204–216.

Hayes, T. B., P. Case, S. Chui, D. Chung, C. Haeffele, K. Haston, M. Lee, V. P. Mai, Y. Marjuoa, J. Parker, and M. Tsui. 2006. Pesticide mixtures, endocrine disruption, and amphibian declines: are we underestimating the impact? Environmental Health Perspectives 114:40–50.

Hayman, P., J. Marchant, and T. Prater. 1986. Shorebirds. Houghton Mifflin, Boston, Massachusetts, USA.

Hayward, B. J., and R. Davis. 1964. Flight speeds in western bats. Journal of Mammalogy 45:236–242.

Hedgal, P. L., and B. A. Colvin. 1988. Potential hazard to eastern screech-owls and other raptors of brodofacoum bait used for vole control in orchards. Environmental Toxicology and Chemistry 7:245–260.

Hejl, S. J., J. A. Holmes, and D. E. Kroodsma. 2002. Winter wren (*Troglodytes troglodytes*). *In* A. Poole and F. Gill, editors. The birds of North America, 623. Birds of North America, Philadelphia, Pennsylvania, USA.

Helm, R. N., D. N. Pashley, and P. J. Zwank. 1987. Notes on the nesting of the common moorhen and purple gallinule in southwestern Louisiana. Journal of Field Ornithology 58:55–61.

Hemesath, L. M. 1998. Iowa's frog and toad survey, 1991–1994. Pages 206–216 *in* M. J. Lannoo, editor. Status and conservation of midwestern amphibians. University of Iowa Press, Iowa City, USA.

Henny, C. J. 1972. An analysis of the population dynamics of selected avian species with special reference to changes during the modern pesticide era. Wildlife Research Report 1, U.S. Fish and Wildlife Service, Washington, D.C., USA.

Henry, P. F. P. 2003. The eastern box turtle at the Patuxent Wildlife Research Center 1940s to the present: another view. Experimental Gerontology 38:773–776.

Henshaw, R. E. 1965. Physiology of hibernation and acclimatization in two species of bats (*Myotis lucifugus* and *M. sodalis*). Dissertation Abstracts 26:2837–2838.

Henshaw, R. E., and G. E. Folk. 1966. Relation of thermoregulation to seasonally changing microclimate in two species of bats (*Myotis lucifugus* and *M. sodalis*). Physiological Zoology 39:223–236.

Herkert, J. R. 1994. The effects of habitat fragmentation on midwestern grassland bird communities. Ecological Applications 4:461–471.

Herkert, J. R., D. E. Kroodsma, and J. P. Gibbs. 2001. Sedge wren (*Cistothorus platensis*). *In* A. Poole and F. Gill, editors. The birds of North America, 582. Birds of North America, Philadelphia, Pennsylvania, USA.

Herkert, J. R., P. D. Vickery, and D. E. Kroodsma. 2002. Henslow's sparrow (*Ammodramus henslowii*. *In* A. Poole and F. Gill, editors. The birds of North America, 672. Academy of Natural Sciences, Philadelphia, Pennsylvania, USA, and American Ornithologists' Union, Washington, D.C., USA.

Hickey, J. J. 1942. Eastern population of the duck hawk. Auk 59:176–204.

Hickey, J. J., editor. 1969. Peregrine falcon populations: their biology and decline. University of Wisconsin Press, Madison, USA.

Hickey, J. J., and D. W. Anderson. 1968. Chlorinated hydrocarbons and eggshell changes in raptorial and fish-eating birds. Science 162:271–273.

Hickey, M. B. C., and M. B. Fenton. 1990. Foraging of red bats (*Lasiurus borealis*): do intraspecific chases mean territoriality? Canadian Journal of Zoology 70:2477–2482.

Hillis, R. E., and E. D. Bellis. 1971. Some aspects of the ecology of the hellbender, *Cryptobranchus alleganiensis alleganiensis* in a Pennsylvania stream. Journal of Herpetology 5:121–126.

Hitch, A. T., and P. L. Leberg. 2007. Breeding distribution of North American bird species moving north as a result of climate change. Conservation Biology 21:534–539.

Hitchcock, H. B. 1955. A summer colony of the least bat, *Myotis subulatus leibii* (Audubon and Bachman). Canadian Field-Naturalist 69:31.

Hitchcock, H. B. 1965. Twenty-three years of bat banding in Ontario and Quebec. Canadian Field-Naturalist 79:4–14.

Hitchcock, H. B., R. Keen, and A. Kurta. 1984. Survival rates of *Myotis leibii* and *Eptesicus fuscus* in southeastern Ontario. Journal of Mammalogy 65:126–130.

Hobson, K. A., and E. Bayne. 2000. Effects of forest fragmentation by agriculture on avian communities in the southern boreal mixedwoods of western Canada. Wilson Bulletin 112:373–387.

Hobson, K. A., and J. Schieck. 1999. Changes in bird communities in boreal mixedwood forest: harvest and wildfire effects over 30 years. Ecological Applications 9:849–863.

Hodges, K. E. 1999. Ecology of snowshoe hare in southern boreal and montane forests. Pages 163–206 *in* L. F. Ruggiero, K. B. Aubry, S. W. Buskirk, G. M. Koehler, C. J. Krebs, K. S. McKelvey, and J. R. Squires, editors. Ecology and conservation of lynx in the United States. General Technical Report RMRS-GTR-30WWW, U.S. Department of Agriculture, Forest Service, Rocky Mountain Research Station, Fort Collins, Colorado, USA.

Hoffman, R. D. 1978. The diets of herons and egrets in southwestern Lake Erie. *In* A. Sprunt IV, J. C. Ogden, and S. Winckler, editors. Wading birds. National Audubon Society Research Report No. 7. National Audubon Society, New York, New York, USA.

Hoffman, R. L. 1980. *Pseudacris brachyphona*. Catalogue of American Amphibians and Reptiles. Society for the Study of Amphibians and Reptiles, Saint Louis, Missouri, USA.

Hogrefe, T. C., R. H. Yahner, and N. Piergallini. 1998. Depredation of artificial ground nests in a suburban versus a rural landscape. Pennsylvania Academy of Science 72:36.

Holdermann, D. A. 1978. The distribution and abundance of eastern cottontail (*Sylvilagus floridanus*) and New England cottontail (*S. transitionalis*) populations on a grouse management area in central Pennsylvania. Thesis, Pennsylvania State University, University Park, USA.

Holmes, R. T. 1986. Foraging patterns of forest birds: male-female differences. Wilson Bulletin 98:196–213.

Holmes, R. T. 1994. Black-throated blue warbler (*Dendroica caerulescens*). *In* A. Poole and F. Gill, editors. The birds of North America, 87. Academy of Natural Science, Philadelphia, Pennsylvania, USA, and American Ornithologists' Union Washington, D.C., USA.

Holmes, R. T., and S. K. Robinson. 1981. Tree species preferences of foraging insectivorous birds in a northern hardwoods forest. Oecologia 48:31–35.

Holmes, R. T., and S. K. Robinson. 1988. Spatial patterns, foraging tactics, and diets of ground-foraging birds in a northern hardwoods forest. Wilson Bulletin 100: 377–394.

Holmes, R. T., N. L. Rodenhouse, and T. S. Sillett. 2005. Black-throated blue warbler (*Dendroica caerulescens*). *In* A. Poole, editor. The birds of North America Online Ithaca. Cornell Laboratory of Ornithology. Retrieved from Birds of North American Online. <http://bna.birds.cornell.edu/bna/species/087doi:10.2173/bna.87>.

Holmes, R. T., T. W. Sherry, P. P. Marra, and K. E. Petit. 1992. Multiple brooding and productivity of a Neotropical migrant, the black-throated blue warbler (*Dendroica caerulescens*), in an unfragmented temperate forest. Auk 109:321–333.

Holmes, S. A., L. M. Curran, and K. R. Hall. 2008. White-tailed deer (*Odocoileus virginianus*) alter herbaceous species richness in the Hiawatha National Forest, Michigan, USA. American Midland Naturalist 159:83–97.

Holt, D. W., and S. M. Leasure. 1993. Short-eared owl. *In* A. Poole, and F. Gill, editors. The birds of North America, 62. Academy of Natural Science, Philadelphia, Pennsylvania, USA, and American Ornithologists Union, Washington, D.C., USA.

Holt, D. W., and S. M. Melvin. 1986. Population dynamics, habitat use, and management needs of the short-eared owl in Massachusetts: summary of 1985 research. Massachusetts Division of Fish and Wildlife, Natural Heritage Program, Boston, USA.

Holub, R. J., and T. J. Bloomer. 1977. The bog turtle, *Clemmys muhlenbergi* . . . a natural history. Bulletin of the New York Herpetological Society 13:9–23.

Holway, D. A. 1991. Nest site selection and the importance of nest concealment in the black throated blue warbler. Condor 93:575–581.

Hooge, P. N., W. Eichenlaub, and E. Solomon. 1999. The animal movement program. U.S. Geological Survey, Alaska Biological Science Center, Anchorage, USA.

Hooper, T. D. 1997. Status of the upland sandpiper in the Chilcotin-Cariboo region, British Columbia. British Columbia Ministry of Environment, Lands and Parks. Williams Lake, British Columbia, Canada.

Hoover, J. P. 2003. Multiple effects of brood parasitism reduce the reproductive success of prothonotary warblers, *Protonotaria citrea*. Animal Behaviour 65:923–934.

Hoover, J. P. 2006. Water depth influences nest predation for a wetland-dependent bird in fragmented bottomland forests. Biological Conservation 127:37–45.

Hoover, J. P., and M. C. Brittingham. 1993. Regional variation in cowbird parasitism of wood thrush. Wilson Bulletin 105:228–238.

Hoover, J. P., and M. C. Brittingham. 1998. Nest-site selection and nesting success of wood thrushes. Wilson Bulletin 110:375–383.

Hoover, J. P., M. C. Brittingham, and L. J. Goodrich. 1995. Effects of forest patch size on nesting success of wood thrushes. Auk 112:146–155.

Horak, G. J. 1970. A comparative study of the foods of the sora and Virginia rail. Wilson Bulletin 82:207–213.

Horn, D. J., R. Fletcher, Jr., and R. Koford. 2000. Detecting area sensitivity: a comment on previous studies. American Midland Naturalist 144:28–35.

Horn, D. J., R. R. Koford, and M. L. Braland. 2002. Effects of field size and landscape composition on grassland birds in south-central Iowa. Journal Iowa Academy of Science 109:1–7.

Horne, M. T., and W. A. Dunson. 1994. Exclusion of the Jefferson salamander from some potential breeding ponds in Pennsylvania: effects of low pH, temperature, and metals on embryonic development. Archives of Environmental Contamination and Toxicology 27:323–330.

Hornocker, M. G., J. P. Messick, and W. E. Melquist. 1983. Spatial strategies in three species of Mustelidae. Acta Zoologica Fennica 174:185–188.

Horsley, S. B., S. L. Stout, and D. S. deCalesta. 2003. White-tailed deer impact on the vegetation dynamics of a northern hardwood forest. Ecological Applications 13:98–118.

Houlahan, J. E., C. S. Findlay, B. R. Schmidt, A. H. Meyer, and S. L. Kuzmin. 2000. Quantitative evidence for global amphibian population declines. Nature 404:752–755.

Houser, C. E., A. R. Pople, and H. P. Possingham. 2006. Should managed populations be monitored every year? Ecological Applications 16:807–819.

Houston, C. S., and D. E. Bowen. 2001. Upland Sandpiper (*Bartramia longicauda*). *In* A. Poole and F. Gill, editors. The birds of North America, 580. Academy of Natural Sciences, Philadelphia, Pennsylvania, USA, and American Ornithologists' Union, Washington, D.C., USA.

Howard, J. M., R. J. Safran, and S. M. Melvin. 1993. Biology and conservation of piping plovers at Breezy Point, New York. Unpublished report. Department of Forestry and Wildlife Management, University of Massachusetts, Amherst, USA.

Howard, W. E., and R. E. Marsh. 1982. Spotted and hog-nosed skunks. Pages 664–673 *in* J. A. Chapman and G. A. Feldhamer, editors. Wild mammals of North America. Johns Hopkins Press, Baltimore, Maryland, USA.

Howell, A. H. 1908. Notes on diurnal migrations of bats. Proceedings of the Biological Society of Washington 21:35–38.

Howell, A. H. 1932. Florida bird life. Florida Department of Game and Fresh Water Fish, and U.S. Department of Agriculture, Bureau of Biological Survey. Coward McCann, New York, New York, USA.

Hoy, P. R. 1883. Catalogue of the cold-blooded vertebrates of Wisconsin. Geological Survey of Wisconsin 1:422–426.

Hubbard, B. 2006. A landscape-level approach for monitoring river otters in Pennsylvania. Thesis, Frostburg State University, Frostburg, Maryland, USA.

Hubbard, B., and T. Serfass. 2005. Assessing the distribution of reintroduced populations of river otters in Pennsylvania (USA): development of a landscape-level approach. International Union for Conservation of Nature. Otter Specialist Group Bulletin 21:63–69.

Hudgins, J. E., G. L. Storm, and J. S. Wakely. 1985. Local movements and diurnal-habitat selection by male American woodcock in Pennsylvania. Journal of Wildlife Management 49:614–619.

Hughes, J. M. 1997. Taxonomic significance of host-egg mimicry by facultative brood parasites of the avian genus *Coccyzus* (Cuculidae). Canadian Journal of Zoology 75:1380–1386.

Hughes, J. M. 1999. Yellow-billed cuckoo (*Coccyzus americanus*). *In* A. Poole and F. Gill, editors. The birds of North America, 418. Birds of North America, Philadelphia, Pennsylvania, USA.

Hughes, J. M. 2001. Black-billed cuckoo (*Coccyzus erythropthalmus*). *In* A. Poole and F. Gill, editors. The birds of North America, 587. Birds of North America, Philadelphia, Pennsylvania, USA.

Hughes, J. P., R. J. Robel, K. E. Kemp, and J. L. Zimmerman. 1999. Effects of habitat on dickcissel abundance and nest success in conservation reserve program fields in Kansas. Journal of Wildlife Management 63:523–529.

Hulse, A. C., C. J. McCoy, and E. J. Censky. 2001. Amphibians and reptiles of Pennsylvania and the Northeast. Cornell University Press, Ithaca, New York, USA.

Humane Society of the United States, Maine Audubon Society, and E. I. Spaulding. 1982. Complaint for declaratory and injunctive relief-preliminary statement. The Humane Society of the United States versus J. G. Watt et al., U.S. District Court for Washington, D.C. Civil Action 82–2689.

Hunt, P. D., and B. C. Eliason. 1999. Blackpoll warbler (*Dendroica striata*). *In* A. Poole and F. Gill, editors. The birds of North American, 431. Academy of Natural Sciences, Philadelphia, Pennsylvania, USA, and American Ornithologists' Union, Washington, D.C., USA.

Hunter, W. C., D. A. Buehler, R. A. Canterbury, J. L. Confer, and P. B. Hamel. 2001. Conservation of disturbance-dependent birds in eastern North America. Wildlife Society Bulletin 29:440–455.

Hussell, D. J. T. 1991. Spring migrations of alder and willow flycatchers in southern Ontario. Journal of Field Ornithology 62:69–77.

Hutchinson, J. T., and M. J. Lacki. 1999. Foraging behavior and habitat use of red bats in mixed mesophytic forests of the Cumberland Plateau, Kentucky. Pages 171–177 *in* J. W. Stringer and D. L. Loftis, editors. Twelfth Central Hardwood Forest Conference. U.S. Forest Service, Southern Research Station, Asheville, North Carolina, USA.

Hutto, R. L., and J. S. Young. 1999. Habitat relationships of landbirds in the Northern Region. USDA Forest Service, General Technical Report RMRS-GTR-32, Rocky Mountain Research Station, Ogden, Utah, USA.

Ickes, R. 1992a. Black-billed cuckoo. Pages 150–151 in D. W. Brauning, editor. Atlas of breeding birds in Pennsylvania. University of Pittsburgh Press, Pittsburgh, Pennsylvania, USA.

Ickes, R. 1992b. Brown thrasher. Pages 280–281 *in* D. W. Brauning, editor. Atlas of breeding birds in Pennsylvania. University of Pittsburgh Press, Pittsburgh, Pennsylvania, USA.

Ickes, R. 1992c. Cerulean warbler. Pages 328–329 *in* D. W. Brauning, editor. Atlas of breeding birds in Pennsylvania. University of Pittsburgh Press, Pittsburgh, Pennsylvania, USA.

Ickes, R. 1992d. Pied-billed grebe. Pages 28–30 *in* D. W. Brauning, editor. Atlas of breeding birds in Pennsylvania. University of Pittsburgh Press, Pittsburgh, Pennsylvania, USA.

Ickes, R. 1992e. Scarlet tanager. Pages 360–361 *in* D. W. Brauning, editor. Atlas of breeding birds in Pennsylvania. University of Pittsburgh Press, Pittsburgh, Pennsylvania, USA.

Ickes, R. 1992f. Summer tanager. Pages 358–359 *in* D. W. Brauning, editor. Atlas of breeding birds in Pennsylvania. University of Pittsburgh Press, Pittsburgh, Pennsylvania, USA.

Ingold, D. J. 1989. Nesting phenology and competition for nest sites among red-headed and red-bellied woodpeckers and European starlings. Auk 106:208–217.

Ingold, D. J., and A. L. Murray. 1998. Habitat use and reproductive success of grassland nesting birds on a reclaimed strip-mine. American Ornithologists Union Abstracts: 1998. Biology Department of Muskingum College, New Concord, Ohio, USA.

Inman, R. L., H. H. Prine, and D. B. Hayes. 2002. Avian communities in forested riparian wetlands of southern Michigan, USA. Wetlands 22:647–660.

International Bird Census Committee. 1970. An international standard for a mapping method in bird census work recommended by the International Bird Census Committee. Audubon Field Notes 24:722–726.

International Union for the Conservation of Nature, Conservation International, and NatureServe. 2004. Global Amphibian Assessment. <www.globalamphibians.org>. Accessed 20 March 2005.

International Union for the Conservation of Nature, Conservation International, and NatureServe. 2008. An analysis of amphibians on the 2008 IUCN Red List. <www.iucnredlist.org/amphibians>. Accessed 2009 May 24.

International Union for the Conservation of Nature. 2006. IUCN red list of threatened species. <www.iucnredlist.org>. Accessed 2006 Oct 6.

Isler, M. L., and P. R. Isler. 1987. The tanagers: natural history, distribution, and identification. Smithsonian Institution Press, Washington, D.C., USA.

Iverson, G. C., G. D. Hayward, K. Titus, E. DeGayner, R. E. Lowell, D. C. Crocker-Bedford, P. F. Schempf, and J. Lindell. 1996. Conservation assessment for the northern goshawk in Southeast Alaska. U.S. Department of Agriculture Forest Service, Portland, Oregon, USA.

Izor, R. J. 1979. Winter range of the silver-haired bat. Journal of Mammalogy 30:641–643.

Jablonski, D. 2004. Extinction: past and present. Nature 427:589–589.

Jackson, H. H. T. 1961. Mammals of Wisconsin. University of Wisconsin Press, Madison, USA.

Jacobs, E. 1999. Nest site fidelity and movements of breeding sharp-shinned hawks in central Wisconsin. Abstract, Raptor Research Foundation annual meeting, 1999. La Paz, Baja, California, USA.

Jacobs, J. W. 1938. The eastern cowbird vs. the Kentucky warbler. Auk 55:260–262.

Jacobs, K. J., J. P. Dunn, and I. Gregg. 2008. Waterfowl population monitoring. Pennsylvania Game Commission Bureau of Wildlife Management Project Annual Job Report. Harrisburg, USA. <www.pgc.state.pa.us>.

James, F. C., R. F. Johnston, N. O. Wamer, G. J. Niemi, and W. J. Boecklen. 1984. The Grinnellian niche of the wood thrush. American Naturalist 124:17–47.

James, R. 1984. Habitat management guidelines for warblers of Ontario's northern coniferous forests, mixed forest or southern hardwood forests. Ontario Ministry of Natural Resources, Royal Ontario Museum, Canada.

James, R. D. 1998. Blue-headed vireo. *In* A. Poole and F. Gill, editors. The birds of North America, 379. Birds of North America, Philadelphia, Pennsylvania, USA.

Jansen, K. P., A. P. Summers, and P. R. Delis. 2001. Spadefoot toads (*Scaphiopus holbrookii holbrookii*) in an urban landscape: effects of nonnatural substrates on burrowing in adults and juveniles. Journal of Herpetology 35:141–145.

Jellen, B. C. 2005. The continued decline of the Eastern Massasauga (*Sistrurus c. catenatus*) in Pennsylvania. Western Pennsylvania Conservancy, Pittsburgh, USA.

Jenkins, D. H., D. A. Devlin, N. C. Johnson, and S. P. Orndorff. 2004. System design and management for restoring Penn's Woods. Journal of Forestry 102:30–36.

Jennings, D. T., and H. S. Crawford. 1983. Pine siskin preys on egg masses of the spruce budworm, *Choristoneura fumiferana* (Lepidoptera: Tortricidae). Canadian Entomologist 115:439–440.

John, R. D. 1991. Observations on soil requirements for nesting bank swallows *Riparai riparia*. Canadian Field-Naturalist 105:251–254.

Johnsgard, P. A. 1988a. The quails, partridges, and francolins of the world. Oxford University Press, New York, New York, USA.

Johnsgard, P. A. 1988b. North American owls: biology and natural history. Smithsonian Institution Press, Washington, D.C., USA.

Johnson, A. J., A. P. Pessier, J. F. X. Wellehan, A. Childress, T. M. Norton, N. L. Stedman, D. C. Bloom, W. Belzer, V. R. Titus, R. Wagner, J. W. Brooks, J. Spratt, and E. R. Jacobson. 2008. *Ranavirus* infection of free-ranging and captive box turtles and tortoises in the United States. Journal of Wildlife Diseases 44:851–863.

Johnson, A. N. 2002. Determining the genetic distances between sub-populations of *Aneides aeneus* in the Westvaco Wildlife and Ecosystem Research Forest. Thesis, Marshall University, Huntington, West Virginia, USA.

Johnson, A. W. 1965. The birds of Chile. Published privately, Buenos Aires, Argentina.

Johnson, C. E. 1932. Notes on a family of red bats in captivity. Journal of Mammalogy 7:35–37.

Johnson, D. H. 1996. Management of northern prairies and wetlands for the conservation of Neotropical migratory birds. Pages 53–67 *in* F. R. Thompson III, editor. Management of midwestern landscapes for the conservation of Neotropical migratory birds. U.S. Department of Agriculture, Forest Service General Technical Report NC 187, Washington, D.C., USA.

Johnson, D. H., and L. D. Igl. 2001. Area requirements of grassland birds: a regional perspective. Auk 118:24–34.

Johnson, G. 1995. Spatial ecology, habitat preferences, and habitat management of the eastern massasauga, *Sistrurus catenatus catenatus*, in a New York transition peatland. Dissertation, State University of New York College of Environmental Science and Forestry, Syracuse, USA.

Johnson, G. 2000. Spatial ecology of the eastern massasauga (*Sistrurus catenatus catenatus*) in a New York peatland. Journal of Herpetology 34:186–192.

Johnson, G., B. Kingsbury, R. King, C. Parent, R. A. Seigel, and J. A. Szymanski. 2000. The eastern massasauga rattlesnake: a handbook for land managers. U.S. Fish and Wildlife Service, Fort Snelling, Minnesota, USA.

Johnson, G., and D. J. Leopold. 1998. Habitat management of the eastern massasauga in a central New York peatland. Journal of Wildlife Management 62:84–97.

Johnson, G. D., W. P. Erickson, M. D. Strickland, M. F. Shepherd, and D. A. Shepherd. 2003. Mortality of bats at a large-scale wind power development at Buffalo Ridge, Minnesota. American Midland Naturalist 150:332–342.

Johnson, K. 1995. Green-winged teal (*Anas crecca*). *In* A. Poole, P. Stettenheim, and F. Gill, editors. The birds of North America, 193. Academy of Natural Sciences, Philadelphia, Pennsylvania, USA, and American Ornithologists' Union, Washington, D.C., USA.

Johnson, K. A. 2003. Abiotic factors influencing the breeding, movement, and foraging of the eastern spadefoot (*Scaphiopus holbrookii*) in West Virginia. Thesis, Marshall University, Huntington, West Virginia, USA.

Johnson, R. R., and J. J. Dinsmore. 1985. Brood-rearing and post breeding habitat use by Virginia rails and soras. Wilson Bulletin 97:551–554.

Johnson, S. A., and K. A. Berkley. 1999. Restoring river otters in Indiana. Wildlife Society Bulletin 27:419–427.

Johnson, S. A., P. E. Hale, W. M. Ford, J. M. Wentworth, J. R. French, O. F. Anderson, and G. B. Pullen. 1995. White-tailed deer foraging in relation to successional stage, overstory type and management of Southern Appalachian forests. American Midland Naturalist 133:18–35.

Johnson, T. R. 1987. The amphibians and reptiles of Missouri. Missouri Department of Conservation, Jefferson City, USA.

Jolley, D. B. 2007. Reproduction and herpetofauna depredation of feral pigs at Fort Benning, Georgia. Thesis, Auburn University, Auburn, Alabama, USA.

Jones, J., J. J. Barg, T. S. Sillett, M. L. Veit, and R. J. Robertson. 2004. Minimum estimates of survival and population growth for cerulean warblers (*Dendroica cerulea*) breeding in Ontario, Canada. Auk 121:15–22.

Jones, J., R. D. DeBruyn, J. J. Barg, and R. J. Robertson. 2001. Assessing the effects of natural disturbance on a Neotropical migrant songbird. Ecology 82:2628–2635.

Jones, J., P. J. Doran, and R. T. Holmes. 2003. Climate and food synchronize regional forest bird abundances. Ecology 84:3024–3032.

Jones, J., and R. J. Robertson. 2001. Territory and nest-site selection of cerulean warblers in eastern Ontario. Auk 118:727–735.

Jones, J. K., E. D. Fleharty, and P. B. Dunnigan. 1967. The distributional status of bats of Kansas. Miscellaneous Publications of the University of Kansas Museum of Natural History 46:1–33.

Jorde, D. G., J. R. Longcore, and P. W. Brown. 1989. Tidal and nontidal wetlands of northern Atlantic states. Pages 1–26 *in* L. M. Smith, R. L. Pederson, and R. M. Kaminiski, editors. Habitat management for migrating and wintering waterfowl in North America. Texas Technical University Press, Lubbock, Texas, USA.

Joy, S. M. 1990. Feeding ecology of sharp-shinned hawks and nest-site characteristics of accipiters in Colorado. Thesis, Colorado State University, Fort Collins, USA.

Joy, S. M., R. T. Reynolds, R. L. Knight, and R. W. Hoffman. 1994. Feeding ecology of sharp-shinned hawks nesting in deciduous and coniferous forests in Colorado. Condor 96:455–467.

Joyal, L. A., M. McCollough, and M. L. Hunter, Jr. 2001. Landscape ecology approaches to wetland species conservation: a case study of two turtle species in southern Maine. Conservation Biology 15:1755–1762.

Just, E. H. 2007. An assessment of bridge-sign surveys for determining the presence of river otters. Thesis, Frostburg State University, Frostburg, Maryland, USA.

Kahl, P. M. 1963. Mortality of common egrets and other herons. Auk 80:295–300.

Kahl, R. B., T. S. Baskett, J. A. Ellis, and J. N. Burroughs. 1985. Characteristics of summer habitats of selected nongame birds in Missouri. Agricultural Experiment Station, University of Missouri–Columbia, Research Bulletin 1056.

Kane, R. 2001. Phragmites use by birds in New Jersey. Records of New Jersey Birds 26(4):122–124.

Kats, L. B., and R. P. Ferrer. 2003. Alien predators and amphibian declines: review of two decades of science and the transition to conservation. Diversity and Distributions 9:99–110.

Kats, L. B., J. W. Petranka, and A. Sih. 1988. Antipredator defenses and the persistence of amphibian larvae with fishes. Ecology 69:1865–1870.

Kaufmann, J. H. 1992a. Habitat use by wood turtles in central Pennsylvania. Journal of Herpetology 26:315–321.

Kaufmann, J. H. 1992b. The social behavior of wood turtles in central Pennsylvania. Herpetological Monographs 6:1–25.

Kaufmann, J. H. 1995. Home ranges and movements of wood turtles, *Clemmys insculpta*, in central Pennsylvania. Copeia 1995:22–27.

Kearney, R. F. 2003. Partners in Flight Landbird Conservation Plan Physiographic Area 10: Mid-Atlantic Piedmont.

Keenlyne K. D., and J. D. Beer. 1973. Food habits of *Sistrurus catenatus catenatus*. Journal of Herpetology 4:381–382.

Keith, L. B. 1963. Wildlife's ten year cycle. University of Wisconsin Press, Madison, USA.

Kelley, J. R., and R. D. Rau. 2006. American woodcock population status. U.S. Fish and Wildlife Service, Laurel, Maryland, USA.

Kelly, G. M. 1977. Fisher (*Martes pennanti*) biology in the White Mountain National Forest and adjacent areas. Dissertation, University of Massachusetts, Amherst, Massachusetts, USA.

Kendall, K. C., L. H. Metzgar, D. A. Patterson, and B. M. Steele. 1992. Power of sign surveys to monitor population trends. Ecological Applications 2:422–430.

Kendeigh, S. C. 1945. Nesting behavior of wood-warblers. Wilson Bulletin 57:145–164.

Kenney, L. P., and M. R. Burne. 2000. A field guide to the animals of vernal pools. Massachusetts Division of Fisheries and Wildlife Natural Heritage and Endangered Species Program, Westborough, USA.

Kent, T. 1951. The least bitterns of Swan Lake. Iowa Bird Life 21:59–61.

Keran, D. 1978. Nest site selection by the broad-winged hawk in north central Minnesota and Wisconsin. Journal of Raptor Research 12:15–20.

Keran, D. 1981. The incidence of man-caused and natural mortalities to raptors. Journal of Raptor Research 15:108–112.

Kerlinger, P. 1992. Sharp-shinned populations in a free-fall. Peregrine Observer 15:1–2.

Kerlinger, P., and C. Doremus. 1981. Habitat disturbance and the decline of dominant avian species in pine barrens of the northeastern United States. American Birds 35:16–20.

Kerns, J., and P. Kerlinger. 2004. A study of bird and bat collision fatalities at the Mountaineer Wind Energy Center, Tucker County, West Virginia: Annual report for 2003. Prepared for FPL Energy and Mountaineer Wind Energy Center Technical Review Committee. Unpublished report by Curry and Kerlinger, LLC, Cape May Point, New Jersey.

Kershner, E. L., and E. K. Bollinger. 1996. Reproductive success of grassland birds at east-central Illinois airports. American Midland Naturalist 136:358–366.

Kibbe, D. P. 1985a. Common moorhen. Pages 100–101 *in* S. B. Laughlin and D. P. Kibbe, editors. The atlas of breeding birds of Vermont. University Press of New England, Hanover, New Hampshire, USA.

Kibbe, D. P. 1985b. Pied-billed grebe *Podilymbus podiceps*. Pages 32–33 *in* S. Laughlin and D. P. Kibbe, editors. The atlas of breeding birds of Vermont. University Press of New England, Hanover, New Hampshire, USA.

Kibbe, D. P. 1985c. Swainson's thrush. Pages 244–245 *in* S. B. Laughlin and D. P. Kibbe, editors. The atlas of breeding birds of Vermont. University Press of New England, Hanover, New Hampshire, USA.

Kibbe, D. P. 1985d. Virginia rail. Pages 96–97 *in* S. B. Laughlin and D. P. Kibbe, editors. The atlas of breeding birds of Vermont. University Press of New England, Hanover, New Hampshire, USA.

Kibbe, D. P. 1995a. Bitterns. Pages 331–334 *in* L. E. Dove and R. M. Nyman, editors. Living Resources of the Delaware Estuary Program. The Delaware Estuary Program, U.S. Environmental Protection Agency National Estuary Program, Washington, D.C., USA.

Kibbe, D. P. 1995b. Draft Pennsylvania Recovery and Management Plan for the least bittern (*Ixobrychus exilis*). Pennsylvania Game Commission, Harrisburg, USA.

Kibbe, D. P. 1995c. Pennsylvania's (Draft) black tern (*Chlidonias niger*) recovery and management plan. Report to Pennsylvania Game Commission, Harrisburg, USA.

Kiester, A. R., C. W. Schwartz, and E. R. Schwartz. 1982. Promotion of gene flow by transient individuals in an otherwise sedentary population of box turtles (*Terrapene carolina triunguis*). Evolution 36:617–619.

Kilpatrick, A. M., S. L. LaDeau, and P. P. Marra. 2007. Ecology of West Nile Virus transmission and its impact on birds in the Western Hemisphere. Auk 124:1121–1136.

Kimmel, J. T., and R. H. Yahner. 1994. The northern goshawk in Pennsylvania: habitat use, survey protocols, and status. Final report. School of Forest Resources, University Park, Pennsylvania, USA.

Kimmel, V. L., and L. H. Fredrickson. 1981. Nesting ecology of the red-shouldered hawk in southeastern Missouri. Academy of Science 15:21–27.

King, D. I., and R. M. DeGraaf. 2000. Bird species diversity and nesting success in mature, clearcut and shelterwood forest in northern New Hampshire, USA. Forest Ecology and Management 129:227–235.

Kinlaw, A. 1995. *Spilogale putorius*. Mammalian Species 511:1–7.

Kinsey, K. P. 1977. Agonistic behavior and social organization in a reproductive population of Allegheny woodrats, *Neotoma floridana magister*. Journal of Mammalogy 58:417–419.

Kinzey, S. 2000. A summary of common nighthawk watches in Pennsylvania, 1999. Pennsylvania Birds 14:11–12.

Kirk, D., and C. Hyslop. 1998. Population status and recent trends in Canadian raptors: a review. Biological Conservation 83:91–118.

Kirkland, G. L., Jr. 1974. Preliminary sampling of small mammals on clearcut and uncut red spruce stands in West Virginia. Proceedings of the West Virginia Academy of Science 46:150–154.

Kirkland, G. L., Jr. 1976. Small mammals of a mine waste

situation in the Central Adirondacks, New York: a case of opportunism by *Peromyscus maniculatus*. American Midland Naturalist 95:103–110.

Kirkland, G. L., Jr. 1977. The rock vole, *Microtus chrotorrhinus* (Miller) (Mammalia: Rodentia) in West Virginia. Annals of the Carnegie Museum 46:45–53.

Kirkland, G. L., Jr. 1981. *Sorex dispar* and *Sorex gaspensis*. Mammalian Species 155:1–4.

Kirkland, G. L., Jr. 1982. Ecology of small mammals on iron and titanium open-pit mine wastes in the Central Adirondack Mountains, New York. National Geographic Society Research Reports 14:371–380.

Kirkland, G. L., Jr. 1985a. New England cottontail. Pages 396–399 *in* H. H. Genoways and F. J. Brenner, editors. Species of Special Concern in Pennsylvania. Carnegie Museum of Natural History Special Publication Number 11, Pittsburgh, Pennsylvania, USA.

Kirkland, G. L., Jr. 1985b. Spotted skunk. Pages 373–375 *in* H. H. Genoways and F. J. Brenner, editors. Species of Special Concern in Pennsylvania. Carnegie Museum of Natural History Special Publication Number 11, Pittsburgh, Pennsylvania, USA.

Kirkland, G. L., Jr. 1986. Small mammal species of Special Concern in Pennsylvania and adjacent states: an overview. Pages 252–267 *in* S. K. Majumdar, F. J. Brenner, and A. F. Rhoads, editors. Endangered and threatened species. Programs in Pennsylvania and other states: causes, issues, and management. Pennsylvania Academy of Science, Easton, USA.

Kirkland, G. L., Jr. 1998. *Microtus chrotorrhinus* (Miller 1894), rock vole. Pages 92–93 *in* D. J. Hafner, E. Yensen, and G. L. Kirkland, Jr. editors. North American rodents, status survey, and conservation action plans. The International Union for Conservation of Nature Rodent Specialist Group, Gland, Switzerland, and Cambridge, United Kingdom.

Kirkland, G. L., Jr., and J. A. Hart. 1999. Recent records for ten species of small mammals in Pennsylvania. Northeastern Naturalist 6:1–18.

Kirkland, G. L., Jr., and F. J. Jannett, Jr. 1982. *Microtus chrotorrhinus*. Mammalian Species 180:1–5.

Kirkland, G. L., Jr., and C. M. Knipe. 1979. The rock vole (*Microtus chrotorrhinus*) as a Transition Zone species. Canadian Field-Naturalist 93:319–321.

Kirkland, G. L., and P. M. Krim. 1990. Survey of the statuses of the mammals of Pennsylvania. Journal of the Pennsylvania Academy of Science 64:33–45.

Kirkland, G. L., Jr., C. R. Schloyer, and D. K. Hull. 1976. A novel habitat for the long-tailed shrew, *Sorex dispar* Batchelder. Proceedings of the West Virginia Academy of Science 48:77–79.

Kirkland, G. L., Jr., and D. F. Schmidt. 1982. Abundance, habitat, reproduction and morphology of forest-dwelling small mammals of Nova Scotia and southeastern New Brunswick. Canadian Field-Naturalist 96:156–162.

Kirkland, G. L., Jr., and H. W. Snoddy. 1999. Biogeography and community ecology of shrews (Mammalia: Soricidae) in the northern Appalachian Mountains. Pages 167–175 *in* R. P. Eckerlin, editor. Proceedings of the Appalachian Biogeography Symposium. Virginia Museum of Natural History, Special Publication 7.

Kirkland, G. L., Jr., and H. M. Van Deusen. 1979. The shrews of the *Sorex dispar* group: *Sorex dispar* Batchelder and *Sorex gaspensis* Anthony and Goodwin. American Museum Novitates 2675:1–21.

Kiser, J. D., and C. L. Elliot. 1996. Foraging habitat, food habits, and roost tree characteristics of the Indiana bat (*Myotis sodalis*) during autumn in Jackson County, Kentucky. Final report E-2. Kentucky Department of Fish and Wildlife Resources, Frankfort, Kentucky, USA.

Klaus, N. A., and D. A. Buehler. 2001. Golden-winged warbler breeding habitat characteristics and nest success in clearcuts in the southern Appalachian Mountains. Wilson Bulletin 113:297–301.

Klem, D., Jr. 1990. Collisions between birds and windows: mortality and prevention. Journal of Field Ornithology 61:120–128.

Klemens, M. K. 1993. Amphibians and reptiles of Connecticut and adjacent regions. State Geological and Natural History Survey of Connecticut, Bulletin Number 112.

Klimstra, W. D., and J. L. Roseberry. 1975. Nesting ecology of the bobwhite in southern Illinois. Wildlife Monographs 41:1–37.

Klingener, D. 1957. A marking study of the short-headed garter snake in Pennsylvania. Herpetologica 13:100.

Kochert, M. N., and K. Steenhof. 2002. Golden eagles in the U.S. and Canada: status, trends, and conservation challenges. Journal of Raptor Research 36(1 Supplement):32–40.

Kochert, M. N., K Steenhof, C. L. McIntyre, and E. H. Craig. 2002. Golden eagle. *In* A. Poole and F. Gill, editors. The birds of North America, 684. Academy of Natural Sciences, Philadelphia, USA, and American Ornithologists' Union, Washington, D.C., USA.

Koenig, W. D. 2001. Eruptions by boreal birds. Condor 103:725–735.

Koford, R. R. 1999. Density and fledging success of grassland birds in Conservation Reserve Program fields in North Dakota and west-central Minnesota. Studies in Avian Biology 19:187–195.

Konig, C., F. Weick, and J. H. Becking. 1999. Owls: a guide to the owls of the world. Yale University Press, New Haven, Connecticut, USA.

Koprowski, J. L. 1994. *Sciurus niger*. Mammalian Species 479:1–9.

Koprowski, J. L., J. L. Roseberry, and W. D. Klimstra. 1988. Longevity records for the fox squirrel. Journal of Mammalogy 69:383–384.

Kramer, D. C. 1973. Movements of western chorus frogs *Pseudacris triseriata triseriata* tagged with Co60. Journal of Herpetology 7:231–235.

Kramer, D. C. 1974. Home range of the western chorus frog *Pseudacris triseriata triseriata*. Journal of Herpetology 8:245–246.

Krebs, C. J., S. Boutin, R. Boonstra, A. R. E. Sinclair, J. N. M. Smith, M. R. T. Dale, K. Martin, and R. Turkington. 1995. Impact of food and predation on the snowshoe hare cycle. Science 269:1112–1115.

Krementz, D. G., and J. J. Jackson. 1999. Woodcock in the Southeast: natural history and management for landowners. University of Georgia College of Agriculture and Environmental Science/Cooperative Extension Service. U.S. Fish and Wildlife Service, Macon, Georgia, USA.

Krichbaum, K., C. G. Mahan, M. A. Steele, G. Turner, and P. J. Hudson. 2010. The potential role of *Strongyloides robustus* on parasite-mediated competition between two species of flying squirrels (*Glaucomys*). Journal of Wildlife Diseases 46:299–235.

Kroodsma, D. E. and J. Verner. 1997. Marsh wren (*Cistothorus palustris*). *In* A. Poole and F. Gill, editors. The birds of North America, 308. Academy of Natural Sciences, Philadelphia, Pennsylvania, USA, and American Ornithologists' Union, Washington, D.C., USA.

Kroodsma, R. L. 1984. Effect of edge on breeding forest bird species. Wilson Bulletin 96:426–436.

Kruse, K. C., and E. L. Smith. 1992. The relationship between land-use and the distribution and abundance of loggerhead shrikes in south-central Illinois. Journal of Ornithology 63:420–427.

Krutzsch, P. H. 1966. Remarks on silver-haired and Leib's bats in eastern United States. Journal of Mammalogy 47:121.

Kubel, J. E. 2005. Breeding ecology of golden-winged warblers in managed habitats of central Pennsylvania. Thesis, Pennsylvania State University, University Park, USA.

Kubel, J. E. 2008. Quality of anthropogenic habitats for golden-winged warblers in central Pennsylvania. Wilson Journal of Ornithology 120:801–812.

Kulba, B., and W. B. McGillivray. 2001. Status of the willow flycatcher (*Empidonax traillii*) in Alberta. Alberta Environment, Fisheries, and Wildlife Management Division, and Alberta Conservation Association, Wildlife Status Report 29, Edmonton, Alberta, Canada.

Kunz, T. H. 1971. Reproduction of some vespertilionid bats in central Iowa. American Midland Naturalist 86:477–486.

Kunz, T. H. 1973. Resource utilization: temporal and spatial components of bat activity in central Iowa. Journal of Mammalogy 54:14–32.

Kunz, T. H. 1982a. *Lasionycteris noctivagans*. Mammalian Species 172:1–5.

Kunz, T. H. 1982b. Roosting ecology. Pages 1–55 *in* T. H. Kunz, editor. Ecology of bats. Plenum Press, New York, New York, USA.

Kunz, T. H., E. B. Arnett, B. M. Cooper, W. P. Erickson, R. P. Larkin, T. Mabee, M. L. Morrison, M. D. Strickland, and J. M. Szewczak. 2007a. Assessing impacts of wind energy development on nocturnally active birds and bats: a guidance document. Journal of Wildlife Management 71:2449–2486.

Kunz, T. H., E. B. Arnett, W. P. Erickson, A. R. Hoar, G. D. Johnson, R. P. Larkin, M. D. Strickland, R. W. Thrasher, and M. D. Tuttle. 2007b. Ecological impacts of wind energy development on bats: questions, research needs, and hypotheses. Frontiers of Ecology and Environment 5:315–324.

Kurta, A. 1995. Mammals of the Great Lakes Region. University of Michigan Press, Ann Arbor, Michigan, USA.

Kurta, A., K. J. Williams, R. Mies. 1996. Ecological, behavioral, and thermal observations of a peripheral population of Indiana bats (*Myotis sodalis*). Pages 102–117 *in* R. M. R. Barclay and R. M. Brigham, editors. Bats and forests symposium. British Columbia Ministry of Forests, Victoria, Canada.

Kus, B. E., and M. K. Sogge. 2003. Status and distribution. Pages 3–4 *in* M. K. Sogge, B. E. Kus, S. J. Sferra, and M. J. Whitfield, editors. Ecology and conservation of the willow flycatcher. Studies in Avian Biology 26:3–4.

Kushlan, J. A. 1973. Black-crowned night-heron diving for prey. Florida Field Naturalist 1:27–28.

Kushlan, J. A. 1978. Feeding ecology of wading birds. Pages 249–298 *in* A. Sprunt, A. C. Ogden, and S. Winkler, editors. Wading birds. National Audubon Society Research Report 7, New York, New York, USA.

Kushlan, J. A., and H. Hafner, editors. 2000. Heron conservation. Academic Press, New York, New York, USA.

Kyle, G. Z., and P. D. Kyle. 2004. Rehabilitation and conservation of chimney swifts (*Chaetura pelagica*). Driftwood Wildlife Association, Austin, Texas, USA.

Labanick, G. M. 1976. Prey availability, consumption and selection in the cricket frog, *Acris crepitans*. Journal of Herpetology 10:293–298.

Lachner, E. A. 1942. An aggregation of snakes and salamanders during hibernation. Copeia 1945:159–162.

LaDeau, S. L., A. M. Kilpatrick, and P. P. Marra. 2007. West Nile virus emergence and large-scale declines of North American bird populations. Nature 447:710–713.

Laerm, J., W. M. Ford, T. S. McCay, M. A. Menzel, L. T. Lepardo, and J. L. Boone. 1999. Soricid communities in the southern Appalachians. Pages 177–193 *in* R. P. Eckerlin, editor. Proceedings of the Appalachian Biogeography Symposium. Virginia Museum of Natural History, Special Publication 7.

Lambert, A., and B. Ratcliff. 1979. A survey of piping plovers in Michigan. Report located at the Michigan Department of Natural Resources, Lansing, USA.

Lambert, A., and B. Ratcliff. 1981. Present status of the piping plover in Michigan. Jack Pine Warbler 59:44–52.

Lambert, J. D. 2005. Mountain birdwatch. Final report to United States Fish and Wildlife Service. Vermont Institute of Natural Science. Technical Report 06–2.

Lanier, J. W., and R. A. Joseph. 1989. Managing human recreational impacts on hacked or free-nesting peregrines. Pages 149–153 *in* M. H. LeFranc, Jr., and M. B. Moss, editors. Proceedings of the Northeast Raptor Management Symposium and Workshop, Scientific and Technical Series 13. National Wildlife Federation, Washington, D.C., USA.

Lanyon, W. E. 1995. Eastern meadowlark (*Sturnella magna*). *In* A. Poole and F. Gill, editors. The birds of North America, 160. Academy of Natural Sciences, Philadelphia, Pennsylvania, USA, and American Ornithologists' Union, Washington, D.C., USA.

Lanzone, M., T. Miller, D. Brandes, D. Ombalski, R. Mulvihill, and T. Katzner. 2007. Golden eagle (*Aquila chrysaetos*) wintering behavior in the Appalachian mountains of Eastern North America using GPS data from satellite telemetry. Raptor Research Foundation and Hawk Migration Association of North America 2007 Joint Meeting.

Lanzone, M. J., and R. S. Mulvihill. 2006. Second Pennsylvania breeding bird atlas owl survey protocols. Powdermill Avian Research Center, Rector, Pennsylvania, USA.

Lariviere, S., and L. R. Walton. 1998. *Lontra canadensis*. Mammalian Species 587:1–8.

Larson, A., D. B. Wake, and K. P. Yanev. 1984. Measuring gene flow among populations having high levels of genetic fragmentation. Genetics 106:293–308.

Latham, R. E. 2003. Shrubland longevity and rare plant species in the northeastern United States. Forest Ecology and Management 185:21–39.

Latham, R. E., J. E. Thompson, S. A. Riley, and A. W. Wibiralske. 1996. The Pocono till barrens: shrub savanna persisting on soils favoring forest. Bulletin of the Torrey Botanical Club 123:330–349.

Latham, R. M. 1950. The food of predaceous animals in northeastern United States. Final Report, Pitmann-Robertson Project 36-R. Pennsylvania Game Commission, Harrisburg, USA.

Latham, R. M., and C. R. Studholme. 1952. The bobwhite quail in Pennsylvania. Pennsylvania Game News Special Issue 4.

Latta, S. C., C. Rimmer, A. Keith, J. Wiley, H. Raffaele, K. McFarland, and E. Fernandez. 2006. Birds of the Dominican Republic and Haiti. Princeton University Press, Princeton, New Jersey, USA.

Laubhan, M. K., and F. A. Reid. 1991. Characteristics of yellow-crowned night-heron nests in lowland hardwood forests of Missouri. Wilson Bulletin 103:272–277.

Laughlin, S. B., and D. P. Kibbe. 1985. The atlas of breeding birds of Vermont. University Press of New England, Hanover, New Hampshire, USA.

LaVal, R. K., R. L. Clawson, M. L. LaVal, and W. Caire. 1977. Foraging behavior and nocturnal activity patterns of Missouri bats, with emphasis on the endangered species *Myotis grisescens* and *Myotis sodalis*. Journal of Mammalogy 58:592–599.

Lawton, J. H. 1999. Are there general laws in ecology? Oikos 84:177–192.

Leberman, R. C. 1976. The birds of the Ligonier Valley. Carnegie Museum of Natural History Special Publication 3, Pittsburgh, Pennsylvania, USA.

Leberman, R. C. 1992a. Bald eagle. Pages 92–93 *in* D. W. Brauning, editor. Atlas of breeding birds in Pennsylvania. University of Pittsburgh Press, Pittsburgh, Pennsylvania, USA.

Leberman, R. C. 1992b. Black tern. Pages 144–145 *in* D. W. Brauning, editor. Atlas of breeding birds in Pennsylvania. University of Pittsburgh Press, Pittsburgh, Pennsylvania, USA.

Leberman, R. C. 1992c. Common moorhen *Gallinula chloropus*. Pages 128–129 *in* D. W. Brauning, editor. Atlas of breeding birds in Pennsylvania. University of Pittsburgh Press, Pittsburgh, Pennsylvania, USA.

Leberman, R. C. 1992d. Prairie warbler. Pages 326–327 *in* D. W. Brauning, editor. Atlas of breeding birds in Pennsylvania. University of Pittsburgh Press, Pittsburgh, Pennsylvania, USA.

Leberman, R. C. 1992e. Prothonotary warbler. Pages 334–335 *in* D. W. Brauning, editor. Atlas of breeding birds in Pennsylvania. University of Pittsburgh Press, Pittsburgh, Pennsylvania, USA.

Leberman, R. C. 1992f. Sedge wren. Pages 256–257 *in* D. W. Brauning, editor. Atlas of breeding birds in Pennsylvania. University of Pittsburgh Press, Pittsburgh, Pennsylvania, USA.

Leberman, R. C. 1992g. Yellow-breasted chat. Pages 356–357 *in* D. W. Brauning, editor. Atlas of breeding birds in Pennsylvania. University of Pittsburgh Press, Pittsburgh, Pennsylvania, USA.

Lee, D. S., and D. W. Herman. 1999. Proposed zoogeographic history of the bog turtle, *Clemmys muhlenbergii*. Pages 31–42 *in* C. W. Swarth, W. R. Roosenberg, and E. Kiviat, editors. Conservation and ecology of turtles of the mid-Atlantic region: a symposium. Bibliomania! Salt Lake City, Utah, USA.

Lee, D. S., and A. W. Norden. 1973. A food study of the green salamander, *Aneides aeneus*. Journal of Herpetology 7:53–54.

Lee, D. S., and A. W. Norden. 1996. The distribution, ecology, and conservation needs of bog turtles, with special emphasis on Maryland. Maryland Naturalist 40:7–46.

Lee, D. S., and W. Spofford. 1990. Nesting of golden eagles in the central and southern Appalachians. Wilson Bulletin 102:693–698.

Lee, Y.-F., and G. F. McCracken. 2004. Flight activity and

food habits of three species of *Myotis* bats (Chiroptera: Vespertilionidae) in sympatry. Zoological Studies 43:589–597.

Lefebvre, G., B. Poulin, and R. McNeil. 1994. Spatial and social behaviour of Nearctic warblers wintering in Venezuelan mangroves. Canadian Journal of Zoology 72:757–764.

LeGrand, H. E., Jr., and S. P. Hall. 1989. Element stewardship abstract—*Contopus borealis*. Nature Conservancy, Arlington, Virginia, USA.

Lehmkuhl, J. F., L. F. Ruggerio, and P. A. Hall. 1991. Landscape-scale patterns of forest fragmentation and wildlife richness and abundance in the southern Washington Cascade Range. USDA Forest Service Report PNW-285:425–442, Portland, Oregon, USA.

Lehnen, S. E., and A. D. Rodewald. 2009. Investigating area-sensitivity in shrubland birds: responses to patch size in a forested landscape. Forest Ecology and Management 257:2308–2316.

Lens, L., S. Van Dongen, and E. Matthysen. 2002. Fluctuating asymmetry as an early warning system in the critically endangered Taita thrush. Conservation Biology 16:479–487.

Leonard, R. D. 1986. Aspects of reproduction of the fisher, *Martes pennanti*, in Manitoba (Canada). Canadian Field-Naturalist 100:32–44.

Leopold, A. 1933. Game management. Charles Scribner's Sons, New York, New York, USA.

Lepage, C., and D. Bordage. 2003. Black duck joint venture. Environment Canada Canadian Wildlife Service, Québec Region. <www.qc.ec.gc.ca/faune/sauvagine/html/bdjv.html>. Accessed Jan 2005.

Lepczyk, C. A., A. G. Mertig, and J. G. Liu. 2004a. Assessing landowner activities related to birds across rural-to-urban landscapes. Environmental Management 33:110–125.

Lepczyk, C. A., A. G. Mertig, and J. Liu. 2004b. Landowners and cat predation across rural-to-urban landscapes. Biological Conservation 115:191–201.

Levell, J. P. 2000. Commercial exploitation of Blanding's turtle, *Emydoidea blandingii*, and the wood turtle, *Clemmys insculpta*, for the live animal trade. Chelonian Conservation and Biology 3:665–674.

Liang, S. Y., and S. W. Seagle. 2002. Browsing and microhabitat effects on riparian forest woody seedling demography. Ecology 83:212–227.

Liers, E. E. 1951. Notes on the river otter (*Lutra canadensis*). Journal of Mammalogy 39:438–439.

Limpert, R. J., and S. L. Earnst. 1994. Tundra swan (*Cygnus columbianus*). *In* A. Poole and F. Gill, editors. The birds of North America, 89. Academy of Natural Sciences, Philadelphia, Pennsylvania, USA, and American Ornithologists' Union, Washington, D.C., USA.

Limpert, R. J., W. J. L. Sladen, and H. A. Allen, Jr. 1991. Winter distribution of tundra swans *Cygnus columbianus columbianus* breeding in Alaska and western Canadian Arctic. Pages 78–83 *in* Proceedings of the Third IWRB International Swan Symposium Wildfowl Supplement Number 1, Oxford, United Kingdom.

Linck, M. H. 2000. Reduction in road mortality in a northern leopard frog population. Journal of the Iowa Academy of Science 107:209–211.

Linck, M. H., J. A. DePari, B. O. Butler, and T. E. Graham. 1989. Nesting behavior of the turtle *Emydoidea blandingi*, in Massachusetts. Journal of Herpetology 23:442–444.

Lindberg, P., Sellström, U., Häggberg, L., and De Wit, C. A. 2004. Higher brominated diphenyl ethers and hexabromocyclododecane found in eggs of peregrine falcons (*Falco peregrinus*) breeding in Sweden. Environmental Science and Technology 38:93–96.

Lindeman, P. V. 1998. Of deadwood and map turtles (*Graptemys*): an analysis of species status for five species in three river drainages using replicated spotting-scope counts of basking turtles. Chelonian Conservation and Biology 3:137–141.

Linder, G., S. K. Krest, and D. W. Sparling, editors. 2003. Amphibian decline: an integrated analysis of multiple stress effects. SETAC Press, Pensacola, Florida, USA.

Linzey, A. V. 1981. Patterns of coexistence in *Microtus pennsylvanicus* and *Synaptomys cooperi*. Dissertation, Virginia Polytechnic Institute and State University, Blacksburg, USA.

Linzey, A. V. 1983. *Synaptomys cooperi*. Mammalian Species 210:1–5.

Linzey, A. V. 1984. Patterns of coexistence in *Microtus pennsylvanicus* and *Synaptomys cooperi*. Ecology 65:382–393.

Linzey, A. V., and J. A. Cranford. 1984. Habitat selection in the southern bog lemming, *Synaptomys cooperi*, and the meadow vole, *Microtus pennsylvanicus*, in Virginia. Canadian Field-Naturalist 98:463–469.

Linzey, D. W. 1967. Food of the leopard frog in central New York. Herpetologica 23:11–17.

Liscinsky, S. A. 1972. The Pennsylvania woodcock management study. Research Bulletin 171. Pennsylvania Game Commission, Harrisburg, Pennsylvania, USA.

Litvaitis, J. A., J. A. Sherburne, and J. A. Bissonette. 1985. Influence of understory characteristics on snowshoe hare habitat use and density. Journal of Wildlife Management 49:866–873.

Litvaitis, J. A., D. L. Wagner, J. L. Confer, M. D. Tarr, and E. J. Snyder. 1999. Early-successional forests and shrub-dominated habitats: land-use artifact or critical community in the northeastern United States. Northeast Wildlife 54:101–118.

Litwin, T. S., and C. R. Smith. 1992. Factors influencing the decline of Neotropical migrants in a northeastern forest fragment: isolation, fragmentation, or mosaic effects? Pages 483–496 *in* J. M. Hagan III and D. W. Johnston, editors. Ecology and conservation of Neotropical migrant landbirds. Smithsonian Institution Press, Washington, D.C., USA.

Litzgus, J. D., and R. J. Brooks. 2000. Habitat and temperature selection of *Clemmys guttata* in a northern population. Journal of Herpetology 34:178–185.

Livo, L. J., D. Chiszar, and H. M. Smith. 1996. Geographic distribution. *Liochlorophis* (=*Opheodrys*) *vernalis*. Herpetological Review 27:154.

Lochmiller, R. L. 1979. Use of beaver ponds by southeastern woodpeckers in winter. Journal of Wildlife Management 43:263–266.

Lockie, J. D. 1955. The breeding habits and food of short-eared owls after a vole plague. Bird Study 2:53–69.

LoGiudice, K. 2003. Trophically transmitted parasites and the conservation of small populations: raccoon roundworm and the imperiled Allegheny woodrat. Conservation Biology 17:258–266.

Long, C. A., C. F. Long, J. Knops, and D. H. Matulionis. 1965. Reproduction in the dickcissel. Wilson Bulletin 77:251–256.

Longcore, J. R., D. G. McAuley, D. A. Clugston, C. M. Bunck, and J. F. Giroux. 2000b. Survival of American black ducks radio-marked in Quebec, Nova Scotia and Vermont. Journal of Wildlife Management 64:238–252.

Longcore, J. R., D. G. McAuley, G. R. Hepp, and J. M. Rhymer. 2000a. American black duck. *In* A. Poole and F. Gill, editors. The Birds of North America, 481. The Academy of Natural Sciences, Philadelphia, Pennsylvania, USA, and American Ornithologists' Union, Washington, D.C., USA.

Lorenzen, A., J. L. Shutt, and S. W. Kennedy. 1997. Sensitivity of common tern (*Sterna hirundo*) embryo hepatocyte cultures to CYPIA induction and porphyry accumulation bihalogenated aromatic hydrocarbons and common tern egg extracts. Archives of Environmental Contamination and Toxicology 32:126–134.

Lorimer, C. G. 2001. Historical and ecological roles of disturbance in eastern North American forests: 9,000 years of change. Wildlife Society Bulletin 29:425–439.

Losey, J. E., and M. Vaughan. 2006. The economic value of ecological services provided by insects. BioScience 56:311–323.

Lott, C. A., and J. P. Smith. 2006. Geographic information system approach to estimating the origin of migrating raptors in North American using stable hydrogen isotope ratios in feathers. Auk 123:822–835.

Lovallo, M. J., and T. S. Hardisky. 2008. Furbearer population and harvest monitoring. Pennsylvania Game Commission, Harrisburg, USA.

Lovette, I. J., S. M. Clegg, and T. B. Smith. 2004. Limited utility of mtDNA markers for determining connectivity among breeding and overwintering locations in three Neotropical migrant birds. Conservation Biology 18:156–166.

Lovich, J. E. 1995. Turtles. Pages 118–121 *in* E. T. Laroe, C. E. Puckett, P. D. Doran, and M. J. Mac, editors. Our living resources: a report to the nation on distribution, abundance and health of U.S. plants, animals, and ecosystems. National Biological Service, Washington, D.C, USA.

Lowther, P. E. 1999. Alder flycatcher (*Empidonax alnorum*). *In* A. Poole and F. Gill, editors. The birds of North America, 446. Academy of Natural Sciences, Philadelphia, Pennsylvania, USA, and American Ornithologists' Union, Washington, D.C., USA.

Luck, M., and J. Wu. 2002. A gradient analysis of urban landscape pattern: a case study from the Phoenix metropolitan region, Arizona, USA. Landscape Ecology 17:327–339.

Lutz, H. J. 1930. Original forest composition in northwestern Pennsylvania as indicated by early land survey notes. Journal of Forestry 28:1098–1103.

Lynch, J. F., and D. L. Whigham. 1984. Effects of forest fragmentation on breeding bird communities in Maryland, USA. Biological Conservation 28:287–324.

Lynn, J. C., T. J. Koronkiewicz, M. J. Whitfield, and M. K. Sogge. 2003. Ecology and conservation of the willow flycatcher. Pages 41–51 *in* M. K. Sogge, B. E. Kus, S. J. Sferra, and M. J. Whitfield, editors. Studies in Avian Biology 26.

MacArthur, D. L., and J. W. T. Dandy. 1982. Physiological aspects of overwintering in the boreal chorus frog (*Pseudacris triseriata maculata*). Comparative Biochemistry and Physiology 72A(1):137–141.

MacCulloch, R. D., and Weller, W. F. 1988. Some aspects of reproduction in a Lake Erie population of the Blanding turtle (*Emydoidea blandingii*). Canadian Journal of Zoology 66:2317–2319.

Mack, D. E., and W. Yong. 2000. Swainson's thrush (*Catharus ustulatus*). *In* A. Poole and F. Gill, editors. The birds of North America, 540. The Birds of North America, Philadelphia, Pennsylvania, USA.

MacWhirter, R. B., and K. L. Bildstein. 1996. Northern harrier (*Circus cyaneus*). *In* A. Poole and F. Gill, editors. The birds of North America, 210. Academy of Natural Sciences, Philadelphia, Pennsylvania, USA, and American Ornithologists' Union, Washington, D.C., USA.

Madge, S., and P. McGowan. 2002. Pheasants, partridges, and grouse: a guide to the pheasants, partridges, quails, grouse, guineafowl, buttonquails, and sandgrouse of the world. Princeton University Press, Princeton, New Jersey, USA.

Magnuson, J. J., D. M. Robertson, B. J. Benson, R. H. Wynne, D. M. Livingstone, T. Arai, R. A. Assel, R. G. Barry, V. Card, E. Kuusisto, N. G. Granin, T. D. Prowse, K. M Stewart, and V. S. Vuglinski. 2000. Historical trends in lake and river ice cover in the Northern Hemisphere. Science 289:1743–1746.

Mahan, C. G., M. A. Steele, M. J. Patrick, and G. L Kirkland, Jr. 1999. The status of the northern flying squirrel (*Glaucomys sabrinus*) in Pennsylvania. Journal of the Pennsylvania Academy of Science 73:15–21.

Manci, K. M., and D. H. Rusch. 1988. Indices to distribution and abundance of some inconspicuous waterbirds on Horicon marsh. Journal of Field Ornithology 59:67–75.

Manuwal, D. A. 1991. Spring bird communities in the southern Washington Cascade Range. USDA Forest Service Report 285:160–174, Portland, Oregon, USA.

Maple, W. T. 1968. The overwintering adaptations of *Sistrurus catenatus catenatus* in northeastern Ohio. Thesis, Kent State University, Kent, Ohio, USA.

Maret, T. J. and J. P. Collins. 1997. Ecological origin of morphological diversity: a study of alternative trophic phenotypes in larval salamanders. Evolution 51:898–905.

Marks, J. S. 1986. Nest-site characteristics and reproductive success of long-eared owls in southwestern Idaho. Wilson Bulletin 98:547–560.

Marks, J. S., D. L. Evan, and D. W. Holt. 1994. Long-eared owl (*Asio otus*). *In* A. Poole and F. Gill, editors. The birds of North America, 133. Birds of North America, Philadelphia, Pennsylvania, USA.

Marquis, R. J., and C. J. Whelan. 1994. Insectivorous birds increase growth of white oak through consumption of leaf-chewing insects. Ecology 75:2007–2014.

Marra, P. P., S. Griffing, and C. Caffrey. 2004. West Nile virus and wildlife. BioScience 54:393–402.

Marshall, J. S., and L. W. Vandruff. 2002. Impact of selective herbicide right-of-way vegetation treatment on birds. Environmental Management 30:801–806.

Marshall, J. T. 1988. Birds lost from a giant sequoia forest during fifty years. Condor 90:359 372.

Marshall, M. R., R. J. Cooper, J. A. DeCecco, J. Strazanac, and L. Butler. 2002. Effects of experimentally reduced prey abundance on the breeding ecology of the red-eyed vireo. Ecological Applications 12:261–280.

Marti, C. D. 1976. A review of prey selection by the long-eared owl. Condor 78:331–336.

Marti, C. D. 1992. Barn Owl. *In* A. Poole, P. Stettenheim and F. Gill, editors. The birds of North America, 1. Academy of Natural Sciences, Philadelphia, Pennsylvania, USA, and American Ornithologists' Union, Washington, D.C., USA.

Marti, C. D., P. W. Wagner, and K. W. Denne. 1979. Nest boxes for the management of barn owls. Wildlife Society Bulletin 7:145–148.

Martin, A. C., H. S. Zim, and A. L. Nelson. 1951. American wildlife and plants a guide to wildlife food habits. McGraw-Hill, New York, New York, USA.

Martin, P. R., F. Bonier, and D. Gibson. 2006. First nest of the yellow-bellied flycatcher for Alaska, with notes on breeding biology. Western Birds 37:8–22.

Martin, R. L., J. T. Pawluk, and T. B. Clancy. 1966. Observations on the hibernation of *Myotis subulatus*. Journal of Mammalogy 47:348–349.

Martin, T. E. 1981. Limitation in small habitat islands: chance or competition? Auk 98:715–734.

Martin, T. E., C. R. Paine, C. J. Conway, W. M. Hochachka, P. Allen, and W. Jenkins. 1997. BBIRD Field Protocol. Montana Cooperative Wildlife Research Unit, University of Montana, Missoula, Montana, USA. Breeding Biology Research and Monitoring Database. <http://pica.wru.umt.edu/BBIRD/>.

Martin, T. G., I. Chadés, P. Arcese, P. P. Marra, H. P. Possingham, and D. R. Norris. 2007. Optimal conservation of migratory species. PLoS ONE 2:e751.

Martin, W. H. 1982. The timber rattlesnake in the Northeast: its range, past, and present. Bulletin of the New York Herpetological Society 17:15–20.

Martin, W. H. 1993. Reproduction of the timber rattlesnake in the Appalachian Mountains. Journal of Herpetology 27:133–143.

Martin, W. H., W. H. Smith, S. H. Harwig, R. O. Magram, and R. Stechert. 1990. Distribution and status of the timber rattlesnake (*Crotalus horridus*) in Pennsylvania. Report to the Carnegie Museum of Natural History and Pennsylvania Fish and Boat Commission, Harrisburg, USA.

Martinez-Vilalta, A., and A. Motis. 1992. Family Ardeidae (herons). Pages 376–429 *in* J. del Hoyo, A. Elliot, and J. Sargatal, editors. Handbook of the birds of the world. Volume 1. Lynx Edicions, Barcelona, Spain.

Marzilli, V. 1989. Up on the roof. Maine Fish and Wildlife 31:25–29.

Massachusetts Division of Fisheries and Wildlife. 2004. Instructions for conducting the Massachusetts whip-poor-will survey. <www.mass.gov/dfwele/dfw/nhesp/whip_instr.htm>. Accessed 8 Dec 2004.

Massemin, S., Y. Le Maho, and Y. Handrich. 1998. Seasonal pattern in age, sex, and body condition of barn owls *Tyto alba* killed on motorways. Ibis 140:70–75.

Master, T. 1992a. Kentucky warbler. Pages 364–365 *in* D. W. Brauning, editor. Atlas of breeding birds in Pennsylvania. University of Pittsburgh Press, Pittsburgh, Pennsylvania, USA.

Master, T. L. 1992b. Composition, structure, and dynamics of mixed-species foraging aggregations in a southern New Jersey salt marsh. Colonial Waterbirds 15:66–74.

Master, T. L. 1992c. Yellow-throated vireo (*Vireo flavifrons*). Pages 292–293 *in* D. W. Brauning, editor. Atlas of breeding birds in Pennsylvania. University of Pittsburgh Press, Pittsburgh, Pennsylvania, USA.

Master, T. L. 2001. Threat assessment and management recommendations for Wade Island. Report to Pennsylvania Game Commission, Harrisburg, USA.

Master, T. L. 2002. Threat assessment and management recommendations for Wade Island. Report to Pennsylvania Game Commission, Harrisburg, USA.

Master, T. L. 2004. Current status and management options for double-crested cormorants on Wade Island. Report to Pennsylvania Game Commission, Harrisburg, USA.

Master, T. L., M. Frankel, and M. Russell. 1993. Benefits of foraging in mixed-species wader aggregations in a southern New Jersey salt marsh. Colonial Waterbirds 16:149-157.

Master, T. L., R. S. Mulvihill, R. C. Leberman, J. Sanchez, and E. Carman. 2005. A preliminary study of riparian songbirds in Costa Rica, with emphasis on wintering Louisiana waterthrushes. Pages 528–532 *in* C. J. Ralph and T. Rich, editors. Bird conservation implementation and integration in the Americas: proceedings of the third international Partners in Flight conference, March 2002, Volume 1. General Technical Report PSW GTR-191. Pacific Southwest Research Station, Forest Service, U.S. Department of Agriculture, Albany, California, USA.

Matray, P. F. 1974. Broad-winged hawk nesting and ecology. Auk 91:307–324.

Matthews, S. N., R. J. O'Connor, L. R. Iverson, and A. M. Prasad. 2004. Atlas of climate change effects in 150 bird species of the eastern United States. General Technical Report NE-318. U.S. Department of Agriculture and Forest Service Northeast Research Station, Newtown Square, Pennsylvania, USA.

Mattice, J. A., D. W. Brauning, and D. R. Diefenbach. 2005. Abundance of grassland sparrows on reclaimed surface mines in western Pennsylvania. Pages 504–510 *in* C. J. Ralph and T. D. Rich, editors. A workshop on bird conservation implementation and integration. U.S. Forest Service General Technical Report PSW-GTR-191, Washington, D.C., USA.

Mattson, T. A., S. W. Buskirk, and N. L. Stanton. 1996. Roost sites of the silver-haired bat (*Lasionycteris noctivagans*) in the Black Hills, South Dakota. Great Basin Naturalist 56:247–253.

Maxted, A. M. 2001. Post-fledging survival, dispersal, and habitat use in two migratory shrubland bird species. Thesis, Purdue University, West Lafayette, Indiana, USA.

Mayer, M., R. Chianese, T. Scudder, J. White, K. Vongpaseuth, and R. Ward. 2002. Thirteen years of monitoring the hemlock woolly adelgid in New Jersey forests. Pages 50–60 *in* B. Onken, R. Reardon, and J. Lashomb, editors. Proceedings, Hemlock woolly adelgid in the Eastern U.S. symposium. 5–7 February 2002. Rutgers University, New Brunswick, New Jersey, USA.

McAuley, D. G., J. R. Longcore, D. A. Clugston, R. B. Allen, A. Weik, S. Williamson, J. Dunn, B. Palmer, K. Evans, W. Staats, G. F. Sepik, and W. Halteman. 2005. Effects of hunting on survival of American woodcock in the Northeast. Journal of Wildlife Management 69:1565 1577.

McCabe, R. A. 1991. The little green bird: ecology of the willow flycatcher. Rusty Rock Press, Madison, Wisconsin, USA.

McCallum, M. L., S. E. Stanley, M. N. Mary, C. McDowell, and B. A. Wheeler. 2004. Fall breeding of the southern leopard frog (*Rana sphenocephala*) in northeastern Arkansas. Southeastern Naturalist 3:401–408.

McCarley, W. H. 1959. An unusually large nest of *Cryptotis parva*. Journal of Mammalogy 40:243.

McClure, H. E. 1942. Summer activities of bats (genus *Lasiurus*) in Iowa. Journal of Mammalogy 30:57–65.

McCoy, C. J. 1982. Amphibians and reptiles of Pennsylvania. Carnegie Museum of Natural History Special Publication Number 6, Pittsburgh, Pennsylvania, USA.

McCoy, C. J. 1989. Amphibians and reptiles. Pages 62–63 *in* D. J. Cugg, W. J. Young, E. K. Muller, W. Zelinsky, and R. F. Abler, editors. The atlas of Pennsylvania. Temple University Press, Philadelphia, Pennsylvania, USA.

McCracken, J. D. 1993. Status report on the cerulean warbler *Dendroica cerulea* in Canada. Committee on the Status of Endangered Wildlife in Canada.

McCrimmon, D. A., Jr. 1978. Nest site characteristics among five species of herons on the North Carolina coast. Auk 95:267–280.

McCrimmon, D. A., Jr., J. C. Ogden, and G. T. Bancroft. 2001. Great egret (*Ardea alba*). *In* A. Poole and F. Gill, editors. The birds of North America, 570. Birds of North America, Philadelphia, Pennsylvania, USA.

McCullough, D. 1996. Metapopulations and Wildlife Conservation. Island Press, Washington, D.C., USA.

McDonald, M. V. 1998. Kentucky warbler (*Oporornis formosus*). *In* A. Poole and F. Gill, editors. The birds of North America, 324. Birds of North America, Philadelphia, Pennsylvania, USA.

McGarigal, K., and W. C. McComb. 1995. Relationships between landscape structure and breeding birds in the Oregon Coast Range. Ecological Monographs 65:235–260.

McGowan, E. M. 1993. Experimental release and fate study of the Allegheny woodrat (*Neotoma magister*). New York Federal Aid Project W-166-E; E-1, Job Number VIII-7.

McGowan, K. J., and K. Corwin. 2008. The second atlas of breeding birds in New York State. Cornell University Press, Ithaca, New York, USA.

McKinney, M. L. 2006. Urbanization as a major cause of biotic homogenization. Biological Conservation 127:247–260.

McLachlin, R. A. 1983. Dispersion of the western winter wren (*Troglodytes troglodytes*) in the coastal western hemlock forest at the University of British Columbia Research Forest in southwestern British Columbia. Dissertation, University of British Columbia, Vancouver, Canada.

McLeod, R. F., and J. E. Gates. 1998. Response of herpetofaunal communities to forest cutting and burning at Chesapeake Farms, Maryland. American Midland Naturalist 139:164–177.

McMorris, F. A., and D. W. Brauning. 2004. Peregrine falcon research and management. Annual report to the Pennsylvania Game Commission, Harrisburg, USA.

McMorris, F. A., and D. W. Brauning. 2005. Peregrine falcon research and management. Annual report to the Pennsylvania Game Commission, Harrisburg, USA.

McMorris, F. A., and D. W. Brauning. 2009. Peregrine falcon research and management. Annual report to the Pennsylvania Game Commission, Harrisburg, USA.

McNair, D. B. 1988a. Breeding attempt of pine siskin on Mt. Mitchell, North Carolina. Migrant 59:49–50.

McNair, D. B. 1988b. Review of breeding records of red crossbill and pine siskin in the southern Appalachian Mountains and adjacent regions. Migrant 59:105–113.

McNaught, B. 2004. Pennsylvania important bird areas 61: Shohola Waterfowl Management Area. Pennsylvania Audubon Society, Harrisburg, USA.

McRae, S. B. 1996. Family values: costs and benefits of communal nesting in the moorhen. Animal Behavior 52:225–245.

McWilliams, G. 1995. Attempted nesting of three species of Laridae at Presque Isle State Park, Erie County. Pennsylvania Birds 9:79–80.

McWilliams, G. M. 1992. Local notes—April through June 1992: Erie County. Pennsylvania Birds 6:79.

McWilliams, G. M. 2005. Local notes—March through May 2005: Erie County. Pennsylvania Birds 19:142.

McWilliams, G. M., and D. W. Brauning. 2000. The birds of Pennsylvania. Cornell University Press, Ithaca, New York, USA.

McWilliams, W. H., C. A. Alerich, D. A. Devlin, J. Lister, T. W. Lister, S. L. Sterner, and J. A. Westfall. 2004. Annual inventory report for Pennsylvania's forests: results from the first three years. Resource Bulletin NE-159. U.S. Forest Service, Northeastern Research Station, Newtown Square, Pennsylvania, USA.

McWilliams, W. H., S. L. Stout, T. W. Bowersox, and L. H. McCormick. 1995. Adequacy of advance tree-seedling regeneration in Pennsylvania's forests. Journal of Applied Forestry 12:187–191.

Meanley, B. 1969. Natural history of the king rail. North American Fauna 67.

Meanley, B. 1992. King rail (*Rallus elegans*). *In* A. Poole, P. Stettenheim, and F. Gill, editors. The birds of North America, 3. Academy of Natural Sciences, Philadelphia, Pennsylvania, USA, and American Ornithologists' Union, Washington, D.C., USA.

Mearns, E. A. 1898. A study of the vertebrate fauna of the Hudson Highlands, with observations on the Mollusca, Crustacea, Lepidoptera, and the flora of the region. Bulletin of the American Museum of Natural History 10:303–352.

Mellon, R. 1990. An ornithological history of the Delaware Valley region. Cassinia 63:36–56.

Melquist, W. E., and M. G. Hornocker. 1983. Ecology of river otters in west central Idaho. Wildlife Monographs 83.

Melquist, W. E., P. J. Polechla, and D. Toweill. 2003. River otter (*Lontra canadensis*). Pages 708–734 *in* G. A. Feldhamer, B. C. Thompson, and J. A. Chapman, editors. Wild mammals of North America: biology, management, and conservation. Second edition. Johns Hopkins University Press, Baltimore, Maryland, USA.

Melvin, S. M., and J. P. Gibbs. 1996. Sora (*Porzana carolina*). *In* A. Poole and F. Gill, editor. The birds of North America, 250. Academy of Natural Sciences, Philadelphia, Pennsylvania, USA, and the Auk, Washington, D.C., USA.

Melvin, S. M., D. G. Smith, D. W. Holt, and G. R. Tate. 1989. Small owls. Pages 88–96 *in* B. G. Pendleton, editor. Proceedings of the northeast raptor management symposium and workshop. National Wildlife Federation, Washington, D.C., USA.

Mendall, H. L., and C. M. Aldous. 1943. The ecology and management of the American Woodcock. Maine Cooperative Wildlife Research Unit, University of Maine, Orono, USA.

Mengak, M. T. 2002. Reproduction, juvenile growth and recapture rates of Allegheny woodrats (*Neotoma magister*) in Virginia. American Midland Naturalist 148:155–162.

Menzel, M. A., T. C. Carter, B. R. Chapman, and J. Laerm. 1998. Quantitative comparison of tree roots used by red bats (*Lasiurus borealis*) and Seminole bats (*L. seminolus*). Canadian Journal of Zoology 76:630–634.

Menzel, M. A., T. C. Carter, W. M. Ford, and B. R. Chapman. 2001. Tree-roost characteristics of subadult and female adult evening bats (*Nycticeius humeralis*) in the Upper Coastal Plain of South Carolina. American Midland Naturalist 145:112–119.

Merrell, D. J. 1977. Life history of the leopard frog, *Rana pipiens*, in Minnesota. Occasional Papers of the Bell Museum of Natural History, University of Minnesota 15:1–23.

Merriam, C. H. 1881. Preliminary list of birds ascertained to occur in the Adirondack region, northeastern New York. Bulletin of the Nuttall Ornithological Club 6:225–235.

Merrill, S. B., F. C. Cuthbert, and G. Oehlert. 1998. Residual patches and their contribution to forest-bird diversity on northern Minnesota aspen clearcuts. Conservation Biology 12:190–199.

Merritt, J. F. 1987. Guide to the mammals of Pennsylvania. University of Pittsburgh Press, Pittsburgh, Pennsylvania, USA.

Merritts, H. V. 1943. The distribution of species of cottontails in Centre County, Pennsylvania. Thesis, Pennsylvania State College, State College, USA.

Meslow, E. C., and L. B. Keith. 1968. Demographic parameters of a snowshoe hare population. Journal of Wildlife Management 32:812–834.

Messenger, S. L., C. E. Rupprecht, and J. S. Smith. 2003. Bats, emerging virus infections, and the rabies paradigm. Pages 622–679 *in* T. H. Kunz and M. B. Fenton, editors. Bat Ecology. University of Chicago Press, Chicago, Illinois, USA.

Messineo, D. J. 1985. The 1985 nesting of pine siskin, red crossbill and white-winged crossbill in Chenango County, New York. Kingbird 35:233–237.

Meyer, K. 1987. Behavioral ecology of breeding and wintering sharp-shinned hawks. Dissertation, University of North Carolina, Chapel Hill, USA.

Meyers, W., J. Bishop, R. Brooks, T. O'Connell, D. Argent, G. Storm, J. Stauffer, and R. Carline. 2000. A Gap of Pennsylvania Analysis: 2001 Final Report. Pennsylvania State University, University Park, USA.

Michener, M. C., and J. D. Lazell. 1989. Distribution and relative abundance of the hognose snake, *Heterodon platirhinos*, in Eastern New England. Journal of Herpetology 23:35–40.

Mikkola, H. 1983. Owls of Europe. Buteo Books, Vermillion, South Dakota, USA.

Milam, J. C., and S. M. Melvin. 2001. Density, habitat use, movements, and conservation of spotted turtles (*Clemmys guttata*) in Massachusetts. Journal of Herpetology 35:418–427.

Miller, D. 2003. Species diversity, reproduction, and sex-ratios of bats in managed pine forest landscapes of Mississippi. Southeastern Naturalist 2:59–72.

Miller, G. S. 1897. Migration of bats on Cape Cod, Massachusetts. Science 5:541–543.

Miller, G. S., Jr., and G. M. Allen. 1928. The American bats of the genera *Myotis* and *Pizonyx*. Bulletin of the U.S. National Museum, Number 144.

Miller, J. C. 1979. Snowy egret nesting in Philadelphia, Pennsylvania. Cassinia 58:22.

Miller, J. C. 1994. December broad-winged hawk in Pennsylvania. Cassinia 66:23.

Miller, J. C., and C. E. Price, Jr. 1959. Birds of Tinicum. Cassinia 44:3–15.

Miller, M., E. Greenstone, W. Greenstone, and K. Bildstein. 2002. Timing and magnitude of broad-winged hawk migration at Montclair Hawk Lookout, New Jersey, and Hawk Mountain Sanctuary Pennsylvania. Wilson Bulletin 14:479–484.

Miller, N. A. 1999. Landscape and habitat predictors of Canada warbler (*Wilsonia canadensis*) and northern waterthrush (*Seiurus noveboracensis*) occurrence in Rhode Island swamps. Thesis, University of Rhode Island, Kingston, USA.

Miller, R. F. 1946. The Florida gallinule: breeding birds of the Philadelphia region. Pt. III. Cassinia 36:1–16.

Miller, T., D. Brandes, M. Lanzone, D. Ombalski, R. Mulvihill, R. Brooks, and T. Katzner. 2007. Flight characteristics of golden eagles (*Aquila chrysaetos*) migrating through Eastern North America as determined by GPS telemetry. Raptor Research Foundation and Hawk Migration Association of North America 2007 Joint Meeting.

Mills, M. A. 2004. Scat-marking by river otters in Pennsylvania and Maryland. Thesis, Frostburg State University, Frostburg, Maryland, USA.

Mills, M. S. 1993. *Heterodon platirhinos* (Eastern hognose snake). Diet. Herpetological Review 24:62.

Milsap, B. A., and S. L. Vana. 1984. Distribution of wintering golden eagles in the eastern United States. Wilson Bulletin 96:692–701.

Minton, S. A., Jr. 1944. Introduction to the study of the reptiles of Indiana. American Midland Naturalist 32:438–477.

Minton, S. A., Jr. 1972. Amphibians and reptiles of Indiana. Indiana Academy of Science, Indianapolis, USA.

Minton, S. A. 1998. Observations on Indiana amphibian populations: a forty-five-year overview. Pages 217–220 *in* M. J. Lannoo, editor. Status and Conservation of Midwestern Amphibians. University of Iowa Press, Iowa City, USA.

Minton, S. A., Jr., J. C. List, and M. J. Lodato. 1983. Recent records and status of amphibians and reptiles in Indiana. Proceedings of the Indiana Academy of Science 92:489–498.

Mitchell, J. C. 1974. Statistics of *Chrysemys rubriventris* hatchlings from Middlesex County, Virginia. Herpetological Review 5:71.

Mitchell, J. C., and M. W. Klemens. 2002. Primary and secondary effects of habitat alteration. Pages 5–32 *in* M. W. Klemens, editor. Turtle conservation. Smithsonian Institution Press, Washington, D.C., USA.

Mitchell, J. C., and T. K. Pauley. 2005. *Pseudacris brachyphona* (Cope, 1889) mountain chorus frog. Pages 465–466 *in* M. Lannoo, editor. Amphibian declines: the conservation status of United States species. University of California Press, Berkeley, USA.

Mitchell, L. C., and B. A. Millsap. 1990. Buteos and golden eagles. Pages 50–62 *in* Proceedings of the Southeast Raptor Management Symposium and Workshop, National Wildlife Federation. Scientific and Technical Series 14, Washington, D.C., USA.

Mitchell, R. C., and J. Brady. 1986. Pennsylvania's bald eagle recovery project. Pages 301–309 *in* S. Majumdar, F. Benner, and A. Rhoads, editors. Endangered and threatened species programs in Pennsylvania and other states: causes, issues, and management. Pennsylvania Academy of Science, Easton, Pennsylvania, USA.

Mitchell, W. A. 1988. Songbird nest boxes. Section 5.1.8, U.S. Army Corp of Engineers, Wildlife Resources Management Manual. Technical Report EL-88-19. Waterways Experiment Station, Vicksburg, Mississippi, USA.

Mohr, C. E. 1932a. *Myotis subulatus leibii* and *Myotis sodalis* in Pennsylvania. Journal of Mammalogy 13:160–161.

Mohr, C. E. 1932b. The seasonal distribution of bats in Pennsylvania. Proceedings of the Pennsylvania Academy of Science 6:189–194.

Mohr, C. E. 1933. Pennsylvania bats of the genus *Myotis*. Proceedings of the Pennsylvania Academy of Science 7:39–43.

Mohr, C. E. 1936. Notes on the least brown bat *Myotis subulatus leibii*. Proceedings of the Pennsylvania Academy of Science 10:62–65.

Mohr, C. E. 1939. Bat tagging in Pennsylvania. Proceedings of the Pennsylvania Academy of Science 13:43–45.

Mohr, C. E. 1942. Bat tagging in Pennsylvania turnpike tunnels. Journal of Mammalogy 23:375–379.

Monroe, B. L., Jr., and C. G. Sibley. 1993. A world checklist of birds. Yale University Press, New London, Connecticut, USA.

Monthey, R. W. 1986. Responses of snowshoe hares, *Lepus americanus*, to timber harvesting in northern Maine. Canadian Field-Naturalist 100:568–570.

Moore, R. P., W. D. Robinson, I. J. Lovette, and T. R. Robinson. 2008. Experimental evidence for extreme dispersal limitation in tropical forest birds. Ecology Letters 11:960–968.

Morgan, R. P., II,, and J. A. Chapman. 1981. The serum proteins of the *Sylvilagus* complex. Pages 64–72 *in* K. Myers, and C. D. MacInnes, editors. Proceedings of the World Lagomorph Conference. University of Guelph, Ontario, Canada.

Morimoto, K., S. M. Patel, S. Corisdeo, D. C. Hooper, Z. F. Fu, C. E. Rupprecht, H. Koprowski, and B. Dietzschold. 1966. Characterization of a unique variant of bat rabies virus responsible for newly emerging human cases in North America. Proceedings of the National Academy of Sciences 93:5653–5658.

Mormann, B., M. Milam, and L. Robbins. 2004. Red bats do it in the dirt. Bats 22:6–8.

Morneau, F., S. Broduer, R. Decarriere, and D. Bird. 1994. Abundance and distribution of nesting golden eagles in Hudson Bay, Quebec. Journal of Raptor Research 28:220–225.

Morrin, H.B., editor. 1991. A guide to the birds of Lancaster County, Pennsylvania. Lancaster County Bird Club. Lancaster, USA.

Morris, B., R. Wiltraut, and F. Brock. 1984. Birds of the Lehigh Valley area. Lehigh Valley Audubon Society, Emmaus, Pennsylvania, USA.

Morris, D. H. 1979. Habitat use by the blackpoll warbler. Wilson Bulletin 91:234–243.

Morrison, R. I. G., B. J. McCaffery, R. E., S. K. Skagen, S. L. Jones, G. W. Page, C. L. Gratto-Trevor, and B. A. Andres. 2006. Population estimates of North American shorebirds. Wader Study Group Bulletin 111:67–85.

Morrison, R. I. G., R. E. Gill, B. A. Harrington, S. Skagen, G. W. Page, C. L. Gratto-Trevor, and S. M. Haig. 2000. Population estimates of nearctic shorebirds. Waterbirds 23:387–552.

Morrow, J. L., J. H. Howard, S. A. Smith, and D. K. Poppel. 2001. Home range and movements of the bog turtle (*Clemmys muhlenbergii*) in Maryland. Journal of Herpetology 35:68–73.

Morse, D. H. 1971. The foraging of warblers isolated on small islands. Ecology 52:216–228.

Morse, D. H. 1976. Variables determining the density and territory size of breeding spruce- woods warblers. Ecology 57:290–301.

Morse, D. H. 1993. Black-throated green warbler (*Dendroica virens*). *In* A. Poole and F. Gill, editors. The birds of North America, 55. Academy of Natural Science, Philadelphia, Pennsylvania, USA, and American Ornithologists' Union, Washington, D.C., USA.

Morse, D. H. 1994. Blackburnian warbler (*Dendroica fusca*). *In* A. Poole, and F. Gill, editors. The birds of North America, 102. Academy of Natural Science, Philadelphia, Pennsylvania, USA, and American Ornithologists' Union, Washington, D.C., USA.

Mosher, J., and P. Matray. 1974. Size-dimorphism: a factor in energy savings for broad-winged hawks. Auk 91:325–341.

Mosher, J. A., M. R. Fuller, and M. R. Kopeny. 1990. Surveying woodland raptors by broadcast of conspecific vocalizations. Journal of Field Ornithology 61:453–461.

Moskoff, W. 1995. Solitary Sandpiper. *In* A. Poole and F. Gill, editors. The birds of North America, 156. Academy of Natural Science, Philadelphia, Pennsylvania, USA, and American Ornithologists' Union, Washington, D.C., USA.

Mossman, M. J., L. M. Hartman, R. Hay, J. R. Sauer, and B. J. Dhuey. 1998a. Monitoring long-term trends in Wisconsin frog and toad populations. Pages 169–198 *in* M. J. Lannoo, editor. Status and Conservation of Midwestern Amphibians. University of Iowa Press, Iowa City, USA.

Mossman, M. J., J. R. Sauer, G. A. Gough, L. M. Hartman, and R. Hay. 1998b. The Wisconsin frog and toad survey home page. Wisconsin Department of Natural Resources (Madison) and U.S. Geological Survey Patuxent Wildlife Research Center, Laurel, Maryland. <www.mbr-pwrc.usgs.gov/wifrog/frog.htm>. Accessed 5 July 2007.

Mott, D. F., and R. D. Flynt. 1995. Evaluation of an electric fence system for excluding wading birds at catfish ponds. Progressive Fish-Culturist 57:88–90.

Mount, R. H. 1975. The reptiles and amphibians of Alabama. University of Alabama Press, Tuscaloosa, USA.

Mousley, H. 1931. A study of the home life of the alder flycatcher (*Empidonax traillii traillii*). Auk 48:547–552.

Mowbray, T. B. 1999. Scarlet tanager. *In* A. Poole and F. Gill, editors. The birds of North America. 479. Birds of North America, Philadelphia, Pennsylvania, USA.

Moyer, B. 2003. Pennsylvania's wildlife and wild places: our outdoor heritage in peril. Pennsylvania Wild Resource Conservation Fund, Harrisburg, USA.

Moyer, B. 2004. Pennsylvania's wildlife and wild places. Pennsylvania Department of Conservation and Natural Resources. Harrisburg, USA.

Mueller, H. 2005. Wilson's snipe (*Gallinago delicata*). *In* A. Poole, editor. The birds of North America Online Ithaca: Cornell Laboratory of Ornithology, Ithaca, New York, USA.

Mueller, H. C., D. D. Berger, and G. Allez. 1977. The periodic invasions of goshawks. Auk 94:652–663.

Muller, M. J., and R. W. Storer. 1999. Pied-billed grebe

(*Podilymbus podiceps*). *In* A. Poole and F. Gill, editors. The birds of North America, 410. Birds of North America, Philadelphia, Pennsylvania, USA.

Mulvihill, R. S. 1999. Effects of stream acidification on the breeding biology of an obligate riparian songbird, the Louisiana waterthrush (*Seiurus motacilla*). Pages 51–61 *in* W. E. Sharpe and J. R. Drohan, editors. The effects of acidic deposition on aquatic ecosystems in Pennsylvania. Proceedings of the 1998 Pennsylvania Acidic Deposition Conference, Volume 2. Environmental Resources Research Institute, Pennsylvania State University, University Park, USA.

Mulvihill, R. S. 1988. The occurrence of dickcissels (*Spiza americana*) in western Pennsylvania during the 1988 nesting season—it's possible bearing on the species' unusual history in Eastern North America. Pennsylvania Birds 2:83–87.

Mulvihill, R. S. 1992a. Dickcissel. Pages 370–371 *in* D. W. Brauning, editor. Atlas of breeding birds in Pennsylvania. University of Pittsburgh Press, Pittsburgh, Pennsylvania, USA.

Mulvihill, R. S. 1992b. Alder flycatcher. Pages 202–203 *in* D. W. Brauning, editor. Atlas of breeding birds in Pennsylvania. University of Pittsburgh Press, Pittsburgh, Pennsylvania, USA.

Mulvihill, R. S. 1992c. Willow flycatcher. Pages 204–205 *in* D.W. Brauning, editor. Atlas of the birds in Pennsylvania. University of Pittsburgh Press, Pittsburgh, Pennsylvania, USA.

Mulvihill, R.S., and D. W. Brauning. 2009. 2nd Pennsylvania breeding bird atlas preliminary results. Cornell Lab of Ornithology for Carnegie Museum of Natural History, Pittsburgh, Pennsylvania, USA. <http://bird.atlasing.org/Atlas/PA>.

Mulvihill, R. S., A. Cunkleman, L. Quattrini, T. O'Connell, and T. L. Master. 2002. Opportunistic polygyny in the Louisiana waterthrush. Wilson Bulletin 114:106–113.

Mulvihill, R. S., S. C. Latta, and F. L. Newell. 2009. Temporal constraints on the incidence of double brooding in the Louisiana waterthrush. Condor 111:341–348.

Mulvihill, R. S., F. L. Newell, and S. C. Latta. 2008. Effects of acidification on the breeding ecology of a stream-dependent songbird, the Louisiana waterthrush (*Seiurus motacilla*). Freshwater Biology 53:2158–2169.

Mumford, R. E. 1964. The breeding biology of the Acadian flycatcher. University of Michigan Museum of Zoology Miscellaneous Publication 175, Ann Arbor, USA.

Mumford, R. E. 1969. The hoary bat in Indiana. Proceedings of the Indiana Academy of Science 78:497–501.

Mumford R. E., and C. E. Keller. 1984. Birds of Indiana. Indiana University Press, Bloomington, USA.

Munyer, E. A. 1967. A parturition date for the hoary bat, *Lasiurus c. cinereus*, in Illinois and notes on the new born young. Transactions of the Illinois Academy of Sciences 60:95–97.

Murdock, N. A. 1994. Rare and endangered plants and animals of southern Appalachian woodlands. Water, Air and Soil Pollution 77:385–405.

Murphy, M. T. 1989. Life history variability in North American breeding tyrant flycatchers. Oikos 54:3–14.

Murphy, M. T., and R. C. Fleischer. 1986. Body size, nest predation, and reproductive patterns in brown thrashers and other mimids. Condor 88:446–455.

Murray, B. G., Jr., and F. B. Gill. 1976. Behavioral interactions of blue-winged and golden winged warblers. Wilson Bulletin 88:231–254.

Murray, D. L. 2000. A geographic analysis of snowshoe hare population demography. Canadian Journal of Zoology 78:1207–1217.

Murray, D. L. 2003. Snowshoe hare and other hares. Pages 147–175 *in* G. A. Feldhamer and B. Thompson, editors. Wild mammals of North America. Volume 2. Johns Hopkins University Press, Baltimore, Maryland, USA.

Murray, S. W., and A. Kurta. 2002. Spatial and temporal variation in diet. Pages 182–192 *in* A. Kurta and J. Kennedy, editors. The Indiana bat: biology and management of an endangered species. Bat Conservation International, Austin, Texas, USA.

Murray, S. W., and A. Kurta. 2004. Nocturnal activity of the endangered Indiana bat (*Myotis sodalis*). Journal of Zoology (London) 262:197–206.

Myers, J. A., M. Vellend, S. Gardescu, and P. L. Marks. 2004. Seed dispersal by white-tailed deer: implications for long-distance dispersal and migration of plants in eastern North America. Oecologia 139:35–44.

Myers, R. F. 1960. *Lasiurus* from Missouri caves. Journal of Mammalogy 41:114–117.

Myers, W., J. Bishop, R. Brooks, T. O'Connell, D. Argent, G. Storm, J. Stauffer, and R. Carline. 2000a. Allegheny woodrat potential habitat map. Pennsylvania GAP Analysis project: Leading Landscapes for Collaborative Conservation. Pennsylvania State University, University Park, USA. <www.orser.psu.edu/PAGAP/Species_Maps/Mammals/Woodrat.jpg>. Accessed 1 Jan 2005.

Myers, W., J. Bishop, R. Brooks, T. O'Connell, D. Argent, G. Storm, J. Stauffer, and R. Carline. 2000b. Pennsylvania Gap Analysis Project: leading landscapes for collaborative conservation. School of Forest Resources and Cooperative Fish and Wildlife Research Unit and Environmental Resources Research Institute. Pennsylvania State University, University Park, Pennsylvania. U.S. Geological Survey, Biological Resources Division, Gap Analysis Program.

Myers, W. L., and J. A. Bishop. 2000. Eastern spotted skunk (*Spilogale putorius*) Potential Habitat Map. <www.orser.psu.edu/PAGAP/Species_Maps/Mammals/SpotSkunk.jpg>. Accessed 22 Feb 2007.

Myers, W. L., and J. A. Bishop. 2000. Pennsylvania conservation gap mammal habitat model for the Eastern red bat (*Lasiurus borealis*). Pennsylvania Spatial Data Access.

<www.pasda.psu.edu/data/gap/species/30meter/mammals/pa30_amacc05010.zip>. Accessed 31 Jan 2006.

Myers, W. L., J. A. Bishop, R. Brooks, T. O'Connell, D. Argent, G. Storm, J. Stauffer, and R. Carline. 2000. The Pennsylvania GAP Analysis Project Final Report. Pennsylvania State University and U.S. Geological Survey, University Park, USA.

Nagle, R. D., C. L. Lutz, and A. L. Pyle. 2004. Overwintering in the nest by hatchling map turtles (*Graptemys geographica*). Canadian Journal of Zoology 82:1211–1218.

Nagle, R. D., C. L. Rowe, and J. D. Congdon. 2001. Accumulation and selective maternal transfer of contaminants in the turtle *Trachemys scripta* associated with coal ash deposition. Archives of Environmental Contamination and Toxicology 40:531–536.

Naidoo, R., A. Balmford, and R. Costanza. 2008. Global mapping of ecosystem services and conservation priorities. Proceedings of the National Academy of Sciences of the United States of America 105:9495–9500.

Nason, E. S. 1948. Morphology of hair of eastern North American bats. American Midland Naturalist 39: 345–361.

National Atmospheric Deposition Program (NRSP-3). 2004. National Atmospheric Deposition Program Office, Illinois State Water Survey, Champaign, IL. <http://nadp.sws.uiuc.edu>. Accessed 28 June 2007.

National Audubon Society. 2005a. The Christmas Bird Count Historical Results. Available <www.audubon.org/bird>. Accessed Jan 2005.

National Audubon Society. 2005b. Partners in flight watch list species nesting in Pennsylvania. <www.audubon.org/watch/watch_list.html>. Accessed 22 Feb 2005.

National Audubon Society. 2008. The 109th Christmas bird count: citizen science in action. <www.audubon.org/bird/cbc/index.html>.

National Research Council. 2007. Environmental impacts of wind-energy projects. National Academies Press, Washington, D.C., USA.

The Nature Conservancy. 1994. The conservation of biological diversity in the Great Lakes ecosystem: issues and opportunities. The Nature Conservancy Great Lakes Program, Chicago, Illinois, USA. NatureServe. 2004. Version 4.1. NatureServe Explorer: an online encyclopedia of life. NatureServe, Arlington, Virginia. USA. <www.natureserve.org/explorer>.

The Nature Conservancy. 1999. Species management abstract: red-shouldered hawk. The Nature Conservancy. Arlington, Virginia, USA.

NatureServe. 2005. Version 4.2. NatureServe Explorer: an online encyclopedia of life. NatureServe, Arlington, Virginia, USA. <www.natureserve.org/explorer>.

NatureServe. 2009. Version 7.1. NatureServe Explorer: an online encyclopedia of life. NatureServe, Arlington, Virginia, USA. <www.natureserve.org/explorer>.

Naughton, G. P., C. B. Henderson, K. R. Foresman, and R. L. McGraw. 2000. Long-toed salamanders in harvested and intact Douglas fir forests of western Montana. Ecological Applications 10:1681–1689.

Neill, W. T. 1948. Hibernation of amphibians and reptiles in Richmond County, Georgia. Herpetologica 4:107–114.

Nelson, B., C. Nowak, S. Reitz, D. deCalesta, and S. Wingate. 1997. Communicating old growth forest management in the Allegheny National Forest. Pages 85–89 *in* Northeastern Forest Experiment Station, editor. Communicating the role of silviculture in managing the national forests: proceedings of the national silviculture workshop (General Technical Report NE-238), U.S. Department of Agriculture, Northeastern Forest Experiment Station, Warren, Pennsylvania, USA.

Nemuras, K. T. 1967. Notes on the natural history of *Clemmys muhlenbergi*. Bulletin of the Maryland Herpetological Society 3:80–96.

Netting, M. G. Blanding's Turtle, *Emys blandingii* (Holbrook) in PA. Copeia 1932:173–175.

Newbury, T. L., and N. P. P. Simon. 2005. The effects of clearcutting on snowshoe hare (*Lepus americanus*) relative abundance in central Labrador. Forest Ecology and Management 210:131–142.

Newell, F. L., and M. S. Kostalos. 2007. Wood thrush nests in dense understory may be vulnerable to predators. Wilson Journal of Ornithology 119:693–702.

Newman, H. H. 1906. The habits of certain tortoises. Journal of Comparative Neurology and Psychology 16:126–152.

Newton, I. 1973. Finches. Taplinger, New York, New York, USA.

Newton, I., I. Wyllie, and P. Freestone. 1990. Rodenticides in British barn owls. Environmental Pollution 68:101–117.

Nickens, T. E. 2004. Small is beautiful. National Wildlife 42:28–31.

Nickerson, M. A., K. L. Krysko, and R. D. Owen. 2003. Habitat differences affecting age class distribution of the hellbender salamander, *Cryptobranchus alleganiensis*. Southeastern Naturalist 2:619–629.

Niederberger, A. J., and M. E. Seidel. 1999. Ecology and status of a wood turtle (*Clemmys insculpta*) population in West Virginia. Chelonian Conservation and Biology 3:414–418.

Niemi, G. J. 1985. Patterns of morphological evolution in the bird genera of New World and Old World peatlands. Ecology 66:1215–1228.

Niemi, G. J., and J. M. Hanowski. 1984. Effects of a transmission line on bird populations in the Red Lake peatland, northern Minnesota. Auk 101:487–498.

Niethammer, K. R., and M. S. Kaiser. 1983. Late summer food habits of three heron species in northeastern Louisiana. Colonial Waterbirds 6:148–153.

Niewiarowski, P. H. 1995. Effects of supplemental feeding and thermal environment on growth rates of eastern fence lizards, *Sceloporus undulatus*. Herpetologica 51:487–496.

Nilsson, G. 1980. River otter research workshop. Florida State Museum, Gainesville, USA.

Nisbet, I. C. T. 1970. Autumn migration of the blackpoll warbler: evidence for long flight provided by regional survey. Bird-Banding 41:207–240.

Nisbet, I. C. T. 2002. Common tern (*Sterna hirundo*). *In* A. Poole and F. Gill, editors. The birds of North America, 618. Birds of North America, Philadelphia, Pennsylvania, USA.

Nisbet, I. C., T. D. B. McNair, W. Post, and T. C. Williams. 1995. Transoceanic migration of the blackpoll warbler: summary of scientific evidence and response to criticisms by Murray. Journal of Field Ornithology 66:612–622.

Nolan, N., Jr., E. D. Ketterson, and C. A. Buerhle. 1999. Prairie warbler (*Dendroica discolor*). *In* A. Poole and F. Gill, editors. The birds of North America, 455. Academy of Natural Sciences, Philadelphia, Pennsylvania, USA, and American Ornithologists' Union, Washington, D.C., USA.

Nolan, V., Jr. 1978. The ecology and behavior of the prairie warbler (*Dendroica discolor*). Ornithology Monographs 26:1–595.

Nolan, V., Jr., and C. F. Thompson. 1975. The occurrence and significance of anomalous reproductive activities in two North American nonparasitic cuckoos, *Coccyzus* species. Ibis 117:496–503.

Norment, C. J., C. D. Ardizzone, and K. Hartman. 1999. Habitat relations and breeding biology of grassland birds in New York. Studies in Avian Biology 19:112–121.

North American Breeding Bird Survey. 2003. NABBS home page. <www.mbr-pwrc.usgs.gov/bbs>. Accessed 1 June 2005.

Norton, R. M., and S. J. Hannon. 1997. Songbird response to partial-cut logging in the boreal mixedwood forest of Alberta. Canadian Journal of Forest Research 27:44–53.

Nott, P., N. Michel, and D. F. DeSante. 2001. Management strategies for reversing declines of inland birds of conservation concern on military institutions. Institute for Bird Populations, Point Reyes Station, California, USA.

Novak, P. G. 1992. Black tern, *Chlidonias niger*. Pages 149–169 *in* K. J. Schneider and Diane M. Pence, editors. Migratory nongame birds of management concern in the Northeast. U.S. Department of the Interior, Fish and Wildlife Service, Newton Corner, Massachusetts, USA.

Novotony, R. J. 1990. Geographic distribution. *Thamnophis brachystoma*. Herpetological Review 21:42.

Nye, P. 2004. New York state 2003 bald eagle report. New York Department of Environmental Conservation, Albany, USA.

Obrist, M. K. 1995. Flexible bat echolocation: the influence of individual, habitat and conspecifics on sonar design. Behavioural Ecology and Sociobiology 36:207–219.

O'Connell, T. J., Brooks, R. P., Laubscher, S. E., Mulvihill, R. S. and T. L. Master. 2003. Using bioindicators to develop a calibrated index of regional ecological integrity for forested headwater ecosystems. U.S. Environmental Protection Agency and Penn State Cooperative Wetlands Center, Report 2003-01, University Park, Pennsylvania, USA.

Odenkirk, J., and S. Owens. 2007. Expansion of a northern snakehead population in the Potomac River system. Transactions of the American Fisheries Society 136:1633–1639.

Ohio Department of Natural Resources. 2004. 2003–2004 Wildlife population status and hunting report. Ohio Department of Natural Resources, Columbus, USA.

Ogden, J. C. 1978. A. Sprunt IV, J. C. Ogden, and S. Winckler, editors. Recent population trends of colonial wading birds on the Atlantic and Gulf Coastal Plain. *In* Wading birds. National Audubon Society Research Report Number 7, National Audubon Society, New York, New York, USA.

Olden, J. D., N. L. Poff, and M. L. McKinney. 2006. Forecasting faunal and floral homogenization associated with human population geography in North America. Biological Conservation 127:261–271.

Oliarnyk, C. J., and R. J. Robertson. 1996. Breeding behavior and reproductive success of Cerulean Warblers in southeastern Ontario. Wilson Bulletin 108:673–684.

Oliver, J. A. 1955. The natural history of North American amphibians and reptiles. D. Van Nostrand, Princeton, New Jersey, USA.

Olnhausen, L. R., and M. R. Gannon. 2004. An evaluation of rabies prevention in the United States, based on an analysis from Pennsylvania. Acta Chiropterologica 6:163–168.

Ontario Birds at Risk Program. 2005. On the road to recovery? The prothonotary warbler in Canada. <www.bsc_eoc.org/prowmain.html>. Accessed 8 May 2005.

Orchard, S. A. 1992. Amphibian population declines in British Columbia. Pages 10–13 *in* C. A. Bishop and K. E. Pettit, editors. Declines in Canadian amphibian populations: designing a national strategy. Canadian Wildlife Service, Occasional Paper 76.

Orwig, D. A., and D. R. Foster. 1998. Forest response to the introduced hemlock woolly adelgid in southern New England, USA. Journal of the Torrey Botanical Society 125:60–73.

O'Shea, T. J., and M. A. Bogan. 2003. Monitoring trends in bat populations of the United States and territories: problems and prospects. U.S. Geological Survey, Biological Resources Discipline, Information and Technology Report USGS/BDR/ITR 2003–0003.

Ostfeld, R. S., and F. Keesing. 2000. Biodiversity and disease risk: the case of Lyme disease. Conservation Biology 14:722–728.

Owen, R. B., Jr., J. M. Anderson, J. W. Artmann, E. R. Clark, T. G. Dilworth, L. E. Gregg, F. W. Martin, J. D. Newsom, and S. R. Pursglove, Jr. 1977. American woodcock (*Philohela minor* = *Scolopax minor* of Edwards 1974). Pages 149–186 *in* G. C. Sanderson, editor. Management of migratory shore and upland game birds in North America. International Association of Fish and Wildlife Agencies, Washington, D.C., USA.

Pack, J. C., and J. I. Cromer. 1981. Reintroduction of the fisher in West Virginia. World Furbearer Conference Proceedings 2:1431–1442.

Packard, G. C., J. T. Tucker, and L. D. Lohmiller. 1998. Distribution of Strecker's chorus frogs (*Pseudacris streckeri*) in relation to their tolerance for freezing. Journal of Herpetology 32:437–440.

Pagels, J. F., L. A. Smock, and S. H. Sklarew. 1998. The water shrew, *Sorex palustris* Richardson (Insectivora: Soricidae) and habitats in Virginia. Brimleyan 25:120–134.

Palen, W. J., D. E. Schindler, M. J. Adams, C. A. Pearl, R. B. Bury, and S. A. Diamond. 2002. Optical characteristics of natural waters protect amphibians from UV-B in the U.S. Pacific Northwest. Ecology 83:2951–2957.

Palis, J. G. 1994. *Rana utricularia* (southern leopard frog) road mortality. Herpetological Review 25:119.

Palisot de Beauvois, A. M. F. J. 1799. Memoir on amphibia: serpents. Transactions of the American Philosophical Society 4:362–381.

Palmer, R. S. editor. 1962. Handbook of North American birds. Volume 1. Yale University Press, New Haven, Connecticut, USA.

Palmer, R. S. 1967. Solitary sandpiper. Pages 196–198 *in* G.D. Stout, editor. The shorebirds of North America. Viking Press, New York, New York, USA.

Palmer, R. S. 1968. *Spinus pinus* (Wilson) pine siskin. Pages 424–447 *in* O. L. Austin, Jr., editor. Life histories of North American cardinals, grosbeaks, buntings, towhees, finches, sparrows, and allies. U.S. National Museum Bulletin 237(1).

Palmer, R. S., editor. 1988a. Red-shouldered hawk. Pages 413–429 *in* Handbook of North American birds. Volume 4. Yale University Press, New Haven, Connecticut, USA.

Palmer, R. S. 1988b. Handbook of North American birds. Volume 4. Yale University Press, New Haven, Connecticut, USA.

Palmer, W. M., and A. L. Braswell. 1976. Communal egg laying and hatchlings of the rough green snake, *Opheodrys aestivus* (Linnaeus) (Reptilia, Serpentes, Colubridae). Journal of Herpetology 10:257–259.

Panko, D. 1990. Decline of migrant sharp-shinned hawks. Hawk Migration Association of North America Newsletter 15:3–13.

Pappas, M. J., and B. J. Brecke. 1992. Habitat selection of juvenile Blanding's turtles, *Emydoidea blandingi*. Journal of Herpetology 26:233–234.

Paragi, T. F. 1990. Reproductive biology of female fisher in southcentral Maine. Thesis, University of Maine, Orono, USA.

Parchman, T. L., C. W. Benkman, and S. C. Britch. 2006. Patterns of genetic variation in the adaptive radiation of new world crossbills (Aves: *Loxia*). Molecular Ecology 115:1873–1887.

Parkhurst, J. A., R. B. Brooks, and D. E. Arnold. 1992. Assessment of predation at trout hatcheries in central Pennsylvania. Wildlife Society Bulletin 20:411–419.

Parmellee, J. R., M. G. Knutson, and J. E. Lyon. 2002. A field guide to amphibian larvae and eggs of Minnesota, Wisconsin, and Iowa. U.S. Geological Survey, Biological Resource Division, Information and Technology Report USGS/BRD/ITR-2002–0004, Washington, D. C. iv + 38 pp.

Parsons, H. J., D. A. Smith, and R. F. Whittam. 1986. Maternity colonies of silver-haired bats, *Lasionycteris noctivagans*, in Ontario and Saskatchewan. Journal of Mammalogy 67:598–600.

Partners in Flight. 2003. Landbird conservation plan. Physiographic area 24. Allegheny plateau. Cornell Lab of Ornithology, Ithaca, New York, USA.

Partners in Flight. 2004. North American landbird conservation plan. Cornell Lab of Ornithology, Ithaca, New York, USA.

Partners in Flight United States. 2006. <www.pwrc.usgs.gov/pif/>.

Partners in Flight Watch List. 2004. Kentucky warbler page. <www.partnersinflight.org/WatchListNeeds/KEWA.htm>. Accessed 1 June 2005.

Patterson, B. D., G. Ceballos, W. Sechrest, M. F. Tognelli, T. Brooks, L. Luna, P. Ortega, I. Salazar, and B. E. Young. 2005. Digital distribution maps of the mammals of the Western Hemisphere, version 2.0. NatureServe, Arlington, Virginia, USA.

Paxton, R. O., W. J. Boyle, Jr., D. A. Cutler, and K. C. Richards. 1980. Hudson-Delaware region. American Birds 34:759.

Payne, J. L., D. R. Young, and J. F. Pagels. 1989. Plant community characteristics associated with the endangered northern flying squirrel, *Glaucomys sabrinus*, in the southern Appalachians. American Midland Naturalist 121:285–292.

Pearson, P. G. 1955. Population ecology of the spadefoot toad, *Scaphiopus h. holbrooki* (Harlan). Ecological Monographs 25:580–267.

Peck, G. K., and R. D. James. 1983. Breeding birds of Ontario: nidiology and distribution. Volume 1. Royal Ontario Museum Life Science, Toronto, Ontario, Canada.

Peck, G. K., and R. D. James. 1987. Breeding birds of Ontario: nidiology and distribution. Volume 2. Royal Ontario Museum Life Science, Toronto, Ontario, Canada.

Pennock, C. J. 1912. Crossbills in Chester County, Pennsylvania, in summer. Auk 29:245–246.

Pennsylvania Biological Survey. 2001. Mammal Technical Committee (PBS-MTC). Comments on Pennsylvania Game Commission Policies. 5205. Reintroductions of Wildlife. One page unpublished document.

Pennsylvania Bureau of Forestry. 2003. What is the wooly

adelgid? Online publication by the Pennsylvania Department of Conservation and Natural Resources. <www.dcnr.state.pa.us/forestry/woollyadelgid/index.htm>.

Pennsylvania Code. 2008. Commonwealth of Pennsylvania, Harrisburg, Pennsylvania. <www.pacode.com/index.html>. Accessed 30 Dec 2008.

Pennsylvania Department of Environmental Protection [PADEP]. 2005. Acid rain isopleth maps. Bureau of Air Quality. <www.dep.state.pa.us/dep/deputate/airwaste/aq/acidrain/isomaps.htm>. Accessed 5 May 2009.

Pennsylvania Department of Environmental Protection [PADEP]. 2006. Wetlands net gain strategy. <www.dep.state.pa.us>. Accessed 8 May 2006.

Pennsylvania Fish and Boat Commission [PFBC]. 2003. Information paper—eastern box turtle. Commonwealth of Pennsylvania, Fish and Boat Commission.

Pennsylvania Fish and Boat Commission [PFBC]. 2008. 2008 Pennsylvania summary of fishing regulations and laws. <www.fish.state.pa.us>. Accessed 30 Dec 2008.

Pennsylvania Game Commission. 1942. Biennial report. Pennsylvania Game Commission, Harrisburg, USA.

Pennsylvania Game Commission. 2002. Great blue heron. Unpublished data, Bureau of Wildlife Management, Pennsylvania Game Commission, Harrisburg, Pennsylvania.

Pennsylvania Game Commission. 2003. Pennsylvania Game Commission, Wildlife Diversity Section, contractor bat trapping/mist netting reports. Harrisburg, Pennsylvania, USA.

Pennsylvania Game Commission. 2006. Pennsylvania's Comprehensive Wildlife Conservation Strategy (PA-CWCS) Pennsylvania Game Commission—State Wildlife Management Agency. <www.pgc.state.pa.us/pgc/cwp/view.asp?a=496&q=162067>. Accessed 31 Jan 2006.

Pennsylvania Natural Heritage Program [PNHP]. 2005. Department of Conservation and Natural Resources, Harrisburg. <www.naturalheritage.state.pa.us>. Accessed 5 May 2009.

Pennsylvania Ornithological Records Committee. 2000. Official list of the birds of Pennsylvania. Pennsylvania Birds 14:109.

Pennsylvania Society for Ornithology Files. 2005. Bird records of the Pennsylvania Society of Ornithology, Franklin Haas, compiler.

Pennsylvania Society for Ornithology Special Areas Project database, Ecology III, Berwick, Pennsylvania, USA.

Perez, J. C., W. C. Haws, and C. Hatch. 1978. Resistance of woodrats to *Crotalus atrox* venom. Toxicon 16:199–211.

Perillo, K. M. 1997. Seasonal movements of and habitat preferences of spotted turtles (*Clemmys guttata*) in north central Connecticut. Chelonian Conservation and Biology 2:445–447.

Peterjohn, B. G., and D. L. Rice. 1991. The Ohio breeding bird atlas. Ohio Department of Natural Resources, Columbus, USA.

Peterjohn, B. G., and J. R. Sauer. 1999. Population status of North American breeding bird survey 1966–1996. Studies in Avian Biology 19:27–44.

Peterjohn, B. G., J. R. Sauer, and C. S. Robbins. 1995. Population trends from the North American breeding bird survey. Pages 3–39 *in* T. E. Martin and D. M. Finch, editors. Ecology and management of Neotropical migratory birds. Oxford University Press, New York, New York, USA.

Peters, S. E. 2002. An evaluation of track-plate surveys for detecting fishers (*Martes pennanti*) in northwestern Pennsylvania. Thesis, Frostburg State University, Frostburg, Maryland, USA.

Petersen, C. E. 1995. Movement patterns and habitat use of the copperhead snake in southeastern Virginia. Thesis, Old Dominion University, Norfolk, Virginia, USA.

Peterson, A. 1986. Habitat suitability index models: bald eagle (breeding season). U.S. Fish and Wildlife Service Biological Report 82(10.126).

Peterson, J. M. C. 1988a. Blackpoll warbler, *Dendoica striata*. Pages 394–395 *in* R. F. Andrle and J. R. Carroll, editors. The atlas of breeding birds in New York state. Cornell University Press, Ithaca, USA.

Peterson, J. M. C. 1988b. Olive-sided flycatcher, *Contopus borealis*. Pages 244–245 *in* R. F. Andrle and J. R. Carroll, editors. The atlas of breeding birds in New York state. Cornell University Press, Ithaca, USA.

Peterson, J. M. C. 1988c. Pine siskin *Carduelis pinus*. Pages 494–495 *in* R. F. Andrle and J. R. Carroll, editors. The atlas of breeding birds in New York state. Cornell University Press, Ithaca, New York, USA.

Peterson, J. M. C. 1988d. Red crossbill (*Loxia curvirostra*). Pages 490–491 *in* R. F. Andrle, and J. R. Carroll, editors. The atlas of breeding birds in New York state. Cornell University Press, Ithaca, New York, USA.

Peterson, J. M. C. 1988e. Swainson's thrush, *Catharus ustulatus*. Pages 322–323 *in* R. F. Andrle, and J. R. Carroll, editors. The atlas of breeding birds in New York state, Cornell University Press, Ithaca, New York, USA.

Peterson, J. M. C. 1988f. Yellow-bellied flycatcher (*Epidonax flaviventris*). Pages 248–249 *in* R. F. Andrle, and J. R. Carroll, editors. The atlas of breeding birds in New York state, Cornell University Press, Ithaca, New York, USA.

Petit, D. R., J. F. Lynch, R. L. Hutto, J. G. Blake, and R. B. Waide. 1993. Management and conservation of migratory landbirds overwintering in the Neotropics. Pages 70–92 *in* D. M. Finch and P. W. Stangel, editors. Status and management of Neotropical migratory birds. U.S. Forest Service, General Technical Report RM-229.

Petit, L. J. 1989. Breeding biology of prothonotary warblers in riverine habitat in Tennessee. Wilson Bulletin 101:51–61.

Petit, L. J. 1999. Prothonotary warbler. *In* A. Poole and F. Gill, editors. The birds of North America, 408. Birds of North America, Philadelphia, Pennsylvania, USA.

Petranka, J. W. 1998. Salamanders of the United States and Canada. Smithsonian Institution Press, Washington, D.C., USA.

Petrie, S. A., S. S. Badzinski, and K. L. Wilcox. 2002. Population trends and habitat use of tundra swans staging at Long Point, Lake Erie. Waterbirds 25:143–149.

Petrie, S. A., and K. L. Wilcox. 2003. Migration chronology of eastern-population tundra swans. Canadian Journal of Zoology 81:861–870.

Petzing, J. E., J. M. Mui, M. J. Dreslik, C. A. Phillips, D. B. Shepard, J. A. Crawford, A. R. Kuhns, M. J. Meyer, T. G. Anton, E. O. Moll, J. G. Palis, and D. Mauger. 2002. Filling in the gaps I: new county records for amphibians and reptiles in Illinois. Herpetological Review 33:327–330.

Phillips, C. A., R. A. Brandon, and E. O. Moll. 1999. Field guide to amphibians and reptiles of Illinois. Illinois Natural History Survey Manual 8.

Pickett, S. T. A., M. L. Cadenasso, J. M. Grove, C. H. Nilon, R. V. Pouyat, W. C. Zipperer, and R. Costanza. 2001. Urban ecological systems: linking terrestrial ecological, physical, and socioeconomic components of metropolitan areas. Annual Review of Ecology and Systematics 32:127–157.

Pike, E. 1985. The piping plover at Waugoshance Point. Jack-Pine Warbler 63:36–41.

Pimentel, D., L. Lach, R. Zuniga, and D. Morrison. 2000. Environmental and economic costs of nonindigenous species in the United States. BioScience 50:53–65.

Pisani, G. R. 1967. Notes on the courtship and mating behavior of *Thamnophis brachystoma* (Cope). Herpetologica 23:112–115.

Pisani, G. R. 1971. An unusually large litter of *Virginia valeriae pulchra*. Journal of Herpetology 5:207–208.

Pisani, G. R., and R. C. Bothner. 1970. The annual reproductive cycle of *Thamnophis brachystoma*. Science Studies 26:15–34.

Pisani, G. R., and J. T. Collins. 1971. The smooth earth snake, *Virginia valeriae* (Baird and Girard), in Kentucky. Transactions of the Kentucky Academy of Science 32:16–25.

Pitelka, F. A., P. Q. Tomich, and G. W. Treichel. 1955. Breeding behavior of jaegers and owls near Barrow, Alaska. Condor 57:3–18.

Pitt, W. C., and M. R. Conover. 1996. Predation at intermountain west fish hatcheries. Journal of Wildlife Management 60:616–624.

Platt, J. B. 1976. Sharp-shinned hawk nesting and nest site selection in Utah. Condor 78:102–103.

Platz, J. E., and A. Lathrop. 1993. Body size and age assessment among advertising male chorus frogs. Journal of Herpetology 27:109–111.

Plissner, J. H., and S. M. Haig. 1997. 1996 International piping plover census. Report located at U.S. Geological Survey-Biological Resources Division, Forest and Rangeland Ecosystem Science Center, Corvallis, Oregon, USA.

Plummer, M. V. 1981. Habitat utilization, diet and movements of a temperate arboreal snake (*Opheodrys aestivus*). Journal of Herpetology 15:425–432.

Plummer, M. V. 1985a. Demography of green snakes (*Opheodrys aestivus*). Herpetologica 41:373–381.

Plummer, M. V. 1985b. Growth and maturity in green snakes (*Opheodrys aestivus*). Herpetologica 41:28–33.

Plummer, M. V. 1987. Geographic variation in body size of green snakes (*Opheodrys aestivus*). Copeia 1987: 483–485.

Plummer, M. V. 1990. Nesting movements, nesting behavior, and nest sites of green snakes (*Opheodrys aestivus*) revealed by radiotelemetry. Herpetologica 46:190–195.

Plummer, M. V. 1996. *Heterodon platirhinos* (Eastern hognose snake). Defensive behavior. Herpetological Review 27:2–4.

Plummer, M. V., and N. E. Mills. 1996. Observations on trailing and mating behaviors in hognose snakes (*Heterodon platirhinos*). Journal of Herpetology 30:80–82.

Plummer, M. V., and N. E. Mills. 2000. Spatial ecology and survivorship of resident and translocated hognose snakes (*Heterodon platirhinos*). Journal of Herpetology 34:565–575.

Plummer, M. V., and H. L. Snell. 1988. Nest site selection and water relations of eggs in the snake, *Opheodrys aestivus*. Copeia 1988:58–64.

Pluto, T. G., and E. D. Bellis. 1986. Habitat utilization by the turtle, *Graptemys geographica*, along a river. Journal of Herpetology 20:22–31.

Pluto, T. G., and E. D. Bellis. 1988. Seasonal and annual movements of riverine map turtles, *Graptemys geographica*. Journal of Herpetology 22:152–158.

Polder, E. 1968. Spotted skunk and weasel populations den and cover usage by northeast Iowa. Iowa Academy of Science 75:142–146.

Polechla, P. 1990. Action plan for North American otters. Pages 74–79 *in* P. Foster-Turley, S. Macdonald, and C. Mason, editors. Otters: an action plan for their conservation. Kelvyn Press, Broadview, Illinois, USA.

Poole, A. F., R. O. Bierregaard, and M. S. Marshall. 2002. Osprey (*Pandion haliaetus*). *In* A. Poole and F. Gill, editors. The birds of North America, 683. Birds of North America, Philadelphia, Pennsylvania, USA.

Poole, E. L. 1932. A survey of mammals of Berks County Pennsylvania. Reading Museum and Art Gallery Bulletin Number 13, Reading, Pennsylvania, USA.

Poole, E. L. 1936. Notes on the young of the Allegheny wood rat. Journal of Mammalogy 17:22–26.

Poole, E. L. 1940. A life history sketch of the Allegheny woodrat. Journal of Mammalogy 21:249–270.

Poole, E. L. 1964. Pennsylvania birds, an annotated list. Livingston Publishing Co., Narberth, Pennsylvania, USA.

Poole, E. L. Manuscript on Pennsylvania birds. Unpublished. Housed in the ornithology department of the

Academy of Natural Sciences of Philadelphia, Pennsylvania, USA.

Pope, C. H. 1928. Some plethodontid salamanders from North Carolina and Kentucky with the description of a new race of *Leurognathus*. American Museum Novitates 306:1–19.

Popotnik, G. J., and W. M. Giuliano. 2000. Response of birds to grazing of riparian zones. Journal of Wildlife Management 64:976–982.

Possessky, S. L., C. E. Williams, and W. J. Moriarity. 2000. Glossy buckthorn, *Rhamnus frangula L.*: a threat to riparian plant communities of the northern Allegheny Plateau (USA). Natural Areas Journal 20:290–292.

Post, W., and C. A. Seals. 2000. Breeding biology of the common moorhen in an impounded cattail marsh. Journal of Field Ornithology 71:437–442.

Poulin, R. G., S. D. Grindal, and R. M. Brigham. 1996. Common nighthawk. *In* A. Poole and F. Gill, editors. The birds of North America, 213. The Birds of North America, Philadelphia, Pennsylvania, USA.

Powell, R. A. 1993. The fisher: life history, ecology, and behavior. University of Minnesota, Minneapolis, USA.

Prescott, D. R. C. 1987b. Territorial responses to song playback in allopatric and sympatric populations of alder (*Empidonax alnorum*) and willow (*E. traillii*) flycatchers. Wilson Bulletin 99:611–619.

Prescott, D. R. C. 1987a. Yellow-bellied flycatcher. Pages 254–255 *in* M. D. Cadman, P. F. J. Eagles, and F. M. Helleiner, editors. Atlas of the breeding birds of Ontario. University of Waterloo Press, Ontario, Canada.

Prescott, D. R. C., and A. L. A. Middleton. 1988. Feeding-time minimization and the territorial behavior of the willow flycatcher (*Empidonax traillii*). Auk 105:17–28.

Price, E. 2002. Piping plover (*Charadrius melodus*) recolonization potential in the Great Lakes: assessment of historic habitat and dispersal events. Thesis, University of Minnesota, Minneapolis, USA.

Price, J., S. Droege, and A. Price. 1995. Summer atlas of North American birds. Academic Press, London.

Price, P. W., M. Westoby, and B. Rice. 1988. Parasite-mediated competition—some predictions and tests. American Naturalist 131:544–555.

Prosser, D. J., and R. P. Brooks. 1998. A verified habitat suitability index for the Louisiana waterthrush. Journal of Field Ornithology 69:288–298.

Proulx, G., K. Aubry, J. Birks, S. Buskirk, C. Fortin, H. Frost, W. Krohn, L. Mayo, V. Monahov, D. Payer, M. Saeki, M. Santos-Reis, R. Weir, and W. Zielnski. 2004. World distribution and status of the genus *Martes*. Pages 21–76 *in* D. J. Harrison, A. K. Fuller, and G. Proulx, editors. Martens and fishers (*Martes*) in human-altered environments: an international perspective. Springer Science and Business Media, New York, New York, USA.

Pruitt, L. 1996. Henslow's sparrow status assessment. Bloomington, Indiana: Bloomington Ecological Services Field Office, U.S. Fish and Wildlife Service.

Punzo, F. 2004. Early-life nutritional environment and spatial navigation in the water shrew, *Sorex palustris* (Insectivora). Journal of Environmental Biology 25:403–411.

Quinn, N. W. S., and D. P. Tate. 1991. Seasonal movement and habitat of wood turtles (*Clemmys insculpta*) in Algonquin Park, Canada. Journal of Herpetology 25:217–220.

Raesly, R. L., and J. E. Gates. 1987. Winter habitat selection by north temperate cave bats. The American Midland Naturalist 118:15–31.

Raffaele, H., J. Wiley, O. Garrido, A. Keith, and J. Raffaele. 1998. A guide to the birds of the West Indies. Princeton University Press, Princeton, New Jersey, USA.

Raine, R. M. 1987. Winter food habits and foraging behaviour of fishers (*Martes pennanti*) and martens (*Martes americana*) in southeastern Manitoba. Canadian Journal of Zoology 65:745–747.

Ralph, C. J., G. R. Geupel, P. Pyle, T. E. Martin, and D. F. DeSante. 1993. Handbook of field methods for monitoring landbirds. General Technical Report PSW-GTW-144. Pacific Southwest Research Station, Forest Service, U.S. Department of Agriculture, Albany, California, USA.

Ralph, C. J., J. R. Sauer, and S. Droege, editors. 1995. Monitoring bird populations by point count. U.S. Forest Service General Technical Report PSW-GTR-149 Albany, California, USA.

Ramos, M. A., and D. W. Warner. 1980. Ecological aspects of migrant bird behavior in Veracruz, Mexico. Pages 353–394 *in* A. Keast and E. S. Morton, editors. Migrant birds in the Neotropics: ecology, behavior, distribution, and conservation. Smithsonian Institute Press, Washington, D.C., USA.

Ramsden, D. J. 1998. Effects of barn conversions on local populations of barn owl *Tyto alba*. Bird Study 45:68–76.

Rand, A. L. 1966. The snipe rediscovered. Audubon Magazine 68:351–354.

Randall, T. E. 1925. Abnormally large clutches of eggs of short-eared owl (*Asio flammeus*). Canadian Field-Naturalist Monographs 39:194.

Rappole, J. H. 2002. Birds of the Mid-Atlantic region and where to find them. John Hopkins University Press, Baltimore, Maryland, USA.

Rappole, J. H., and M. V. McDonald. 1994. Cause and effect in population declines of migratory birds. Auk 111:652–660.

Rappole, J. H., M. A. Ramos, R. J. Oehlenschlager, D. W. Warner, and C. P. Barkan. 1979. Timing of migration and route selection in North American songbirds. Pages 199–214 *in* D. L. Drawe, editor. Proceedings of First Welder Wildlife Foundation Symposium, Sinton, Texas, USA.

Rappole, J. H., and D. W. Warner. 1980. Ecological aspects of migrant bird behavior in Veracruz, Mexico. Pages

353–393 *in* A. Keast and E. S. Morton, editors. Migrant birds in the Neotropics: ecology, behavior, distribution, and conservation. Smithsonian Institution Press, Washington, D.C., USA.

Rastall, K., C. Kondo, and J. S. Strazanac. 2003. Lethal effects of biological insecticide applications on nontarget Lepidopterans in two Appalachian forests. Environmental Entomology 32:1364–1369.

Ratcliffe, D. 1980. The peregrine falcon. Buteo Books, Vermillion, South Dakota, USA.

Ratcliffe, D. 1993. The peregrine falcon. Second edition. Calton: T & A D Poyser, London, United Kingdom.

Rattner, B. A., and P. C. McGowan. 2007. Potential hazards of environmental contaminants to avifauna residing in the Chesapeake Bay Estuary. Waterbirds 30:63–81.

Raub, M.W. 1892. The golden eagle in Pennsylvania. Auk 9:200.

Record, J. 1995. An amphibian and reptile survey of Great Swamp National Wildlife Refuge emphasizing endangered and threatened species with management recommendations. Thesis, State University of New York, Syracuse, USA.

Reed, J. M. 1992. A system of ranking conservation priorities for Neotropical migrant birds based on relative susceptibility to extinction. Pages 524–536 *in* J. M. Hagan III and D. W. Johnston, editors. Ecology and conservation of Neotropical migrant landbirds. Smithsonian Institution Press, Washington, D.C., USA.

Reed, R. N., and J. W. Gibbons. 2002. Conservation status of live United States nonmarine turtles in domestic and international trade. Report to Division of Scientific Authority, United States Fish and Wildlife Service, Arlington, Virginia, USA.

Reeder, D. M., and G. R. Turner. 2008. Working together to combat white-nose syndrome: a report of a meeting on 9–11 June 2008 in Albany, New York. Bat Research News 49:75–78.

Reese, J. G. 1972. A Chesapeake barn owl population. Auk 89:106–104.

Reese, J. G. 1996. Whip-poor-will. Pages 194–195 *in* C. S. Robbins, editor. Atlas of the breeding birds of Maryland and the District of Columbia. University of Pittsburgh Press, Pittsburgh, Pennsylvania, USA.

Reid, D. G., T. E. Code, A. C. H. Reid, and S. M. Herrero. 1994. Spacing, movements, and habitat selection of the river otter in boreal Alberta. Canadian Journal of Zoology 72:1314–1324.

Reid, F. A., B. Meanley, and L. H. Fredrickson. 1994. King rail. Pages 180–191 *in* T. C. Tacha and C. E. Braun, editors. Migratory shore and game bird management in North America. Allen Press, Lawrence, Kansas, USA.

Reid, W. 1992. Marsh wren. Pages 258–259 *in* D. W. Brauning, editor. Atlas of breeding birds in Pennsylvania. University of Pittsburgh Press, Pittsburgh, Pennsylvania, USA.

Reimann, E. J. 1947. Summer birds of Tamarack Swamp, 1900 and 1947. Cassinia 36:17–24.

Reinert, H. K. 1975. Another winter record of a snake (*Natrix septemvittata*). Bulletin of the Philadelphia Herpetological Society 23:7.

Reinert, H. K. 1978. The ecology and morphological variation of the Massasauga rattlesnake, *Sistrurus catenatus*. Thesis, Clarion State College, Clarion, Pennsylvania, USA.

Reinert, H. K. 1981. Reproduction by the massasauga, *Sistrurus catenatus catenatus*. American Midland Naturalist 105:393–395.

Reinert, H. K. 1984a. Habitat separation between sympatric snake populations. Ecology 65:478–486.

Reinert, H. K. 1984b. Habitat variation within sympatric snake populations. Ecology 65:1673–1682.

Reinert, H. K. 1985. Eastern massasauga, *Sistrurus catenatus catenatus* (Rafinesque). Special Publication of the Carnegie Museum of Natural History 11:273–275.

Reinert, H. K. 1990. A profile and impact assessment of organized rattlesnake hunts in Pennsylvania. Journal of the Pennsylvania Academy of Science 64:136–144.

Reinert, H. K. 1991. The spatial ecology of timber rattlesnakes (*Crotalus horridus*). Program of the Joint Annual Meeting of the Society for the Study of Amphibians and Reptiles/Herpetologists League, Pennsylvania State University, USA. (Abstract)

Reinert, H. K. 1993. Habitat selection in snakes. Pages 201–240 *in* R. A. Seigel and J. T. Collins, editors. Snakes: ecology and behavior. McGraw-Hill, New York, New York, USA.

Reinert, H. K. 2005. A telemetric study of the survivorship, behavior, and spatial ecology of neonatal timber rattlesnakes, *Crotalus horridus*. Program of the Biology of Rattlesnakes Symposium, Loma Linda University, California. (Abstract.)

Reinert, H. K., and L. M. Bushar. 1992. The massasauga rattlesnake in Pennsylvania: continuing habitat loss and population isolation. Pages 55–59 *in* B. Johnson and V. Menzies, editors. International Symposium and Workshop on the Conservation of the Eastern Massasauga Rattlesnake, *Sistrurus catenatus catenatus*. Metro Toronto Zoo, Ontario, Canada.

Reinert, H. K., D. Cundall, and L. M. Bushar. 1984. Foraging behavior of the timber rattlesnake, *Crotalus horridus*. Copeia 1984:976–981.

Reinert, H. K., and W. R. Kodrich. 1982. Movements and habitat utilization by the Massasauga, *Sistrurus catenatus catenatus*. Journal of Herpetology 16:162–171.

Reinert, H. K., and R. R. Rupert, Jr. 1999. Impacts of translocation on behavior and survival of timber rattlesnakes, *Crotalus horridus*. Journal of Herpetology 33:45–61.

Reinert, H. K., and R. T. Zappalorti. 1988a. Timber rattlesnakes (*Crotalus horridus*) of the pine barrens: their movement patterns and habitat preference. Copeia 1988:964–978.

Reinert, H. K., and R. T. Zappalorti. 1988b. Field observation of the association of adult and neonatal timber rattlesnakes, *Crotalus horridus*, with possible evidence for conspecific trailing. Copeia 1988:1057–1059.

Relyea, R. A. 2004. Growth and survival of five amphibian species exposed to combinations of pesticides. Environmental Toxicology and Chemistry 23:1737–1742.

Relyea, R., and N. Diecks. 2008. An unforeseen chain of events for amphibians: lethal effects of pesticides at sublethal concentrations. Ecological Applications 18:1728–1742.

Reynolds, K. D., T. R. Rainwater, E. J. Scollon, S. S. Sathe, B. M. Adair, K. R. Dixon, G. P. Cobb, and S. T. McMurry. 2001. Accumulation of DDT and mercury in prothonotary warblers (*Protonotaria citrea*) foraging in a heterogeneously contaminated environment. Environmental Toxicology and Chemistry 20:2903–2909.

Reynolds, R. T., E. C. Meslow, and H. M. Wight. 1982. Nesting habitat of coexisting accipiter in Oregon. Journal of Wildlife Management 46:124–138.

Reynolds, R. T., and H. Wight. 1978. Distribution, density and productivity of accipiter hawks breeding in Oregon. Wilson Bulletin 90:182–196.

Rhoads, S. 1903. Mammals of Pennsylvania and New Jersey. Privately printed, Philadelphia, Pennsylvania, USA.

Rhoads, S. N. 1903. Exit the dickcissel—a remarkable case of local extinction. Proceedings of the Delaware Valley Ornithological Club 7:17–28.

Rice, J. N. 1969. The decline of the peregrine population in Pennsylvania. Pages 155–163 *in* J. J. Hickey, editor. Peregrine falcon population: their biology and decline. University of Wisconsin Press, Madison, USA.

Rich, A. C., D. S. Dobkin, and L. J. Niles. 1994. Defining forest fragmentation by corridor width: the influence of narrow forest dividing corridors on forest-nesting birds in southern New Jersey. Conservation Biology 8:1109–1121.

Rich, T. D., C. J. Beardmore, H. Berlanga, P. J. Blancher, M. S. W. Bradstreet, G. S. Butcher, D. W. Demarest, E. H. Dunn, W. C. Hunter, E. E. Inogo-Elias, J. A. Kennedy, A. M. Martell, A. O. Panjabi, D. N. Pashley, K. V. Rosenberg, C. M. Rustay, J. S. Wendt, and T. C. Will. 2004. Partners in Flight North American landbird conservation plan. Cornell Laboratory of Ornithology, Ithaca, New York, USA.

Richards, K. C. 1976. Some declining bird species of southeastern Pennsylvania. Cassinia 55:33–36.

Richmond, N. D. 1952. First record of the green salamander in Pennsylvania, and other range extensions in Pennsylvania, Virginia and West Virginia. Annals of Carnegie Museum 32:313–318.

Richmond, N. D. 1954. The ground snake, *Haldea valeriae* in Pennsylvania and West Virginia with description of new subspecies. Annals of Carnegie Museum 33:251–260.

Richmond, N. D., and H. R. Roslund. 1949. Mammal survey of northwestern Pennsylvania. Final Report, Pittman-Robertson Project 20-R, Pennsylvania Game Commission, Harrisburg, Pennsylvania, USA.

Richter, A. R., S. R. Humphrey, J. B. Cope, and V. Brack. 1993. Modified cave entrances: thermal effect on body mass and resulting decline of the endangered Indiana bats (*Myotis sodalis*). Conservation Biology 7:407–415.

Ridgely, R. S., and J. A. Gwynne, Jr. 1989. A guide to the birds of Panama, with Costa Rica, Nicaragua and Honduras. Princeton University Press, Princeton, New Jersey, USA.

Ridgely, R. S., and G. Tudor. 1989. The birds of South America. Volume 1. The oscine passerines. University of Texas Press, Austin, USA.

Ridgely, R. S., and G. Tudor. 1994. The birds of South America. Volume 2. The suboscine passerines. University of Texas, Austin, USA.

Riegner, M. F. 1982. The diet of yellow-crowned night-herons in the eastern and southern United States. Colonial Waterbirds 5:173–176.

Rimmer, C. C., K. P. McFarland, D. C. Evers, E. K. Miller, Y. Aubry, D. Busby, and R. J. Taylor. 2005. Mercury concentrations in Bicknell's thrush and other insectivorous passerines in montane forests of Northeastern North America. Ecotoxicology 14:223–240.

Ringleman, J. K., and J. R. Longcore. 1982. Survival of juvenile black ducks during brood rearing. Journal of Wildlife Management 46:622–628.

Rising, J. D. 1996. The sparrows of the United States and Canada. Academic Press, San Diego, California, USA.

Rittman, S. E., E. Muths, and D. E. Green. 2003. *Pseudacris triseriata* (Western Chorus Frog) and *Rana sylvatica* (Wood Frog). Chytridiomycosis. Herpetological Review 34:53.

Robbins, A. E. 1994. Olive-sided flycatcher, *Contopus borealis*. Pages 144–145 *in* C. R. Foss, editor. Atlas of breeding birds in New Hampshire. Audubon Society of New Hampshire, Dover, USA.

Robbins, C. S. 1980. Effect of forest fragmentation on breeding bird populations in the Piedmont of the mid-Atlantic region. Atlantic Naturalist 33:31–36.

Robbins, C. S. 1986. Conservation of migratory raptors: an overview based on fifty years of raptor banding. Pages 26–34 *in* S. E. Senner, C. M. White, and J. R. Parrish, editors. Raptor conservation in the next 50 years. Raptor Research Reports, 5, Raptor Research Foundation, Hastings, Minnesota, USA, and Proceedings of Raptor Conference at Hawk Mountain Sanctuary, Kempton, Pennsylvania, USA.

Robbins, C. S., D. Bystrak, and P. H. Geissler. 1986. The breeding bird survey: its first fifteen years, 1965–1979. U.S. Department of Interior Fish and Wildlife Service Research Publication 157, U.S. Fish and Wildlife Service, Washington, D.C., USA.

Robbins, C. S., D. K. Dawson, and B. A. Dowell. 1989. Habi-

tat area requirements of breeding forest birds of the Middle Atlantic States. Wildlife Monographs 103:1–34.
Robbins, C. S., J. W. Fitzpatrick, and P. Hamel. 1992. A warbler in trouble: *Dendroica cerulea*. Pages 549–562 *in* J. M. Hagan III and D. W. Johnston, editors. Ecology and conservation of Neotropical migrant landbirds. Smithsonian Institute Press, Washington, D.C., USA.
Robbins, L. W., J. G. Boyles, B. M. Mormann, and M. B. Milam. 2004. Fall, winter and spring roosting behavior of eastern red bats and evening bats in Missouri. Bat Research News 45:69.
Robbins, M. B., and D. A. Easterla. 1992. Birds of Missouri: their distribution and abundance. University of Missouri Press, Columbia, USA.
Robbins, S. D., Jr. 1991. Wisconsin birdlife: population and distribution past and present. University of Wisconsin Press, Madison, USA.
Robel, R. J. 1993. Symposium wrap-up: what is missing? Pages 156–158 *in* K. E. Church and T. V. Dailey, editors. Quail III: National Quail Symposium. Kansas Department of Wildlife and Parks, Pratt, USA.
Roberts, C. and C. J. Norment. 1999. Effects of plot size and habitat characteristics on breeding success of scarlet tanagers. Auk 116:73–82.
Roberts, H. A., and R. C. Early. 1952. Mammal survey of southeastern Pennsylvania. Final Report, Pittman-Robertson Project 43-R, Pennsylvania Game Commission, Harrisburg, Pennsylvania, USA.
Robertson, B., and K. V. Rosenberg. 2003. Partners in Flight landbird conservation plan: Physiographic Area 24: Allegheny Plateau. Version 1.1. <www.blm.gov/wildlife/pl_24sum.htm>.
Robertson, B., and K. V. Rosenberg. 2003. Partners in Flight Bird Conservation Plan, Allegheny Plateau (Physiographic Area 24). Version 1.1. Cornell Laboratory of Ornithology, Ithaca, New York, USA.
Robinson, S. K. 1992. Population dynamics of breeding Neotropical migrants in a fragmented Illinois landscape. Pages 408–418 *in* J. M. Hagan III and D. W. Johnson, editors. Ecology and conservation of Neotropical migrant landbirds. Smithsonian Institution Press, Washington, D.C., USA.
Robinson, S. K., and W. D. Robinson. 2001. Avian nesting success in a selectively harvested north temperate deciduous forest. Conservation Biology 15:1763–1771.
Robinson, S. K, F. R. Thompson III, T. M. Donovan, D. R. Whitehead, and J. Faaborg. 1995. Regional forest fragmentation and the nesting success of migratory birds. Science 267:1987–1990.
Robinson, W. D. 1995. Louisiana waterthrush (*Seiurus motacilla*). *In* A. Poole and F. Gill, editors. Birds of North America, 151. Academy of Natural Sciences, Philadelphia, Pennsylvania, USA, and the American Ornithologists' Union, Washington, D.C., USA.
Robinson, W. D. 1996. Summer tanager (*Piranga rubra*). Page 248 *in* A. Poole and F. Gill, editors. The birds of North America. Birds of North America, Philadelphia, Pennsylvania, USA.
Robinson, W. D., and S. K. Robinson. 1999. Effects of selective logging on forest bird populations in a fragmented landscape. Conservation Biology 13:58–66.
Roche, B. 2002. Committee on the status of endangered wildlife in Canada status report on the northern map turtle, *Graptemys geographica* in Canada. Committee on the Status of Endangered Wildlife in Canada, Ottawa.
Roddy, H. J. 1928. Reptiles of Lancaster County and the state of Pennsylvania. Science Press, Lancaster, Pennsylvania, USA.
Rodewald, A. D. 2004. Landscape and local level influences of forest management on cerulean warblers in Pennsylvania. Pages 472–477 *in* D. Yaussy, D. N. Hix, R. P. Long, and P. C. Goebel, editors. General Technical Report NE-316, Proceedings of the 14th Central Hardwoods Forest Conference. U.S. Forest Service, Northeastern Research Station, Newtown Square, Pennsylvania, USA.
Rodewald, A. D., and R. H. Yahner. 2000. Bird communities associated with harvested hardwood stands containing residual trees. Journal of Wildlife Management 64:924–932.
Rodewald, P. G. 1997. Two new breeding species of wood-warblers (Parulinae) in Arkansas. Southwestern Naturalist 42:106.
Rodewald, P. G., and M. C. Brittingham. 2004. Stopover habitats of landbirds during fall use of edge-dominated and early-successional forests. Auk 121:1040–1055.
Rodewald, P. G., and R. D. James. 1996. Yellow-throated Vireo (*Vireo flavifrons*). *In* A. Poole and F. Gill, editors. The birds of North America, 247. Academy of Natural Sciences, Philadelphia, Pennsylvania, USA, and American Ornithologists' Union, Washington, D.C, USA.
Rogers, C. M. 2006. Nesting success and breeding biology of cerulean warblers in Michigan. Wilson Journal of Ornithology 118:145–151.
Rogers, J. A., Jr. 1983. Foraging behavior of seven species of herons in Tampa Bay, Florida. Colonial Waterbirds 6:11–23.
Rohrbaugh, R. W., Jr., D. L. Reinking, D. H. Wolfe, S. K. Sherrod, and M. A. Jenkins. 1999. Effects of prescribed burning and grazing on nesting and reproductive success of three grassland passerine species in tallgrass prairie. Studies in Avian Biology 19:165–170.
Romano, W. B. 2007. Habitat selection and foraging behavior of great egrets (*Ardea alba*) in a riparian environment on the Susquehanna River in Harrisburg, Pennsylvania. Masters Thesis, East Stroudsburg University, East Stroudsburg, Pennsylvania, USA.
Rose, R. K. 1981. *Synaptomys* not extinct in the Dismal Swamp. Journal of Mammalogy 62:844–845.
Rose, R. K. 2005. The small mammals of Isle of Wight County, Virginia, as revealed by pitfall trapping. Virginia Journal of Science 56:83–92.
Rose, R. K., and A. M. Spevak. 1978. Aggressive behavior in

two sympatric microtine rodents. Journal of Mammalogy 59:213–216.

Roseberry, J. L., and W. D. Klimstra. 1984. Population ecology of the bobwhite. Southern Illinois University Press, Carbondale, USA.

Rosenberg, K. V. 2003a. Partners in Flight Landbird Conservation Plan: Physiographic Area 17: Northern Ridge and Valley. Version 1.0. June 2003.

Rosenberg, K. V. 2003b. Partners in Flight Landbird Conservation Plan: Physiographic Area 22: Ohio Hills. Draft.

Rosenberg, K. V. 2004. Partners in Flight continental priorities and objectives defined at the state and bird conservation region level—Pennsylvania. Cornell University Laboratory of Ornithology, Ithaca, New York, USA.

Rosenberg, K. V., S. E. Barker, and R. W. Rohrbaugh. 2000. An atlas of cerulean warbler populations. Final report to USFWS: 1997–2000 breeding seasons. Cornell Laboratory of Ornithology, Ithaca, New York, USA.

Rosenberg, K. V., and R. Dettmers. 2004. Partners in Flight Landbird Conservation Plan: Physiographic Area 22: Ohio Hills. American Bird Conservancy, Washington, D.C., USA.

Rosenberg, K. V., R. S. Hames, R. W. Rohrgaugh, Jr., S. Barker Swarthout, J. D. Lowe, and A. A. Dhondt. 2003. A land manager's guide to improving habitat for forest thrushes. The Cornell Lab of Ornithology, Ithaca, New York, USA.

Rosenberg, K. V., J. D. Lowe, and A. A, Dhondt. 1999a. Effects of forest fragmentation on breeding tanagers: a continental perspective. Conservation Biology 13:568–583.

Rosenberg, K. V., R. D. Ohmart, W. C. Hunter, and B. W. Anderson. 1991. Birds of the lower Colorado River Valley. University of Arizona Press, Tucson, USA.

Rosenberg, K. V., and M. G. Raphael. 1986. Effects of forest fragmentation on vertebrates in Douglas-fir forests. Pages 263–272 *in* J. Verner, M. L. Morrison, and C. J. Ralph, editors. Wildlife 2000: modeling habitat relationships of terrestrial vertebrates. University of Wisconsin Press, Madison, USA.

Rosenberg, K. V., R. W., Rohrbaugh, Jr., S. E. Barker, J. D. Lowe, R. S. Hames, and A. A. Dhondt. 1999b. A land manager's guide to improving habitat for scarlet tanagers and other forest interior birds. The Cornell Lab of Ornithology, Ithaca, New York, USA.

Rosenberg, K. V., and J. V. Wells. 1995. Importance of geographic areas to Neotropical migrant birds in the Northeast. Final report to the U.S. Fish and Wildlife Service, Region 5, Hadley, Massachusetts, USA.

Rosenberg, K. V., and J. V. Wells. 2000. Global perspectives on Neotropical migratory bird conservation in the Northeast: long-term responsibility vs. immediate concern. *In* R. Bonney, D. N. Pashley, R. J. Cooper, and L. Niles, editors. Strategies for bird conservation: the Partners in Flight planning process. Cornell Lab of Ornithology, Ithaca, New York, USA.

Rosenburg, C. 1992. Barn owl, *Tyro alba*. Pages 253–279 *in* K. J. Schneider and D. M. Pence, editors. Migratory non-game birds of Management Concern in the Northeast. U.S. Department of the Interior, Fish and Wildlife Service, Newton Center, Massachusetts, USA.

Rosene, W. 1969. The bobwhite quail, its life and management. Rutgers University Press, New Brunswick, New Jersey, USA.

Rosenfield, R. N. 1984. Nesting biology of broad-winged hawks in Wisconsin. Journal of Raptor Research 18:6–9.

Rosenfield, R. N., J. Bielefeldt, and R. K. Anderson. 1988. Effectiveness of broadcast calls for detecting breeding Cooper's hawks. Wildlife Society Bulletin 16:210–212.

Rosenfield, R. N., J. Bielefeldt, R. K. Anderson and J. M. Papp. 1991. Accipiters. Pages 42–49 *in* B. G. Pendleton, editor. Proceedings of the Midwest Raptor Management Symposium and Workshop. Scientific and Technical Series, 15. National Wildlife Federation, Washington, D.C., USA.

Rosenfield, R. N., M. W. Gratson, and L. B. Carson. 1984. Food brought by broad-winged hawks to a Wisconsin nest. Journal of Field Ornithology 55:246–247.

Roslund, H. R. 1951. Mammal survey of north central Pennsylvania. Final report. Pittman Robertson Project 37-R, Pennsylvania Game Commission, Harrisburg, USA.

Ross, B. D., M. L. Morrison, W. Hoffman, T. S. Fredericksen, R. J. Sawicki, E. Ross, M. B. Lester, J. Beyea, and B. N. Johnson. 2001. Bird relationships to habitat characteristics created by timber harvesting in Pennsylvania. Journal of the Pennsylvania Academy of Sciences 74:71–84.

Ross, D. A., and R. K. Anderson. 1990. Habitat use, movements, and nesting of *Emydoidea blandingi* in central Wisconsin. Journal of Herpetology 24:6–12.

Ross, D. A., K. N. Brewster, R. K. Anderson, N. Ratner, and C. M. Brewster. 1991. Aspects of the ecology of wood turtles, *Clemmys insculpta*, in Wisconsin. Canadian Field-Naturalist 105:363–367.

Ross, R. M. 1990. Saving a heron rookery. Bird Watcher's Digest 12:62–67.

Ross, R. M., L. A. Redell, R. M. Bennett, and J. A. Young. 2004. Mesohabitat use of threatened hemlock forests by breeding birds of the Delaware River Basin in northeastern United States. Natural Areas Journal 24:307–315.

Rossman, D. A. 1963. The colubrid snake genus *Thamnophis:* a revision of the *sauritus* group. Bulletin of the Florida State Museum 7:99–178.

Rossman, D. A., N. B. Ford, and R. A. Seigel. 1996. The garter snakes evolution and ecology. University of Oklahoma Press, Norman, USA.

Roth, A. M., and S. Lutz. 2004. Relationship between territorial male golden-winged warblers in managed aspen stands in northern Wisconsin, USA. Forest Science 50:153–161.

Roth, R. R., M. S. Johnson, and T. J. Underwood. 1996.

Wood thrush (*Hylocichla mustelina*). *In* A. Poole and F. Gill, editors. The birds of North America, 246. Academy of Natural Sciences, Philadelphia, Pennsylvania, USA, and American Ornithologists Union, Washington, D.C., USA.

Roth, R. R., and R. K. Johnson. 1993. Long-term dynamics of a wood thrush population breeding in a forest fragment. Auk 110:37–48.

Rothstein, S. I. 1975. An experimental and teleonomic investigation of avian brood parasitism. Condor 77:250–271.

Roux, K., and P. Marra. 2007. The presence and impact of environmental lead in passerine birds along an urban to rural land use gradient. Archives of Environmental Contamination and Toxicology 53:261–268.

Rowe, C. L., and W. A. Dunson. 1993. Relationships among abiotic parameters and breeding effort by three amphibians in temporary wetlands of central Pennsylvania. Wetlands 13:7–10.

Rowe, C. L., W. J. Sadinski, and W. A. Dunson. 1992. Effects of acute and chronic acidifications on three larval amphibians that breed in temporary ponds. Archives of Environmental Contamination and Toxicology 23:339–350.

Rowe, J. W. 1992. Dietary habits of the Blanding's turtle (*Emydoidea blandingi*) in northeastern Illinois. Journal of Herpetology 26:111–114.

Rowe, J. W., and E. D. Moll. 1991. A radiotelemetric study of activity and movements of the Blanding's turtle (*Emydoidea blandingi*) in northeastern Illinois. Journal of Herpetology 25:178–185.

Roy, K. D. 1991. Ecology of reintroduced fishers in the Cabinet Mountains of northwestern Montana. Thesis, University of Montana, Missoula, USA.

Rozanski, B., S. Gerst, and M. Ezdebski. 1996. Northern water shrews as indicators of habitat quality in sensitive ecosystems of northeastern Pennsylvania. Unpublished Undergraduate Thesis, Wilkes University, Wilkes-Barre, Pennsylvania, USA.

Rubbo, M. J., and J. M. Kiesecker. 2005. Amphibian breeding distribution in an urbanized landscape. Conservation Biology 19:504–511.

Ruedas, L. A., R. C. Dowler, and E. Aita. 1989. Chromosomal variation in the New England cottontail, *Sylvilagus transitionalis*. Journal of Mammalogy 70:860–864.

Ruelas-Inzunza, E. 2005. Raptor migration through Veracruz, Mexico. Dissertation, University of Missouri–Columbia, Columbia, USA.

Ruiz, G., M. Rosemann, F. F. Novoa, and P. Sabat. 2002. Hematological parameters and stress index in rufous-collared sparrows dwelling in urban environments. Condor 104:162–166.

Rumsey, R. L. 1970. Woodpecker nest failures in creosoted utility poles. Auk 87:367–369.

Rusch, D. H., and P. D. Doerr. 1972. Broad-winged hawk nesting and food habits. Auk 89:139–145.

Russell, A. L., C. M. Butchkoski, L. Saidak, and G. F. McCracken. 2008. Road-killed bats, highway design, and the commuting ecology of bats. Endangered Species Research, esr 00121.

Ryan, T. J., and C. T. Winne. 2001. Effects of hydroperiod on metamorphosis in *Rana sphenocephala*. American Midland Naturalist 145:46–53.

Sabo, S. R. 1980. Niche and habitat relations in subalpine bird communities of the White Mountains of New Hampshire. Ecological Monographs. 50:241–259.

Sabo, S. R., and R. T. Holmes. 1983. Foraging niches and the structure of forest bird communities in contrasting montane habitats. Condor 85:121–138.

Sadinski, W. J., and W. A. Dunson. 1992. A multilevel study of effects of low pH on amphibians of temporary ponds. Journal of Herpetology 26:413–422.

Sallabanks, R. 1993. Species management abstract: prothonotary warbler. The Nature Conservancy. <http://conserveonline.org/2001/05/m/en/prow.doc>. Accessed 8 May 2005.

Sample, B. E., R. J. Cooper, and R. C. Whitmore. 1993. Dietary shifts among songbirds from a diflubenzuron-treated forest. Condor 95:616–624.

Samuels, M., and D. L. Elliot. 2000. The costs of sprawl in Pennsylvania. 10,000 Friends of Pennsylvania, Philadelphia, USA.

Sandeen, J. L., D. M. Tollini, C. Skinner, and D. W. Brauning. 1999. Project annual job report: black tern nesting ecology and wetland delineation. Unpublished report. Pennsylvania Game Commission, Harrisburg, USA.

Sanders, J. S., and J. S. Jacob. 1981. Thermal ecology of the copperhead (*Agkistrodon contortrix*). Herpetologica 37:264–270.

Sandilands, A. P., and C. A. Campbell. 1988. Status report on the least bittern, *Ixobrychus exilis*. Committee on the Status of Endangered Wildlife in Canada. Ottawa, Ontario, Canada.

Santner, S. 1992a. Long-eared owl. Pages 162–163 *in* D. W. Brauning, editor. Atlas of breeding birds in Pennsylvania. University of Pittsburgh Press, Pittsburgh, Pennsylvania, USA.

Santner, S. 1992b. Whip-poor-will. Pages 172–173 *in* D. W. Brauning, editor. Atlas of breeding birds in Pennsylvania. University of Pittsburgh Press, Pittsburgh, Pennsylvania, USA.

Sauer, J. R., and J. B. Bortner. 1991. Population trends from the American woodcock singing-ground survey, 1970–1988. Journal of Wildlife Management 55:300–312.

Sauer, J. R., and J. F. Fallon. 2008. The North American breeding bird survey, results and analysis 1966–2007. Version 5.15.2008. U.S. Geological Survey Patuxent Wildlife Research Center, Laurel, Maryland, USA.

Sauer, J. R., J. E. Hines, and J. Fallon. 2005. The North American breeding bird survey, results and analysis 1966–2004. Version 6.2.2006. U.S. Geological Survey Patuxent Wildlife Research Center, Laurel, Maryland, USA.

Sauer, J. R., J. E. Hines, and J. Fallon. 2006. The North American breeding bird survey, results and analysis 1966–2005. Version 6.2.2006. U.S. Geological Survey Patuxent Wildlife Research Center, Laurel, Maryland, USA. <www.mbr-wrc.usgs.gov/bbs/bbs.html>. Accessed 6 Oct 2006.

Sauer, J. R., J. E. Hines, and J. Fallon. 2008. The North American breeding bird survey, results and analysis 1966–2007. Version 5.15.2008. U.S. Geological Survey Patuxent Wildlife Research Center, Laurel, Maryland, USA.

Saumure, R. A., and J. R. Bider. 1998. Impact of agricultural development on a population of wood turtles (*Clemmys insculpta*) in southern Quebec, Canada. Chelonian Conservation and Biology 3:43–47.

Saunders, A. A. 1913. A study of the nesting of the marsh hawk. Condor 15:99–104.

Saunders, A. A. 1927. The summer birds of the northern Adirondack Mountains. Roosevelt Wild Live Bulletin 5:319–504.

Saunders, A. A. 1938. Studies of breeding birds of the Allegany State Park. New York State Museum Bulletin, 318, Albany, New York, USA.

Sayre, M. W., and W. D. Rundle. 1984. Comparison of habitat use by migrant soras and Virginia rails. Journal of Wildlife Management 48:599–600.

Schaadt, C. P., and L. M. Rymon. 1983. The restoration of ospreys by hacking. Pages 299–305 *in* D. M. Bird, N. R. Seymour, and J. M. Gerrards, editors. Biology and management of bald eagles and ospreys. Harpell Press, Ste. Anne de Bellevue, Quebec, Canada.

Scheffer, V. B. 1992. Mammals of the Olympic National Park and vicinity (1949). Northwest Fauna 2:1–133.

Scheuhammer, A. M., C. A. Rogers, and D. Bond. 1999. Elevated lead exposure in American woodcock (*Scolopax minor*) in eastern Canada. Archives of Environmental Contamination and Toxicology 36:334–340.

Schmaljohann, H., F. Liechti, and E. Bachler. 2008. Quantification of bird migration by radar—a detection probability problem. Ibis 150:342–355.

Schmidt, K. A. 2003. Nest predation and population declines in Illinois songbirds: a case for mesopredator effects. Conservation Biology 17:1141–1150.

Schmidt, K. A., S. A. Rush, and R. S. Ostfeld. 2008. Wood thrush nest success and post-fledging survival across a temporal pulse of small mammal abundance in an oak forest. Journal of Animal Ecology 77:830–837.

Schmidt, K. A., and C. J. Whelan. 1999. The relative impacts of nest predation and brood parasitism on seasonal fecundity in songbirds. Conservation Biology 13:46–57.

Schuett, G. W. 1982. A copperhead (*Agkistrodon contortrix*) brood produced from autumn copulations. Copeia 1982:700–702.

Schuett, G. W., and J. C. Gillingham. 1986. Sperm storage and multiple paternity in the copperhead, *Agkistrodon contortrix*. Copeia 1986:807–811.

Schutsky, R. M. 1992a. Great egret. Pages 52–53 *in* D. W. Brauning, editor. Atlas of breeding birds in Pennsylvania. University of Pittsburgh Press, Pittsburgh, Pennsylvania, USA.

Schutsky, R. M. 1992b. Red-headed woodpecker. Pages 180–181 *in* D. W. Brauning, editor. Atlas of breeding birds in Pennsylvania. University of Pittsburgh Press, Pittsburgh, Pennsylvania, USA.

Schutsky, R. M. 1992c. Yellow-crowned night-heron. Pages 62–63 *in* D. W. Brauning, editor. Atlas of breeding birds in Pennsylvania. University of Pittsburgh Press, Pittsburgh, Pennsylvania, USA.

Schwalbe, G., and P. Schwalbe. 1993. Breeding season record of a short-eared owl. Pennsylvania Birds 7:53.

Schwalbe, P. W. 1992. Virginia rail. Pages 124–125 *in* D. W. Brauning, editor. Atlas of breeding birds in Pennsylvania. University of Pittsburgh Press, Pittsburgh, Pennsylvania, USA.

Schwalbe, P. W., and R. M. Ross. 1992. Great blue heron. Pages 50–51 *in* D. W. Brauning, editor. Atlas of breeding birds in Pennsylvania. University of Pittsburgh Press, Pittsburgh, Pennsylvania, USA.

Schwartz, M. W., L. R. Iverson, and A. M. Prasad. 2006. Predicting extinctions as a result of climate change. Ecology 87:1611–1615.

Sciascia, J. C., and E. Pehek. 1995. Small mammal and amphibian populations and their microhabitat preferences within selected hemlock ecosystems in the Delaware Water Gap National Recreation Area. Draft final report, U.S. Department of the Interior, National Park Service, Mid-Atlantic Region, Washington, D.C., USA.

Scott, D. 1997. The long-eared owl. Hawk and Owl Trust, London.

Scott, D. P., and R. H. Yahner. 1989. Winter habitat and browse use by snowshoe hares, *Lepus americanus*, in a marginal habitat in Pennsylvania. Canadian Field-Naturalist 103:560–563.

Scott, P. E., T. L. Devault, R. A. Bajema, and S. L. Lima. 2002. Grassland vegetation and bird abundances on reclaimed mid-western coal mines. Wildlife Society Bulletin 30:1006–1014.

Scott, S. L., editor. 1987. Field guide to birds of North America. Second edition. National Geographic Society, Washington, D.C., USA.

Sechler, F. 2004. Field surveys to Tioga State Forest, old possessions Road Bog, Tioga County, Pennsylvania, USA. Natural Heritage Program, The Nature Conservancy, F04SEC25.

Sedgwick, J. A. 2000. Willow flycatcher. The birds of North America, Volume 14:533. American Ornithologists' Union, Washington, D.C., USA, and Academy of Natural Sciences of Philadelphia, Pennsylvania, USA.

Sedgwick, J. A., and W. M. Iko. 1999. Costs of brown-headed cowbird parasitism to willow flycatchers. Studies in Avian Biology 18:167–181.

Sedgwick, J. A., and F. L. Knopf. 1992. Describing willow

flycatcher habitats: scale perspectives and gender differences. Condor 94:720–733.

Seigel, R. A. 1986. Ecology and conservation of an endangered rattlesnake, *Sistrurus catenatus*, in Missouri, USA. Biological Conservation 35:333–346.

Seigel, R. A., and H. S. Fitch. 1985. Annual variation in reproduction in snakes in a fluctuating environment. Journal of Animal Ecology 54:497–505.

Seigel, R. A., and C. A. Sheil. 1999. Population viability analysis: applications for the conservation of massasaugas. Pages 17–22 *in* B. Johnson and M. Wright, editors. Second International Symposium and Workshop on the Conservation of the Eastern Massasauga Rattlesnake, *Sistrurus catenatus catenatus*. Toronto Zoo, Toronto, Ontario.

Sekercioglu, C. H. 2002. Impacts of birdwatching on human and avian communities. Environmental Conservation 29:282–289.

Sekercioglu, C. H. 2006. Increasing awareness of avian ecological function. Trends in Ecology and Evolution 21:464–471.

Sekercioglu, C. H., S. H. Schneider, J. P. Fay, and S. R. Loarie. 2008. Climate change, elevational range shifts, and bird extinctions. Conservation Biology 22:140–150.

Semlitsch, R. D. 1998. Biological delineation of terrestrial buffer zones for pond breeding salamanders. Conservation Biology 12:1113–1119.

Semlitsch, R. D. 2003. Conservation of pond-breeding amphibians. Pages 8–23 *in* R. D. Semlitsch, editor. Amphibian conservation. Smithsonian, Washington, D.C., USA.

Semlitsch, R. D., and J. R. Bodie. 2003. Biological criteria for buffer zones around wetlands and riparian habitats for amphibians and reptiles. Conservation Biology 17:1219–1228.

Senner, S. E., and M. R. Fuller. 1989. Status and conservation of North American raptors migrating to Neotropics. Pages 53–58 *in* B. U. Meyburg and R. D. Chancellor, editors. Raptors in the modern world. Proceedings of the Third World Conference on Birds of Prey, Eilat, Israel, World Work Group, and Birds of Prey, Berlin, Germany.

Senner, S. E., and L. J. Goodrich. 1992. The broad-winged hawk. Pages 104–105 *in* D. Brauning, editor. Atlas of breeding birds in Pennsylvania. University of Pittsburgh Press, Pittsburgh, Pennsylvania, USA.

Sepik, G. F., and E. L. Derleth. 1993. Habitat use, home range size, and patterns of moves of the American woodcock in Maine. Pages 41–49 *in* J. R. Longcore and G. F. Sepik, editors. Proceedings of the Eighth American Woodcock Symposium. U.S. Fish and Wildlife Service Biological Report 16, Washington, D.C., USA.

Sepik, G. F., R. B. Owen, Jr., and T. J. Dwyer. 1983. The effect of drought on a local woodcock Population. Transactions of the Northeast Section of the Wildlife Society 40:1–8.

Serfass, T. L. 1991. Up from the bottom. Pennsylvania Game News 62:6–11.

Serfass, T. L. 1994. Conservation genetics and reintroduction strategies for river otters. Dissertation, Pennsylvania State University, University Park, USA.

Serfass, T. L. 2001. Expectations for the North American region from the Otter Action Plan. Pages 24–27 *in* C. Reuther and C. Santiapillai, editors. How to implement an otter action plan? Proceedings from Workshop IUCN/SSC Otter Specialist Group, Hankensbuttel, Germany.

Serfass, T. L., R. P. Brooks, and L. M. Rymon. 1993. Evidence of long-term survival and reproduction by translocated river otters, *Lutra canadensis*. Canadian Field-Naturalist 107:59–63.

Serfass, T. L., R. P. Brooks, L. M. Rymon, and O. E. Rhodes, Jr. 2003. River otters in Pennsylvania, USA: lessons for predator reintroduction. *In* J. W. H. Conroy, A. C. Gutleb, J. Ruiz-Olmo, and G. M. Yoxon, editors. Proceedings of the European Otter Conference "Return of the Otter in Europe Where and How." Journal of the International Otter Survival Fund, Number 2. International Otter Survival Fund, Broadford, Isle of Skye, Scotland (CD-Rom).

Serfass, T. L., R. P. Brooks, W. M. Tzilkowski, D. H. Mitcheltree, M. R. Dzialak, and T. J. Swimley. 2001. Fisher reintroduction in Pennsylvania: final report. Department of Biology, Frostburg State University, Frostburg, Maryland, USA.

Serfass, T. L., M. J. Lovallo, R. P. Brooks, A. H. Hayden, and D. H. Mitcheltree. 1999. Status and distribution of river otters in Pennsylvania following a reintroduction project. Journal of the Pennsylvania Academy of Science 73:10–14.

Serfass, T. L., and L. M. Rymon. 1985. Success of river otter reintroduced into Pine Creek drainage in northcentral Pennsylvania. Transactions Northeast Section of the Wildlife Society 24:28–40.

Serfass, T. L., L. M. Rymon, and R. P. Brooks. 1990. Feeding relationships of river otters in northeastern Pennsylvania. Transactions of the Northeast Section of the Wildlife Society 47:43–53.

Serfass, T. L., L. M. Rymon, and J. D. Hassinger. 1986. Development and progress of Pennsylvania's river otter reintroduction program. Pages 322–344 *in* S. K. Majumdar, F. J. Brenner, and A. F. Rhoads, editors. Endangered and threatened species programs in Pennsylvania and other states: causes, issues, and management. Pennsylvania Academy of Science, Easton, USA.

Serfass, T. L., M. T. Whary, R. L. Peper, R. P. Brooks, T. J. Swimley, W. R. Lawrence, and C. E. Rupprecht. 1995. Rabies in a river otter (*Lutra canadensis*) intended for reintroduction. Journal of Zoo and Wildlife Medicine 26:311–314.

Serie, J., and R. Raftovich. 2004. Atlantic Flyway waterfowl harvest and population survey data. U.S. Fish and Wildlife Service, Laurel, Maryland, USA.

Serie, J., and R. Raftovich. 2005. Atlantic Flyway waterfowl and harvest population survey data. U.S. Fish and Wildlife Service, Laurel, Maryland, USA.

Serie, J. R., D. Luszcz, and R. V. Raftovich. 2002. Population trends, productivity, and harvest of EP tundra swans. Waterbirds 25 (Special Publication) 1:32–36.

Serrao, J. 2000. The reptiles and amphibians of the Poconos and northeastern Pennsylvania. Llewellyn and McKane, Wilkes-Barre, Pennsylvania, USA.

Shaffer L. L. 1991. Pennsylvania amphibians and reptiles. Pennsylvania Fish and Boat Commission.

Shapiro, L. H., R. A. Canterbury, D. M. Stover, and R. C. Fleischer. 2004. Reciprocal introgression between golden-winged warblers (*Vermivora chrysoptera*) and blue-winged warblers (*V. pinus*) in eastern North America. Auk 121:1019–1030.

Sharp, B. L. 1934. Dickcissel (*Spiza americana*) and prairie horned lark (*Otocoris a. praticola*) again in Lancaster County, Pennsylvania. Auk 51:531.

Sharpe, W. E. 2002. Acid deposition explains sugar maple decline in the east. BioScience 52:45.

Sharpe, W. E., and J. R. Drohan, editors. 1999. The effects of acidic deposition on Pennsylvania's forests. Proceedings of the 1998 Acidic Deposition Conference. Environmental Resources Research Institute, University Park, Pennsylvania, USA.

Sheehan, J. 2003. Habitat selection in the Acadian flycatcher: the potential impact of hemlock woolly adelgid infestations and other anthropogenic stressors. Thesis, East Stroudsburg University, East Stroudsburg, Pennsylvania, USA.

Sheffield, S. R. 1999. Owls as biomonitors of environmental contamination. Pages 383–398 *in* J. R. Duncan, D. H. Johnson, T. H. Nicholls, editors. Biology and conservation of owls of the Northern Hemisphere. U.S. Department of Agriculture Forest Service General Technical Report NC-190. North Central Research Station, Saint Paul, Minnesota, USA.

Sheldon, W. G. 1971. The book of the American woodcock. Second edition. University of Massachusetts Press, Amherst, USA.

Shelly, L. O. 1929. Twig gathering of the chimney swift. Auk 46:116.

Shepard, D. B., M. J. Dreslik, C. A. Phillips, and B. C. Jellen. 2003. *Sistrurus catenatus catenatus* (eastern massasauga). Male-male aggression. Herpetological Review 34:155–156.

Shiffer, C. N., C. J. McCoy, and C. W. Bier. 1987. Memorandum: management of state game lands number 51 for the protection of the green salamander and the timber rattlesnake. Pennsylvania Fish and Boat Commission.

Shire, G. G., K. Brown, and G. Winegrad. 2000. Communication towers: a deadly hazard to birds. American Bird Conservancy, Washington, D.C., USA.

Shump, K. A., Jr., and A. U. Shump. 1982. *Lasiurus cinereus*. Mammalian Species 185:1–5.

Sibley, D. 1997. The birds of Cape May. Second edition. New Jersey Audubon Society, Cape May Point, New Jersey, USA.

Sibley, D. A. 2000. The Sibley guide to birds. Alfred A. Knopf, New York, New York, USA.

Siebenheller, B., and N. Siebenheller. 1987. Possible effects of artificial feeding on nest-site selection by pine siskins. Chat 51:57–58.

Simberloff, D. 1998. Flagships, umbrellas, and keystones: is single species management passé in the landscape era? Biological Conservation 83:247–257.

Simons, T. R., G. L. Farnsworth, and S. A. Shriner. 2000. Evaluating Great Smoky Mountains National Park as a population source for the wood thrush. Conservation Biology 14:1133–1144.

Simpson, M. R., Jr. 1992. Birds of the Blue Ridge Mountains. University of North Carolina, Chapel Hill, USA.

Simpson, R. B. 1909. Letter on the birds of Warren County. Oologist 26:25–26.

Sims, V., K. L. Evans, S. E. Newson, J. A. Tratalos, and K. J. Gaston. 2008. Avian assemblage structure and domestic cat densities in urban environments. Diversity and Distributions 14:387–399.

Sinclair, K. E., G. R. Hess, C. E. Moorman, and J. H. Mason. 2005. Mammalian nest predators respond to greenway width, landscape context and habitat structure. Landscape and Urban Planning 71:277–293.

Skelly, D. K. 1995. A behavioral trade-off and its consequences for the distribution of *Pseudacris* treefrog larvae. Ecology 76:150–164.

Skelly, D. K. 1996. Pond drying, predators, and the distribution of *Pseudacris* tadpoles. Copeia 1996:599–605.

Smith, A. G. 1945. *Agkistrodon mokeson mokeson* (Daudin) in Pennsylvania. Proceedings of the Pennsylvania Academy of Science 19:69–72.

Smith, D. C. 1983. Factors controlling tadpole populations of the chorus frog (*Pseudacris triseriata*) on Isle Royale, Michigan. Ecology 64:501–510.

Smith, D. C. 1987. Adult recruitment in chorus frogs: effects of size and date at metamorphosis. Ecology 68:344–350.

Smith, D. F. 1997. Foraging strategies of sympatric lagomorphs: implications of habitat fragmentation. Dissertation, University of New Hampshire, Durham, USA.

Smith, D. G., C. R. Wilson, and H. H. Frost. 1974. History and ecology of a colony of barn owls in Utah. Condor 76:131–163.

Smith, H. M. 1967. The handbook of lizards. Fourth edition. Comstock, Binghamton, New York, USA.

Smith, K. G., J. H. Withgott, and P. G. Rodewald. 2000. Red-headed woodpecker (*Melanerpes erythrocephalus*). *In* A. Poole and F. Gill, editors. The birds of North America, 518. Birds of North America, Philadelphia, Pennsylvania, USA.

Smith, L., and K. E. Clark. 2006. New Jersey bald eagle management project, 2006. New Jersey Department of

Environmental Protection, Division of Fish, Game, and Wildlife, Trenton, New Jersey, USA.
Smith, P. W. 1961. The amphibians and reptiles of Illinois. Illinois Natural History Survey Bulletin 28:1–298.
Smith, T. L. 1991. Natural ecological communities of Pennsylvania. First revision. Unpublished Report, Pennsylvania Science Office of the Nature Conservancy, Middletown, USA.
Smithers, B. L., C. W. Boal, and D. E. Anderson. 2005. Northern goshawk diet in Minnesota: an analysis using video recording systems. Journal of Raptor Research 39:264–273.
Smyers, S. D., M. J. Rubbio, R. V. Townsend, and C. C. Swart. 2002. Intra- and interspecific characterizations of burrow use and defense by juvenile ambystomatid salamanders. Herpetologica 58:422–429.
Snider, A. T., and J. K. Bowler. 1992. Longevity of reptiles and amphibians in North American Collections. Society for the Study of Amphibians and Reptiles, Oxford, Ohio, USA.
Snyder, L. L., and C. E. Hope. 1938. A predator-prey relationship between the short-eared owl and the meadow mouse. Wilson Bulletin 50:110–112.
Snyder, N. F. R., H. A. Snyder, J. Lincer, and R. T. Reynolds. 1973. Organochlorines, heavy metals, and the biology of North American accipiters. BioScience 23:300–305.
Sogge, M. K., R. M. Marshall, S .J. Sferra, and T. J. Tibbitts. 1997. A Southwestern willow flycatcher natural history summary and survey protocol. U.S. National Park Service Technical Report, NPS/NAUCPRS/NRTR-97/12.
Sole, J. D. 1999. Distribution and habitat of Appalachian cottontails in Kentucky. Proceedings of the Annual Conference of the Southeastern Association of Fish and Wildlife Agencies 53:444–448.
Solymár, B. D., and J. D. McCracken. 2002. Draft National Recovery Plan for the barn owl and its habitat (*Tyto alba*): Ontario population. Wildlife Ministers' Council, Recovery of Nationally Endangered Wildlife (RENEW) Committee, Ottawa, Ontario, Canada.
Somers, A. B., K. A. Bridle, D. W. Herman, and A. B. Nelson. 2000. The restoration and management of small wetlands of the mountains and Piedmont in the Southeast: a manual emphasizing endangered and threatened species habitat with a focus on bog turtles. Natural Resource Conservation Service. <www.wsi.nrcs.usda.gov/products/w2q/strm_rst/docs/pied/fm.ppd>. Accessed 13 May 2009.
Speare R., and L. Berger. 2004. Global distribution of chytridiomycosis in amphibians. <www.jcu.edu.au/school/phtm/PHTM/frogs/chyglob.htm>. Accessed 24 May 2009.
Speiser, R., and T. Bosakowski. 1987. Nest site selection by northern goshawks in northern New Jersey and southeastern New York. Condor 89:387–394.
Spencer, R. K., and J. A. Chapman. 1986. Seasonal feeding habits of New England and eastern cottontails. Proceedings of the Pennsylvania Academy of Science 60:157–160.
Spencer, S. J. 1962. A study of the physical characteristics of nesting sites used by Bank Swallows. Dissertation, Pennsylvania State University, State College, USA.
Sperring, C. 2001. Long-eared owl (*Asio otus*) Mendip biodiversity action plan. Environmental Action Fund.
Spielman, D., B. W. Brook, and R. Frankham. 2004. Most species are not driven to extinction before genetic factors impact them. Proceedings of the National Academy of Sciences of the United States of America 101:15261–15264.
Spinola, R. M. 2003. Spatio-temporal ecology of river otters translocated to western New York. Dissertation, Pennsylvania State University, University Park, USA.
Spitzer, P. R. 1989. Osprey. Pages 22–29 *in* B. G. Pendleton, editor. Proceedings of the Northeast Raptor Management Symposium and Workshop. National Wildlife Federation Science and Technical Series 13, Washington, D.C., USA.
Spofford, W. R. 1971. The breeding status of the golden eagle in the Appalachians. American Birds 25:3–7.
Sprunt, A. 1954. Florida birdlife. Coward-McCann, New York, New York, USA.
Spurgeon, D. J., and S. P. Hopkin. 1999. Seasonal variation in the abundance, biomass and biodiversity of earthworms in soils contaminated with metal emissions from a primary smelting works. Journal of Applied Ecology 36:173–183.
Squires, J. R., and R. T. Reynolds. 1997. Northern goshawk (*Accipiter gentilis*). *In* A. Poole and F. Gill, editors. The birds of North America, 298. Academy of Natural Sciences, Philadelphia, Pennsylvania, USA, and American Ornithologists' Union, Washington, D.C., USA.
Stallmaster, M. V. 1987. The bald eagle. Universe Books, New York, New York, USA.
Stapanian, M. A., and C. C. Smith. 1978. A model for seed scatterhoarding: coevolution of fox squirrels and black walnuts. Ecology 59:884–896.
Stapanian, M. A., and C. C. Smith. 1984. Density-dependent survival of scatterhoarded nuts: an experimental approach. Ecology 65:1387–1396.
State of Connecticut. 2004. Connecticut's endangered, threatened and special concern species. <http://dep.state.ct.us/burnatr/wildlife/pdf/ETS04.pdf>. Accessed 31 Jan 2006.
Stebbins, R. C., and N. W. Cohen. 1995. A natural history of amphibians. Princeton University Press, Princeton, New Jersey, USA.
Steele, B. B. 1993. Selection of foraging and nesting sites by black-throated blue warblers: their relative influence on habitat choice. Condor 95:568–579.
Steele, M., and G. G. Kwiecinski. 1994. Mammals of Pennsylvania: a checklist with notes on status, habitat, and behavior. Pennsylvania Game Commission, Harrisburg, USA.

Steele, M., C. G. Mahan, and G. Turner. 2004. A manual for long-term monitoring and management of the threatened northern flying squirrel (*Glaucomys sabrinus*) in Pennsylvania. Final Report for State Wildlife Grants, Wilkes Barre, Pennsylvania Game Commission, Harrisburg, Pennsylvania, USA.

Steele, M. A., and J. L. Koprowski. 2001. North American tree squirrels. Smithsonian Institution Press, Washington, D.C., USA.

Steele, M. A., N. Lichti, and R. K. Swihart. 2010. Avian-mediated seed dispersal: an overview and synthesis with an emphasis on temperate forests of Central and Eastern U.S. Pages 28–43 *in* S. K. Majumdar, T. L. Masters, M. C. Brittingham, R, M. Ross, R. S. Mulvihill, and J. E. Huffman, editors. Avian ecology and conservation: a Pennsylvania focus with national implications. Pennsylvania Academy of Science, Easton, Pennsylvania, USA.

Steele, M. A., and R. Powell. 1999. Biogeography of small mammals in the southern Appalachians: patterns of local and regional abundance. Pages 155–165 *in* R. Eckerlin, editor. Appalachian biogeography. Virginia Museum of Natural History, Martinsville, USA.

Steele, M. A., and P. D. Smallwood. 2002. Acorn dispersal by birds and mammals. Pages 182–195 *in* W. McShea and W. M. Healy, editors. Oak forest ecosystems: ecology and management for wildlife. Johns Hopkins University Press, Baltimore, Maryland, USA.

Steele, M. A., L. Wauters, and K. Larsen. 2005. Selection, predation, and dispersal of seeds by tree squirrels in temperate and boreal forests: Are tree squirrels keystone granivores? Pages 205–221 *in* P. M. Forget, S. Vander Wall. J. Lambert, and P. Hulme, editors. Seed fate: predation, dispersal and seedling establishment. CAB International, Wallingford, United Kingdom.

Steen, D. A., M. J. Aresco, S. G. Beilke, B. W. Compton, E. P. Condon, C. Kenneth Dodd, Jr., H. Forrester, J. W. Gibbons, J. L. Greene, G. Johnson, T. A. Langen, M. J. Oldham, D. N. Oxier, R. A. Saumure, F. W. Schueler, J. M. Sleeman, L. L. Smith, J. K. Tucker, and J. P. Gibbs. 2006. Relative vulnerability of female turtles to road mortality. Animal Conservation 9:269–273.

Stein, R. C. 1963. Isolating mechanisms between populations of Traill's flycatchers. Proceedings of the American Philosophical Society 107:21–50.

Stevens, M. A., and R. E. Barry. 2002. Selection, size, and use of home range of the Appalachian cottontail, *Sylvilagus obscurus*. Canadian Field-Naturalist 116: 529–535.

Stevens, S. 2005. Visitation patterns and behavior of river otters at latrine sites in Pennsylvania and Maryland. Thesis, Frostburg State University, Frostburg, Maryland, USA.

Stevens, S. S., and T. Serfass. 2008. Visitation patterns and behavior of nearctic river otters (*Lontra canadensis*) at their latrines. Northeastern Naturalist 15:1–12.

Stevenson, H. M., and B. H. Anderson. 1994. The birdlife of Florida. University Press of Florida, Gainesville, USA.

Stickel, L. F. 1978. Changes in a box turtle population during three decades. Copeia 1978:221–225.

Stickel, L. F. 1989. Home range behavior among box turtles (*Terrapene c. carolina*) of a bottomland forest in Maryland. Journal of Herpetology 23:40–44.

Stiles, F. G., and A. F. Skutch. 1989. A guide to the birds of Costa Rica. Cornell University Press, Ithaca, New York, USA.

Stoleson, S. H. 2004. Cerulean warbler habitat use in an oak-northern hardwood transition zone: implications for management. Page 535 *in* D. Yaussy, D. N. Hix, R. P. Long, and P. C. Goebel, editors. General Technical Report NE-316, Proceedings of the Fourteenth Central Hardwoods Forest Conference. U.S. Forest Service, Northeastern Research Station, Newtown Square, Pennsylvania, USA.

Stone, W. 1928. Dickcissel (*Spiza americana*) in Delaware County, Pennsylvania. Auk 45:507–508.

Stone, W. 1894. Birds of eastern Pennsylvania and New Jersey. Delaware Valley Ornithological Club, Philadelphia, Pennsylvania, USA.

Stone, W. 1900. The summer birds of the higher parts of Sullivan and Wyoming counties, Pennsylvania. Abstract Proceedings of the Delaware Valley Ornithological Club 3:20–23.

Stone, W. 1906. Notes on reptiles and batrachians of Pennsylvania, New Jersey and Delaware. American Naturalist 40:159–170.

Storey, K. B., and J. M. Storey. 1986. Freeze tolerance and intolerance as strategies of winter survival in terrestrially-hibernating amphibians. Comparative Biochemistry and Physiology 83A:613–617.

Storey, K. B., and J. M. Storey. 1987. Persistence of freeze tolerance in terrestrially hibernating frog after spring emergence. Copeia 1987:720–726.

Storm, G. L., W. L. Palmer, and D. R. Diefenbach. 2003. Ruffed grouse responses to management of mixed oak and aspen communities in central Pennsylvania. Grouse Research Bulletin Number one, Pennsylvania Game Commission, Harrisburg, USA.

Storm, G. L., W. K. Shope, and W. M. Tzilkowski. 1993. Cottontail population response to forest management using clearcutting and short rotations. Northeast Wildlife 50:91–99.

Stotz, D. F., J. W. Fitzpatrick, T. A. Parker, and D. K. Moskovits. 1996. Neotropical birds: ecology and conservation with ecological and distributional databases. University of Chicago Press, Chicago, Illinois, USA.

Strang, C. A. 1983. Spatial and temporal activity patterns in two terrestrial turtles. Journal of Herpetology 17:43–47.

Stratford, J. A. 2010. The effect of environmental contaminants on avian populations. Pages 340–358 *in* S. K. Majumdar, T. L. Masters, M. C. Brittingham, R. M. Ross, R. S. Mulvihill, and J. E. Huffman, editors. Avian ecology and conservation: a Pennsylvania focus with national implications. Pennsylvania Academy of Science, Easton, Pennsylvania, USA.

Stratford, J. A., and W. D. Robinson. 2005. Distribution of Neotropical migratory bird species across an urbanizing landscape. Urban Ecosystems 8:59–77.

Straw, J. A., D. G. Krementz, M. W. Olinde, and G. F. Sepik. 1994. American woodcock. Pages 97–114 *in* T. C. Tacha and C. E. Braun, editors. Migratory shore and upland game bird management in North America. International Association of Fish and Wildlife Agencies.

Street, J. F. 1923. On the nesting grounds of the solitary sandpiper and the lesser yellowlegs. Auk 40:577–583.

Street, P. B. 1954. Birds of the Pocono Mountains. Cassinia 41:3–76.

Strickland, M. A., C. W. Douglas, M. Novak, and N. P. Hunziger. 1982. Fisher. Pages 586–598 *in* J. A. Chapman and G. A. Feldhamer, editors. Wild mammals of North America: biology, management, and economics. Johns Hopkins University Press, Baltimore, Maryland, USA.

Strong, A. M., and T. W. Sherry. 2000. Habitat-specific effects of food abundance on the condition of ovenbirds wintering in Jamaica. Journal of Animal Ecology 69:883–895.

Stuart, J. N. 2002. *Liochlorophis* (=*Opheodrys*) *vernalis* (smooth green snake). Reproduction. Herpetological Review 33:140–141.

Stuart, S. N., J. S. Chanson, N. A. Cox, B. E. Young, A. S. L. Rodrigues, D. L. Fischman, and R. W. Waller. 2004. Status and trends of amphibian declines and extinctions worldwide. Science 306:1783–1786.

Stucker, J. H., and F. J. Cuthbert. 2005. Piping plover breeding biology and management in the state of Michigan, 2005. Report submitted to Michigan Department of Natural Resources, Endangered Species Office, Lansing, USA.

Stull, J., J. A. Stull, and G. M. McWilliams. 1985. Birds of Erie County Pennsylvania, including Presque Isle. Allegheny Press, Eglin, Pennsylvania, USA.

Stutchbury, B. J. M., S. A. Tarof, T. Done, E. Gow, P. M. Kramer, J. Tautin, J. W. Fox, and V. Afanasyev. 2009. Tracking long-distance songbird migration by using geolocators. Science 323:896–896.

Surdick, J. 1995. The Louisiana waterthrush in southeastern Minnesota. Loon 62:201–206.

Surface, H. A. 1906. The serpents of Pennsylvania. Pennsylvania State Department of Agriculture Monthly Bulletin of the Division of Zoology 4:114–208.

Sutherland, G. D., A. S. Harestad, K. Price, and K. P. Lertzman. 2000. Scaling of natal dispersal distances in terrestrial birds and mammals. Conservation Ecology 4:44.

Sutton, C., and P. Sutton. 1994. How to spot an owl. Chapters Publishing Limited, Shelburne, Vermont, USA.

Sutton, G. M. 1928a. The birds of Pymatuning Swamp and Conneaut Lake, Crawford County, Pennsylvania. Annals of the Carnegie Museum 18:19–239.

Sutton, G. M. 1928b. An introduction to the birds of Pennsylvania. J. Horace McFarland Company, Harrisburg, Pennsylvania, USA.

Sutton, G. M. 1967. Oklahoma birds. University of Oklahoma Press, Norman, USA.

Swaddle, J. P., and S. E. Calos. 2008. Increased avian diversity is associated with lower incidence of human West Nile infection: observation of the dilution effect. PLoS ONE 3:e2488.

Swanson, P. L. 1930. Notes on the massasauga. Bulletin of the Antivenin Institute of America 4:70–71.

Swanson, P. L. 1948. Notes on the amphibians of Venango County, Pennsylvania. American Midland Naturalist 40:362–371.

Swanson, P. L. 1952. The reptiles of Venango County, Pennsylvania. American Midland Naturalist 47:161–182.

Swarth, C. W. 2004. Natural history and reproductive biology of the red-bellied turtle (*Pseudemys rubriventris*). Pages 73–83 *in* C. W. Swarth, W. M. Roosenburg, and E. Kiviat, editors. Conservation and ecology of turtles of the mid-Atlantic Region, a Symposium. Bibliomania!, Salt Lake City, Utah, USA.

Swartzentruber, B. A., and T. L. Master. 2003. Habitat use of blue-headed vireos. Journal of the Pennsylvania Academy of Science 76:137.

Sweeney, J. R., J. M. Sweeney, and S. W. Sweeney. 2003. Feral hog (*Sus scrofa*). Pages 1164–1179 *in* G. A. Feldhamer, B. C. Thompson, and J. A. Chapman, editors. Wild mammals of North America: biology, management, and conservation. Johns Hopkins University Press, Baltimore, Maryland, USA.

Sweitzer, R. A., and D. Van Vuren. 2002. Rooting and foraging effects of wild pigs on tree regeneration and acorn survival in California's oak woodland ecosystems. Proceedings of the 5th symposium on oak woodlands: oaks in California's changing landscape. General Technical Report PSW-GTR-184.

Swift, B. L. 1987. An analysis of avian breeding habitats in Hudson River tidal marshes. Unpublished report. New York State Department of Environmental Conservation, Division of Fish and Wildlife, Delmar, New York, USA.

Swift, B. L., S. R. Orman, and J. W. Ozard. 1988. Response of least bitterns to tape-recorded calls. Wilson Bulletin 100:496–499.

Swimley, T. J., R. P. Brooks, and T. L. Serfass. 1999. Otter and beaver interactions in the Delaware Water Gap National Recreation Area. Journal of the Pennsylvania Academy of Science 72:97–101.

Swimley, T. J., T. L. Serfass, R. P. Brooks, and W. M. Tzilkowski. 1998. Predicting river otter latrine sites in Pennsylvania. Wildlife Society Bulletin 26: 836–845.

Szaro, R. C., and R. P. Balda. 1986. Relationships among weather, habitat structure, and ponderosa pine forests. Journal of Wildlife Management 50:253–260.

Szep, T. 1995. Relationship between West-African rainfall and the survival of Central-European sand martins *Riparia riparia*. Ibis 137:162–168.

Szep, T. 1991. Number and distribution of the Hungarian sand martin population breeding along the Hungarian reaches of the River Tisza. Aquila 98:111–124.

Szymanski, J. A. 1998. Range-wide status assessment for the eastern massasauga (*Sistrurus c. catenatus*). U.S. Fish and Wildlife Service, Fort Snelling, Minnesota, USA.

Taber, T. T. 1970. Ghost lumber towns of central Pennsylvania: Laquin, Masten, Ricketts, Grays Run. Published by the author, Muncy, Pennsylvania, USA.

Takats, D. L., C. M. Francis, G. L. Holroyd, J. R. Duncan, K. M. Mazur, R. J. Cannings, W. Harris, and D. Holt. 2001. Guidelines for nocturnal owl monitoring in North America. Beaverhill Bird Observatory and Bird Studies Canada, Edmonton, Alberta.

Tallamy, D. W. 2007. Bringing nature home. Timber Press, Portland, Oregon, USA.

Talliaferro, E. H., R. T. Holmes, and J. D. Blum. 2001. Eggshell characteristics and calcium demands of a migratory songbird breeding in two New England forests. Wilson Bulletin 113:94–100.

Tate, G. R. 1991. Population dynamics, reproductive ecology, and conservation of two insular populations of the short-eared owl in Massachusetts. Thesis, University of Massachusetts, Boston, USA.

Tate, G. R. 1992. Short-eared owl, *Asio flammeus*. Pages 171–189 *in* K. J. Schneider and D. M. Pence, editors. Migratory nongame birds of management concern in the Northeast. U.S. Fish and Wildlife Service, Newton Corner, Massachusetts, USA.

Tate, J., Jr. 1986. The blue list for 1986. American Birds 40:227–236.

Taylor, G. J. 1973. Present status and habitat survey of the Delmarva fox squirrel (*Sciurus niger cinereus*) with a discussion of reasons for its decline. Proceedings of the Annual Conference of the Southeastern Association of Game and Fish Commissions 27:278–289.

Taylor, G. J. 1976. Range determination and habitat description of the Delmarva fox squirrel in Maryland. Thesis, University of Maryland, College Park, USA.

Taylor, J. S., K. E. Church, D. H. Rusch, and J. R. Cary. 1999. Macrohabitat effects on summer survival, movements, and clutch success of northern bobwhites in Kansas. Journal of Wildlife Management 63:675–685.

Teats, P. J., K. W. Anderson, and D. W. Brauning. 1998. Project annual job report: study of the breeding habits of the black tern (*Chlidonias niger*) in Erie and Crawford counties: habitat assessment and management recommendations. Unpublished report, Pennsylvania Game Commission, Harrisburg, USA.

Tefft, B. C., and J. A. Chapman. 1983. Growth and development of nestling New England cottontails, *Sylvilagus transitionalis*. Acta Theriologica 28:317–377.

Tefft, B. C., and J. A. Chapman. 1987. Social behavior of the New England cottontail, *Sylvilagus transitionalis*, with a review of social behavior of the New World rabbits. Revue d'Ecologie: La Terra et la Vie 42:235–276.

Temple, S. A. 2002. Dickcissel (*Spiza americana*). *In* A. Poole and F. Gill, editors. The birds of North America, 703. Birds of North America, Philadelphia, Pennsylvania, USA.

Temple, S. A. 1986. Predicting impacts of habitat fragmentation on forest birds: a comparison of two models. Pages 301–304 *in* J. Verner, M. L. Morrison, and C. J. Ralph, editors. Wildlife 2000: Modeling habitat relationships of terrestrial Vertebrates. University of Wisconsin Press, Madison, USA.

Terres, J. K. 1980. The Audubon Society encyclopedia of North American birds. Alfred A. Knopf, New York, New York, USA.

Therres, G. D. 1999. Wildlife species of regional conservation concern in the northeastern United States. Northeast Wildlife 54:93–100.

Therres, G. D., and G. W. Willey, Sr. 2002. Reintroductions of the endangered Delamarva fox squirrel in Maryland. Proceedings of the Annual Conference of the Southeastern Fish and Wildlife Agencies 56:265–274.

Thomas, C. D., A. Cameron, R. E. Green, M. Bakkenes, L. J. Beaumont, Y. C. Collingham, B. F. N. Erasmus, M. Ferreira de Siqueira, A. Grainger, L. Hannah, L. Hughes, B. Huntley, A. S. van Jaarsveld, G. F. Midgley, L. Miles, M. A. Ortega-Huerta, A. T. Peterson, O. L. Phillips, and S. E. Williams. 2004. Extinction risk from climate change. Nature 427:145–148.

Thompson, C. F. 1977. Experimental removal and replacement of territorial male yellow-breasted chats. Auk 94:107–113.

Thompson, C. F., and V. Nolan, Jr. 1973. Population biology of the yellow-breasted chat (*Icteria virens* L.) in southern Indiana. Ecological Monographs 43:145–171.

Thompson, E. L., and G. J. Taylor. 1985. Notes on the green salamander, *Aneides aeneus* in Maryland. Bulletin of Maryland Herpetological Society 21:107–114.

Thompson, F. R., III, and D. E. Burhans. 2003. Predation of songbird nests differs by predator and between field and forest habitats. Journal of Wildlife Management 67:408–416.

Thompson, F. R., T. M. Donovan, R. M. DeGraaf, J. Faaborg, and S. K. Robinson. 2002. A multi-scale perspective of the effects of forest fragmentation on birds in eastern forests. Studies in Avian Biology 25:8–19.

Thompson, W. L. 2002. Towards reliable bird surveys: accounting for individuals present but not detected. Auk 119:18–25.

Thomson, C. E. 1982. *Myotis sodalis*. Mammalian Species 163:1–5.

Thorne, S., K. C. Kim, K. C. Steiner, and B. J. McGuinness. 1995. A heritage for the 21st century: conserving Pennsylvania's native biological diversity. Pennsylvania Fish and Boat Commission, Harrisburg, USA.

Thurow, G. R. 1993. *Clonophis kirtlandii*, diet. Herpetological Review 24:34–35.

Tiner, R. W. 1989. Current status and recent trends in Pennsylvania's wetlands. Pages 368–378 *in* S. K. Majumdar, R. P. Brooks, F. J. Brenner, and R. W. Tiner, editors. Wetlands ecology and conservation: emphasis in Pennsylvania. Pennsylvania Academy of Science, Easton, Pennsylvania, USA.

Tiner, R. W. 1990. Pennsylvania's wetlands: current status and recent trends. U.S. Fish and Wildlife Service, Newton Corner, Massachusetts, USA.

Tingley, M. W., D. A. Orwig, R. Field, and G. Motzkin. 2002. Avian response to removal of a forest dominant: consequences of hemlock woolly adelgid infestations. Journal of Biogeography 29:1505–1516.

Tinkle, D. W., and R. E. Ballinger. 1972. *Sceloporus undulatus*: a study of the intraspecific comparative demography of a lizard. Ecology 53:570–584.

Tittler, R., L. Fahrig, M. Villard. 2006. Evidence of large scale source-sink dynamics and long-distance dispersal among Wood Thrush populations. Ecology 87:3029–3086.

Titus, K. 1984. Uniformity in relative habitat selection by red-shouldered and broad-winged hawks in two temperate forest regions. Dissertation, University of Maryland, Catonsville, Maryland, USA.

Titus, K., and J. A. Mosher. 1981. Nest site habitat selected by woodland hawks in the central Appalachians. Auk 98:270–281.

Titus, K., M. R. Fuller, D. F. Stauffer, and J. R. Sauer. 1989. Buteos. Pages 53–64 *in* Proceedings of the Northeast Raptor Management Symposium and Workshop, May 1988, Syracuse, New York. National Wildlife Federation Scientific and Technical Series 13.

Todd, W. E. C., editor. 1940a. Birds of western Pennsylvania. University of Pittsburgh Press, Pittsburgh, Pennsylvania, USA.

Todd, W. E. C. 1940b. Eastern winter wren. Pages 414–417 *in* W. E. C. Todd, editor. Birds of western Pennsylvania. University of Pittsburgh Press, Pittsburgh, Pennsylvania, USA.

Todd, W. E. C. 1940c. Golden eagle. Pages 149–150 *in* Birds of Western Pennsylvania. University of Pittsburgh Press, Pittsburgh, Pennsylvania, USA.

Todd, W. E. C. 1963. Birds of the Labrador Peninsula and adjacent areas. University of Toronto Press, Ontario.

Tome, R., and J. Valkama. 2001. Seasonal variation in the abundance and habitat use of barn owls, *Tyto alba,* on lowland farmland. Ornis Fennica 78:109–118.

Toms, J. D., F. K. A. Schmiegelow, S. J. Hannon, and M.-A. Villard. 2006. Are point counts of boreal songbirds reliable proxies for more intensive abundance estimators? Auk 123:438–454.

Trani, M. K., R. T. Brooks, T. L. Schmidt, V. A. Rudis, and C. M. Gabbard. 2001. Patterns and trends of early successional forests in the eastern United States. Wildlife Society Bulletin 29:413–424.

Trautman, M. B. 1940. The birds of Buckeye Lake. Miscellaneous Publications Museum of Zoology, 44, University of Michigan, Ann Arbor, USA.

Trenham, P. C., and H. B. Shaffer. 2005. Amphibian upland habitat use and its consequences for population viability. Ecological Applications 15:1158–1168.

Trent, T. T., and O. S. Rongstad. 1974. Home range and survival of cottontail rabbits in southwestern Wisconsin. Journal of Wildlife Management 38:459–472.

Trexel, D. R., R. N. Rosenfield, J. Bielefeldt, and E. A. Jacobs. 1999. Comparative nest site habitats in sharp-shinned and Cooper's hawks in Wisconsin. Wilson Bulletin 111:7–14.

Trimble, R. 1940. Changes in bird life at Pymatuning Lake, Pennsylvania. Annals of the Carnegie Museum 28:83–132.

Trine, C. L. 1998. Wood thrush population sinks and implications for the scale of regional conservation strategies. Conservation Biology 12:576–585.

Tuck, L. M. 1972. The snipes: a study of the genus *Capella*. Canadian Wildlife Service Monographs Serial Number 5. Ottawa, Ontario, Canada.

Tucker, J. K. 1976. Observations on the birth of a brood of Kirtland's water snake, *Clonophis kirtlandii*. Journal of Herpetology 10:53–54.

Tucker, J. K. 1977. Notes on the food habits of Kirtland's water snake, *Clonophis kirtlandii*. Bulletin of the Maryland Herpetological Society 13:193–195.

Tucker, J. K. 1994. A laboratory investigation of fossorial behavior in Kirtland's snake *Clonophis kirtlandii* (Kennicott) Serpentes: Colubridae, with some comments on management of the species. Bulletin of the Chicago Herpetological Society 29:93–94.

Turnbull, W. P. 1869. The birds of east Pennsylvania and New Jersey. Henry Grambo and Company, Philadelphia, Pennsylvania, USA.

Turner, G. 2007. Indiana bat hibernacula surveys. Annual job report 17110-9797. Pennsylvania Game Commission, Harrisburg, USA.

Turner, W. R., K. Brandon, T. M. Brooks, R. Costanza, G. A. B. da Fonseca, and R. Portela. 2007. Global conservation of biodiversity and ecosystem services. BioScience 57:868–873.

Tuttle, S. E., and D. M. Carroll. 2003. Home range and seasonal movements of the wood turtle (*Glyptemys insculpta*) in southern New Hampshire. Chelonian Conservation and Biology 4:655–663.

Tuttle, M. D. 1964. *Myotis subulatus* in Tennessee. Journal of Mammalogy 45:148–149.

Tuttle, M. D., and J. Kennedy. 1999. Indiana bat hibernation roost evaluation: phase II—results from the first annual cycle. Bat Conservation International, Austin, Texas, USA.

Twedt, D. J., and J. L. Henne-Kerr. 2001. Artificial cavities enhance breeding bird densities in managed cottonwood forests. Wildlife Society Bulletin 29:680–687.

Ulrich, W. D., editor. 1997. A century of bird life in Berks County, Pennsylvania. Reading Public Museum, Reading, Pennsylvania, USA.

Ultsch, G. R., and J. T. Duke. 1990. Gas exchange and habitat selection in the aquatic salamanders *Necturus maculosus* and *Cryptobranchus alleganiensis*. Oecologia 83:250–258.

Underwood T. J., and R. R. Roth. 2002. Demographic variables are poor indicators of wood thrush productivity. Condor 104:92–102.

U.S. Department of Agriculture (USDA). 2002. National Resources Inventory 2002 Annual NRI. <www.nrcs.usda.gov/technical/land/nri02/>. Accessed 19 Sep 2006.

U.S. Department of Agriculture. 2004. The state of the forest: a snapshot of Pennsylvania updated forest inventory 2004. U.S. Department of Agriculture Forest Service website. Information on hemlock woolly adelgid. Morgantown, West Virginia, USA. <www.fs.fed.us/na/morgantown/fhp/hwa/hwasite.html>.

U.S. Department of Agriculture Forest Service. Undated. Pest alert: elongate hemlock scale. Northeastern Area NA-PR-01-02.

U.S. Department of Agriculture Forest Service. 2002. Conservation assessment for the red-shouldered hawk. Eastern Region.

U.S. Department of Agriculture Forest Service. 2009. Forest stewardship program. <www.fs.fed.us/spf/coop/programs/loa/fsp.shtml>.

U.S. Department of Interior. 1990. American woodcock management plan. U.S. Fish and Wildlife Service, Washington, D.C., USA.

U.S. Department of Interior. 1996. American woodcock management plan. U.S. Fish and Wildlife Service, Region 5, Hadley, Massachusetts, USA.

U.S. Department of Interior. United States Geological Survey. 2001. A gap analysis of Pennsylvania. Final report. Pennsylvania State University, University Park, USA.

U.S. Fish and Wildlife Service. 1985. Determination of endangered and threatened status for the piping plover. Federal Register 50(238):50720–34.

U.S. Fish and Wildlife Service. 1987. Migratory nongame birds of management concern in the United States: the 1987 list. U.S. Fish and Wildlife Service Office of Migratory Bird and Management, Washington, D.C., USA.

U.S. Fish and Wildlife Service. 1988. Great Lakes and Northern Great Plains piping plover recovery plan. Twin Cities, Minnesota, USA.

U.S. Fish and Wildlife Service. 1993. Delmarva fox squirrel (*Sciurus niger cinereus*) recovery plan. Second edition. U.S. Fish and Wildlife Service, Northeast Region, Hadley, Massachusetts, USA.

U.S. Fish and Wildlife Service. 1994. Plymouth redbelly turtle (*Pseudemys rubriventris*) recovery plan, second revision. Hadley, Massachusetts, USA.

U.S. Fish and Wildlife Service. 1995. Final rule determining endangered status for the southwestern willow flycatcher. Federal Register 60(38):10694–10715

U.S. Fish and Wildlife Service. 1997. Final rule to list the northern population of the bog turtle as threatened and the southern population as threatened due to similarity of appearance. Federal Register 62(213):59605–59623.

U.S. Fish and Wildlife Service. 1998. Endangered and threatened wildlife and plants; proposed rule to remove the peregrine falcon in North America from the list of endangered and threatened wildlife. Federal Register 63(165):45446–45463.

U.S. Fish and Wildlife Service. 1999a. Agency draft. Indiana Bat (*Myotis sodalis*) revised recovery plan. U.S. Fish and Wildlife Service, Fort Snelling, Minnesota, USA.

U.S. Fish and Wildlife Service. 1999b. Endangered and threatened wildlife and plants; final rule to remove the peregrine falcon in North America from the list of endangered and threatened wildlife, and to remove the similarity of appearance provision for free-flying peregrines in the coterminous United States. Federal Register 64(164):46542–46558.

U.S. Fish and Wildlife Service. 2001a. Bog turtle (*Clemmys muhlenbergii*), northern population, recovery plan. Hadley, Massachusetts, USA.

U.S. Fish and Wildlife Service. 2001b. Final determination of critical habitat for the Great Lakes breeding population of the piping plover. Federal Register 66(88):22938–22969.

U.S. Fish and Wildlife Service. 2001c. Final determination of critical habitat for wintering piping plovers. Federal Register 66(132):36038–30143.

U.S. Fish and Wildlife Service. 2002a. 90-day finding on a petition to list the cerulean warbler as threatened with critical habitat. Federal Register 67:65083–65086.

U.S. Fish and Wildlife Service. 2002b. National wetlands inventory: a strategy for the Twenty-first century. <www.nwi.fws.gov/Pubs_Reports/NWI121StatFNL.pdf>. Accessed 19 Sep 2006.

U.S. Fish and Wildlife Service. 2002c. Recovery data call—East Lansing Ecological Services Field Office. Threatened and Endangered Species System (TESS), East Lansing, Michigan, USA.

U.S. Fish and Wildlife Service. 2003a. Interim guidelines to avoid and minimize wildlife impacts from wind turbines. U.S. Department of the Interior, Fish and Wildlife Service, Washington, D.C., USA.

U.S. Fish and Wildlife Service. 2003b. Recovery Plan for the Great Lakes piping plover (*Charadrius melodus*). Fort Snelling, Minnesota, USA.

U.S. Fish and Wildlife Service. 2004a. Waterfowl Population Status, 2004. U.S. Department of the Interior, Washington, D.C., USA.

U.S. Fish and Wildlife Service. 2004b. North American Waterfowl Management Plan: 2004 strategic guidance. U.S. Fish and Wildlife Service, Washington, D.C., USA.

U.S. Fish and Wildlife Service. 2005. Migratory bird harvest information, 2004: preliminary estimates. U.S. Department of the Interior, Washington, D.C., USA.

U.S. Fish and Wildlife Service. 2007a. Endangered and threatened wildlife and plants; removing the bald eagle in the lower 48 states from the list of endangered and threatened wildlife; final rule Department of the Interior, Fish and Wildlife Service, 50 CFR Part 17. Federal Register 37346–37372.

U.S. Fish and Wildlife Service. 2007b. Indiana bat (*Myotis sodalis*) draft recovery plan: first revision. U.S. Fish and Wildlife Service, Fort Snelling, Minnesota, USA.

U.S. Fish and Wildlife Service. 2007c. National Bald Eagle Management Guidelines, Anchorage, Alaska, USA.

U.S. Fish and Wildlife Service. 2008. Birds of conservation concern 2008. United States Department of Interior, Fish and Wildlife Service, Division of Migratory Bird Management, Arlington, Virginia, USA. Online version available at <www.fws.gov/migratorybirds/>.

U.S. Fish and Wildlife Service. 2009. The White-Nose Syndrome mystery: something is killing our bats. <www.fws.gov/northeast/white_nose.html>. Accessed 24 May, 2009.

U.S. Fish and Wildlife Service, Environment Canada an Secretaria de Medio Ambiente Recursos Naturales, Mexico. 2004. North American management plan. 2004 Strategic Guidance. North American Waterfowl Management Plan Implementation Office, Gatineau, Québec, Canada.

Valentini, A., F. Pompanon, and P. Taberlet. In press. DNA barcoding for ecologists. Trends in Ecology and Evolution.

Vana-Miller, S. L. 1987. Habitat suitability index models: osprey. U.S. Department of the Interior Fish and Wildlife Society Biological Report 82(10.154):1–46.

Van Fleet. 1884. Notes from Dubois, Pennsylvania. Ornithologist and Oologist 9:108.

Van Fleet, K. 2001. Pages 23–41 *in* K. L. Bildstein and D. Klem, Jr., editors. Hawkwatching in the Americas: geography of diurnal raptors migrating through the valley-and-ridge province of central Pennsylvania 1991–1994. Hawk Migration Association of North America, North Whales, Pennsylvania, USA.

Van Gelder, R. G. 1959. A taxonomic revision of the spotted skunk (genus *Spilogale*). Bulletin of the American Museum of Natural History 117:229–392.

Van Horne, B. 1983. Density as a misleading indicator of habitat quality. Journal of Wildlife Management 47:893–901.

Van Horne, B. 1995. Assessing vocal variety in the winter wren, a bird with a complex repertoire. Condor 97:413–420.

Van Zyll de Jong, C. G. 1979. Distribution and systematic relationships of long-eared *Myotis* in western Canada. Canadian Journal of Zoology 57:987–994.

Van Zyll de Jong, C. G. 1983. Handbook of Canadian mammals: marsupials and insectivores. National Museum of Natural Science, National Museum. Canada, Ottawa 1:1–210.

Van Zyll de Jong, C. G. 1984. Taxonomic relationships of Nearctic small-footed bats of the *Myotis leibii* group (Chiroptera: Vespertilionidae). Canadian Journal of Zoology 62:2519–2526.

Vega Rivera, J. H., C. A. Haas, J. H. Rappole, and W. J. McShea. 2000. Parental care of fledgling wood thrushes. Wilson Bulletin 112:233–237.

Vega Rivera, J. H., W. J. McShea, J. H. Rappole, and C. H. Haas. 1999. Post breeding movements and habitat use of adult wood thrushes in northern Virginia. Auk 116:458–466.

Vega Rivera, J. H., J. H. Rappole, W. J. McShea, and C. H. Haas. 1998. Wood thrush postfledging movement and habitat use in northern Virginia. Condor 100:69–78.

Vermont Institute of Natural Science. 2005. Mountain Birdwatch Project. <www.vinsweb.org/cbd/mtn_bird watch.html>.

Vickery, P. D. 1996. Grasshopper sparrow (*Ammodramus savannarum*). *In* A. Poole and F. Gill, editors. The birds of North America, 239. Academy of Natural Sciences, Philadelphia, Pennsylvania, USA, and American Ornithologists' Union, Washington, D.C., USA.

Vickery, P. D., H. E. Casanas, and A. S. Di Giacomo. 2003. Effects of altitude on the distribution of Nearctic and resident grasslands birds in Cordoba province, Argentina. Journal of Field Ornithology 74:172–178.

Vickery, P. D., P. L. Tubaro, J. M. Cardoso Da Silva, B. G. Peterjohn, J. R. Herkert, and R. B. Cavalcanti. 1999. Conservation of grassland birds in the Western Hemisphere. Studies in Avian Biology 19:2–26.

Villard, M., K. Freemark, and G. Merriam. 1992. Metapopulation theory and Neotropical migrant birds in temperate forests: an empirical investigation. Pages 474–482 *in* J. M. Hagan III and D. W. Johnston, editors. Ecology and conservation of Neotropical migrant landbirds. Smithsonian Institution Press, Washington, D.C., USA.

Vitt, L. J. 1986. Foraging and diet of a diurnal predator (*Eumeces laticeps*) feeding on hidden prey. Journal of Herpetology 20:408–415.

Vitt, L. J., J. P. Caldwell, H. M. Wilbur, and D. C. Smith. 1990. Amphibians as harbingers of decay. BioScience 40:418.

Vitt, L. J., and W. E. Cooper, Jr. 1985. The evolution of sexual dimorphism in the skink *Eumeces laticeps*: an example of sexual selection. Canadian Journal of Zoology 63:995–1002.

Viverette, C. B., L. J. Goodrich, and M. Pokras. 1990. Levels of DDE in eastern flyway populations of migrating sharp-shinned hawks and the question of recent declines in numbers sighted. Journal of the Hawk Migration Association of North America 20:5–7.

Viverette, C. B., S. Struve, L. J. Goodrich, and K. L. Bildstein. 1996. Decreases in migrating sharp-shinned hawks

(*Accipiter striatus*) at traditional raptor-migration watch sites in eastern North America. Auk 113:32–40.

Vogel, W. G. 1981. A guide for revegetating coal minespoils in the eastern United States. General Technical Report NE-68. Northeast Forest Experiment Station, Forest Service, U.S. Department of Agriculture, Broomall, Pennsylvania, USA.

Vogt, R. C. 1980. Natural history of the map turtles *Graptemys pseudogeographica* and *G. ouachitensis* in Wisconsin. Tulane Studies in Zoology and Botany 22:17–48.

Vogt, R. C. 1981. Food partitioning in three sympatric species of map turtle, genus *Graptemys* (Testudinata, Emydidae). American Midland Naturalist 105:102–111.

Vonhof, M. J. 1996. Roost site preferences of big brown bats (*Eptesicus fuscus*) and silver haired bats (*Lasionycteris noctivagans*) in Pend d'Oreille Valley in southern British Colombia. Pages 62–79 *in* R. M. R. Barclay and R. M. Brigham, editors. Bats and forests Symposium. Ministry of Forests Research Program, University of British Columbia, Victoria, British Columbia, Canada.

Von Steen, D. A. 1965. A study of nesting dickcissels in Nebraska. Nebraska Bird Review 33:22–24.

Wakeley, J. S., and T. H. Roberts. 1996. Bird distributions in forest zonation in a bottomland hardwood wetland. Wetlands 16:296–308.

Walde, A. D., J. R. Bider, C. Daigle, D. Masse, J. C. Bourgeois, J. Jutras, and R. D. Titman. 2003. Ecological aspects of a wood turtle, *Glyptemys insculpta*, population at the northern limit of its range in Quebec. Canadian Field-Naturalist 117:377–388.

Walk, J. W., K. Wentworth, E. L. Kershner, E. K. Bollinger, and R. E. Warner. 2004. Renesting decisions and annual fecundity of female dickcissels (*Spiza americana*) in Illinois. Auk 121:1250–1261.

Walker, C. F., and W. Goodpaster. 1941. The green salamander, *Aneides aeneus*, in Ohio. Copeia 1941:178.

Walker, D. J. 1963. Notes on broods of *Virginia v. valeriae* Baird and Girard in Ohio. Journal of the Ohio Herpetological Society 4:54.

Walker, E. P. 1964. Mammals of the world. Volume 2. Johns Hopkins University Press, Baltimore, Maryland, USA.

Walkinshaw, L. R. 1935. Studies of the short-billed marsh wren (*Cistothorus stellaris*) in Michigan. Wilson Bulletin 52:361–368.

Walkinshaw, L. H. 1953. Life-history of the prothonotary warbler. Wilson Bulletin 65:152–168.

Walkinshaw, L. H. 1966. Studies of the Acadian flycatcher in Michigan. Bird-Banding 37:227–257.

Walley, W. J. 1989. Breeding blackpoll warblers, *Dendoica striata*, in Duck Mountain Provincial Park, Manitoba. Canada Field-Naturalist 103:396–397.

Walsh, J., V. Elia, R. Kane, and T. Halliwell. 1999. Birds of New Jersey. New Jersey Audubon Society, Bernardsville, USA.

Walters, J. R. 1998. The ecological basis of avian sensitivity to habitat fragmentation. Pages 181–192 *in* J. M. Marzlaff and R. Sallabanks, editors. Avian conservation—research and management. Island Press, Washington, D.C., USA.

Ward, F. P., C. J. Hohmann, J. F. Ulrich, and S. E. Hill. 1976. Seasonal microhabitat selections of spotted turtles (*Clemmys guttata*) in Maryland elucidated by radioisotope tracking. Herpetologica 32:60–64.

Ward, J. S., M. E. Montgomery, C. A. Cheah, B. P. Onken, and R. S. Cowles. 2004. Eastern hemlock forests: guidelines to minimize the impacts of hemlock woolly adelgid. U.S. Forest Service, Northeastern Area State and Private Forestry, Newtown Square, Pennsylvania, USA.

Warner, E. D. 2003. Pennsylvania land cover, 2000. Pennsylvania State University, University Park, USA. <www.pasda.psu.edu/data/orser/psu-palulc_2000.zip>. Accessed 19 Sep 19 2006.

Warner, R. E. 1994. Agricultural land-use and grassland habitat in Illinois—future-shock for midwestern birds. Conservation Biology 8:147–156.

Warren, B. H. 1888. Report on the birds of Pennsylvania. State Board of Agriculture, Harrisburg, Pennsylvania, USA.

Warren, B. H. 1890. Birds of Pennsylvania. Commonwealth of Pennsylvania, Harrisburg, USA.

Warren, B. H. 1890. Report on the birds of Pennsylvania. Second edition. State Board of Agriculture, Harrisburg, Pennsylvania, USA.

Watmough, B. R. 1978. Observations on nocturnal feeding by night herons *Nycticorax nycticorax*. Ibis 120:356–358.

Watson, J. 1997. The golden eagle. T & AD Poyser, Ltd., London.

Watts, B. D. 1987. The effects of mortality and time constraints on productivity in yellow-crowned night-herons. Condor 90:860–865.

Watts, B. D. 1989. Nest site characteristics of yellow-crowned night-herons in Virginia. Condor 91:979–983.

Watts, B. D. 1995. Yellow-crowned night-heron. *In* A. Poole and F. Gill, editors. The birds of North America, 74. Academy of Natural Sciences, Philadelphia, Pennsylvania, USA, and American Ornithologists' Union, Washington, D.C., USA.

Watts, B. D., and D. S. Bradshaw. 1994. The influence of human disturbance on the location of great blue heron colonies in the lower Chesapeake Bay. Colonial Waterbirds 17:184–186.

Weakland, C. A. 2000. Effects of diameter-limit and two-age timber harvesting on songbird populations on an industrial forest in central West Virginia. Dissertation, West Virginia University, Morgantown, USA.

Weakland, C. A., and P. B. Wood. 2005. Cerulean warbler (*Dendroica cerulea*) microhabitat and landscape-level habitat characteristics in southern West Virginia. Auk 122:497–508.

Weatherhead, P. J., and K. A. Prior. 1992. Preliminary observations of habitat use and movements of the Eastern Massasauga rattlesnake (*Sistrurus c. catenatus*). Journal of Herpetology 26:447–452.

Webb, W. L., D. F. Behrend, and B. Sainsorn. 1977. Effect of logging on songbird populations in a northern hardwood forest. Wildlife Monographs 55:1–34.

Webster, W. D., J. F. Parnell , and W. C. Biggs, Jr. 1985. Mammals of the Carolinas, Virginia, and Maryland. University of North Carolina Press, Chapel Hill, North Carolina, USA.

Weigl, P. D. 1969. The distribution of the flying squirrels (*Glaucomys volans* and *G. sabrinus*): an evaluation of the competitive exclusion idea. Dissertation, Duke University, Durham, North Carolina, USA.

Weigl, P. D. 1977. Status of the northern flying squirrel, *Glaucomys sabrinus coloratus*, in North Carolina. Pages 398–400 *in* J. E. Cooper, S. S. Robinson, and J. B. Funderburg, editors. Endangered and threatened plants and animals of North Carolina. North Carolina State Museum of Natural History, Raleigh, USA.

Weigl, P. D. 1978. Resource overlap, interspecific interactions and the distribution of the flying squirrels, *Glaucomys volans* and *G. sabrinus*. American Midland Naturalist 100:83–96.

Weigl, P. D., T. W. Knowles, and A. C. Boynton. 1999. The distribution and ecology of the northern flying squirrel, *Glaucomys sabrinus coloratus*, in the southern Appalachians. North Carolina Wildlife Commission Nongame and Endangered Wildlife Program Division of Wildlife Management, Raleigh, USA.

Weigl, P. D., L. J. Sherman, A. I. Williams, M. A. Steele, and D. S. Weaver. 1998. Geographic variation in the fox squirrel (*Sciurus niger*): a consideration of size clines, habitat vegetation, food habits and historical biogeography. Pages 171–184 *in* M. A. Steele, J. F. Merritt, and D. A. Zegers, editors. Ecology and evolutionary biology of tree squirrels. Virginia Museum of Natural History, Special Publication 6, Martinsburg, Virginia, USA.

Weinberg, H. J., and R. P. Roth. 1998. Forest area and habitat quality for nesting wood thrush. Auk 115:879–889.

Weir, B. S. 1996. Intraspecific differentiation. Pages 385–405 *in* D. M. Hillis, C. Moritz, and B. K. Mable, editors. Molecular systematics. Second edition. Sinauer Associates, Sunderland, Massachusetts, USA.

Weller, M. W. 1958. Observations of the incubation behavior of a common nighthawk. Auk 75:48–59.

Weller, M. W. 1961. Breeding biology of the least bittern. Wilson Bulletin 73:11–35.

Wells-Gosling, N., and L. R. Heaney. 1984. *Glaucomys sabrinus*. Mammalian Species 247:1–8.

Wemmer, L. C. 2000. Conservation of the piping plover (*Charadrius melodus*) in the Great Lakes region: a landscape-ecosystem approach. Dissertation, University of Minnesota, Twin Cities, USA.

Wentworth, K. 2001. Renesting of the dickcissel (*Spiza americana*) at Prairie Ridge State Natural Area. Thesis, Eastern Illinois University, Charleston, USA.

Wentworth, K. L., M. C. Brittingham, and A. M. Wilson. 2010. Conservation Reserve Enhancement program fields: benefits for grassland and shrub-scrub species. Journal of Soil and Water Conservation 65:50–60.

Wetzel, R. M. 1955. Speciation and dispersal of the southern bog lemming, *Synaptomys cooperi* (Baird). Journal of Mammalogy 36:1–20.

Wheeler, B. A., E. Prosen, A. Mathis, and R. F. Wilkinson. 2003. Population declines of a long-lived salamander: a 20 year + study of hellbenders, *Cryptobranchus alleganiensis*. Biological Conservation 109:151–156.

Wheeler, B. K. 2003. Raptors of eastern North America. Princeton University Press, Princeton, New Jersey, USA.

Whitaker, J. O., Jr. 1971. A study of the western chorus frog, *Pseudacris triseriata*, in Vigo County, Indiana. Journal of Herpetology 5:127–150.

Whitaker, J. O., Jr. 1974. *Cryptotis parva*. Mammalian Species 43:1–8.

Whitaker, J. O. 2004. Prey selection in a temperate zone insectivorous bat community. Journal of Mammalogy 85:460–469.

Whitaker, J. O., V. Brack, and J. B. Cope. 2002. Are bats in Indiana declining? Proceedings of the Indiana Academy of Sciences 111:95–106.

Whitaker, J. O., Jr., and W. J. Hamilton, Jr. 1998. Mammals of the eastern United States. Comstock Publishing Associates, Cornell University Press, Ithaca, New York, USA.

Whitaker, J. O., Jr., and R. E. Martin. 1977. Food habits of *Microtus chrotorrhinus* from New Hampshire, New York, Labrador and Quebec. Journal of Mammalogy 58:99–100.

Whitaker, J. O., Jr., C. Maser, and L. E. Keller. 1977. Food habits of bats of western Oregon. Northwest Science 52:46–55.

Whitcomb, R. F., C. S. Robbins, J. F. Lynch, M. K. Klimkiewicz, B. L. Whitcomb, and D. Bystrak. 1981. Effects of forest fragmentation on avifauna of the eastern deciduous forest. Pages 125–206 *in* R. L. Burgess and D. M. Sharpe, editors. Forest island dynamics in man-dominated landscapes. Springer-Verlag, New York, New York, USA.

White, D., and D. Moll. 1991. Clutch size and annual reproductive potential of the turtle *Graptemys geographica* in a Missouri stream. Journal of Herpetology 25:493–494.

Whitehead, D. R. 1992. Factors influencing the reproductive success of Neotropical migrant landbirds in south-central Indiana: the effect of landscape pattern and wildlife management activities. Report submitted to the National Fish and Wildlife Foundation, Washington, D.C., USA.

Whitehead, D. R., and T. Taylor. 2002. Acadian flycatcher (*Empidonax virescens*). *In* A. Poole and F. Gill, editors. The birds of North America, 614. Academy of Natural

Sciences, Philadelphia, Pennsylvania, USA, and the Auk, Washington, D.C., USA.

Whitlock, A. L. 2002. Ecology and status of the bog turtle (*Clemmys muhlenbergii*) in New England. Dissertation, University of Massachusetts, Amherst, USA.

Whitmore, R. C., and G. A. Hall. 1978. The response of passerine species to a new resource: reclaimed strip mines in West Virginia. American Birds 32:6–9.

Whitney, G. G. 1990. The history and status of the hemlock-hardwood forests of the Allegheny Plateau. Journal of Ecology 78:443–458.

Widrlechner, M. P., and S. K. Dragula. 1984. Relation of cone-crop size to irruptions of four seed-eating birds in California. American Birds 38:840–846.

Wiggers, E. P., and K. J. Kritz. 1994. Productivity and nesting chronology of the Cooper's hawk and sharp-shinned hawk in Missouri. Journal of Raptor Research 28:1–3.

Wikelski, M., R. W. Kays, and N. J. Kasdin. 2007. Going wild: what a global small-animal tracking system could do for experimental biologists. Journal of Experimental Biology 210:181–186.

Wilcove, D. S., D. Rothstein, and J. Dubow. 1998. Quantifying threats to imperiled species in the United States. BioScience 48:607–615.

Wilcox, J. T., and D. H. Van Vuren. 2009. Wild pigs as predators in oak woodlands of California. Journal of Mammalogy 90:114–118.

Wiley, R. W. 1980. *Neotoma floridana*. Mammalian Species 139:1–7.

Wilhelm, G. 1993. King Rail breeding in western Pennsylvania. Pennsylvania Birds 7(3):89–90.

Wilhelm, G. 1994. Breeding dickcissel behavior. Pennsylvania Birds 8:139–140.

Wilhelm, G. 1995. Scenario of the upland sandpiper in western Pennsylvania. Pennsylvania Birds 8:204–205.

Wilkins, K. A. 2007. Movement, survival rate estimation, and population modeling of eastern tundra swans, *Cygnus columbianus columbianus*. Dissertation, Cornell University, Ithaca, New York, USA.

Will, T. C. 1986. The behavioral ecology of species replacement: blue-winged and golden-winged warblers in Michigan. Dissertation, University of Michigan, Ann Arbor, USA.

Willard, D. E. 1977. The feeding ecology and behavior of five species of herons in southeastern New Jersey. Condor 79:462–470.

Williams, A. B. 2000. The effects of gypsy moth treatment applications of *Bacillus thuringiensis* on worm-eating warblers in middle Appalachia. Thesis, University of Georgia, Athens, USA.

Williams, B. 1979. Black-crowned night-heron. Pages 448–449 *in* D.W. Linzey, editor. Proceedings of the Symposium of Endangered and Threatened Plants and Animals of Virginia, Virginia Polytechnic Institute, Blacksburg, USA.

Williams, D. F., and J. S. Findley. 1979. Sexual size dimorphism in vespertilionid bats. American Midland Naturalist 102:113–126.

Williams, E. C., Jr., and W. S. Parker. 1987. A long-term study of a box turtle (*Terrapene carolina*) population at Allee Memorial Woods, Indiana, with emphasis on survivorship. Herpetologica 43:328–335.

Williams, G. E, and P. B. Wood. 2002. Are traditional methods of determining nest predators and nest fates reliable? an experiment with wood thrushes (*Hylocichla mustelina*) using miniature video cameras. Auk 119:1126–1132.

Williams, L. M. 2007. Pennsylvania's Comprehensive Wildlife Conservation Strategy. <www.pgc.state.pa.us/pgc/cwp/view.asp?a=496&q=162067>.

Williams, W. 2004. When blade meets bat. Scientific American February 2004:20–21.

Wilsmann, L. A., and M. A. Sellers, Jr. 1988. *Clonophis kirtlandii* rangewide survey. U.S. Fish and Wildlife Service Region 3, Office of Endangered Species, Twin Cities, Minnesota, USA.

Wilson, A. M. 2009. Bird population responses to conservation grasslands in Pennsylvania. Dissertation, Pennsylvania State University, University Park, USA.

Wilson, C. 2001. Green Salamander, *Aneides aeneus*. Chattooga Quarterly, Chattooga Conservancy. Spring/Summer 2001. <www.chattoogariver.org/Articles/2001SS/GreenSalamander.htm>. Accessed 25 March 2005.

Wilson, C. R. 2003. Woody and arboreal habitats of the green salamander (*Aneides aeneus*) in the Blue Ridge Mountains. Contemporary Herpetology 2003(2). <www.contemporaryherpetology.org/ch/2003/2/index.htm>. Accessed 5 May 2009.

Wilson, R. R., and R. J. Cooper. 1998. Acadian flycatcher nest placement: does placement influence reproductive success? Condor 100:673–679.

Wiltraut, R. 1994. Courtship display of the yellow-crowned night-heron. Pennsylvania Birds 8:95.

Wingate, D. B. 1982. Successful reintroduction of the yellow-crowned night-heron as a nesting resident in Bermuda. Colonial Waterbirds 5:104–115.

Winter, M. 1999. Nesting biology of dickcissels and Henslow's sparrows in Missouri prairie fragments. Wilson Bulletin 111:515–527.

Winter, M., and J. Faaborg. 1998. Patterns of area sensitivity in grassland-nesting birds. Conservation Biology 13:1424–1436.

Witmer, G. W., R. B. Sanders, A. C. Taft. 2003. Feral swine—are they a disease threat to livestock in the United States? Pages 316–325 *in* K. A. Fagerstone and G. W. Witmer, editors. Proceedings of the 10th Wildlife Damage Management Conference. USDA National Wildlife Research Center, Fort Collins, Colorado, USA.

Woleslagle, B. A. 1994. Survey of small mammals in northeastern Pennsylvania with an emphasis on six target species. Thesis, Shippensburg University, Shippensburg, Pennsylvania, USA.

Wolfe, M. L., N. V. Debyle, C. S. Winchell, T. R. McCabe. 1982. Snowshoe hare cover relationships in northern Utah. Journal of Wildlife Management 46:662–670.

Wolford, J. W., and D. A. Boag. 1971. Food habits of black-crowned night-herons in southern Alberta. Auk 88:435–437.

Wood, J. T. 1949. Observations on *Natrix septemvittata* (Say) in southeastern Ohio. American Midland Naturalist 42:744–750.

Wood, M. 1973. Birds of Pennsylvania. Pennsylvania State University, University Park, USA.

Wood, M. 1979. Birds of Pennsylvania, when and where to find them. Pennsylvania State University, University Park, USA.

Wood, N. A. 1974. Breeding behavior and biology of the moorhen. British Birds 67:104–115, 137–158.

Wood, P. B., C. B. Viverette, L. J. Goodrich, M. Pokras, and C. Tibbott. 1996. Environmental contaminant levels in sharp-shinned hawks from the eastern United States. Journal of Raptor Research 30:136–144.

Woodward, A. A., A. D. Fink, and F. R. Thompson. 2001. Edge effects and ecological traps: effects on shrubland birds in Missouri. Journal of Wildlife Management 65:668–675.

World Conservation Union (International Union for the Conservation of Nature). 2007. Species Survival Commission. Global Invasive Species Database. <www.issg.org/database>. Accessed 24 May 2009.

World Owl Trust. 2005. Long-eared owl conservation project, including the month of the long-eared owl 2005. <www.owls.org/Whatis/long_eared_project.htm>.

Worthington, R. D. 1973. Remarks on the distribution of the smooth green snake, *Opheodrys vernalis blanchardi* (Grobman) in Texas. Southwestern Naturalist 18: 341–357.

Wright, A. H., and A. A. Wright. 1949. Handbook of frogs and toads. Comstock, Ithaca, New York, USA.

Wright, A. H., and A. A. Wright. 1957. Handbook of snakes. Comstock, Ithaca, New York, USA.

Wright, J., and G. L. Kirkland. 2000. A possible role for chestnut blight in the decline of the Allegheny woodrat. Journal of the American Chestnut Foundation 8:30–35.

Yahner, R. H., R. J. Hutnik, and S. A. Liscinsky. 2002. Bird populations associated with an electric transmission right-of-way. Journal of Arboriculture 28:123–130.

Yahner, R. H., and R. W. Rohrbaugh, Jr. 1996. Birds on reclaimed surface mines. Northeast Wildlife 53:11–18.

Yamasaki, M., R. M. DeGraaf, and J. W. Lanier. 2000. Wildlife habitat associations in eastern hemlock birds, smaller mammals, and forest carnivores. Pages 135–143 *in* K. A., McManus, K. S. Shields, and D. R. Souto, editors. Proceedings: Symposium on sustainable management of hemlock ecosystems in eastern North America. U.S. Forest Service, Northeastern Research Station, General Technical Report NE-267.

Yoakum, J., W. P. Dasmann, H. R. Sanderson, C. M. Nixon, and H. S. Crawford. 1980. Habitat improvement techniques. Pages 329–403 *in* S. D. Schemnitz, editor. Wildlife management techniques manual. Fourth Edition, Revised. The Wildlife Society, Washington, D.C., USA.

Yong, W., and F. R. Moore. 1994. Flight morphology, energetic condition, and the stopover biology of migrating thrushes. Auk 111:683–692.

Yosef, R. 1996. Loggerhead shrike (*Lanius ludovicianus*). *In* A. Poole and F. Gill, editors. The birds of North America, 231. Academy of Natural Sciences, Philadelphia, Pennsylvania, USA, and American Ornithologists' Union, Washington, D.C., USA.

Young, R. T. 1896. Summer birds of the anthracite coal regions of Pennsylvania. Auk 13:278–285.

Yunick, R. P. 1981. Some observations on the breeding status of the pine siskin. Kingbird 31:219–225.

Zappalorti, R. T., G. Rocco, and P. J. Drake. 1995. Results of a two phase bog turtle (*Clemmys muhlenbergii*) study within Pennsylvania, with special notes on nesting. February 25, 1995. Herpetological Associates. File Numbers 94.18 and 94.19.

Zegers, D. A. 1985. Eastern fox squirrel. Pages 399–402 *in* H. H. Genoways and F. J. Brenner, editors. Species of special concern in Pennsylvania. Carnegie Museum of Natural History, Special Publication 11, Pittsburgh, Pennsylvania, USA.

Zeller, N. S. 1997. The Nature Conservancy species management abstract for yellow throated vireo (*Vireo flavifrons*). Arlington, Virginia, USA.

Zielinski, W. J., N. P. Duncan, E. C. Farmer, R. L. Truex, A. P. Clevenger, and R. H. Barrett. 1999. Diet of fishers (*Martes pennanti*) at the southernmost extent of their range. Journal of Mammalogy 80:961–971.

Zielinski, W. J., and T. E. Kucera. 1995. Survey methods for the detection of wolverines, lynx, fishers, and martens. U.S. Department of Agriculture Forest Service General Technical Report PSW-157.

Zielinski, W. J., and H. B. Stauffer. 1996. Monitoring *Martes* populations in California: survey design and power analysis. Ecological Applications 6:1254–1267.

Zimmerman, A. L., J. A. Dechant, D. H. Johnson, C. M. Goldade, J. O. Church, and B. R. Euliss. 2002a. Effects of management practices on wetland birds: marsh wren. Northern Prairie Wildlife Research Center, Jamestown, North Dakota USA.

Zimmerman, A. L., J. A. Dechant, D. H. Johnson, C. M. Goldade, B. E. Jamison, and B. R. Euliss. 2002b. Effects of management practices on wetland birds: black tern. Northern Prairie Wildlife Research Center, Jamestown, North Dakota, USA.

Zimmerman, A. L., B. E. Jamison, J. A. Dechant, D. H. Johnson, C. M. Goldade, J. O. Church, and

B. R. Enliss. 2003. Effects of management practices on wetland birds. U.S. Department of the Interior, and U.S. Geological Survey, Northern Prairie Wildlife Research Center (NPWRC), Jamestown, North Dakota, USA.

Zimmerman, J. L. 1966. Polygyny in the dickcissel. Auk 83:534–546.

Zimmerman, J. L. 1971. The territory and its density dependent effect in *Spiza americana*. Auk 88:591–612.

Zimmerman, J. L. 1977. Virginia Rail (*Rallus limicola*). Pages 46–56 *in* G. C. Sanderson, editor. Management of migratory shore and upland game birds in North America. Washington, D.C. International Association of Fish and Wildlife Agencies, USA.

Zimmerman, J. L. 1982. Nesting success of dickcissels (*Spiza americana*) in preferred and less preferred habitat. Auk 99:292–98.

Zimmerman, J. L., and E. J. Finck. 1989. Philopatry and correlates of territorial fidelity in male dickcissels. North American Bird Bander 14:83–85.

Zink, R. M. 1996. Species concepts, speciation, and sexual selection. Journal of Avian Biology 27:1–6.

Zollner, P. A. 2000. Comparing the landscape level perceptual abilities of forest sciurids in fragmented agricultural landscapes. Landscape Ecology 15:523–533.

Zorn, T. 1998. Highway mortality of barn owls in northeastern France. Journal of Raptor Research 32:229–232.

# Index

Page numbers in *italics* indicate figures and tables.